Stimulating Concepts in Chemistry

Other Titles of interest

G. R. Newkome/
C. N. Moorefield/F. Vögtle

Dendritic Molecules
Concepts, Syntheses,
Perspectives

1996. XI. 261 pages
with 218 figures and
4 tables.
Hardcover.
ISBN 3–527-29325–6

H. Hopf

**Classics in Hydrocarbon
Chemistry**

2000. XIII. 547 pages
with 434 figures.
Softcover.
ISBN 3–527-29606–9

K. C. Nicolaou/
E. J. Sorensen

Classics in Total Synthesis

1996. XXIII. 798 pages
with 444 figures.
Softcover.
ISBN 3–527-29231–4

F. Diederich/P. J. Stang

**Templated Organic
Synthesis**

1999. XX. 410 pages
with ca 340 figures and
10 tables.
Hardcover.
ISBN 3–527-29666–2

Stimulating Concepts in Chemistry

Editors: Fritz Vögtle,
J. Fraser Stoddart,
Masakatsu Shibasaki

WILEY-VCH

Weinheim · New York · Chichester ·
Brisbane · Singapore · Toronto

Prof. Dr. Fritz Vögtle
Kekulé-Institut für
Organische Chemie und
Biochemie
Universität Bonn
Gerhard-Domagk-Straße 1
53121 Bonn
Germany

Prof. Dr. J. Fraser Stoddart
Department of Chemistry
and Biochemistry
UCLA
405 Hilgard Avenue
Los Angeles CA
90095–1569
USA

Prof. Dr.
Masakatsu Shibasaki
Graduate School of
Pharmaceutical Sciences
University of Tokyo
7–3–1 Hongo, Bunkyo-ku
Tokyo 113–0033
Japan

Library of Congress Card No. applied for.

British Library Cataloguing-in-Publication Data: A catalogue record for this book is available from the British Library.

Die Deutsche Bibliothek – CIP Cataloguing-in-Publication-Data
A catalogue record for this publication is available from Die Deutsche Bibliothek

ISBN 3-527-29978-5

Composition: Mitterweger & Partner GmbH, Plankstadt
Printing: Hans Rappold Offsetdruck GmbH, Speyer
Bookbinding: Wilhelm Osswald & Co., Neustadt (Weinstraße)
Cover Design: Günther Schulz, Fussgönheim
Printed in the Federal Republic of Germany.

Table of Contents

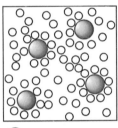
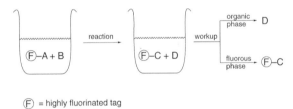

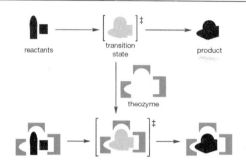
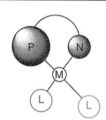
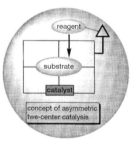

Metal-catalyzed enantiose-lective reactions are used in the synthesis of complex organic molecules. These transformations offer cost-effective, highly selective alternatives to the more classical methods. Additional levels of efficiency are achieved when various cata-lytic reactions are used successively.

The hypothetical fullerene – acetylene hybrid at right symbolizes the exciting prospects that have arisen from the discovery of fuller-enes, and the profound changes they have induced in our view of acetylene and carbon allotrope chemistry. The preparation of molecu-lar and polymeric acetylenic carbon allotropes, as well as carbon-rich nanometer-sized structures, has opened up new avenues in fundamen-tal and technological research at the interface of chemistry with materials science.

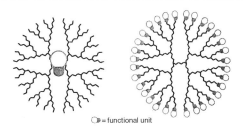

In dendrimer chemistry there are different ways to display functionalities. On the one hand, a central functional unit can be den-drylated to change some of its properties. On the other hand, it is possible to attach functional groups to the periphery of a dendrimer skeleton, i.e., the functional-ity can be multiplied.

= functional unit

Self-assembling capsules are about to leave the play-ground of basic research and begin their career in materials sciences. The long way from the principles of self-organization through noncovalent interactions and the filling of space inside host molecules *via* chiral recognition and catal-ysis in reaction vessels to hydrogen-bonded liquid crystals and fibers forms the backbone of this review.

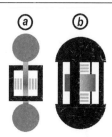
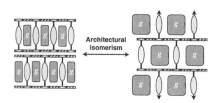
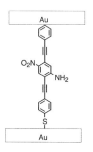
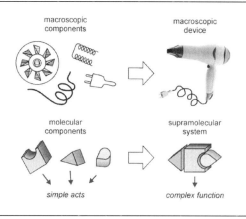

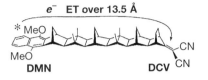

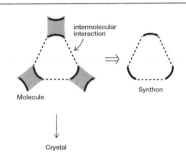
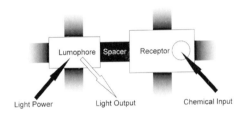
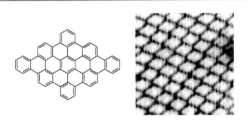
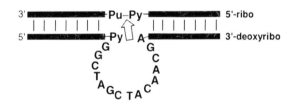
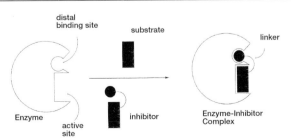

This chapter highlights the **interplay between organic synthesis, biophysics and cell biology** in the study of protein lipidation and its relevance to targeting proteins towards the plasma membrane of cells in precise molecular detail.

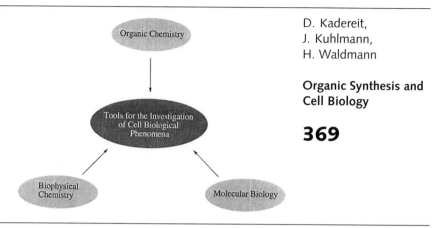

D. Kadereit,
J. Kuhlmann,
H. Waldmann

Organic Synthesis and Cell Biology

369

Preface

Dwelling momentarily on the words *Stimulating Concepts in Chemistry* conjures up more than one line of thought. One is "Concepts in Chemistry" that are stimulating in themselves. Another is how "Concepts in Chemistry" can be stimulated at the interfaces with the life sciences and with materials science. These are powerful undercurrents that provide the changing circumstances for the creative developments in chemistry through – in many instances – the dreams and imaginations of intelligent and well-trained young men and women entering the profession as researchers at the graduate student and post-doctoral levels. Those academic, government, and industrial scientists in mid-career and beyond, who have demonstrated their creativity in articles, communications, papers, and reviews in the scientific literature – and from the lecterns in giving research seminars and named lectures in a variety of different settings, as well as from the podiums in presenting plenary talks and award lectures at society meetings and international symposia – are certainly in a position to commit their accomplishments and experiences to print, often with the direct involvement of their industrious and enthusiastic young colleagues.

By focusing the spotlight on concepts that are particularly stimulating to chemists, a collection of essays, that address some of the most exciting conceptual developments in chemistry in recent years, have been gathered together in the pages that follow herein. The exercise was initiated by inviting over 30 highly renowned chemists, all of them pioneers in their field, to contribute an essay to this treatise. In the event, some two dozen world leaders in their research areas responded somewhat differently to a fairly prescriptive format: some followed the rubric closely, others veered away from the recommendations somewhat, and the remainder did what they felt most comfortable with doing – and the outcome is what you have in your hands to select from and savor, we hope.

There was some discussion by the editors about the title of this book, especially with regard to the problem that some areas of "chemistry" (for example inorganic chemistry) are not represented to the same degree of detail as others. A consensus was then achieved that this volume, by touching, illustrating and interconnecting a selection of undoubt-edly important branches of chemistry, should stimulate, challenge and initiate further monographs to include other subdisciplines.

Chemistry receives a lot of its current stimulation from its two-way trafficking with the biological world. Three essays under the heading **Biological Aspects** in **Section IV** (p 339) serve to illustrate this point extremely well.

In an up-date of a recent Angewandte review, *Kirby* discusses **enzyme mimics** (p 341) under five banners – namely, mimics based (1) on natural enzymes, (2) on other proteins, (3) on other biopolymers, (4) on synthetic macromolecules, and (5) on small-molecule host-guest interactions. He is at pains to point out that, although a lot of progress has been made toward 'explaining' enzyme catalysis, our current level of understanding is nowhere like sufficient to design and synthesize enzyme analogs that even begin to vie with enzymes, at least as far as catalysis is concerned. So far, supramolecular chemists have by and large failed the practical test of delivering artificial enzymes that actually do a respectable job: their attempts to date to translate the sophisticated molecular recognition of their (ground state) systems into effective catalysis is judged to have been 'generally disappointing'. Consequently, there is still a big challenge out there for chemists to meet and it is our hunch that it will take a combination of some form of supramolecular or dynamic covalent chemistry, probably relying heavily on computational methods and virtual/dynamic combinatorial libraries for its ultimate success. Even then, by its very nature, substrate specific catalysis with artificial enzymes could end up being one job at a time.

As far as enzymes are concerned, the search for inhibitors of their mode of actions, it has to be said, has been much more rewarding than the quest for artificial enzyme catalysis. The prospect of being able to treat a lot of human diseases with **enzyme inhibitors** (p 355) has provided a considerable fillip to the understanding of this concept. It is one that *Kreutter* and *Wong* examine from the traditional angles of (1) transition state analogs, (2) suicide substrates, (3) 'quiescent' affinity labels, and (4) bisub-strate analogs, before extolling the virtues of (5) distal binding analogs – an approach to drug design that employs binding sites that have no catalytic function at all and can often be located quite distant from the enzyme's active site. It is an approach which has already yielded clinically approved drugs. Moreover, it is rendered extremely powerful as a means of uncovering potent enzyme inhibitors (1) when it is augmented by computer-aided molecular modeling at the design stage, (2) when parallel and combinatorial approaches to synthesis are employed, and (3) when high resolution X-ray crystal structures are performed on enzyme-inhibitor complexes. It is clear that we are on the threshold of a new era when this latest approach to enzyme inhibition will be put severely to the test in the drive to discover new drugs. The reason lies in the simple fact that the development of functional genomics and proteomics will uncover many new enzymes as targets for drug discovery.

The importance of the interplay between three core disciplines – **organic synthesis**, biophysics, **and cell biology** (p 369) is highlighted in the essay written by *Kadereit, Kuhlmann,* and *Waldmann.* These authors advocate the use of any structural information that is known about a particular biological phenomenon to identify chemical problems that lead to the development of new synthetic pathways and methods. This kind of conceptual thinking characterizes the field that has come to be known as either 'Bioorganic Chemistry' or 'Chemical Biology'. The authors make the point most succinctly at the end of their article with the words – 'In this field, research has to be carried out in both chemistry and biology. The researcher has to cross the barriers and bridge the undoubtedly existing gap in research culture between the two disciplines. The researcher will be rewarded by experiencing the excitement that is created in both disciplines by the ability to describe a biological phenomenon in the precise molecular language of chemistry and by gaining insights that could not have been obtained by employing either discipline alone.' Our wish to paraphrase what is contained in this collection of essays forsook us when we read the above passage – fittingly, the very last one in the book! It says it all wonderfully well and in so doing outlines the challenge that goes out nowadays to all scientists who consider themselves to be either a bioorganic chemist or a chemical biologist. There is no mistaking the take-home message in this forthright statement. The authors put on record a pretty convincing case history in support of their thesis by calling attention to some proteins – the so-called Ras proteins – that influence numerous signal transduction processes. The Ras signal transduction cascade is central to the regulation of cell growth and differentiation:

it occupies a position of great physiological importance that has been probed successfully using synthetic Ras peptides and proteins in experimental cell biology.

What has been stated in the preceding paragraph about the relationship between chemistry and biology can equally well be said about the interface between chemistry and materials. Consequently, this collection contains some half-dozen essays in **Section III** (p 235) on **Molecular Devices and Materials Properties** that relate to materials in general and molecular devices in particular. A moment's reflection reminds us that chemists brought dyes into being in the 19th century and, in the process, brightened up our lives considerably. In the 20th century, it has been drugs with their ability to save and prolong our lives, as well as to make them considerably more bearable, that chemists will be remembered for above all else. The question which is now at the forefront of many chemists' minds is – will a device industry be spawned on the back of chemistry in the 21st century that will be as successful and central to our advanced societies as are the dye-stuffs and pharmaceutical industries of today? The answer to this question is not so much whether, as when – and when a device industry based on functioning molecules does evolve, it will be revolutionary in its consequences for the whole human race. There is no doubting that exciting times lie ahead of the materials chemist.

Computers are here to stay and we would all like them to get smaller, faster, and cheaper. However, there is a serious problem looming upon the horizon for the conventional silicon-based technology that has served us well over the last 35 years or so: if the current rate of computer miniaturization continues, it will reach limits of a fundamental nature in about a decade from now. Although there is no certainty that it will be **molecular wires and devices** (p 237) that will come to the rescue, *Tour* tackles this emotive issue in a well-argued and logical manner in his essay. He points out that chemical synthesis which brought dyes and drugs into our lives is equally capable of delivering devices for an electronics industry that could begin to run out of steam otherwise. He sees the route to the continued success of this industry as involving hybrid technologies where 'molecules work in concert with silicon … as we develop ultradense and ultrafast computational systems'. The challenge presently is to be able to construct nanoelectronic components – e.g., diodes, switches, transistors, wires, etc. – by using single molecules or small packets of them in the context of computer circuitry. In this essay, we learn, amongst many other things, how the Rice group has fashioned many well-characterized, conducting poly(phenylene-ethynylene) and poly(thiophene-ethynylene) polymers, as well as poly(phenylene-ethynylene)/poly(thiophene-ethynylene) block copolymers that are precise in their lengths using ingenious iterative synthetic routes. What is more – their conductivities can be interrogated one molecule at a time. Speculation as to a new technology aside, it is impressive science in the making.

In their essay on **molecular-level devices and machines** (p 255), *Balzani, Credi,* and *Venturi* emphasize how the extension of the concept of a device from the macroscopic world to the molecular level is important for the growth of nanoscience and the development of nanotechnology – not least of all the construction of molecular-based (chemical) computers! They draw attention to the fact that molecular-level devices, which operate by means of electronic and/or nuclear rearrangements are characterized by exactly the same features that we associate with macroscopic devices. They define a molecular-level device ' as an assembly of a discrete number of molecular components designed to achieve a specific function' and a molecular-level machine as 'a device in which the component parts can display changes in their relative positions as a result of some external stimulus'. The stimulus is commonly supplied as chemical energy, electrical energy, or light. The essay draws for proof of principle on some of the spectacular examples demonstrated of late in the Bologna laboratories. Molecular devices based on the transfer of electrons – modeling antennas for light harvesting – are formed in suitably designed metallodendrimers where electronic energy can be channeled toward a specific region of the dendrimer. Devices based on nuclear motions – reminiscent of a piston moving up and down a cylinder or of beads being slid along an abacus – are described that are based (1) on the complexation/decomplexation equilibria of pseudorotaxanes and (2) on the relative linear and rotary motions of the components of rotaxanes and catenanes. Systems, at both the supramolecular and molecular levels are highlighted and it is demonstrated that they can be driven chemically, electrochemically, or photochemically. Aside from molecular-level devices that work as a result of either electron or nuclear movements, there are other more sophisticated devices (e.g., plug-in-socket and three-pole systems) whose functions are based on both electronic and nuclear rearrangements. The 'secrets' to all these machine and device-like processes are exactly the same as those utilized in living systems and include molecular recognition processes and switching controlled by electrons or protons or metal ions.

Remarkable progress has been made in our fundamental understanding of the nature of long-range electron transfer processes during the last 15 years. In his essay on **electron and energy transfer** (p 267), *Paddon-Row* reminds us that the characteristics of the through-bond mechanism for electron transfer have only begun to be understood in recent times. It is now clear from the research done by the essayist and others that the distance and orientation dependence of the dynamics of long-range electron transfer between a donor and acceptor chromophore is markedly facilitated by the nature and composition of the surrounding medium. He goes on to point out that it is 'this aspect of the long-range electron transfer processes that has captured the attention of a broad cross section of the chemical community, partly because a deeper understanding of long-range electron transfer is essential to the successful design of molecular photovoltaic systems and other nanoscale electronic devices.' Although rapid electron transfer is expected in conjugated systems, it was much more intriguing for chemists to find out if an intervening non-conjugated medium comprised of, for example, an oligopeptide, or a saturated hydrocarbon bridge or p-p stacks of unsaturated (aromatic) molecules, or indeed simply solvent molecules, could facilitate electron transfer. In fact, it does! Electron transfer mediated by saturated norbornylogous bridges greater than 12 bonds in length can take place on a subnanoscale timescale over distances exceeding 10 Å by a through-bond mechanism. The dependence of the dynamics of electron transfer on the constitution and configurational stereochemistry of the bridge, as well as on orbital symmetry constraints, is consistent with a through-bond mechanism. The fact that electron transfer can also be mediated by hydrogen bonding networks and solvent molecules means that noncovalent synthesis can be used for the construction of photovoltaic supramolecular assemblies. Exciting also are the recent investigations of electron transfer mechanisms in DNA molecules where superexchange and hole-hopping mechanisms – both mediated apparently by the p-stacked base pairs – are being invoked to explain experimental observations which could have a profound impact on the field of molecular electronics. Also, let us not lose sight of the fact that electron transfer plays a key role in many essential biological processes.

Chemical synthesis does not begin and end with the chemistry of the covalent bond. The art of non-covalent synthesis has been evolving rapidly over the last 20 years and now the time is ripe to entertain the concept of **the supramolecular synthon in crystal engineering** (p 293), introduced onto the scene a few years ago by *Desiraju*. In his essay, he argues that the 'synthon is the supramolecular equivalent of the functional group and is as useful in supramolecular chemistry as the functional group is in molecular chemistry.' However, he also points out that the functional group – a molecular phenomenon – cannot be the centerpiece of crystal packing information since this is a property that is supramolecular in nature. He views supramolecular synthons as multimolecular units that take into account the complementary – electronic as well as steric – association of functional groups from different molecules. He sees them as much closer to actual crystal structures than are molecules, functional groups, or even single interactions. Just as in covalent synthesis, a retrosynthetic analysis of a crystal structure can be used to define crystal engineering strategies. Provided the level of recognition insulation between molecules is high in a crystal structure, then the noncovalent synthesis of this structure is possible. Examples of this approach to crystal engineering are discussed in the essay. It is stressed that the supramolecular synthon concept is implicated in all the stages (molecules → recognition → nucleation → crystallization → growth → crystal) through which molecules progress as they form crystals. One cannot help thinking that, if this is the case, then the concept has utility in the solution as well as the solid state and is most likely being employed instinctively and intuitively by supramolecular chemists whenever they embark upon a noncovalent synthesis whether they end up with crystals or not. However, there is also no doubt that crystal engineers need to explore any concept at all that might help to bring some rhyme or reason into the design of solid state structures. Nowhere is the concept of supramolecular synthons more welcome!

In their essay on **luminescent logic and sensing** (p 307), the Belfast group of *de Silva*, *Fox*, and *Moody* highlight how light-powered molecular-level devices, given a set of chemical inputs, can generate a particular output. The patterns associated with these outputs correspond to different members in the logic vocabulary that can be classified according to truth tables. Luminescent molecular sensors and switches, based on molecules that contain a receptor separated by a rigid spacer from a lumophore, have been designed and synthesized that exhibit operations associated with YES (PASS), NOT, AND, XOR, and INHIBIT logic. 'Since logic gates drive the information technology revolution .....

rudimentary computation at the molecular-scale emerges over the horizon.' The molecular devices – some of them designed and made in Belfast – depend on photoinduced electron transfer for their operation and are formatted in terms of (1) lumophores for photonic communication, (2) receptors for chemical sensing, and (3) spacers to segregate these different functions. The Belfast team sees the increasing demands from biologists, clinicians, and engineers for monitoring and imaging particular analytes as driving the development of luminescent logic and sensing down to the single molecule level – and you can't do any better than that in chemical terms! Nanotechnology is one of the more emotive words these days that serious-minded scientists have to come to terms with in the context of their science.

In their essay on **nanochemistry – architecture at the mesoscale** (p 317) – *Becker* and *Müllen*, while stating their belief that it is not unreasonable to assert that 'we will be able to achieve viable nanotechnologies in the not too distant future', issue the cautionary note of warning – 'Nanochemistry has not yet advanced to a level where it can meet crucial technological requirements.' Historically, one of the constraints that chemists have imposed on themselves has been the methods of characterization and analyses they have employed: until recent times, they have been preoccupied by compound purity and so have relied on methods of characterization and analyses that examine ensembles and measures their averaged properties. Old ways die hard! The Mainz group, in recommending that chemists should be 'taking each molecule seriously' make the case strongly for doing chemistry at the level of the single molecule, pointing out that ultimately it should lead to the most efficient use of materials possible. The nanochemistry, ushered in by this conceptual approach, is envisaged to involve four steps – (1) design and synthesis, (2) immobilization using self-assembly principles, (3) visualization by scanning probe microscopies, for example, and (4) manipulation and function. They track their own remarkable contributions through molecular structures – where the benzene ring always plays a significant role – from rods to discs to spheres with electronic properties in mind. The take-home message to chemists reading this essay is – think small!

It is evident that the essays in **Sections III** and **IV** have drawn heavily on the theme of **Section II** (p 161) – namely, **Architecture, Organization, and Assembly**. Simply by designing and synthesizing – in both the molecular and supramolecular sense – architectures from tens to thousands of nanometers, chemists can now start to investigate structure and function on a scale that holds a lot of promise since it is the one that living systems have evolved around and operate on with awe-inspiring efficiency and elegance.

The essay by *Rubin* and *Diederich* points to the **exciting prospects for organic synthesis** on the way **from fullerenes to novel carbon allotropes** (p 163). They highlight the on-going search for stable molecular and polymeric acetylenic carbon allotropes which has fostered the development of advanced synthetic methodologies in a remarkable manner during the last 15 years. The impetus for this research is two-fold – one is the sheer elegance of the acetylenic all-carbon and carbon-rich networks that can be constructed and the other is the promise that they and their fullerene cousins hold as providers of novel functional materials for emerging technologies. The Bonn group of *Gestermann, Hesse, Windisch*, and *Vögtle* feature another kind of novel molecular architecture – **dendritic architectures** (p 187) – that has blossomed in an amazing fashion during the same time-frame of the last 15 years. These novel architectures, not only give scientists access to homogenous macromolecules that span the nanometer scale almost effortlessly, but they are also amenable to endless compositional, structural, and functional variations. The dendrimer concept is one of the most profound and stimulating ideas to have entered the chemical arena in quarter of a century. Moreover, the implications for the technological scene are as broad as one's imagination can stretch. Another concept that has flourished over the last decade is **chemical encapsulation in self-assembling capsules** (p 199). In their essay on this topic, *Schalley* and *Rebek* extol the virtues of assembling capsules from curved subunits in a dynamic manner using noncovalent (hydrogen) bonding. Not only are guests bound reversibly but the host subunits are also exchanging in solution in these kinetically labile supermolecules. Their unique properties, which are characteristic of the complete assembly. have been expressed in terms of chiral recognition, catalysis, and the formation of novel hydrogen-bonded polymers and fibers. Mechanically-interlocked molecules – the catenanes and rotaxanes – form the basis for molecular switches and shuttles that could find applications in molecular electronics and other nanoscale devices. In their essay on **slippage and constrictive binding** (p 211), *Fyfe, Raymo*, and *Stoddart* draw an analogy between slipping appropriately-sized rings over dumbbells to make rotaxanes and the ingression of a guest into the cavity of a hemicarcerand to form a hemicarceplex. They are both 'examples of kinetically stable species that can be synthesized

noncovalently through the combined action of noncovalent bonding and mechanical coercion.' In their essay, *Holman* and *Ward* reveal how **crystal engineering with soft and topologically adaptable molecular host frameworks** (p 221) allows a crystalline lattice 'to achieve dense packing through low-energy deformations while retaining their inherent dimensionality and supramolecular connectivity.' Crystalline inclusion compounds based on two-dimensional hydrogen-bonded sheets comprising guanidinium cations and anionic organosulfonates exemplify the concept that could lead to a new generation of designer materials which could impact technologies such as optoelectronics, magnetics and chemical storage.

Design and Synthesis – which constitutes **Section I** (p 1) and pervades all the other (**II-IV**) sections – has always been and will continue to reside at the very heart of chemistry. After all, as many who have written on the subject previously have pointed out, it is what singles chemistry out from the other sciences – the opportunity to be creative in an abstract medium – and it is extremely adaptive to human needs. Take, for example, the challenge to decrease the use of harmful organic solvents in chemical processes, leading to environmentally friendly 'green chemistry' taken up by *Kobayashi* and *Manabe* in their essay entitled, **Lewis acid catalysis in aqueous media** (p 3). They record how aldol and Diels-Alder reactions – amongst others – can now be catalyzed by water-stable (rare earth) metal salts in water or water-containing solvents. Even asymmetric catalysis with moderately good enantiomeric excesses can be performed successfully in aqueous solution. A growing awareness of the need for 'green chemistry' is also one of the reasons why essayists *Noyori* and *Ikariya* extol the virtues of using **supercritical fluids for organic synthesis** (p 13). They remind us that 'solvent molecules associate with reactants, intermediates, and products as well as transition state structures in chemical reactions'. The unique properties (high density, low viscosity and large diffusivity) of supercritical fluids, such as supercritical carbon dioxide, means that they form a single phase mixture with gaseous reactants like hydrogen and carbon monoxide. The outcome is that the mass transfer rate-limiting step in reactions, such as catalytic (asymmetric) hydrogenation, hydroformylation, and carbonylation is avoided and the efficiencies of many reactions can be greatly enhanced under conditions that are rather simple to operate. On account of their unique solubility properties in supercritical carbon dioxide, fluorous compounds also have a special role to play in 'green chemistry'. In his essay on **fluorous tech-**

niques for the synthesis of organic molecules (p 25), *Curran* outlines **a unified strategy for reaction and separation**. The rendering of organic compounds fluorous by attaching a highly fluorinated domain to them is already transforming the science of separating the products of some reactions into another realm of perfection. Instead of time-consuming chromatography, liquid-liquid and solid-liquid extraction techniques can be used to separate efficiently fluorous compounds from other organic compounds. The strategy is also ideal for the production of combinatorial libraries solution phase parallel synthesis. It will have a major impact on both chemical process development and chemical discovery research. In their essay, *Tietze* and *Haunert* argue the case for the **domino reaction in organic synthesis** (p 39) as **an approach to efficiency, elegance, ecological benefit, economic advantage and preservation of our resources in chemical transformations**. The challenge to make complex molecules in as few steps as possible from simple precursors has led to domino reactions – 'combining bond-forming steps in one process under identical or nearly identical reaction conditions' – that have been employed in efficient synthesis of carbocycles and heterocycles, as well as of natural products. These domino reactions, which can involve combinations of pericyclic, radical, cationic, anionic and transition-metal catalyzed processes, are especially useful for doing combinatorial chemistry in solution. 'In the field of drug discovery, combinatorial chemistry has played an increasingly important role for identification and optimization of drug leads which target therapeutically important biomolecules' – so claim *Lee, Szewczyk*, and *Ellman* in their essay on **combinatorial libraries for drug development** (p 65). Historically, compounds with desired properties were stumbled upon serendipitously for the most part. Nowadays, the solution-phase and solid-phase synthesis of large collections of libraries of compounds from increasing well-chosen sets of precursors is changing, not only how we discover new drugs, but also how new materials and catalysts are uncovered. Speaking of catalysts, their rational design for organic reactions has often been based on the mimicry of biological catalysts. In their essay on **theozymes and catalyst design** (p 79), *Tantillo* and *Houk* advocate an alternative approach that identifies theoretical enzymes (theozymes) as catalyst models consisting of arrays of functional groups that can be optimized computationally so as to maximize transition state stabilities for catalytic reactions. Theozymes can then be used (1) to provide quantitative information on mechanisms of catalysis, (2) to improve the preferences of known catalysts

and (3) to design new synthetic catalysts (chemzymes). Imagine going one step further and devising automated procedures for sampling all possible arrangements of catalytic groups by a combinatorial approach. The catalytic array derived from such an approach that would identify the best of a family of theozymes has been called a compuzyme! **Enantioselective catalysis using sterically and electronically unsymmetrical ligands** (p 89) is the subject of an essay by *Humphries* and *Pfaltz* that reveals how enantioselective variants of allylic alkylation, the Heck reaction, and the hydrogenation of unfunctionalized olefins benefit significantly in terms of enantioselectivities from dispensing with C_2 symmetry in chiral bidentate ligands used in asymmetric catalysis. For example, when the classical chelating P/P- and N/N-ligands are replaced by chelating P/N ligands, asymmetric metal catalysts emerge that are considerably superior to their symmetric counterparts. Moreover, their modular architectures allow them to be fine tuned to perfection – or near to it (> 99 % enantiomeric excesses). In an essay on **asymmetric two-center catalysis** (p 105), *Shibasaki* draws attention to the fact that most synthetic asymmetric catalysts show limited activity in terms of either enantioselectivities or chemical yields – and sometimes both. The problem with these catalysts is that they activate only a small portion of the substrate in an intermolecular interaction or reaction. Taking a leaf out of the book of nature's catalysts suggests that, if chemists were to activate a substrate from both sides and control its orientation at the same time, there would be advantages to be gained in terms of enantioselectivities. Indeed, this concept has been exploited successfully in a wide range of important catalytic asymmetric reactions using, for example, heterobimetallic complexes which function simultaneously as a Lewis acid and a Brønsted base. Another successful approach to enhancing enantioselectivities during asymmetric catalysis is highlighted by *Shiori* and *Arai* in their essay on **asymmetric phase transfer catalysts** (p 123). Chiral onium (ammonium, phosphonium and arsonium) salts, including quaternary ones and chiral crown ethers (which form complexes with hard metal cations) can distribute themselves between heterogeneously different phases and mediate, in catalytic fashion, reactions producing new stereogenic centers. Despite the fact that 'Diels-Alder cycloaddition, aldol reaction, olefin epoxidation, olefin dihydroxylation, olefin metathesis and cyclopropanation are among the myriad of important transformations that now have catalytic enantioselective variants, ... the science of stereo- and regioselective organic synthesis is far from mature'. So writes

Hoveyda in his essay on **asymmetric catalysis in target-oriented synthesis** (p 145). The concept of catalysis-based total synthesis, in which a series of catalytic enantioselective reactions are employed in combination with other catalytic reactions, is emerging as the desirable way to make complex natural products and medicinally-important target compounds.

It is earnest hope of the editors of this collection of essays that it will be the first of many such collections. For one thing, this first collection has drawn almost entirely on the organic aspects of chemistry, reflecting the research interests of the trio of editors. Other areas of chemistry are equally rich in stimulating concepts. We have been most fortunate to have been supported throughout this venture by a group of highly professional technical editors in Weinheim. Anette Eckerle, Roland Kessinger, and Peter Biel deserve a hearty vote of thanks from all three of us. This project would not have gotten off the ground without the enthusiastic backing of Peter Gölitz. When it comes to putting chemistry into words and pictures, his creativity and imagination is only matched by his magic and experience. Finally, we would like to express our personal appreciation to the 45 authors who joined with us in a very positive manner to produce this collection of essays on *Stimulating Concepts in Chemistry*. We leave it to the readers to have the last word – and hopefully some new ideas.

Masakatsu Shibasaki/Tokyo
Fraser Stoddart/Los Angeles
Fritz Vögtle/Bonn

I Design and Synthesis

Lewis Acid Catalysis in Aqueous Media

Shū Kobayashi and Kei Manabe

Graduate School of
Pharmaceutical Sciences,
The University of Tokyo,
Hongo, Bunkyo-ku,
Tokyo 113–0033, Japan

Phone: 81-3-5841-4790
Fax: 81-3-5684-0634
Email: skobayas@mol.f.u
-tokyo.ac.jp

Keywords ■ Lewis Acids ■ Rare Earth Metal Triflate ■ Aldol Reactions ■ Aqueous Media ■ Chiral Catalysts ■ Surfactants ■ Colloidal Dispersions ■ Asymmetric Reactions ■ Mannich-Type Reactions ■ Environmentally Friendly Processes ■ Green Chemistry

Concept: Conventional Lewis acids are moisture-sensitive and easily deactivated by water. On the other hand, water-stable Lewis acid catalysts have been developed recently.

These Lewis acids function as catalysts in the presence of moisture and even in water. Asymmetric catalysis has also been performed successfully in aqueous solution.

Abstract: New types of Lewis acids as water-stable catalysts have been developed. Metal salts, such as rare earth metal triflates, can be used in aldol reactions of aldehydes with silyl enolates in aqueous media. These salts can be recovered after the reactions and reused. Furthermore, surfactant-aided Lewis acid catalysis, which can be used for aldol reactions in water without using any organic solvents, has been also developed. These reaction systems have been applied successfully to catalytic asymmetric aldol reactions in aqueous media. In addition, the surfactant-aided Lewis acid catalysis for Mannich-type reactions in water has been disclosed. These investigations are expected to contribute to the decrease of the use of harmful organic solvents in chemical processes, leading to environmentally friendly green chemistry.

Prologue

Lewis acid catalysis has been and continues to be of great interest in organic synthesis.[1] While various kinds of Lewis acid-promoted reactions have been developed and many have been applied in industry, these reactions must generally be carried out under strictly anhydrous conditions. The presence of even a small amount of water stops the reaction because most conventional Lewis acids react immediately with water, rather than with the substrates, and decompose. This destructive reaction has restricted the use of Lewis acids in organic synthesis. From a viewpoint of today's environmental consciousness, however, it is desirable to use water instead of organic solvents as a reaction solvent.[2,3]

In the course of our investigations to develop new synthetic methods, we have found that lanthanide triflates ($Ln(OTf)_3$) can be used as water-stable Lewis acids in water-containing solvents. Lanthanide compounds were expected to act as strong Lewis acids because of their hard character and to have strong affinities toward carbonyl oxygens. Furthermore, their hydrolysis was postulated to be slow, based on their hydration energies and hydrolysis constants.[4–6] In fact, while most metal triflates are prepared under strictly anhydrous conditions, $Ln(OTf)_3$ compounds are reported to have been made in aqueous solution.[7] The usefulness of these Lewis acids was first demonstrated by us in the hydroxymethylation of silyl enol ethers using commercial aqueous formaldehyde solution.[8] Among the $Ln(OTf)_3$ compounds tested for the reaction, ytterbium triflate ($Yb(OTf)_3$) was found to be the most effective catalyst (Eq. 1). It should be noted that only a catalytic amount of $Yb(OTf)_3$ was required to complete the reaction. These findings prompted us to investigate Lewis acid catalysis in aqueous media. Here, we present a brief overview of some of the recent research carried out in this area.

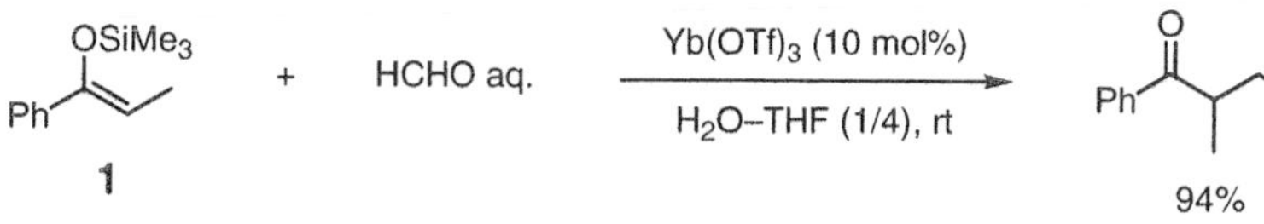

$$\text{(1)} \quad \underset{Ph}{\overset{OSiMe_3}{\diagup}} \quad + \quad \text{HCHO aq.} \quad \xrightarrow[\text{H}_2\text{O–THF (1/4), rt}]{\text{Yb(OTf)}_3 \text{ (10 mol\%)}} \quad \underset{94\%}{Ph\overset{O}{\diagdown}\diagup\diagdown OH}$$

Eq. 1.

Aldol Reactions in Water-Containing Solvents

Titanium tetrachloride-mediated aldol reactions of aldehydes with silyl enol ethers were first reported[9] in 1973. The reactions (Mukaiyama aldol reactions) are notable because they can be distinguished from the conventional aldol reactions carried out under basic conditions. Mukaiyama aldol reactions proceed in a highly regioselective manner to afford cross aldols in high yields.[10] Since this pioneering work, several efficient activators have been developed to realize high yields and selectivities, and now the reaction is considered to be one of the most important carbon–carbon bond-forming reactions in organic synthesis.[11] However, for the reaction, anhydrous solvents have to be used because of water-labile nature of the Lewis acids.

On the other hand, Ln(OTf)$_3$ compounds, which were found to be effective catalysts for the hydroxymethylation in aqueous media, also activate aldehydes other than formaldehyde in aldol reactions with silyl enol ethers in aqueous solvents.[12] One feature of the present reactions is that water-soluble

thane. Almost 100 % of Ln(OTf)$_3$ compounds are quite easily recovered from the aqueous layers after the reactions are complete and they can be reused. The reactions are usually quenched with water and the products are extracted with an organic solvent. The catalyst resides in aqueous layer and only removal of water gives the catalyst, which can be used in the next reaction (Scheme 1). No loss of the catalytic activity was observed, even after using the catalyst several times. It is noteworthy that Ln(OTf)$_3$ compounds are expected to solve some severe environmental problems induced by Lewis acid-promoted reactions in industrial chemistry.

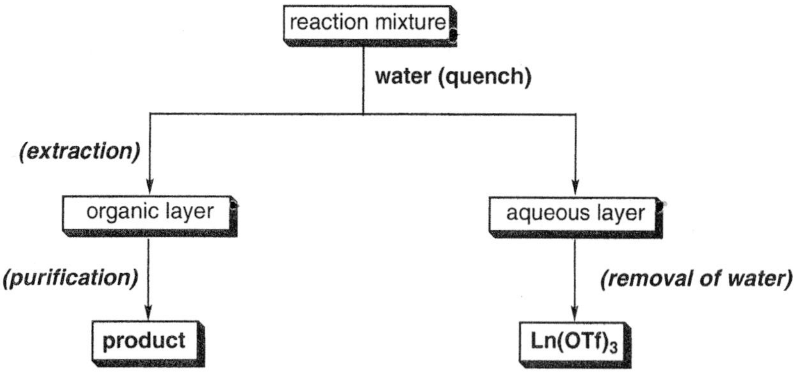

In the present Ln(OTf)$_3$-catalyzed aldol reactions in aqueous media, the amount of water strongly influences the yields of the aldol adducts. The effects of the amount of water on the yields in the model reaction of benzaldehyde with the silyl enol ether **2** in the presence of 10 mol% Yb(OTf)$_3$ in THF were investigated (Eq. 2). The best yields are obtained when the amount of water present in THF is in the range 10–20 %. When the amount of water is increased, the yield begins to decrease. The reaction system becomes a two phase one when the

Scheme 1. Recovery of the catalyst

aldehydes – for instance, acetaldehyde, acrolein, and chloroacetaldehyde – can be used directly in aqueous solutions for reactions with silyl enol ethers to afford the corresponding cross aldol adducts in high yields. Another striking feature of Ln(OTf)$_3$ compounds is that it is very easy to recover them from the reaction mixture. Ln(OTf)$_3$ are more soluble in water than in organic solvents, such as dichlorome-

amount of water is increased, and the yield decreases. Only 18 % of product was isolated from pure water. On the other hand, when water is not added or 1–5 eq. of water are added to Yb(OTf)$_3$, the yield of the desired aldol adduct is also low (ca. 10 % yield). The yield improves as the amount of water is increased to 6–10 eq., and when more than 50 eq. of water are added, the yield improves to more than

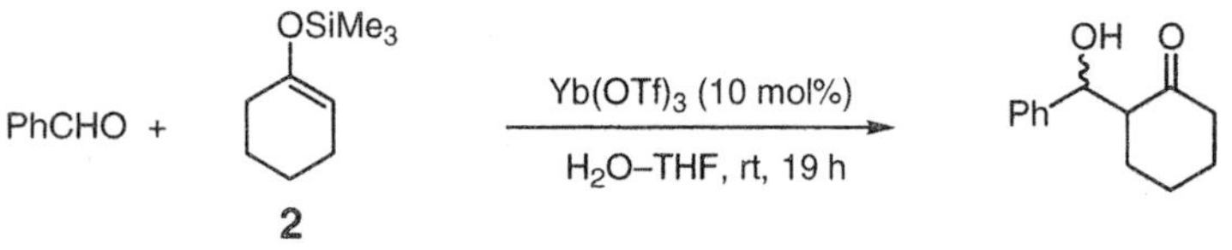

$$\text{PhCHO} \quad + \quad \underset{\textbf{2}}{\overset{OSiMe_3}{\bigcirc}} \quad \xrightarrow[\text{H}_2\text{O–THF, rt, 19 h}]{\text{Yb(OTf)}_3 \text{ (10 mol\%)}} \quad Ph\underset{}{\overset{OH \quad O}{\diagup\diagdown}}\bigcirc$$

Eq. 2.

80 %. These results indicate that water acts as, not only a co-solvent, but also a kind of activator in the reaction.

Scandium triflate ($Sc(OTf)_3$) was also found to be an effective catalyst in aldol reactions in aqueous media.[13] In many cases, $Sc(OTf)_3$ is more active than $Yb(OTf)_3$, as expected from the smaller ionic radius of Sc(III).

Stable Lewis Acids in Aqueous Media

In order to identify other Lewis acids stable in aqueous solvents, group 1–15 metal chlorides, perchlorates, and triflates were screened.[14] As a model, the reaction of benzaldehyde with the silyl enol ether **1** in H_2O–THF was selected. In the screening exercise, the salts of Fe(II), Cu(II), Zn(II), Cd(II), and Pb(II), as well as the rare earth metals [Sc(III), Y(III), Ln(III), all gave good yields. When the salts of B(III), Si(IV), and Ti(IV), etc. were used, decomposition of the silyl enol ether occurred rapidly and no aldol adduct was obtained. This happens because hydrolysis of such metal salts is very fast and the silyl enol ether is protonated then hydrolyzed to afford the corresponding ketone. On the other hand, no product or only trace amounts of the product was detected using the metal of Li(I), Na(I), and Mg(II), etc. Although these salts are stable in water, they have low catalytic activities. We noticed a correlation of their catalytic activity in aqueous media with hydrolysis constants (Kh) and exchange rate constants for substitution of inner-sphere water ligands [water exchange rate constant (WERC)].[4–6] Table 1

lists these constants. Elements surrounded by red squares represent effective cations for the aldol reaction. The metal compounds which were active in the reaction have pKh values in the range 4.3–10.08 and WERC greater than 3.2×10^6 $M^{-1}s^{-1}$. Cations are generally difficult to hydrolyze when their pKh values are large. In the case that pKh values are less than 4.3, cations are easy to hydrolyze and oxonium ions are formed. Under these conditions, silyl enol ethers decompose rapidly. On the other hand, in the cases when the pKh values are more than 10.08, the Lewis acidity of the cations are too low to catalyze the aldol reaction. WERC values should be large to act as effective Lewis acids in aqueous media, because large WERC values secure fast exchange between hydrated water molecules and an aldehyde which must coordinate to the metal cation to be activated.

Judging from these findings, the mechanism of Lewis acid catalysis in water (for example, aldol reactions of aldehydes with silyl enol ethers) can be assumed to be as follows. When metal compounds are added to water, the metals dissociate and hydration occurs immediately. At this stage, the intramolecular and intermolecular exchange reactions of water molecules frequently occur. If an aldehyde exists in the system, there is a chance that it will coordinate to the metal cations instead of the water molecules and the aldehyde is then activated. A silyl enol ether attacks this activated aldehyde to produce the aldol adduct. According to this mechanism, it is expected that many Lewis acid-catalyzed reactions should be successful in aqueous solutions. Although the precise activity as Lewis acids in aqueous media cannot be predicted quantitatively

Table 1. Hydrolysis constants[a] and exchange rate constants for substitution of inner-sphere water ligands[b]

1	2	3	4	5	6	7	8	9	10	11	12	13	14	15
Li^{+1} 13.64 4.7×10^7	*Be* — —											B^{+3} — —	*C* — —	*N* — —
Na^{+1} 14.18 1.9×10^8	Mg^{+2} 11.44 5.3×10^5											Al^{+3} 1.14 1.6×10^0	Si^{+4} — —	P^{+5} — —
K^{+1} 14.46 1.5×10^8	Ca^{+2} 12.85 5×10^7	Sc^{+3} 4.3 4.8×10^7	Ti^{+4} ≤2.3 —	V^{+3} 2.26 1×10^3	Cr^{+3} 4.0 5.8×10^{-7}	Mn^{+2} **10.59** 3.1×10^7	Fe^{+2} **9.5** 3.2×10^6	Co^{+2} 9.65 2×10^5	Ni^{+2} 9.86 2.7×10^4	Cu^{+2} 7.53 2×10^8	Zn^{+2} 8.96 5×10^8	Ga^{+3} 2.6 7.6×10^2	Ge^{+4} — —	*As* — —
Rb — —	*Sr* — —	Y^{+3} 7.7 1.3×10^7	Zr^{+4} 0.22 —	Nb^{+5} (0.6) —	Mo^{+5} — —	*Tc* — —	Ru^{+3} — —	Rh^{+3} 3.4 3×10^{-8}	Pd^{+2} 2.3 —	Ag^{+1} 12 $>5\times10^6$	Cd^{+2} 10.08 $>1\times10^8$	In^{+3} 4.00 4.0×10^4	Sn^{+4} — —	Sb^{+5} — —
Cs — —	Ba^{+2} 13.47 $>6\times10^7$	Ln^{+3} 7.6-8.5 10^6-10^8	Hf^{+4} 0.25 —	Ta^{+5} (-1) —	W^{+6} — —	Re^{+5} — —	Os^{+3} — —	Ir^{+3} — —	Pt^{+2} 4.8 —	Au^{+1} — —	Hg^{+2} 3.40 2×10^9	Tl^{+3} 0.62 7×10^5	Pb^{+2} 7.71 7.5×10^9	Bi^{+3} 1.09 —

La^{+3}	Ce^{+3}	Pr^{+3}	Nd^{+3}	*Fm*	Sm^{+3}	Eu^{+3}	Gd^{+3}	Tb^{+3}	Dy^{+3}	Ho^{+3}	Er^{+3}	Tm^{+3}	Yb^{+3}	Lu^{+3}
8.5	8.3	8.1	8.0	—	7.9	7.8	8.0	7.9	8.0	8.0	7.9	7.7	7.7	7.6
2.1×10^8	2.7×10^8	3.1×10^8	3.9×10^8	—	5.9×10^8	6.5×10^8	6.3×10^7	7.8×10^7	6.3×10^7	6.1×10^7	1.4×10^8	6.4×10^6	8×10^7	6×10^7

[a] $pKh = -\log K_{xy}$.

$$x M^{z+} + y H_2O \rightleftharpoons M_x(OH)_y^{(xz-y)+} + y H^+$$

$$K_{xy} = \frac{[M_x(OH)_y^{(xz-y)+}][H^+]^y}{[M^{z+}]^x} \cdot \frac{g_{xy}\, g_{H^+}^{\,y}}{g_{M^{z+}}^{\,x}\, a_{H_2O}^{\,y}}$$

[b] Measured by NMR, sound absorption, or multidentate ligand method.

by pKh and WERC values, these results have demonstrated the possibility of using several promising metal compounds as Lewis acid catalysts in water.

Aldol Reactions in Water Without Using Organic Solvents

While the Lewis acid-catalyzed aldol reactions in aqueous solvents described above are catalyzed smoothly by several metal salts, a certain amount of an organic solvent such as THF had still to be combined with water to promote the reactions efficiently. This requirement is probably because most substrates are not soluble in water. To avoid the use of the organic solvents, we have developed a new reaction system in which metal triflates catalyze aldol reactions in water with the aid of a small amount of a surfactant, such as sodium dodecyl sulfate (SDS).

The surfactant-aided Lewis acid catalysis was first noted[15] in the model reaction shown in Table 2. While the reaction proceeded sluggishly in the presence of 20 mol% Yb(OTf)$_3$ in water, remarkable enhancement of the reactivity was observed when the reaction was carried out in the presence of 20 mol% Yb(OTf)$_3$ in an aqueous solution of SDS (20 mol%, 35 mM). The corresponding aldol adduct was obtained in 50 % yield. The yield was improved when Sc(OTf)$_3$ was used as the Lewis acid catalyst. It was found that different kinds of surfactants influenced the product yield, and that TritonX-100, a neutral surfactant, was effective in the aldol reaction (but required long reaction time), while only a

trace amount of the adduct was detected when using a representative cationic surfactant, cetyltrimethylammonium bromide (CTAB).

A surprising feature of the Sc(OTf)$_3$–SDS system is that the ketene silyl acetal **3**, which is known to be hydrolyzed very easily in the presence of water, reacts with an aldehyde to afford the corresponding aldol adduct in a high yield (Eq. 3).

With these results in hand, we have next introduced new types of Lewis acids, e.g., scandium tris(-dodecyl sulfate) (**4a**) and scandium trisdodecanesulfonate (**5a**) (Chart 1).[16] These "Lewis acid–surfactant-combined catalysts (LASCs)" were found to form stable colloidal dispersions with organic substrates in water and to catalyze efficiently aldol reactions of aldehydes with very water-labile silyl enol ethers.

Chart 1.

$M(O_3SO\ C_{12}H_{25})_n$ $M(O_3S\ C_{12}H_{25})_n$

4a: M = Sc, n = 3	**5a:** M = Sc, n = 3	**5e:** M = Cu, n = 2
4b: M = Cu, n = 2	**5b:** M = Yb, n = 3	**5f:** M = Zn, n = 2
	5c: M = Mn, n = 2	**5g:** M = Na, n = 1
	5d: M = Co, n = 2	**5h:** M = Ag, n = 1

To investigate this LASC system in detail, we have synthesized dodecyl sulfate and dodecanesulfonate salts with various metal cations (Chart 1) and studied the effects of the metal cations on the catalysis of aldol reactions in water.[17,18] The catalysts were used in the aldol reactions of benzaldehyde (1 eq) with thioketene silyl acetal **6** (1.5 eq) in water (Eq. 4). Remarkable effects of the metal cations on catalytic activity can be observed from the reaction profiles (Fig. 1). The order of catalytic activity at the initial stage of the reaction is as follows: Cu (**5e**) > Zn (**5f**), Ag (**5h**) > Sc (**5a**), Yb (**5b**) > Na (**5g**) > Mn (**5c**), Co (**5d**). The Cu salt (**5e**) has the highest ability to catalyze the aldol reaction among the catalysts tested. However, the yield of **7** did not exceed 70 %, because **5e** accelerated not only the aldol reaction but also the hydrolysis of the thioketene silyl acetal **6**. The same trend was observed for the Zn and Ag salts (**5f, 5h**). On the other hand, the Sc and Yb salts (**5a, 5b**) afforded the aldol product **7** in > 90 % final yields, although the catalytic activities of **5a** and **5b** in the initial stages of the reaction were slightly lower

Table 2. Effect of M(OTf)$_3$ and surfactants

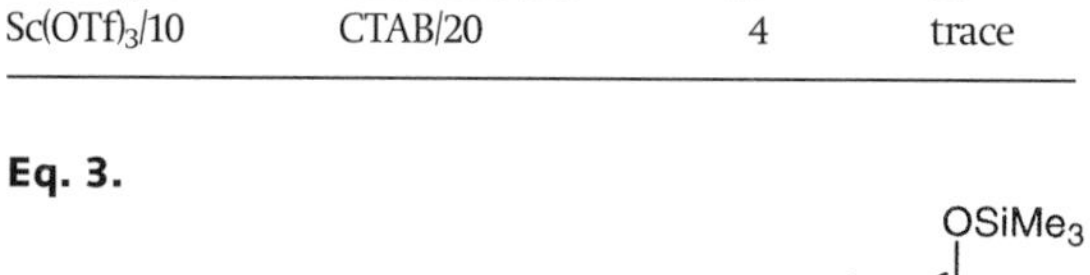

PhCHO + $\overset{\displaystyle OSiMe_3}{\underset{\displaystyle Ph}{\diagup}}$ **1** $\xrightarrow[\text{H}_2\text{O, rt}]{\text{cat. M(OTf)}_3 \text{ surfactant}}$ Ph $\overset{O\ \ OH}{\diagup}$ Ph

M(OTf)$_3$/mol%	Surfactant/mol%	Time/h	Yield/%
Yb(OTf)$_3$/20	–	48	17
Yb(OTf)$_3$/20	SDS/20	48	50
Sc(OTf)$_3$/10	SDS/20	4	88
Sc(OTf)$_3$/10	TritonX-100/20	60	89
Sc(OTf)$_3$/10	CTAB/20	4	trace

Eq. 3.

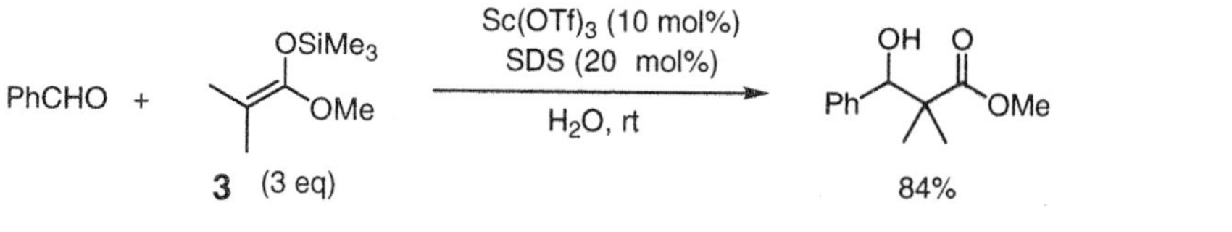

PhCHO + $\overset{\displaystyle OSiMe_3}{\underset{\displaystyle}{\diagup}}$ OMe $\xrightarrow[\text{H}_2\text{O, rt}]{\text{Sc(OTf)}_3 \text{ (10 mol\%)} \\ \text{SDS (20 mol\%)}}$ Ph $\overset{OH\ \ O}{\diagup}$ OMe

3 (3 eq)   84%

Eq. 4.

PhCHO + $\overset{\displaystyle OSiMe_3}{\underset{\displaystyle}{\diagup}}$ SEt $\xrightarrow[\text{H}_2\text{O, 30 °C}]{\text{LASC (10 mol\%)}}$ Ph $\overset{OH\ \ O}{\diagup}$ SEt

6   **7**

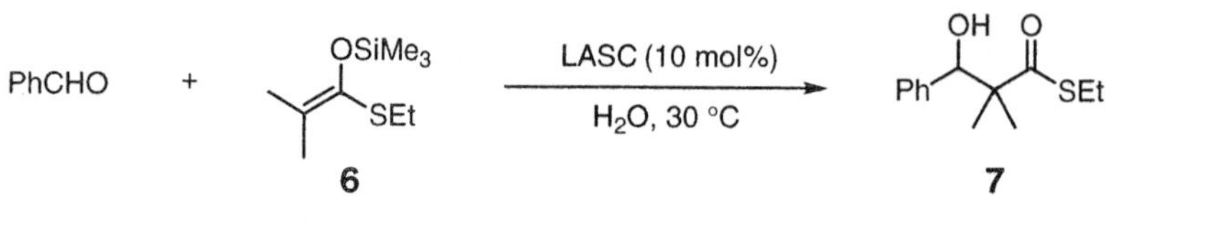

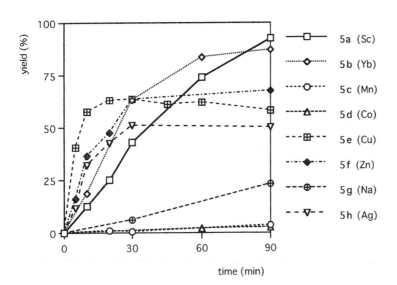

Figure 1. Plot of yield versus time for the aldol reactions in the presence of the dodecanesulfonate salts

than those of **5e**, **5f**, and **5h**. It should be noted that, in the dispersion system derived from **5a** and **5b**, the hydrolysis of the thioketene silyl acetal **6** was attenuated. Especially in the case of **5a**, a small amount of **6** still remained when the aldol reaction was complete. When the Na, Mn, and Co salts (**5g**, **5c**, **5d**) were used as catalysts, the aldol reactions proceeded very slowly, and the hydrolysis of **6** competed with the desired reaction, resulting in low yields of **7**.

During our investigations of the reactions mediated by LASCs, we have found that addition of a small amount of a Brønsted acid dramatically increased the rate of the aldol reaction (Eq. 5).[19] This cooperative effect of a LASC and an added Brønsted acid was also observed in the allylation of benzaldehyde with tetraallyltin in water.[20] Although, from a mechanistic point of view, little is known about the real catalytic function of scandium and proton, this cooperative effect of a Lewis acid and a Brønsted acid provides a new methodology for efficient catalytic systems in synthetic chemistry.

Catalytic Asymmetric Aldol Reactions in Aqueous Media

Catalytic asymmetric aldol reactions provide one of the most powerful carbon-carbon bond-forming processes affording synthetically useful, optically active β-hydroxy ketones and esters.[21,22] Chiral Lewis acid-catalyzed reactions of aldehydes with silyl enol ethers are the most convenient and promising methods, and several successful examples have been reported since the first chiral tin(II)-catalyzed reactions appeared in 1990.[23,24] Common characteristics of these catalytic asymmetric reactions include (1) the use of aprotic anhydrous solvents such as dichloromethane, toluene, and propionitrile, and (2) low reaction temperatures (–78 °C), which are also employed in many other catalytic asymmetric reactions.

In the course of our investigations to develop new chiral catalysts and catalytic asymmetric reactions in water, we focused on several elements whose salts are stable and behave as Lewis acids in water. In addition to the findings of the stability and activity of Lewis acids in water related to hydration constants and exchange rate constants for substitution of inner-sphere water ligands of elements (cations) (see above), it was expected that undesired achiral side reactions would be suppressed in aqueous media and that desired enantioselective reactions would be accelerated in the presence of water. Moreover, besides metal chelations, other factors such as hydrogen bonds, specific solvation, and hydrophobic interactions are anticipated to increase enantioselectivities in such media.

After the screening of chiral Lewis acids which could be used in aqueous solvents, a combination of

Eq. 5.

PhCHO + (structure with OSiMe$_3$, SEt)

LASC (10 mol%)
+
Brønsted Acid (10 mol%)
H$_2$O
23 °C, 5 min

→ Ph (structure with OH, O, SEt)

LASC = **5a**; Brønsted acid = none: 10% yield
LASC = none; Brønsted acid = HCl: 0% yield
LASC = **5a**; Brønsted acid = HCl: 67% yield

Table 3. Catalytic asymmetric aldol reactions in aqueous media

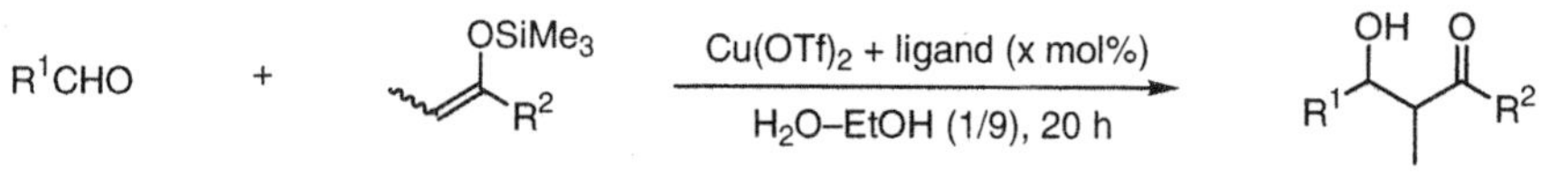

$$R^1CHO + \overset{OSiMe_3}{\underset{R^2}{\diagdown}} \xrightarrow[\text{H}_2\text{O–EtOH (1/9), 20 h}]{\text{Cu(OTf)}_2 + \text{ligand (x mol\%)}} \overset{OH\ O}{R^1\diagup\diagdown R^2}$$

R^1	R^2	E/Z	Ligand/mol%	Temp/°C	Yield/%	*syn/anti*	ee/% (syn)
Ph	Ph	Z^a	20	−10	74	3.2/1	67[b]
Ph	Et	Z^a	20	−15	81	3.5/1	81
Ph	Et	E^c	20	−10	32	1.6/1	32
Ph	i-Pr	Z^d	20	0	17	4.0/1	85
Ph	i-Pr	Z^d	20	5	95	4.0/1	77
p-ClPh	Et	Z^a	10	−10	88	2.6/1	76
2-naphthyl	i-Pr	Z^d	20	−10	97	4.0/1	81
2-furyl	Et	Z^a	20	−10	86	4.0/1	76

[a] E/Z = <1/99. [b] (2S, 3S). [c] E/Z = 77/23. [d] E/Z = 2/98.

ligand =

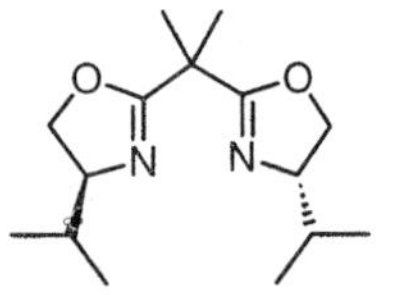

copper triflate [Cu(OTf)$_2$] and bis(oxazoline) ligands was found to give good enantioselectivities. Several examples of the catalytic asymmetric aldol reactions of aldehydes with silyl enol ethers are summarized in Table 3.[25]

It should be noted that the reaction of benzaldehyde with (Z)-3-trimethylsiloxy-2-pentene in ethanol or dichloromethane in the presence of the chiral catalyst resulted in a much lower yield and selectivity. On the basis of these results, we propose the catalytic cycle shown in Scheme 2. The catalyst **A** formed from Cu(OTf)$_2$ and a bis(oxazoline) ligand accelerates the aldol reaction to generate the intermediate **B**. In aqueous solvents, **B** is rapidly hydrolyzed to produce the aldol product **C** and regenerate A. On the other hand, **B** is converted to the corresponding silyl ether **D** in anhydrous organic solvents. If this catalyst regeneration step is rather slow, the silyl group of **B** and/or other Lewis acidic silyl species such as silyl triflate, which is generated form **B** and triflate anion, catalyze the aldol reaction to afford an almost racemic adduct and, as a result, reduce the enantioselectivity. In fact, it has been crucial for the success of the reported catalytic asymmetric aldol reactions in aprotic organic solvents to suppress this undesired pathway.[22] The rapid hydrolysis of **B** in the present aqueous system is, therefore, crucial, not only for the high yields but also for the high enantioselectivities.

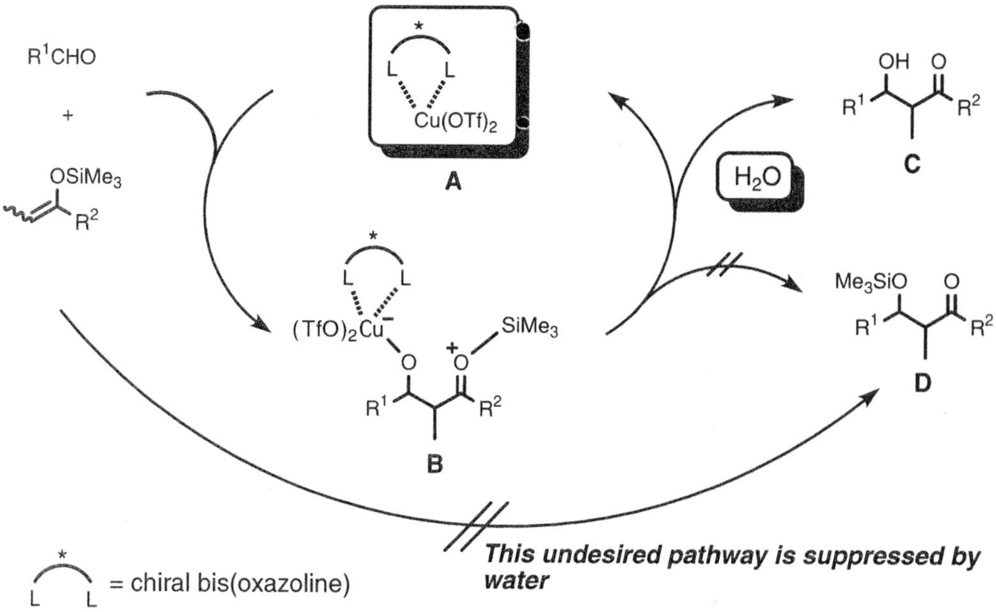

Scheme 2. Assumed catalytic cycle of aldol reactions in water

Eq. 6.

76% yield (*syn/anti* = 2.8/1), 69% ee (*syn*)

The catalytic asymmetric aldol reaction has been applied to the LASC system, which uses copper bis(-dodecyl sulfate) (**4b**) instead of Cu(OTf)$_2$.[26] An example is shown in Eq. 6. In this case, a Brønsted acid, such as lauric acid, is necessary to obtain a good yield and enantioselectivity. This example is the first one involving Lewis acid-catalyzed asymmetric aldol reactions in water without using organic solvents. Although the yield and the selectivity are still not yet optimized, it should be noted that this appreciable enantioselectivity has been attained at ambient temperature in water.

Mannich-Type Reactions in Water

Mannich and related reactions provide one of the most fundamental and useful methods for the synthesis of β-amino carbonyl compounds, which constitute various pharmaceuticals, natural products, and versatile synthetic intermediates.[27] Conventional protocols for three-component Mannich-type reactions of aldehydes, amines, and ketones in organic solvents include some severe side reactions and have some substrate limitations, especially for enolizable aliphatic aldehydes. The direct synthesis of β-amino ketones from aldehydes, amines, and silyl enolates under mild conditions is desirable from a synthetic point of view. Our working hypothesis was that aldehydes could react with amines in a hydrophobic reaction field created in water in the presence of a catalytic amount of a metal triflate and a surfactant, to produce imines, which could then react with hydrophobic silyl enolates.

A model reaction of benzaldehyde, *o*-methoxyaniline, and 1-phenyl-1-trimethylsiloxyethene was performed in the presence of 5 mol% of Sc(OTf)$_3$ in an aqueous solution of SDS (20 mol%).[28] The reaction proceeded smoothly at room temperature to afford the corresponding β-amino ketone derivative in 87% yield (Eq. 7). It should be noted that the dehydration (imine formation) and the coupling reaction between two water-unstable substrates, the imine and the silyl enolate, occur successfully in water, and that only a trace amount of the product is obtained without SDS under the same reaction conditions. Other catalysts such as Yb(OTf)$_3$ and Cu(OTf)$_2$ were also found to be effective in this reaction. Side reaction adducts, such as deamination and aldol products, were not obtained at all in aqueous media. On the other hand, when the same reaction was carried out in dichloromethane, the yield of the desired product decreased to 66% and the deamination product (chalcone) was obtained in 22% yield. The product was readily converted into the corresponding β-amino ketone. Thus, treatment of the products with cerium ammonium nitrate (CAN) in acetonitrile–water (9:1) at room temperature induced smooth deprotection of the 2-methoxyphenylamino group.[29]

Quite recently, not only Lewis acids, but also Brønsted acids were found to be effective catalysts for the three-component Mannich-type reactions in water with the aid of a surfactant. For example, Akiyama and co-workers[30] have reported that a combination of HBF$_4$ and SDS is effective for the reactions of aldehydes, amines, and silyl enolates. We have found that dodecylbenzenesulfonic acid (DBSA), a Brønsted acid with a surfactant moiety, also catalyzes the reactions in water.[31] Furthermore, DBSA can be used for the "direct" Mannich-type reactions of aldehydes, amines, and ketones, without using silyl enolates as nucleophilic components (Eq. 8).[32]

Eq. 7.

(1 eq)   (1 eq)   (1.5 eq)   87%

$$\text{PhCHO} + \text{PhNH}_2 + \underset{(5\ eq)}{\text{cyclohexanone}} \xrightarrow[\text{H}_2\text{O, 23 °C, 1 h}]{\text{DBSA (1 mol\%)}} \text{product}$$

Eq. 8.

PhCHO + PhNH₂ + (cyclohexanone)

(1 eq)   (1 eq)   (5 eq)

97%

Strecker-Type Reactions in Water

Strecker reactions provide one of the most efficient methods for the synthesis of α-amino nitriles, which are useful intermediates in the synthesis of amino acids and nitrogen-containing heterocycles. Although classical Strecker reactions have some limitations, use of trimethylsilyl cyanide (TMSCN) as a cyano anion source provides promising and safer routes to these compounds.[33–35] Consequently, we focused our attention on tributyltin cyanide (Bu₃SnCN), because Bu₃SnCN is stable in water and is also a potential cyano anion source. Indeed, the Strecker-type reactions of aldehydes, amines, and Bu₃SnCN proceeded smoothly in water (Eq. 9).[36] It should be noted that no surfactants are required in this reaction. Furthermore, Complete recovery of the toxic tin compounds is also possible in the form of bis(tributyltin) oxide after the reaction is over. Since conversion of bis(tributyltin) oxide to tributyltin cyanide is known in the literature, this procedure provides a solution to the problem associated with toxicity of tin compounds.

$$\text{PhCHO} + \text{PhCH}_2\text{NH}_2 + \text{Bu}_3\text{SnCN}$$

$$\xrightarrow[\text{H}_2\text{O, rt, 20 h}]{\text{Sc(OTf)}_3\ (10\ mol\%)}$$

Eq. 9.

88%

Epilogue

Since our first paper[8] on Lewis acid catalysis in aqueous media appeared, many investigations and results in this area have been reported. Water-stable Lewis acids are now becoming common and useful catalysts in organic synthesis. These catalysts have been applied to various types of Lewis acid-catalyzed reactions.

Diels–Alder reactions are one of the most famous examples which are accelerated by a Lewis acid. Various water-stable Lewis acids such as Ln(OTf)₃,[37] methylrhenium trioxide,[38] copper nitrate,[39] copper bis(dodecyl sulfate) (4b),[40] indium chloride,[41] and bismuth triflate[42] have been used for Diels-Alder and aza-Diels–Alder reactions in water. Furthermore, a catalytic asymmetric Diels–Alder reaction in water using a copper complex of an amino

acid has been reported recently.[43] Michael reactions of β-keto esters[44] and also of α-nitro esters[45] with α,β-unsaturated carbonyl compounds were found to be catalyzed by Yb(OTf)₃ in water. The usefulness of organometallic reagents are now well-recognized in organic synthesis. Recently, much attention has been focused on the reactions of allyl organometallics with carbonyl compounds in water-containing solvents.[46] By using a combination of Sc(OTf)₃ and SDS, we have attained Lewis acid-catalyzed allylations in water.[47] The LASC/Brønsted acid system can be also used for allylations.[20] Allylations of aldehydes with tetraallylgermane are catalyzed efficiently by Sc(OTf)₃ in aqueous nitromethane.[48]

Various metal salts can function as Lewis acids in aqueous media. The Lewis acids described in this essay are expected to lead to new types of catalysts and hopefully provide some solutions to environmental problems. At a fundamental level, investigations on reactions in aqueous media are providing a much fuller understanding of the role of water in chemical reactions. At a practical level, these investigations are contributing to the development of "greener" reaction processes. Chemistry in water is also "green" chemistry.

References and Notes

1. *Selectivities in Lewis Acid Promoted Reactions* (Ed. D. Schinzer), Kluwer Academic Publishers, Dordrecht, **1989**.
2. *Organic Synthesis in Water* (Ed. P. A. Grieco), Blackie Academic and Professional, London, **1998**.
3. C.-J. Li, "Organic Reactions in Aqueous Media – With a Focus on Carbon–Carbon Bond Formation" *Chem. Rev.* **1993**, *93*, 2023–2035.
4. C. F. Baes, Jr., R. E. Mesmer, *The Hydrolysis of Cations*, John Wiley & Sons, New York, **1976**.
5. K. B. Yatsimirksii, V. P. Vasil'ev, *Instability Constants of Complex Compounds*, Pergamon, New York, **1960**.
6. *Coordination Chemistry* (Ed. A. E. Martell), ACS Monograph 168, American Chemical Society, Washington, DC, **1978**.
7. J. H. Forsberg, V. T. Spaziano, T. M. Balasubramanian, G. K. Liu, S. A. Kinsley, C. A. Duckworth, J. J. Poteruca, P. S. Brown, J. L. Miller, "Use of Lanthanide(III) Ions as Catalysts for the Reactions of Amines with Nitriles" *J. Org. Chem.* **1987**, *52*, 1017–1021.
8. S. Kobayashi, "Lanthanide Trifluoromethanesulfonates as Stable Lewis Acids in Aqueous Media. Yb(OTf)₃ Catalyzed Hydroxymethylation Reaction of Silyl Enol Ethers with Commercial Formaldehyde Solution" *Chem. Lett.* **1991**, 2187–2190.

9. T. Mukaiyama, K. Narasaka, T. Banno, "New Aldol Type Reaction" *Chem Lett.* **1973**, 1011–1014.

10. T. Mukaiyama, "The Directed Aldol Reaction" *Org. React.* **1982**, *28*, 203–331.

11. T.-H. Chan, "Formation and Addition Reactions of Enol Ethers" In *Comprehensive Organic Synthesis* (Ed. B. M. Trost), Pergamon Press, New York, **1991**, vol. 2, pp. 595–628.

12. S. Kobayashi, I. Hachiya, "The Aldol Reaction of Silyl Enol Ethers with Aldehydes in Aqueous Media" *Tetrahedron Lett.* **1992**, 1625–1628.

13. S. Kobayashi, "Scandium Triflate in Organic Synthesis" *Eur. J. Org. Chem.* **1999**, 15–27.

14. S. Kobayashi, S. Nagayama, T. Busujima, "Lewis Acid Catalysts Stable in Water. Correlation between Catalytic Activity in Water and Hydrolysis Constants and Exchange Rate Constants for Substitution of Inner-Sphere Water Ligands" *J. Am. Chem. Soc.* **1998**, *120*, 8287–8288.

15. S. Kobayashi, T. Wakabayashi, S. Nagayama, H. Oyamada, "Lewis Acid Catalysis in Micellar Systems. Sc(OTf)$_3$-Catalyzed Aqueous Aldol Reactions of Silyl Enol Ethers with Aldehydes in the Presence of a Surfactant" *Tetrahedron Lett.* **1997**, *38*, 4559–4562.

16. S. Kobayashi, T. Wakabayashi, "Scandium Tridodecylsulfate (STDS). A New Type of Lewis Acid That Forms Stable Dispersion Systems with Organic Substrates in Water and Accelerates Aldol Reactions Much Faster in Water Than in Organic Solvents" *Tetrahedron Lett.* **1998**, *39*, 5389–5392.

17. K. Manabe, S. Kobayashi, "Effects of Metal Cations in Lewis Acid–Surfactant-Combined Catalyst-Mediated Aldol Reactions in Water" *Synlett* **1999**, 547–548.

18. K. Manabe, Y. Mori, S. Kobayashi, "Effects of Lewis Acid–Surfactant-Combined Catalysts on Aldol and Diels–Alder Reactions in Water" *Tetrahedron* **1999**, *55*, 11203–11203.

19. K. Manabe, S. Kobayashi, "Remarkable Enhancement of Reactivity by Brønsted Acids in Aldol Reactions Mediated by Lewis Acids–Surfactant-Combined Catalysts in Water" *Tetrahedron Lett.* **1999**, *40*, 3773–3776.

20. K. Manabe, Y. Mori, S. Nagayama, K. Odashima, S. Kobayashi, "Synthetic Reactions Using Organometallics in Water. Aldol and Allylation Reactions Catalyzed by Lewis Acid–Surfactant-Combined Catalysts/ Brønsted Acids Systems" *Inorg. Chim. Acta* in press.

21. H. Gröger, E. M. Vogl, M. Shibasaki, "New Catalytic Concepts for the Asymmetric Aldol Reaction" *Chem. Eur. J.* **1998**, *4*, 1137–1141.

22. S. G. Nelson, "Catalyzed Enantioselective Aldol Additions of Latent Enolate Equivalents" *Tetrahedron: Asymmetry.* **1998**, *9*, 357–389.

23. T. Mukaiyama, S. Kobayashi, H. Uchiro, I. Shiina, "Catalytic Asymmetric Aldol Reaction of Silyl Enol Ethers with Aldehydes by the Use of Chiral Diamine Coordinated Tin(II) Triflate" *Chem. Lett.* **1990**, 129–132.

24. S. Kobayashi, Y. Fujishita, T. Mukaiyama, "The Efficient Catalytic Asymmetric Aldol-Type Reaction" *Chem. Lett.* **1990**, 1455–1458.

25. S. Kobayashi, S. Nagayama, T. Busujima, "Chiral Lewis Acid Catalysis in Aqueous Media. Catalytic Asymmetric Aldol Reactions of Silyl Enol Ethers with Aldehydes in a Protic Solvent Including Water" *Chem. Lett.* **1999**, 71–72.

26. S. Kobayashi, Y. Mori, S. Nagayama, K. Manabe, "Catalytic Asymmetric Aldol Reactions in Water Using a Chiral Lewis Acid–Surfactant-Combined Catalyst" *Green Chem.* **1999**, *1*, 175–177.

27. E. F. Kleinman, "The Bimolecular Aliphatic Mannich and Related Reactions" In *Comprehensive Organic Synthesis* (Ed. B. M. Trost), Pergamon Press, New York, **1991**, vol. 2, pp. 893–951.

28. S. Kobayashi, T. Busujima, S. Nagayama, "Ln(OTf)$_3$- or Cu(OTf)$_2$-Catalyzed Mannich-Type Reactions of Aldehydes, Amines, and Silyl Enolates in Micellar Systems. Facile Synthesis of β-Amino Ketones and Esters in Water" *Synlett.* **1999**, 545–546.

29. D. R. Kronenthal, C. Y. Han, M. K. Taylor, "Oxidative *N*-Dearylation of 2-Azetidinones. *p*-Anisidine as a Source of Azetidine Nitrogen" *J. Org. Chem.* **1982**, *47*, 2765–2768.

30. T. Akiyama, J. Takaya, H. Kagoshima, "One-Pot Mannich-Type Reaction in Water: HBF$_4$ Catalyzed Condensation of Aldehydes, Amines, and Silyl Enolates for the Synthesis of β-Amino Carbonyl Compounds" *Synlett.* **1999**, 1426–1428.

31. K. Manabe, Y. Mori, S. Kobayashi, "A Brønsted Acid–Surfactant-Combined Catalyst for Mannich-Type Reactions of Aldehydes, Amines, and Silyl Enolates in Water" *Synlett.* **1999**, 1401–1402.

32. K. Manabe, S. Kobayashi, "Mannich-Type Reactions of Aldehydes, Amines, and Ketones in a Colloidal Dispersion System Created by a Brønsted Acid–Surfactant-Combined Catalyst in Water" *Org. Lett.* **1999**, *1*, 1965–1967.

33. I. Ojima, S. Inaba, K. Nakatsugawa, "A New Route to Aminonitriles via Cyanosilylation of Schiff Bases and Oximes" *Chem. Lett.* **1975**, 331–334.

34. K. Mai, G. Patil, "Facile Synthesis of α-Aminonitriles" *Tetrahedron Lett.* **1984**, *25*, 4583–4586.

35. S. Kobayashi, H. Ishitani, M. Ueno, "Facile Synthesis of α-Amino Nitriles Using Lanthanide Triflate as a Lewis Acid Catalyst" *Synlett.* **1997**, 115–116.

36. S. Kobayashi, T. Busujima, S. Nagayama, "Scandium Triflate-Catalyzed Strecker-Type Reactions of Aldehydes, Amines, and Tributyltin Cyanide in Both Organic and Aqueous Solutions. Achievement of Complete Recovery of the Tin Compounds toward Environmentally Friendly Chemical Processes" *Chem. Commun.* **1998**, 981–982.

37. L. Yu, D. Chen, P. G. Wang, "Aqueous Aza Diels–Alder Reactions Catalyzed by Lanthanide(III) Trifluoromethanesulfonates" *Tetrahedron Lett.* **1996**, *37*, 2169–2172.

38. Z. Zhu, J. H. Espenson, "Aqueous Catalysis: Methylrhenium Trioxide (MTO) as a Homogeneous Catalyst for the Diels–Alder Reaction" *J. Am. Chem. Soc.* **1997**, *119*, 3507–3512.

39. S. Otto, J. B. F. N. Engberts, "Lewis-Acid Catalysis of a Diels–Alder Reaction in Water" *Tetrahedron Lett.* **1995**, *36*, 2645–2648.

40. S. Otto, J. B. F. N. Engberts, J. C. T. Kwak, "Million-Fold Acceleration of a Diels–Alder Reaction due to Combined Lewis Acid and Micellar Catalysis in Water" *J. Am. Chem. Soc.* **1998**, *120*, 9517–9525.

41. T.-P. Loh, J. Pei, M. Lin, "Indium Trichloride (InCl$_3$) Catalyzed Diels–Alder Reaction in Water" *Chem. Commun.* **1996**, 2315–2316.

42. H. Laurent-Robert, C. Le Roux, J. Dubac, "Enhancement of Dienophilic and Enophilic Reactivity of the Glyoxylic Acid by Bismuth(III) Triflate in the Presence of Water" *Synlett.* **1998**, 1138–1140.

43. S. Otto, G. Boccaletti, J. B. F. N. Engberts, "A Chiral Lewis-Acid-Catalyzed Diels–Alder Reaction. Water-Enhanced Enantioselectivity" *J. Am. Chem. Soc.* **1998**, *120*, 4238–4239.

44. E. Keller, B. L. Feringa, "Ytterbium Triflate Catalyzed Michael Additions of β-Ketoesters in Water" *Tetrahedron Lett.* **1996**, *37*, 1879–1882.

45. E. Keller, B. L. Feringa, "Highly Efficient Ytterbium Triflate Catalyzed Michael Additions of α-Nitroesters in Water" *Synlett.* **1997**, 842–844.

46. A. Lubineau, J. Auge, Y. Queneau, "Carbonyl Additions and Organometallic Chemistry in Water" in *Organic Synthesis in Water* (Ed. P. A. Grieco), Blacky Academic and Professional, London, **1998**, pp.102–140.

47. S. Kobayashi, T. Wakabayashi, H. Oyamada, "Use of an Organometallic Reagent in Water: $Sc(OTf)_3$-Catalyzed Allylation Reactions of Aldehydes in Micellar Systems" *Chem. Lett.* **1997**, 831–832.

48. T. Akiyama, J. Iwai, "Scandium Trifluoromethanesulfonate-Catalyzed Chemoselective Allylation Reactions of Carbonyl Compounds with Tetraallylgermane in Aqueous Media" *Tetrahedron Lett.* **1997**, *38*, 853–856.

Supercritical Fluids for Organic Synthesis

Ryoji Noyori[a] and Takao Ikariya[b]

[a] Department of Chemistry
and Research Center for
Material Science, Nagoya
University, Chikusa, Nagoya
464–8602, Japan
Phone: + 81 52–789 2956,
Fax: + 81 52 783 4177,
Email: noyori@chem3.chem.
nagoya-u.ac.jp

[b] Graduate School of Science
and Technology, Tokyo
Institute of Technology, and
CREST, Meguro-ku, Tokyo
152–8552, Japan
Phone: + 81 3 5734 2636
Fax: + 81 3 5734 2637
e-mail: tikariya@o.cc.titech.
ac.jp

Concept: Solvent molecules associate with reactants, intermediates, and products as well as transition structures in chemical reactions. A high reaction rate and desired selectivity are accessible only in a suitable environment that facilitates the selective formation of a product-determining molecular assembly and its transformation. Practical synthetic reactions must be effective and operationally simple. Supercritical fluids are emerging reaction media of scientific interest and with technical merits.

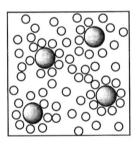

○ : Solute molecule

○ : Solvent molecule

Abstract: Any pure compound becomes supercritical when compressed to a pressure higher than the critical pressure above the critical temperature. The unique properties of supercritical fluids such as high density, low viscosity, and large diffusivity can be utilized for homogeneous catalysis. A small amount of a functional additive can dramatically change the attributes. Supercritical fluids form a single-phase mixture with gaseous reactants including H_2 and CO, avoiding a mass-transfer rate-limiting step in reactions such as hydrogenation, hydroformylation, or carbonylation. Thus, reaction rate and selectivity are readily tunable by a subtle change in the additives or conditions, such as pressure and temperature. Supercritical carbon dioxide (scCO$_2$) is a nontoxic, nonflammable alternative to hazardous organic solvents for certain reactions. Examples include highly productive hydrogenation of CO$_2$, asymmetric hydrogenation of functionalized olefins, and carbonylation of aryl halides.

Keywords: ■ Asymmetric Hydrogenation ■ Carbon Dioxide ■ Carbonylation ■ Dimethylformamide ■ Enantioselectivity ■ Formic Acid ■ Homogeneous Hydrogenation ■ Palladium Catalysts ■ Radical Reactions ■ Ruthenium Catalysts ■ Supercritical Fluids ■ Solvent Replacement

Prologue

Why are synthetic reactions performed largely in liquid solutions? The reason is that reactants dissolve in solvents to achieve homogeneity and to facilitate mass transport and molecular (ionic) collision.[1] However, certain chemical reactions also occur in a gas phase and in a solid state at high rates and with unique selectivities. Changes in the reaction phase sometimes increase the efficiency enormously. There exists a diverse array of chemical reactions. Figure 1 schematically illustrates the reaction phases: liquid, supercritical, and gas. In the liquid phase, solvent molecules are packed densely to form a solvation sphere, where there is continuous rotation or exchange of the solvent shell molecules. In the gas phase, molecules have high mobility and interact without the formation of rigid aggregation of these molecules. Supercritical fluid (SCF) with the beneficial effects of both liquid- and gas-phase chemistry is an emerging reaction medium for many scientific and technical reasons.[2] The appropriate combination of microscopic molecular science and macroscopic chemical engineering exploits the benefits in each chemical process. Thus, SCFs provide a great opportunity to enhance chemical reactivity and to attain desired selectivity.[3–6]

Why Supercritical Fluids?

All stable pure compounds have a triple point and a critical point. The critical point is the endpoint of the liquid–gas line in the phase diagram and the point where the liquid and gas phases become indistinguishable. Any gaseous compound becomes supercritical when compressed to a pressure higher than the critical pressure (P_c) above the critical temperature (T_c). Figure 2 shows photographs

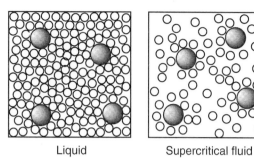

Liquid   Supercritical fluid   Gas

Figure 1. Schematic reaction phases

 : Solute molecule   O : Solvent molecule

of the interior of a window-equipped vessel under various conditions; the left photograph shows gently boiling liquid and gaseous CO_2 below the critical point and the right one shows only the one fluid phase of $scCO_2$, where the meniscus separating liquid and gas disappears above the critical point. The critical points of some selected compounds are illustrated in Figure 3. Carbon dioxide has a T_c of 31 °C and a P_c of 73 atm, and hence, its supercritical state ($scCO_2$) is readily accessible. Water with T_c = 374.0 °C and P_c = 217.8 atm and methane with T_c = –82.6 °C and P_c = 45.4 atm are offscale.

Physical properties of SCFs are very different from those of ordinary gases and liquids and are strongly and uniquely influenced simply by changing the pressure and temperature. Table 1 compares the density, viscosity, and diffusivity of SCFs with those of typical liquids and gases. Notably, the density varies by a factor of 10^3 at conditions close to the critical point from 1 (liquid) to 10^{-3} g cm^{-3} (gas), implying that the average molecular distance can change by a factor of 10. An increase in the pressure leads to the continuous increase in density as visualized in Figure 4. Thus intermediate densities,

Table 1. Typical liquid, supercritical fluid (SCF), and gas properties.

	Liquid	SCF	Gas
Density (g cm^{-3})	1	0.1–0.5	10^{-3}
Viscosity (Pa · s)	10^{-3}	10^{-4}–10^{-5}	10^{-5}
Diffusivity (cm^2 s^{-1})	10^{-5}	10^{-3}	10^{-3}

which are impossible below the T_c, can be obtained in the supercritical region. The density dependent properties including the dielectric constant (ε), viscosity, and overall solvent strength are also tunable through changes in the pressure and temperature.

Most SCFs with T_c <100 °C are very nonpolar. The few exceptions are CHF_3, CH_3F, CO, and PF_3. Figure 5 shows the dielectric constant of $scCHF_3$; the value of the polar compound is variable and is far from "constant".[8] Such large changes in the dielectric constant cannot be attained by using a single conventional liquid solvent. The viscosity of SCFs (10^{-4} to 10^{-5} Pa·s) is between gas (10^{-5} Pa·s) and liquid (10^{-3} Pa·s) viscosities, and then the diffusivity is generally much higher than those of liquids (10^{-3} vs 10^{-5} cm^2 s^{-1}). Consequently, the reduction of the cage effect

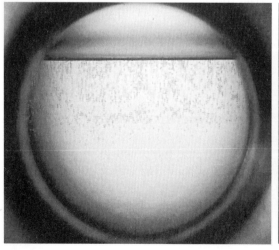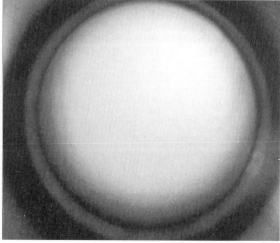

Figure 2. Photographs indicating gently boiling liquid CO_2 (left) and supercritical CO_2 (right)

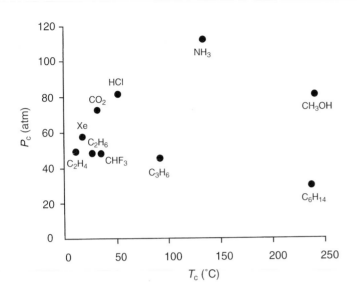

Figure 3. The critical points of selected pure compounds. Water (374.0 °C and 217.8 atm) and methane (−82.6 °C and 45.4 atm) are offscale

could be important in radical reactions involving paired radicals or diffusion-limited reactions.[9] Moreover, the high compressibility near the critical point often results in very large negative activation volumes. These unusual pressure tunable properties of SCFs certainly affect the profile of certain chemical reactions leading to higher reaction rates and/or selectivity with potentially small changes in the pressure and temperature.[2–7, 9–11]

The solvating power of SCFs is much smaller than that of conventional liquid solvents, but this in turn can be utilized to generate unique molecular clusters or assemblies in a homogenous phase.[12,13] The local inhomogeneity of SCFs is among the most noteworthy, although scientific knowledge of this aspect remains limited. Figure 6 shows schematic illustration of the inhomogeneity, clustering molecules, where the solvation shell is formed around a solute molecule in an SCF. The number of the mole-

cules in the shell is strongly dependent on the density of the SCF.[13] Such phenomena have already been recognized in spectroscopic studies, and the same effects are expected to change chemical reactivity and selectivity. Experimentally, the solvent effects, molecular interactions, and phase behavior in ground-state chemistry are evaluated quantitatively under high-pressure and high-temperature SCF conditions by IR, near-IR, Raman, UV-visible, and NMR spectroscopy.[14] Solvatochromism is useful to empirically evaluate solvent polarity.

Solvent molecules interact with reactants, reactive intermediates, and products as well as transition structures. The efficiency of intermolecular reactions is controlled by various thermodynamic and kinetic factors, particularly the concentration of a product-determining molecular complex and its reactivity. However, the requisite molecular association is often prevented by the strong association

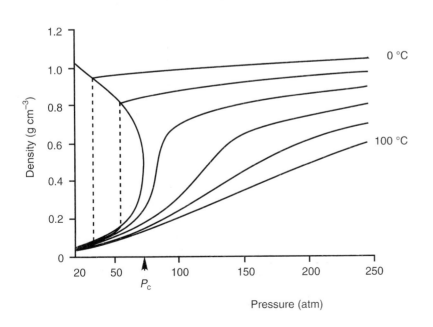

Figure 4. Carbon dioxide density depending on pressure and temperature

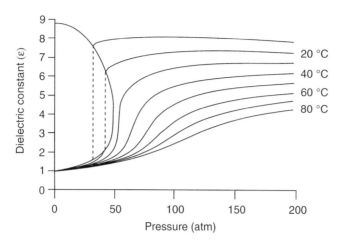

Figure 5. The variable dielectric constant of CHF$_3$ with changes in pressure and temperature

of solvent molecules to a reactant or catalyst molecule. A high rate is accessible by stabilizing the transition state rather than the ground state, but the solvent effect in an ordinary homogenous solution is often opposite. This is the origin of the relatively low reaction rate in comparison to that of gas-phase reactions. Many enzymatic reactions as well as chemical reactions in zeolite matrices proceed rapidly owing to the lack of solvation to substrates. This weakening the solvation process may be better achieved in SCFs because of their tunable physical properties.[9] Reactant/substrate complexes might be formed locally in high concentration because of the local clustering or the weak solvation. Alternatively, if one can stabilize the transition state selectively, the reactivity should be greatly increased.

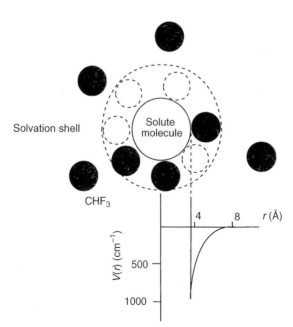

Figure 6. A simple model for the aggregation of supercritical fluid molecules around a polar solute molecule. The potential value, $V(r)$, of interaction between benzonitrile and CHF$_3$ molecules is also given.[13]

Recent progress in theoretical chemistry allows for the reliable prediction of transition state structures in vacuum-phase reactions. The precise knowledge of properties of the loose molecular assembly allows experimentalists to rationally select suitable additives that stabilize the structure by forming molecular clusters. Such molecular manipulation using non-covalent interactions such as hydrogen bonding, electrostatic interaction, or π-stacking is much easier in SCFs than in a liquid phase where solvent molecules are packed more densely to form a solvation sphere.[12,13] Here, for example, a calculated electron-density map would help visually to find the way to accelerate the reaction.[12] In SCF, as in the gas phase, molecules have high mobility and diffusivity and interact without the formation of rigid molecular aggregates. The deliberate, external control of transition state stability is a dream of synthetic chemists.

In principle, the combination of a bulk SCF and a small amount of a functional additive(s) acting as reaction promoter can create a range of unique reaction media. The permutability of this approach is unlimited. The performance of molecular catalyses can also be improved in SCFs by knowing the detailed mechanistic pathway including the structures of the catalytic species as well as the turnover-limiting and product-determining steps.[6] Many gaseous reactants such as H$_2$ and CO are sparingly soluble in liquid solvents. However, SCFs form a single-phase mixture with such reactants, sometimes avoiding a mass-transfer rate-limiting step and thus enhancing reaction rates. Selectivity of some reactions is strongly affected by solvent-cage, which is also adjustable with SCFs.[9]

Characteristic properties of SCFs include variable density, dielectric constant, viscosity, and cage strength, local inhomogeneity, high diffusion rate, high miscibility with gaseous substances, and high sensitivity of all properties to added substances. All of these properties influence the thermodyanamic and kinetic parameters of reaction. The reaction rate and selectivity are tunable with very small changes in operating conditions.

Furthermore, many SCFs possess economical, technical, environmental, and health advantages. The high volatility of CO_2, for example, allows it to be completely and easily removed from the product, resulting in an overall "solvent-free" reaction. $scCO_2$ is a promising alternative to hazardous organic solvents.

What is Supercritical Carbon Dioxide?

$scCO_2$ is the most commonly used SCF. A large body of phase and solubility data has been published for binary mixtures of $scCO_2$ and organic compounds. The dipolarity and polarizability of CO_2, as measured using solvatochromic dyes by the Kamlet–Taft π^* parameter, ranges from –0.52 to –0.60 for $scCO_2$ and –0.34 to –0.46 for liquid CO_2, slightly lower than that of alkanes, 0.0, and fluorocarbons, –0.33 to –0.41. The polarity for liquid and $scCO_2$, measured on the $E_T(30)$ scale, is 31–33, which is comparable to the value of 31 for alkanes.

Because CO_2 is a nonpolar molecule, it dissolves a wide variety of nonpolar organic compounds having high volatility, low polarity, and no protic functional groups. In addition, fluoropolymers which are soluble only in chlorofluorocarbons dissolve in $scCO_2$ probably due to a specific interaction with the fluorinated groups.[15] This implies that $scCO_2$ can replace environmentally hazardous halogenated solvents. On the other hand, its ability to dissolve polar, ionic, or polymeric compounds is exceedingly limited. However, a small quantity of a polar entrainer or an appropriate surfactant drastically changes the microenvironment and, consequently, greatly increases the solubility of such substances.[6] Normally volatile polar or protic compounds such as small alcohols, acetone, or amines are used. Perfluorinated compounds are also effective for this purpose. Water in $scCO_2$ forms microemulsions stabilized by an ammonium carboxylate perfluoropolyether surfactant.[16]

For organometallic catalysis, the solubility of transition metal complexes is often increased by ligand modifications. Figure 7 summarizes typical examples of the ligand modifications that have been reported. Replacement of aryl substituents in organic ligands by alkyl or alkoxy groups and introduction of "CO_2-philic" polyfluoroalkyl or silicone groups significantly increases the solubility in $scCO_2$. Although ionic metal complexes are generally insoluble in $scCO_2$, a suitable choice of counter ions can greatly improve the solubility. $[(3,5-(CF_3)_2C_6H_3)_4B]^-$ and $CF_3SO_3^-$ are the typical counter anions for cationic Rh and Ir complexes.[6, 7, 17–20] Because the solubility of catalyst molecules or reactants is crucial for homogeneous reactions, solubility considerations set a lower limit on the pressures which can be used. Figure 8 illustrates the solubility of $Fe_3(CO)_{12}$ in $scCO_2$ showing the importance of pressure. The solubility is obviously affected by pressure and temperature. Below a "crossover pressure",[6] increasing the temperature decreases the solubility of compounds because the SCF density decreases, whereas, above this pressure, increasing the temperature enhances the solubility of the solute because the volatility of the solute increases. Catalysts and products can be separated or recovered by changes in the temperature as well as the pressure. Experimental data concerning phase behavior of organic compounds have been accumulating. However, complete phase diagrams are available for only a few binary mixtures of solutes and $scCO_2$ or scH_2O. At this moment therefore, one must determine the phase behavior at operating conditions to elucidate the SCF effects.

$scCO_2$ offers a range of technical advantages. It is an inexpensive, stable, nonflammable, and nontoxic alternative to hazardous organic solvents for certain reactions. Large-scale chemical manufacturing is facing a serious solvent problem in connection with environmental concerns. Regulation of the use of hazardous organic solvents such as chlorinated hydrocarbons or benzene is becoming increasingly stringent and spurs the development of environmentally conscious, economical reaction media, that are "greener" media for chemical synthesis.[18] $scCO_2$ has excellent potential for achieving this urgent goal. The use of $scCO_2$ allows facile separation of reactants, catalysts, and products. Venting volatile CO_2 after a reaction avoids the presence of solvent residues. $scCO_2$ can also be used to efficiently recover nonvolatile organic compounds from a room-temperature ionic liquid such as 1-butyl-2-methylimidazolium hexafluorophosphate.[21] Thus it may eventually be used as a substitute for environmentally unacceptable solvents.

Reactions in Supercritical Carbon Dioxide

Performance of synthetic reactions is highly sensitive to reaction media. However, the general importance of SCFs in homogeneous catalysis was

- Alkyl or partially hydrogenated arylphosphine ligands

- Alkoxy- or (dialkylamino)phosphines ligands

$PdCl_2[P(OR)_3]_2$ $PdCl_2[P(NR_2)_3]_2$

R = C_2H_5, C_6H_5 R = CH_3, C_2H_5

- Polyfluorinated ligands

Figure 7. Typical examples of ligand modifications to improve the solubility of complexes in scCO$_2$

ignored until the early 1990s. Since the first use of scCO$_2$ by Rathke for hydroformylation of olefins,[22] a number of researchers became interested in this unique reaction media. We describe here some examples taken from our research repertoire which could spur the further development of this important field.

Hydrogenation of CO$_2$ to Formic Acid Derivatives

CO$_2$ is the most abundant and cheapest C$_1$ resource. Now a considerable amount of CO$_2$ is being converted to organic bulk chemicals directly or indirectly via CO. Hydrogenation of CO$_2$ to produce useful organic substances could be a promising synthetic method in the 21st century, only if H$_2$ becomes accessible economically from water by solar or hydrothermally powered electrolysis. We found that the hydrogenation of scCO$_2$ proceeds

with a high turnover number (TON, mole of product per mole of catalyst) and a high turnover frequency (TOF, defined as TON h^{-1}).[23] Figure 9 shows a schematic diagram for the equipment used for scCO$_2$ hydrogenation. Reaction of H$_2$ (85 atm) and CO$_2$ (125 atm, total gas pressure 210 atm) in the presence of triethylamine (5.0 mmol) and RuH$_2$[P(CH$_3$)$_3$]$_4$ (0.3 µmol) at 50 °C resulted in rapid production of a formic acid/triethylamine adduct with a TOF of 680 h^{-1} (Figure 10). RuCl$_2$[P(CH$_3$)$_3$]$_4$ had an induction period but was very active overall. Addition of a trace amount of water (0.1 mmol) or methanol (13 mmol) greatly increased the TOF to 1400 and >4000 h^{-1}, respectively, although addition of too much protic compounds resulted in drastic rate decreases due to the formation of a separate liquid phase.[24] The marked enhancement of the reaction rate is possibly attributed to the operation of a metal–ligand bifunctional catalysis in which the transition state is stabilized by a hydrogen bond as shown in Figure 10. The reaction does not involve

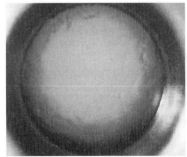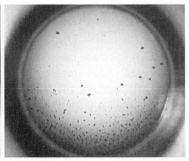

Figure 8. Fe$_3$(CO)$_{12}$ is soluble in scCO$_2$ at 35 °C and 100 atm, providing a green-colored fluid (left photograph). The green color of the fluid deepens as the pressure is raised to 200 atm, (middle photograph). When the pressure is decreased to atmospheric pressure, black green crystals are deposited on the walls and the windows of the reaction vessel (right photograph)

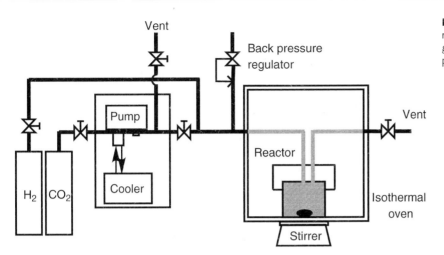

Figure 9. Reaction equipment used for CO_2 hydrogenation in the supercritical phase

ligation of the CO_2 molecule to the Ru center but proceeds via a possible six-membered pericyclic mechanism using the other coordination sphere.

When the hydrogenation was performed with methanol at an elevated temperature, 80 °C, a mixture of methyl formate and formic acid was produced with TON as high as 3500.[24] The use of biphasic conditions, viz., liquid methanol, increased the rate of hydrogenation of $scCO_2$. Use of ammonia and primary or secondary amines in place of triethylamine gave the corresponding formamides in high yield. For example, reaction of $scCO_2$, H_2, and dimethylamine gave DMF with up to 420 000 TON and 99 % selectivity (Figure 10).[24] The formamide synthesis differs from the synthesis of formic acid and methyl formate in that two phases are present from the start. The dimethylamine and CO_2 form an insoluble liquid dimethylammonium dimethylcarbamate. Because the Ru catalyst is insoluble in

this liquid salt and also in the aqueous product phase that forms in a later stage of the reaction, we suppose that the hydrogenation occurs in the supercritical phase. The high rate and productivity are probably due to the high concentration of H_2 in $scCO_2$, rapid mass transfer between phases, and high reactivity of the catalyst in the supercritical phase as well as favorable phase control. The high TON demonstrates the long catalyst lifetime possible under these conditions. Later, Baiker showed that $RuCl_2[P(CH_3)_3]_4$ heterogenized on a sol-gel-derived silica matrix was very active for the production of DMF, giving a TON of 110 800. $RuCl_2[(C_6H_5)_2PCH_2CH_2P(C_6H_5)_2]$ has proved to be a practical catalyst giving a TON of 740 000 at the initial 18 % conversion after 2 h.[25]

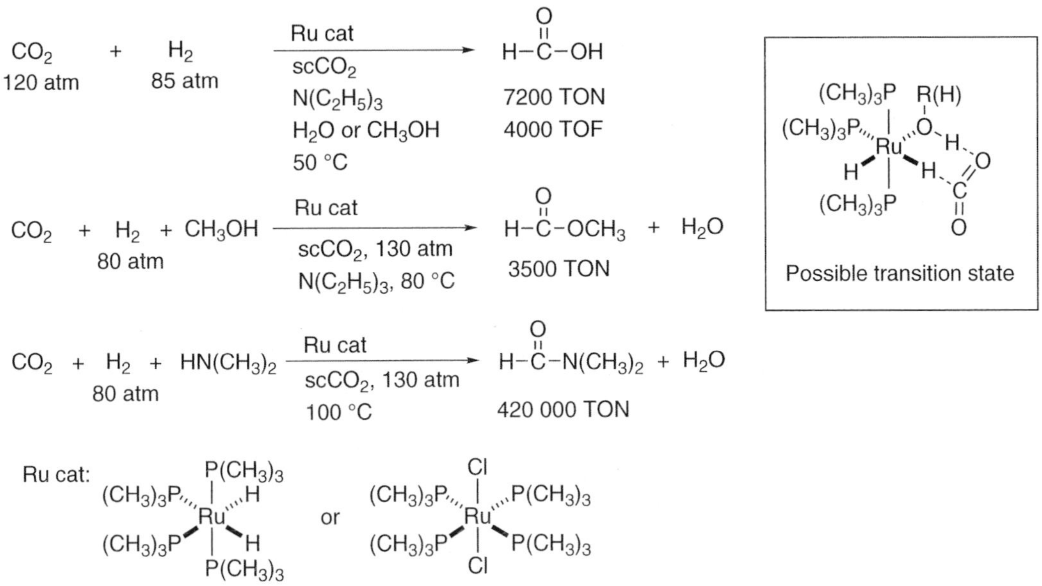

Figure 10. Syntheses of formic acid and its derivatives via CO_2 hydrogenation catalyzed by a Ru–$P(CH_3)_3$ complex in the supercritical phase. TON = mol product mol catalyst, TOF = TON h

Asymmetric Hydrogenation

Certain olefinic substrates can be hydrogenated in an enantioselective manner in $scCO_2$. Because the extent of stereoselectivity of asymmetric hydrogenation is often influenced by hydrogen concentration,[26] the high miscibility of H_2 in $scCO_2$ might affect the stereochemical outcome. We found that tiglic acid was hydrogenated with the CO_2-soluble $Ru(OCOCH_3)_2[(S)-H_8$-binap] catalyst at 30 atm H_2 and 50 °C to give the (S)-product in 81% ee (Figure 11). The ee value is similar to that obtained in methanol, 82%, and higher than that in hexane. Upon reduction of the hydrogen pressure, the ee increased in methanol but not in $scCO_2$ probably because the hydrogen concentration in $scCO_2$ is already very high. Addition of $CF_3(CF_2)_6CH_2OH$ to the reaction in $scCO_2$ increased the product ee to 89%.[27]

The enantioselectivity of asymmetric hydrogenation of tiglic acid in $scCHF_3$ is influenced not only by the H_2 partial pressure but also by the pressure of the SCF, as shown in Figure 11. The product ee decreased from 90 to 84% with a decrease in the pressure of CHF_3 from 188 to 82 atm, while the H_2 pressure was maintained at 12 atm. This change is well correlated to the change of the dielectric constant of the solvent as seen in conventional liquid solvents.[6] The dielectric constant at high pressures

is around six, but three near the critical pressure. The marked H_2 pressure dependency of enantioselectivity in methanol is interpreted in terms of the coexistence of a highly stereoselective monohydride mechanism and a less selective polyhydride mechanism. The increase in enantioselectivity with the increase of polarity of the SCF phase is ascribable to the enhanced contribution of the monohydride mechanism which requires heterolytic cleavage of H_2 by the Ru catalyst.[26]

Asymmetric hydrogenation of α-(acylamino) acrylic esters in $scCO_2$ has been studied by Tumas and Burk (Figure 12). The reaction was effected with a cationic Et-DuPHOS–Rh catalyst under a partial H_2 pressure of 14 atm (total pressure 340 atm) to give the chiral amino acid derivatives. The enantioselectivity is normally comparable to that obtained in conventional alcoholic solvents, while some substrates are hydrogenated with a better stereoselectivity.[19] The key of the success is the use of the phosphine ligand having alkyl substituents and BArF⁻ or $CF_3SO_3^-$ as a counter anion of the Rh cation, both of which solubilize the transition metal complex in $scCO_2$.

Pfaltz and Leitner found that some imines are hydrogenated enantioselectively at a partial H_2 pressure of 30 atm and 40 °C in $scCO_2$ (d = 0.75 g mL^{-1}) containing a chiral phosphinodihydrooxazole–Ir complex to give optically active amines in up to 81%

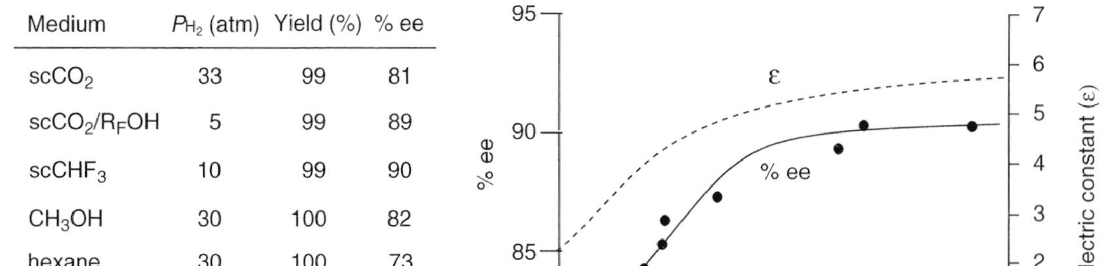

Medium	P_{H_2} (atm)	Yield (%)	% ee
$scCO_2$	33	99	81
$scCO_2/R_FOH$	5	99	89
$scCHF_3$	10	99	90
CH_3OH	30	100	82
hexane	30	100	73

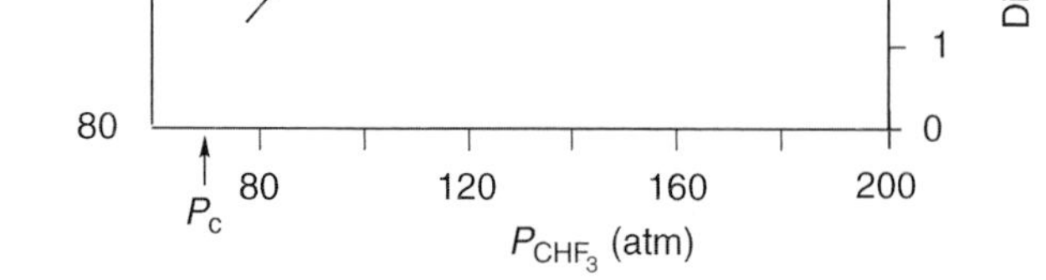

Figure 11. Asymmetric hydrogenation of tiglic acid with (S)-H_8-BINAP–Ru in $scCO_2$ and other solvents. The ee values depend on the pressure of $scCHF_3$. Reaction conditions: cat 4.4–4.7 μmol, S/C = 150–160, P_{CHF_3} = 170–180 atm, P_{CHF_2} = 82–188 atm, time 12–15 h, R_FOH = $CF_3(CF_2)_6CH_2OH$ 1.5 mmol

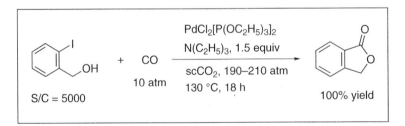

Figure 12. Asymmetric hydrogenation of α-enamides catalyzed by chiral Rh complexes in scCO₂

Chiral Rh cat:

$[(R,R)$-Et-DuPHOS–Rh](BArF)
$[(R,R)$-Et-DuPHOS–Rh](CF$_3$SO$_3$)

ee. The cationic catalyst again had BArF⁻ as a counter anion. The ee value was independent of hydrogen pressure in a range of 10 to 40 atm. The enantioselectivity is similar to that observed in dichloromethane, while the initial TOF exceeded 2800 h⁻¹ which is ca. two times higher than that attained in organic solvents. Therefore, cleaner scCO₂ serves as a promising alternative to dichloromethane which is a useful but environmentally unfavorable chlorinated solvent.[20]

Carbonylation of Aryl Halides

scCO₂ is a suitable medium for Pd-catalyzed carbonylation of aryl halides, where the use of CO₂-soluble PdCl₂[P(OC₂H₅)₃]₂ is crucial to obtain high

efficiency.[28] The solubility test showed that 5.1 mg of the Pd complex dissolves in 300 atm of scCO₂ in a 50-mL reactor at 130 °C. The carbonylative cyclization of 2-iodobenzyl alcohol in a supercritical mixture of CO₂ (200 atm) and CO (10 atm) containing the Pd catalyst and triethylamine at 130 °C gave the phthalide product with a TON of 5000 after 18 (Figure 13). The TON is independent of the CO pressure because of the high solubility of CO in scCO₂. The reaction occurs even with 1 atm of CO pressure (CO:substrate = 1.5:1) giving the product in near 100 % yield (TON = 4650). The reaction is faster than in toluene solution and is completed within 3–4 h after a ca. 2-h induction period. The carbonylation of iodobenzene with a substrate/catalyst molar ratio of 500 in scCO₂ containing methanol gave methyl benzoate with TON = 260.

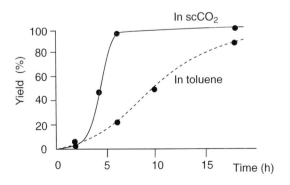

Figure 13. Carbonylation of 2-iodobenzyl alcohol catalyzed by PdCl₂[P(OC₂H₅)₃]₂ in scCO₂ and in toluene. Reaction conditions: S/C = 5000, Pd cat 1.0 × 10⁻² M DMF solution, P_{CO} = 10 atm, P_{CO_2} = 200 atm, 130 °C, toluene 50 mL

Supercritical Fluid as Mechanistic Probe

In some cases, reactions in scCO$_2$ can shed light on mechanisms. Reaction of a cyclopropene derivative with MnH(CO)$_5$ is performed either catalytically in the presence of H$_2$ and CO or stoichiometrically in the absence of H$_2$ and CO. As illustrated in Figure 14, both hydrogenation and hydroformylation had been thought to proceed by a radical mechanism via cage escape and cage collapse.[29] According to this mechanism, a solvent with a weak cage effect should increase selectivity for hydrogenation over hydroformylation. In fact, however, the reaction in scCO$_2$ at 200 atm and 60 °C gave the hydrogenation product with 66 % selectivity, which is close to the selectivity in pentane or hexane, 63–66 %, but is much higher than that obtained in a micellar solution in water 8 % at 50 °C. The result suggests that the solvent cage strength of scCO$_2$ is comparable to that of liquid alkanes or, more likely, that the hydroformylation occurs by a nonradical pathway which is independent of solvent viscosity.

Free-Radical Carbonylation of Alkyl Halides in scCO$_2$

The utility of SCFs is not limited to organotransition metal chemistry. scCO$_2$ is a promising medium for free-radical carbonylation of organic halides to ketones or aldehydes. As shown in Figure 15A, the silane-mediated reaction of 1-iodooctane, acrylonitrile, and CO in scCO$_2$ (total CO$_2$ 300 atm) containing 2,2'-azobis(isobutyronitrile) (AIBN) as a radical initiator affords ketones in an excellent yield.[30] The reaction involves formation of *n*-octyl radical which is followed by carbonylation, addition of the acyl radical to acrylonitrile, and quenching by the silane. The yields are comparable to or higher than those attained in benzene. The carbonylation occurs even at 2 atm of CO pressure with CO:halide = ca. 3:1, giving the product in 40 % yield because of the high solubility of CO in scCO$_2$. The reaction of alkyl halides without olefins under otherwise identical conditions gives aldehydes in comparable yield to those attained in benzene. Thus scCO$_2$ can replace environmentally unacceptable benzene which is one of the most useful solvents for this reaction.

When 4-hexenyl iodide was used as a substrate for the reductive carbonylation with [(CH$_3$)$_3$Si]$_3$SiH in scCO$_2$ (CO 50 atm, total CO$_2$ 330 atm, 80 °C), the five-membered and six-membered cyclic ketones were produced in a 1:2.2 molar ratio (Figure 15B). Decreasing the total CO$_2$ pressure to 180 atm resulted in a significant increase of the five-membered product (cyclopentanone:cyclohexanone = 1:1.1) possibly because of weakening of the cage effects at lower densities of CO$_2$. The unsaturated acyl radical intermediate cyclizes to give the exocyclic and endocyclic β-keto radicals. The 5-*exo-trig* process is kinetically more favored over the 6-*endo-trig* cyclization which gives the more stable six-membered rad-

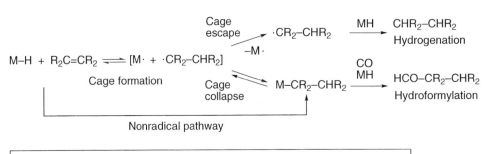

Figure 14. Reaction of 3,3-dimethyl-1,2-diphenylcyclopropene with MnH(CO)$_5$ in scCO$_2$

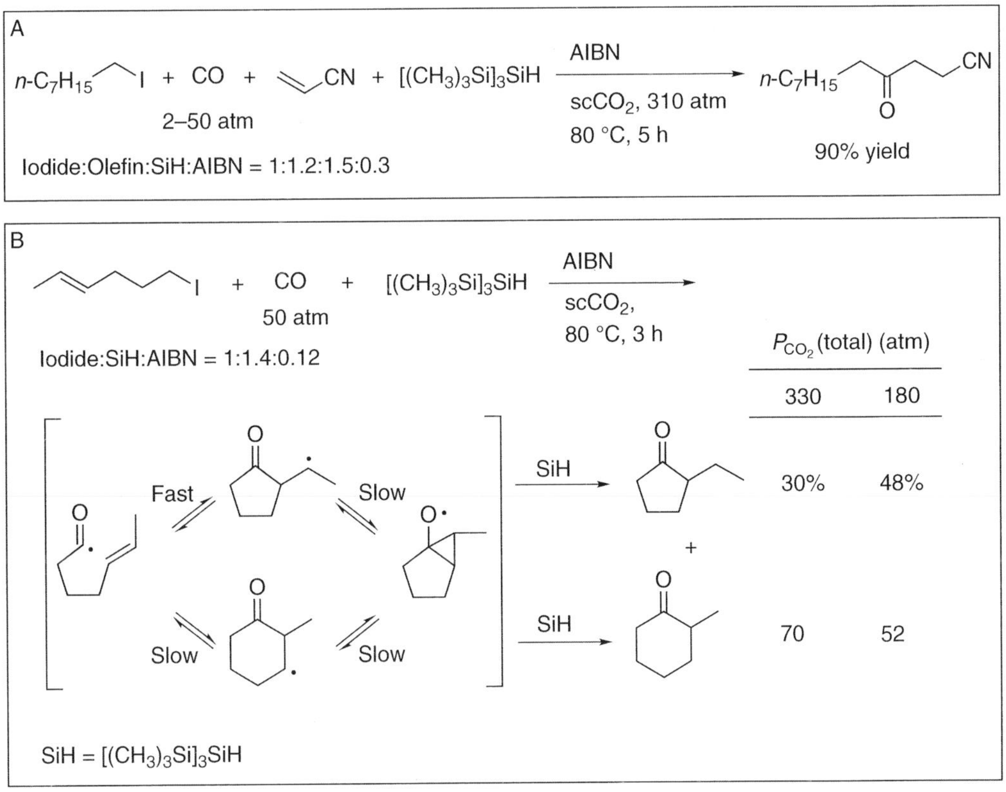

Figure 15. Silane-mediated free-radical carbonylations of alkyl halides in scCO$_2$

ical. Prior to reaction with a silane, the initially formed exocyclic radical slowly isomerizes to the ring-expanded isomer in a solvent cage. Under a lower CO$_2$ pressure condition with a weaker cage effect, the contribution of the kinetically favored exocyclic radical is increased.

Epilogue

Utilizing the adjustable solvent power of SCFs, the decaffeination of coffee and tea, the extraction of hops, spices, and drugs in extraction technology, and chromatography technology are all performed on an industrial scale. In the last five years, research on catalysis and synthetic reactions in scCO$_2$ has demonstrated that SCFs present a series of opportunities for the control of reactivity and selectivity of chemical processes. The ideal catalytic reactions would have high rates because of the high solubility of reactants gases in SCFs, large diffusivity, and weak catalyst or substrate solvation. The advances have so far been pronounced for reactions involving gaseous reagents, such as hydrogenation, hydroformylation, and carbonylation.[6, 7, 17] In fact, controlled polymerization of supercritical ethylene has been an important commercial process for about 60 years.[7]

Although SCFs can affect the outcome of many homogeneously and heterogeneously catalyzed reactions as well as stoichiometric organic reactions in numerous ways, their potential benefits have not been fully realized. Future research in catalysis is warranted in a number of areas. For example, a variety of stereocontrolled reactions including asymmetric catalysis stand to potentially benefit from tunable SCF properties. The unique properties of SCFs affect enzyme activity, specificity, and stability in biocatalytic reactors as well. Compared to liquid water, scH$_2$O (T_c = 374 °C and P_c = 218 atm) is much less polar, can dissolve organic compounds, and has a low dielectric constant (ε = 6 at the critical point but ε = 90 at the freezing point) and a large dissociation constant (pK_w = 8 at the critical point but pK_w = 15 at the freezing point). Although this phase requires safety precautions, it can be used for the chemical synthesis of useful compounds and retrosynthesis of chemical wastes.[31, 32]

Note that, compared to conventional liquid solvents, SCFs are not always a panacea. They have both merits and disadvantages. Many chemical reactions are better performed in ordinary fluid solutions. However, chemistry of the reaction in SCFs still is a young and fully unexplored scientific field. We need deeper understanding of the microscopic and macroscopic properties of SCFs. The industrial outlook, particularly for the use of scCO₂, is bright largely because of the growing awareness of the need for "green chemistry".[18, 19, 33]

References and Notes

1. C. Reichardt, *Solvents and Solvent Effects in Organic Chemistry*, VCH, Weinheim, **1990**.
2. P. G. Jessop, T. Ikariya, R. Noyori, "Homogeneous Catalysis in Supercritical Fluids", *Science* **1995**, *269*, 1065–1069.
3. T. Ikariya, R. Noyori, "Organic Reactions in Supercritical Fluids", In *Transition Metal Catalyzed Reactions*, (Eds. S.-I. Murahashi and S. G. Davies), IUPAC, Blackwell Science, New York, **1999**, pp. 1–28.
4. A. A. Clifford, " In *Supercritical Fluids Fundamentals for Application of the NATO ASI Series E*, (Eds. E. Kiran, J. M. H. L. Sengers), Dordrecht, Kluwer Academic, **1994**, 449–479.
5. P. E. Savage, S. Gopalan, T. I. Mizan, C. J. Martino, E. E. Brock, "Reactions at Supercritical Conditions: Applications and Fundamentals", *AIChE J.* **1995**, *41*, 1723–1778.
6. P. G. Jessop, T. Ikariya, R. Noyori, "Homogeneous Catalysis in Supercritical Fluids", *Chem. Rev.* **1999**, *99*, 475–493.
7. P. G. Jessop, W. Leitner (Eds.), *Chemical Synthesis Using Supercritical Fluids*, Wiley–VCH, Weinheim, **1999**.
8. T. A. Rhodes, K. O'Shea, G. Bennett, K. P. Johnston, M. A. Fox, "Effect of Solvent–Solute and Solute–Solute Interactions on the Rate of a Michael Addition in Supercritical Fluoroform and Ethane", *J. Phys. Chem.* **1995**, *99*, 9903–9908.
9. J. F. Brennecke, J. E. Chateauneuf, "Homogeneous Organic Reactions as Mechanistic Probes in Supercritical Fluids", *Chem. Rev.* **1999**, *99*, 433–452, and references cited therein.
10. C. A. Eckert, B. L. Knutson, P. G. Debenedetti, "Supercritical Fluids as Solvents for Chemical and Materials Processing", *Nature* **1996**, *383*, 313–318.
11. T. Clifford, K. Bartle, "Chemical Reactions in Supercritical Fluids", *Chem. Ind.* **1996**, 449–452.
12. S. C. Tucker, "Solvent Density Inhomogeneities in Supercritical Fluids", *Chem. Rev.* **1999**, *99*, 391–418, and references cited therein.
13. O. Kajimoto, "Solvation in Supercritical Fluids: Its Effects on Energy Transfer and Chemical Reactions", *Chem. Rev.* **1999**, *99*, 355–389, and references cited therein.
14. FT-IR: S. M. Howdle, M. W. George, M. Poliakoff, "Vibrational Spectroscopy"; NMR: J. W. Rathke, R. J. Klingler, R. E. Gerald H, D. E. Fremgen, K. Woelk, S. Gaemers, C. J. Elsevier, NMR Spectroscopy"; UV, EPR, X-ray, etc.: C. R. Yonker, J. C. Linehan, J. L. Fulton, "UV, EPR, X-ray, and Related Spectroscopic Techniques", (Eds. P. G. Jessop, W. Leitner), In *Chemical Synthesis Using Supercritical Fluids*, Wiley–VCH, Weinheim, Chapter 3, **1999**, pp. 147–212.
15. J. M. DeSimone, Z. Guan, C. S. Elsbernd, "Synthesis of Fluoropolymers in Supercritical Carbon Dioxide", *Science* **1992**, *257*, 945–947.
16. K. P. Johnston, K. L. Harrison, M. J. Clarke, S. M. Howdle, M. P. Heitz, F. V. Bright, C. Carlier, T. W. Randolph, "Water-in-Carbon Dioxide Microemulsions: An Environment for Hydrophiles Including Proteins", *Science* **1996**, *271*, 624–626.
17. J. A. Darr, M. Poliakoff, "New Direction in Inorganic and Metal-Organic Coordination Chemistry in Super-critical Fluids", *Chem. Rev.* **1999**, *99*, 495–541, and references cited therein.
18. R. T. Baker, W. Tumas, "Toward Greener Chemistry", *Science* **1999**, *284*, 1477–1479.
19. S. Buelow, P. Dell'Orco, D. K. Morita, D. Pesiri, E. Birnbaum, S. L. Borkowsky, G. H. Brown, S. Feng, L. Luan, D. A. Morgenstern, W. Tumas, "Recent Advances in Chemistry and Chemical Processing in Dense Phase Carbon Dioxide at Los Alamos", In *Green Chemistry: Frontiers in Benign Chemical Syntheses and Processes*, (Eds. P. T. Anastas and T. C. Williamson), Oxford University Press, Oxford, **1998**, pp. 265–285.
20. S. Kainz, A. Brinkmann, W. Leitner, A. Pfaltz, "Iridium-Catalyzed Enantioselective Hydrogenation of Imines in Supercritical Carbon Dioxide", *J. Am. Chem. Soc.* **1999**, *121*, 6421–6429.
21. L. A. Blanchard, D. Hancu, E. J. Beckman, J. F. Brennecke, "Green Processing Using Ionic Liquids and CO₂", *Nature* **1999**, *399*, 28–29.
22. J. W. Rathke, R. J Klingler, T. R. Krausem, "Propylene Hydroformylation in Supercritical Carbon Dioxide", *Organometallics* **1991**, *10*, 1350–1355.
23. P. G. Jessop, T. Ikariya, R. Noyori, "Homogeneous Catalytic Hydrogenation of Supercritical Carbon Dioxide", *Nature* **1994**, *368*, 231–233.
24. P. G. Jessop, Y. Hsiao, T. Ikariya, R. Noyori, "Homogeneous Catalysis in Supercritical Fluids: Hydrogenation of Supercritical Carbon Dioxide to Formic Acid, Alkyl Formates, and Formamides", *J. Am. Chem. Soc.* **1996**, *118*, 344–355.
25. A. Baiker, "Supercritical Fluids in Heterogeneous Catalysis", *Chem. Rev.* **1999**, *99*, 453–473, and references cited therein.
26. R. Noyori, *Asymmetric Catalysis in Organic Synthesis*, John Wiley and Sons, New York, **1994**, pp. 28–33.
27. J.-L. Xiao, S. C. A. Nefkens, P. G. Jessop, T. Ikariya, R. Noyori, "Asymmetric Hydrogenation of α,β-Unsaturated Carboxylic Acids in Supercritical Carbon Dioxide", *Tetrahedron Lett.* **1996**, *37*, 2813–2816.
28. Y. Kayaki, Y. Noguchi, S. Iwasa, T. Ikariya, R. Noyori, "An Efficient Carbonylation of Aryl Halides Catalysed by Palladium Complexes with Phosphite Ligands in Supercritical Carbon Dioxide", *Chem. Commun.* **1999**, 1235–1236.
29. P. G. Jessop, T. Ikariya, R. Noyori, "Selectivity for Hydrogenation or Hydroformylation of Olefins by Hydridopentacarbonylmanganese(I) in Supercritical Carbon Dioxide", *Organometallics* **1995**, *14*, 1510–1513.
30. Y. Kishimoto, T. Ikariya, submitted for publication.
31. R. W. Shaw, T. B. Brill, A. A. Clifford, C. A. Eckert, C. A. Franck, "Supercritical Water: A Medium for Chemistry", *Chem. Eng. News* **1991**, *69 (51)*, 26–39.
32. P. E. Savage, "Organic Reactions in Supercritical Water", *Chem. Rev.* **1999**, *99*, 603–621, and references cited therein.
33. P. T. Anastas, J. C. Warner, *Green Chemistry: Theory and Practice*, Oxford University Press, Oxford, **1998**.

Fluorous Techniques for the Synthesis of Organic Molecules:
A Unified Strategy for Reaction and Separation

Dennis P. Curran

*Department of Chemistry,
University of Pittsburgh,
Pittsburgh, PA 15260,
USA*

*Phone: +1 412 624 8240
Fax: +1 412 624 9861
e-mail: curran@pitt.edu*

Keywords ■ *Fluorous Biphasic Catalysis* ■ *Fluorous Synthesis* ■ *Combinatorial Chemistry*
■ *Parallel Synthesis* ■ *Separation* ■ *Fluorocarbon Bonded Silica Gel* ■ *Green Chemistry*
■ *Supercritical Carbon Dioxide*

Concept: Organic molecules can be rendered fluorous[1] by attachment of a suitable highly fluorinated domain ("F" in the graphic below). The resulting fluorous-tagged molecules resemble their organic parents in terms of reactivity, yet they can readily be separated from essentially all standard organic molecules that lack a fluorous domain. Originally introduced as a catalyst immobilization method, fluorous techniques have quickly spread and they stand poised to impact the field of organic synthesis on many fronts.

Traditional organic reaction and separation

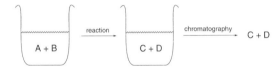

Fluorous Variant

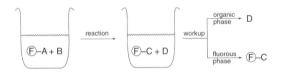

Ⓕ = highly fluorinated tag

Prologue

It is ironic that organic synthesis and separation science are separate disciplines because synthesis and separation are inseparable. The vast majority of organic reactions involve the combination of a substrate with other organic molecules (reagents, reactants, catalysts) to make a new organic product. The synthesis exercise is not complete until the desired product of the reaction has been separated from everything else in the final reaction mixture. Accordingly, the yield of every chemical reaction is limited by both the efficiency of the reaction and the efficiency of the separation.

Process chemists have long sought to couple reaction and separation chemistry when possible because it is often the separation steps that account for much of the cost of a chemical synthesis (material, labor, time, energy).[2] The chromatographic techniques that are so powerful for separation of small quantities of mixtures of organic molecules become less and less attractive as scales of reactions increase.

Abstract: Recently introduced fluorous techniques provide strategic new options for conducting organic reactions and for separating the resulting reaction mixtures. Fluorous molecules typically contain at least one highly fluorinated domain attached to an organic domain. The fluorinated domain can be an integral part of the molecule (permanent attachment) if the intended use is as a reagent, reactant or catalyst. A temporary attachment of a removable fluorous group is required to render a reaction substrate or product fluorous. Fluorous compounds can be separated from standard organic compounds by simple workup techniques of liquid-liquid extraction (two- or three-phase) or solid--liquid extraction. The original fluorous technique, biphasic catalysis, is becoming increasingly well established and holds significant potential for future development in chemical processes. More recently, techniques that use fluorous phases only at the separation stage and not at the reaction stage have been introduced. These have the potential to impact on both chemical process development and chemical discovery research. Many of these new techniques are especially suited to the preparation of combinatorial libraries by solution phase parallel synthesis. This chapter provides a brief introduction to the concepts of strategy level purification, and then introduces fluorous chemistry with representative examples of reactions, reagents and techniques.

Synthetic chemists have responded to the separation problem by providing increasingly more efficient[3] and selective reactions.[4] This response is necessary – indeed even vital. But it is not sufficient for two reasons. First, catalytic methods are the most efficient types of transformations conceivable for many types of organic reactions.[5] So even when many of today's stoichiometric reactions are replaced by powerful catalytic alternatives, it will still be necessary to separate the catalyst from the product. Second, chemists involved in discovery-oriented synthesis research cannot be constrained by process concerns. To discover as many useful molecules as quickly as possible, they must have available the whole range of synthetic reactions and techniques. An illustrative example is the preparation of large libraries of organic compounds by solid phase synthesis.[6] This technique miserably fails any test of synthetic efficiency or atom economy. But it passes the speed test with flying colors. It is difficult to match solid phase synthesis for the rapid preparation of large libraries.

In keeping with the division between the two fields, separation scientists have responded to the separation problem by providing ever more powerful means to separate small organic molecules. Again, this response is necessary but not sufficient.

The popularity of solid phase synthesis makes obvious the answer to the separation problem: do not mix difficult-to-separate things in the first place! But experience teaches us that this answer is deceptively simple. Solid phase synthesis techniques render separation trivial but at the same time make reactions more difficult.

Accordingly, the best methods to use for a given organic synthesis may vary depending on whether discovery or production is the goal, and they must address not only the efficiency of the chemical reactions but also the ease and efficiency of the separation. Finally, while reaction and separation are the main concerns, all useful methods must also accommodate requirements to identify and analyze products.

Strategy Level Separations

Issues that arise in the reactions of a synthetic scheme have long been treated at a strategy level.[7] In other words, prior to conducting any reactions, a plan is formulated on how to build the skeleton of the target molecule, when to introduce functional groups, and how to control chemo-, regio- and stereoselectivity. It follows that the best laid synthetic schemes will also have a strategy for separation.[8]

We have suggested that separation strategies be planned with the following goal in mind: the target product(s) in a final reaction mixture should partition into a phase that is different from all the other components of the mixture.[8] When this goal is met, reactions can be purified simply by workup, which involves simple phase separation techniques such as evaporation, extraction, and filtration.

The problem with this goal is that there are too many organic compounds in organic synthesis. The large majority of things added to or formed during a typical organic reaction end up in the organic liquid phase after workup. At a conceptual (strategic) level, the solution to the problem is to "tag" organic molecules with a domain that will force them into another phase during a standard workup while at the same time not compromising their reactivity properties. Aside from the organic liquid phase, the three commonly used phases in workup of an organic reaction mixture are the gas phase, the aqueous phase and the solid phase.[9] Since tags invariably add molecular weight, it is difficult to envision tagging strategies for the gas phase. But the addition of ionizable groups can divert organic molecules to the aqueous phase under appropriate conditions (acid/base extractions),[10] and the covalent or ionic attachment of small organic molecules to polymers or other insoluble materials is a very general way to tag molecules for the solid phase.[11–13] It is within this context that the "fluorous phase" presents itself as a fifth phase, that (in simple terms) is mutually orthogonal to the other four.

The Fluorous Phase[14]

The fluorous liquid phase consists of perfluorinated or very highly fluorinated organic liquids. Members of 3M's "FC™" family of "fluorinert fluids" are becoming popular fluorous solvents. For example, FC-72™ is a solvent mixture consisting mostly of perfluorohexanes (C_6F_{14}), and boils at about 56 °C. Although more expensive, perfluoromethylcyclohexane ($CF_3C_6F_{11}$) has the advantage of being a pure compound, and it is also popular in some circles. A number of other related solvents are available. These low boiling compounds are immiscible both in water and in typical organic solvents at ambient temperatures, and effectively form a separate phase for use in two-phase or three-phase liquid–liquid extractions.

Organic compounds can be rendered soluble in fluorous solvents by attaching them to highly fluorinated tags, and some representative fluorous molecules are shown in Figure 1. The tags are often long

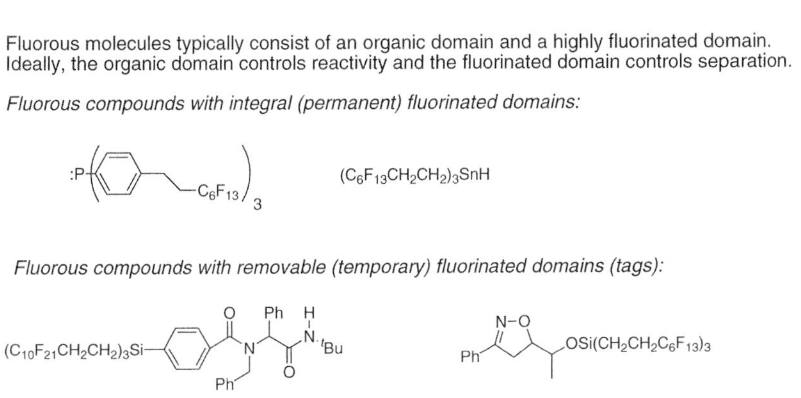

Fluorous molecules typically consist of an organic domain and a highly fluorinated domain. Ideally, the organic domain controls reactivity and the fluorinated domain controls separation.

Figure 1. Representative fluorous molecules

Fluorous compounds with integral (permanent) fluorinated domains:

Fluorous compounds with removable (temporary) fluorinated domains (tags):

perfluoroalkyl chains that are sometimes called "pony tails". If the resulting fluorous compounds are to be used as reagents, reactants, catalysts, etc., then the tags can be an integral part of the molecule since they will never be removed. But if the resulting fluorous compounds are substrates or products, then the tags must be removable. Experience has shown that partitioning of the resulting compounds between fluorous and organic solvents depends heavily on the choice of organic solvent, with polar or aromatic solvents often being favored. High partition coefficients can be obtained by building tagged molecules that have 60 % or more fluorine by molecular weight. Since the tags themselves are typically only about 75 % fluorine by molecular weight, it becomes increasingly problematic to tag larger and larger organic molecules with enough fluorines to provide high partition coefficients.

Increasing size (more fluorine content), easier separation, but lower solubility in common reaction solvents

Decreasing size (less fluorine content), improved solubility in common reaction solvents, but more difficult separation

A = the organic domain

F = the fluorinated domain

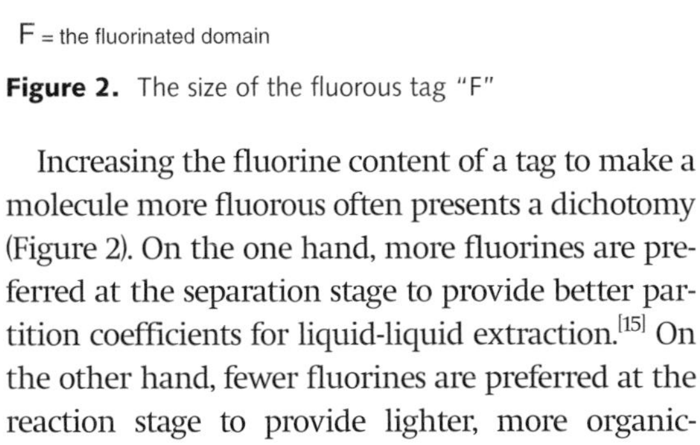

Figure 2. The size of the fluorous tag "F"

Increasing the fluorine content of a tag to make a molecule more fluorous often presents a dichotomy (Figure 2). On the one hand, more fluorines are preferred at the separation stage to provide better partition coefficients for liquid-liquid extraction.[15] On the other hand, fewer fluorines are preferred at the reaction stage to provide lighter, more organic-soluble molecules. To some extent, decreasing the size of a fluorous tag can be accommodated by changing the extraction procedure. The organic solvent can be cut with water to force the fluorous-tagged components out,[16] or a continuous extrac-

tion can be conducted.[17] Both of these techniques rely on the extraordinarily low solubility of organic compounds in fluorous solvents. However, at some point, the fluorine content becomes too low and liquid-liquid extraction techniques begin to break down. At this point, solid–liquid extractions come to the rescue.

We introduced the technique of fluorous solid phase extraction in 1997.[18] This technique uses silica gel with a fluorocarbon bonded phase[19] (fluorous reverse phase silica gel), although the use of other fluorinated solid phases is also conceivable. Fluorous reverse phase silica gel has a very low retention capacity for organic molecules, and it separates primarily by fluorine content: the more fluorines a molecule has, the better it is retained by the silica gel. While the separation of molecules with similar numbers of fluorine requires a chromatography, the separation of untagged organic molecules from fluorous-tagged molecules containing 15–20 or so fluorines becomes a simple solid phase extraction. Effectively, this is a two-stage filtration that is very easy to conduct either on individual reaction mixtures or on multiple mixtures in parallel (Figure 3). A reaction mixture is charged to the top of a short column containing fluorous silica gel, and the column is eluted with a fluorophobic solvent to remove organic (non-tagged) compounds. A second, more fluorophilic solvent, is then applied to elute the fluorous-tagged compound(s). Aside from the practical advantage of simplicity, the solid phase extraction is far superior to the liquid-liquid extraction in separating tagged molecules with lower fluorine content. Indeed, we have even separated tagged molecules bearing so few fluorines that they are not soluble in fluorous solvents!

Reactions in Fluorous Solvents

In their original incarnations,[20] fluorous techniques were designed with the notion that fluorous solvents should be used in both the reaction stage

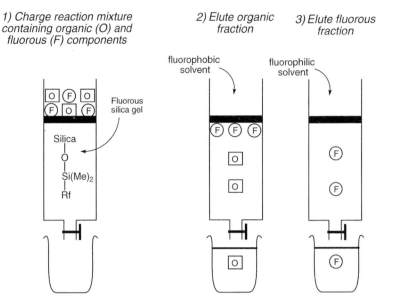

1) Charge reaction mixture containing organic (O) and fluorous (F) components

2) Elute organic fraction

3) Elute fluorous fraction

Figure 3. Solid phase extraction over fluorous reverse phase silica

and the separation stage. In a far-reaching 1994 paper, Horváth and Rábai laid the conceptual foundations for "fluorous biphasic catalysis".[21] This catalyst immobilization technique is potentially very general and may offer advantages over traditional methods such as using water soluble or insoluble catalysts. The rapid growth of publications in this area suggests that this technique is poised to broadly impact the catalysis field.

Several reviews and overviews survey the development and current state of fluorous biphasic catalysis,[22–25] and a recent application by Gladysz and

Dinh is illustrative of the methods and their potential (Figure 4).[26] Hydrosilylation of enones like cyclohexenone was conducted in two different ways. Under biphasic reaction conditions, an organic solvent like toluene containing the enone and phenyldimethyl silane was heated with a fluorocarbon solvent (perfluoromethylcyclohexane) containing a fluorous rhodium catalyst. The hydrosilylation reaction occurred over 10 h, and the reaction mixture was cooled and the phases were separated. Distillation of the residue from the organic phase gave the hydrosilylated products in excellent isolated yields

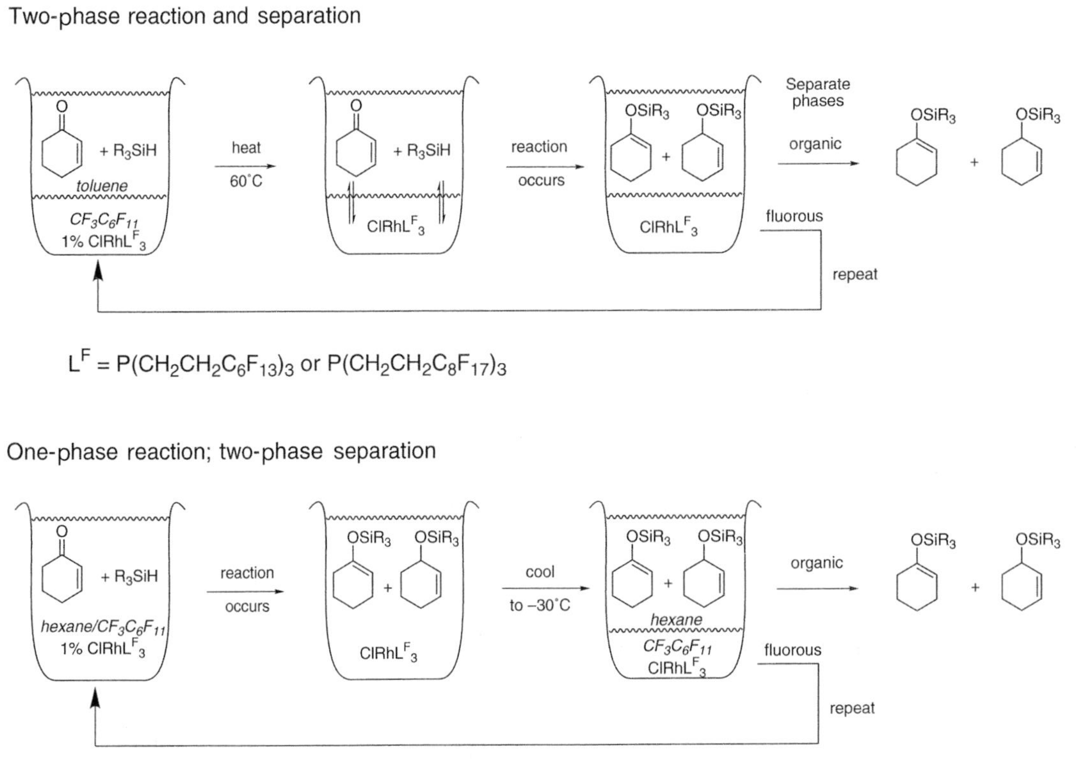

Two-phase reaction and separation

L^F = P(CH$_2$CH$_2$C$_6$F$_{13}$)$_3$ or P(CH$_2$CH$_2$C$_8$F$_{17}$)$_3$

One-phase reaction; two-phase separation

Figure 4. Fluorous biphasic catalysis

(85–90 %). The fluorous phase contained the active catalyst, and it was reused two times to repeat the reaction. Yields in runs two and three were comparable to run one and turnover numbers were 100–115 for each run, although each successive run was slower than its predecessor.

The same reaction can also be run under homogenous conditions by changing the organic solvent and relying on the strong temperature dependence on the miscibility of (some) organic and fluorous solvents. In this example, hexane was used as the organic solvent, and the reaction mixture at 60 °C was homogeneous. Hexane is one of the rare organic solvents that is miscible with (some) fluorous solvents at room temperature, so the reaction mixture was cooled to –30 °C prior to phase separation. As might be expected, the homogenous reactions occurred considerably faster than their heterogeneous analogs, but the yields, turnover numbers, and recylability of the catalyst were comparable.

The recently reported asymmetric alkylation of aromatic aldehydes with diethylzinc provides a nice example of the potential use of fluorous catalysts in sophisticated processes (Figure 5).[27] Addition

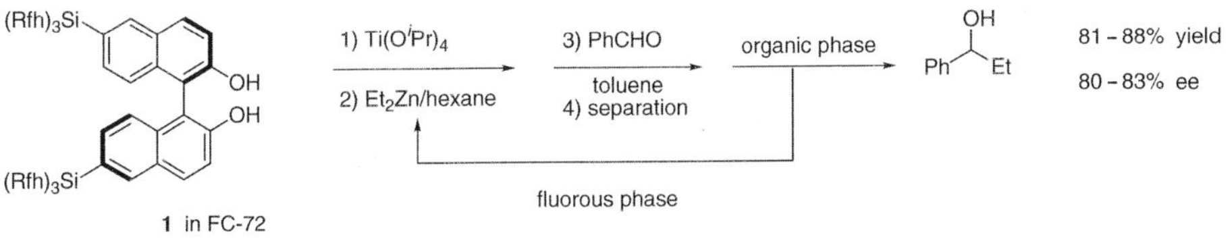

vents are relatively non-toxic, and the fluorous phase can be reused directly without concentration or re-isolation of the original catalyst. There is no water involved, so water sensitive reactions can be conducted.[28] The very low solubility of most organic compounds in fluorous solvents ensures a high recovery of the organic product. On the flip side, loss of the catalyst to the organic phase can be more problematic. The technique generally requires a large number of fluorines to provide acceptably high partition coefficients, and even then, leaching of toxic or precious catalysts can be a problem.

Reactions in Hybrid Solvents and Organic Solvents

Despite their vast untapped potential, techniques like fluorous biphasic catalysis still have limitations. These limitations center on solubility and emanate directly from the strengths of the technique: heavily fluorous compounds tend to be highly insoluble in organic solvents while organic compounds (especially polar ones) tend to be highly

Figure 5. Catalytic addition of Et_2Zn with a fluorous binol ligand

of $Ti(O^ipr)_4$ to a solution of fluorous BINOL (FBINOL) **1** in FC-72 generated an active catalyst that effected nucleophilic addition of diethylzinc to benzaldehyde at 0 °C under biphasic conditions. Separation of the phases gave a fluorous phase which could be reused to catalyze additional reactions with identical yields and enantioselectivities. Unfortunately, about 10 % of the FBINOL ligand was lost to the organic phase in each cycle, so the process was limited in practice to about 5 cycles of the fluorous solution. However, the organic product was readily separated from the FBINOL derivative by fluorous solid phase extraction, so the leached ligand could be recovered and reused.

These examples reveal the attractive features of fluorous biphasic catalysis methods for chemical processes. Reactions occur in the liquid phase and can be either homogeneous or biphasic. In either case, biphasic conditions are established at the end of the reaction so the separation is easy. Fluorous sol-

insoluble in fluorous solvents. Clearly, neither fluorous nor organic solvents will be ideal for broad application with heavily fluorous compounds. In short, the features that create such a favorable separation scenario at the same time create a reaction solvent problem (Figure 2).

We immediately encountered this reaction solvent problem in our initial foray into the fluorous field early in 1995. Dr. Sabine Hadida synthesized a fluorous tin hydride **2** that was designed as an analog of the popular reagent tributyltin hydride (Figure 6). However, while separation of this tin reagent from prospective organic products was indeed trivial, the initial conditions surveyed to effect reactions were highly unsatisfactory. Attempts to run reductions of a simple organic compound like adamantyl iodide were not very successful in organic solvents, fluorous solvents, or biphasic solvent mixtures.[29] We believe that radical chains will not propagate under these conditions because neither phase has a critical concentration of all the needed components to ensure that chain propagation is faster than termination. Adamantyl iodide can be reduced (albeit slowly) in hexane alone or in a hexane/fluorous mixture, but these solvents are only useful for the most non-polar types of organic compounds.

In simple terms, one can envision three general solutions to the reaction solvent problem: 1) modify the organic compounds to make them soluble in fluorous solvents, 2) modify the fluorous compounds to make them soluble in organic solvents, or 3) find solvents that have good power to dissolve both organic and fluorous compounds. Solution one has little value since organic compounds can only be soluble in fluorous solvents by fluorinating them (and thereby defeating the fluorous/organic separation). In contrast, we have developed very general solutions to dramatically extend the utility of fluorous chemistry based on the other two approaches.

Partially fluorinated ("hybrid") solvents have the unique ability to dissolve both highly fluorous and non-fluorinated organic compounds. In 1996,[30] we reported the first fluorous reactions that took place in a non-fluorous reaction solvent (Figure 6). Radical reduction with fluorous tin hydride **2** occurs very smoothly in the solvent benzotrifluoride (BTF, trifluoromethylbenzene, $C_6H_5CF_3$).[31] Benzotrifluoride[32] is not a fluorous solvent because it is fully miscible in all common organic solvents. However, it is fully miscible in fluorous solvents as well, and by implication (and experience) has the ability to dissolve both fluorous and organic compounds. Even better than the stoichiometric procedure shown in Figure 6 is a catalytic procedure that uses 5–10 % tin hydride **2** along with sodium cyanoborohydride in 1/1 *tert*-butanol/BTF. To meet the strategic goal of isolation of pure product by workup, the reaction mixture is purified by a three-phase liquid–liquid

extraction with water (to remove the inorganic product), an organic solvent (to isolate the desired product), and a fluorous solvent (to recover the fluorous tin hydride).

The separation of organotin residues is a general problem in tin chemistry, and we have developed several other fluorous variants of important organotin reactions including Stille couplings,[33] allylations,[34] and tin azide cycloadditions.[35] In many of these, benzotrifluoride is used as a solvent or cosolvent. Occasionally, we have also used other "hybrid" solvents such as 1,1,2-trichloro-1,2,2-trifluoroethane ($CF_2ClCFCl_2$) and 2,2,2-trifluoroethanol (CF_3CH_2OH). All of these reactions can be separated by liquid–liquid extractions, although partial concentration of the "hybrid" reaction solvent may be needed on larger scales prior to extraction to prevent it from homogenizing the organic and fluorous extraction solvents. In most of these tin reactions, the fluorous tin products are recovered from the fluorous phase in very high yield and are routinely reused. In short, the simple expedient of using partially fluorinated reaction solvents or cosolvents very dramatically extends the applicability of fluorous techniques with respect to both the kinds of organic reactions that can be conducted and the types of organic substrates that can be reacted.

Reducing the number of fluorines on the fluorous tag also provides a general solution to the reaction solvent problem; as the number of fluorines is reduced, the solubility in organic solvents tends to go up. Of course, the solubility in fluorous solvents tends to go down at the same time, and the residual tag must strike a balance between too many fluorines (low solubility in organic solvents) and too few fluorines (cannot easily be separated from organic compounds). However, thanks to the technique of fluorous solid-liquid extraction, there is surprisingly broad latitude here.

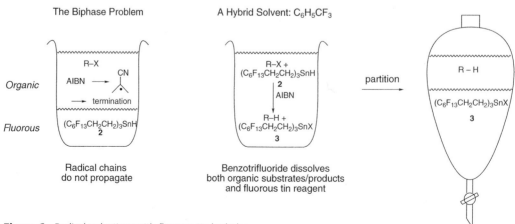

Figure 6. Radical reductions with fluorous tin hydrides

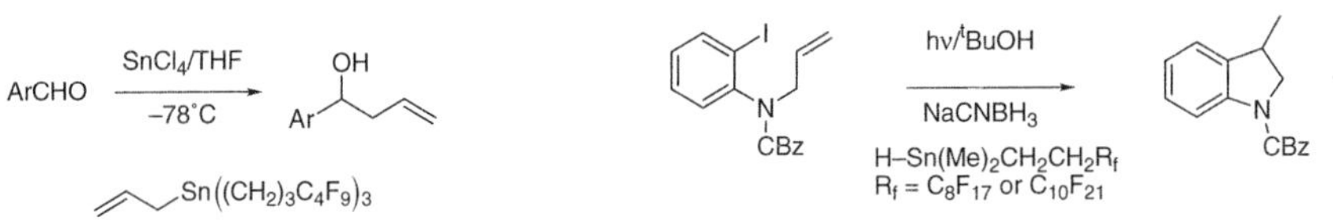

Figure 7. Fluorous tin reactions with separation by solid-liquid extraction

Our initial work used fluorous solid phase extraction mainly as a convenience. For example, tin reagents **4** bearing a longer spacer component [(CH₂)₃ rather than (CH₂)₂][36] and a shorter fluorocarbon tail (C₄F₉ rather than C₆F₁₃) can be readily separated from organic reaction products by solid phase extraction (Figure 7).[37] Likewise, tin hydride reagents bearing a single fluorous chain rather than three can also be used.[38] All these reagents are sufficiently soluble in organic solvents, so no fluorous or hybrid solvents or cosolvents are needed to conduct reactions. However, due to the low fluorous/organic partition coefficients of these reagents, repeated extractions (as many as 8–20) can be needed in liquid-liquid extractive workups. This inconvenience is neatly bypassed with a single solid phase extraction.[39] Solid phase extractions are unbeatable in parallel synthesis applications because they are so easy to conduct in parallel.[40]

Unfortunately, the appeal of solid phase extractions on small scale fades as the scale increases due to the cost and inconvenience of using large amounts of fluorous silica gel. Here, modified techniques to reduce the tedium of repeated extractions are attractive. For example, Crich has recently introduced the minimally fluorous selenide C₆F₁₃CH₂CH₂C₆H₄SeH[17]. This selenol is added in catalytic quantities to tin hydride reductions of reactive aryl and vinyl radicals. The high reducing capacity of the aryl selenide suppresses undesired reactions of product radicals without suppressing the reactions of the aryl and vinyl radicals themselves. After the reaction is complete, the selenol can be recovered by a modified continuous extraction procedure.

All of these methods capture the favorable features of fluorous biphasic catalysis including the ability to recover and reuse the reagent or catalyst. However, the scope of these processes should be considerably broader than processes that employ fluorous reaction solvents. There are a few types of reactions that capitalize on unique features of fluorous solvents (such as high solubility of gases), and fluorous biphasic catalysis procedures are also attractive for large scale chemical production. But reactions in organic or hybrid solvents appear generally more appealing for the types of small scale organic reactions that are routinely run in chemical discovery research.

Supercritical carbon dioxide is a non-traditional solvent that is increasingly touted as the green organic reaction solvent of the future.[41] Supercritical CO_2 is a relatively poor solvent for many types of organic compounds at low pressures but its dissolving capability improves as the pressure increases. However, it readily dissolves fluorinated compounds even at low pressures. In 1997, we reported radical reductions with fluorous tin hydride **2** in supercritical CO_2, and in the process pointed out that supercritical CO_2 is effectively the equivalent of a hybrid solvent. Under suitable conditions, it can dissolve both organic and fluorous reaction components.[42] Accordingly, the union of supercritical CO_2 as a reaction solvent with fluorous/organic separation techniques offers bright prospects for future green chemical reaction and separation processes.[43]

Fluorous Synthesis Techniques[44]

The work in organotin chemistry discussed above is unified by the fact that the substrates and the products are always small organic molecules, while the reagents, reactants or catalysts are fluorous. The other major area of fluorous research reverses this approach and renders the substrates and/or reaction products fluorous. This requires a "phase tagging" strategy wherein small organic molecules are tagged with fluorous groups. The approach is conceptually analogous to solid phase synthesis (where a small organic molecule is tagged with a polymer) but operationally very different. Fluorous tags are usually, though not always, modified protecting groups and thereby serve a dual role. Many of the concepts and the initial demonstration of fluorous synthesis appeared in our 1997 *Science* paper,[45] and some of these will be highlighted below.

The first demonstration of fluorous synthesis was in the preparation of small (8–12 members) isoxazoline and isoxazole libraries by the three-step procedure outlined in Figure 8.[46] All reactions were purified by three-phase liquid-liquid extraction. The starting substrates were simple allylic alcohols which were tagged with the fluorous silyl halide **5** to make substrates **6** for an ensuing dipolar cycloaddition. This was conducted by the Mukaiyama method with a large excess of nitro compound and

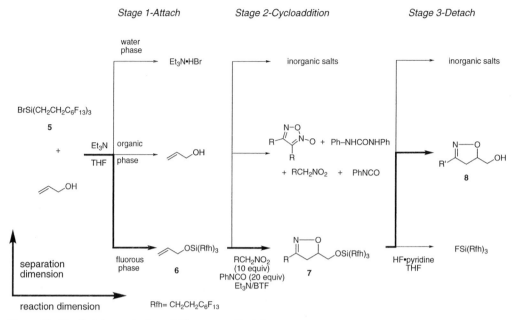

Figure 8. The first demonstration of "fluorous synthesis"

phenylisocyanate. Three-phase extraction provided the pure fluorous isoxazolines **7** in the fluorous phase, while the organic phase contained all the byproducts and unreacted reagents. Detagging (deprotection) of **7** with fluoride gave the recovered tag in the fluorous phase along with the pure products **8** in the organic phase. The sequence allows the isolation of pure organic isoxazolines without the need for chromatographic separation from a number of other organic products.

This simple series of experiments gave us a handle on some of the features of fluorous synthesis and also identified some crucial problems. The fluorous phase tagging strategy is a powerful tool for separating organic reaction mixtures. Unlike polymers, fluorous products are single entities that can be characterized by all standard small molecule chromatographic (TLC, HPLC) and spectroscopic techniques (MS, NMR, IR). The goal is to purify the fluorous-tagged products by workup, but if this goal is not met then the products can simply be

purified by chromatography (not an option in solid phase synthesis). The "perfluorinated" part of the tag is very stable, although this particular silyl tag is probably too sensitive to hydrolysis to be generally useful (it has a reactivity towards nucleophiles that is qualitatively similar to the TMS group). Limitations were also quickly identified. The silyl group **5** bearing 39 fluorines was only large enough to tag compounds in the MW 100–200 range. Larger tagged compounds resisted extraction into the fluorous liquid phase. Also, partially fluorinated solvents like BTF were essential for some reactions.

To be of practical use, fluorous synthesis must be applicable to larger organic molecules in the molecular weight range of at least 350–500. To make molecules of this size, we initially took the simple approach of adding more fluorines, which can be done either by making longer tags or attaching more tags. Tags bearing 63 fluorines were first made and then used successfully in both fluorous Ugi[47] and fluorous Biginelli reactions.[48] Figure 9 shows

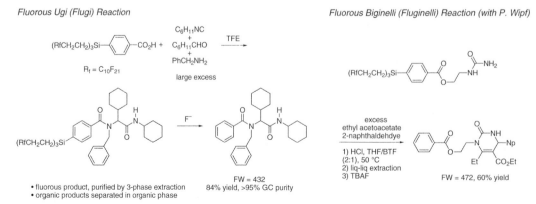

Figure 9. Examples of fluorous Ugi and Biginelli reactions

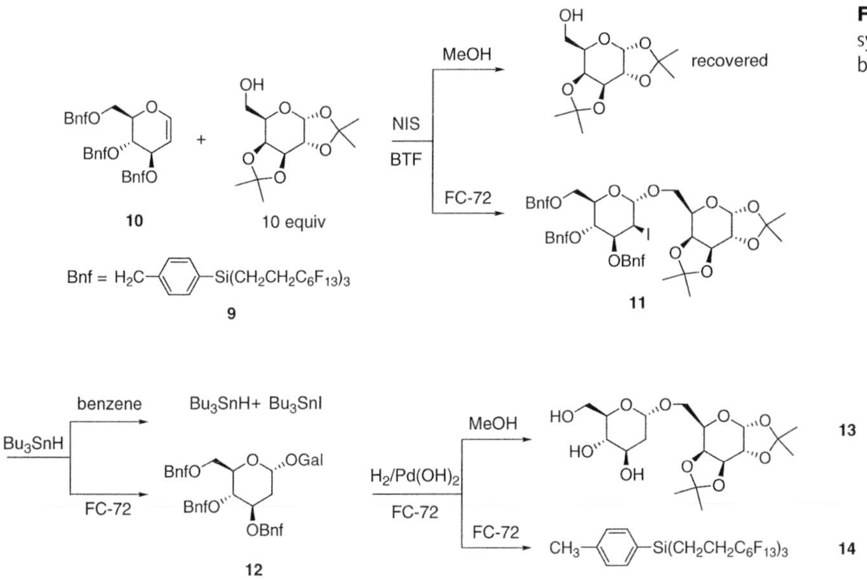

Figure 10. Disaccharide synthesis with a fluorous benzyl group

one from among about a dozen examples of each of these reactions. The Biginelli work was conducted in collaboration with Professor Peter Wipf and Dr. Patrick Jäger.

In these examples, the tag behaved as a "traceless" unit rather than a protecting group, and molecules with polar functionalities and MWs >400 were successfully passed through the fluorous liquid phase in the extractive workup. However, we experienced general insolubility problems with substrates bearing multiple $C_{10}F_{21}$ tags. The molecular weights of the tagged compounds are rather high (2000–2500), and their low solubility in organic solvents caused problems at the reaction stage. For example, the Ugi reaction completely failed in MeOH and EtOH, but succeeded in trifluoroethanol (TFE). A couple of the Biginelli reactions failed entirely, possibly due to the insolubility of the substrates.

We investigated the use of more fluorous tags in disaccharide synthesis as shown in Figure 10.[49] A fluorous benzyl group **9** was prepared and attached to glucal to give tribenzyl derivative **10** bearing 117 fluorines (39 per benzyl group). This highly fluorous molecule was coupled with excess diacetone galactose (NIS/BTF) followed by liquid-liquid extraction. The resulting fluorous disaccharide **11** was reductively deiodinated with tributyltin hydride in a procedure that is essentially the reverse of that described above (fluorous substrate with organic tin hydride rather than organic substrate with fluorous tin hydride). Finally, reductive debenzylation of **12** with a standard heterogeneous catalyst gave the organic disaccharide **13** along with the fluorous toluene **14** (which was rebrominated with NBS to recycle the benzyl group). Molecules **10–12** with nine fluorous chains are highly soluble in FC-72 and reasonably soluble in BTF. However, their very high

molecular weights (~3500) and complete lack of solubility in organic solvents are detractions.

We now refer to the techniques in Figures 8–10 as "heavy fluorous" synthesis. While these methods have value that could be extended by future work, the results served to convince us that a version of "light fluorous" synthesis would be considerably more valuable. Cutting back on the number of fluorines in the tag is expected to increase the organic solubility of tagged molecules (useful for reaction chemistry) while decreasing the fluorous solubility. The later effect poses a problem when using liquid-liquid extractions, so light fluorous synthesis uses solid-liquid extractions instead. The left side of Figure 11 shows the dramatic effect of fluorine substitution on retention times of a series of tagged amides **15a–i** on a commercial "Fluofix™" column.[50] The control amide **15a** (bearing an alkyl instead of a fluoroalkyl group) comes off with the solvent front, as do most other non-fluorinated organic compounds under these conditions (80% MeOH). As the eluting power of the solvent is increased from 80% MeOH to 100% MeOH, the amides bearing fluorinated tags come off in order of fluorine content. The separation must be expedited by a gradient; under isocratic conditions, the retention time of **15h** is close to 1 h and **15i** does not appear to come off at all (up to 2 h). Most importantly, all tagged compounds **15** are soluble in dichloromethane and insoluble in FC-72. In other words, even **15i** with 21 fluorines does not have nearly enough fluorines to be useful in liquid-liquid extractions.

Despite the lack of solubility of **15** in fluorocarbon solvents, the huge retention differences between organic compounds (solvent front) and fluorous-tagged compounds (>30 min for the longer tags) translate into simple separations by solid

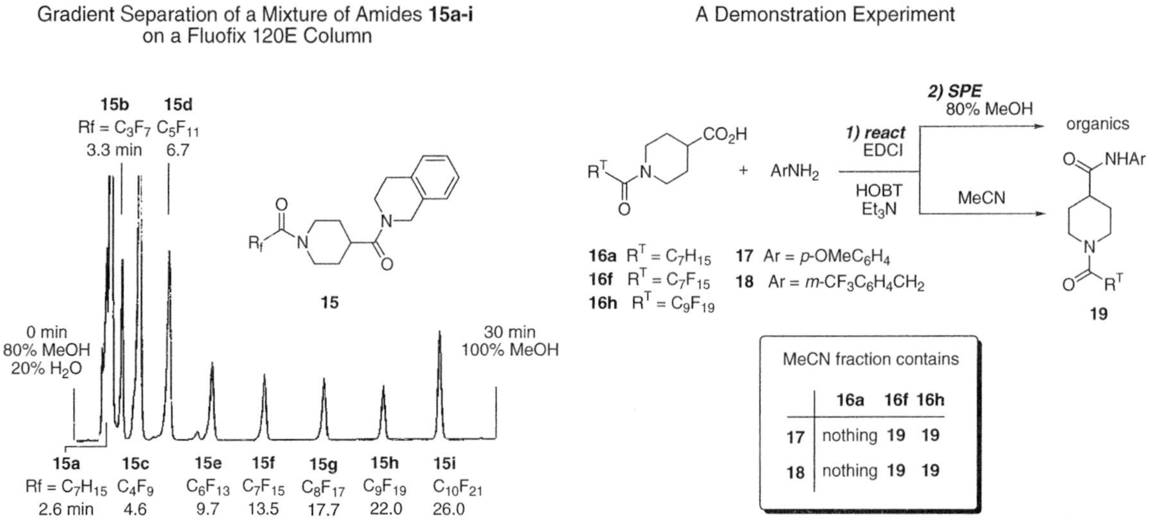

Gradient Separation of a Mixture of Amides 15a-i on a Fluofix 120E Column
15b 15d
Rf = C3F7 C5F11
3.3 min 6.7
15
0 min
80% MeOH
20% H2O
30 min
100% MeOH
15a 15c 15e 15f 15g 15h 15i
Rf = C7H15 C4F9 C6F13 C7F15 C8F17 C9F19 C10F21
2.6 min 4.6 9.7 13.5 17.7 22.0 26.0
A Demonstration Experiment
CO2H
+ ArNH2
1) react
EDCI
HOBT
Et3N
2) SPE
80% MeOH
organics
MeCN
NHAr
19
16a RT = C7H15
16f RT = C7F15
16h RT = C9F19
17 Ar = p-OMeC6H4
18 Ar = m-CF3C6H4CH2
MeCN fraction contains
16a 16f 16h
17 nothing 19 19
18 nothing 19 19

Figure 11. Light fluorous synthesis

phase extractions, as shown in the demonstration experiment on the right side of Figure 11.

One control (**16a**) and two fluorous-tagged amides (**16f,h**) were coupled in a small parallel synthesis with two anilines (**17** and **18**) by using standard coupling conditions. The crude reaction mixtures (~200 mg) were eluted through 2 g of fluorous silica first with 4 mL 80% MeOH (to collect the organic fraction) and then with 4 mL acetonitrile (to collect the "fluorous" fraction). (Note that no fluorous solvent is needed to collect the fluorous fraction!) Evaporation of the fluorous fractions provided pure products **19** in the case of the fluorous-tagged substrates, but there was nothing in the fluorous fractions of the control substrate **16a** (the coupling succeeded but the product eluted with all of the other components in the organic fraction). As a follow up, a 16-compound amide library was made by parallel synthesis.[50] Based on these experiments, we are now confident that this type of light fluorous synthesis (tagged substrate have MWs ~750) can rapidly be moved into the mainstream of both parallel and non-parallel synthesis.

Finally, in our 1997 *Science* paper,[45] we introduced a number of fluorous phase switching techniques, and one of these is illustrated in Figure 12. Cycloaddition of nitriles with fluorous tin azide **20** provides fluorous products **21** from organic substrates. In this fluorous phase switch, the reaction product **22** can be isolated in pure form even when the starting material is not completely consumed. Perhaps even more impressive, the use of starting materials doped with organic impurities still gives pure products (provided that the impurities do not react with the tin azide). This type of phase switch will be very useful in rapid, chromatography-free multistep synthesis. When "one-phase" methods (solid phase synthesis, fluorous synthesis) are used exclusively, then pure products can only be obtained in reactions that go to 100% yield and 100% conversion based on the tagged substrate (because tagged products cannot be separated easily from tagged substrates or side products). This fundamental problem is solved by phase switching. While the illustration is in the heavy fluorous mode, there is no reason why light fluorous applications cannot be developed. Furthermore, it is also possible to switch undesired products by using quenching techniques. For example, the fluorous tin hydride **2** can be used to remove unreacted alkenes or alkynes by hydrostannation, and a fluorous amine can be used to quench and remove electrophiles like isocyanates.[51]

(Rfh)3SnN3 + R-CN
20
Rfh = CH2CH2C6F13
BTF
benzene
Unreacted R-CN
FC-72
R
Sn(Rfh)3
21
+ other fluorous products
CH3CN
Ether/HCl
FC-72
22
(Rfh)3SnCl

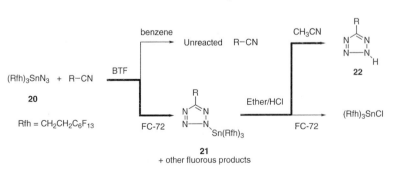

Figure 12. An example of fluorous phase switch

Epilogue

Although the field of fluorous chemistry is very young and much additional discovery and development are needed, the strategies and techniques that have been introduced have the potential to impact across the field of organic synthesis and beyond. In the production area, chemists have long been concerned about separation, and fluorous techniques provide fundamental new options. The use of fluorous catalysts is especially attractive, whether under biphasic conditions with fluorous solvents or under monophasic conditions with organic solvents or cosolvents, or with hybrid solvents. The ability to recover and reuse expensive catalysts or reagents will serve process chemists of the future well. Fluorous compounds have a special role to play in the green chemistry future of supercritical CO_2 due to their unique solubility properties in this solvent.

At the discovery end of the synthesis continuum, organic chemists have only recently begun to think about the coupling of reaction and separation chemistry to replace the traditional reaction/chromatography approach to research. In this area, fluorous reagents, reactants and catalysts are useful for expediting traditional solution phase organic synthesis, especially parallel synthesis. In addition, the techniques of fluorous synthesis (fluorous protecting groups, phase switches, quenchers, etc.) offer fundamentally new options for both traditional synthesis and parallel synthesis.

Although many of the key principles and concepts of fluorous chemistry have been introduced, there is still much interesting and important work to be done. There are too few fluorous reagents, reactants, ligands, catalysts, quenchers, protecting groups, tags, etc., and more are needed. Even the fluorous compounds that exist now may not be optimized for reaction or separation chemistry. For example, some fluorous reagents may benefit from different spacers and others may now have too many fluorines considering the recent improvements in separation techniques. Indeed, one of the most interesting features of fluorous research is that the exercise of "converting" an organic compound to a fluorous one does not result in a single compound but instead provides a family of compounds with differing properties. In other words, there is an ability to tune the fluorous compounds to meet the needs of a given transformation. Research and development in the fluorous field will keep researchers busy well into the 21st century.

Acknowledgements

I warmly thank the enthusiastic group of co-workers cited in the references for their intellectual and experimental contributions to this project. I also thank the National Science Foundation, the National Institutes of Health, Warner-Lambert, Combichem, and Merck for funding our work in this area.

References and Notes

1. The word "fluorous" was introduced in 1994 by Horváth and Rábai. It is coined by analogy to "aqueous" and is intended to convey the notion that the fluorous phase is unique when compared to aqueous and organic phases. See reference[21].
2. T. Laird, "Working Up to Scratch", *Chem. Brit.* **1996**, *32*, 43.
3. B. M. Trost, "Atom Economy: A Search for Synthetic Efficiency.", *Science* **1991**, *254*, 1471.
4. K. C. Nicolaou, E. J. Sorenson, *Classics in Total Synthesis*; VCH: Weinheim, **1995**.
5. B. M. Trost, "Atom Economy – A Challenge for Organic Synthesis: Homogeneous Catalysis Leads the Way", *Angew. Chem., Int. Ed. Eng.* **1995**, *34*, 259.
6. F. Balkenhohl, C. von dem Büssche-Hunnefeld, A. Lansky, C. Zechel, "Combinatorial Synthesis of Small Organic Molecules", *Ang. Chem., Int. Ed. Eng.* **1996**, *35*, 2289.
7. E. J. Corey, X.-M. Cheng, *The Logic of Chemical Synthesis*; Wiley: New York, **1989**.
8. D. P. Curran, "Strategy-Level Separations in Organic Synthesis: From Planning to Practice", *Angew. Chem., Int. Ed. Engl.* **1998**, *37*, 1175.
9. D. P. Curran, "Combinatorial Organic Synthesis and Phase Separation: Back to the Future", *Chemtracts – Org. Chem.* **1996**, *9*, 75.
10. D. L. Boger, C. M. Tarby, P. L. Myers, L. H. Caporale, "Generalized Dipeptidomimetic Template: Solution Phase Parallel Synthesis of Combinatorial Libraries", *J. Am. Chem. Soc.* **1996**, *118*, 2109.
11. S. W. Kaldor, M. G. Siegel, "Combinatorial Chemistry using Polymer supported Reagents", *Curr. Opin. Chem. Bio.* **1997**, *1*, 101.
12. R. J. Booth, J. C. Hodges, "Solid-Supported Reagent Strategies for Rapid Purification of Combinatorial Synthesis Products", *Acc. Chem. Res.* **1999**, *32*, 18.
13. L. A. Thompson, J. A. Ellman, "Synthesis and Applications of small Molecule Libraries", *Chem. Rev.* **1996**, *96*, 555.
14. This review focuses on highlighting concepts with illustrative examples. Experimental information and practical advice on fluorous techniques can be found in the following two sources: 1) L. P. Barthel-Rosa, J. A. Gladysz, "Chemistry in fluorous media: a user's guide to practical considerations in the application of fluorous catalysts and reagents", *Coord. Chem. Rev.* **1999**, *192*, 587, 2) D. P. Curran, S. Hadida, A. Studer, M. He, S.-Y. Kim, Z. Luo, M. Larhed, A. Hallberg, B. Linclau, Experimental Techniques in Fluorous Synthesis: A user's Guide, in "Combinatorial Chemistry: A Practical Approach", H. Fenniri, ed., Oxford Univ. Press, in press.

15. The discussion is oversimplified since increasing the number of fluorines does not always increase the partition coefficient into a fluorous phase.

16. P. Wipf, J.-L. Methot, "Silver(I)-Catalyzed Addition of Zirconocenes to Epoxy Esters: A New Entry to 1,4-Dicarbonyl Compounds and Pyridazinones", *Org. Lett.* **1999**, 1253.

17. D. Crich, X. Hao, M. A. Lucas, "Design, Application and Recovery of a Minimally Fluorous Diaryl Diselenide for Catalysis of Stannane-Mediated Radical Chain Reductions", *Org. Lett.* **1999**, *1*, 1426.

18. D. P. Curran, s. Hadida, M. He, "Thermal Allylations of Aldehydes with a Fluorous Allylstannane. Separation of Organic and Fluorous Products by Solid Phase Extraction with Fluorous Reverse Phase Silica Gel", *J. Org. Chem.* **1997**, *62*, 6715.

19. N. D. Danielson, L. G. Beaver, J. Wangsa, "Fluoropolymers and Fluorocarbon Bonded Phases as Column Packings for Liquid Chromatography", *J. Chromat.* **1991**, *544*, 187.

20. The technique now called "fluorous biphasic catalysis" was apparently first described in the Ph.D. thesis of M. Vogt in 1991; however, these studies did not become known to the community until sometime later. W. Keim, M. Vogt, P. Wasserscheid, B. Driessen-Holscher, "Perfluorinated polyethers for the immobilization of homogeneous nickel catalysts", *J. Mol. Catal. A Chem.* **1999**, *139*, 171.

21. I. T. Horváth, J. Rábai, "Facile Catalyst Separation without Water: Fluorous Biphase Hydroformylation of Olefins", *Science* **1994**, *266*, 72.

22. I. T. Horváth, "Fluorous Biphase Chemistry", *Acc. Chem. Res.* **1998**, *31*, 641.

23. M. Cavazzini, F. Montanari, G. Pozzi, S. Quici, "Perfluorocarbon-Soluble Catalysts and Reagents and the Application of FBS (Fluorous Biphase System) to Organic Synthesis", *J. Fluorine Chem.* **1999**, *94*, 183.

24. R. H. Fish, "Fluorous Biphasic Catalysis: A New Paradigm for the Separation of Homogeneous Catalysts from their Reaction Substrates and products", *Chem. Eur. J.* **1999**, *5*, 1677.

25. E. de Wolf, G. van Koten, B. J. Deelman, "Fluorous Phase Separation Techniques in Catalysis", *Chem. Soc. Rev.* **1999**, *28*, 37.

26. L. V. Dinh, J. Gladysz, "Transition Metal Catalysis in Fluorous Media: Extension of a New Immobilization Principle to Biphasic and Monophasic Rhodium-Catalyzed Hydrosilylations of Ketones and Enones", *Tetrahedron Lett.* **1999**, *40*, 8995.

27. Y. Nakamura, S. Takeuchi, Y. Ohgo, D. P. Curran, *Tetrahedron Lett.*, **2000**, *41*, 57.

28. This provides new options for synthesis and purification of sensitive organometallic complexes: N. Spetseris, S. Hadida, D. P. Curran, T. Y. Meyer, "Organic/fluorous phase extraction: A new tool for the isolation of organometallic complexes", *Organometal.* **1998**, *17*, 1458.

29. However, microwave irradiation can induce otherwise unsuccessful reactions to occur: K. Olofsson, S. Y. Kim, M. Larhed, D. P. Curran, A. Hallberg, "High-speed, highly fluorous organic reactions", *J. Org. Chem.* **1999**, *64*, 4539.

30. D. P. Curran, S. Hadida, "Tris(2-(Perfluorohexyl)tin Hybride: A New Fluorous Reagent for Use in Traditional Organic Synthesis and Liquid Phase Combinatorial Synthesis", *J. Am. Chem. Soc.* **1996**, *118*, 2531.

31. J. J. Maul, P. J. Ostrowski, G. A. Ublacker, B. Linclau, D. P. Curran, In *Tropics in Current chemistry, "Modern Solvents in Organic Synthesis"*; P. Knochel, Ed.; Springer-Verlag: Berlin, 1999; Vol. 206; p 80.

32. As an aside, benzotrifluoride is a slightly polar, non-Lewis basic solvent that has favorable properties for many kinds of organic reactions. A. Ogawa, D. P. Curran, "Benzotrifluoride: A useful alternative solvent for organic reactions currently conducted in dichloromethane and related solvents", *J. Org. Chem.* **1997**, *62*, 450.

33. M. Hoshino, P. Degenkolb, D. P. Curran, "Palladium-Catalyzed Stille Couplings with Fluorous Tin Reactants", *J. Org. Chem.* **1997**, *62*, 8341.

34. D. P. Curran, Z. Luo, P. Degenkolb, "Propylene Spaced" Allyl Tin Reagents: A New Class of Fluorous Tin Reagents for Allylations under Radical and Metal-Catalyzed Conditions", *Bioorg. Med. Chem. Lett.* **1998**, *8*, 2403.

35. D. P. Curran, S. Hadida, S. Y. Kim, "Tris(2-Perfluorohexylethyl)tin Azide: A New Reagent for Preparation of 5-Substituted Tetrazoles from Nitriles with Purification by Fluorous Organic Liquid-Liquid Extraction", *Tetrahedron* **1999**, *55*, 8997.

36. The length and nature of the spacer can have major effects on certain reactions. For example, this ionic reaction does not succeed with the reagent bearing an ethylene spacer.

37. D. P. Curran, Z. Luo, "Rapid, Parallel Synthesis of Homoallylic Alcohols by Lewis Acid Mediated Allylations of Aldehydes with New Fluorous Allyl Stannanes", *Med. Chem. Res.* **1998**, *8*, 261.

38. D. P. Curran, S. Hadida, S. Y. Kim, Z. Y. Luo, "Fluorous Tin Hybrides: A New Family of Reagents for Use and Reuse in Radical Reactions", *J. Am. Chem. Soc.* **1999**, *121*, 6607.

39. We usually conduct solid phase extractions on small scale and have not focused on the recovery of the tin reagents. However, in principle this should be possible for many types of reactions.

40. For a short review highlighting the attractive features for parallel synthesis, see: D. P. Curran, "Parallel Synthesis with Fluorous Reagents and Reactants", *Med. Res. Rev.* **1999**, *19*, 432.

41. R. Noyori, "Supercritical Fluids: Introduction", *Chem. Rev.* **1999**, *99*, 353.

42. S. Hadida, M. S. Super, E. J. Beckman, D. P. Curran, "Radical Reactions with Alkyl and Fluoroalkyl (Fluorous) Tin Hybride Reagents in Supercritical CO_2", *J. Am. Chem. Soc.* **1997**, *119*, 7406.

43. D. Koch, W. Leitner, "Rhodium-Catalyzed Hydroformylation in Supercritical Carbon Dioxide", *J. Am. Chem. Soc.* **1998**, *120*, 13398.

44. By analogy to the usual usage of "solid phase synthesis", we intend "fluorous synthesis" to mean techniques in which the substrate or product is fluorous. However, the term "fluorous synthesis" is also sometimes used more broadly to encompass essentially any fluorous technique.

45. A. Studer, S. Hadida, R. Ferritto, S. Y. Kim, P. Jeger, P. Wipf, D. P. Curran, "Fluorous Synthesis: A Fluorous-Phase Strategy for Improving Separation Efficiency in Organic Synthesis", *Science* **1997**, *275*, 823.

46. A. Studer, D. P. Curran, "A Strategic Alternative to Solid Phase Synthesis: Preparation of A Small Isocazoline Library by "Fluorous Synthesis", *Tetrahedron* **1997**, *53*, 6681.

47. New fluorous reactions also present entertaining naming opportunities. For example, we call the fluorous Ugi reaction the "flugi" reaction. The "fluginelli" reactions follows accordingly.

48. A. Studer, P. Jeger, P. Wipf, D. P. Curran, "Fluorous Synthesis: Fluorous Protocols for the Ugi and Biginelli Multicomponent Condensations", *J. Org. Chem.* **1997**, *62*, 2917.

49. D. P. Curran, R. Ferritto, Y. Hua, "Preparation of a Fluorous Benzyl Protecting Group and its Use in a Fluorous Synthesis Approach to a Disaccharide", *Tetrazhedron Lett.* **1998**, *39*, 4937.

50. D. P. Curran, Z. Y. Luo, "Fluorous Synthesis with Fewer Fluorines (Light Fluorous Synthesis): Separation of Tagged from Untagged Products by Solid-Phase Extraction with Fluorous Reverse-Phase Silica Gel", *J. Am. Chem. Soc.* **1999**, *121*, 9069.

51. B. Linclau, A. K. Singh, D. P. Curran, "Organic-Fluorous Phase Switches: A Fluorous Amine Scavenger for Purification in Solution Phase Parallal Synthesis", *J. Org. Chem.* **1999**, *64*, 2835.

Domino Reaction in Organic Synthesis.
An Approach to Efficiency, Elegance, Ecological Benefit, Economic Advantage and Preservation of Our Resources in Chemical Transformations

Lutz F. Tietze and Frank Haunert

Institute of Organic Chemistry of the Georg-August-Universität, Tammannstraße 2, D-37075 Göttingen, Germany

Phone: +49 551 39 3271
Fax: (internat.)
+49 (0) 551 39 9476
e-mail: ltietze@gwdg.de

Dedicated to our respected colleague and friend Prof. Jürgen Troe on the occasion of his 60th birthday

Keyword ■ Carbocycles ■ Cascade Reactions ■ Cycloadditions ■ Combinatorial Chemistry ■ Domino Reactions ■ Enantioselective Transformations ■ Ene Reactions ■ Heterocycles ■ Natural products ■ Preservation of Resources and Environment ■ Sigmatropic Rearrangements ■ Tandem Reactions ■ Transition Metal-Catalyzed Transformations

Concept: Organic synthesis is one of the centers of chemistry; its domain is not only the preparation of new materials and bioactive compounds but also the production of goods in industry. Its future lies in the development of new efficient methods which allow the formation of complex molecules in a few steps, starting from simple substrates. This is the strength of domino reactions, combining several bond-forming steps in one process under identical or nearly identical reaction conditions. This type of reaction will revolutionize organic synthesis, especially if it is performed in a catalytic way. It is not only economically highly efficient since it saves time and money but is also ecologically beneficial since it reduces the amount of waste formed in a synthesis and preserves our resources.

Abstract: In this chapter different types of domino-processes are described which consist of the combination of cationic, anionic, radical, pericyclic and transition metal-catalyzed as well other reactions. The methodology is used for the highly effective synthesis of carbocycles and heterocycles as well as of natural products and other interesting materials. It is also employed as an efficient tool in combinatorial chemistry.

Prologue

Synthesis is one of the hearts of organic chemistry, it is the foundation of the chemical and pharmaceutical industry and it is the basis of many interdisciplinary work with medicine and biology as well as it supplies the test probes for analytical investigations. Thus, the synthesis of relevant organic compounds such as natural products and analogues, drugs, diagnostics, agrochemicals and any kind of material is a main topic in academic and industrial chemistry. Whereas the race for more selectivity was one of the driving forces in the past – it is still going on – the main goal is now efficiency, the compatibility with our environment, the preservation of our resources and also the economical advantage. This new view is clearly a change of a paradigm in synthesis.[1] The proportion of the numbers of steps and the increase of complexity is now an important standard for the quality of a synthesis. Multi-step syntheses with much more than 20 steps are clearly not state of the art since they are neither economically nor ecologically justifiable. In addition, the use of toxic reagents and solvents should be avoided, the amount of waste produced in a process must be reduced and finally one must deal carefully with our resources and our time.

In recent years, synthetic methodology has been developed to allow the synthesis of diversified substance libraries mainly for pharmaceutical testing in an automated way using either solid phase or solution chemistry.[2] Here, efficiency is also an important goal.

A general way to improve synthetic efficiency and also address the mentioned criteria as well as give access to a multitude of diversified molecules is the development of domino processes.[1a, 1d, 3] This methodology allows the formation of complex

Table 1. Classification of domino processes

1. Step	2. Step etc.
1a cationic	2a cationic
1b anionic	2b anionic
1c radical	2c radical
1d pericyclic	2d pericyclic
1e photochemical	2e photochemical
1f transition metal-catalyzed	2f transition metal-catalyzed
1g oxidative/reductive	2g oxidative/reductive

compounds starting from simple substrates in very few steps. Domino reactions are defined as processes of two or more bond forming transformations under widely identical conditions in which the subsequent reactions take place at the functionalities obtained in the former transformation. A way to judge the quality of a domino process is its bond forming efficiency that means the number of bonds which are formed in one process, the increase in structural complexity of the product compared to the substrates and finally its suitability for a general application especially in combinatorial chemistry. Domino reactions can be classified according to the mechanism of the single steps. Combination of the different reaction types as in Table 1 allows the creation of a multitude of domino processes. Many of these permutations are already known, but there is plenty of space for the development of new combinations. They can consist of the same but also of different reaction types. Most of the so far developed domino processes

belong to the first category and may include two or more cationic, anionic, radical, pericyclic or transition metal-catalyzed transformations.

An example for the combination of mechanistically different reactions is the anionic-pericyclic process such as the domino-Knoevenagel-hetero-Diels–Alder reaction. In the inter-intramolecular version of this process an aldehyde **1** containing a dienophile moiety and a 1,3-dicarbonyl compound **2** can be mixed together to give unusual heterocycles of diversified structures such as **4** via the intermediate formation of an 1-oxa-1,3-butadiene **3** (scheme 1).[4]

Several multi-component domino reactions for the preparation of libraries have been developed in the last years.[4] One of the first examples in solution is a four-component reaction described by Ugi et al. in which a carbocyclic acid **5**, an aldehyde **6** and an isocyanide **7** are condensed in presence of methanol to give amino acids **8** (scheme 2).[5]

A three-component domino reaction of isobutyraldehyde, N-benzylidene aniline and pyridine-2-thiol as nucleophile gives access to 1,2,3,4-tetrahydroquinolines as shown by Annunziata et al..[6]

As the first multi-component domino reaction on solid phase the above mentioned domino-Knoevenagel-hetero-Diels–Alder reaction was performed using a 1,3-dicarbonyl compound bound to a modified Merrifield resin.[7]

The principle of a domino process is also found in nature quite often as the efficient cyclotetrameriza-

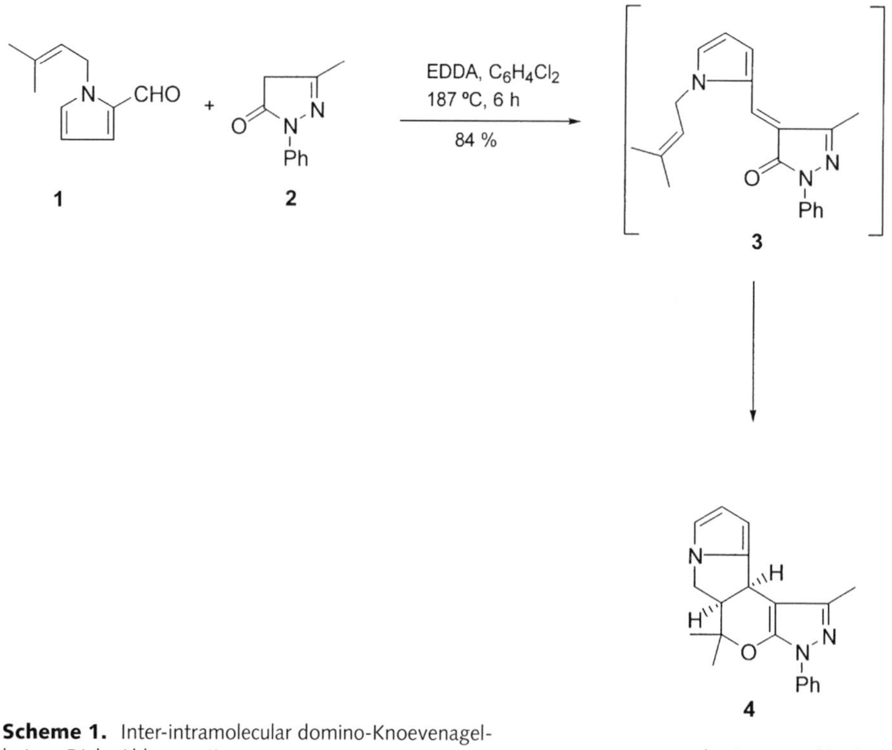

Scheme 1. Inter-intramolecular domino-Knoevenagel-hetero-Diels–Alder reaction

cis : trans = > 99 : 1

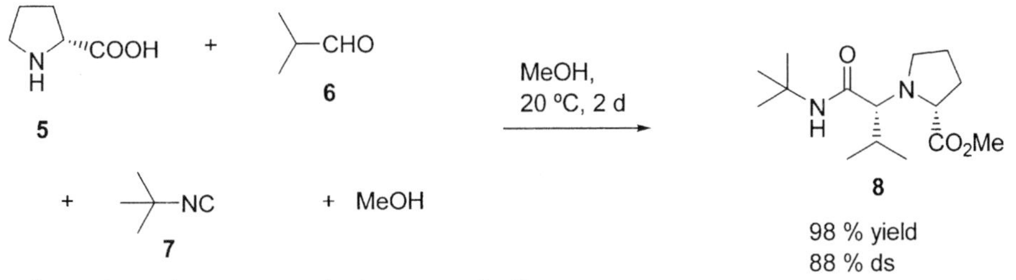

Scheme 2. Multi-component domino process for the synthesis of amino acids

tion of porphobilinogen **9** to give uroporphyrinogen **10**[8] or the cyclization of (*S*)-2,3-oxidosqualene **11** to form lanosterol **12** (scheme 3).[9]

An impressive example of a multi-step biomimetic domino process is the synthesis of codaphniphyllin by Heathcock et al. (see scheme 14). Another example is the highly efficient biomimetic synthesis of (+)-hirsutine by my group (see scheme 13). In addition several other total syntheses of natural products have been developed using domino processes; these are described in the different sections of this chapter.

Certain examples are known for asymmetric induction in domino reactions using either chiral substrates or educts with removable chiral auxiliaries. In contrast, only a few enantioselective domino reactions have been developed so far. The first example was described by us using a titanium complex of glucose diacetonide with 88 % ee.[10] Quite

recently, an enantioselective domino-inter-intranitroaldol reaction was presented by Shibasaki (see scheme 9) and an enantioselective domino-Heck–Tsuji–Trost reaction by Helmchen (see scheme 26).

In this article novel developments of the different types of domino processes are presented which are subdivided according to our classification. Since our first review on this topic and the book of Ho several overviews on special domino reactions have been published.[1a, 1d, 11]

Cationic Domino Processes

In the cationic domino process which is a synonym for an electrophilic reaction a carbocation is formed first, either formally or in reality, which under bond formation reacts with a nucleophile to form a new carbocation. In most of the known

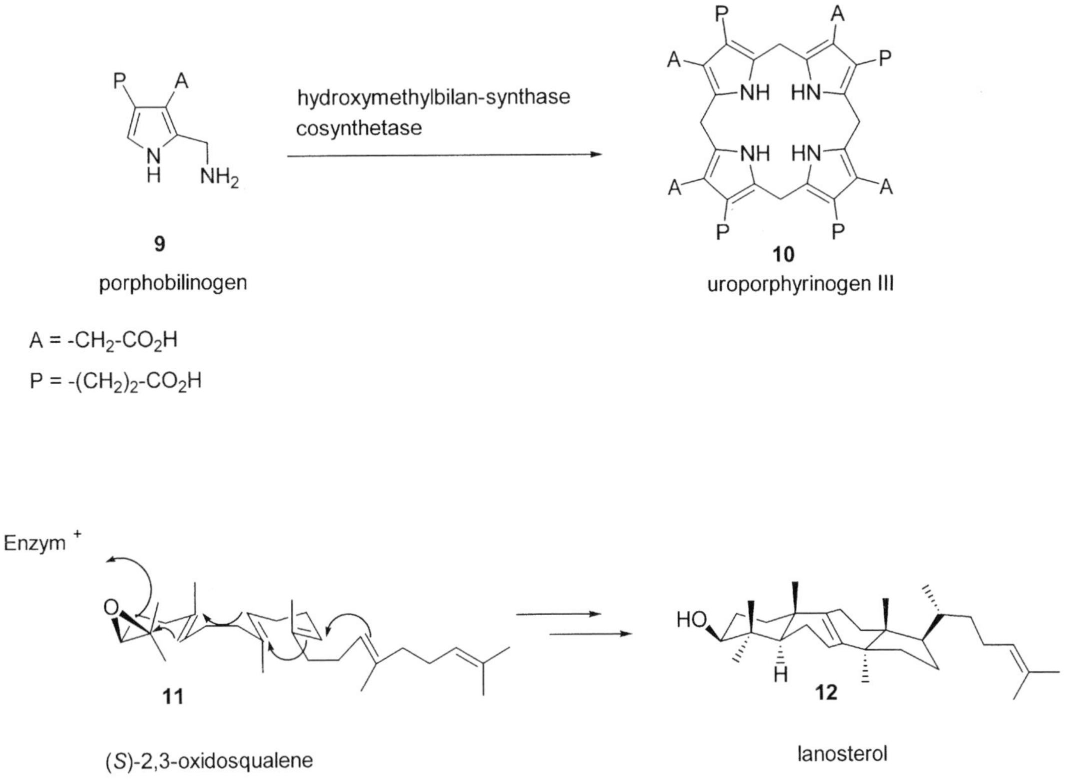

A = -CH₂-CO₂H

$A = -CH_2-CO_2H$

$P = -(CH_2)_2-CO_2H$

Scheme 3. Biosynthesis of uroporphyrinogen III **10** and lanosterol **12**

domino processes of this type another cationic process follows where the final carbocation is either stabilized by elimination of a proton or by addition of another nucleophile to give the product. However, a succession of a cationic by a pericyclic step is also possible as shown in scheme 7.

The formation of the primary carbocation can be achieved by treatment of an alkene or an epoxide with a Brønsted or a Lewis acid, by elimination of water from an alcohol or an alcohol from an acetal and by reaction of enones and imines with Lewis acids. The two latter reactions may also be classified under anionic domino reactions depending on the following steps.

Cationic–Cationic Domino Processes

The biomimetic cationic domino cyclization of an acyclic unsaturated substrate to give the tetracyclic scaffold of triterpenes and steroids is intensively described in the literature.[12] The concept has recently been used by Corey et al. to prepare enantiopure (+)-dammarenediol II **18** in an exceptional short way.[13] The synthesis demonstrates the power of the combination of cation-olefin polyannulation with the aldol cyclization for tetraannulation (scheme 4). Successive treatment of the acylsilane **13** with 2-propenyllithium **14** and the iodoalkane **15** efficiently yields the epoxytriene **16**. The Lewis acid

mediated tricyclization of **16** followed by removal of the silyl group and oxidative cleavage of the thioacetal subunit leads to the tricyclic diketone **17**. Only six further steps are necessary to get (+)-dammarenediol II **18**.

Another cationic–cationic domino reaction which follows the way of biosynthesis is used by Koert et al. for the synthesis of the whole left part of the polyether etheromycin.[14] The starting step is an acid catalyzed cyclization of a diepoxide to give a bistetrahydrofuran with creation of three new oxacycles in over 30 % yield.

A cationic–cationic domino process of a 3-hydroxytetrahydro-β-carboline which leads to a stereoselective preparation of (±)-ulein was described by Blechert et al.[15] Although the intermediates of this fast proceeding sequence could not be identified definitely, it is acceptable to assume that the process is initiated by the formation of a 2-indolyl carbocation, which cyclizes after tautomerization to a more stable iminium ion.

Recently, we have shown that iminium ions can induce a hydride shift to form a new carbocation which then reacts with a nucleophile. By this way the novel unusual bridged steroid alkaloids **25** were prepared from the secoestron derivative **19** (scheme 5).[16] Treatment of **20** obtained from **19** by hydrogenation with aniline or an aniline derivative **21** containing an electron-withdrawing group in the presence of the Lewis acid BF$_3$·OEt$_2$ leads to the iminium

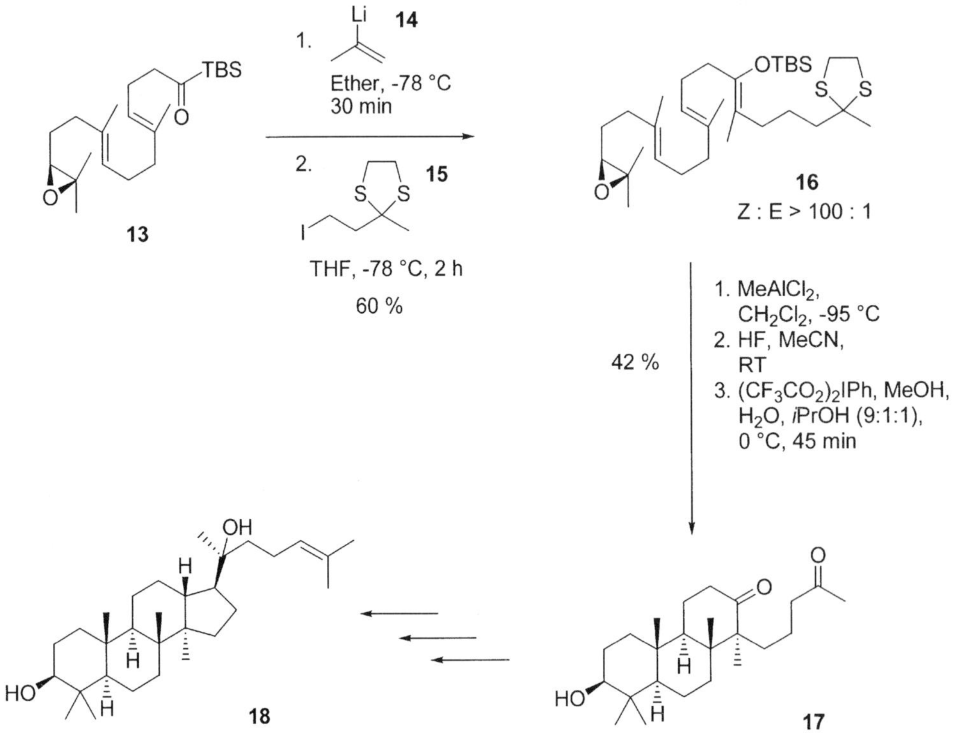

Scheme 4. Cationic domino polycyclization to the triterpene dammarenediol II **18**

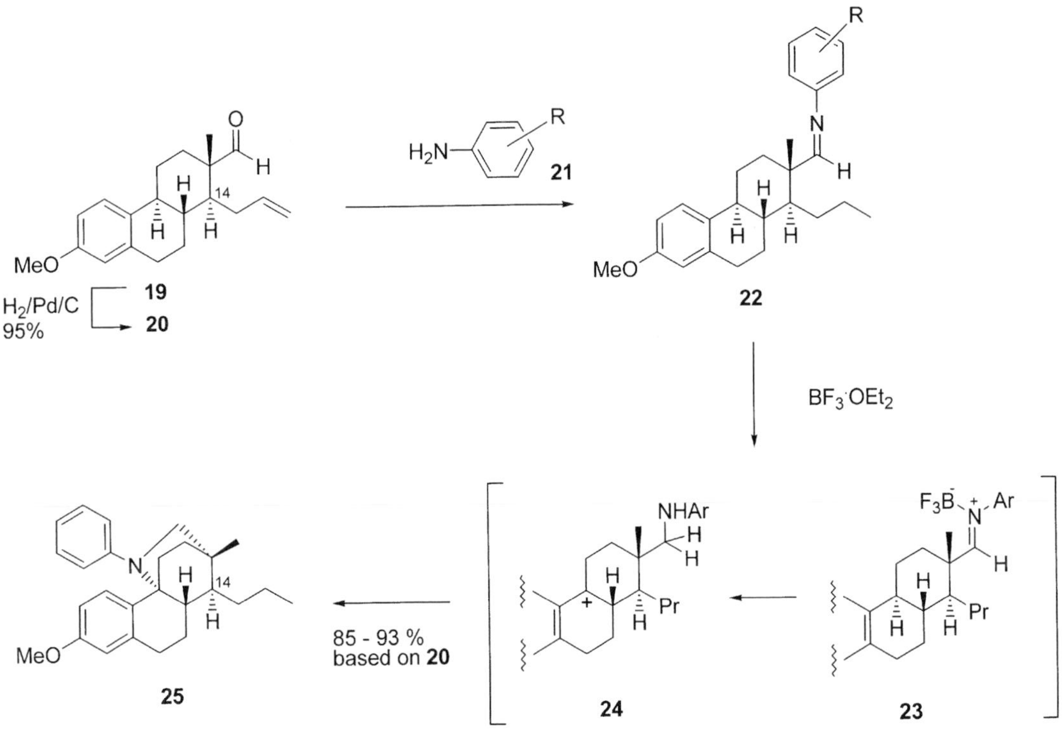

Scheme 5. Synthesis of bridged steroid alkaloids by a cationic domino 1,5-H-shift of a benzylic hydride

ion **23**. This undergoes a 1,5-hydride shift to give **24**, which contains a secondary amine moiety and a carbocation. Finally, addition of the amino group to the carbocationic center in **24** yields **25** in an excellent yield as single diastereomer.

Barluenga et al. developed a novel exo-endo-cyclization of α,ω-diynes by use of bis(pyridyl) iodonium(I)-tetrafluoroborate (IPy$_2$BF$_4$) as catalyst.[17] The reaction follows a cationic-cationic pathway (scheme 6). First the electrophilic iodo ion reacts with the triple bond of **26** to give the relative stable vinyl cation **28**. Ring closure leads to a seven-

membered ring containing another vinyl cation **29**, which then cyclizes to give the final tricycle **27**.

Cationic–Pericyclic Processes

An attractive strategy to functionalized cycloheptenes was developed by West al.[18] As illustrated in scheme 7, the substrate **30** is exposed to FeCl$_3$ as Lewis acid at −30 °C to induce a Nazarov electrocyclization to give the carbocation **31**. Under the reaction conditions **31** undergoes a [4+3]-cycloaddition

Scheme 6. Intramolecular cyclization of alkynylsulfides

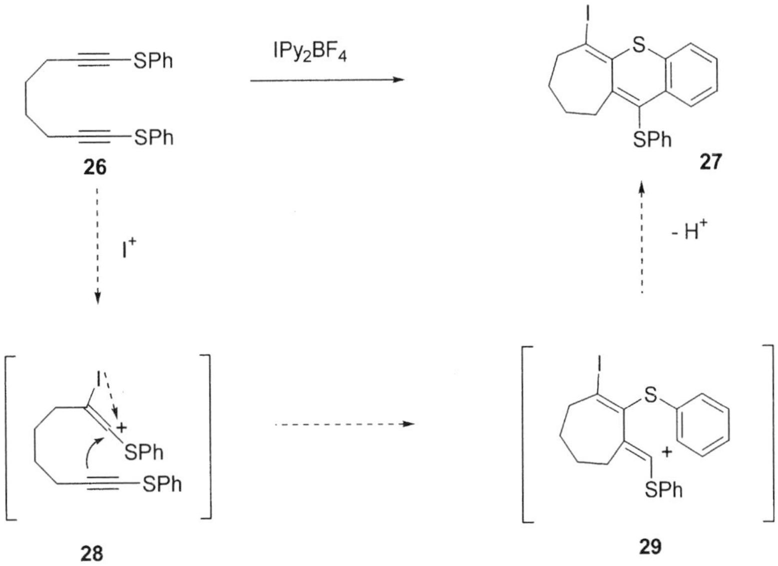

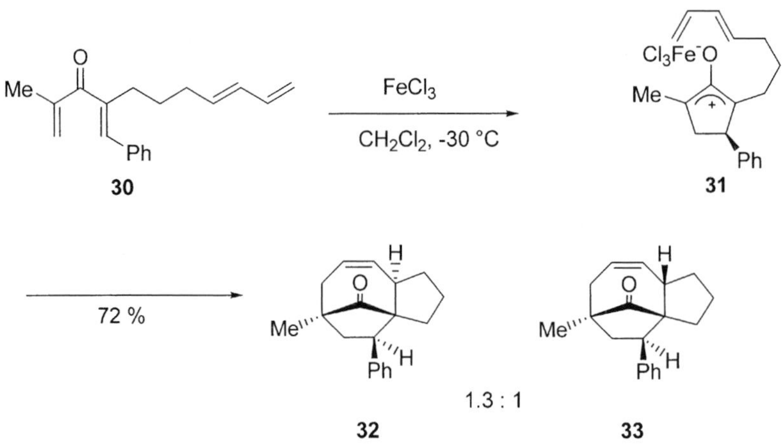

Scheme 7. Cationic domino cycloisomerization of tetraenones

to yield a 1.3 : 1 mixture of the two diastereomers **32** and **33** in fairly good yield. The two other possible diastereomers are not formed. The diastereoselectivity can be rationalized by a facial differentiation in the cycloaddition with preferred approach from the less hindered site of the cyclic oxyallyl cation. However, the formation of **32** and **33** indicates that there is nearly no preference of the endo- over the exo-orientation of the diene and the oxyallyl cation in the transition state.

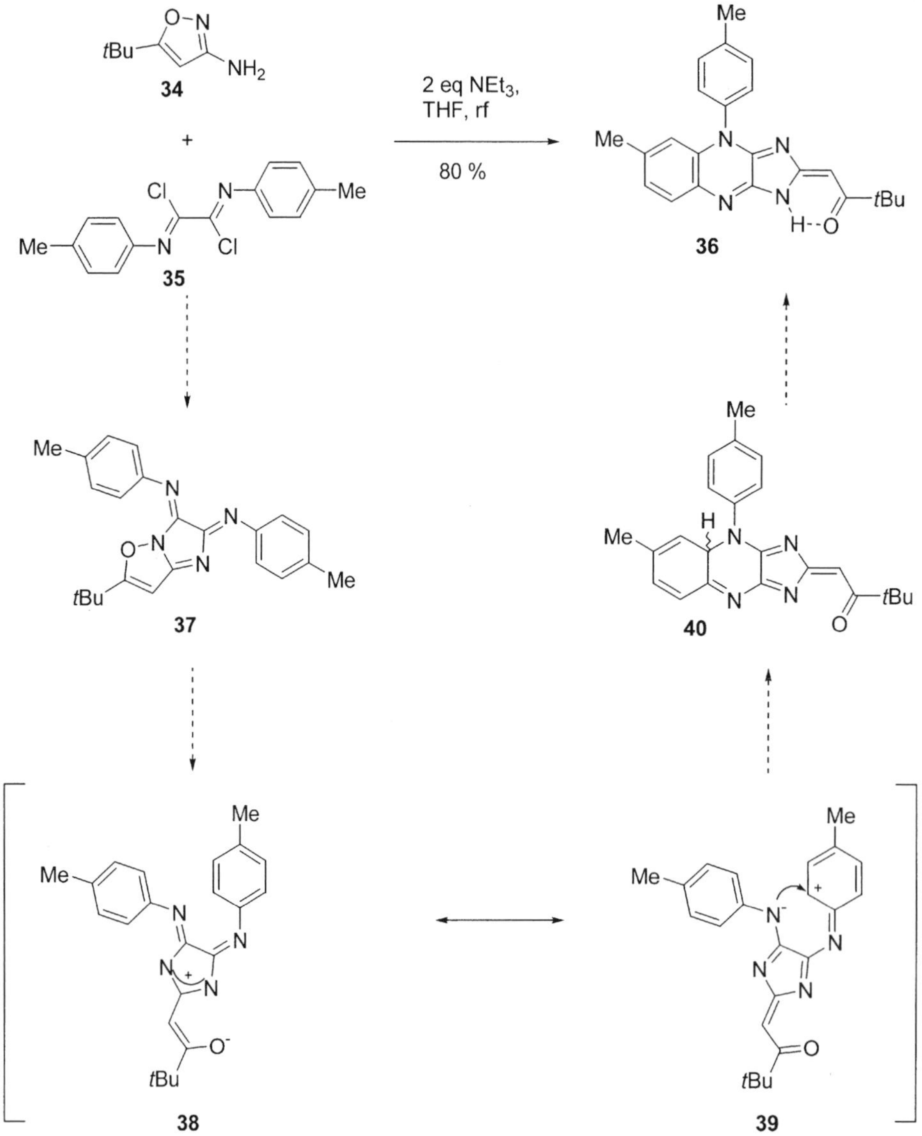

Scheme 8. Synthesis of 1,4-dihydro-imidazo [4,5-b]quinoxalines

Anionic Domino Processes

The anionic domino reaction is the most often encountered domino reaction in literature.

In this process the primary step is the formation of an anion, which is a synonym for a nucleophile, mostly by deprotonation using a base. It follows a reaction with an electrophile to give a new anion which in the anionic-anionic process again reacts with an electrophile. The reaction is then completed either by addition of another electrophile as a proton or by elimination of an X⁻ group. Besides the anionic–anionic process there are several examples of anionic-pericyclic domino reactions as the domino-Knoevenagel-hetero-Diels–Alder reaction in which after the first step an 1-oxa-1,3-butadiene is formed.

Anionic–Anionic Domino Processes

An efficient approach to biologically relevant 1,4-dihydro-imidazo[4,5-b]quinoxalines **36** is described by Langer et al. employing a five-step anionic domino reaction of 3-aminoisoxazol **34** with bis(imidoyl)chlorides **35** in the presence of triethylamine in high yield (scheme 8).[19] Initially, the isoxazo[2,1-b]imidazoline **37** is formed, which leads to the zwitterionic intermediate **38/39** by cleavage of the N–O bond. Subsequent nucleophilic attack at the aryl moiety by the nitrogen atom gives **40**, which tautomerizes to the final heterocycle **36**. The stereoselective formation of the semicyclic double bond with an (E)-configuration can be explained by the existence of a stabilizing intramolecular N–H···O bond.

A novel example of a catalytic enantioselective domino process[20] is the inter-intramolecular nitro-aldol reaction described by Shibasaki et al. which generates substituted indanones. As catalyst a praseodym-heterobimetallic complex with binaphthol as chiral ligand is employed. Treatment of keto-aldehyde **41** with nitromethane in the presence of the catalyst **46** at –40 °C and successive warming to room temperature affords directly the product **42** in an overall yield of 41 % and 96 % ee after several recrystallizations (scheme 9). As intermediates the nitromethane adduct **43** and the hemiacetal **44** can be proposed. In a second aldol reaction **44** leads to **45** which isomerizes to the thermodynamically more stable epimer **42**.

An interesting example of a facile anionic multi-step one-pot reaction is the condensation of 2-ethylpropenal with substituted cyclopentanones using DBU as base to give functionalized cyclohep-

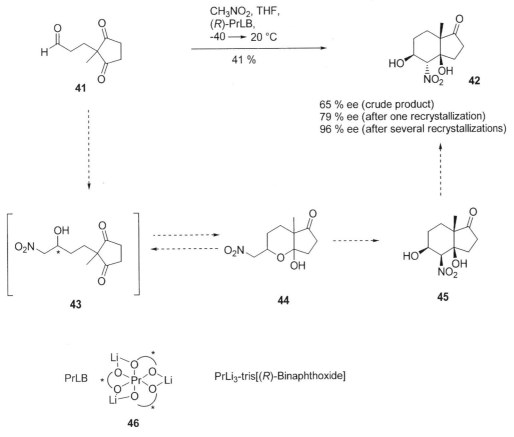

Scheme 9. Enantioselective domino reaction for the formation of hydrindanones with heterobimetallic catalysts

Scheme 10. Three-component domino reaction of nitroenones

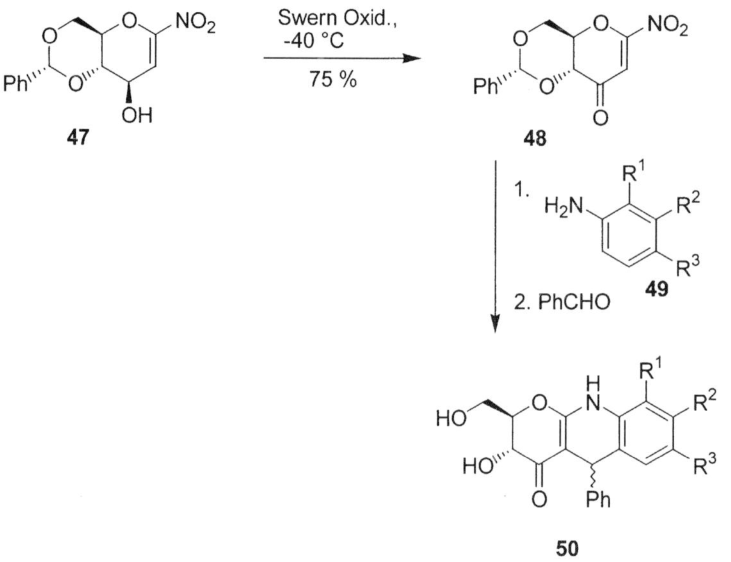

tenedicarboxylic acid stereoselectively as single diastereomers. Rodriguez et al.[21] described the overall transformation as a sequence involving a Michael addition, an intramolecular aldol condensation, a retro-Dieckmann reaction followed by dehydration and a chemoselective ester saponification. The described sequence has also been applied to several other 2-substituted α,β-unsaturated aldehydes including compounds with propargylic and functionalized alkyl chains. In the synthesis of methyl (R)-4-methylene pipecolate developed by Agami et al., the formation of an iminium ion by condensation of a β-aminoalcohol with glyoxal is followed by an intramolecular reaction with an allylsilane to give a bicyclic intermediate.[22]

Domino reactions increasingly gain importance in the search for new drugs. Especially appropriate is the use of multi-component reactions in solution combinatorial chemistry. In such a process described by Wessel et al.[23] an alkoxy-nitroenone 48 was treated with different anilines 49 to give ketene-NO-acetals which in the presence of aromatic aldehydes and TfOH are transformed into 50 (scheme 10). The substrate 48 is readily available by oxidation of the nitrosugar 47.

An impressive example for the successful use of domino reactions for the synthesis of pharmacological lead structures was described by Paulsen et al.[24] Recently, the difluoro compound 57 has been identified as highly potent inhibitor of the cholesterin-ester-transferprotein (CETP), which is responsible for a transfer of cholesterin from high-density lipoprotein (HDL) to low-density lipoprotein (LDL). This clearly results in an increase of LDL and a decrease of HDL which raise the risk of coronary heart deseases. The core structure of 57 is now accessible efficiently by a combination of a Mukaiyama–Michael,

a Michael and an aldol reaction (scheme 11). Thus, treatment of the silyl enol ether 51, the cyclohexanone 52 and the enone 53 in the presence of the Lewis acid TiCl$_4$/Ti(OiPr)$_4$ first leads to the adduct 54 which reacts with the enone 53 to give 55. The end of the domino process is an intramolecular aldol reaction which yields 56. 57 can be obtained from 56 by elimination, aromatization, enantioselective reduction of one of the carbonyl groups, OH protection, reduction of the second carbonyl group, treatment with DAST (Et$_2$NSF$_3$) for the introduction of the fluoro atom and deprotection.

Anionic–Radical Processes

Novel anionic–radical domino processes have been developed by Molander et al. using one-electron donor reagents like SmI$_2$ (scheme 12).[25] In this reaction an intramolecular acyl substitution and an intramolecular ketyl-olefin coupling take place. The process can be extended by a third C–C bond forming reaction if one uses ketones as solvents for trapping the generated organosamarium species. Reaction of the carboxylic acid ester 58, bearing an iodoalkyl group and an alkene moiety, with 4 equivalents of samarium diiodide in presence of acetone yields the diol 60. As the first steps an iodide-samarium exchange followed by an anionic cyclization to give a cyclohexanone takes place. This is then reduced to give a ketyl-radical, which leads to the organosamarium-intermediate 59 by a radical cyclization. Addition to acetone yields product 60.

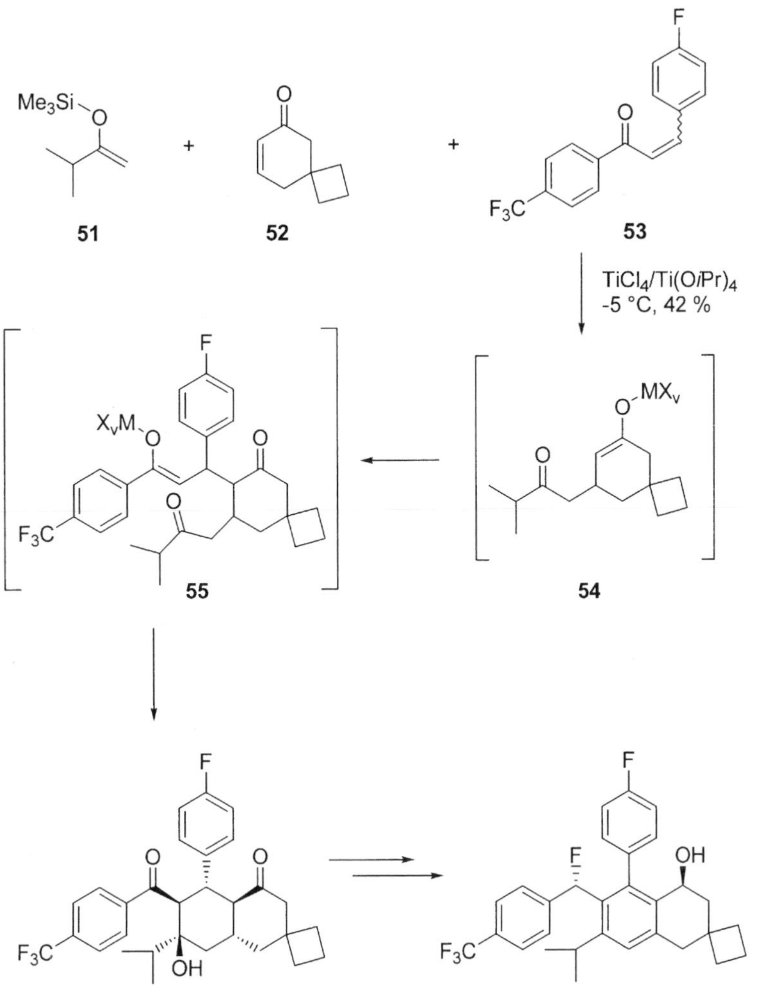

Scheme 11. Anionic–anionic multi-component domino process

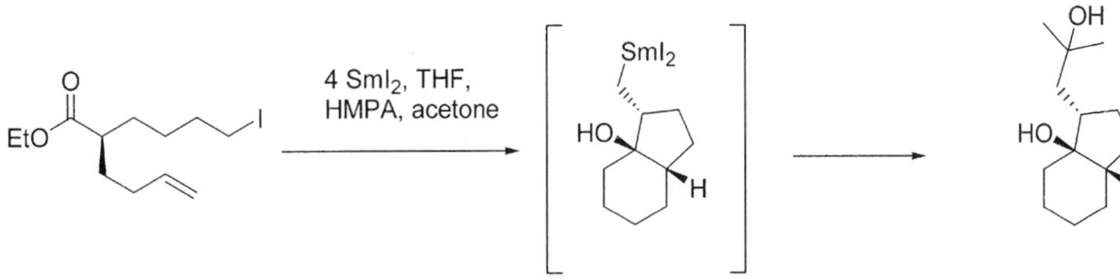

Scheme 12. Domino anionic-radical polycyclizations with one-electron-donor reagents

Anionic–Pericyclic Processes

An efficient and also elegant synthesis of the active anti-influenza A virus indole alkaloid hirsutine **67** is performed by an inter-intermolecular anionic-pericyclic three- component domino reaction followed by solvolysis and hydrogenation (scheme 13).[26] The synthetic sequence developed by us contains first a Knoevenagel condensation of enantiopure **61** and **62** with the formation of the 1,3-oxabutadiene **64**, which then undergoes a facial-differentiating hetero-Diels–Alder reaction with the vinyl ether **63** with an excellent 1,3-induction of >25:1. In a one-pot process the crude product is treated with methanol in presence of potassium carbonate and immediately afterwards hydrogenated using palladium on charcoal as catalyst. In the solvolysis the lactone moiety in **65** is transformed into a methyl ester with concomitant release of an aldehyde group. Under the hydrogena-

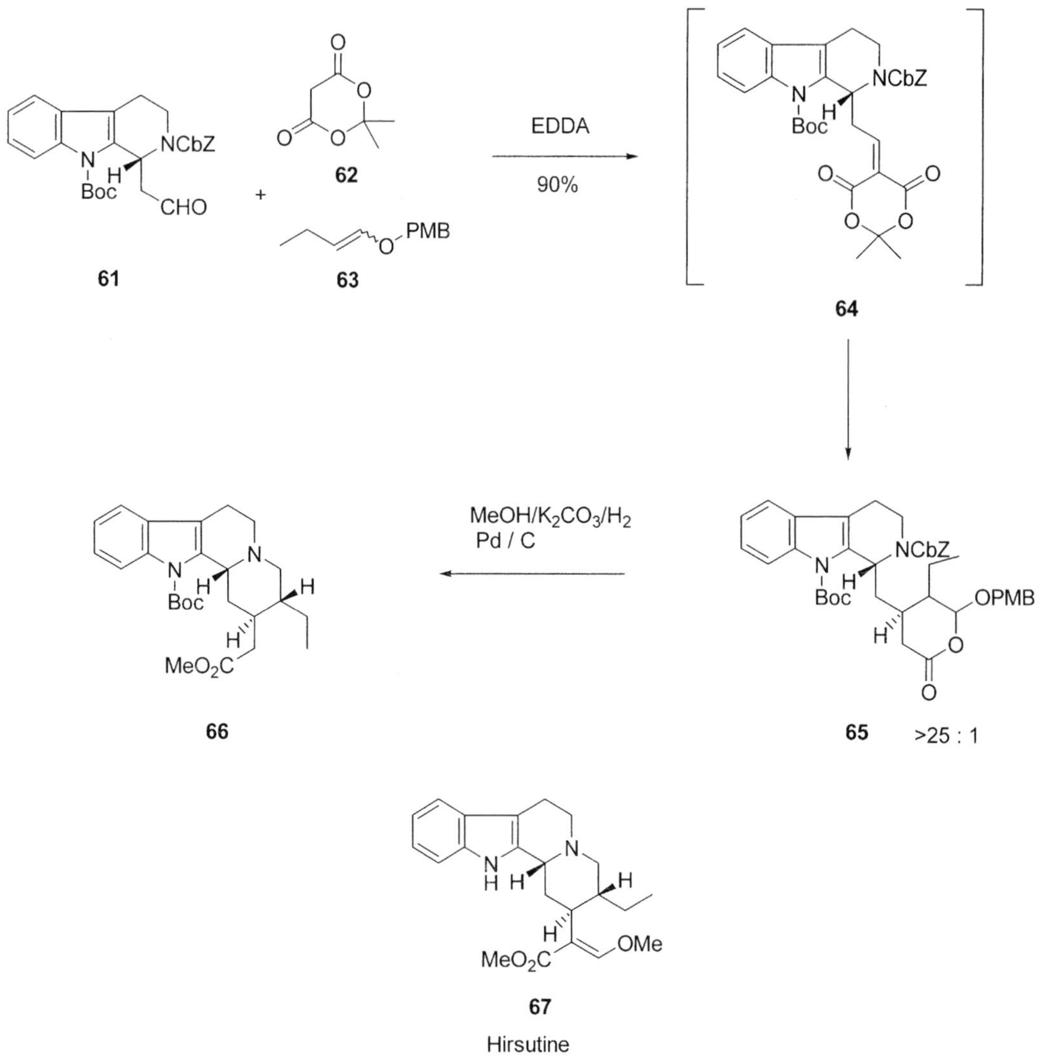

Scheme 13. Enantioselective synthesis of hirsutine **67** by a Knoevenagel-hetero-Diels–Alder solvolysis hydrogenation process

tion conditions the CbZ-group is then cleaved and the resulting secondary amine forms an enamine with the aldehyde which is finally hydrogenated selectively under stereoelectronic control to give the desired product **66**. Removal of the *tert*-butoxycarbonyl group, condensation with methyl formate and treatment with diazomethane leads to the enantiopure indole alkaloid hirsutine **67**.

Another beautiful anionic-pericyclic domino process is the already mentioned biomimetic synthesis of the daphniphyllum alkaloid (+)-codaphniphyllin by Heathcock et al.[27] The diol **68** is employed as the starting material which is obtained from oxocyclopentanecarboxylic acid methyl ester via an enantioselective hydrogenation using [RuCl(C$_6$H$_6$)-(*R*)-(+)BINAP]Cl as catalyst with 93 % ee. Oxidation of the diol **68** leads to the dialdehyde **70** which by condensation with methylamine yields the unsaturated iminium salt **71** (scheme 14). This 2-aza-1,3-butadiene reacts in an intramolecular

hetero-Diels–Alder reaction and successive intramolecular aza-Prins reaction to give the hexacyclic carbocation **73** via **72**. **73** is stabilized by a hydride shift from the N-methyl group and not as anticipated by loss of a proton in α-position to the carbocation. In the reaction an isopropyl group and an iminium ion are formed, which after hydrolysis yields the desired product **69** with a secondary amino functionality. A few additional manipulations allow the synthesis of (+)-codaphniphyllin **75** from **69**.

Mixed Anionic Domino Processes

An elegant approach to polyquinanes has recently been described by Paquette et al.[28] By means of an anionic–pericyclic domino reaction four new C–C bonds are formed in one process. First, a two-fold addition of a vinyl lithium reagent, e.g. cyclopentyl lithium to diisopropyl squarate takes

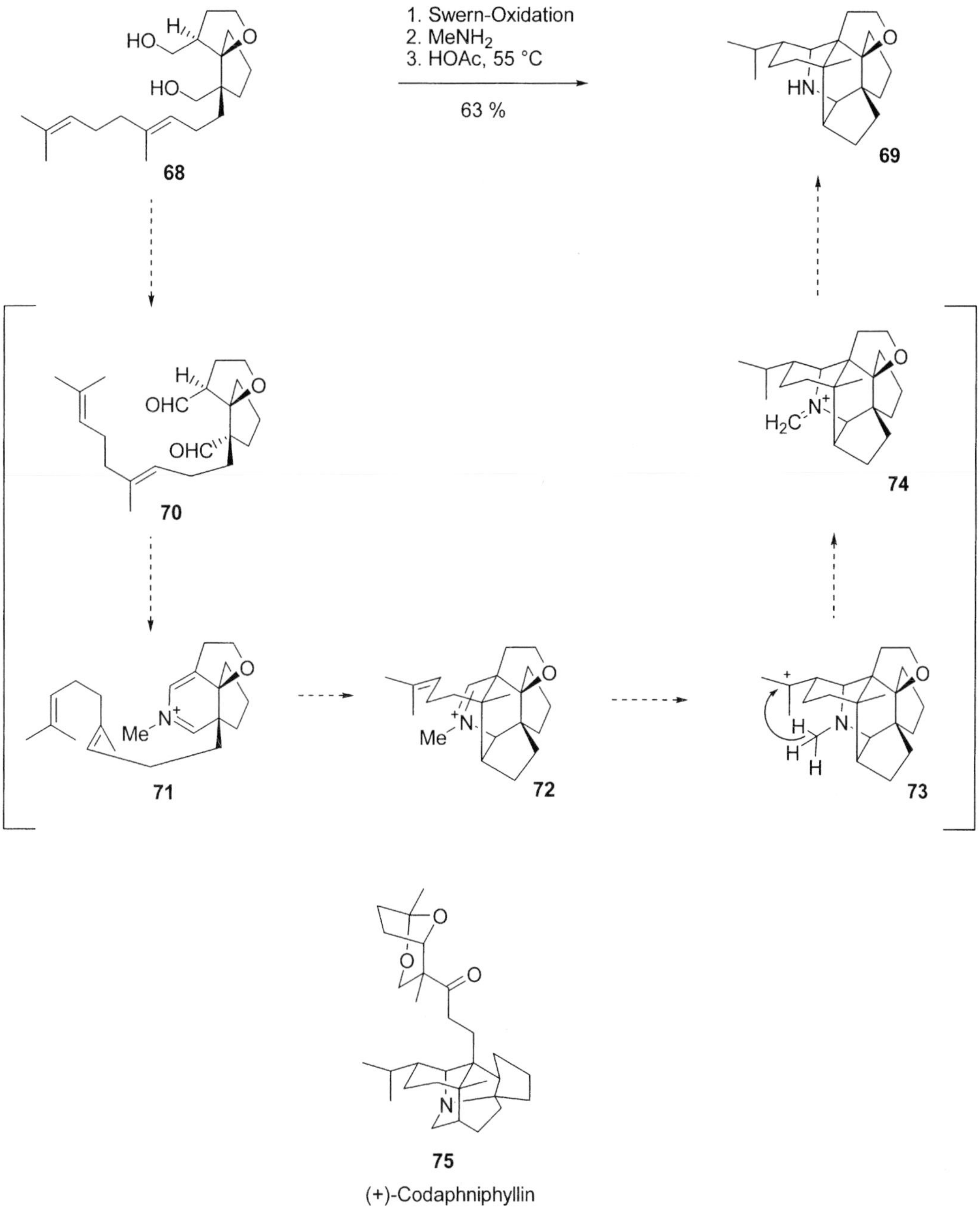

Scheme 14. Biomimetic synthesis of (+)-codaphniphyllin **75**

place. Depending on the stereochemistry of the addition two different reaction pathways are found for the subsequent transformations. In case of a trans 1,2-addition a conrotatoric 4π-cycloreversion and a 8π-cyclization is observed and in case of a cis 1,2-addition a dianionic oxy-Cope rearrangement follows with highly substituted cyclooctadienones as the products. In both cases successive protonation irreversibly yields the final tetraquinanes by a trans-annular aldol reaction. The value of the described method has raised recently due to the commercial availability of different vinyl lithium reagents.

Risch et al. have developed a new multi-step dom-ino reaction in which benzaldehyde reacts with a secondary amine to give an azomethine ylide.[29] Reaction with two more moles of the aldehyde leads to pyrrole derivatives by cyclization and aromatiza-tion. Also an anionic domino reaction is introduced by Langer et al. to prepare substituted imidazole and pyrimidine derivatives.[30] Initial reaction of three molecules of a nitrile with an allene dianion and subsequent loss of one and attack of two more mole-cules of the nitrile leads to the heterocycles. In gen-eral, anionic domino reactions are highly feasible for the preparation of functionalized heterocycles, as isoxazolidines by Grigg et al.,[31] isoxazolines by Rosini et al.,[32] tetrahydrofurans by Hassner et al.[33] and enantiopure oxetanones by Schick et al.[34] Kobayashi et al. prepared various γ-acyl-δ-lactam derivatives on the basis of three- and four-component coupling reactions by Lewis acid-catalyzed Michael-imino aldol reactions.[35] In the presence of a Lewis acid, silyl enolates react with α,β-unsaturated carbonyl compounds to afford the corresponding Michael adducts as silyl enolates. Subsequent aldol reaction with imines followed by cyclization leads to the lactams. Takeda et al. developed an interesting route to functionalized carbocycles via a domino Brook-rearrangement and successive intramolecular Michael reaction.[36] Addition of phenyl lithium to acylsilanes in THF results in the formation of four- to six-membered carbocycles.

Michael additions as part of a domino process are quite common. Another interesting example is the twofold Michael reaction of chiral α-bromo-α,β-unsaturated esters with cyclic dienolates followed by a γ-elimination to give tricyclo[3.2.1.0[2,7]] octanes by Spitzner et al.[37] Further processes of this type are the construction of functionalized naphthalenes by Krohn et al.,[38] the synthesis of α,β-unsaturated esters using a stereoselective thio-Michael-aldol-domino reaction by Kamimura et al.[39] as well as the preparation of substituted cyclopentenols via a secondary amine mediated Baylis-Hillman reaction by Murphy et al.[40] Spitzner et al. recently introduced also a resin-bound anionically induced domino reaction.[41] Furthermore anionic domino reactions are used in the efficient syntheses of natural products. For example Parsons et al. developed a short route to (±)-anatoxin-a via an anionically induced small ring opening/ring closure and ring opening process.[42] Indole alkaloids of the vallesiachotamine group are efficiently synthesized by our group also employing an anionic domino process.[43] Schmittel et al. presented a novel domino process consisting of two aldol reactions followed by a hemiacetal formation which leads to tetrahydropyran-2,4-diols containing five stereogenic centers.[44] In the course of the reaction an aldehyde as benzaldehyde or butanal reacts with a titanium bis-(Z)-enolate of an aromatic ketone like propiophenone (acetophenone is not a suitable substrate) to yield the anti-selective aldol product which undergoes a second aldol reaction to form a titanium bound hemiacetal. Aqueous work-up leads stereoselectively to the products. Beyond the described anionic domino processes several other types of combinations exist. One example is a domino Wittig-Diels–Alder reaction described by Jarosz which allows an easy entry to enantiopure highly oxygenated decalins.[45] Another interesting example is the synthetic elaboration of sultones by Metz et al. using a domino alkylation-desulfurization process.[46] Hiemstra et al. describe in their enanti-

oselective synthesis of (+)-gelsedin a new iodide induced intramolecular addition of an allene to a monocyclic iminium ion to give a bicyclic vinyl iodide.[47]

Radical Domino Processes

The combination of several radical reactions in one process has gained considerable importance for the fast construction of complex and sterically demanding compounds. However, also examples for a combination of a primary radical and an anionic step exist.

The use of free radicals allows great versatility and tolerance. Usually, the primary free radicals are generated by reaction of halides as well as phenyl-thio and phenylselenium compounds with stannanes such as nBu$_3$SnH, silanes and germanes. Redox processes e.g. using Mn^{III} oxidation may also be employed. Since tin compounds are quite toxic there was a search going to use other methods for the generation of radicals.

Recently, an elegant radical domino reaction was employed by Pattenden et al. for the synthesis of the taxane skeleton starting from **76** which first undergoes a radical macrocyclization to provide **77**; it follows a cyclization to give **78** as a 3 : 1 mixture of two diastereomers (scheme 15). Although the yield is not very high, this is a straightforward approach to the taxane ring system.[48]

A new effective metal-free radical approach by Murphy et al. generates the free radical by treatment with tetrathiafulvalene (TTF).[49] As depicted in scheme 16 the aromatic amine **79** is transformed into the diazonium salt **81** which on treatment with TTF leads to the radical **82**. The following stereoselective cyclization gives the hexahydrocarbazole scaffold **80**, a substructure of alkaloids like aspidospermidin, strychnin and vinblastin. Also the non-toxic tris(trimethylsilyl)silane was employed for domino reactions, e.g. for the preparation of the alkaloid aspidospermidin.

Takasu et al. found that at high temperature a free radical reaction of polyolefinic vinyl iodides leads to phenanthrene derivatives via a 6-endo-6-exo-polycyclization while at low temperature a mono-cyclized product is obtained through a 5-exo mode.[50]

An unexpected C-arylation was observed by Sherburn et al. in an attempted radical deoxygenation of a homoallylic secondary alcohol **83** via its thiocarbonyl-derivative **84** (scheme 17).[51] Thus, reaction of **84**, obtained from **83** with ArOC(=S)Cl, with nBu$_3$SnH yielded the 10-aryl compound **87**. As inter-

Scheme 15. Domino radical cyclization for the synthesis of the taxane skeleton

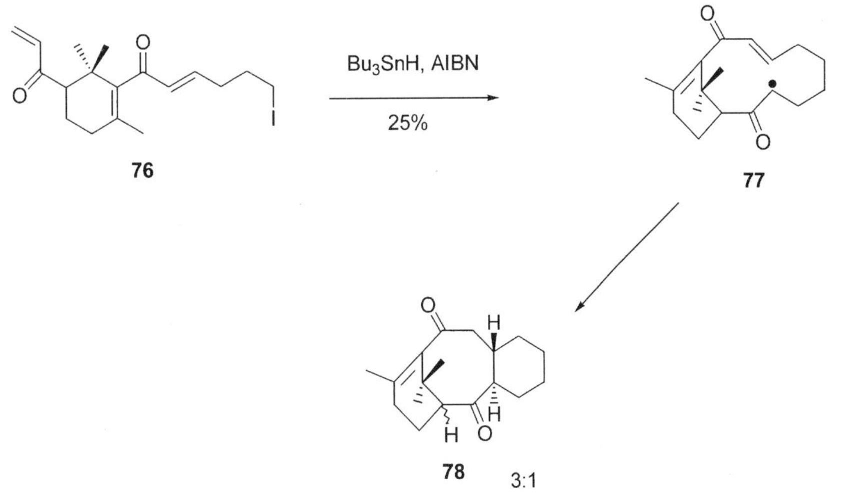

Scheme 16. Metal-free radical domino reaction to a hexahydrocarbazole scaffold **80**

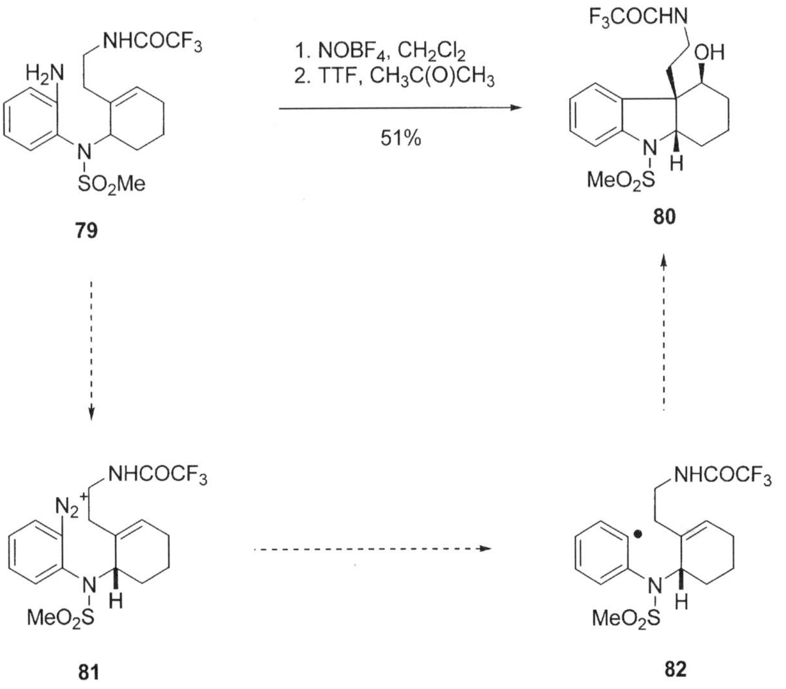

mediate the radical **85** can be assumed which attacks the double bond to give **86**. Further reaction with the aromatic moiety and rearomatization leads to **87**.

In general, the homo radical domino reaction is a powerful tool for the construction of complex molecules. Further examples are the synthesis of the BCD-Ring system of progesterone by Takahashi et al.,[52] the construction of bicyclic octanol derivatives by Sonoda et al.[53] and the total synthesis of (+)-claantholide and (−)-estafiatin by Lee et al.[54]

Domino reactions are not only useful for the construction of molecules, but also for their degradation. This concept is often encountered in nature. Thus, ribonucleotide reductases (RNRs) are enzymes that catalyze the formation of DNA monomers from ribonucleotides by radical mediated 2'-deoxygenation. This process has also been studied

in vitro using a chemical approach. Giese et al.[55] used a photochemical degradation of a selenoester model and Robins et al. performed studies on 6'-O-nitro esters of homonucleosides **88** containing a 2'-chloro substituent with tributylstannane and AIBN to give a 6'-oxygen radical **89**. It follows an abstraction of H-3' in a [1,5]-hydrogen shift to give a C-3' radical which looses the 2'-substituent to give **90** as postulated for the action of ribonucleotide reductase (scheme 18).[56] The reaction could be accompanied by an additional loss of the base to form **91**.

Radical–Anionic Processes

The formation of carbon–carbon bonds adjacent to nitrogen is of current interest in organic synthesis. Normally, the cyclization of an α-aminoalkyl

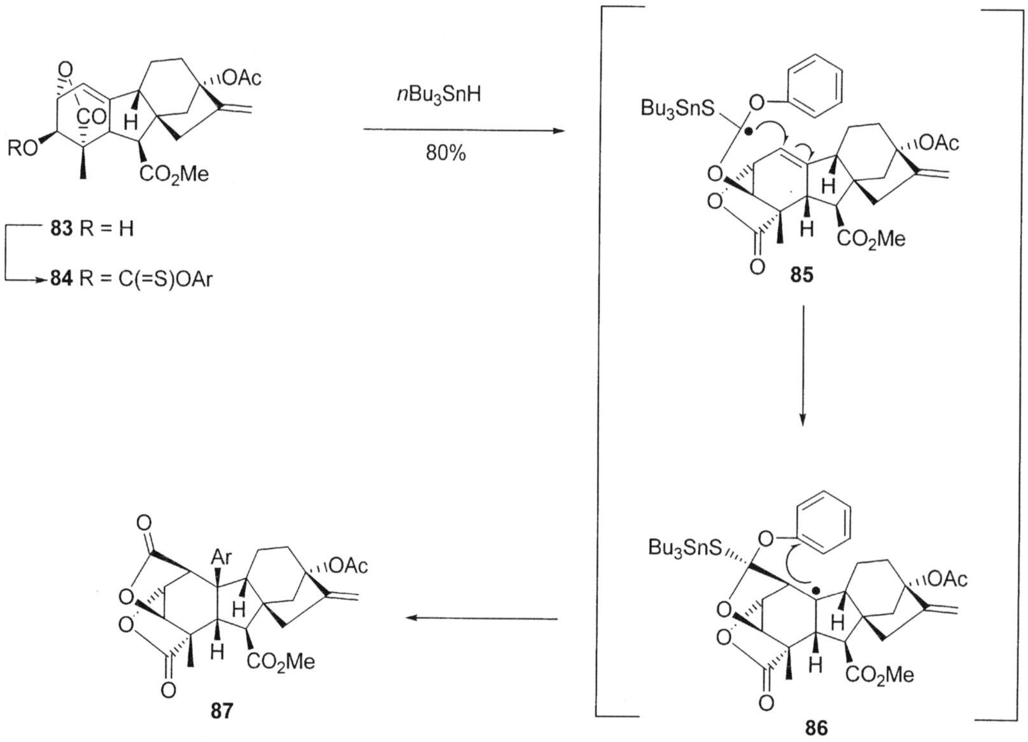

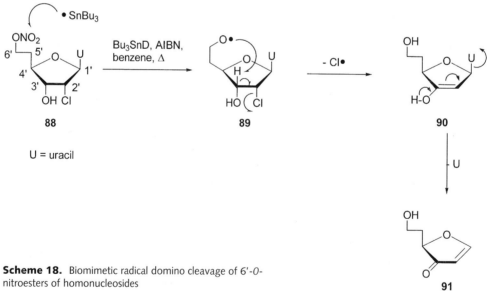

Scheme 17. Unexpected C-arylation by a radical domino process

Scheme 18. Biomimetic radical domino cleavage of 6'-O-nitroesters of homonucleosides

radical **94** onto an unactivated carbon–carbon double bond is rather difficult, since in general, acyclic radicals readily undergo reduction or dimerization. However, in presence of SmI$_2$ as shown by Katritzky et al. a radical initiated domino process of N-(α-aminobutyl)-benzotriazole **93**, obtained from **92**, is possible, which leads to the pyrrolidine derivative **97** (scheme 19).[57] The suggested mechanism comprises a site specific formation of an α-aminoalkyl radical, regioselective addition of the radical to the double

bond, reduction to a carbanion and final carbanion trapping by an electrophile. An equilibrium between the radical intermediates **94** and **95** is proposed.

Pericyclic Domino Processes

Pericyclic reactions such as cycloadditions, electrocyclic, sigmatropic, as well as ene reactions can easily be combined among one another. However,

Scheme 19. Radical domino reaction with SmI$_2$

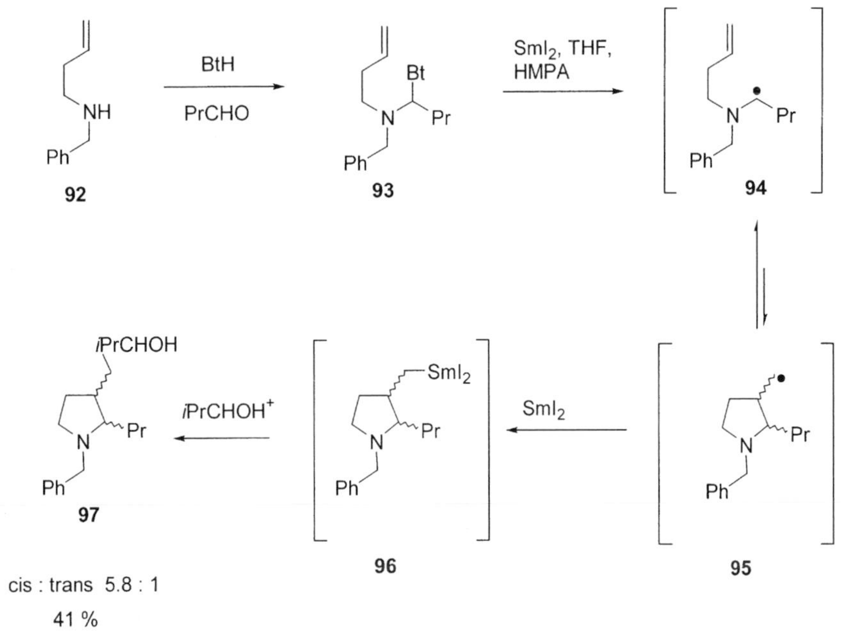

cis : trans 5.8 : 1

41 %

most of the known examples deal with sequences consisting of two or more Diels–Alder reactions, which is clearly one of the most powerful tools in organic synthesis. But there are also domino processes in which a pericyclic reaction is coupled with an anionic reaction.

An interesting pericyclic-anionic-pericyclic domino reaction showing a high stereoselectivity is the cycloaddition-aldol-retro-ene process depicted in scheme 20.[58] The procedure presumably starts with a [4+2]-cycloaddition of diene **98** and SO$_2$ in presence of a Lewis acid. After opening of the formed adduct reaction with (Z)-silyl vinyl ether **99** leads to a mixture of alk-2-enesulfinic acids **101**. It follows a retro-ene reaction which affords a 7 : 3 mixture of the products **102** and **103**. The reaction described by Vogel et al. is a nice example for the efficient generation of polypropionate chains with the stereoselective formation of three stereogenic centers and one (E)-double bond in a three-component domino reaction in its strict definition.

Lallemand et al. have found a pericyclic-anionic domino three-component reaction to prepare highly functionalized alcohols.[59a] This reaction was originally developed by Vaultier, Hoffmann et al.[59b] Diels–Alder reaction of a 1,3-dienylboronate with an acrylate yields a mixture of endo and exo diastereomers of the coupled allylboronate, which in the presence of an aldehyde such as 4-phenoxybutyraldehyde undergoes an allylation reaction. After hydrolysis the resulting diastereomeric alcohols are obtained in about 50 % yield, whereby two new stereogenic centers are formed in a stereoselective fashion.

In the course of the total synthesis of pseudomonic acid C analogues a novel domino ene reaction is employed by Markó et al. (scheme 21).[60] This methodology consists of an ene-reaction which is followed by a concomitant intramolecular silyl-mediated Sakurai (ISMS) cyclization. Using two equivalents of the same aldehyde the reaction proceeds in the presence of BF$_3 \cdot$OEt$_2$. However, the syn-

Scheme 20. Stereoselective synthesis of polypropionate fragments by a three-component pericyclic domino reaction

7 : 3

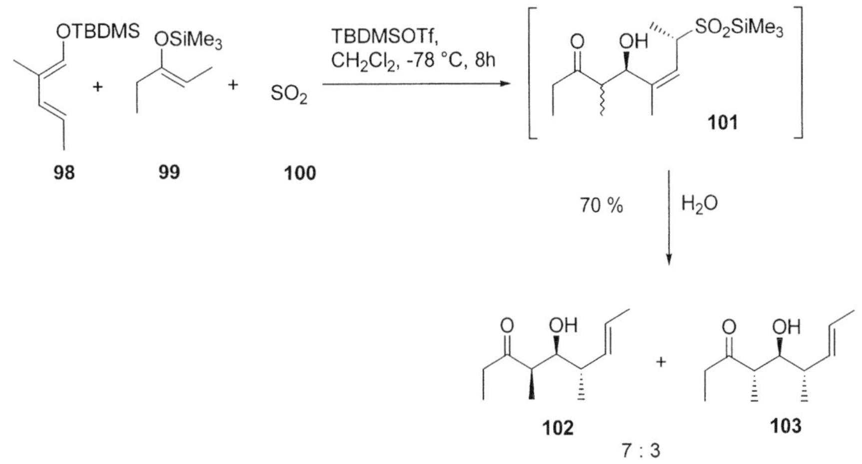

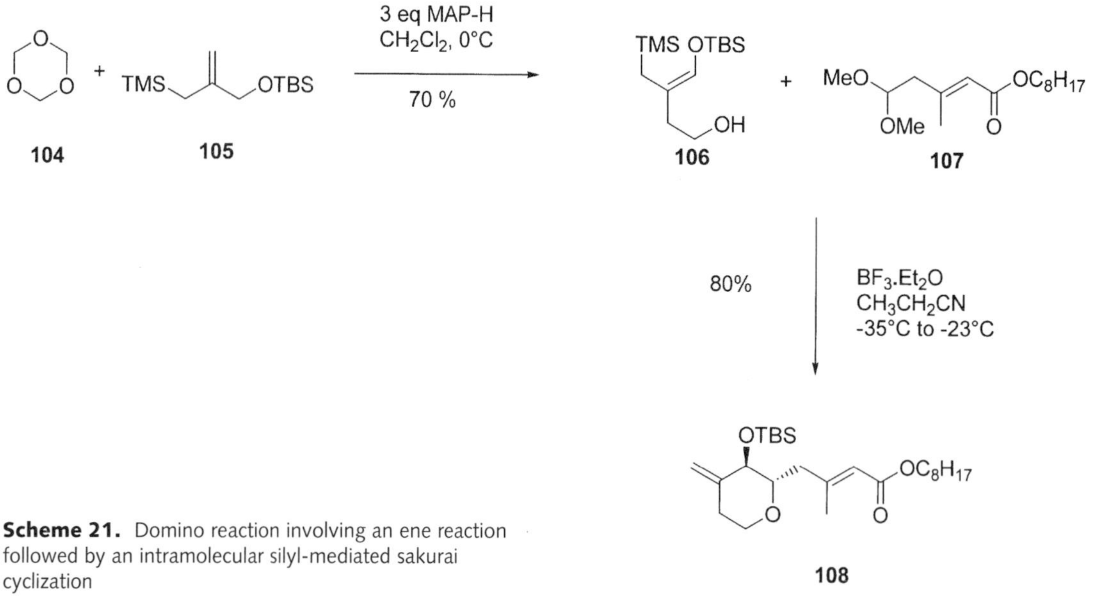

Scheme 21. Domino reaction involving an ene reaction followed by an intramolecular silyl-mediated sakurai cyclization

thesis of **108** requires two different aldehydes. In this case the initial ene-reaction between **105** and trioxane **104** is catalyzed by Yamamoto's aluminium reagent, MAP-H,[61] affording the homoallylic alcohol **106** in 70 % yield. Subsequent ISMS condensation between (*E*)-acetal **107** and alcohol **106** in propionitrile using BF$_3$·OEt$_2$ yields the exo-methylene tetrahydropyran **108** in 80 yield as a single diastereomer, possessing the 2,3-trans-stereochemistry.

A combination of an intramolecular [5+2]-cycloaddition of the pyrone **110** followed by an intermolecular [4 + 2]-cycloaddition with dimethylbutadiene leads to the fused 6,7,5-tricyclocarboxylic system **112** via **111** as described by Rodriguez et al. (scheme 22).[62] The prepared skeleton is found in nature in several terpenes. The starting material for this domino reaction is obtained from maltol **109**

by a sequence of three steps which includes the protection of the enol moiety, allylic bromination and in situ alkylation of the sodium salt of allylmalonitrile. The two cycloadditions take place at 160 °C using dimethylbutadiene in a fivefold excess to give **112** as a single diastereomer in 81 % yield. Also other butadienes have been employed in this process.

Holmes et al. used another interesting pericyclic domino reaction in the synthesis of (–)-histrionicotoxine **116** (scheme 23).[63] In a not fully understood sequence **113** at 190 °C generates the cycloadduct **115** in 80 % yield with loss of styrene. It can be assumed that the process consists of a retro-1,3-dipolar and a 1,3-dipolar cycloaddition with intermediary formation of the nitrone **114**. Overall, three new stereogenic centers, necessary for the natural product, have been efficiently created. An

Scheme 22. Intra-inter-domino-[5+2]-/[4+2]-cycloaddition

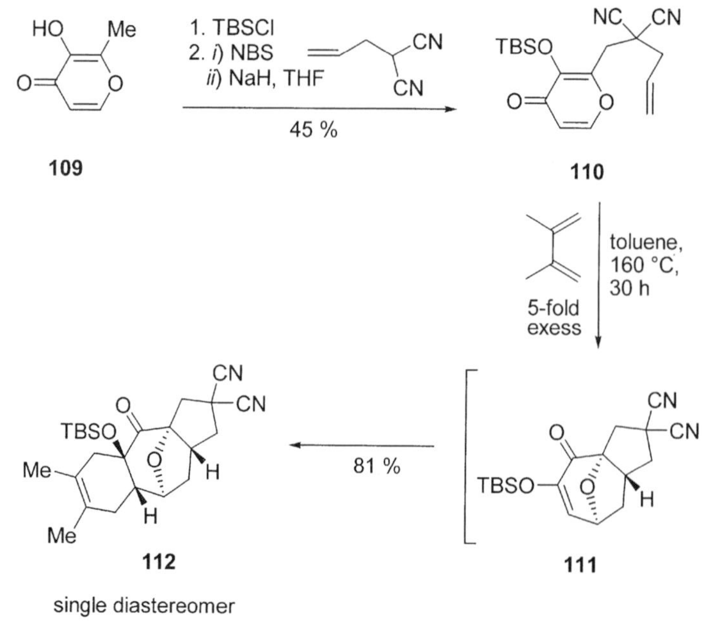

single diastereomer

Scheme 23. Synthesis of (–)-histrionicotoxine by a pericyclic domino process

113

toluene, sealed tube, 190 °C, 3.5 h

80%

114

115

116

Histrionicotoxin

Transition Metal-Catalyzed Domino Processes

Transition metal-catalyzed transformations are of increasing importance in synthetic organic chemistry. Therefore the application of this type of transformation as part of a domino process will be of growing interest. Certain transition metals are known to be suitable for domino type reactions and several combinations have already been developed.[11e] In these processes, first the formation of an organometallic compound takes place, e.g. by an oxidative addition to an alkenyl or aryl halide or triflate as in the Heck reaction or by oxidative coupling or by hydrometallation which is followed by an addition of the organometallic compound to an alkene or alkyne with the formation of a new organometallic compound which can again add to an alkene or alkyne. Negishi, Overman and Trost have disclosed this type of domino process using palladium as catalyst which can also include a mono and double carbonylation.[11e] In scheme 24 some examples for Pd-catalyzed intra-intra-domino processes are shown which either allow the synthesis of fused products as in the reaction of **117** and **123** to give **118** and **124**, respectively or alkylidencycloalkenes and spiro compounds as in the reaction of **119** and **121** to give **120** and **122**, respectively. Naturally, there are also inter-intra and intra-intra domino processes possible.

Elegant contributions to this field have also been made by Grigg et al.; especially interesting are his palladium-anion capture domino processes. By this way a direct route to protected pseudoargiopinine III **127**, a toxine from *Argiope lobata*, has been developed using a three-component process of allene **125**

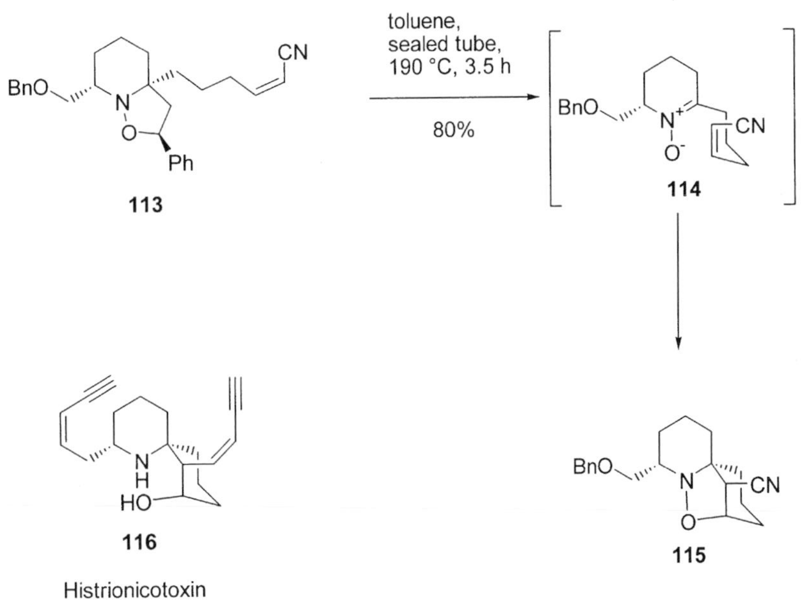

explanation for the observed regiochemistry which is contrary to a number of related examples is still under investigation.

Mixed Pericyclic Domino Processes

Pericyclic domino processes involving at least one cycloaddition have frequently been used for the synthesis of complex polycyclic ring systems. Examples are the synthesis of cyclohexene derivatives by Neier et al. combining a Diels–Alder reaction and a Ireland–Claisen rearrangement,[64] the rapid construction of bridged polyoxycyclic ring systems by Lautens et al.[65] and the formation of azapropellanes by Denmark et al. using a domino-inter-[4 + 2] -intra-[3 + 2] -cycloaddition of nitroalkenes with high stereoselectivity.[66] Avalos et al. demonstrated the use of asymmetric domino reactions based on nitroalkenes. Initially, an inverse electronic demand [4+2]-cycloaddition occurs with ethyl vinyl ether and a nitroalkene, the resulting nitronate reacts in a [3 + 2]-cycloaddition with the electron deficient vinyl ether to give nitrosoacetals as precursors for homologated carbohydrates.[67] Notably in the case of the pericyclic domino reaction many different transformations can be employed. A domino Mukaiyama-aldol-addition/aza-cope rearrangement of oxazinones has been used by Obrecht et al. to synthesize bicyclic pyrrolidine derivatives,[68] Engler et al.[69] employed domino-[5 + 2]-[3 + 2]- and [5 + 2]-[3 + 3]-cycloadditions to give tricyclic substrates with up to eight stereogenic centers from substituted benzoquinones and Padwa et al. developed a strategy for the construction of the erythrinane scaffold employing a domino Pummerer/Diels–Alder/N-acylium ion cyclization.[70]

Scheme 24. Various modes of Pd-catalyzed domino Heck processes

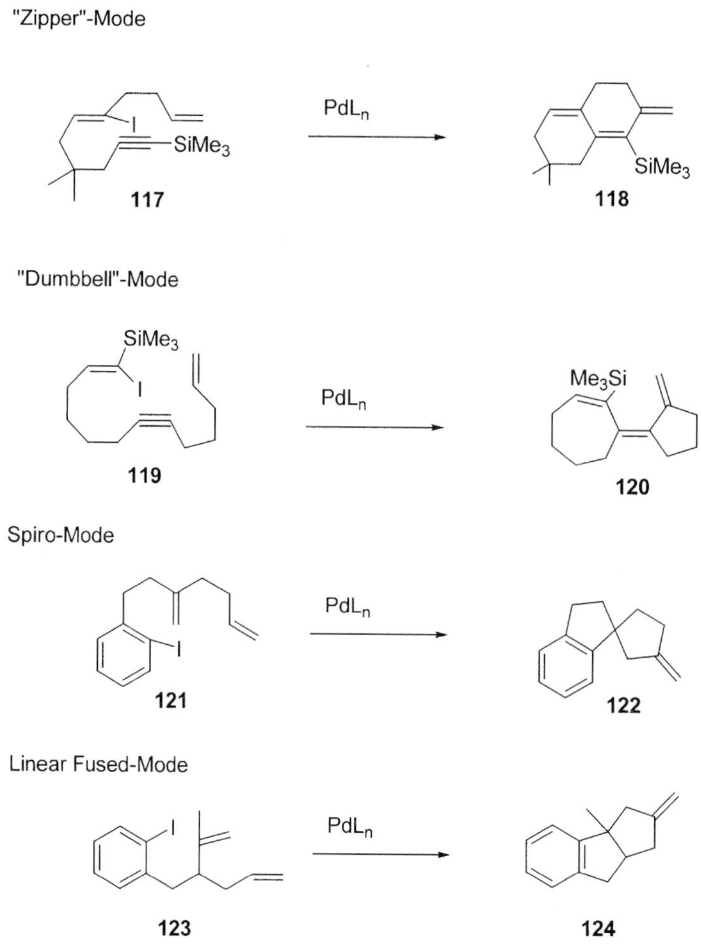

"Zipper"-Mode

117   PdL$_n$ **118**

"Dumbbell"-Mode

119   PdL$_n$ **120**

Spiro-Mode

121   PdL$_n$ **122**

Linear Fused-Mode

123   PdL$_n$ **124**

and amine **126** in the presence of carbon monoxide (scheme 25).[71]

Heck reactions have also been used by Helmchen et al. for a two-component domino process of α,ω-amino-1,3-dienes.[72] By using palladium complexes with chiral phosphino-oxazolines as catalysts an enantiomeric excess of up to 80 % is achieved. In a

typical experiment as illustrated in scheme 26 a suspension of Pd(OAc)$_2$, the chiral ligand L*, the aminodiene **128** and the arene trifluorosulfonate **129** in DMF are heated at 100 °C for 10 h. Via the chiral palladium complex **130** the resulting cyclic amine derivative **131** is obtained in 47 % yield and 80 % ee.

Scheme 25. Palladium anion capture domino process

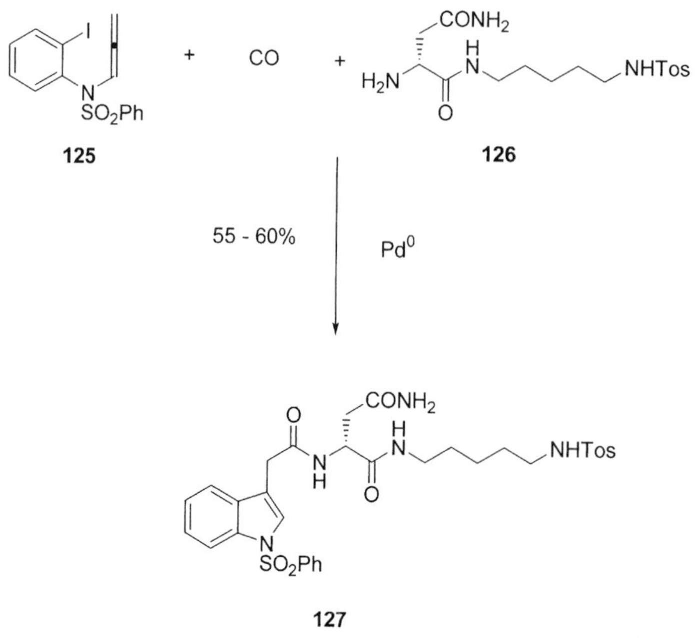

125   +   CO   +   **126**

55 - 60 %   Pd^0

127

Scheme 26. Enantio-
selective domino
Heck-allyllation process

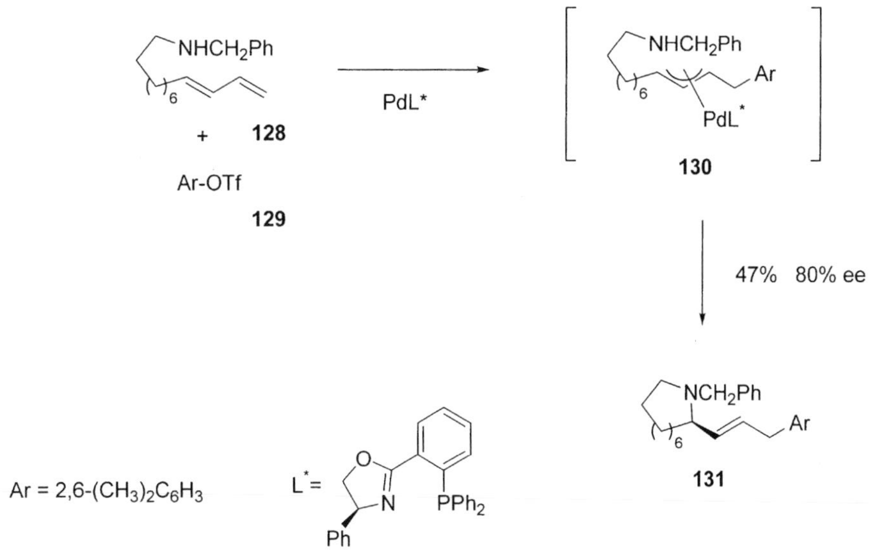

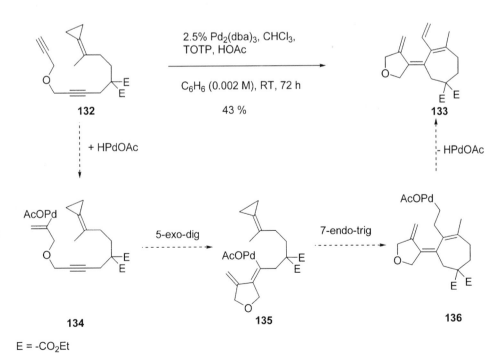

Ar = 2,6-(CH$_3$)$_2$C$_6$H$_3$

Reaction cascades consisting of several inter- and intramolecular Heck-type couplings followed by thermal reactions such as Diels–Alder reactions or 6π-electrocyclization have been developed by de Meijere et al.[11b, 73] Dieneynes proceed smoothly to form a tetracyclic array by a domino-Heck–Diels–Alder reaction in the presence of Pd(OAc)$_2$ at 130 °C. In general, cross-conjugated trienes, so-called dendralenes, containing two new cycles with a central cyclohexa- or cyclohepta-1,3-diene moiety can be obtained by two successive Heck reactions. An interesting new example is the facile cycloisomerization of enediyne **132** (scheme 27). After addition of the palladium catalyst a 5-exo-dig and a subsequent 7-endo-trig ring closure followed by β-hydride

elimination takes place to give **133** via the proposed intermediates **134–136**.

A domino Heck reaction of **137** has recently been used by Overman et al. for the total synthesis of (+)- and (–)-scopadulcic acid A **139**. The process leads to **138** which can be transformed into **139** (scheme 28).[74]

A combination of two palladium catalyzed reactions has also been used for the synthesis of estradiol[75a] **144** and cephalotaxine **145** by us;[75b, c] however, here the step-wise procedures give better yields. Reaction of **140** and enantiopure **141** leads via **142** to **143** in 55% yield which can be transformed into estradiol in three steps (scheme 29). For the transformation of **142** into **143** the palladacene **146** was used.

E = -CO$_2$Et

Scheme 27. Domino Heck-Diels–Alder reaction and formation of dendralenes by cycloisomerization of enediynes

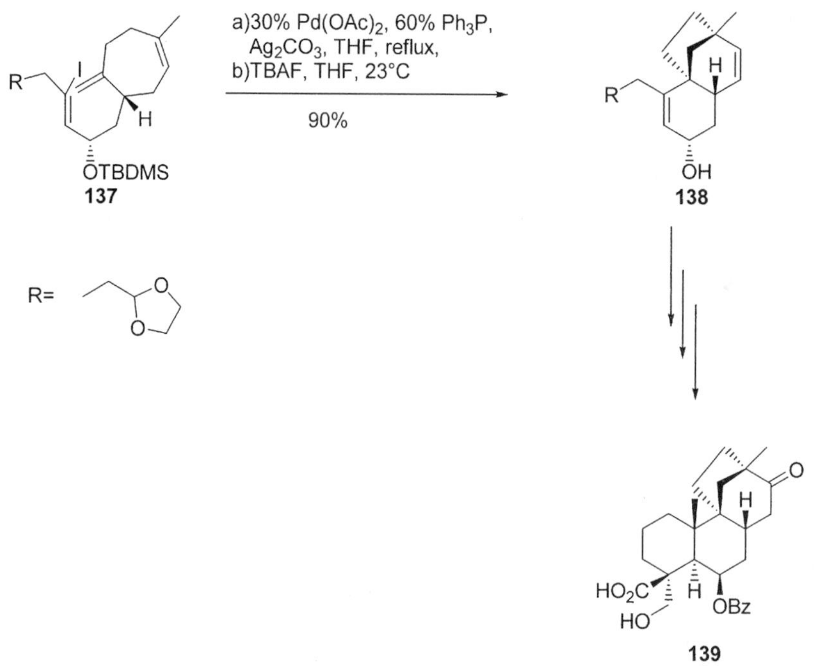

Scheme 28. Domino Heck reaction for the total synthesis of (–)-scopadulcic acid A **129**

a)30% Pd(OAc)$_2$, 60% Ph$_3$P,
   Ag$_2$CO$_3$, THF, reflux,
b)TBAF, THF, 23°C

90%

R=

137   **138**   **139**

Matsuda et al. recently described a domino reaction of 1,6-enyne derivatives with a hydrosilane and carbon monoxide in the presence of a catalytic amount of a rhodium complex to give a five-membered ring product containing a silylmethylene group.[76]

A domino-cyclopropanation/Cope rearrangement was used by Davies et al. for the enantioselective

synthesis of functionalized tropanes.[77] Thus, reaction of various N-Boc-protected pyrroles with vinyl-diazomethanes under rhodium(II)-carboxylate catalysis leads to functionalized tropanes. Best asymmetric induction was obtained using either (S)-lactate or (R)-pantolactone, respectively as chiral auxiliary on the vinyldiazomethane. The use of chiral catalysts was less successful.

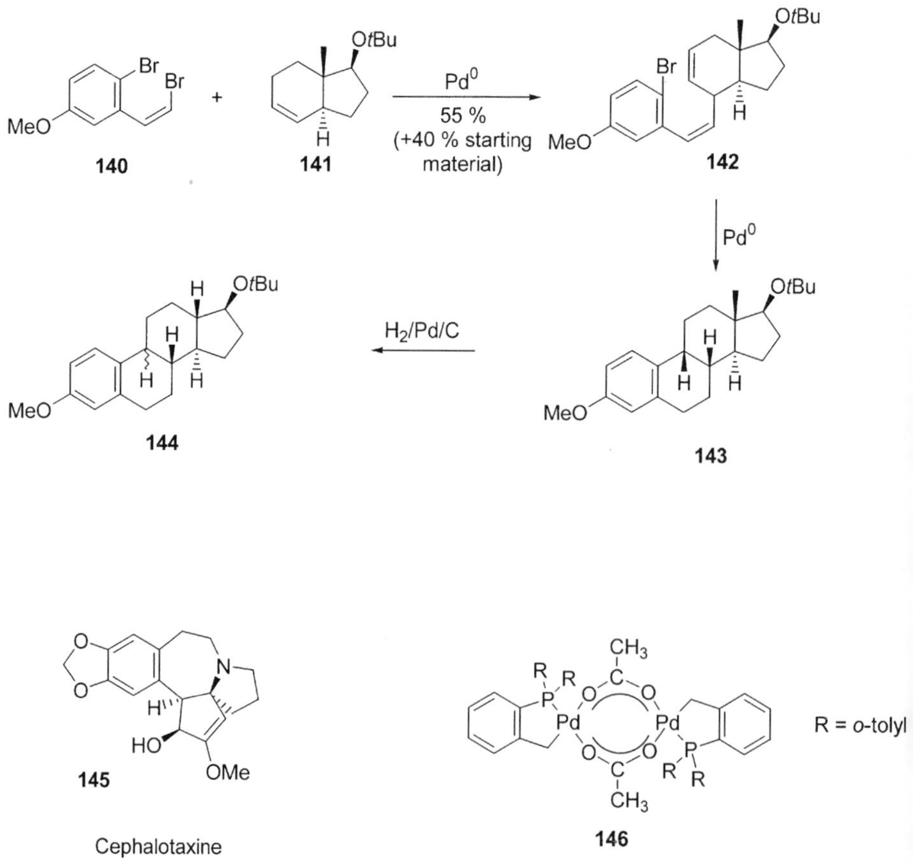

Scheme 29. Synthesis of estradiol and cephalotaxine by two subsequent palladium-catalyzed domino reactions

140   +   **141** $\xrightarrow{Pd^0}$ 55 %   (+40 % starting material)   **142**

$\downarrow Pd^0$

144 $\xleftarrow{H_2/Pd/C}$ **143**

145

Cephalotaxine

R = o-tolyl

146

Another rhodium-catalyzed domino process was developed by Padwa.[78] In the presence of rhodium(II)-perfluorobutyrate the diaza-1,3-dicarbonyl compound **147** gives a carbonylylide which undergoes a 1,3-dipolar cycloaddition to provide a 3 : 2 mixture of the diastereomeric oxido-bridged cyclohexanone derivatives **148** and **149** in 97% yield (scheme 30). Exposure of the mixture to BF$_3$· 2 AcOH generates the tetracyclic amides **151** as a 4 : 1 mixture of diastereomers about the tertiary hydroxyl center in 71% yield. The stereoselection of this reaction may be explained in terms of an anti arrangement of the bridgehead hydrogen atom and the tethered aromatic ring in the cyclohexylidene ring of a bicyclic iminium ion formed from **148**. A cyclization of the diastereomeric N-acylium salt, in which the bridgehead hydrogen atom and the tethered aromatic ring are in a syn arrangement, to give the epi-derivative of **151** is not observed. In this case the cyclohexylidene ring must adopt an unfavorable boat conformation. Consequently, the initially formed iminium ion derived from **148** can easily cyclize to give **151**, whereas the isomeric iminium ion obtained from **149** must first undergo a proton loss to give the corresponding enamide **150**. This is followed by reprotonation and a subsequent π-cyclization to give **151**. In a few steps **151** can be converted into the key intermediate previously used by Stork for the synthesis of (±)-lycopodine **152**.

A domino Pauson-Khand-Reaction was developed by Keese et al. starting from enediyne **155** leading to the shortest synthesis of a fenestrane **157** (scheme 31).[79]

The process is initiated by a double Grignard reaction of 4-pentynoic acid **153**, first with 3-butenyl magnesium bromide and subsequently magnesium acetylide followed by silylation of the formed tertiary hydroxyl function. The cobalt induced polycyclization leads directly to the fenestrane **157**; interestingly, the reaction halts at the stage of **156** when employing the unprotected alcohol.

Another interesting domino Pauson-Khand reaction was presented by Cook et al. generating six carbon–carbon bonds in a one-pot process in the synthesis of dicyclopenta[a,e]pentalene derivatives.[80]

The transition metal-catalyzed domino reactions will undoubtedly have a splendid future which is underlined by the increasing number of publications in this area. Steglich et al. presented an approach to arcyriacyanin A via a domino Heck reaction between a bromo(indolyl)maleiimide and 4-bromoindole.[81] The synthesis of 3,3'-bifurans by Ling et al. was achieved using a novel palladium-catalyzed domino dimerization and subsequent cyclization of acetylenic ketones.[82] Other applications of a combination of Heck reactions for domino processes are the syntheses of aza-heterocycles developed by our group.[83]

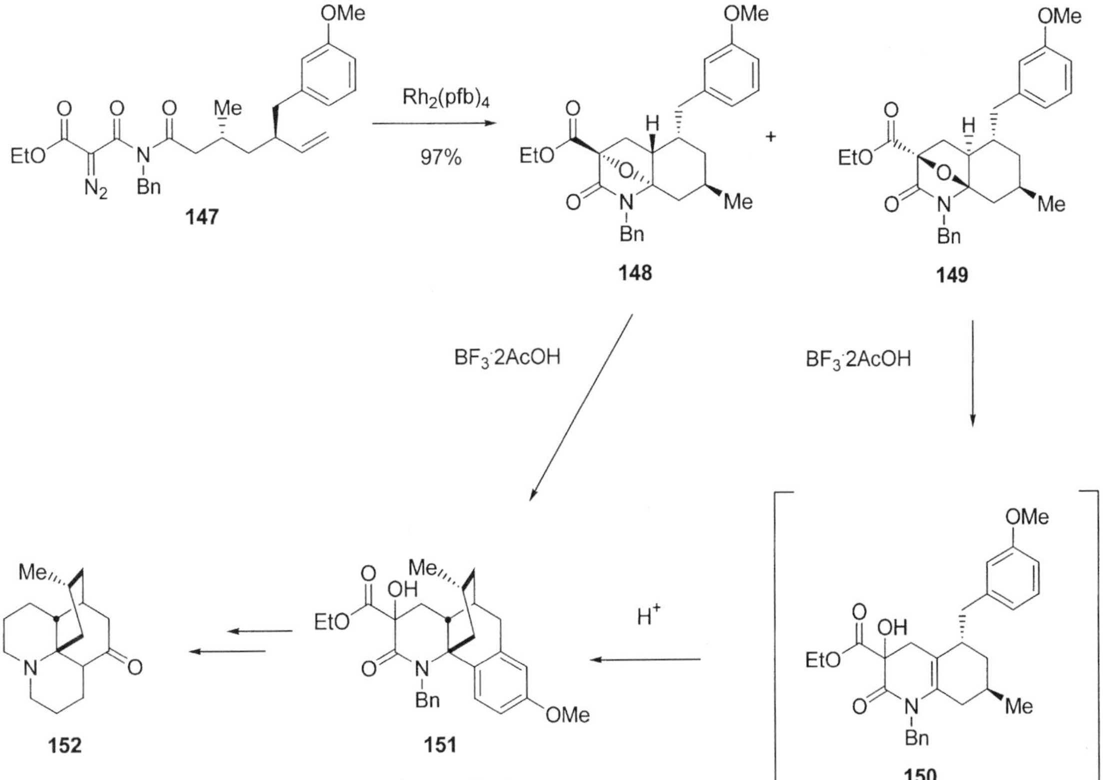

Scheme 30. Rhodium-initiated domino cyclization for the synthesis of lycopodine **152**

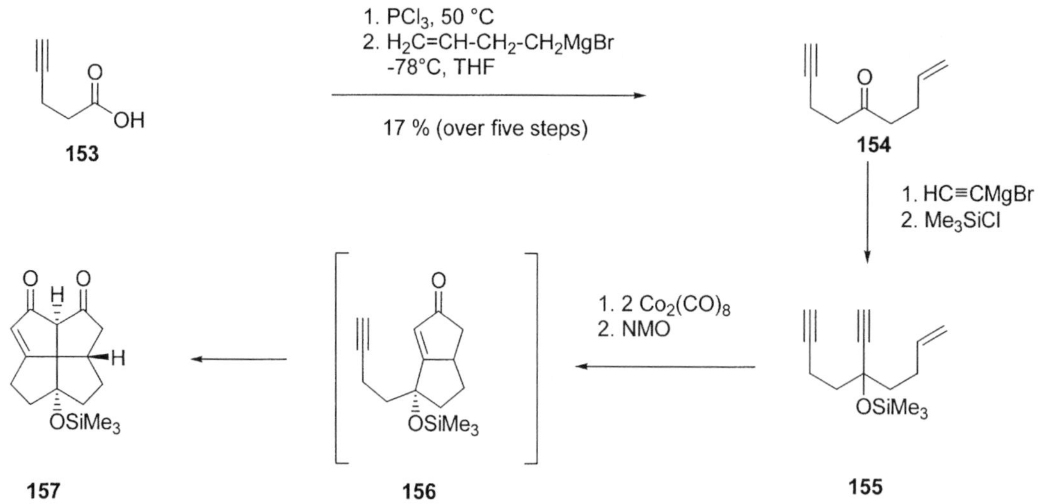

Scheme 31. Synthesis of fenestranes by domino Pauson–Khand reaction

Beyond the numerous applications of palladium in transition metal-catalyzed domino reactions there are a lot of other metals inducing domino processes. Ihara et al. found a strategy for the enantioselective synthesis of (+)-equilenin catalyzed by mangan and palladium complexes,[84] Whitby et al.[85] initiated domino cyclizations on a zirconocene template and furthermore Scherf et al. generated phenantrones by a nickel-mediated one-pot domino reaction.[86]

Enzymatic Domino Processes

A novel interesting approach is the use of multienzyme cocktails described by Scott et al. for the synthesis of precorrin-5 **159** starting from δ-amino levulinic acid **158** (scheme 32).[87] For the process, a multienzyme cocktail of eight different enzymes including the ALA-dehydratase to form porphobilinogen as well as PBG deaminase and cosynthetase to give the tetracyclic uroporphyrinogen III (**10**) was employed.

Waldmann et al. used tyrosinase which is obtained from *Agaricus bisporus* for the oxidation of phenols to give *ortho*-quinones via the corresponding catechols in the presence of oxygen (scheme 33).[88] A combination of this enzymatic-initiated domino process with a Diels–Alder reaction yields the functionalized bicyclic components **164** and **165** as a 33 : 1 mixture starting from simple *p*-methylphenol **160** in the presence of ethyl vinyl ether **163** as an electron rich dienophile via the intermediates **161** and **162** in an overall yield of 77%.

Photochemical Domino Processes

In a combination of photochemical cyclization and a radical reaction Yoshimatsu et al. synthesized 2-azabicyclo[3.3.0]octa-3,7-diene **169** from the trienal hydrazone **166**.[89] The domino process was initiated by irradiation of **166** at 400–500 nm in benzene. The transformation may include an intermolecular [2+2]-cyclization, followed by ring opening to give

Scheme 32. Multienzyme cocktail for the domino synthesis of precorrin-5 **159**

1. ALA Dehydratase
2. PBG Deaminase
3. Cosynthetase
4. M - 1
5. M - 2
6. CobG/O_2
7. CobJ
8. CobM
   SAM

30 %

A = -CH_2-CO_2H

P = -$(CH_2)_2$-CO_2

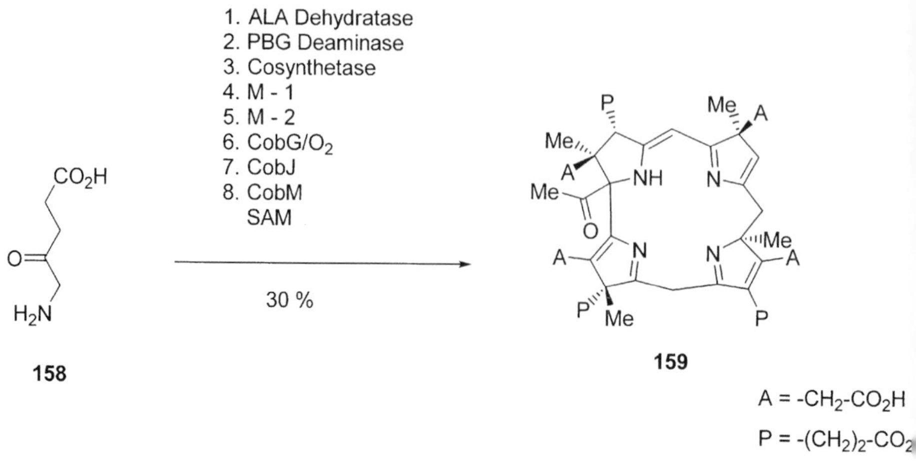

Scheme 33. Enzyme-initiated domino reaction to give bicyclo[2.2.2]octenones **164** and **165**

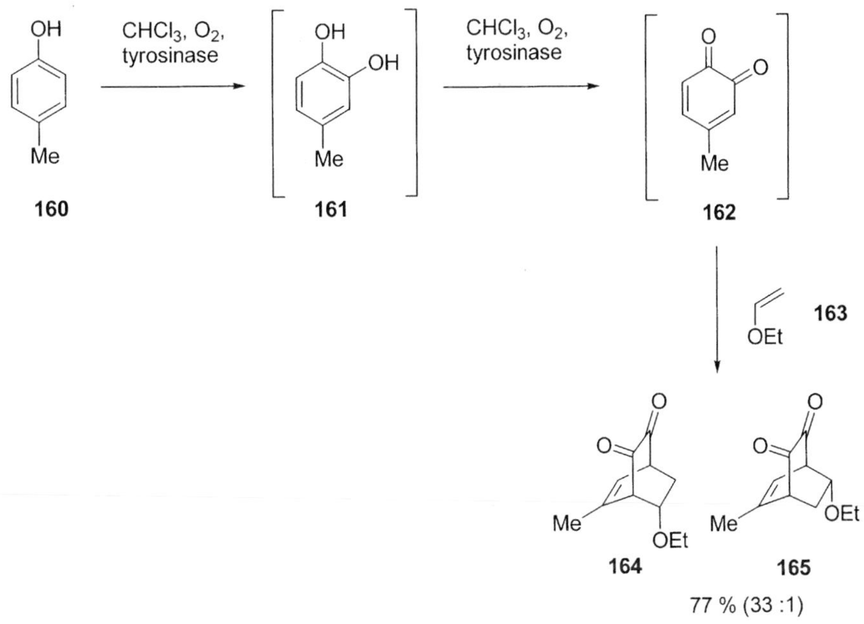

77 % (33 :1)

the 1,4-biradical **167** which isomerizes to give **168**, subsequent cyclization affords **169** in 41% yield (scheme 34).

Another interesting example of a photochemically induced domino process is the combination of the photocyclization of aryl vinyl sulfides with an intramolecular addition as described by Dittami et al.;[90] as intermediate a thiocarbonyl ylide can be assumed. The domino-Norrish I-Knoevenagel-allylsilane cyclization developed by us allows the efficient stereoselective formation of 1,2-trans-subsituted five- and six-membered carbocycles.[91] A photochemical cycloaddition of enamino-aldehydes and enamino-ketones with the intermediate formation of an iminium salt followed by addition to allylsilanes gives access to novel bicyclic heterocycles. New examples of photochemically induced

domino processes are also the photocyclization of 1,2-diketones conjugated with an eneyne moiety and the photochemical rearrangement of 1,3-diaryl-1,2-dihydropentalenes. The first procedure described by Nakatani et al. leads to bifuran derivatives [92] and the second described by Nair et al.[93] to 1,5-isomers of the pentalenes. However on the whole, the number of photochemically induced domino-processes is still rather small and there is plenty of room for new developments.

Electrochemical Domino Processes

Olivero et al. have found a new electrochemical domino process.[94] In an electrochemical overall two-electron-process in an one-compartment cell

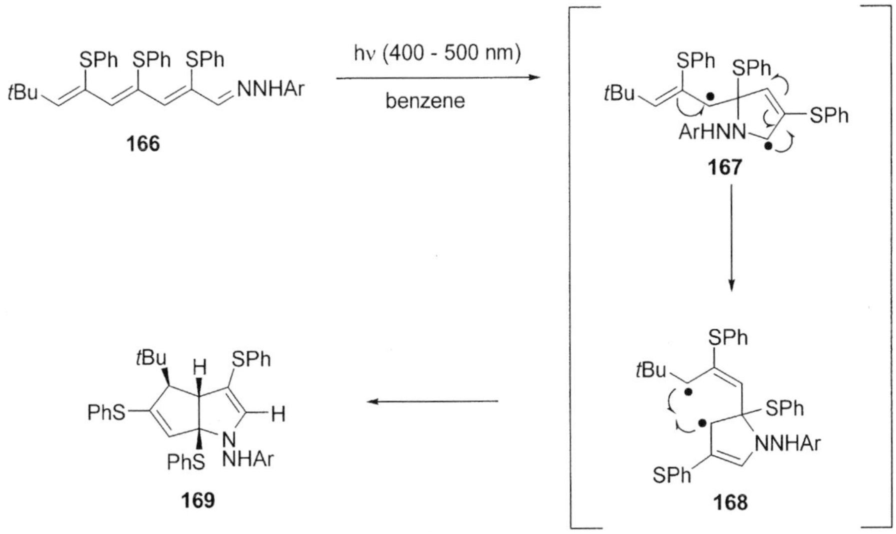

Scheme 34. Domino intramolecular photocyclization

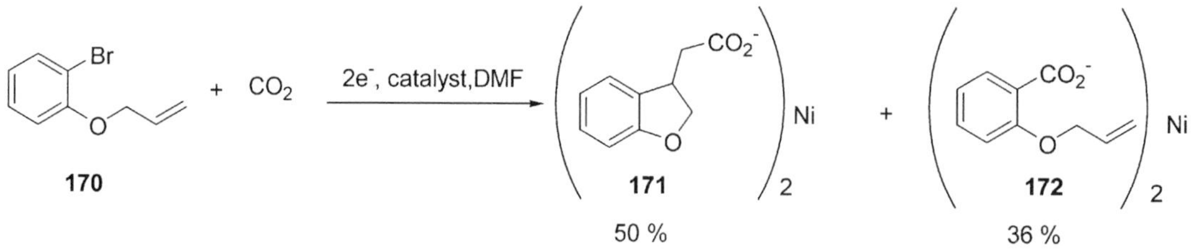

Scheme 35. Electrochemically induced domino process

with magnesium anode unsaturated haloaryl allyl ether as **170** are converted into benzofuranacetic acid derivatives **171** in the presence of carbon dioxide (scheme 35). As byproducts the non-cyclized material **172** as well as a small amount of the non-carboxylated compound are formed.

As a catalyst Ni^{II}(cyclam)Br_2 is employed which in a proposed cycle is reduced to Ni^{I}(cyclam)$^+$ coordinated to CO_2. In the following, the catalyst undergoes an oxidative addition with the bromoaryl compound to form a Ni^{III} species which in a radical like reaction inserts into the double bond. It follows another $1e^-$ reduction to give a Ni^{II} species which undergoes CO_2 uptake to form the nickel carboxylate **171** as the final product.

Epilogue

The development and use of domino reactions is an appropriate answer to the need for processes that allow the highly efficient synthesis of complex molecules starting from simple substrates. The point of these reactions is the formation of a bond – either C–C or C–O or C–N etc. – with the creation of a new functionality which can undergo a second bond formation with a repeated formation of another functionality which again undergo a further bond formation. In an optimized process the single steps should take place under identical conditions. There is actually no limitation in the combination of different types of transformation. Thus, reactions of the same type like cationic–cationic domino processes but also of different type like aniomic–pericyclic- or radical–anionic domino processes are known. There are even examples where five different steps are combined in one domino process. It is obvious that such processes are not only highly elegant and efficient but they also reduce the amount of waste produced in a synthesis, save our resources and they clearly improve the product/time factor. These reaction are therefore especially useful in combinatorial chemistry in solution.

We thank the Deutsche Forschungsgemeinschaft (SFB 416), the VW-Foundation, the German Ministry of Education and Research, the state Lower Saxony and the Fonds der chemischen Industrie for financial support for the domino processes described in this article which were developed by our group. We are also indebted to *BASF AG Ludwigshafen, Bayer AG Leverkusen, Degussa AG Frankfurt, Schering AG Berlin* and *Wacker Chemie GmbH, München* for generous gifts of chemicals.

References and Notes

1. (a) L. F. Tietze, U. Beifuss, Angew. Chem. **1993**, *105*, 137; Angew. Chem. Int. Ed. Engl. **1993**, *32*, 131. (b) B. M. Trost, Angew. Chem. **1995**, *107*, 285; Angew. Chem. Int. Ed. Engl. **1995**, *34*, 259. (c) T. Hudlicky, Chem. Rev. **1996**, *96*, 3. (d) L. F. Tietze, Chem. Rev. **1996**, *96*, 115.

2. (a) F. Balkenhohl, C. Bussche-Hünnefeld, A. Lansky, C. Zechel, Angew. Chem. **1996**, *108*, 2436; Angew. Chem. Int. Ed. Engl. **1996**, *35*, 2288. (b) L. A. Thompson, J. A. Ellman, Chem. Rev. **1996**, *96*, 555. (c) J. S. Früchtel, G. Jung, Angew. Chem. **1996**, *108*, 19; Angew. Chem. Int. Ed. Engl. **1996**, *35*, 17. (d) A. Nefzi , J. M. Ostresh, R. A. Houghton, Chem. Rev. **1997**, *97*, 449. (e) R. C. D. Brown, J. Chem. Soc., Perkin Trans. 1 **1998**, 3293. (f) A. R. Brown, P. H. H. Hermkens, H. C. J. Ottenheijm, D. C. Rees, Synlett **1998**, 817. (g) L. F. Tietze, M. Lieb, Curr. Op. Chem. Biol. **1998**, *2*, 363. (h) C. Watson, Angew. Chem. **1999**, *111*, 2025; Angew. Chem. Int. Ed. Engl. **1999**, *38*, 1903. (i) C. James, Tetrahedron **1999**, *55*, 4855.

3. In this chapter the combinations of different chemical steps in one process are always called domino process although some authors may have used the word tandem or cascade. We prefer the word domino over tandem and cascade since domino is not used in any other context in chemistry and thus facilitates the search for this type of transformation in the literature. On the other hand, the word tandem does not describe these time-resolved processes in a proper way, since it means two at the same time.

4. L. F. Tietze, G. Kettschau, J. A. Gewert, A. Schuffenhauer, Curr. Org. Chem. **1998**, *2*, 19.

5. I. Ugi, J. prakt. Chem. **1997**, *339*, 499.

6. R. Annunziata, M. Cinquini, F. Cozzi, V. Molteni, O. Schupp, Tetrahedron **1997**, *53*, 9715.

7. L. F. Tietze, A. Steinmetz, Angew. Chem. **1996**, *108*, 682; Angew. Chem. Int. Ed. Engl. **1996**, *35*, 651.

8. (a) D. J. Chadwick, K. Ackrill (eds.), *Biosynthesis of the Tetrapyrrole Pigments* (Series: Ciba Foundation Symposia) **1994**, *180*. (b) L. F. Tietze, G. Schulz, Chem. Eur. J. **1997**, *3*, 523.

9. C. H. Baker, S. P. T. Matsuda, D. R. Liu, E. J. Corey, Biochem. Biophys. Res. Commun. **1995**, *213*, 154.

10. L. F. Tietze, P. Saling, Chirality **1993**, *5*, 329.

11. T.-L. Ho, *Tandem Organic Reactions*; J. Wiley, New York, **1992**. (b) A. de Meijere, S. Bräse, J. Organomet. Chem. **1999**, *576*, 88. (c) K. Neuschütz, J. Velker, R. Neier, Synthesis **1998**, 227. (d) R. Grigg, V. Sridharan, J. Organomet. Chem **1999**, *576*, 65. (e) E. Negishi, C. Copéret, S. Ma, S.-H. Liou, F. Liu, Chem. Rev. **1996**, *96*, 365. (f) J. Streith, A. Defoin, Synlett **1996**, 189.

12. W. S. Johnson, M. S. Plummer, S. Pulla Reddy, W. R. Barlett, J. Am. Chem. Soc. **1993**, *115*, 515.

13. E. J. Corey, S. Lin, J. Am. Chem. Soc. **1996**, *118*, 8765.

14. U. Koert, Synthesis **1995**, 115.

15. M. H. Schmitt, S. Blechert, Angew. Chem. **1997**, *109*, 1516; Angew. Chem. Int. Ed. Engl. **1997**, *36*, 1474.

16. J. Wölfling, E. Frank, G. Schneider, L. F. Tietze, Angew. Chem. **1999**, *111*,151; Angew. Chem. Int. Ed. Engl. **1999**, *38*, 200.

17. J. Barluenga, G. P. Romanelli, L. J. Alvarez-García, I. Llorente, J. M. González, E. García-Rodríguez, S. García-Granda, Angew. Chem. **1998**, *110*, 3332; Angew. Chem. Int. Ed. Engl. **1998**, *37*, 3136.

18. Y. Wang, A. M. Arif, F. G. West, J. Am. Chem. Soc. **1999**, *121*, 876.

19. J. Wuckelt, M. Döring, P. Langer, R. Beckert, Synlett **1999**, 468.

20. M. Shibasaki, H. Sasai, T. Arai, T. Iida, Pure Appl. Chem. **1998**, *70*, 1027.

21. J. Rodriguez, Synlett **1999**, 505.

22. C. Agami, D. Bihan, R. Morgentin, C. Puchot-Kadouri, Synlett **1997**, 799.

23. G. Scheffer, M. Justus, H. Vasella, H. P. Wessel, Tetrahedron Lett. **1999**, *40*, 5845.

24. H. Paulsen, S. Antons, A. Brandes, M. Lögers, S. N. Müller, P. Naab, C. Schmeck, S. Schneider, J. Stoltefuß, Angew. Chem. **1999**, *111*, 3574; Angew. Chem. Int. Ed. Engl. **1999**, *38*, 3373.

25. G. A. Molander, C. R. Harris, J. Am. Chem. Soc. **1996**, *118*, 4059.

26. L. F. Tietze, Y. Zhou, Angew. Chem. **1999**, *111*, 2076; Angew. Chem. Int. Ed. Engl. **1999**, *38*, 2045.

27. C. H. Heathcock, J. C. Kath, R. B. Ruggeri, J. Org. Chem. **1995**, *60*, 1120.

28. L. A. Paquette, T. M. Morwick, J. Am. Chem. Soc. **1997**, *119*, 1230.

29. C. Wittland, N. Risch, Heterocycles **1998**, *48*, 2631.

30. P. Langer, M. Döring, D. Seyferth, Chem. Commun. **1998**, 1927.

31. J. Markandu, H. A. Dondas, M. Frederickson, R. Grigg, Tetrahedron, **1997**, *53*, 13165.

32. P. Righi, E. Marotta, G. Rosini, Chem. Eur. J. **1998**, *4*, 2501.

33. A. Hassner, O. Friedman, W. Dehaen, Liebigs Ann./Receuil **1997**, 587.

34. C. Wedler, B. Costisella, H. Schick, J. Org. Chem. **1999**, *64*, 5301.

35. S. Kobayashi, R. Akiyama, M. Moriwaki, Tetrahdron Lett. **1997**, *38*, 4819.

36. K. Takeda, T. Tanaka, Synlett, **1999**, 705.

37. N. A. Braun, U. Bürkle, M. P. Feth, I. Klein, D. Spitzner, Eur. J. Org. Chem. **1998**, 1569.

38. K. Krohn, C. Freund, U. Flörke, Eur. J. Org. Chem. **1998**, 2713.

39. A. Kamimura, H. Mitsudera, S. Asano, S. Kidera, A. Kakehi, J. Org. Chem. **1999**, *64*, 6353.

40. F. Dinon, E. Richards, P. J. Murphy, Tetrahedron Lett. **1999**, *40*, 3279.

41. J. Gutke, D. Spitzner, Tetrahedron **1999**, *55*, 3931.

42. J. Parsons, N. P. Camp, J. M. Underwood, D. M. Harvey, Tetrahedron **1996**, *52*, 11637.

43. L. F. Tietze, J. Bachmann, J. Wichmann, Y. Zhou, T. Raschke, Liebigs Ann./Recueil **1997**, 881.

44. M. Schmittel, M. K. Ghorai, A. Haeuseler, W. Henn, T. Koy, R. Söllner, Eur. J. Org. Chem. **1999**, 2007.

45. S. Jarosz, J. Chem. Soc., Perkin Trans. 1 **1997**, 3579.

46. B. Plietker, P. Metz, Tetrahedron Lett. **1998**, *39*, 7827.

47. W. G. Beyersbergen van Henegouwen, R. M. Fieseler, F. P. J. T. Rutjes, H. Hiemstra, Angew. Chem **1999**, *111*, 2351; Angew. Chem. Int. Ed. Engl. **1999**, *38*, 2214.

48. S. A. Hitchcock, G. Pattenden, Tetrahedron Lett. **1992**, *33*, 4843.

49. R. J. Fletcher, D. E. Hibbs, M. Hursthouse, C. Lampard, J. A. Murphy, S. J. Roome, Chem. Commun. **1996**, 739.

50. K. Takasu, J. Kuroyanagi, A. Katsumata, M. Ihara, Tetrahedron Lett. **1999**, *40*, 6277.

51. L. N. Mander, M. S. Sherburn, Tetrahedron Lett. **1996**, *37*, 4255.

52. S. Tomida, T. Doi, T. Takahashi, Tetrahedron Lett. **1999**, *40*, 2363.

53. S. Tsunoi, I. Ryu, S. Yamasaki, M. Tanaka, N. Sonoda, M. Komatsu, Chem. Commun. **1997**, 1889.

54. E. Lee, J. W. Lim, C. H. Yoon, Y. Sung, Y. K. Kim, J. Am. Chem. Soc. **1997**, *119*, 8391.

55. R. Lenz, B. Giese, J. Am. Chem. Soc. **1997**, *119*, 2784.

56. M. J. Robins, Z. Guo, M. C. Samano, S. F. Wnuk, J. Am. Chem. Soc. **1999**, *121*, 1425.

57. A. R. Katritzky, D. Feng, M. Qi, J. M. Aurrecoechea, R. Suero, N. Aurrekoetxea, J. Org. Chem. **1999**, *64*, 3335.

58. J.-M. Roulet, G. Puhr, P. Vogel, Tetrahedron Lett. **1997**, *38*, 6201.

59. (a) Y. Six, J.-Y. Lallemand, Tetrahedron Lett. **1999**, *40*, 1295. (b) M. Vaultier, F. Truchet, B. Carboni, R. W. Hoffmann, I. Denne, Tetrahedron Lett. **1987**, *28*, 4169.

60. I. E. Markó, J.-M. Plancher, Tetrahedron Lett. **1999**, *40*, 5259.

61. S. Saito, H. Yamamoto, J. Chem. Soc., Chem. Commun. **1997**, 1585.

62. J. R. Rodríguez, A. Rumbo, L. Castedo, J. L. Mascareñas, J. Org. Chem. **1999**, *64*, 966.

63. G. M. Williams, S. D. Roughley, J. E. Davies, A. B. Holmes, J. Am. Chem. Soc. **1999**, *121*, 4900.

64. J. Velker, J.-P. Roblin, A. Neels, A. Tesouro, H. Stoeckli-Evans, F.-G. Klaerner, J.-S. Gehrke, R. Neier, Synlett, **1999**, 925.

65. M. Lautens, E. Fillion, J. Org. Chem. **1997**, *62*, 4418.

66. (a) S. E. Denmark, D. S. Middleton, J. Org. Chem. **1998**, *63*, 1604. (b) S. E. Denmark, J. A. Dixon, J. Org. Chem. **1998**, *63*, 6167.

67. M. Avalos, R. Babiano, P. Cintas, J. L. Jiménez, J. C. Palacios, M. A. Silva, Chem. Commun. **1998**, 459.

68. D. Obrecht, C. Zumbrunn, K. Müller, J. Org. Chem. **1999**, *64*, 6891.

69. T. A. Engler, C. M. Scheibe, R. Iyengar, J. Org. Chem. **1997**, *62*, 8274.

70. A. Padwa, R. Hennig, C. O. Kappe, T. S. Reger, J. Org. Chem. **1998**, *63*, 1144.

71. R. Grigg, B. Putnikovic, D. Sykes, V. Savic, unpublished results

72. D. Flubacher, G. Helmchen, Tetrahedron Lett. **1999**, *40*, 3867.

73. S. Schweizer, Z. Z. Song, F. E. Meyer, P. J. Parsons, A. de Meijere, Angew. Chem. **1999**, *111*, 1550; Angew. Chem. Int. Ed. Engl. **1999**, *38*, 1452.

74. M. E. Fox, C. Li, J. P. Marino, Jr., L. E. Overman, J. Am. Chem. Soc. **1999**, *121*, 5467.

75a). L. F. Tietze, T. Nöbel, M. Spescha. J. Am. Chem. Soc. **1998**, *120*, 8971.

75b). L. F. Tietze, H. Schirok, J. Am. Chem. Soc. **1999**, *121*, 10264.

75c). L. F. Tietze, H. Schirok, M. Wöhrmann, Chem. Eur. J. **2000**, *6*, 510.

76. Y. Fukuta, I. Matsuda, K. Itoh, Tetrahedron Lett. **1999**, *40*, 4703.

77. H. M. L. Davies, J. L. Matasi, M. Hodges, N. J. S. Huby, C. Thornley, N. Kong, J. H. Houser, J. Org. Chem. **1997**, *62*, 1095.

78. A. Padwa, Chem. Commun. **1998**, 1417.

79. M. Thommen, R. Keese, Synlett **1997**, 231.

80. S. G. Van Ornum, J. M. Cook, Tetrahedron Lett. **1997**, *38*, 3657.

81. M. Brenner. G. Mayer, A. Terpin, W. Steglich, Chem. Eur. J. **1997**, *3*, 70.

82. A. Jeevanandam, K. Narkunan, C. Cartwright, Y.-C. Ling, Tetrahedron Lett. **1999**, *40*, 4841.

83. L. F. Tietze, R. Ferracioli, Synlett **1998**, 145.

84. H. Nemoto, M. Yoshida, K. Fukumoto, M. Ihara, Tetrahedron Lett. **1999**, *40*, 907.

85. S. F. Fillery, G. J. Gordon, T. Luker, R. J. Whitby, Pure Appl. Chem. **1997**, *69*, 633.

86. H. A. Reisch, V. Enkelmann, U. Scherf, J. Org. Chem. **1999**, *64*, 655.

87. A. I. Scott, Synlett, **1994**, 871.

88. G. H. Müller, A. Lang, D. R. Seithel, H. Waldmann, Chem. Eur. J. **1998**, *4*, 2513.

89. M. Yoshimatsu, S. Gotoh, G. Tanabe, O. Muraoka, Chem. Commun. **1999**, 909.

90. J. P. Dittami, X. Y. Nie, H. Nie, H. Ramanthan, C. Buntel, S, Rigatti, J. Bordner, D. L. Decosta, P. Williard, J. Org. Chem. **1992**, *57*, 1151.

91. L. F. Tietze, J. R. Wünsch, M. Noltemeyer, Tetrahedron **1992**, *48*, 2081.

92. K. Nakatani, K. Adachi, K. Tanabe, I. Saito, J. Am. Chem. Soc. **1999**, *121*, 8221.

93. V. Nair, G. Anilkumar, C. N. Jayan, N. P. Rath, Tetahedron Lett. **1998**, *39*, 2437.

94. S. Olivero, E. Duñach, Eur. J. Org. Chem. **1999**, 1885.

Combinatorial Libraries for Drug Development

Alice Lee,
Jason W. Szewczyk,
Jonathan A. Ellman

Department of Chemistry,
University of California,
Berkeley, CA 94720–1460,
USA

Phone: +510 642 4488,
Fax: +510 642 8369,
e-mail: jellman@uclink4.berkeley.edu

Keywords ■ *Combinatorial Chemistry* ■ *Solid-Phase Synthesis* ■ *Solid-Phase Scavenger* ■ *Linker* ■ *Targeted Library* ■ *Small Molecule Libraries*

Concept: A major endeavor in science is the identification of compounds having a desired property. Historically, these compounds were found by the serendipitous testing of compound collections. More recently, the rational design of compounds has gained increased emphasis. Usually many iterations in the design, synthesis, and testing of compounds are required to iden-tify a molecule that has the sought-after property. A rate-limiting step in both approaches is the preparation of compounds for evaluation. Combinatorial chemistry, which may be generally defined as the synthesis of large collections (libraries) of compounds from sets of precursor building blocks (Figure 1), provides rapid access to the compounds that drive these efforts.

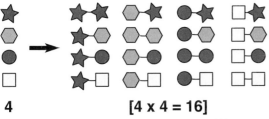

Figure 1. General schematic of a combinatorial library prepared from a small set of building blocks.

4   **[4 x 4 = 16]**

Abstract: Combinatorial approaches increasingly impact all areas of chemistry and biology.[1] These methods now pervade all aspects of drug development and have growing influence in the fields of molecular recognition, catalysis, and materials science. The area first impacted by combinatorial methods and where the greatest investment of time, resources, and effort has been placed is for the identification and opti-mization of low molecular weight (< 600–700 amu) molecules that bind to target biomolecules of therapeutic relevance.[2] To enable the preparation of compound libraries, many innovative methods have been developed for solution-phase and solid-phase syn-thesis, including the application of support-bound scavengers and reagents. Compound selection for preparation and testing continues to be one of the most difficult and thought-provoking aspects of combinatorial chemistry. Libraries have been designed based on a weakly-binding lead molecule or on the mechanism and/or structure of the target biomolecule. For lead identification, libraries have been designed for the maximal display of diverse functionality upon novel structural templates. In addition, strategies have recently emerged for lead identification where the target biomolecule determines which building blocks are selected for library preparation.

Prologue

The preparation of a combinatorial library requires the simultaneous manipulation and isolation of many different compounds. A uniform sequence of operations is required to efficiently prepare and isolate each member of the library. In contrast, the traditional synthesis of compounds utilizes conditions tailored specifically to the compound desired. A key challenge for library preparation is the development of a robust synthesis sequence that cleanly incorporates chemical building blocks containing a diverse range of chemical functionality in high yields. Equally important are the development of uniform and efficient methods to isolate intermediates and products from solvents, reagents, and byproducts.

Solution-Phase Synthesis

Conceptually, parallel solution-phase synthesis is the most straightforward method for library preparation due to the close resemblance to traditional synthesis. However, in a solution-phase format, uniform methods for the separation of intermediates and products from reagents and byproducts can be difficult to achieve, because liquid-liquid extraction is frequently required. Partitioning of compounds between organic solvents and water depends upon the hydrophilicity and charge of compounds; thus extractions often produce emulsions or loss of the hydrophilic and/or charged library members. Additionally, reagents and byproducts may partition with the library members resulting in contamination. To minimize byproducts, reaction sequences have been developed that rely on reagents known to cleanly partition under extractive conditions. McCarthy and coworkers reported an early and ele-

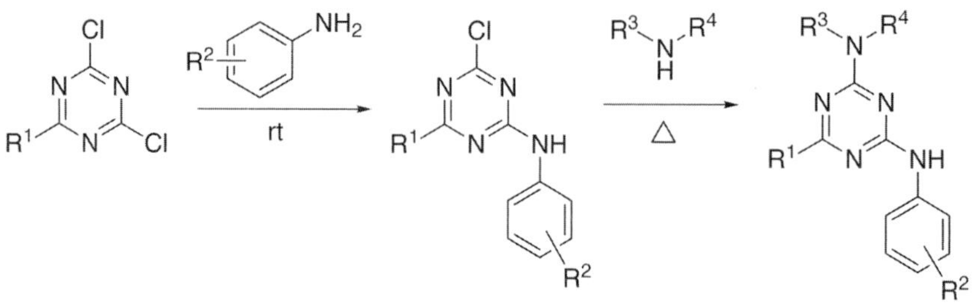

Scheme 1. A triazine library prepared by solution phase synthesis.

gant example for the synthesis of triazine-based corticotropin-releasing factor₁ receptor antagonists by the sequential addition of anilines and amines to dichlorotriazine (Scheme 1).[3] The symmetrical nature of the triazine starting material combined with the differential nucleophilicity of the aniline and amine reagents yields products that are of high purity following acid extraction to remove excess amine reagents. Clearly, many reactions are not amenable to straightforward isolation procedures, placing increased emphasis on high-throughput purification methods.

Solid-Phase Synthesis

The power of solid-phase synthesis has been amply demonstrated by the rapid and automated parallel synthesis of peptides and oligonucleotides. Likewise for library synthesis, the two most significant advantages of solid-phase over solution-phase synthesis are the ability to drive reactions to completion by the use of excess reagents and the straightforward isolation of intermediates and products by simple filtration away from reagents and byproducts. By definition, solid-phase synthesis requires the attachment of the starting material to support. The linkage must be stable throughout the synthesis sequence but be cleavable upon synthesis completion to liberate the final product.[4,5] A common strategy for library synthesis employs the linkage to solid support as a protecting group for an invariant functionality present in both the starting material and final product. Examples of this strategy are the attachment of alcohols to support using silyl or tetrahydropyranyl linkages as well as the attachment of amines to support via a carbamate linkage (Figure 2).

Traceless linkages leave no obvious site of attachment to support upon synthesis completion and are

frequently employed when an invariant functional group is not present in the target molecule. A popular approach uses protodesilylation to release aromatic and heteroaromatic compounds from support (Scheme 2),[6] although many innovative methods could also be considered, as illustrated by the traceless synthesis of isoquinolines by Kurth and coworkers.[7]

Cyclative cleavage strategies release the final compound into solution following intramolecular attack of a nucleophile or electrophile upon the linkage site. Synthesis byproducts and intermediates do not incorporate the necessary nucleophile or electrophile; therefore only the desired products are released into solution to yield high purity materials. Seminal examples of this approach are the library syntheses of benzodiazepines and hydantoins (Scheme 3).

The development of diversification linkers allows introduction of an additional element of diversity. Upon completion of the synthesis sequence, the linker is activated facilitating nucleophilic release of the library members from support. In the ideal case, as implemented with the acylsulfonamide linker (Scheme 4a), the activated linker is sufficiently reactive that *limiting* amounts of nucleophile may be added to provide pure product after resin filtration.[8] Diversification linkers have been developed for the preparation of carboxylic acid derivatives (Scheme 4a), amines (Scheme 4b),[9] aromatic (Scheme 4c) and even heteroaromatic compounds (Scheme 4d).[10]

Support-Bound Reagents and Scavengers

Support-bound reagents and scavengers have become increasingly popular in the preparation of compound libraries.[11] When polymer-supported

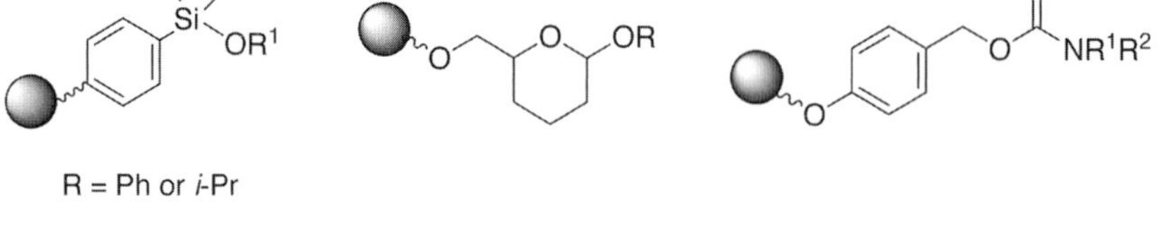

Figure 2. Examples of linkers which protect a specific functional group by attachment to solid support.

R = Ph or *i*-Pr

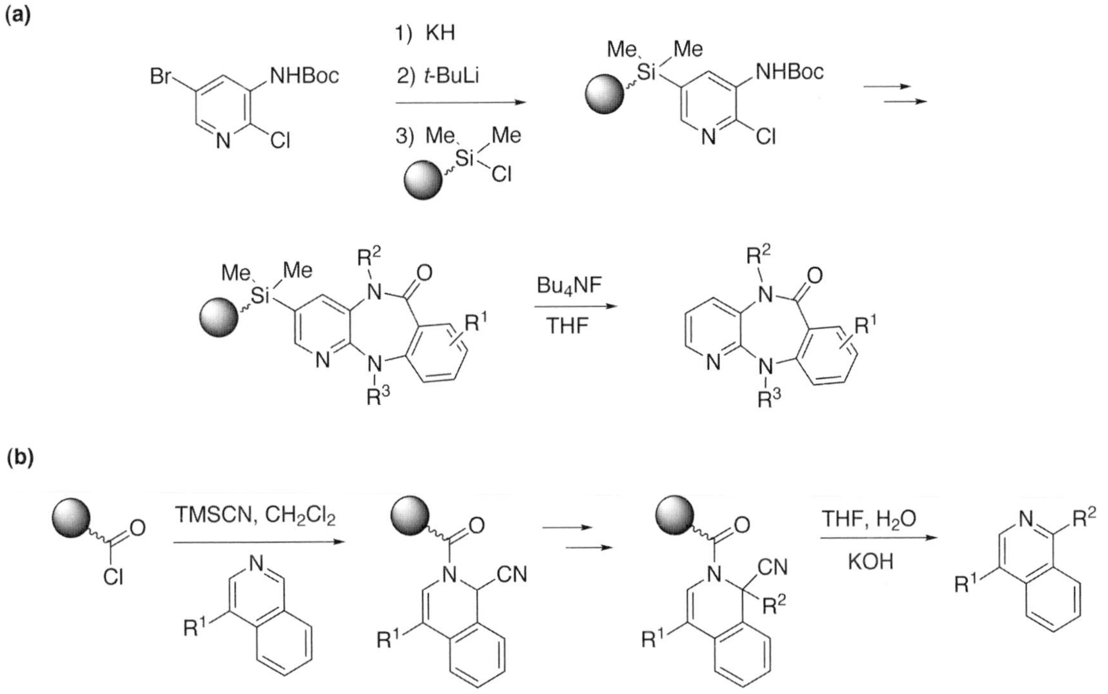

(a)

(b)

Scheme 2. Traceless linkers for the solid-phase synthesis of (**a**) pyridine-based tricycles and (**b**) isoquinolines.

reagents and catalysts are used (Figures 3a, 3b), the product is isolated by simple filtration from the support-bound materials. Alternatively, polymer-supported scavengers (Figure 3c) may be added upon reaction completion to sequester and remove impurities (e.g., reagents and byproducts) either through covalent bond formation or ionic interactions. These new methods greatly simplify product isolation and offer the additional advantage of traditional solution-phase reaction monitoring methods. Kaldor and coworkers provided an early demonstration of the use of polymer-bound reagents and scavengers (Scheme 5).[12]

Reductive amination of an aldehyde with excess primary amine, using a support-bound borohydride, provides the desired secondary amine contaminated with the primary amine precursor. Covalent capture of the primary amine with a support-bound aldehyde provides the pure secondary amine. Treatment with excess isocyanate yields the final urea product, which is purified by reaction with a support-bound amine to remove unreacted isocyanate. For the full potential of this method to be realized, further development of support-bound reagents and scavengers for most of the important chemical transformations will be necessary. Al-

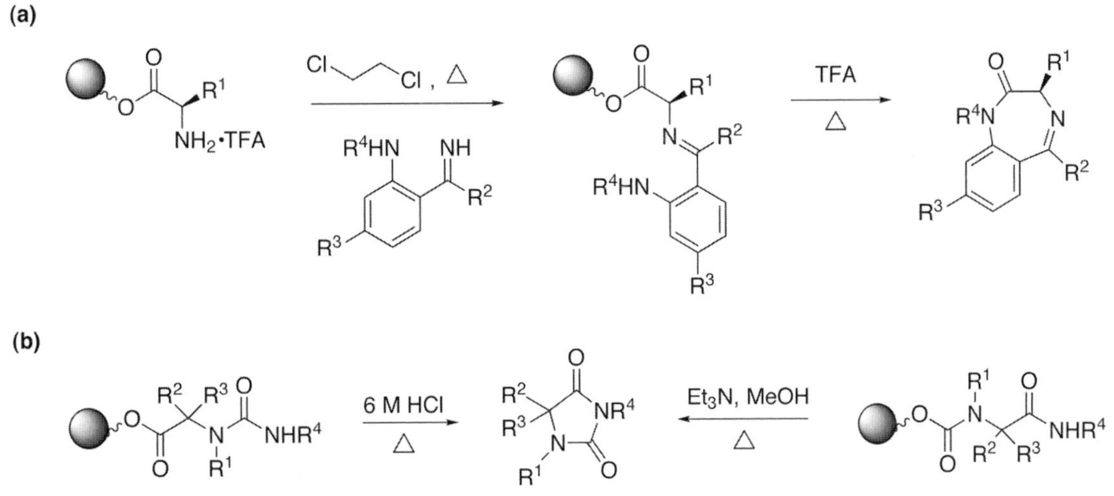

(a)

(b)

Scheme 3. Cyclative cleavage strategies for the synthesis of (**a**) benzodiazepines and (**b**) hydantoins.

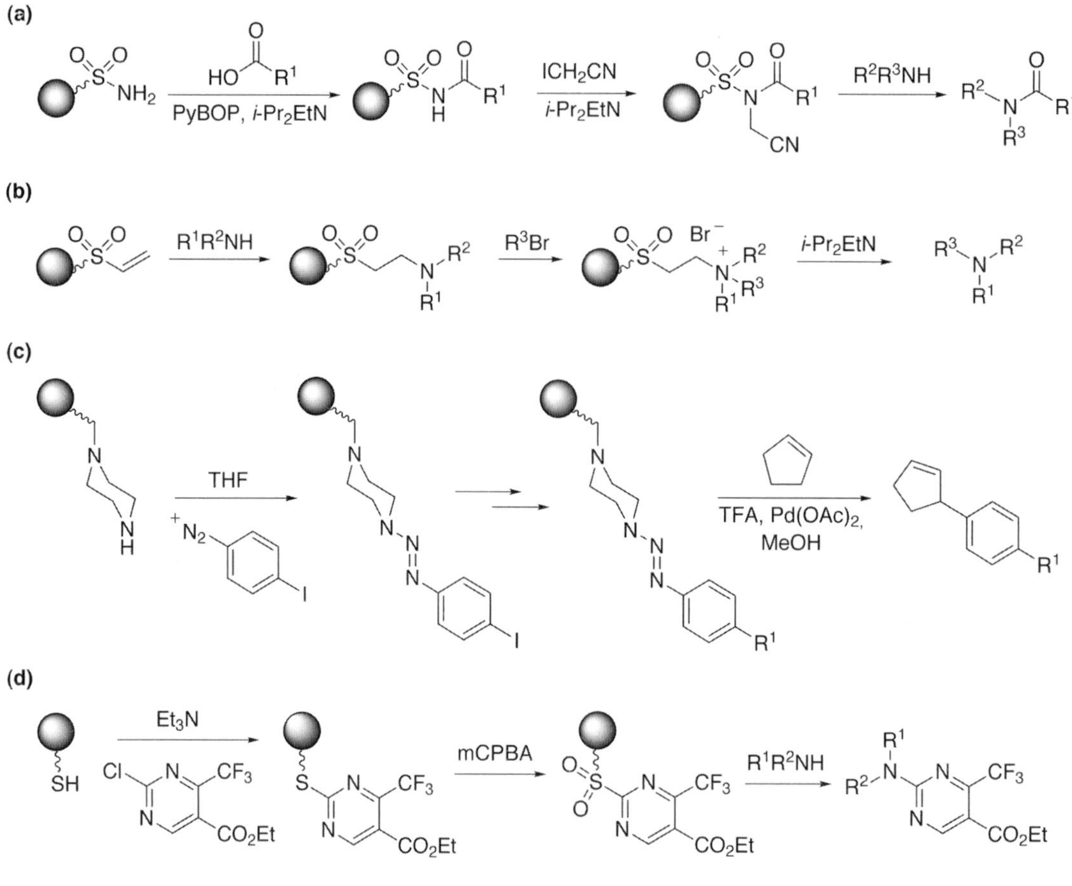

(a)

(b)

(c)

(d)

Scheme 4. Use of diversification linkers for the synthesis of (**a**) amides, (**b**) amines, (**c**) aromatic, and (**d**) heteroaromatic compounds.

though significant progress has been made, as illustrated by Ley's multi-step synthesis of tetrasubstituted pyrroles (Scheme 6),[13] the development of additional low cost, high-loading supports is critical to enhance the efficiency of multi-step sequences that rely on solid-phase scavengers, since multiple supports are often used for each reaction.

Discrete Compound versus Mixture Libraries

Combinatorial libraries have been prepared as discrete compounds or as mixtures with each approach having obvious advantages and disadvantages. Establishing the presence and purity of library members is straightforward when discrete compounds are synthesized. In addition, biological evaluation of discrete compounds directly corre-

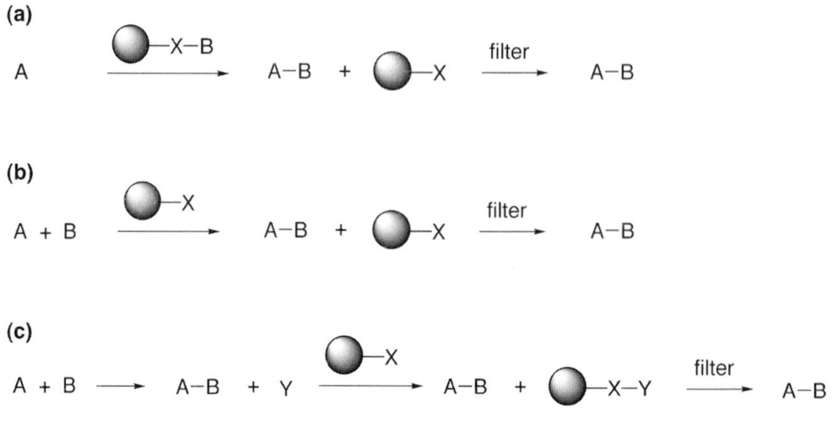

Figure 3. General examples of polymer-supported (**a**) reagents, (**b**) catalysts, and (**c**) scavengers.

(a)

(b)

(c)

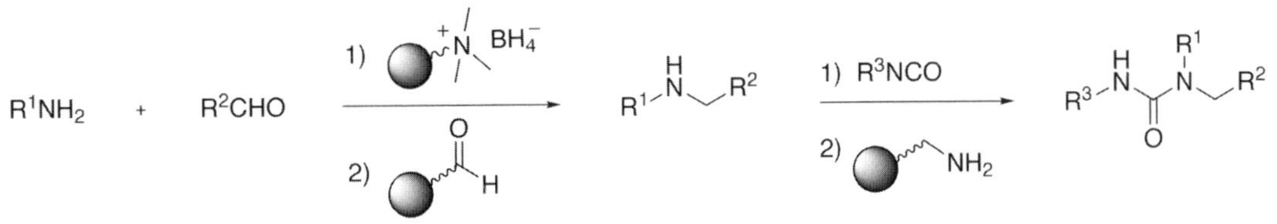

Scheme 5. A library of functionalized ureas synthesized using support-bound reagents and scavengers.

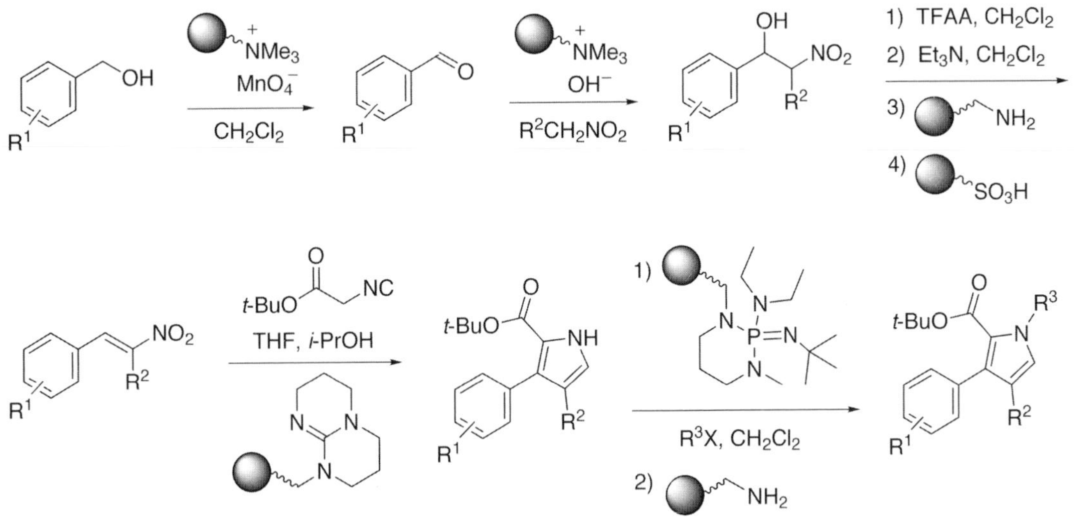

Scheme 6. Multi-step synthesis of pyrroles employing support-bound reagents and scavengers.

lates structure with activity. In contrast, compound mixtures offer less precise information; however this approach enhances the efficiency of library production enabling the preparation of very large libraries (millions to billions of compounds). Inherent limitations in the use of compound mixtures include: the difficulty of assessing the presence and purity of library members, the possibility of additive or opposing activities of compounds, and the necessity for resynthesis of the putative active members. While the preparation and testing of libraries of discrete compounds predominate in the pharmaceutical industry, mixture-based methods continue to be used and have resulted in the successful identification of bioactive compounds towards a number of therapeutic targets.[14–15]

The solid-phase technique of split and mix synthesis relies on the efficiency of mixture-based synthesis to provide very large libraries (millions) of discrete compounds (Figure 4).[16] In this approach, each resin bead is treated with a single building block for each synthesis step. Thus any single resin bead possesses identical copies of one library member, but the identity of the library member on any bead is lost due to the "mix" step of the process. Elegant strategies have been developed to chemically encode the syn-

thesis history of library members by introducing chemical tags at each synthesis step.[17] Upon completion of the synthesis, release and structural elucidation of the tags from any library bead provides the chemical equivalent of a UPC bar code for easy compound identification. Significant advances have also been made for the structural determination of the small amount of compound present per bead (~100 pmol/bead up to 1µmol/bead for newer bead designs). An increasingly popular variation of the split and mix method relies on radio-frequency (rf) memory chips, which are physically associated with the solid support during library synthesis. The synthesis history is encoded through rf signals that can be received and transmitted by the memory chip.[18]

Library Design

From first principles, we can only prepare and test an infinitesimally small fraction of the greater than 10^{60} small molecules that could theoretically be prepared. Therefore, the selection of the compounds to be prepared is critical for success of any combinatorial effort. Libraries have been designed upon: lead compounds which bind a target biomol-

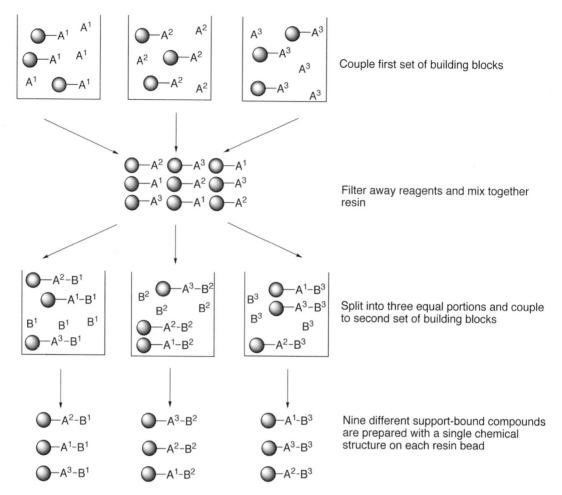

Figure 4. A general schematic depicting the split and mix synthesis concept.

ecule, the structure or mechanism of the target biomolecule, or the maximal display of diverse functionality for use towards a broad range of targets.[19]

Lead Optimization Libraries

A proven approach for the design of a combinatorial library employs a lead structure or compound class that has previously been identified to bind to a biomolecule target. Several combinatorial efforts for lead optimization have contributed compounds that have entered clinical trials.[20] At Merck, Chapman and coworkers provided an impressive example of this strategy with the solid-phase preparation of large mixture-based libraries of hundreds of thousands of compounds based on a lead compound that bound to the somatostatin receptors (Figure 5).[21] Library screening identified several highly selective and potent agonists to four of the five somatostatin receptor sub-types. These ligands have served as powerful pharmacological tools for understanding receptor sub-type function.[21]

Targeted Libraries

The design of libraries based upon the mechanism and/or structural motif of a target biomolecule has also proven to be highly successful. Since proteins are classified into families according to mechanism and structure, a sufficiently versatile and robust library synthesis approach enables *any* member of a protein family to be potentially targeted. This strategy will be of increasing importance because genome-sequencing efforts are rapidly expanding the number of protein targets within each protein family.

Targeted libraries have been most effective when based upon the display of diverse functionality about a minimal mechanism-based pharmacophore targeting an enzyme family. Early successes with this approach were first achieved with proteases.[22] Our own efforts to design libraries which target enzyme families require that the minimal pharmacophore serves as the site for attachment to solid support.[23] The pharmacophore is the only invariant part of the inhibitor structure, which

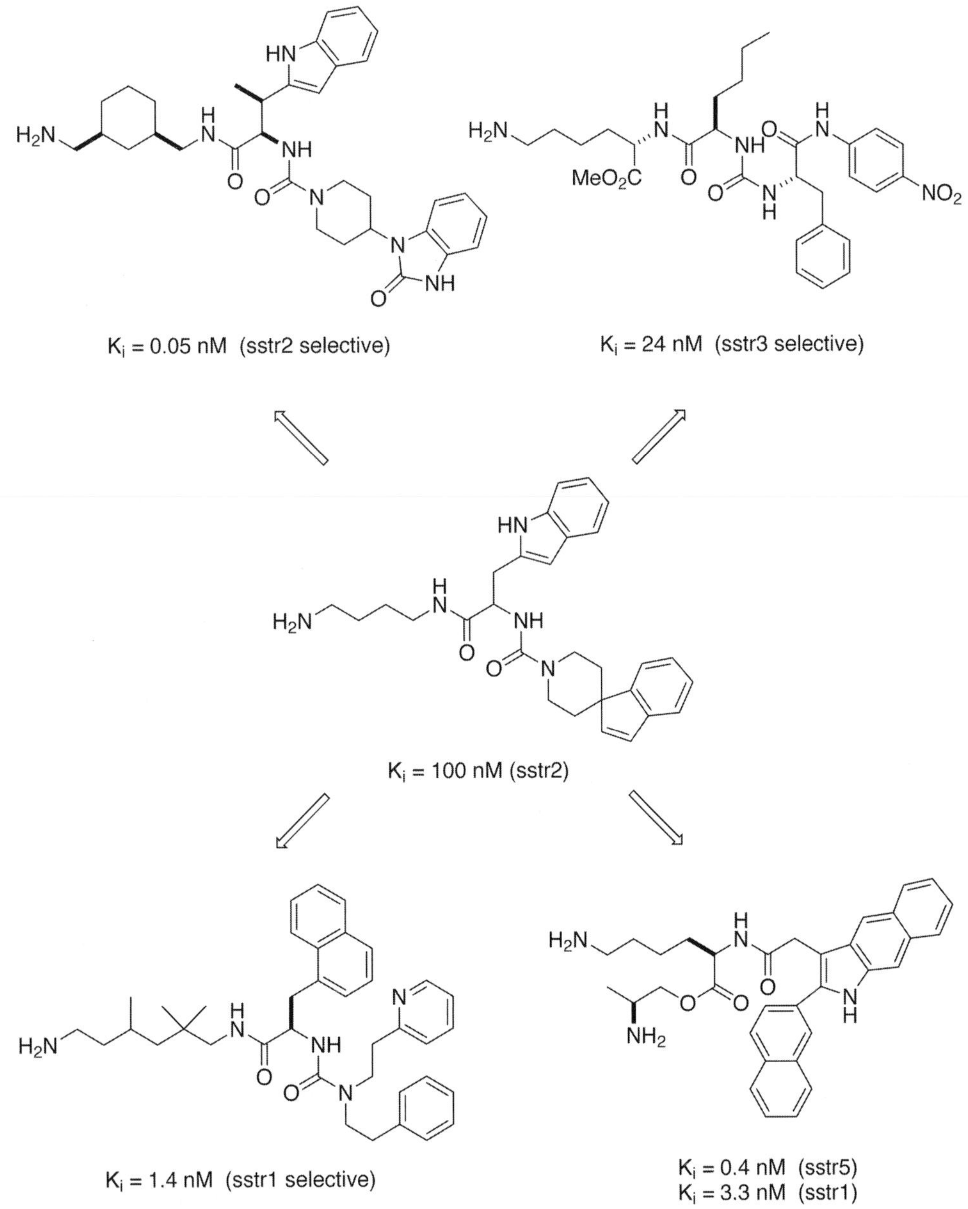

Figure 5. Potent somatostatin inhibitors produced by combinatorial lead optimization.

allows diverse functionality to be displayed at all variable sites of the inhibitor.

In the first demonstration of this conceptual strategy, we targeted the aspartyl proteases, which are a ubiquitous class of enzymes playing an important role in mammals, plants, fungi, and retroviruses. Potent inhibitors of the aspartyl proteases

have been developed that utilize a secondary alcohol as the minimal pharmacophore, which serves as a stable mimetic of the tetrahedral intermediate formed in aspartyl protease-catalyzed hydrolysis (Figure 6a). We specifically chose to display functionality about the hydroxyethylamine-based isostere (Figure 6b), since this isostere is amenable to the

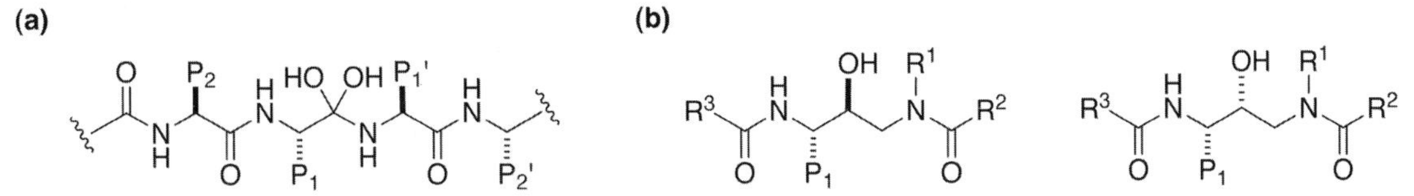

Figure 6. (a) Tetrahedral intermediate formed during aspartyl protease-catalyzed hydrolysis and **(b)** the(S)- and (R)-hydroxyethylamine-based isosteres.

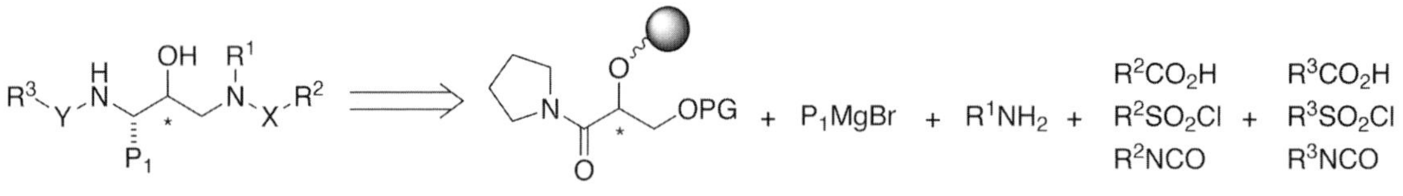

1a: (*S*) epimer
1b: (*R*) epimer

2a: (*S*) epimer
2b: (*R*) epimer

$X = CO, CONH, SO_2$
$Y = CO, NHCO, SO_2$

Figure 7. Potent aspartyl protease inhibitors synthesized from simple Grignard, amine, acid, sulfonyl chloride, and isocyanate building blocks.

introduction of a wide variety of side chains about both sides of the secondary alcohol. The aspartyl protease inhibitors are prepared by assembly of four readily available building blocks upon the minimal scaffolds **2a** and **2b** (Figure 7). Notably, a stereoselective synthesis was designed to access both the *S* and *R* epimeric alcohols since the preferred alcohol stereochemistry depends on both the targeted aspartyl protease and the overall inhibitor structure.

Libraries of hundreds to thousands of spatially separate inhibitors have been prepared and screened to identify small molecule inhibitors of the human protease cathepsin D and the essential malarial proteases, plasmepsins I and II. The best inhibitors do not incorporate any amino acids and possess high affinity ($K_i < 5$ nM).[24] Furthermore, these lead compounds were optimized by combinatorial methods for good physicochemical properties and minimal binding to human serum albumin. The optimized inhibitors effectively block cathepsin D-mediated proteolysis in human hippocampyl slices and are currently being used to evaluate the therapeutic potential of cathepsin D inhibition in the treatment of Alzheimer's disease. Additionally, the plasmepsin inhibitors serve as promising leads for the treatment of malaria.

We have also developed targeted library approaches towards cysteine proteases, which are important pharmaceutical targets due to their role in the pathogenesis of many diseases.[25] A common feature of virtually all cysteine protease inhibitors is an electrophilic functionality, such as a carbonyl or a Michael acceptor, which can react with the nucleophilic active site cysteine residue. We specifi-

cally chose to employ the ketone carbonyl as the minimal pharmacophore since functionality may be displayed on both sides of the carbonyl to achieve protease specificity.[26] The initial chloromethyl ketone introduces the P_1 side chain and provides sites for further functionalization (Scheme 7). Linking to support through the ketone carbonyl is ideal since the carbonyl pharmacophore is invariant regardless of the cysteine protease targeted. The hydrazone linkage allows nucleophilic substitution at the α-position while simultaneously preventing nucleophilic attack upon the carbonyl. The hydrazone also eliminates racemization, which is problematic for the corresponding enolizable α-acylamino substituted chiral ketone. As outlined in Scheme 7, a wide variety of different classes of cysteine protease inhibitors may be accessed by this method.

Prospecting Libraries

Combinatorial approaches have been most successful when information about the target biomolecule has been considered in the design of the library. However, for many biomolecules, structural or mechanistic information is not available or does not provide sufficient insight to enable productive library design. Also, lead compounds are not available for many targets, and in some cases, novel motifs for binding are desired. Under these circumstances, it is no surprise that the successful application of combinatorial chemistry has been less fre-

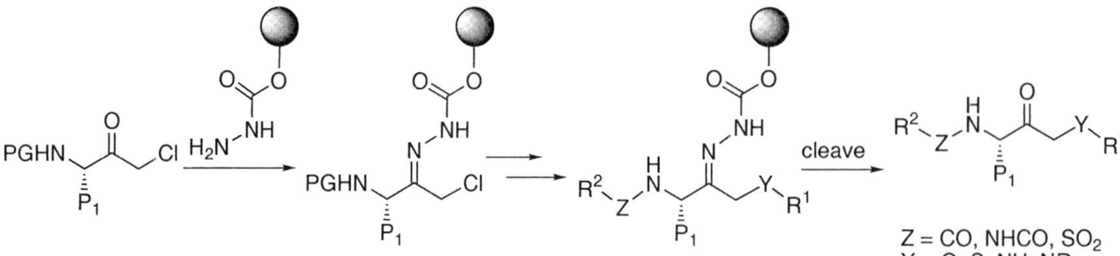

$Z = CO, NHCO, SO_2$
$Y = O, S, NH, NR$

Scheme 7. Synthesis strategy for ketone-based cysteine protease inhibitors.

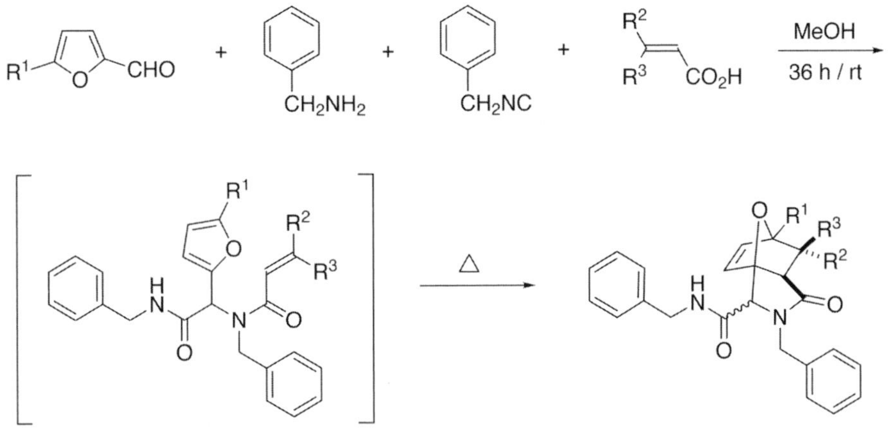

Scheme 8. Tandem Ugi-Diels–Alder reaction.

quent. Prospecting libraries, i.e., lead identification libraries, are driven by the selection of readily available input materials with a goal towards creating novel structures possessing the maximal display of diverse functionality.[27] To produce compounds in high yield and purity, a premium is placed on reducing the number of synthesis transformations. Equally essential is the development of a synthesis sequence that is compatible with a broad range of functional groups to maximize the diverse display of functionality. Multi-component condensation reactions, where three or more building blocks are introduced in a single step, are particularly efficient. The Passerini, Biginelli, and many variations of the Ugi multi-component reactions have been employed extensively. For example, Paulvannan uses simple building blocks in an Ugi reaction fol-

lowed by an intramolecular Diels-Alder reaction to prepare tricyclic heterocycles containing multiple sites of diversity (Scheme 8).[28] Cycloadditions and cyclocondensations enable rapid display of diversity upon rigid scaffold structures. These approaches are nicely represented by the azomethine ylide cycloaddition reaction wherein three readily available inputs (amine, aldehyde, and dipolarophile) are introduced in two sequential steps (Scheme 9a).[27b] Interesting multicyclic structures may be further accessed by tethering two of the three components (Scheme 9b).[29]

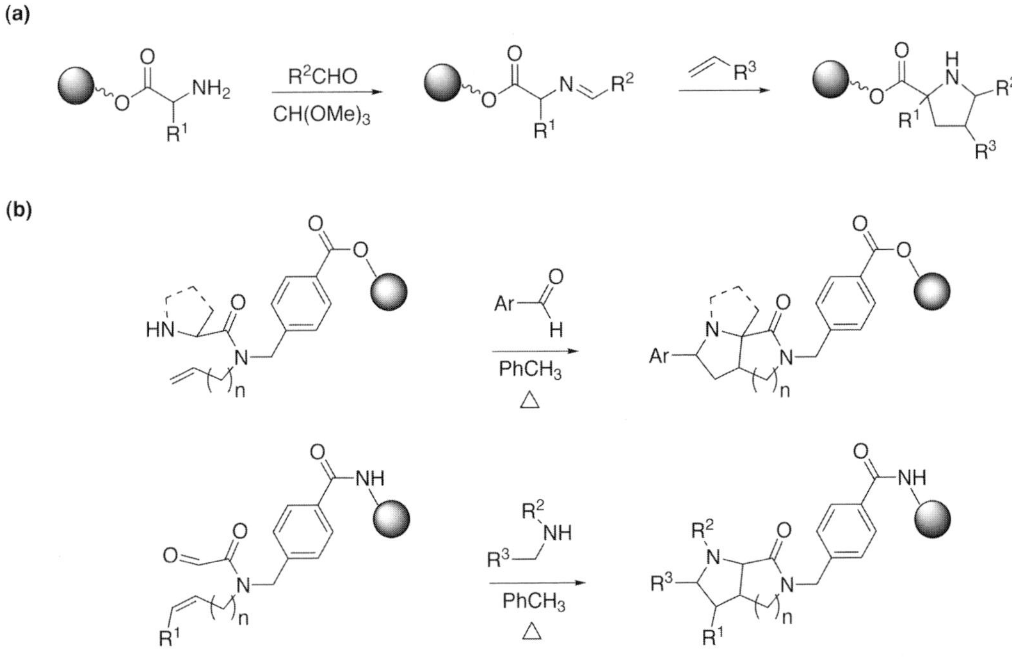

(a)

(b)

Scheme 9. Examples of azomethine ylide cycloaddition by (a) Affymax and (b) Bartlett and coworkers.

Targeted-Guided Compound Selection

Emerging approaches for identifying new leads to target biomolecules rely on feedback from the biomolecule target to select the building blocks for library preparation. One of the first approaches to be explored relies on genetic algorithms,[30] which are mathematical optimization techniques modeled upon natural evolution. Rather than preparing and testing a library that incorporates all possible combinations with the greatest number of building blocks, a smaller library of compounds is prepared using a subset of building blocks in only a few combinations. In subsequent generations of synthesis and testing, the building blocks that provide the greatest activity are recombined in analogy to genetic recombination. In addition, new building blocks may be introduced corresponding to mutation. Weber and coworkers elegantly demonstrated this approach by identifying a submicromolar small molecule inhibitor of thrombin (Figure 8). The inhibitor was identified after 20 generations of synthesis and testing, which required the preparation of only 400 compounds out of 160 000 possible combinations from $10 \times 40 \times 10 \times 40$ pre-selected Ugi reaction building blocks.

Lehn and coworkers have also introduced an innovative approach for the target-assisted selection of compounds.[31] In this approach, termed dynamic combinatorial chemistry, the target biomolecule selects the preferred combination of building blocks from a set capable of reversible self-assembly. Lehn first presented dynamic combinatorial chemistry for recognition induced assembly of inhibitors to carbonic anhydrase. Reductive amination trapped the equilibrium mixture of twelve possible imines obtained by combining three aldehydes and four amines (Figure 9). In the presence of carbonic anhydrase, the relative proportion of imine **3** almost doubled. From the twelve imines, imine **3** is the closest in structure to a known strong inhibitor of the enzyme. Although this approach is still in its infancy, increasing numbers of researchers are working to apply the approach to a range of applications.

We have recently demonstrated an alternative method "Combinatorial Target-Guided Ligand Assembly," which enables the rapid and efficient identification of ligands to predetermined biomolecule targets (Figure 10).[32] The combinatorial method does not require prior structural or mechanistic information. Instead, a collection of low molecular weight compounds, each with a common chemical linkage group, are screened against a target biomolecule to select building blocks for library preparation that bind to a target biomolecule. A library is prepared by connecting all combinations of the selected building blocks with a set of flexible scaffolds through the common chemical linkage group. Library screening against the target biomolecule identifies the combinations of scaffold and building blocks that provide the most productive binding. This approach was demonstrated by identifying new, potent and sub-type selective inhibitors of the tyrosine kinase c-Src. Specifically, 305 *O*-methyl oximes were screened to select 37 weakly-binding compounds. A library of 3515 members was then prepared from the aldehyde precursors of the weakly-binding oximes and five flexible *O,O*-bisaminoalkanediol scaffolds. The most active library member **4** showed a >500-fold increase in binding affinity relative to the initial binding elements from which it was composed (Figure 11). Furthermore, a profound linker dependence was observed with compound **4** showing 30-fold greater affinity than an analog containing a single additional methylene in the alkane scaffold. In contrast, traditional combinatorial methods would have required the preparation and testing of a library of greater than 230 000 members resulting from all possible combinations of the 305 aldehydes and five linkers.

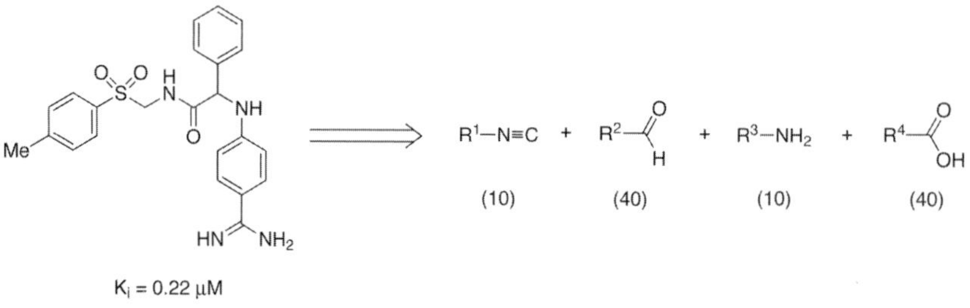

$R^1-N{\equiv}C$ + $R^2-\overset{O}{\underset{H}{\cdots}}$ + R^3-NH_2 + $R^4-\overset{O}{\underset{OH}{\cdots}}$

(10)   (40)   (10)   (40)

$K_i = 0.22\ \mu M$

Figure 8. Thrombin inhibitors identified by a genetic algorithm selection method.

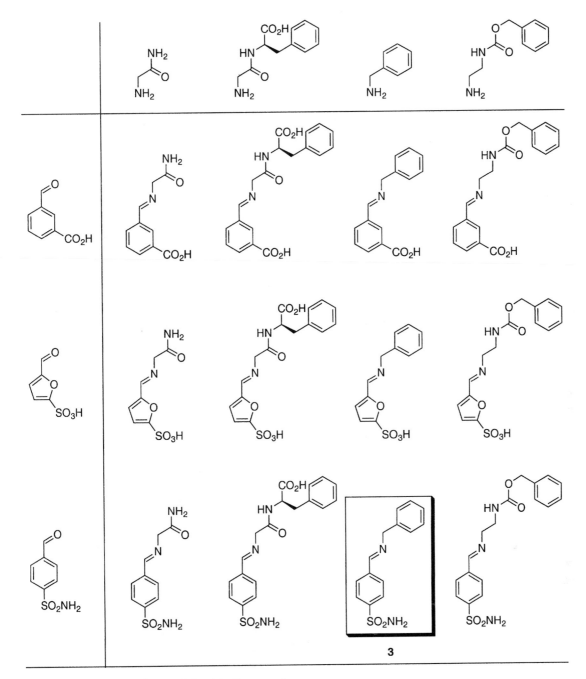

Figure 9. A dynamic combinatorial chemistry library used to identify a carbonic anhydrase inhibitor.

1. Prepare a set of potential binding elements with a common chemical linkage group X

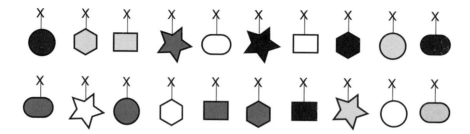

2. Screen potential binding elements to identify elements that bind to target

3. Prepare library of all possible combinations of linked binding elements

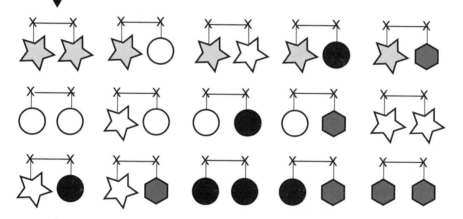

4. Screen library of linked binding elements to identify the tightest binding ligands

Figure 10. "Combinatorial Target-Guided Ligand Assembly."

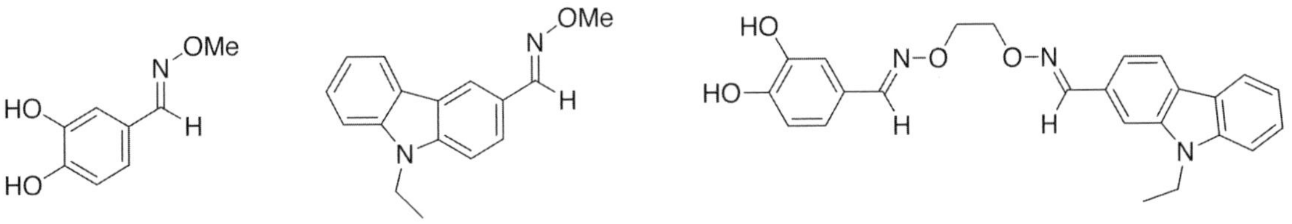

Figure 11. Potent tyrosine kinase inhibitors discovered through "Combinatorial Target-Guided Ligand Assembly."

Epilogue

The identification of compounds with a desired property is a central pursuit in science. In the field of drug discovery combinatorial chemistry has played an increasingly important role for identification and optimization of drug leads which target therapeutically important biomolecules. For the successful implementation of combinatorial methods, new and innovative synthesis methods have been developed. Additionally, novel conceptual approaches to the design of compounds have been pursued to more efficiently generate libraries of small molecules.

References and Notes

1. J. A. Ellman, M. A. Gallop, "Combinatorial Chemistry", *Curr. Opin. Chem. Biol.* **1998**, *2*, 305–428.

2. R. E. Dolle, K. H. Nelson, Jr., "Comprehensive Survey of Combinatorial Library Synthesis: 1998", *J. Comb. Chem.* **1999**, *1*, 235–282.

3. J. P. Whitten, Y. F. Xie, P. E. Erickson, T. R. Webb, E. B. De Souza, D. E. Grigoriadis, J. R. McCarthy, "Rapid Microscale Synthesis, a New Method for Lead Optimization Using Robotics and Solution Phase Chemistry: Application to the Synthesis and Optimization of Corticotropin-Releasing Factor₁ Receptor Antagonists", *J. Med. Chem.* **1996**, *39*, 4354–4357.

4. a) B. J. Backes, J. A. Ellman, "Solid Support Linker Strategies", *Curr. Opin. Chem. Biol.* **1997**, *1*, 86–93; b) I. W. James, "Linkers for Solid Phase Organic Synthesis", *Tetrahedron* **1999**, *55*, 4855–4946.

5. For the vast majority of drug discovery efforts compound libraries have been screened following cleavage from support. For other applications, support-bound libraries have been employed much more extensively.[1]

6. F. X. Woolard, J. Paetsch, J. A. Ellman, "A Silicon Linker for Direct Loading of Aromatic Compounds to Supports. Traceless Synthesis of Pyridine-Based Tricyclics", *J. Org. Chem.* **1997**, *62*, 6102–6103.

7. B. A. Lorsbach, R. B. Miller, M. J. Kurth, "Reissert-Based 'Traceless' Solid-Phase Synthesis: Isoquinoline, and Isoxazoline-Containing Heterocycles", *J. Org. Chem.* **1996**, *61*, 8716–8717.

8. B. J. Backes, A. A. Virgilio, J. A. Ellman, "Activation Method to Prepare a Highly Reactive Acylsulfonamide 'Safety-Catch' Linker for Solid-Phase Synthesis", *J. Am. Chem. Soc.* **1996**, *118*, 3055–3056.

9. F. E. K. Kroll, R. Morphy, D. Rees, D. Gani, "Resin-Immobilized Benzyl and Aryl Vinyl Sulfones: New Versatile Traceless Linkers for Solid-Phase Organic Synthesis", *Tetrahedron Lett.* **1997**, *38*, 8573–8576.

10. A. de Meijere, H. Nüske, M. Es-Sayed, T. Labahn, M. Schroen, S. Bräse, "New Efficient Multicomponent Reactions with C-C Coupling for Combinatorial Application in Liquid and on Solid Phase", *Angew. Chem. Int. Ed.* **1999**, *38*, 3669–3672.

11 R. J. Booth, J. C. Hodges, "Solid-Supported Reagent Strategies for Rapid Purification of Combinatorial Synthesis Products", *Acc. Chem. Res.* **1999**, *32*, 18–26.

12. S. W. Kaldor, M. G. Siegel, J. E. Fritz, B. A. Dressman, P. J. Hahn, "Use of Solid Supported Nucleophiles and Electrophiles for the Purification of Non-Peptide Small Molecule Libraries", *Tetrahedron Lett.* **1996**, *37*, 7193–7196.

13. M. Calderelli, J. Habermann, S. V. Ley, "Clean Five-Step Synthesis of an Array of 1, 2, 3, 4-Tetra-substituted Pyrroles using Polymer-Supported Reagents", *J. Chem. Soc., Perkin Trans. 1* **1999**, 107–110.

14. R. A. Houghten, C. Pinilla, J. R. Appel, S. E. Blondelle, C. T. Dooley, J. Eichler, A. Nefzi, J. M. Ostresh, "Mixture-Based Synthetic Combinatorial Libraries", *J. Med. Chem.* **1999**, *42*, 3743–3778.

15. A number of very innovative methods have been developed to minimize the number of compounds that must be resynthesized in order to establish the identity of active library members present in mixtures.

16. K. S. Lam, M. Lebl, V. Krchnák, "The 'One-Bead-One-Compound' Combinatorial Library Method", *Chem. Rev.* **1997**, *97*, 411–448.

17. A. W. Czarnik, "Encoding Methods for Combinatorial Chemistry", *Curr. Opin. Chem. Biol.* **1997**, *1*, 60–66.

18. Due to the statistical nature of split synthesis, an average 20 beads per member must be used to ensure that all library members are prepared successfully. This requirement results in increased reagent consumption and extra effort in screening and structural determination. One of the most powerful aspects of the rf memory chips is the ability to perform automated *sorting* of library members. Thus a complete library may be prepared with each member synthesized only once.

19. In this review we have chosen to focus on strategies for designing the templates and scaffolds used in library synthesis. Equally important is the selection of building blocks for library preparation. A number of powerful computational methods have been developed to select the minimal set of building blocks that maxi-

mize the functional group diversity. Other computational approaches base building block selection on structure-based methods. Finally, physicochemical properties of building blocks and their potential impact upon the pharmacokinetics of library members are increasingly being considered.

20. a) S. Borman, "Combinatorial Chemists Focus on Small Molecules, Molecular Recognition, and Automation", *Chem. Eng. News* **1996**, *74*, 29–54; b) A. M. Thayer, "Combinatorial Chemistry Becoming Core Technology at Drug Discovery Companies", *Chem. Eng. News* **1996**, *74*, 57–64.

21. S. P. Rohrer, E. T. Birzin, R. T. Mosley, S. C. Berk, S. M. Hutchins, D.-M. Shen, Y. Xiong, E. C. Hayes, R. M. Parmar, F. Foor, S. W. Mitra, S. J. Degrado, M. Shu, J. M. Klopp, S.-J. Cai, A. Blake, W. W. S. Chan, A. Pasternak, L. Yang, A. A. Patchett, R. G. Smith, K. T. Chapman, J. M. Schaeffer, "Rapid Identification of Subtype-Selective Agonists of the Somatostatin Receptor Through Combinatorial Chemistry", *Science* **1998**, *282*, 737–740.

22. M. Whittaker, "Discovery of Protease Inhibitors Using Targeted Libraries", *Curr. Opin. Chem. Biol.* **1998**, *2*, 386–396.

23. C. E. Lee, E. K. Kick, J. A. Ellman, "General Solid-Phase Synthesis Approach to Prepare Mechanism-Based Aspartyl Protease Inhibitor Libraries. Identification of Potent Cathepsin D Inhibitors." *J. Am. Chem. Soc.* **1998**, *120*, 9735–9747.

24. T. S. Haque, A. G. Skillman, C. E. Lee, H. Habashita, I. Y. Gluzman, T. J. A. Ewing, D. E. Goldberg, I. D. Kuntz, J. A. Ellman, "Potent, Low-Molecular-Weight Non-Peptide Inhibitors of Malarial Aspartyl Protease Plasmepsin II", *J. Med. Chem.* **1999**, *42*, 1428–1440.

25. H.-H. Otto, T. Schirmeister, "Cysteine Proteases and Their Inhibitors", *Chem. Rev.* **1997**, *97*, 133–171, and references cited therein.

26. A. Lee, L. Huang, J. A. Ellman, "General Solid-Phase Method for the Preparation of Mechanism-Based Cysteine Protease Inhibitors", *J. Am. Chem. Soc.* **1999**, *121*, 9907–9914.

27. a) M. R. Spaller, M. T. Burger, M. Fardis, P. A. Bartlett, "Synthetic Strategies in Combinatorial Chemistry", *Curr. Opin. Chem. Biol.* **1997**, *1*, 47–53; b) E. M. Gordon, M. A. Gallop, D. V. Patel, "Strategy and Tactics in Combinatorial Organic Synthesis. Applications in Drug Discovery", *Acc. Chem. Res.* **1996**, *29*, 144–154.

28. K. Paulvannan, "Preparation of Tricyclic Nitrogen Heterocycles via Tandem Four-Component Condensation/ Intramolecular Diels–Alder Reaction", *Tetrahedron Lett.* **1999**, *40*, 1851–1854.

29. M. A. Marx, A.-L. Grillot, C. T. Louer, K. A. Beaver, P. A. Bartlett, "Synthetic Design for Combinatorial Chemistry. Solution and Polymer-Supported Synthesis of Polycyclic Lactams by Intramolecular Cyclization of Azomethine Ylides", *J. Am. Chem. Soc.* **1997**, *119*, 6153–6167.

30. L. Weber, "Applications of Genetic Algorithms in Molecular Diversity", *Curr. Opin. Chem. Biol.* **1998**, *2*, 381–385.

31. J.-M. Lehn, "Dynamic Combinatorial Chemistry and Virtual Combinatorial Libraries", *Chem. Eur. J.* **1999**, *5*, 2455–2463.

32. D. J. Maly, I. C. Choong, J. A. Ellman, "Combinatorial Target-Guided Ligand Assembly. Identification of Potent, Sub-type Selective c-Src Inhibitors", *Proc. Natl. Acad. Sci., USA,* **2000**, *97*, 2419–2424.

Theozymes and Catalyst Design

**Dean J. Tantillo
and K. N. Houk**

*Department of Chemistry
and Biochemistry, University
of California at Los Angeles,
Los Angeles, CA
90095–1569, USA*

*Phone: +310 206 0515,
Fax: +310 206 1843,
e-mail:
houk@chem.ucla.edu*

Keywords ■ Theozyme ■ Compuzyme ■ Ezyme ■ Abzyme ■ Chemzyme ■ Catalytic Antibodies ■ Catalysis ■ Rational Design ■ Preorganization ■ Combinatorial ■ Theoretical

Concept: A theozyme ("theoretical enzyme") is a theoretical catalyst model consisting of an array of functional groups optimized so as to maximize stabilization of a transition state (see cartoon below). Theozymes may be used to provide quantitative information about mechanisms of catalysis, to design improvements to known catalysts, and to design new synthetic catalysts ("chemzymes").

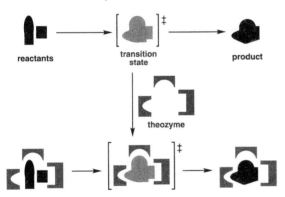

Abstract: Rational catalyst design involves understanding how specific intermolecular interactions present in transition state · catalyst complexes lower the activation barrier for a given reaction. Theozymes are a tool that can provide this information. Several examples that demonstrate the potential of theozymes are discussed in this essay. The first example involves the use of theozyme calculations to understand and predict the mechanism of an organic reaction promoted by a biological catalyst. The second case involves an application of theozymes to the rational design of synthetic catalysts.

Prologue

Rational design of catalysts for organic reactions has frequently been based on the mimicry of biological catalysts.[1] In order for this strategy to succeed, a quantitative understanding of the factors contributing to biological catalysis is necessary. Although there is little doubt that the free energy of activation for the rate-determining step of a catalyzed reaction must be less than that of the uncatalyzed process, the source of this barrier lowering is still not well understood.[2] Many studies of catalysis concentrate on structure – usually obtained from X-ray crystallography or spectroscopic techniques – yet structure does not reveal energetics; structural data cannot reveal the magnitudes of individual contributions to catalysis. The fundamental function of a theozyme is to quantitate the effects of specific interactions between a catalyst and transition state and to compare these with the effects of the corresponding interactions between the catalyst and substrates or products.

Constructing a theozyme involves the following steps: (1) The structure of a transition state for a particular reaction is obtained using quantum mechanical methods. (2) The positions of functional groups of interest ("residues") are then optimized around the transition state. The geometry of the transition state is often held static during this process but may be allowed to relax in the presence of the theozyme. (3) The reaction energetics in the presence of the theozyme are computed and are compared with those in the gas phase or in solution, which involves the computation of the interaction of the theozyme with reactant(s), transition state, and product(s). This procedure gives quantitative information about the influence of each catalyst functional group studied on rate. Applications of this technique are far-reaching, ranging from the study of known biocatalysts to the design of synthetic catalysts for which no counterparts exist in nature.

Quantitating the Importance of Noncovalent Interactions in Enzymatic Catalysis: A Brief Survey of Theozymes

To date, most theozymes have been used to quantitate the effects of noncovalent interactions on catalysis by enzymes and catalytic antibodies (abzymes).[3] Theozymes have provided insight into the biological catalysis of substitution, cyclization, reduction, elimination, decarboxylation, hydrolysis, radical, and pericyclic reactions. A selection of theozymes and the reactions they catalyze are shown in Figure 1; the reaction is shown in the box surrounded by the normal line and the catalyst is given over the arrow. In the box marked with a double dagger, the transition state of the model system, surrounded by the theozyme in bold print, are shown. More detailed descriptions of these and other theozymes may be found in reference.[3]

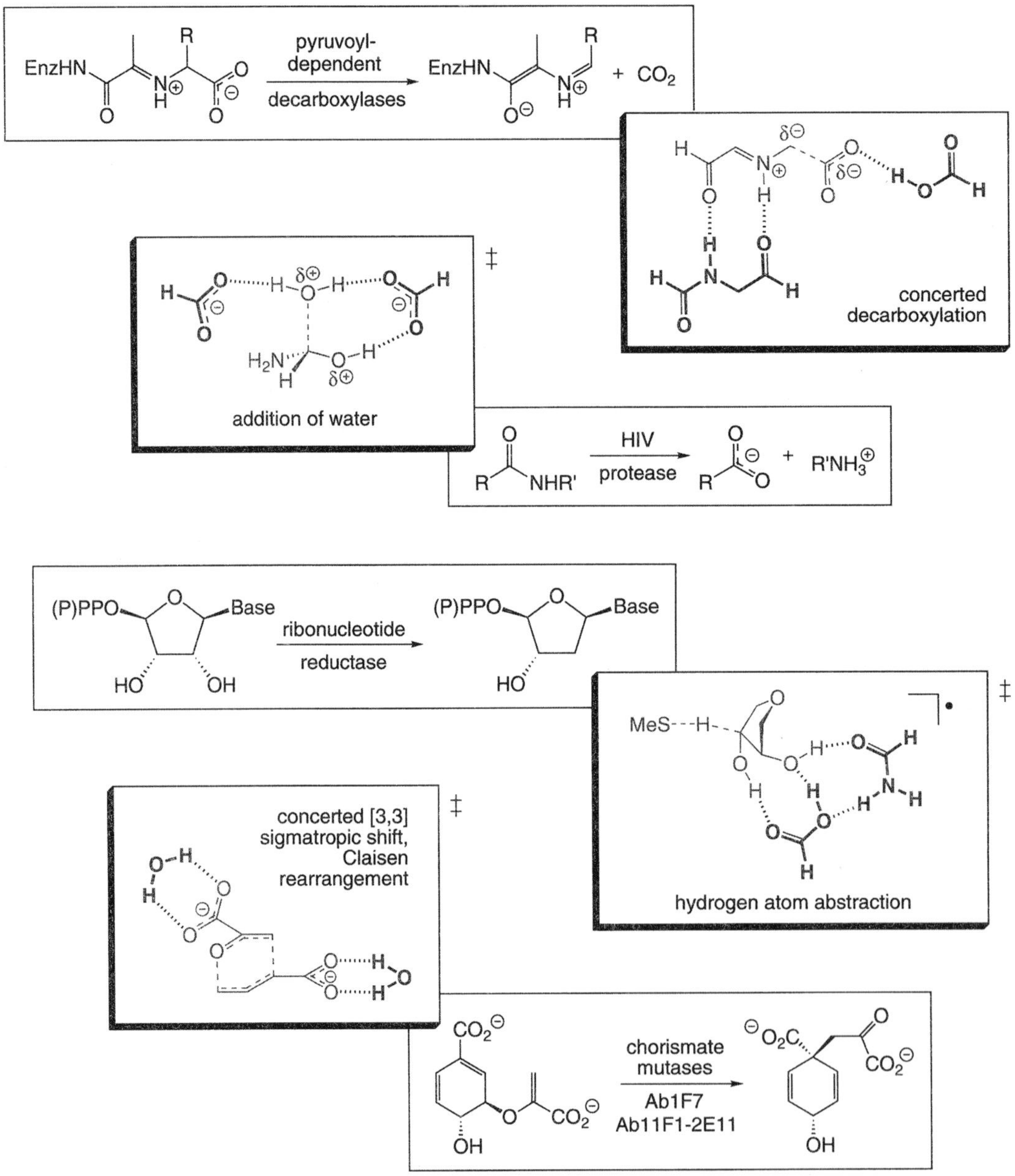

Figure 1. Selected theozymes (bold), the reactions they catalyze, and computed transition states complexed to them.[3]

Understanding and Predicting Biological Catalysis: Antibody Catalysis of Disfavored Cyclization Reactions

The theozyme concept was first fully developed by Na and Houk in their investigation of abzyme-catalyzed intramolecular cyclizations of hydroxyepoxides. These investigations focused on antibody 26D9, an antibody that selectively catalyzes 6-endo over 5-exo ring closure of substrate **1** (*n* = 1, Figure 2) to form tetrahydopyran derivatives **2** (*n* = 1).[4] In contrast, the acid- or base-catalyzed process in solution selectively forms tetrahydrofuran derivivatives **3** (*n* = 1), as expected based on Baldwin's Rules for cyclization reactions.[5] This situation points to a major success of antibody catalysis: alteration of the selectivity (chemo-, regio-, or stereo-) of a reaction from its normal course in solution. While many catalytic antibodies accentuate biases inherent in solution reactions, 26D9 counteracts a solution bias.

Initial theoretical studies on the hydroxyepoxide system[6] were aimed at elucidating and comparing the structures of the competing 5-exo and 6-endo transition states for cyclization, in pursuit of key differences that could be exploited by the antibody binding site. The gas phase reactions of protonated **1** (*n* = 1) to form **2** and **3** were examined with ab initio calculations.[7] It was predicted that the 5-exo process is favored over the 6-endo process by 1.9 kcal/mol in the gas phase, in accord with Baldwin's Rules. This preference was shown to arise primarily from the inability of the nucleophilic hydroxyl group to obtain an ideal angle of attack on the epoxide in the 6-endo transition state. Additional calculations including a continuum dielectric envi-

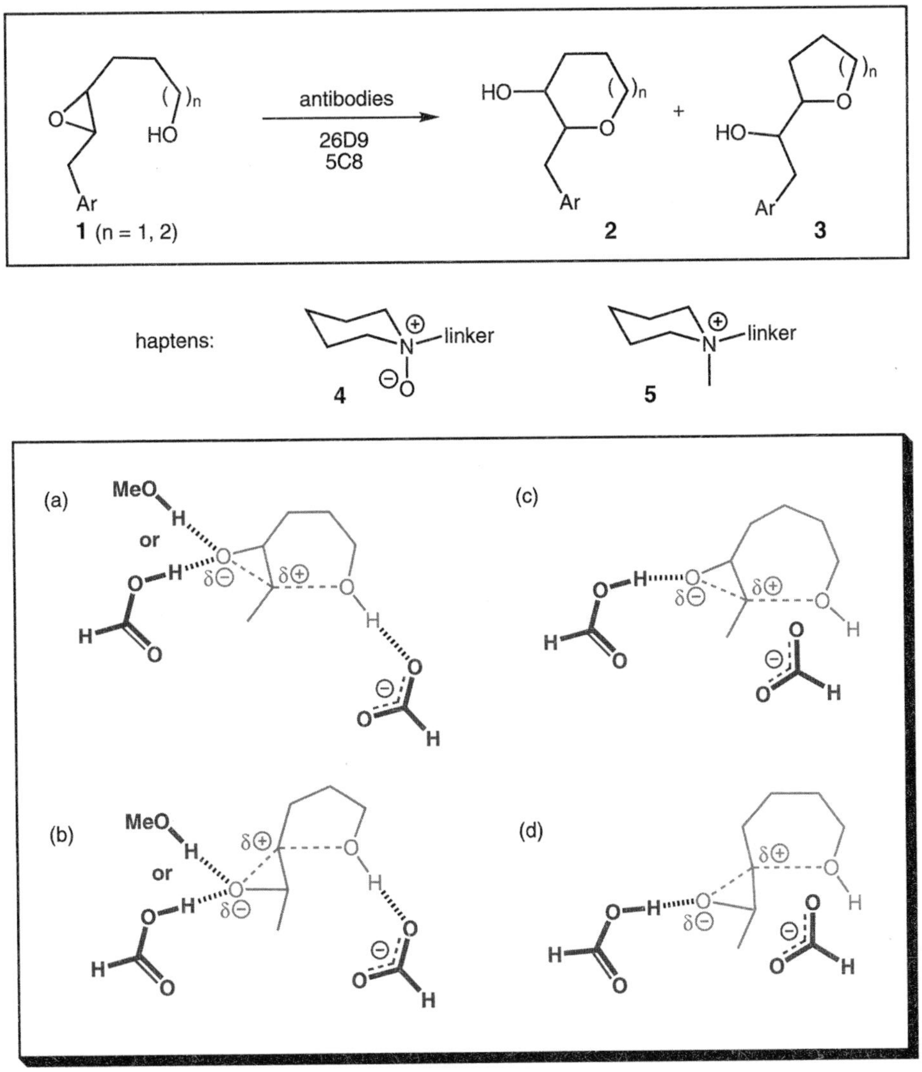

Figure 2. Theozymes for hydroxyepoxide cyclizations. Antibody-catalyzed cyclization reactions are shown schematically in the box. Antibodies 26D9 and 5C8 were raised against haptens **4** and **5**, respectively. (**a**) Theozymes (bold) complexed to the 6-endo cyclization transition state model. (**b**) Theozymes (bold) complexed to the 5-exo cyclization transition state model. (**c**) Theozyme (bold) complexed to the 7-endo cyclization transition state model. (**d**) Theozyme (bold) complexed to the 6-exo cyclization transition state model.

ronment to represent aqueous solvation[8] revealed that this difference in activation energy does not change in a homogenous dielectric environment. It was therefore suggested that the antibody alters the cyclization selectivity through specific noncovalent interactions that selectively stabilize the 6-endo transition state (selective binding by approximately 4 kcal/mol is necessary in order for no 5-exo product to be observed in this case).[9] The primary difference between the two transition state structures is that the 6-endo transition state has longer breaking and forming C–O bonds and more positive charge on the electrophilic carbon than does the 5-exo transition state (ie. the 6-endo process is more S_N1-like). As a result, it was proposed that antibody 26D9 favors the anti-Baldwin process by selective stabilization of this partial positive charge.

In order to test this proposal, several theozymes were constructed.[10] Antibody 26D9 was originally elicited in response to a piperidine-N-oxide hapten (**4**, Figure 2), in which the polarized N–O bond was meant to mimic the breaking C–O bond of the epoxide in the cyclization transition state, and the 6-membered piperidine ring was used to represent the size and shape of the 6-endo transition state. Hapten **4** was used as a template to create several

theozymes to test how anti-Baldwin selectivity is produced. Since no structural information on the combining site of this antibody was available, formate was chosen as a model for potential anionic groups (aspartate or glutamate), and formic acid or methanol was selected as a model for hydrogen bond donor or acidic groups (aspartic acid, glutamic acid, tyrosine, serine, or threonine).

The construction of a theozyme for the cyclization reaction is outlined in Figure 3 for the formate/formic acid system. First, the geometries of a truncated model of hapten **4** (N-methylpiperidinium-N-oxide), formate, formic acid, and the transition state models for endo- and exo-cyclization were optimized independently. Based on the idea that the microevolved antibody binding pocket should be optimized to bind the hapten, the positions of formate and formic acid were then optimized around the hapten model.[11] A hydrogen bond between the acidic proton of formic acid and the anionic oxygen of the N-oxide and an ion-pairing interaction between the formate and the cationic nitrogen of the N-oxide were observed in the resulting complex. The location of the formate and formic acid, fixed in space, constitute the candidate theozyme.

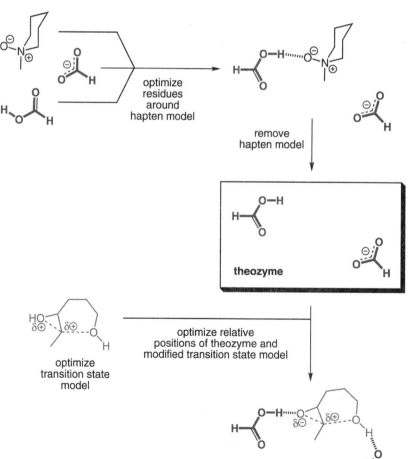

Figure 3. The procedure used in constructing the formate/formic acid theozyme. First, the hapten and theozyme residues were optimized independently. Second, the positions of the residues around the hapten were optimized. Third, the hapten was removed, resulting in the theozyme. Subsequently, the interaction of this theozyme with model cyclization transition states was optimized.

Calculations were then performed to see whether this arrangement of functional groups can stabilize one cyclization transition state more than the other. The 5-exo and 6-endo transition state models were first modified by removal of the proton on the epoxide oxygen to allow a theozyme residue to function as a general acid. The position of each these modified transition state models within the theozyme was then optimized,[12] leading to the complexes shown schematically in Figure 2a and b. For both the 5-exo and 6-endo transition states, hydrogen bonds to the epoxide oxygen and the nucleophilic hydroxyl group are observed. In addition, the partially positive epoxide carbon atoms interact with the formate residue at distances of approximately 4–5 Å. The 6-endo transition state is selectively stabilized by the formate/formic acid theozyme over that for 5-exo cyclization by 4 kcal/mol, demonstrating that selective electrostatic stabilization is sufficient to explain the observed abzyme selectivity; selective binding of the nonpolar alkyl tether is not necessary.[13] This suggests that the antibody binding site may function by having appropriately preorganized acidic and anionic residues.

Given the success of the electrostatic argument in explaining the known abzyme selectivity, a further prediction was made[10] for the homologous 6-exo versus 7-endo competition for substrate **1** (n = 2); because the theozyme had no recognition elements

for ring size, it was considered possible that the theozyme could stabilize other cyclization transition states. As in the 5-exo versus 6-endo case, the transition states were located and allowed to interact with the formate/formic acid theozyme described above (Figure 2c and d). It was predicted that these properly oriented residues could again promote the anti-Baldwin process (leading to the product of endo cyclization) by providing 5 kcal/mol of selective stabilization and overcoming a 2.6 kcal/mol preference in the absence of the antibody. Subsequently, experiments on hydroxyepoxide **1** (n = 2) by Janda, Shevlin, and Lerner at Scripps[14] revealed that antibody 26D9 does in fact favor 7-endo cyclization, producing the oxepane **2** (n = 2) in 98 % yield!

The Scripps group also produced a different antibody for hydroxyepoxide cyclization from the piperidinium hapten **5**.[15] In this hapten, the piperidinium nitrogen bears a larger positive charge than in hapten **4**, and the hydrogen bond-accepting anionic oxygen of the N-oxide is replaced by a methyl group. A new abzyme, 5C8, was produced in response to hapten **5**. 5C8 was found to catalyze the selective 6-endo cyclization of **1** (n = 1) with a kinetic profile similar to that of 26D9.

The X-ray crystal structures of antibody 5C8 in complex with piperidine-N-oxide hapten **4** and piperidinium hapten **5** were determined at 2.0 Å resolution.[15] In the 5C8 · **5** complex, two carboxylate residues (AspH95 and AspH101) flank the cationic

(a)

Antibody 5C8

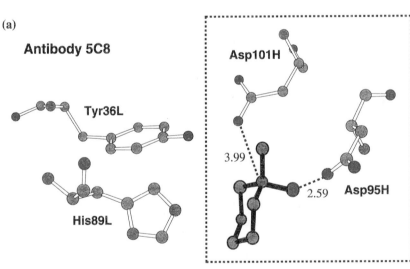

Tyr36L

His89L

Asp101H

3.99

2.59 **Asp95H**

(b)

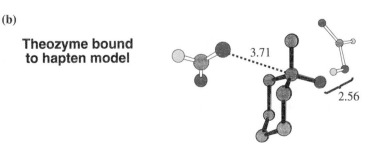

Theozyme bound to hapten model

3.71

2.56

Figure 4. (**a**) Selected residues from the combining site of antibody 5C8 complexed to piperidine-*N*-oxide hapten **4**, as determined by X-ray crystallography.[15] The linker portion of the hapten has been truncated to a methyl for comparison with the theozyme complex. (**b**) The formate/formic acid theozyme complexed to the model of hapten **4**, as optimized through quantum mechanical calculations.[6]

piperidinium nitrogen at distances of 3.7 and 4.0 Å, respectively. In the 5C8 · **4** complex, the piperidinium ring is rotated slightly, bringing the oxygen of the *N*-oxide in the vicinity of AspH95 (Figure 4), which is presumably protonated in this case. The similarity of the aspartate/aspartic acid pair in 5C8 to the formate/formic acid theozyme is undeniable!

The 5-exo and 6-endo transition states were then docked into the 5C8 combining site using the Auto-Dock suite of programs,[16] and the noncovalent interactions between the abzyme and transition states were analyzed. In both transition states, the attacking hydroxyl group is near to both HisL89 and TyrL36 (see Figure 4), either of which may activate it towards nucleophilic attack by acting as a general base, and the epoxide oxygen is near to AspH95, which presumably functions as a general acid. These interactions contribute to catalysis, but as shown by the theozyme studies described above, the regioselectivity of cyclization is likely controlled by the selective stabilization of the developing positive charge on the electrophilic carbon in the 6-endo transition state. Nearby AspH101 may serve exactly this function. Sequence comparison of 5C8 and 26D9 reveals that AspH95 in 5C8 is replaced by a tyrosine in 26D9, which may also function as a general acid, while AspH101, HisL89, and TyrL36 are present in both 5C8 and 26D9. The congruence between the natural and theoretical solutions to the problem of selective stabilization of the anti-Baldwin transition state is striking!

These theozyme studies emphasize the importance of specific noncovalent interactions between polar groups in the active sites of biological catalysts and the transition states to which they bind. The use of theozymes to predict previously unknown structures and reactivity patterns of biological catalysts is certainly risky, yet both sorts of prediction have been validated by experiment in this case.

Catalyst Design from Theoretical Principles: Chemzymes for Peptide Synthesis from Theozyme Blueprints

Theozymes can also be used to guide the rational design of synthetic catalysts or "chemzymes."[17] In contrast to theozymes that consist of only certain key functional groups present in biological catalysts, theozyme calculations on synthetic catalysts often include entire catalyst molecules. One reaction for which theozymes have been used both to rationalize the performance of known organic catalysts and to design new catalysts is the aminolysis of

esters. The development of efficient, selective, well-defined catalysts for peptide synthesis has been a long-standing goal of organic chemists and biochemists alike, and represents a rich area of research on the rapidly blurring frontier between chemistry and biology.

The potential of theozyme-aided catalyst design was explored in detail by Zipse and Houk[18] for polyether-catalyzed aminolysis of aryl esters (Figure 5). Their general plan for catalyst development was: "(1) Find the lowest energy reaction pathway for the *uncatalyzed* reaction under study. (2) Determine the structure of the highest transition state on this pathway. (3) Design a catalyst that recognizes this transition state better than any other structure on the pathway. (4) Synthesize the catalyst and study its catalytic properties."[18]

Initial theoretical studies focused on steps (1) and (2). Several model systems were examined with ab initio calculations.[19] For the reaction of methyl amine with methyl acetate, it was shown that the addition/elimination (through a neutral tetrahedral intermediate) and the direct displacement (through a transition state similar to that shown in Figure 5a) mechanisms for aminolysis had comparable activation barriers. However, in the case of methyl amine addition to phenyl acetate, it was shown that the direct displacement pathway is favored by approximately 5 kcal/mol.[20] Noncovalent stabilization of the direct displacement transition state was therefore the focus of the subsequent catalyst design process.

The effects of noncovalent catalyst–transition state interactions were explored through theozyme calculations.[18] Known polyethers were first examined, and the transition state for methyl amine attack on methyl acetate was used due to its moderate size. Optimization of the position of a dimethyl ether molecule with respect to this transition state led to a hydrogen bond between the ether oxygen and the amine proton that is not transferred in the reaction (Figure 5a).[21] This hydrogen bond activates the amine towards nucleophilic attack, and promotes formation of the N–C bond and breaking of the C–O bond. Inclusion of the dimethyl ether molecule lowers the activation barrier for aminolysis by almost 9 kcal/mol. Calculations on a dimethoxyethane (DME) theozyme led to a complex (Figure 5b) in which both ether oxygens interact with the amine proton. The effects on the transition state structure and the activation barrier in the presence of DME were comparable to those with dimethyl ether. These calculations suggest that selective stabilization of the aminolysis transition state by hydrogen bonding to the amine proton is an important factor in polyether catalysis.

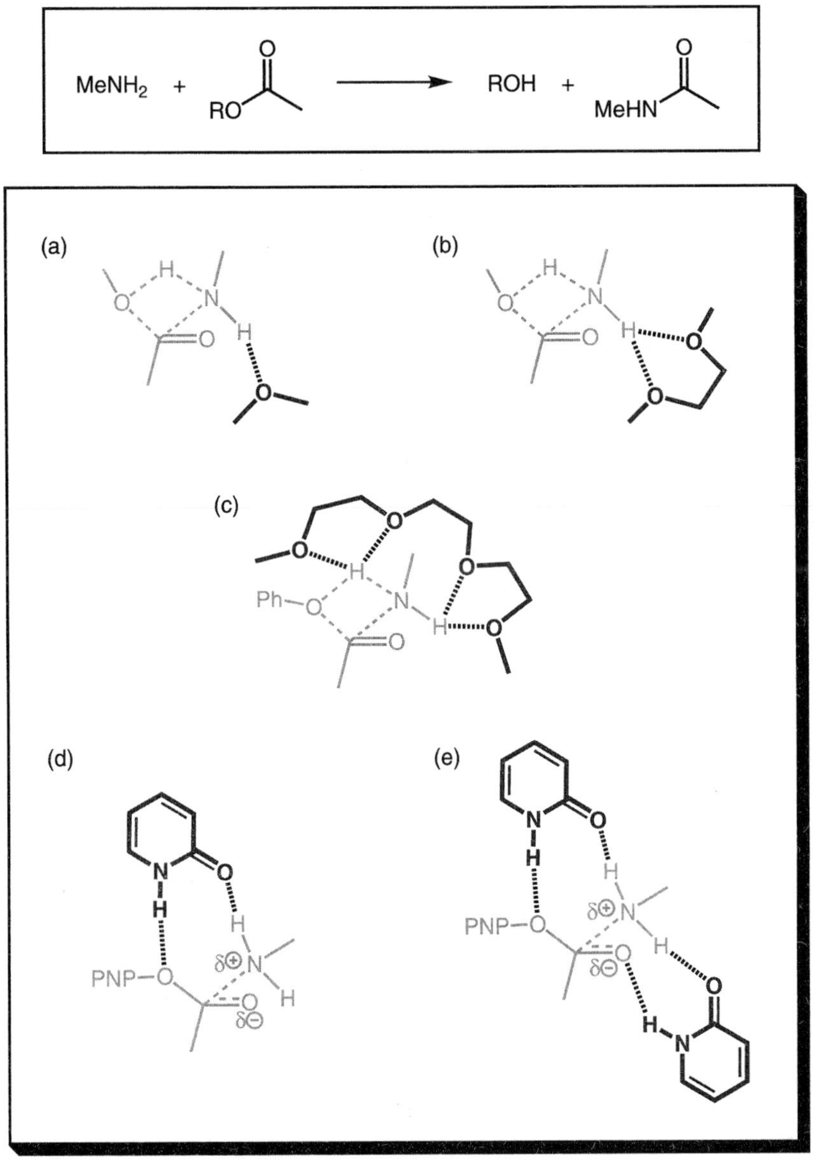

Figure 5. Theozymes for ester aminolysis. (**a**) Ether theozyme (bold) complexed to the direct displacement transition state for aminolysis. (**b**) DME theozyme (bold) complexed to the direct displacement transition state for aminolysis. (**c**) Triglyme theozyme (bold) complexed to the direct displacement transition state for aminolysis. (**d**) 2-Pyridone theozyme (bold) complexed to the transition state for addition of methyl amine to *para*-nitrophenyl acetate. (**e**) Theozyme consisting of two 2-pyridone molecules (bold) complexed to the transition state for addition of methyl amine to *para*-nitrophenyl acetate.

The study of larger systems by fully quantum mechanical methods was impractical, so a method combining quantum mechanical calculations of transition state geometries and force field calculations of transition state·catalyst interaction energies was developed.[18] The geometry of the direct displacement transition state in the case of methyl amine attack on phenyl acetate was used in these studies; the transition state was not allowed to relax in the presence of catalyst molecules. Various polyethers were placed around this transition state and their positions (including internal conformational flexibility) were optimized by force field calculations. The effect of complexation on activation energies was calculated from force field derived interaction energies between catalyst molecules and reactants or transition states. An example of the transition state complexes produced by this methodology is shown schematically in Figure 5c

for the polyether triglyme. Catalysis by triglyme is predicted to involve multiple hydrogen bonds to both amine protons. It is also predicted that triglyme should lower the activation barrier for aminolysis by almost 2 kcal/mol more than DME. This prediction is consistent with the relative rate constants for aminolysis of *para*-nitrophenyl acetate with *n*-butyl amine in chlorobenzene in the presence of triglyme or DME : 0.32 and 0.02 $M^{-2} s^{-1}$, respectively.

Based on the triglyme theozyme, additional catalysts were designed.[18] Of the potential catalysts examined, **6** was predicted to provide the best differential stabilization of the reactants and transition state. This molecule contains a polyether substructure as well as an additional hydrogen bond donor (the carbamate NH) which may further promote departure of the aryloxide leaving group. In addition, this catalyst is preorganized such that its polyether array is predisposed towards efficient transition state binding. Experimental exploration of this catalyst is planned.

6

Subsequent studies on catalysis of ester aminolysis by Zipse focused on bifunctional catalysts that could function as both general acids and bases.[22–23] Detailed examination of the reaction coordinate for direct displacement in the aminolysis of *para*-nitrophenyl acetate revealed that this process, although concerted, is asynchronous: C–N bond-making and C–O bond-breaking processes commence before the transition state is reached, while proton transfer occurs primarily after the transition state is reached, and through a relatively strained bent geometry.[24] In order to improve the geometry of proton transfer, a catalyst molecule was sought that would allow for a 6-membered rather than 4-membered cyclic transition state.

Ultimately, the effect of 2-pyridone complexation on the transition state was examined.[25] The lowest energy pathway for aminolysis of *para*-nitrophenyl acetate in the presence of 2-pyridone is no longer concerted, but instead involves formation of a zwitterionic tetrahedral intermediate, proceeding through a rate determining addition transition state stabilized by two hydrogen bonds as shown in Figure 5d. Theozyme calculations including two 2-pyridone molecules (Figure 5e) suggest that catalysis can be improved further by additional hydrogen bonding interactions. This sort of catalysis probably does not occur in solution due to the highly unfavorable entropy penalty associated with attaining the transition state and the tendency of 2-pyridones to form dimeric hydrogen bonded complexes. Consequently, catalysts are currently being designed by Zipse and coworkers that rigidly display two 2-pyridone groups in an arrangement that is preorganized for transition state binding rather than dimerization.

Similar catalysts, although not designed computationally, have been used successfully for related addition reactions. In an approach to asymmetric

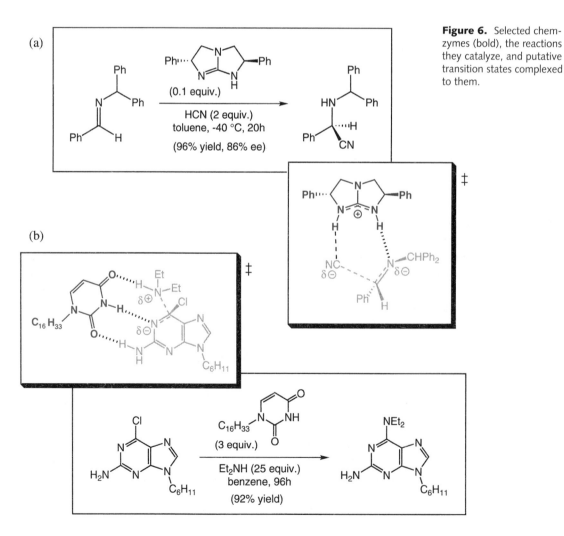

Figure 6. Selected chemzymes (bold), the reactions they catalyze, and putative transition states complexed to them.

Strecker synthesis of α-amino acids, Corey and Grogan used a guanidinium-based chemzyme to stabilize the putative transition state for cyanide attack on imines through hydrogen bonding and charge pairing interactions (Figure 6a).[26] This reaction proceeded in excellent yield with 10 mol% of catalyst. Konishi et al. utilized uracil derivatives to provide noncovalent stabilization of the transition state for the aminolysis of 6-chloropurine derivatives (Figure 6b).[27] Although this reaction uses stoichiometric quantities of the uracil derivatives, an order of magnitude rate acceleration is observed compared to the background reaction. The formation of a preorganized ternary complex is proposed which may lead to rate acceleration through transition state stablization by a rigid array of hydrogen bonds (Figure 6b). Theozyme calculations would be ideally suited to validate the proposed transition state assemblies for these two reactions and to quantitate the importance of the various noncovalent interactions involved in catalysis.[28]

Combinatorial Theoretical Catalyst Design: Compuzymes

Theozymes are theoretical enzymes whose creation involves a choice by the chemist of possible catalytic functionalities, followed by computer optimization of geometries providing stabilization. One might imagine alternative processes of theozyme generation that do not rely on chemist's choices, but instead utilize automated procedures that sample all possible arrangements of catalytic groups taken from a library of residues. The catalytic array derived from such a combinatorial approach – the best of a family of theozymes – has been called a "compuzyme".[3] A compuzyme could be found for a given reaction by screening a library of theozymes against a single transition state, or alternatively, the compuzyme could be "evolved" using a genetic algorithm approach. Given enormous computational resources, this process could use an exhaustive library of theozyme residues and should generate an optimal catalytic array for a particular reaction. Comparison of a compuzyme with the catalytic array optimized by nature through evolution (or microevolution in the case of catalytic antibodies) for the same reaction would lend insight into the limits of nature's catalyst development process. This blueprint could then be employed in the service of chemzyme creation, guiding the positioning of key functional groups to create highly preorganized catalysts.

Epilogue

Computational studies using theozymes have revealed the importance of selective stabilization of transition states over ground states by arrays of preorganized noncovalent interactions. The potential of these interactions in catalysis has been demonstrated in both biological and synthetic systems. Theozyme calculations are performed with relatively high levels of theory, and are therefore only suitable for relatively small systems, yet calculations on simple model systems have lead to predictions of catalyst structure and function that have been validated by experiment. The use of theozymes to understand and improve known catalysts and to design new synthetic catalysts is a promising area of research that has implications in the realms of organic synthesis, biochemistry, and supramolecular science.

References and Notes

1. See for example the pioneering work of Breslow: Breslow, R.; Dong, S. D. "Biomimetic Reactions Catalyzed by Cyclodextrins and their Derivatives" *Chem. Rev.* **1998**, *98*, 1997–2011 and Breslow, R. "Biomimetic Chemistry and Artificial Enzymes – Catalysis by Design" *Acc. Chem. Res.* **1995**, *28*, 146–153.

2. Much discussion has transpired on this issue. See, for example, the following series of mini-reviews: Cleland, W. W.; Frey, P. A.; Gerlt, J. A. "The Low Barrier Hydrogen Bond in Enzymatic Catalysis" *J. Biol. Chem.* **1998**, *273*, 25529–25532, Cannon, W. R.; Benkovic, S. J. "Solvation, Reorganization Energy, and Biological Catalysis" *J. Biol. Chem.* **1998**, *273*, 26257–26260, and Warshel, A. "Electrostatic Origin of the Catalytic Power of Enzymes and the Role of Preorganized Active Sites" *J. Biol. Chem.* **1998**, *273*, 27035–27038. Among the factors commonly invoked to explain enzymatic catalysis are entropy effects, medium effects, selective transition state binding interactions, and dynamic effects (for discussions on dynamic effects in enzyme catalysis, see: Cannon, W. R.; Singleton, S. F.; Benkovic, S. J. "A Perspective on Biological Catalysis" *Nat. Struc. Biol.* **1996**, *3*, 821–833 and Alhambra, C.; Gao, J. L.; Corchado, J. C.; Villa, J.; Truhlar, D. G. "Quantum Mechanical Dynamical Effects in an Enzyme-Catalyzed Proton Transfer Reaction" *J. Am. Chem. Soc.* **1999**, *121*, 2253–2258).

3. Tantillo, D. J.; Chen, J. G.; Houk, K. N. "Theozymes and Compuzymes: Theoretical Models for Biological Catalysis" *Curr. Op. Chem. Biol.* **1998**, *2*, 743–750.

4. Janda, K. D.; Shevlin, C. G.; Lerner, R. A. "Antibody Catalysis of a Disfavored Chemical Transformation" *Science* **1993**, *259*, 490–493.

5. Baldwin, J. E. "Rules for Ring Closure" *J. Chem. Soc., Chem. Commun.* **1976**, 734–736.

6. Na, J.; Houk, K. N.; Shevlin, C. G.; Janda, K. D.; Lerner, R. A. "The Energetic Advantage of 5-Exo Versus 6-Endo Epoxide Openings: A Preference Overwhelmed by Antibody Catalysis" *J. Am. Chem. Soc.* **1993**, *115*, 8453–8454.

7. Geometries were fully optimized at the HF/6–31G* level of theory, and single point energies were evaluated at the MP2/6–31G* level to include the effects of electron correlation. Transition states were characterized by harmonic frequency analysis.

8. A homogenous dielectric environment with the dielctric constant of water was modeled using SCRF calculations (see Wong, M. W.; Frisch, M. J.; Wiberg, K. B. "Solvent Effects. 1. The Mediation of Electrostatic Effects by Solvents" *J. Am. Chem. Soc.* **1991**, *113*, 4776–4782).

9. Another possibility is that the antibody alters the mechanism of the cyclization reaction. This is unlikely, however, in light of subsequent experimental evidence discussed later.

10. Na, J.; Houk, K. N. "Predicting Antibody Catalyst Selectivity from Optimum Binding of Catalytic Groups to a Hapten" *J. Am. Chem. Soc.* **1996**, *118*, 9204–9205.

11. Internal bond lengths and angles of the hapten model and theozyme residues were not allowed to relax in these calculations.

12. In these calculations, the geometries and relative position of the two residues in the theozyme were held fixed to their values as optimized previously in construction of the theozyme around the hapten model. The geometries of each transition state (previously optimized in the absence of the theozyme) were also held fixed in these calculations.

13. Coxon and Thorpe have reported additional calculations on the cyclization of protonated and BF$_3$-complexed epoxy alcohols (Coxon, J. M.; Thorpe, A. J. "Ab Initio Study of Intramolecular Ring Cyclization of Protonated and BF$_3$-Coordinated *trans*- and *cis*-4,5-Epoxyhexan-1-ol" *J. Org. Chem.* **1999**, *64*, 5530–5541 and Coxon, J. M.; Thorpe, A. J. "Theozymes for Intramolecular Ring Cyclization Reactions" *J. Am. Chem. Soc.* **1999**, *121*, 10955–10957), and Gruber et al. have shown,[15] using a simple theozyme consisting of an ammonia molecule (as a model of potential combining site lysine, histidine, or alcohol residues), that basic activation of the hydroxyl nucleophile lowers the activation barrier for cyclization but does not effectively differentiate between the exo and endo transition states.

14. Janda, K. D.; Shevlin, C. G.; Lerner, R. A. "Oxepane Synthesis Along a Disfavored Pathway: The Rerouting of a Chemical Reaction Using a Catalytic Antibody" *J. Am. Chem. Soc.* **1995**, *117*, 2659–2660.

15. Gruber, K.; Zhou, B.; Houk, K. N.; Lerner, R. A.; Shevlin, C. G.; Wilson, I. A. "Structural Basis for Antibody Catalysis of a Disfavored Ring Closure Reaction" *Biochemistry* **1999**, *38*, 7062–7074.

16. AutoDock V.2.4, a suite of programs that dock flexible guests into rigid host molecules, was used in these calculations (Morris, G. M.; Goodsell, D. S.; Huey, R.; Olson, A. J. *AutoDock V. 2.4*, The Scripps Research Institute, 10666 North Torrey Pines Road, La Jolla, CA 92037–5025, 1995). This version of AutoDock utilizes a Monte Carlo simulated annealing technique to sample potential binding modes. The interaction energy (consisting of van der Waals plus electrostatic interactions) of guest and host in each binding orientation is calculated internally by AutoDock utilizing grid-based affinity potentials to describe the host and atomic charges supplied by the user (and generally derived from ab initio or semiempirical calculations) for the guest.

17. Corey has discussed the similarities and differences between synthetic and biological catalysts, coined the term "chemzymes" for the former, and discussed the connections between these microscopic catalysts and macroscopic robots. See: Corey, E. J. "New Enantioselective Routes to Biologically Interesting Compounds" *Pure Appl. Chem.* **1990**, *62*, 1209–1216.

18. Zipse, H.; Wang, L.-H.; Houk, K. N. "Polyether Catalysis of Ester Aminolysis – A Computational and Experimental Study" *Liebigs Ann.* **1996**, 1511–1522.

19. All stationary point geometries were fully optimized at the HF/6–31G** level of theory and characterized by harmonic frequency analysis. Single point energies were evaluated at the MP2/6–31G** level to account for the effects of electron correlation. Since experiments were carried out in a relatively low dielectric environment (chlorobenzene solvent), it is likely that the shape of the potential energy surface in the gas phase and solution would be comparable.

20. This mechanism is also supported by kinetic isotope effects, Brønsted correlations, and other evidence as described in reference[18].

21. In these calculations the transition state geometry was allowed to relax in the presence of the catalyst. Starting the geometry optimization with the ether molecule near to the hydrogen that is being transferred in the transition state led to the same complex.

22. Wang, L.-H.; Zipse, H. "Bifunctional Catalysis of Ester Aminolysis – A Computational and Experimental Study" *Liebigs Ann.* **1996**, 1501–1509.

23. Müller, C.; Wang, L.-H.; Zipse, H. "Enzymes, Abzymes, Chemzymes–Theozymes?" In *Transition State Modeling for Catalysis*; Truhlar, D. G.; Morokuma, K., Eds.; ACS Symposium Series 721; American Chemical Society: Washington, DC, 1999; pp 61–73.

24. These calculations were performed at the HF/3–21G level of theory.

25. In this case, single point energies on HF/3–21G geometries were evaluated at the B3LYP/6–31G* level, a density functional theory method.

26. Corey, E. J.; Grogan, M. J. "Enantioselective Synthesis of α-Amino Nitriles from *N*-Benzhydryl Imines and HCN with a Chiral Bicyclic Guanidine as Catalyst" *Org. Lett.* **1999**, *1*, 157–160.

27. Tominaga, M.; Konishi, K.; Aida, T. "Catalysis of Nucleobase via Multiple Hydrogen-Bonding Interactions: Acceleration of Aminolysis of 6-Chloropurine Derivatives by Uracils" *J. Am. Chem. Soc.* **1999**, *121*, 7704–7705.

28. Both reactions are carried out in nonpolar solvents which should not only accentuate polar interactions, but should also make gas phase calculations relevant to these solution processes.

Enantioselective Catalysis using Sterically and Electronically Unsymmetrical Ligands

Alexander C. Humphries and Andreas Pfaltz   *Department of Chemistry, University of Basel, St. Johanns-Ring 19, CH-4056 Basel, Switzerland*   *Phone: +41 61 267 1108, Fax: +41 61 267 1103, e-mail: andreas.pfaltz@unibas.ch*

Concept: Classical C_2-symmetric, chelating P,P- and N,N-ligands can be sterically and electronically de-symmetrized by mixing the two types of coordinating unit, i.e. a phosphine group with a nitrogen-containing heterocycle, furnishing a chelating P,N-ligand. These hybrid ligands have been found to afford superior results in various metal-catalyzed processes. In addition, their modular architecture enables systematic 'fine tuning' of a generalized structure to suit a specific process.

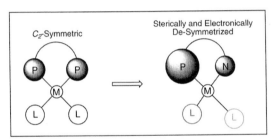

Abstract: Until recently, transition metal-mediated asymmetric catalysis has been dominated by the use of bidentate ligands possessing C_2-symmetry. However, the concept of 'electronic-differentiation', first introduced by Faller and further developed by Åkermark, together with the 'respective control concept' proposed by Achiwa, have contributed to the emergence of sterically and electronically unsymmetrical ligands for asymmetric catalysis. An important family of these is the bidentate P,N-ligands, of which the phosphino-oxazolines (PHOX) are particularly significant. These chelating ligands incorporate one P-coordinating and one N-coordinating unit and are constructed on a modular design, thus facilitating the rapid, individual tailoring of a specific ligand for a spe-cific process. Amongst other applications, the enantiose-lective variants of allylic alkylation, the Heck reaction and the hydrogenation of unfunctionalised olefins have benefited significantly from the use of these P,N-ligands. Given their relative immaturity, therefore, there can be little doubt that these ligands, as well as the concepts upon which they are designed, will be of importance to the further development of asymmetric catalysis.

Prologue

Transition metal-mediated asymmetric catalysis is one of the frontiers of development in modern synthetic methodology, with applications ranging from bench-top preparations to large scale industrial production. Arguably, one of the seminal events in the development of this fascinating field was the demonstration, in the mid-1960s, that RhCl(PPh₃)₃ **1** catalysed the hydrogenation of alkenes in homogeneous solution (Scheme 1).[1,2] It was soon established that two of the Rh-bound phosphine units were retained throughout the catalytic cycle and, in addition, the catalyst activity displayed a dependence on the substituent arrangement around the C=C bond, suggesting that substrate-ligand steric interactions were occurring.

Scheme 1. The mechanism of hydrogenation of simple alkenes using Wilkinson's catalyst RhCl(PPh₃)₃ (S = solvent, P = PPh₃).[2]

Keywords ■ Allylic Alkylation ■ Asymmetric Catalysis ■ C_2-Symmetry ■ De-symmetrization ■ Electronic Differentiation ■ Heck Reaction ■ Hydrogenations ■ Metal Complexes ■ Modular Design ■ P,N-Ligands ■ Steric Differentiation ■ Respective Control

Hence, the concept that the process might be rendered enantioselective by simply incorporating chiral ligands in place of the achiral ones was relatively straightforward, and early results with chiral phosphines showed promise.[3] However, the enantiomeric excesses obtained in this fashion were low, and it was clear that considerable improvements were required.

In 1971, Kagan published his ground-breaking research in the field.[4] He demonstrated that high enantioselectivities could be obtained in the Rh-catalyzed hydrogenation of functionalized olefins, such as the dehydroamino acid derivatives **3** and **4**, using a novel diphosphine ligand which he called DIOP **2** (Scheme 2).

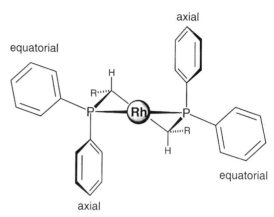

Figure 1. The transmission of 'backbone-chirality' to the reaction environment by way of the differing spatial orientation of the P-phenyl substituents in a C_2-symmetric ligand.

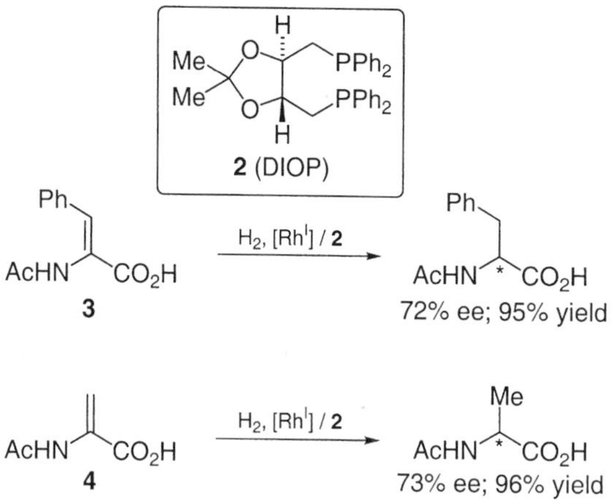

Scheme 2. Enantioselective reduction of dehydroamino acid derivatives **3** and **4** using Kagan's Rh-DIOP system.

The design of this ligand introduced three important concepts:

1. Bidenticity – Kagan realised that the enantiofacial differentiation by the phosphine could be enhanced by using a bidentate ligand, since this would display greater conformational rigidity. In addition, the chelate effect promotes the formation of a 1:1 Rh-diphosphine complex, whereas monodentate phosphines can produce mixtures of *cis*- and *trans*-diphosphines, together with mono-, tris- and tetra-phosphine complexes.

2. Backbone chirality – a chiral backbone unit was used instead of stereogenic phosphorus atoms. At first sight, it might appear that this places the chiral unit in a remote situation from the reacting centre, however, the effect is actually to force the ligand into a single conformation; the stereochemical information can then be transferred by way of differing interactions with the fixed equatorial and axial *P*-phenyl substituents (e.g. see Figure 1). This approach also demonstrated the viability of using naturally occurring, functionalized *C*-chiral units for ligand construction, thus simplifying the ligand synthesis.

3. C_2-Axis – by invoking this symmetry element in the ligand, the number of undesirable, competing diastereomeric transition states in the catalytic cycle is effectively halved, when compared with a C_1-symmetric bidentate ligand. Further advantages can also be imparted by using C_2-symmetry, for example: a potential simplification of the ligand synthesis by enabling methods such as dimerization of a monomeric chiral unit, or derivatization of a naturally occurring C_2-symmetric chiral building block; a reduction in the complexity of the NMR spectra of both the ligand and its complexes; and a simplification of the mechanistic analysis and the rationalization of an enantioselective outcome.[5]

Given these strong arguments, it is not surprising that following the introduction of DIOP **2**, a plethora of other C_2-symmetric and *pseudo* C_2-symmetric[6] diphosphine ligands soon emerged (Figure 2)[7] – some with chiral backbones, as in **5**, **7** and **9**, and others with chiral phosphorus centres, e.g. **6**

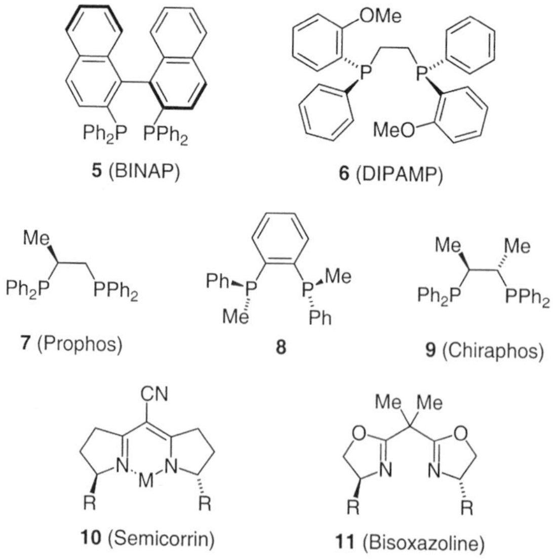

Figure 2. Examples of C_2-symmetric ligands used in asymmetric catalysis (R = alkyl or aryl).

and **8**. C$_2$-Symmetric ligands possessing alternative coordinating atoms were also introduced, for example the semicorrins **10** and the bisoxazolines **11**.[8] Concurrently, an array of other useful enantioselective, metal-catalysed reactions was developed: from oxidation processes, e.g. epoxidation and dihydroxylation, to C-C bond forming reactions such as allylic alkylation, cycloadditions and the Heck reaction.[9]

Despite the many successes obtained with C$_2$-symmetric ligands, there is no fundamental reason why C$_2$-symmetry should necessarily be better than C$_1$-symmetry; indeed, nature has developed highly enantioselective, catalytic processes which occur at enzyme sites that are devoid of symmetry. It may be that the organic chemist's reluctance to abandon C$_2$-symmetry has been based on a reluctance to discard the relative simplicity which it imparts, particularly when considering reaction intermediates and mechanistic pathways. However, a number of reasoned studies, together with the sometimes serendipitous discovery of successful C$_1$-symmetric ligands, suggested that, for certain applications, non-symmetric ligands might well prove superior.[10]

Breaking from C$_2$-Symmetry

In the early 1980s, Achiwa and co-workers conducted a study with the aim of designing improved P,P-ligands for the well-known Rh-catalyzed hydrogenation of functionalised olefins.[11] The detailed mechanism of this process had already been determined for dehydroamino acid-derived substrates (Scheme 3);[2, 12] initial coordination of the active Rh-species **12** to the olefin substrate is followed by turnover-limiting, oxidative addition of H$_2$, giving **13**; the coordinated C=C then immediately inserts into the periplanar M–H bond forming **14**, prior to reductive elimination of the desired product, regenerating the Rh-species **12**.

The turnover-limiting hydrogen addition step is irreversible since the subsequent insertion is so rapid. Therefore, hydrogen addition also controls the enantioselective outcome. Achiwa postulated that in this crucial step, the two phosphine groups must perform different functions as a result of their non-equivalent spatial relationships with the η^2-coordinated alkene and the incoming H$_2$ molecule. More specifically, it was argued intuitively that the *cis*-phosphine (P$_{cis}$) would be best positioned for steric interaction with the olefin, thus controlling the enantiomeric outcome, whilst the *trans*-phosphine (P$_{trans}$) is better situated to interact electronically with the olefin (cf. the *trans* effect), making it primarily responsible for activity. Accordingly, an ideal ligand would contain two different phosphine groups, with each one being optimized for its specific function. Achiwa soon demonstrated the potential of what he termed the 'respective control concept' in the Rh-catalyzed hydrogenation of dimethyl itaconic acid **17** and ketopantolactone **18** (Scheme 4). The chosen sterically and electronically

Scheme 3. The mechanism of Rh-catalyzed hydrogenation of functionalized olefins in the presence of a bidentate P,P-ligand (S = solvent, P = phosphine unit, R = alkyl).

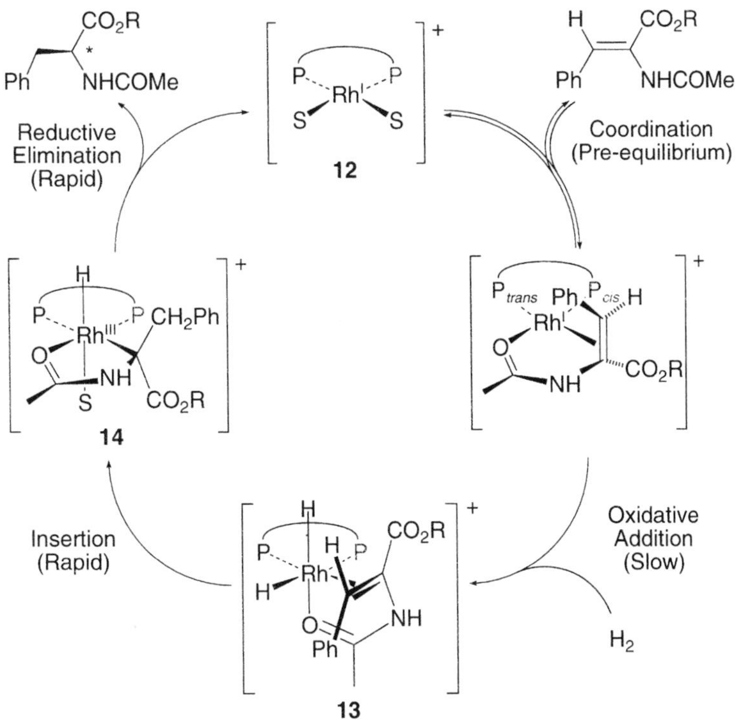

Scheme 4. Rh-catalysed hydrogenation of functionalised double-bonds using sterically and electronically unsymmetrical *P,P*-ligands.

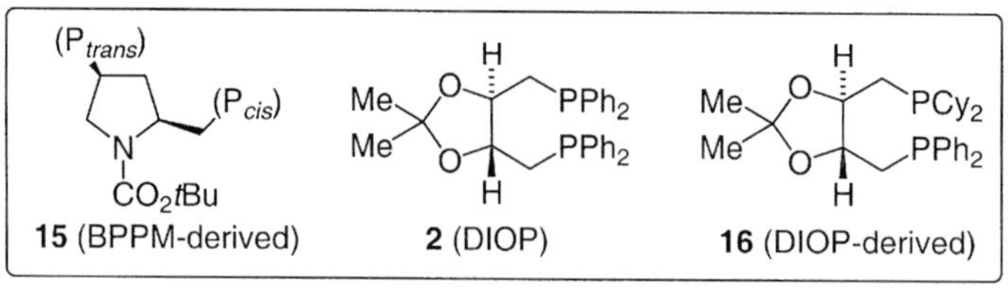

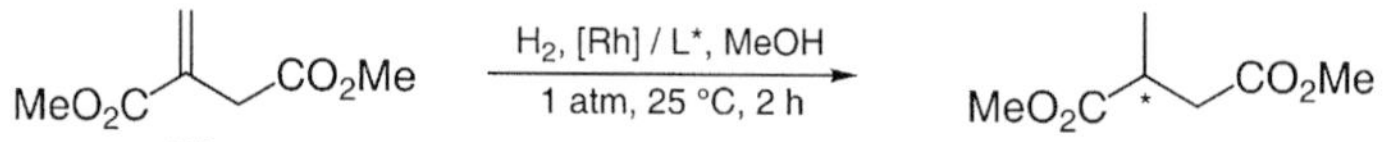

P_{trans} / P_{cis}

15 (BPPM-derived)   **2** (DIOP)   **16** (DIOP-derived)

$$\text{MeO}_2\text{C} \diagup \diagdown \text{CO}_2\text{Me} \xrightarrow[\text{1 atm, 25 °C, 2 h}]{\text{H}_2, \text{[Rh] / L*, MeOH}} \text{MeO}_2\text{C} \diagdown^* \diagup \text{CO}_2\text{Me}$$

17

L*	P_{cis}	P_{trans}	[Substrate]/[Rh]	Yield %	ee %
15	PPh_2	PPh_2	10^3	36	5 (S)
15	$P(p\text{-}Me_2NPh)_2$	PPh_2	10^3	55	68 (S)
15	PPh_2	$P(p\text{-}Me_2NPh)_2$	10^3	100	93 (S)

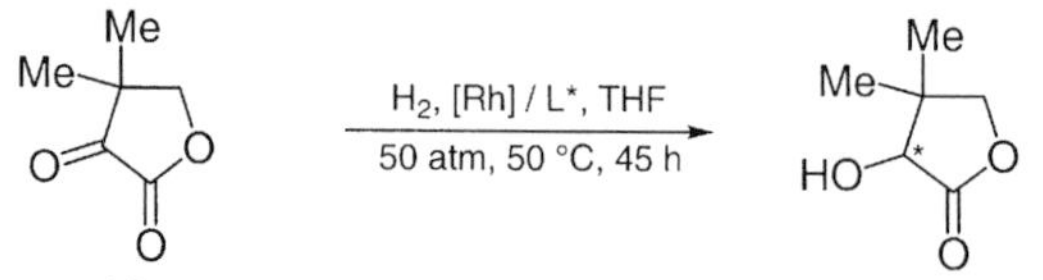

$$\xrightarrow[\text{50 atm, 50 °C, 45 h}]{\text{H}_2, \text{[Rh] / L*, THF}}$$

18

L*	P_{cis}	P_{trans}	[Substrate]/[Rh]	Yield %	ee %
15	PPh_2	PPh_2	10^3	44	72 (R)
15	PCy_2	PPh_2	10^3	75	9 (R)
15	PPh_2	PCy_2	10^3	100	91 (R)
15	PCy_2	PCy_2	10^4	100	61 (S)
2	-	-	10^3	45	37 (R)
16	-	-	10^3	100	72 (R)

unsymmetrical ligands were based on either the C_2-symmetric DIOP structure **2**, or the non-C_2-symmetric BPPM ligand **15**. [13]

The results illustrate the potential of bidentate ligands which differentiate the two *trans* disposed coordination sites, since independent variation of the electronic and steric influences of either phosphine unit clearly has a profound effect on both the activity and enantioselectivity of the hydrogenation process, as Achiwa had predicted.

Another striking example of this principle came from the former Ciba-Geigy central research services.[14] During investigation into the discovery of a feasible protocol for the enantioselective hydrogenation of MEA-imine **20**, variation of the steric and electronic influences of each phosphine unit in the

bidentate ligand led to the selection of the unsymmetric XYLIPHOS structure **19** (Scheme 5). After optimisation, exceptionally high turnover frequencies and numbers, as well as enantioselectivities, were achieved and, ultimately, the process was applied on an industrial scale.

It is relatively straightforward to recognize that a more effective approach to de-symmetrization would be to switch from a *P,P*-ligand to one which uses two different coordinating heteroatoms as in, for example, a *P,N*-ligand. Not surprisingly, a number of groups, including our own, pursued this line of thought in the early 1990s. However, despite the fact that the original studies using differentiating ligands had been based on catalytic hydrogenation, the mechanistically-reasoned evolution of *P,N*-

Scheme 5. Ciba-Geigy's
Ir-catalyzed hydrogenation
of MEA-imine **20**.

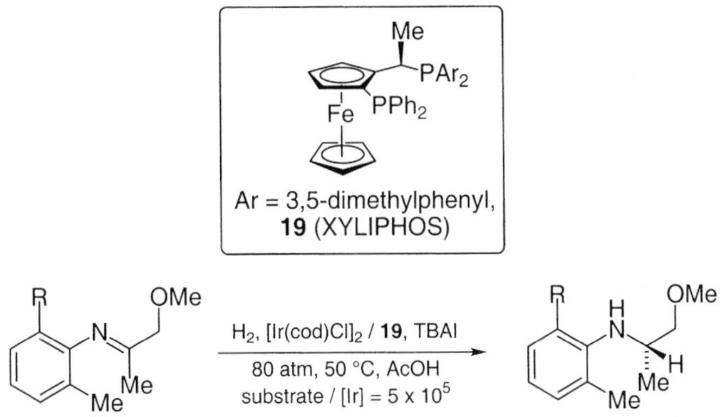

Ar = 3,5-dimethylphenyl,
19 (XYLIPHOS)

H_2, [Ir(cod)Cl]$_2$ / **19**, TBAI
80 atm, 50 °C, AcOH
substrate / [Ir] = 5 x 10^5

20

100% conversion;
79% ee

ligands ultimately took place in conjunction with a fundamentally distinct process, that of Pd-catalyzed allylic alkylation.

Pd-Catalyzed Allylic Alkylation and the Advent of *P,N*-Ligands

Pd-catalyzed allylic alkylation with 'soft' nucleophiles[15] is probably the most important category within the more general area of transition metal-catalyzed allylic substitution (Scheme 6).[16,17]

Originally, *P,P*-ligands which had been previously developed for hydrogenation (e.g. DIOP **2**, *vide supra*)

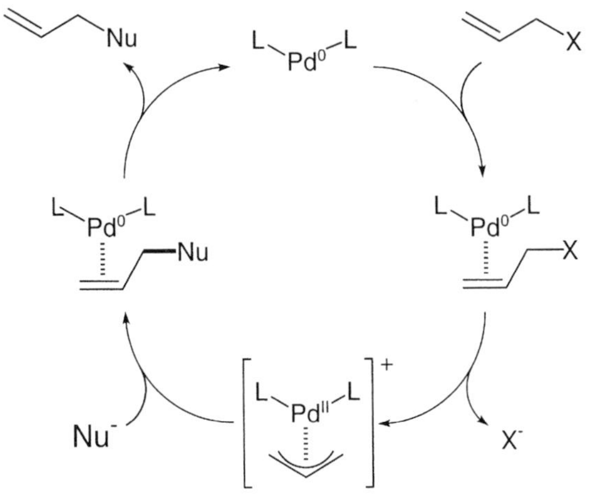

Scheme 6: The catalytic cycle for Pd-catalyzed allylic substitution.

were tested for induction of enantioselectivity in allylic substitution. However, in general, these ligands afforded only moderate success. More recently, certain other C_2-symmetric *P,P*-and *N,N*-ligands have been developed, for example **21** and **22** (Figure 3), and these have enabled the attainment of good to excellent enantioselectivities in a number of cases. However, with particular classes of substrate, arguments existed to suggest that unsymmetrical ligands might prove superior to C_2-symmetric ligands. These arguments are set out below.

The simplest case of an enantioselective allylic substitution involves the reaction of a racemic substrate containing identical substituents at the C1 and C3 positions (i.e. **23**, Scheme 7). The reaction commences with metal displacement of the leaving group (X$^-$) from **23**, producing, in the situation with a C_2-symmetric ligand, a single η^3-allyl intermediate **24**, in which the allyl moiety adopts a W-shaped *syn, syn*-conformation.[18] Formation of the product then follows from nucleophilic addition to either the C1 or C3 positions, furnishing **25** or *ent*-**25** after decomplexation. Consequently, the problem of governing the *enantioselectivity* of the overall reaction is equivalent to controlling the *regioselectivity* of nucleophilic attack at the allylic termini of **24**.

The origin of the regioselectivity of nucleophilic addition to the η^3-intermediate can be understood with reference to the interactions which take place in the intermediate itself. In the situation with a

Figure 3. C_2-Symmetric
P,P- and *N,N*-ligands developed specifically for allylic substitution.

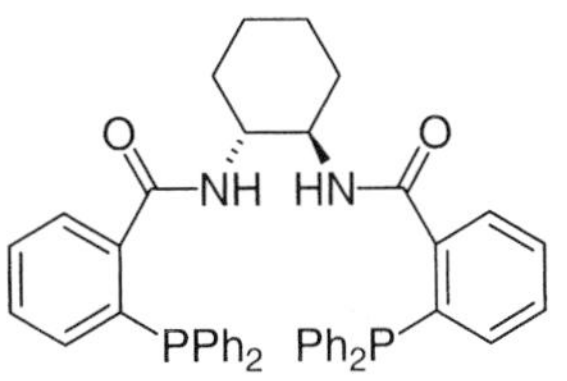

21 (DPPBA-derived)

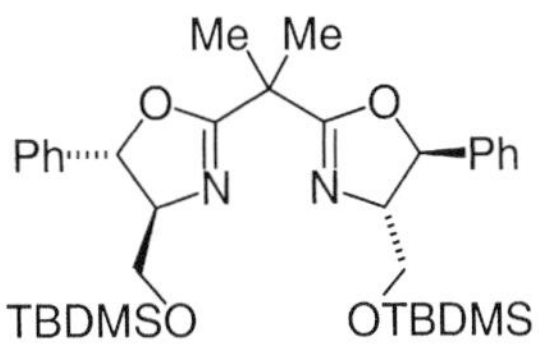

22 (Bisoxazoline)

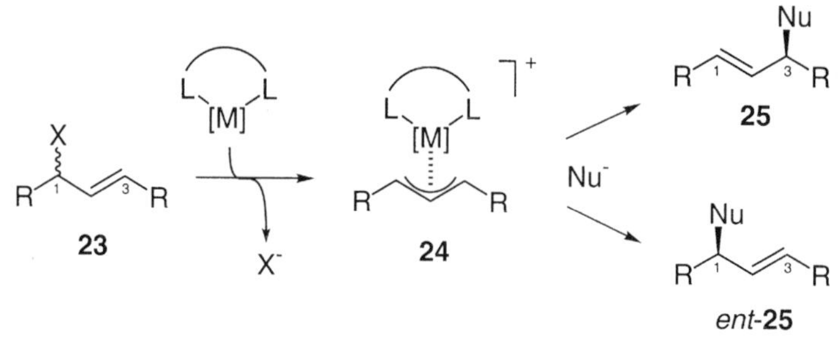

Scheme 7. The simplest type of enantioselective allylic alkylation which occurs on treatment of an allylic substrate with a metal derivative, together with a stabilized nucleophile (R = H, alkyl or aryl; X⁻ = leaving group; [M] = metal catalyst; Nu⁻ = nucleophile; L = coordinating atom).

complex derived from a C_2-symmetric ligand – for example **26** (Figure 4), incorporating a bisoxazoline – these interactions have been determined by both X-ray crystallographic and NMR structural analyses.

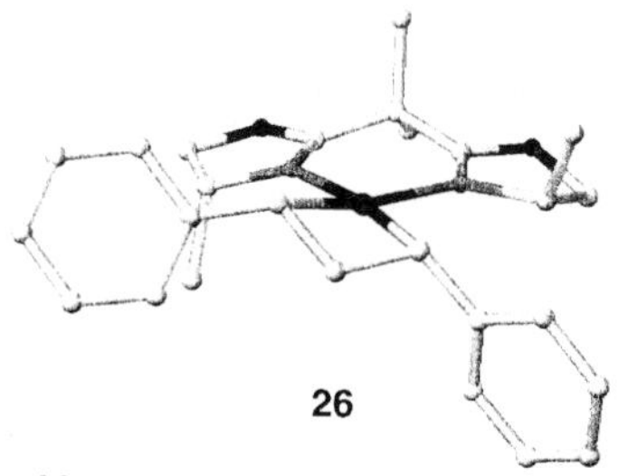

Figure 4. Structural data obtained for the cationic η^3-allyl complexes **26** (R = Me or CH$_2$Ph), illustrating the differing Pd-C bond lengths which result from unfavorable steric interactions.

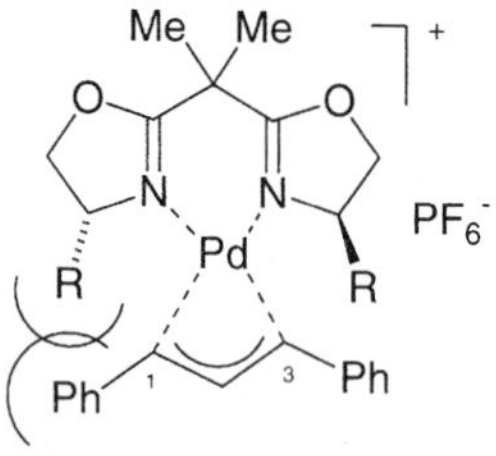

R	Pd–C(1)	Pd–C(3)
Me	2.20 Å	2.16 Å
CH$_2$Ph	2.17 Å	2.12 Å
	(±0.005 Å)	

It is evident that steric repulsion occurs between one of the allylic substituents and the *syn*-disposed oxazoline substituent, causing a significant lengthening of the corresponding Pd-C bond and a slight rotation (ca. 15°) of the allyl moiety. Comparison of the η^3-intermediate **26** with the absolute configuration of the product indicated that nucleophilic

attack occurs *trans* to the elongated Pd-C bond, a property which is most simplistically explained by invoking the Hammond postulate for an early transition state.[19] However, a better estimation of the transition state can be gained by considering the changing interactions which occur on proceeding along the reaction coordinate towards the transient η^2-intermediates **27** and **28** (Scheme 8). By this approach, it is apparent that addition of the nucleophile *trans* to the elongated N–Pd bond results in a favourable rotation of the allyl unit, forming **27**, in which the steric compression is released; whilst by contrast, addition to the opposite allylic terminus causes rotation in the opposite sense, resulting in a significant increase in steric interaction.

In this situation, it is evident that it is not crucial to determine whether the reaction proceeds via an early or a late transition state, since the outcome would be the same in both cases. This property is a result of the close similarity between the initial and final structures; indeed, the allyl moiety undergoes a rotation of only 30° from its idealised initial geometry to form the η^2-coordinated alkene complex.

It is important to highlight that the regioselectivity of nucleophilic addition in the above situation is entirely a result of unfavorable steric interactions between the bisoxazoline ligand and the η^3-allyl moiety. However, this is not the only means by which the regioselectivity may be influenced. As early as 1976, Faller and co-workers demonstrated

Scheme 8. Estimation of transition state interactions by consideration of the η^2-coordinated intermediates.

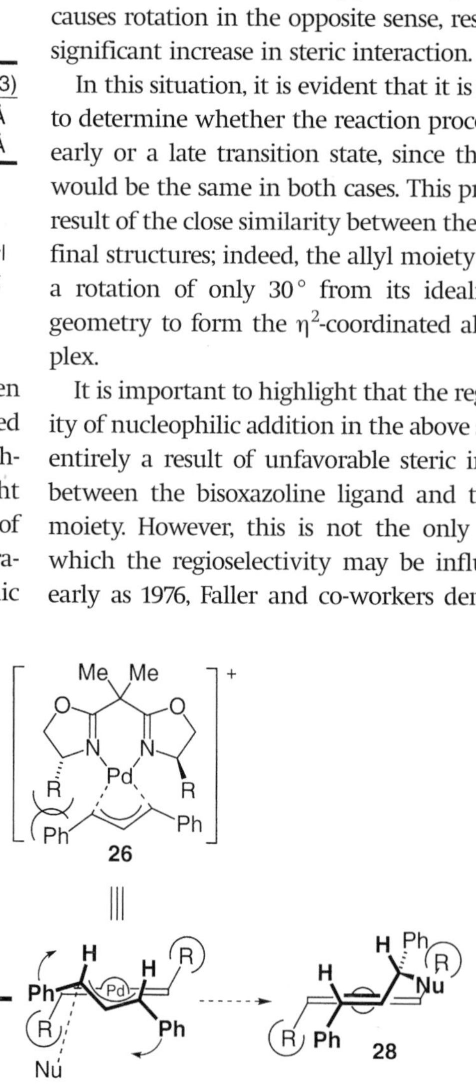

that it was also possible to discriminate the allylic termini by electronic means.[20] The system studied was the η^5-CpMo(NO)(CO)(η^3-allyl)$^+$ cation, which exists as a pair of rapidly interconverting isomers, termed *exo* and *endo* (**29** and **30** respectively, Scheme 9), with the former dominating at equilibrium.[21]

Scheme 9. Faller's demonstration of electronic differentiation in stoichiometric allyl alkylation.

Exo, **29**   *Endo,* **30**

Although the steric environment around each terminus of the allyl unit is virtually identical, Faller showed that addition of a 'soft' nucleophile occurs exclusively *trans* to the carbonyl ligand. It was clear that electronic rather than steric factors were responsible for this remarkable selectivity, and, even though the precise origin of the effect was not immediately understood, the principle of electronic differentiation had been experimentally demonstrated.[22]

Åkermark further developed the idea of electronic differentiation in the early 1980s, using the more familiar Pd-system.[23] He postulated that for a given allylic adduct, e.g. **31** (Figure 5), the relative positive charge residing on the allylic termini could be determined by comparing the ^{13}C NMR chemical shifts, the relevance of which he later confirmed experimentally.[24] In his initial studies, Åkermark used identical ligands (L^1=L^2), enabling him to categorise some of the more common ligand types into an order which could be related to their π-acceptor properties (Figure 5). He thus concluded that a greater positive charge is induced at the allylic termini by strong π-acceptor ligands, such as phosphite (P(OPh)$_3$), than by pure σ-donors, such as amines (NR$_3$).

Åkermark subsequently turned his attention to the situation with a non-symmetrical ligand

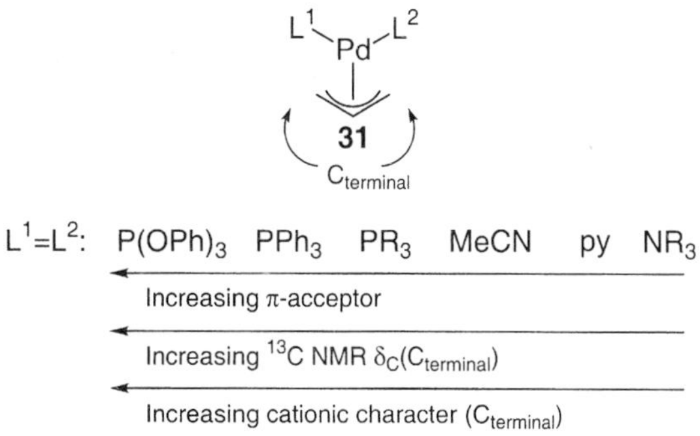

Figure 5. Åkermark's results with symmetrically ligated Pd-complexes.

L^1=L^2:	P(OPh)$_3$	PPh$_3$	PR$_3$	MeCN	py	NR$_3$

Increasing π-acceptor

Increasing ^{13}C NMR δ_C(C$_{terminal}$)

Increasing cationic character (C$_{terminal}$)

arrangement, i.e. $L^1 \neq L^2$. He confirmed that the allylic termini in a *P,N*-ligated compound such as **33** exhibit different ^{13}C NMR chemical shifts (Figure 6), as expected. However, by comparing the values for the *P,N*-ligated compound **33** with those for the *P,P*- and *N,N*-situations **32** and **34**, Åkermark demonstrated that the electronic environment – or, at least the ^{13}C NMR chemical shift – of each allylic terminus is affected essentially only by the *trans*-disposed ligand, in analogy to the well known *trans* kinetic effect for ligand substitution at a metal site.[23]

The application of these concepts to catalytic asymmetric synthesis led to the emergence of the *P,N*-ligands. In order to gain both steric and electronic differentiation, units from previously-developed, C_2-symmetric, *P*-chelate (-PPh$_2$, cf. DIOP, BINAP etc.) and *N*-chelate (*N*-heterocycle, cf. bisoxazoline, bipyridyl etc.) structures were mixed, producing the hybrid ligands. In the event, the generalized phosphino-oxazoline (PHOX) ligand **35** was developed simultaneously, and independently, by the groups of Helmchen[25] and Williams,[26] as well as in our own laboratories (Figure 7).[8] In addition, a further ligand, (quinylnaphthyl)phosphine **36** (QUINAP), was reported at about the same time by Brown and co-workers.[27]

The anticipated effect of a sterically and electronically differentiating *P,N*-ligand, e.g. PHOX **35**, on a

Figure 6. Comparison of the δ_C values for non-symmetrically ligated and symmetrically ligated complexes.

Ph$_3$P$\diagdown$Pd$\diagup$PPh$_3$ 78.6 ⟵ ⟶ 78.6   **32**

Ph$_3$P$\diagdown$Pd$\diagup$py   58.8 ⟵ ⟶ 80.0   **33**

py$\diagdown$Pd$\diagup$py   61.4 ⟵ ⟶ 61.4   **34**

Figure 7. The original *P,N*-ligands PHOX **35** and QUINAP **36**.

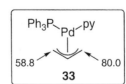

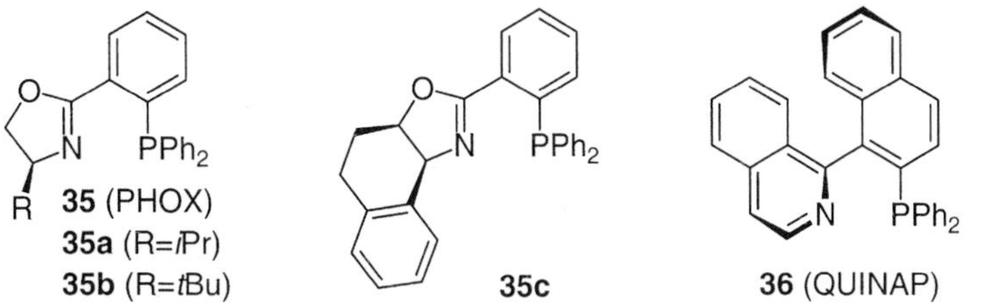

35 (PHOX)
35a (R=*i*Pr)
35b (R=*t*Bu)

35c

36 (QUINAP)

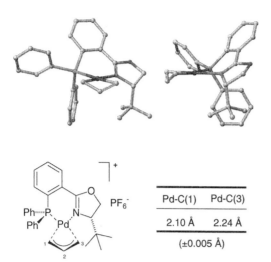

Figure 8. X-ray structure of a Pd-PHOX η³-allyl complex.

Pd-C(1)	Pd-C(3)
2.10 Å	2.24 Å
(±0.005 Å)	

η³-allyl Pd-complex is clearly illustrated by its X-ray crystal structure (Figure 8).[16, 28] This highlights two important features:

1. Steric differentiation – the chelate ring is strongly puckered, forcing one of the *P*-phenyl substituents into an equatorial orientation and the other into an axial one (cf. Figure 1, *vide supra*), with the oxazoline substituent bent 'backwards' away from the metal. Consequently, the *N*- and *P*-coordinating units are sterically unsymmetrical and it is in fact the equatorial *P*-phenyl group, and not the oxazoline substituent, which is spatially orientated to transmit the stereochemical information from the ligand to the η³-allyl moiety.

2. Electronic differentiation – the expected electronic influence of the unsymmetrical ligand is clearly displayed in the differing lengths of the two Pd-C bonds, with Pd-C(1) (*trans* to phosphorus) being considerably longer than Pd-C(3) (*trans* to nitrogen).

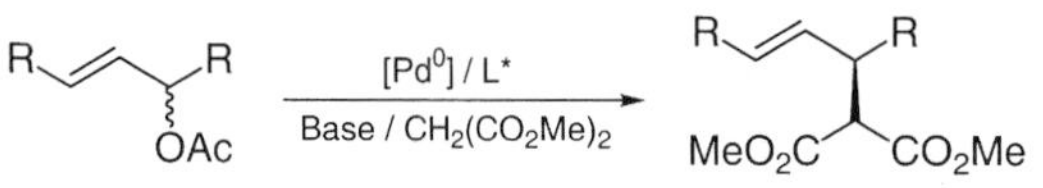

A selection of results from Pd-catalyzed allylic alkylation reactions of 'symmetrically substituted' substrates using PHOX ligands, together with some using C_2-symmetric ligands, are displayed below (Scheme 10).[16] These indicate that the PHOX ligands **35** are able to induce high enantioselectivities for both 1,3-diaryl and 1,3-dialkyl allylic derivatives, with only minor variations to the basic structure. In comparison, the bisoxazoline *N,N*-ligands, e.g. **22**, are less reactive and their use is generally restricted to the more reactive 1,3-diaryl substrates; whilst the DPPBA-derived ligand **21**, although particularly successful in conjunction with 'slim' substrates such as the 1,3-dimethyl allyl acetate, displays poor reactivity with allyl substrates bearing more bulky substituents.[29]

The origin of enantioselectivity with *P,N*-ligated complexes is more complex than the previously discussed situation using a C_2-symmetric bisoxazoline. With the *P,N*-ligand, initial displacement of the allylic leaving group affords a pair of rapidly equilibrating isomers, i.e. *endo* **37** and *exo* **38** (Scheme 11, cf. **26**, Scheme 8), with the former dominating at equilibrium. As illustrated earlier (Figure 8), the ligand's equatorial *P*-phenyl substituent alone interacts significantly with the η³-coordinated allyl moiety. Hence, the relative stability of the *endo* isomer **37** can be readily understood, since the *syn*-orientated terminal allyl substituent bisects the two *P*-phenyl substituents, whereas in the *exo* isomer **38**, the same allyl substituent eclipses the equatorial *P*-phenyl substituent.

It is important to recall that it is the relative transition state energies and not the relative ground state energies which determine the outcome of nucleophilic addition (Curtin–Hammett principle).[30] Therefore, the substrate-ligand interactions in the four possible η²-alkenyl products **39**–**42** must be considered as well as those in the two η³-allyl intermediates. Nucleophilic addition *trans* to nitrogen in the *endo* isomer **37**, or *trans* to phosphorus in the *exo* isomer **38**, results in η²-coordinated compounds **39** and **42**, respectively, both of which possess significant steric congestion about the equatorial *P*-phenyl substituent; this is in contrast to nucleophilic addition *trans* to nitrogen in the *exo* isomer **38**, or *trans* to phosphorus in the *endo* isomer **37**, furnishing **41** and **40**, respectively, in which the unfavorable steric interaction is absent. When these steric factors are taken in conjunction with the strong electronic preference for nucleophilic attack *trans* to the phosphine group, the latter-most of the four pathways is clearly preferred, forming **40**, in accordance with the experimentally observed

R	L*	yield %	ee %
Ph	**35a** (PHOX)	98	99
Ph	**22** (Bisoxazoline)	97	97
Ph	**5** (BINAP)	85	90
Ph	**21** (DPPBA-derived)	9	52
*i*Pr	**35b** (PHOX)	88	96
Me	**35b** (PHOX)	96	71
Me	**35c** (PHOX)	97	85
Me	**21** (DPPBA-derived)	98	92

Scheme 10: Selected results illustrating the success of PHOX ligands in allylic alkylation.

Scheme 11. The detailed mechanism of allylic alkylation in the presence of a PHOX ligand, illustrating the four possible pathways and η^2-coordinated products.

Scheme 11. The detailed mechanism of allylic alkylation in the presence of a PHOX ligand, illustrating the four possible pathways and η^2-coordinated products.

39

Addition *trans* to N

Endo, **37**

40

Addition *trans* to P

41

Exo, **38**

42

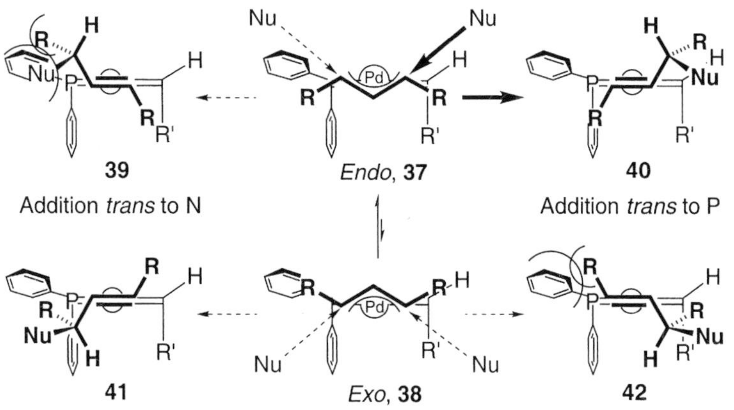

outcome. Recently, Helmchen has provided evidence for this pathway in the course of an elegant NMR study, during which the intermediacy of the η^2-alkenyl complex **40** was observed.[28]

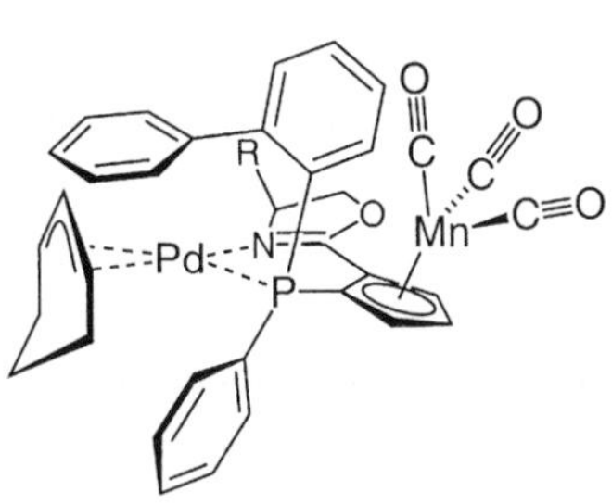

Figure 9. Helmchen's modified PHOX ligand for Pd-catalyzed allylic alkylation of cyclic substrates.

An essential requirement for the success of the above mechanism is the adoption of the *syn, syn*-(W) conformation by the η^3-allyl moiety. For acyclic substrates this is generally the case, however, with small- and medium-ring cyclic substrates, this conformation is precluded and poor results are furnished when using the conventional PHOX ligands (e.g. 0 % ee with cyclohexenyl acetate). Nevertheless, by rational argument, Helmchen has succeeded in designing a special PHOX ligand which affords high enantioselectivities and yields in reactions with cyclic substrates.[25] The ligand, shown above as its corresponding Pd-complex (Figure 9), possesses an additional phenyl substituent at the *ortho*-position of the axial *P*-phenyl group. η^5-Complexation of a Mn(CO)₃ unit onto the cyclopentadienyl backbone is then used to 'lock' the additional phenyl substituent

so that it completely blocks one of the faces of the Pd, thus providing enhanced control over the orientation of the allyl substrate.

In the examples presented so far, only two enantiomeric products have been possible in each case, since the substrates have all contained identical substituents on the C1 and C3 positions. However, a more complex situation occurs when the allyl system is unsymmetrically-substituted, as in **43a** or **43b** (Scheme 12).[16] Here, nucleophilic addition to the corresponding η^3-allyl intermediate **44** may afford an achiral, linear product **45**, in addition to the pair of enantiomeric, branched products **46**.

The regioselectivity of nucleophilic addition to the Pd-complex **44** is generally controlled by the steric preference to add to the least hindered allylic terminus. When $R^1 = H$, the expected product is therefore the linear, achiral compound **45**. However, in a continuation of his studies into the relative ^{13}C NMR shifts of Pd-bound η^3-allyl termini, Åkermark examined the effect of taking an unsymmetrical ligand arrangement together with an unsymmetrically-substituted allyl unit.[23, 24] As established previously, if one of the ligands is a powerful acceptor and the other a strong donor, the former will induce greater carbocationic character in the η^3-allyl terminus *trans* to it, compared with the latter. Åkermark discovered that when the allylic termini are also dissimilar, the greater positive charge prefers to reside on the more substituted terminus, so that the isomer **47** is electronically favoured over **48** (Scheme 13). In addition, the electronic influence

Scheme 12. Allylic alkylation of a non-symmetrically substituted allyl system (R^1 = H, aryl, alkyl; R^2 = alkyl, aryl; X = leaving group; [M] = metal catalyst; Nu = nucleophile).

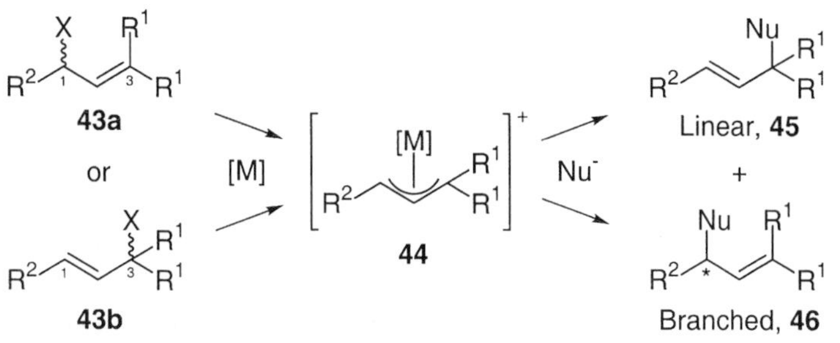

43a

or

43b

[M]

44

Linear, **45**

+

Branched, **46**

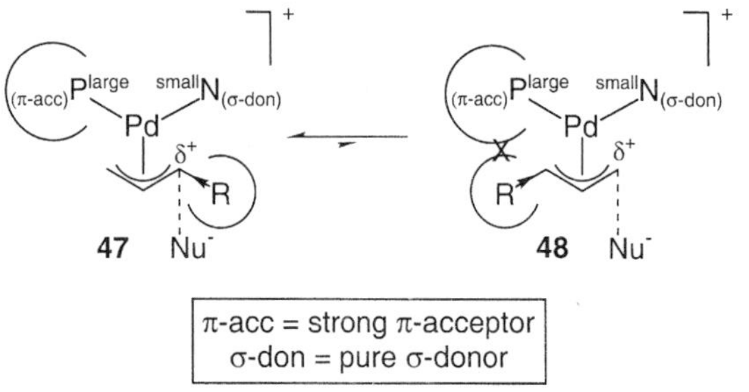

Scheme 13. Methods of favouring nucleophilic addition to the least-substituted terminus in allylic alkylations of mono-substituted substrates.

may be combined with a steric preference to form the intermediate **47**, by increasing the bulk around the more π-accepting heteroatom, thereby sterically forcing the more substituted allyl terminus to a *trans* position with respect to phosphorus. Nucleophilic addition *trans* to phosphorus would then afford the desired branched, chiral product.[31]

These ideas prompted the development of chiral *P,N*-ligands such as **49** and **50** (Scheme 14),[32] in which the phosphorus units display both greater electron-acceptor properties and increased steric bulk, compared to the earlier PHOX structure **35**. Pleasingly, in certain cases, these new ligands are able to completely outweigh the inherent steric preference for nucleophilic addition to the least-substituted terminus.[33]

Further Applications of Chiral *P,N*-Ligands

One of the beauties of transition metal-catalysed allylic alkylation is the existence of a relatively high degree of mechanistic understanding, enabling fea-

sible explanations for the observed outcomes. More importantly, this understanding has provided a limited ability to rationally design a specific ligand for a particular situation. When considered alongside the results obtained with unsymmetrically substituted *P,P*-ligands (*vide supra*), it is not surprising that these early successes prompted the evaluation of *P,N*-ligands in a variety of other metal-catalyzed processes, some of which are detailed below.[28, 34] In many applications, these ligands have made a valuable contribution and often display superior results to the more conventional, C_2-symmetric *P,P*- and *N,N*-varieties.

The enantioselective hydrogenation of alkenes is one of the most important applications in asymmetric catalysis. However, the vast majority of studies have used functionalized olefins as substrates (*vide supra*).[2] By contrast, the enantioselective hydrogenation of unfunctionalized olefins is not well developed and few successful catalyst systems exist.[35] Prior to the introduction of the PHOX ligand, the most promising system was that reported by Buchwald, using the titanium catalyst **51** (Figure 10) which is based on Brintzinger's titanocene complex. With this catalyst, enantioselectivities in excess of 90% have been obtained in the enantioselective hydrogenation of unfunctionalized olefins, e.g. **52** and **53**. However, the methodology is seriously limited by the low productivity of the catalyst, with typical conditions involving a high catalyst loading (substrate/catalyst = 10–20), high hydrogen pressure (up to 140 atm) and long reaction times (1–8 days).[36]

In the late 1970s, Crabtree and Morris introduced the Ir-complex [Ir(cod)PCy$_3$(py)]PF$_6$ which they dem-

Scheme 14. Allylic substitution of mono-substituted substrates using *P,N*-ligands.

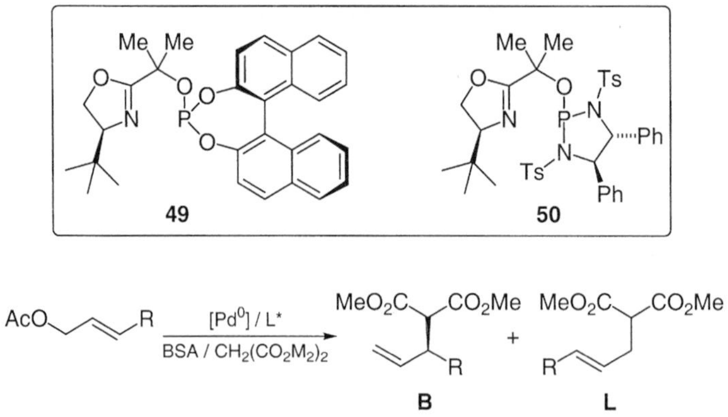

R	L*	yield %	ee %	B:L
Ph	49	86	90	76:24
Ph	50	92	94	84:16
1-Naphthyl	50	91	98	98:2

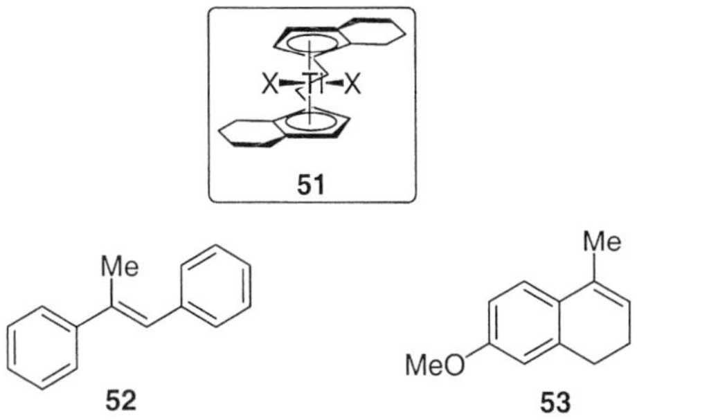

Figure 10. Buchwald's catalyst system for the enantio-selective hydrogenation of unfunctionalized olefins such as **52** and **53**.

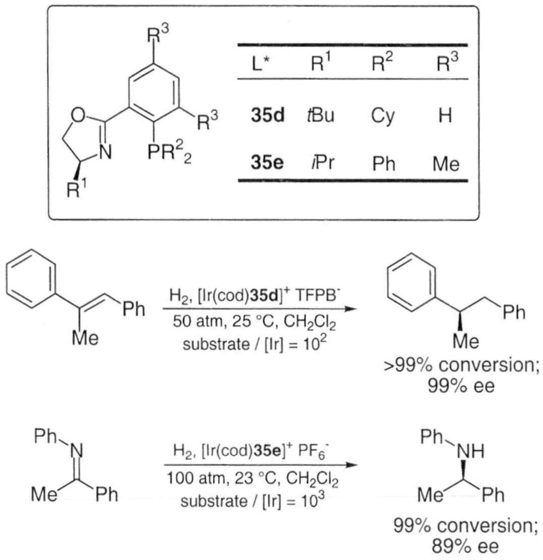

Scheme 15. Ir-Catalysed hydrogenation of unfunctional-ized olefins and imines using PHOX-ligands.

onstrated to have exceptionally high activity in the homogeneous hydrogenation of tri- and tetra-substituted, unfunctionalized olefins.[37–38] The mixed heteroatom ligand arrangement in Crab-tree's catalyst prompted speculation that an Ir-PHOX complex might act as a chiral analogue. Pleasingly, after optimization of the PHOX ligand structure, along with other external conditions,[$39] promising results have been obtained in the hydro-genation of trisubstituted unfunctionalized ole-fins,[40] as well as imines (Scheme 15).[41] For unfunc-tionalized olefins, such high enantioselectivities and catalyst activities have not been obtained with any other catalyst system.

P,N-Ligands have also enhanced the scope of the Pd-catalyzed enantioselective Heck reaction.[42] Tra-ditionally, the process has been carried out using the C_2-symmetric BINAP ligand **5** (Figure 2), which often affords excellent results.[43] However, in many cases (*e.g.* the reaction of **54** with **55**, Scheme 16), the Pd-BINAP system catalyzes a C=C bond migration after the initial coupling reaction, i.e. converting **56** to **57**, so that the major compound isolated is not the actual Heck reaction product. If this rearrange-ment leads to the desired product, then BINAP is

the ligand of choice. However, in situations where the rearrangement is undesirable, or where there is no thermodynamic incentive to drive it to a single product, as in the reaction of **58** with **55**, the Pd-BINAP system fails, in the latter case affording a complex mixture of isomeric products. This prob-lem has been overcome by using the chiral PHOX ligand **35b** (Figure 7), which gives excellent enantio-selectivities and high conversions in the initial Heck coupling, without any significant C=C bond migration (Scheme 16).

In total, over the past six years, the chelating P,N-ligands have shown considerable promise in a vari-ety of enantioselective processes, including transfer-hydrogenation and hydrosilylation of ketones, hydroboration of alkenes, conjugate addition to enones and Lewis-acid catalysed Diels-Alder reac-tions, in addition to those described above.[28, 34] It is anticipated that this list will continue to grow, and

Scheme 16. Results from the enantioselective Heck reaction using BINAP and PHOX.

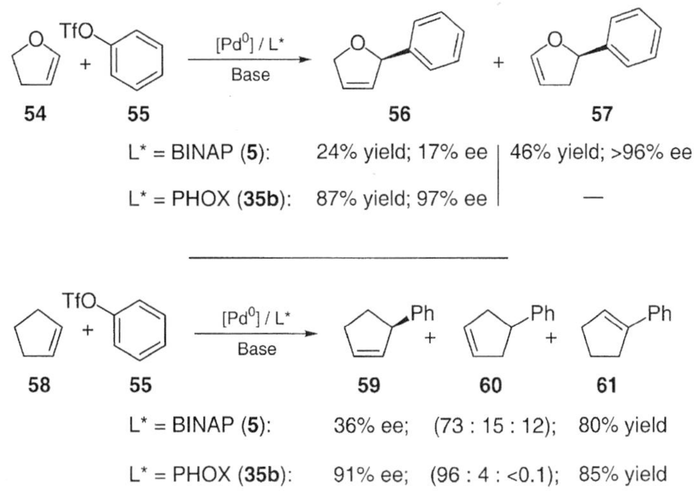

L* = BINAP (**5**):   24% yield; 17% ee | 46% yield; >96% ee

L* = PHOX (**35b**):   87% yield; 97% ee | —

L* = BINAP (**5**):   36% ee;   (73 : 15 : 12);   80% yield

L* = PHOX (**35b**):   91% ee;   (96 : 4 : <0.1);   85% yield

it is hoped that with it will come a greater understanding of the mode by which the unsymmetrical *P,N*-ligands are operating.

The Modular Nature of Chiral *P,N*-Ligands

The processes discussed in the preceding sections illustrate how fine adjustments in either the electronic or steric environment imposed by the ligand may have a profound effect on the reaction product distribution; it follows that a specific ligand structure may only be useful for a single reaction category, and possibly with only a single substrate. It is paramount, therefore, that ligands are not designed individually, but rather as classes, using a systematic approach which can enable facile and independent variation of the various structural units, so as to 'fine tune' them for each particular application.

A bidentate *P,N*-ligand can be visualized as a conglomerate of three distinct modules: one phosphorus-containing ('P') and one nitrogen-containing ('N'), separated by a 'spacer' backbone ('B') which usually achieves a 1,5-disposition of the coordinating atoms (Figure 11).[44] The chiral variants may then incorporate a single chiral unit in any one of the modules, or multiple chiral units in any combination.

To date, the most popular and successful designs for *P,N*-ligands have incorporated an achiral phosphine with a substituted non-aromatic heterocycle as the primary chiral unit, largely due to the ease by which the 'P' and 'N' modules may be altered, and the ready availability of enantiopure, amino acid-derived starting materials.[28] Considerable variation of the precise steric and electronic environment imposed on a chelated metal centre has been achieved by the synthesis of a substantial library of

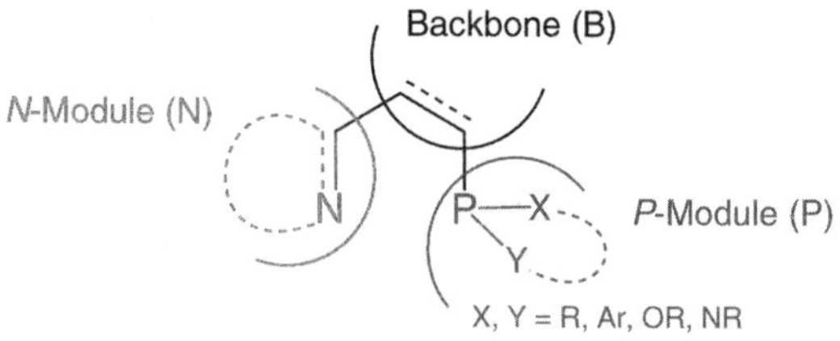

Figure 11. The generalised modular design of chiral *P,N*-ligands.

ligands. Although related, these *P,N*-ligands make use of a number of different combinations of chiral and achiral units (Figure 12): the archetypal PHOX ligand design **35** (chiral 'N' module) has been widely exploited in many processes and remains the 'work horse' standard; phosphite and related oxazolines, e.g. **49** and **50**, (chiral 'N' and 'P' modules) have found application in allylic substitution of non-symmetrical allyl substrates and in conjugate addi-

Figure 12. Some examples of *P,N*-ligands used in asymmetric catalysis.

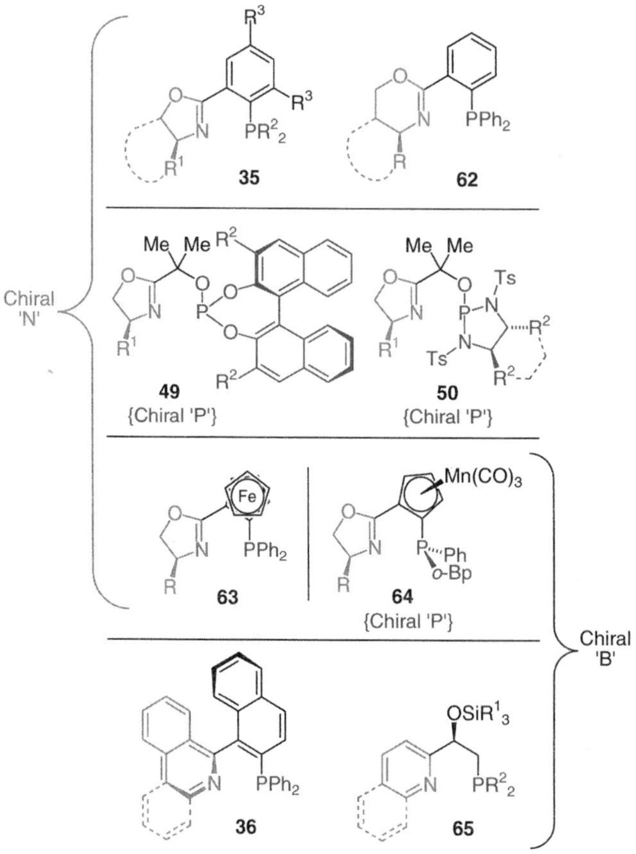

tion reactions to enones; backbone-ferrocenyl phosphino-oxazolines such as **63** (chiral 'N' and 'B' modules) are versatile ligands for ketone hydrogenation; and finally, the highly functionalized η^5-cyclopentadienyl-Mn(CO)$_3$ ligand **64** (chiral 'P', 'N' and 'B' modules) has been developed specifically for allylic substitution of cyclic substrates. *P,N*-Ligands which incorporate alternative heterocyles – most commonly pyridine-type ones – generally possess a chiral backbone.[34] Despite the early advent of QUINAP **36**, this type of ligand architecture has been much less widely exploited. However, recent studies conducted in our laboratories have resulted in the development of the related phosphino-pyridine and phosphino-quinoline ligands **65**.[42] Interestingly, these possess similar properties to the original PHOX ligands, whilst offering complimentary scope for 'fine tuning'.

Epilogue

Until recently, transition metal-mediated asymmetric catalysis has been dominated by the use of bidentate ligands possessing C_2-symmetry.[5] However, the concept of 'electronic-differentiation', which was first introduced by Faller[22] and further developed by Åkermark,[23] together with the 'respective control concept' proposed by Achiwa,[11] have contributed to the emergence of electronically and sterically unsymmetrical ligands. When complexed to a metal, these ligands differentiate the opposite two equatorial coordination sites of the metal, which, in certain situations, enhances both the catalyst activity and the enantioselectivity. A particularly versatile type of differentiating ligands are the *P,N*-ligands, most notable of which are the phosphino-oxazolines (PHOX).[28] The modular construction of these ligands has facilitated the rapid, individual tailoring of a specific ligand for a specific function. Indeed, even though *P,N*-ligands in general are still in their infancy, they already rival the best C_2-symmetric varieties, and are in some situations the ligands of choice. Related differentiating ligands have also emerged, such as *P,S*-, and *N,S*-varieties, which have been used in a number of applications.[17, 34, 45] In the broader spectrum, the concept of combined electronic and steric differentiation is manifested by any '*X,Y*'-ligand and consequently the possibilities are immense. There can be little doubt, therefore, that *P,N*-ligands, together with other related *X,Y*-ligands, will play an important rôle in the future development of metal-mediated asymmetric catalysis.

References and Notes

1. J. A. Osborne, F. J. Jardine, J. F. Young, G. Wilkinson, The Preparation and Properties of Tris(triphenylphosphine)halogenorhodium(I) and some Reactions thereof including Catalytic Homogeneous Hydrogenation of Olefins and Acetylenes and their Derivatives, *J. Chem. Soc. A.* **1966**, 1711–1732.
2. E. N. Jacobson, A. Pfaltz, H. Yamamoto, Hydrogenation of Functionalised Carbon-Carbon Double Bonds, J. M. Brown in Comprehensive Asymmetric Catalysis, Vol. I (Eds.), Springer, Berlin, **1999**, pp. 121–182.
3. W. S. Knowles, M. J. Sabacky, Catalytic Asymmetric Hydrogenation employing a Soluble, Optically Active, Rhodium Complex, *J. Chem. Soc., Chem. Commun.* **1968**, 1445–1446.
4. H. B. Kagan, T.-P. Dang, Asymmetric Catalytic Reduction with Transition Metal Complexes. I. A Catalytic System of Rhodium(I) with (–)-2,3-*O*-Isopropylidene-2,3-dihydroxy-1,4-bis(diphenylphosphino)butane, a New Chiral Diphosphine, *J. Am. Chem. Soc.* **1972**, 94, 6429–6433.
5. J. K. Whitesell, C_2-Symmetry and Asymmetric Induction, *Chem. Rev.* **1989**, 89, 1581–1590.
6. Although many ligands possess a C_2-symmetric backbone, this is actually remote from the coordination sphere of the metal. Therefore, it is possible to have a non-C_2-symmetric backbone, as in **7**, and still have a C_2 symmetric arrangement of the *P*-phenyl groups about the metal. This situation is termed *pseudo* C_2 symmetric.
7. H. B. Kagan, Chiral Ligands for Asymmetric Catalysis, in *Asymmetric Synthesis, Vol. 5* (Ed. J. D. Morrison), Academic Press, London, **1985**, pp. 1–40.
8. A. Pfaltz, From Corrin Chemistry to Asymmetric Catalysis – A Personal Account, *Synlett*, **1999**, 835–842.
9. I Ojima (Ed.), Catalytic Asymmetric Synthesis VCH, New York, **1993**, pp. 1–476.
10. In 1979, Rauchfuss and Jeffrey postulated that ligands possessing one 'hard' and one 'soft' coordinting atom would act as 'hemilabile' ligands. Thus, essentially monodentate coordination via the soft atom (usually P) could be enhanced by additional coordination of the hard atom (N or O) to a vacant metal site at appropriate opportunities in the catalytic cycle, thus enhancing the reactivity of the catalysts.[9] However, in certain cases, ligands designed to be hemiliable actually turned out to be chelating, with both heteroatoms bound to the metal throughout the catlytic cycle. J. C. Jeffrey, T. B. Rauchfuss, Metal Complexes of Hemilabile Ligands, *Inorg. Chem.* **1979**, 2658–2666.
11. K. Inoguchi, S. Sakuraba, K. Achiwa, Design Concept for Developing Highly Efficient Chiral Bisphosphine Ligands in Rhodium-Catalysed Asymmetric Hydrogenations, *Synlett*, **1992**, 169–178.
12. J. Halpern, Asymmetric Catalytic Hydrogenation: Mechanism and Origin of Enantioselection, in *Asymmetric Synthesis, Vol. 5* (Ed. J. D. Morrison), Academic Press, London, **1985**, pp. 41–69.
13. An advantage of using the BPPM-derived ligands was that, as a result of the lack of symmetry, Achiwa was able to determine which phosphine group was orientated *cis* and which one was *trans* to the η^2-coordinated alkene (cf. Scheme 3).

14. F. Spindler, B. Pugin, H.-P. Jalett, H.-P. Buser, U. Pittelkow, H.-U Blaser, A Technically Useful Catalyst for the Homogeneous Enantioselective Hydrogenation of N-Aryl Imines: A Case Study, in *Catalysis of Organic Reactions* (Ed. R. E. Maltz), Dekker, New York, **1996**, pp. 153–168.

15. In general, Pd-catalyzed allylic substitutions with 'soft' nucleophiles involve nucleophilic attack directly on the allyl unit, on the opposite face to that occupied by the metal. This is contrasted with the situation for 'hard' nucleophiles where the initial attack occurs at the metal, with subsequent migration of the nucleophile to the allyl moiety – the addition to the allyl unit therefore occurring from the same face as the metal. Obviously, this has profound implications on the stereochemical outcome.

16. A. Pfaltz and M. Lautens, Allylic Substitution Reactions, in *Comprehensive Asymmetric Catalysis, Vol. II* (Eds. E. N. Jacobson, A. Pfaltz, H. Yamamoto), Springer, Berlin, **1999**, pp. 883–886.

17. B. M. Trost, D. L. Van Vranken, Asymmetric Transition Metal-Catalysed Allylic Alkylations, *Chem. Rev.* **1996**, *96*, 395–422.

18. The *syn*-descriptor refers to the orientation of the substituent with respect to the single proton on the central carbon atom (C2) of the η^3-allyl unit. Although interconversion to the *anti,syn* or even the *anti,anti* isomers may occur via the 'π-σ-π' mechanism,[16] these higher-energy species generally do not play a rôle in the catalytic cycle, so they can be ignored in the mechanistic discussion.

19. The Hammond postulate can be used to estimate the interactions in a limiting early transition state by simply examining the interactions in the initial η^3-complex.

20. R. D. Adams, D. F. Chodosh, J. W. Faller, A. M. Rosan, Stereospecificity in Reactions of Activated η^3-Allyl Complexes of Molybdenum, *J. Am. Chem. Soc.* **1979**, *101*, 2570–2578.

21. The mechanism of this interconversion is known as 'apparent allyl rotation' and results in a switching of the allylic termini in addition to the movement of the central allyl carbon from one side of the coordination plane to the other; although, for situations with symmetrically substituted allyl units, only this latter change is noticeable. As the name suggests, the process is non-trivial and, in reality, the mechanism is dependent on the specific conditions.[16] Nevertheless, in most cases, the interchange is rapid compared with the rate of nucleophilic attack and, therefore, the product outcome does not necessarily depend on the intermediate isomer ratio (Curtin–Hammett principle).

22. J. W. Faller, K- H- Chao, H. H. Murray, Controlling the Regioselectivity of Nucleophilic Addition on Unsaturated Ligands Bound to Molybdenum. The Cationic [CpMo(NO)(CO)(cyclooctenyl)]⁺ System, *Organometallics*, **1984**, *3*, 1231–1240.

23. B. Åkermark, B. Krakenberger, S. Hansson, Ligand Effects and Nucleophilic Addition to (η^3-Allyl)palladium Complexes. A Carbon-13 Nuclear Magnetic Resonance Study, *Organometallics*, **1987**, *6*, 620–628.

24. B. Åkermark, K. Zetterberg, S. Hansson, B. Krakenberger, The Mechanism of Nucleophilic Addition to η^3-Allylpalladium Complexes: the Influence of Ligands on Rates and Regiochemistry, *J. Organometallic Chem.* **1987**, *335*, 133–142.

25. G. Helmchen, S. Kudis, H. Steinhagen, Enantioselective Catalysis with Complexes of Asymmetric P,N-Ligands, *Pure & Appl. Chem.* **1997**, *69*, 513–518.

26. J. M. J. Williams, The Ups and Downs of Allylpalladium Complexes in Catalysis, *Synlett*, **1996**, 705–710.

27. J. M. Brown, D. I. Hulmes, P. J. Guiry, Mechanistic and Synthetic Studies in Catalytic Allylic Alkylation with Palladium Complexes of 1-(2-Diphenylphosphino-1-naphthyl)isoquinoline, *Tetrahedron*, **1994**, *50*, 4493–4506.

28. G. Helmchen, A. Pfaltz, Phosphinooxazolines – A New Class of Versatile, Modular P,N-Ligands for Asymmetric Catalysis, *Acc. Chem. Res.* **2000**, *33*, 336–345.

29. It should be noted that Trost's DPPBA-derived ligand **21** possesses considerable utility in many other situations of allylic substitution, for example with substrates which are cyclic or which possess enantiotopic leaving groups, or with prochiral nucleophiles.[17]

30. This is underlined by the observation that *endo/exo* ratios do not correlate with the observed enantiomeric excess.

31. It should be noted that in cases when R¹ and R² ≠ H (Scheme 12), the steric propensity to add to the least substituted terminus is again the overriding factor, giving rise to formation of a chiral product. This, more trivial situation has been successfully accomplished using PHOX ligand **35**, as well as conventional C_2-symmetric ligands.[16, 26]

32. R. Hilgraf, A. Pfaltz, Chiral Bis(N-tosylamino)phosphine- and TADDOL-Phosphite-Oxazolines as Ligands in Asymmetric Catalysis, *Synlett*, **1999**, *11*, 1814–1816.

33. An alternative to ligand-based regiocontrol is to use a different, more electropositive metal, e.g. W, Ir or Mo. The success of this approach is assumed to arise from the induction of a greater positive charge on the allyl moiety, promoting nucleophilic attack at the more substituted terminus as before. Finally, good results have also been obtained by Hayashi, using a Pd-catalyst with the monophosphine 'MOP' ligand.[16, 17]

34. P. Espinet, K. Soulantica, Phosphine-Pyridyl and Related Ligands in Synthesis and Catalysis, *Coord. Chem. Rev.* **1999**, *195*, 499–556.

35. R. L. Halterman, Hydrogenation of Non-Functionalised Carbon–Carbon Double Bonds, in *Comprehensive Asymmetric Catalysis, Vol. I* (Eds. E. N. Jacobson, A. Pfaltz, H. Yamamoto), Springer, Berlin, **1999**, pp. 183–189.

36. Recently, Buchwald has developed an analogous, cationic zirconium system which is, to date, the most effective catalyst for the hydrogenation of tetrasubstituted, unfunctionalized olefins.[46]

37. By comparison, the diphosphine analogue [Ir(cod)(P-MePh₂)₂]PF₆ was less active by a factor of two for tri-substituted alkenes and a factor of eight for tetrasubstituted ones; whilst the analogous bispyridine complex showed no activity at all, apparently due to a failure to add H₂.

38. R. Crabtree, Iridium Compounds in Catalysis, *Acc. Chem. Res.* **1979**, *12*, 331–337.

39. Interestingly, it was found that the catalyst activity is highly dependent on the nature of the counter ion. The best results in the hydrogenation of unfunctionalised olefins were obtained using the tetrakis[2,6-bis(trifluoromethyl)phenyl]borate (TFPB or BARF) anion instead of the more common non-coordinating anions such as hexafluorophosphate.

40. A. Lightfoot, P. Schnider, A. Pfaltz, Enantioselective Hydrogenation of Olefins with Iridium-Phosphano-phanodihydrooxazole Catalysts, *Angew. Chem. Int. Ed.* **1998**, *37*, 2897–2899.

41. P. Schnider, G. Koch, R. Préfot, G. Wang, F. M. Bohnen, C. Krüger, A. Pfaltz, Enantioselective Hydrogenation of Imines with Chiral (Phospanodihydrooxazole)iridium Catalysts, *Chem. Eur. J.* **1997**, *3*, 887–892.

42. O. Loiseleur, M. Hayashi, M. Keenan, N. Schmees, A. Pfaltz, Enantioselective Heck Reactions using Chiral P,N-Ligands, *J. Organomet. Chem.* **1999**, *576*, 16–22.

43. Intermolecular, enantioselective Heck reactions require a cyclic olefin as substrate, since 'syn' carbopalladation of a cyclic olefin results in a geometrically defined σ-alkyl-palladium compound. By necessity, the subsequent 'syn' dehydropalladation must take place away from the newly formed chiral centre, thereby affording a chiral product.

44. Although a 1,5-disposition of the coordinating atoms is by far the most common situation, there are a number of successful ligands which possess alternative arrangements, such as a 1,4- or a 1,6-disposition (e.g. PYDIPHOS).[34]

45. Q.-L. Zhou, A. Pfaltz, Chiral Mercaptoaryl-oxazolines as Ligands in Enantioselective Copper-Catalyzed 1,4-Additions of Grignard Reagents to Enones, *Tetrahedron*, **1994**, *50*, 4467–4478.

46. M. V. Troutman, D. H. Appella, S. L. Buchwald, Asymmetric Hydrogenation of Unfunctionalised Tetrasubstituted Olefins with a Cationic Zirconocene Catalyst, *J. Am. Chem. Soc.* **1999**, *121*, 4916–4917.

Asymmetric Two-Center Catalysis

Masakatsu Shibasaki

*Graduate School of Pharma-
ceutical Sciences, The
University of Tokyo, 7-3-1,
Hongo, Bunkyo-ku,
Tokyo 113-0033, Japan,*

*Phone: +81–3 5684 0651
Fax: +81 3 5684 5206
e-mail: mshibasa@mol.f.u-
tokyo.ac.jp*

Keywords ■ *Heterobimetallic Asymmetric Catalyst* ■ *Rare Earth Metal*
■ *Aluminum* ■ *Gallium* ■ *Alkali Metal* ■ *1,1'-Bi-2-naphthol*
■ *Lewis Acid* ■ *Lewis Base* ■ *Brønsted Base*

Concept: Most of the synthetic asymmetric catalysts show limited activity in terms of either enantioselectivity or chemical yields. The major difference between synthetic asymmetric catalysts and enzymes is that the former activate only one side of the substrate in an intermolecular reaction, whereas the latter can not only activate both sides of the substrate but can also control the orientation of the substrate. If this kind of synergistic cooperation can be realized in synthetic asymmetric catalysis, the concept will open up a new field in asymmetric synthesis, and a wide range of applications may well ensure.

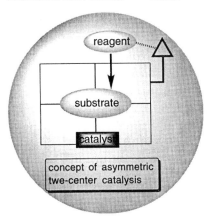

Abstract: In the first part of this mini review a variety of efficient asymmetric catalysis using heterobimetallic complexes is discussed. Since these complexes function at the same time as both a Lewis acid and a Brønsted base, similar to enzymes, they make possible many catalytic asymmetric reactions such as nitroaldol, aldol, Michael, Michael-aldol, hydrophosphonylation, hydrophosphination, protonation, epoxide opening, Diels–Alder and epoxidation reaction of α,β-unsaturated ketones. In the second part catalytic asymmetric reactions such as cyanosilylations of aldehydes and Strecker-type reactions promoted by complexes displaying a Lewis acidity and a Lewis basicity is described.

Prologue

The development of catalytic asymmetric reactions is one of the major areas of research in the field of organic chemistry. So far, a number of chiral catalysts have been reported, and some of them have exhibited a much higher catalytic efficiency than enzymes, which are natural catalysts.[1] Most of the synthetic asymmetric catalysts, however, show limited activity in terms of either enantioselectivity or chemical yields. The major difference between synthetic asymmetric catalysts and enzymes is that the former activate only one side of the substrate in an intermolecular reaction, whereas the latter can not only activate both sides of the substrate but can also control the orientation of the substrate. If this kind of synergistic cooperation can be realized in synthetic asymmetric catalysis, the concept will open up a new field in asymmetric synthesis, and a wide range of applications may well ensure. In this review we would like to discuss two types of asymmetric two-center catalysis promoted by complexes showing Lewis acidity and Brønsted basicity and/or Lewis acidity and Lewis basicity.[2]

Heterobimetallic Asymmetric Catalysis

Our preliminary attempts to obtain a basic chiral rare earth complex have led us to create several new chiral heterobimetallic complexes which catalyze various types of asymmetric reactions. The rare earth-alkali metal-tris(1,1'-bi-2-naphthoxide) complexes (LnMB, where Ln = rare earth, M = alkali metal, and B = 1,1'-bi-2-naphthoxide) have been efficiently synthesized from the corresponding metal chloride and/or alkoxide,[3, 4] and the structures of the LnMB complexes have been unequivocally

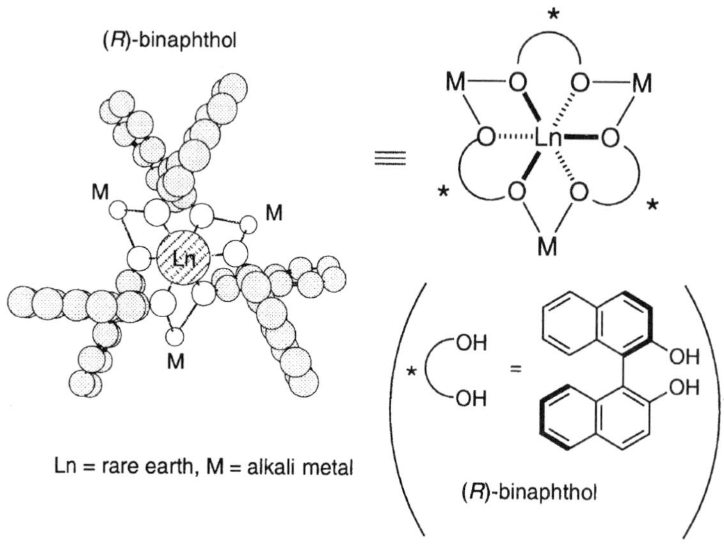

(R)-binaphthol

Ln = rare earth, M = alkali metal

(R)-binaphthol

Figure 1. The structure of rare earth-alkali metal binaphthoxide complexes (LnMB).

determined by a combination of X-ray crystallography and LDI-TOF-mass spectroscopy as shown in Figure 1.[5, 6]

For example, the effective procedure for the synthesis of LLB (where L = lanthanum and lithium, respectively) is the treatment of LaCl$_3\cdot$ 7H$_2$O with 2.7 mol equiv of BINOL dilithium salt, and NaO-t-Bu (0.3 mol equiv) in THF at 50 °C for 50 h. Alternatively we established another efficient procedure for the preparation of LLB, this time starting from La(O-i-Pr)$_3$, [7] the exposure of which to 3 mol equiv of BINOL in THF is followed by the addition of butyllithium (3 mol equiv) at 0 °C. It is noteworthy that these heterobimetallic asymmetric complexes including LLB are stable in organic solvents such as THF, CH$_2$Cl$_2$ and toluene, which contain small amounts of water, and are also insensitive to oxygen. These heterobimetallic complexes can promote a variety of efficient asymmetric reactions such as nitroaldol, aldol, Michael, hydrophosphonylation, hydrophosphination, protonation and Diels–Alder reactions by choosing suitable rare earth metals and alkali metals, respectively. Herein, a catalytic asymmetric nitroaldol reaction and a direct catalytic asymmetric aldol reaction are discussed in detail.

The nitroaldol (Henry) reaction has been recognized as a powerful synthetic tool and has also been utilized in the construction of numerous natural products and other useful compounds. As shown in Scheme 1, we succeeded in realizing the first example of a catalytic asymmetric nitroaldol reaction by the use of a catalytic amount of LLB.

The rare earth metals are generally regarded as a group of 17 elements with similar properties, especially with respect to their chemical reactivity. However, in the case of the above-mentioned catalytic asymmetric nitroaldol reaction, we observed pronounced differences both in reactivity and enantioselectivity of various rare earth metals used. For example, when benzaldehyde and nitromethane were used as starting materials, the corresponding Eu complex gave the nitroaldol in 72 % ee (91 %) in contrast to 37 % ee (81 %) in the case of the LLB (–40 °C, 40 h). These results suggest that small changes in the structure of the catalyst (ca. 0.1Å in the ionic radius of the rare earth cation) cause a drastic change in the optical purity of the nitroaldols produced. Although in general nitroaldol reactions are regarded as equilibrium processes, no detectable retronitroaldol reactions were observed in the Ln-BINOL complex-catalyzed asymmetric nitroaldol reactions. Having succeeded in obtaining the first results from a catalytic asymmetric nitroaldol reaction, we then attempted to apply the method to catalytic asymmetric synthesis of biologically important compounds. The nitroaldol products were readily converted into β-amino alcohols and/or α-hydroxy carbonyl compounds. Thus, convenient syntheses of three kinds of optically active β-blockers are presented in Scheme 2.[8] Interestingly, the nitroaldol products **8**, **11** and **14** were

Scheme 1. Catalytic asymmetric nitroaldol reaction

$$\text{RCHO} + \text{CH}_3\text{NO}_2 \xrightarrow[\text{THF, -42 °C, 18 h}]{\text{LLB (3.3 mol \%)}} \text{R} \overset{\text{OH}}{\wedge} \text{NO}_2$$

(10 equiv)

1: R = PhCH$_2$CH$_2$ **2:** 79% (73% ee), R = PhCH$_2$CH$_2$
3: R = i-Pr   **4:** 80% (85% ee), R = i-Pr
5: R = cyclohexyl   **6:** 91% (90% ee), R = cyclohexyl

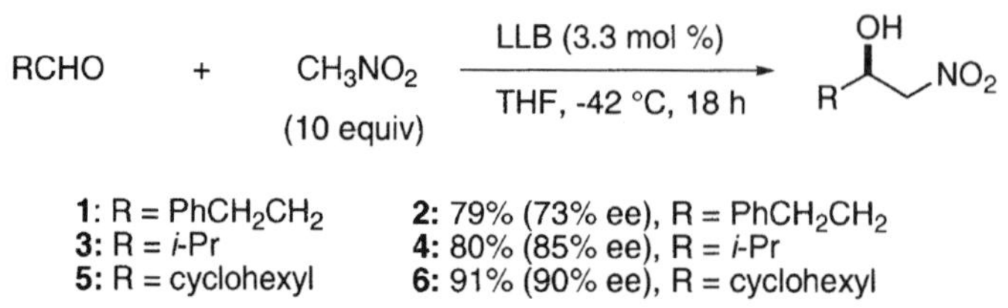

Scheme 2. Catalytic asymmetric synthesis of β-blockers.

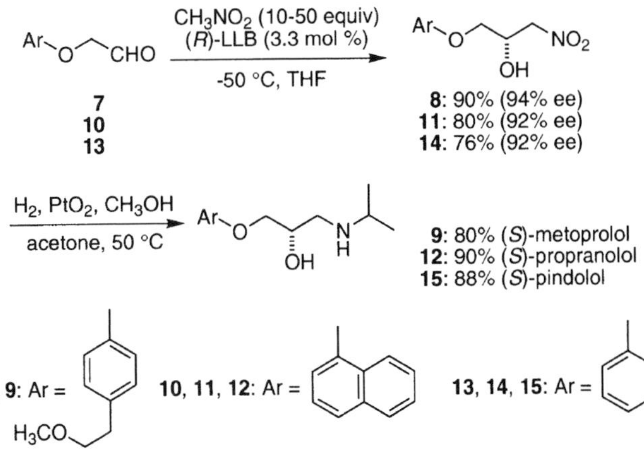

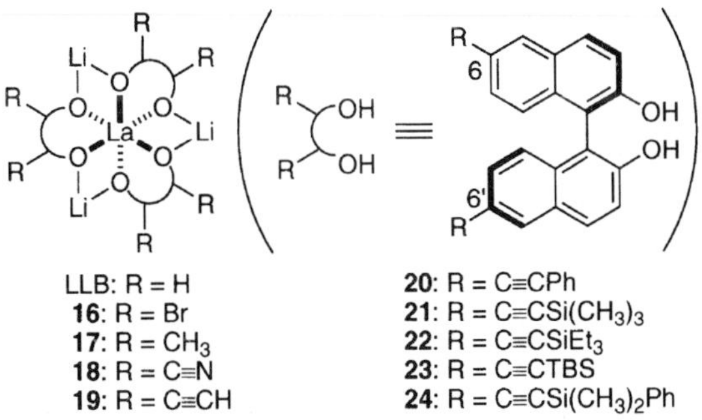

$$Ar\text{-O-CH}_2\text{CHO} \xrightarrow[\text{-50 °C, THF}]{\substack{\text{CH}_3\text{NO}_2 \text{ (10-50 equiv)} \\ (R)\text{-LLB (3.3 mol %)}}} Ar\text{-O-CH}_2\text{-CH(OH)-CH}_2\text{-NO}_2$$

7
10
13

8: 90% (94% ee)
11: 80% (92% ee)
14: 76% (92% ee)

$$\xrightarrow[\text{acetone, 50 °C}]{\text{H}_2, \text{PtO}_2, \text{CH}_3\text{OH}} Ar\text{-O-CH}_2\text{-CH(OH)-CH}_2\text{-NH-CH(CH}_3)_2$$

9: 80% (S)-metoprolol
12: 90% (S)-propranolol
15: 88% (S)-pindolol

7, 8, 9: Ar = (4-(CH₃OCH₂CH₂)phenyl)   **10, 11, 12**: Ar = (naphthyl)   **13, 14, 15**: Ar = (indolyl)

Figure 2. Structural modification of LLB.

LLB: R = H
16: R = Br
17: R = CH₃
18: R = C≡N
19: R = C≡CH

20: R = C≡CPh
21: R = C≡CSi(CH₃)₃
22: R = C≡CSiEt₃
23: R = C≡CTBS
24: R = C≡CSi(CH₃)₂Ph

found to have (S)-absolute configuration when (R)-LLB was used. The nitronates thus appear to react preferentially with the *si* face of the aldehydes, in contrast to the enantiofacial selectivity which might have been expected on the basis of previous results (cf. Scheme 1). These results suggest that the presence of an oxygen atom at the β-position greatly influences the enantiofacial selectivity. LLB-type catalysts were also able to promote diastereoselective and enantioselective nitroaldol reactions starting from prochiral materials. However, limited enantioselectivities (<78 % ee) and diastereoselectivities (ca. 2:1–3:1) were obtained using LLB. In order to obtain both high enantio- and diastereoselectivity, we focused our attention to the preparation of a novel asymmetric catalyst. Among many catalysts prepared, catalysts **16–24** were first found to give higher enantioselectivity in the catalytic asymmetric nitroaldol reaction of hydrocinnamaldehyde **1** using nitromethane. With more effective asymmetric catalysts in hand, we next applied the most efficient catalysts **21** and/or **22** to diastereoselective nitroaldol reactions. We were very pleased to find that, in all cases, high *syn* selectivity and enantioselectivity were obtained using 3.3 mol % of the catalysts.[9] Representative results are shown in Table 1. It appears that

Table 1. *syn*-Selective catalytic asymmetric nitroaldol reaction.

$$RCHO + R'CH_2NO_2 \xrightarrow[\text{THF}]{\substack{\text{catalyst} \\ \text{(3.3 mol %)}}} R\text{-CH(OH)-CH(R')-NO}_2 \text{ (syn)} + R\text{-CH(OH)-CH(R')-NO}_2 \text{ (anti)}$$

1: R = PhCH₂CH₂
28: R = CH₃(CH₂)₄

25: R' = CH₃
29: R' = Et
32: R' = CH₂OH

26 (syn), **27** (anti): R = PhCH₂CH₂, R' = CH₃
30 (syn), **31** (anti): R = PhCH₂CH₂, R' = Et
33 (syn), **34** (anti): R = PhCH₂CH₂, R' = CH₂OH
35 (syn), **36** (anti): R = CH₃(CH₂)₄, R' = CH₂OH

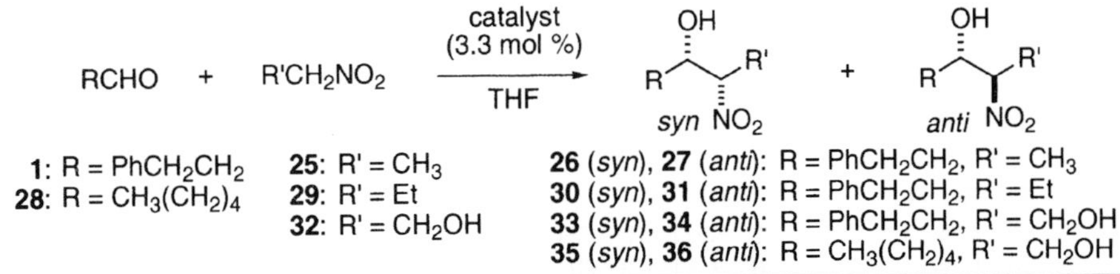

entry	aldehyde	nitroalkane	catalyst	time (h)	temp (°C)	nitro-aldols	yield (%)	*syn/anti*	ee of *syn* (%)
1	1	25	LLB	75	-20	26 + 27	79	74:26	66
2	1	25	21	75	-20	26 + 27	72	85:15	92
3	1	25	22	75	-20	26 + 27	70	89:11	93
4	1	29	LLB	138	-40	30 + 31	89	85:15	87
5	1	29	22	138	-40	30 + 31	85	93:7	95
6	1	32	LLB	111	-40	33 + 34	62	84:16	66
7	1	32	22	111	-40	33 + 34	97	92:8	97
8	28	32	LLB	93	-40	35 + 36	79	87:13	78
9	28	32	22	93	-40	35 + 36	96	92:8	95

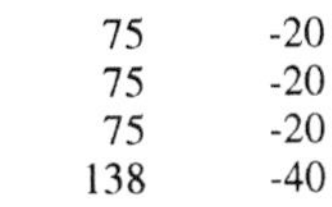

Ph-CH₂-CH(OH)-CH(NH₂)-COOH   **37**

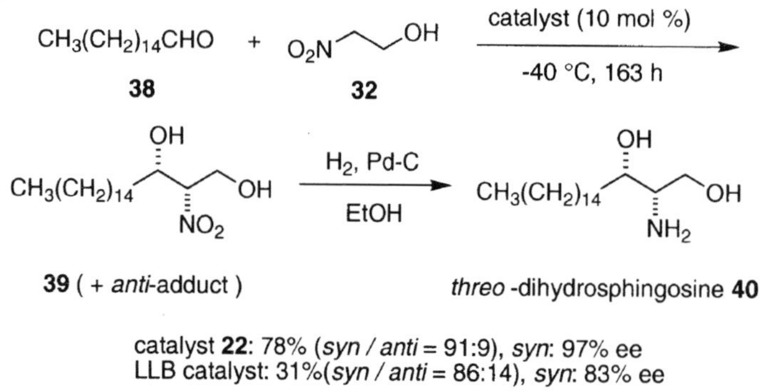

Scheme 3. Catalytic asymmetric synthesis of *threo*-dihydrosphingosine.

$$CH_3(CH_2)_{14}CHO \quad + \quad O_2N\diagdown\diagup OH \quad \xrightarrow[\text{-40 °C, 163 h}]{\text{catalyst (10 mol \%)}}$$

38   **32**

39 (+ *anti*-adduct)   *threo*-dihydrosphingosine **40**

catalyst **22**: 78% (*syn / anti* = 91:9), *syn*: 97% ee
LLB catalyst: 31%(*syn / anti* = 86:14), *syn*: 83% ee

the *syn* selectivity in the nitroaldol reaction can best be explained as arising from steric hindrance in the bicyclic transition state, and higher stereoselectivities obtained by using catalysts **21** and **22** seems to be ascribed to the increase of catalyst stability in the presence of excess nitroalkanes. The *syn*-selective asymmetric nitroaldol reaction was successfully applied to the catalytic asymmetric synthesis of *threo*-dihydrosphingosine **40**, which elicits a variety of cellular responses by inhibiting protein kinase C. Moreover, an efficient synthesis of *erythro*-AHPA **37** from L-phenylalanine was achieved using LLB. Catalytic asymmetric nitroaldol reactions promoted by LLB or its derivatives require at least 3.3 mol% of asymmetric catalysts for efficient conversion. Moreover, even in the case of 3.3 mol% of catalysts, the reactions are rather slow. In order to enhance the activity of the catalyst, a consideration of the possible mechanism for catalytic asymmetric nitroaldol reactions is clearly a necessary prerequisite to formulation of an effective strategy. One possible mechanism for the catalytic asymmetric nitroaldol reactions is shown at the top of Scheme 4. We strove to detect the postulated intermediate I using

various methods. These attempts, however, proved to be unsuccessful, probably owing to the low concentrations of the intermediate, which we thought might be ascribable to the presence of an acidic OH group in close proximity.

In order to remove a proton from I, we added almost 1 equiv of base to the LLB catalyst. After many attempts, we were finally pleased to find that 1 mol% of second-generation LLB (LLB-II), prepared from LLB, 1 mol equiv of H_2O, and 0.9 mol equiv of butyllithium efficiently promoted the catalytic asymmetric nitroaldol reactions. Moreover, we also found that the use of LLB-II (3.3 mol%) accelerated these reactions. The use of other bases such as NaO-*t*-Bu, KO-*t*-Bu and Ca(O-*i*-Pr)$_2$ gave less satisfactory results. The results are shown in Table 2. The structure of LLB-II has not yet been unequivocally determined. We propose here, however, that it is a complex of LLB and LiOH: a proposed reaction course for its use in an improved catalytic asymmetric nitroaldol reaction is shown at the bottom of Scheme 4. Industrial application of a catalytic asymmetric nitroaldol reaction is being examined.

Table 2. Comparisons of catalytic activity between either LLB and second-generation LLB (LLB-II) or **22** and **22-II**.

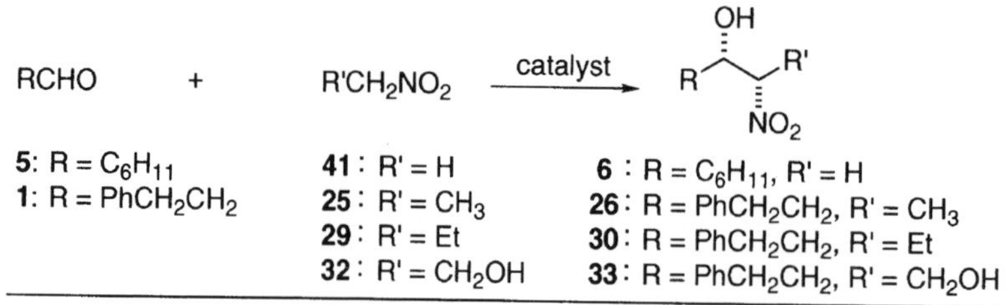

$$RCHO \quad + \quad R'CH_2NO_2 \quad \xrightarrow{\text{catalyst}} \quad R\diagup\overset{OH}{\diagdown}R' \,\,\overset{}{\underset{NO_2}{}}$$

5: R = C$_6$H$_{11}$ **41**: R' = H   **6** : R = C$_6$H$_{11}$, R' = H
1: R = PhCH$_2$CH$_2$ **25**: R' = CH$_3$ **26**: R = PhCH$_2$CH$_2$, R' = CH$_3$
                                  29: R' = Et   **30**: R = PhCH$_2$CH$_2$, R' = Et
                                  32: R' = CH$_2$OH   **33**: R = PhCH$_2$CH$_2$, R' = CH$_2$OH

entry	substrate	catalyst[a] (mol %)	time (h)	temp (°C)	product	yield (%) (*syn/anti*)	ee (%) of *syn*
1	**5** + **41**	LLB (1)	24	-50	**6**	5.6	88
2	**5** + **41**	LLB-II (1)	24	-50	**6**	73	89
3	**5** + **41**	LLB-II (3.3)	4	-50	**6**	70	90
4	**1** + **25**	**22** (1)	113	-30	**26**	25 (70/30)	62
5	**1** + **25**	**22-II** (1)	113	-30	**26**	83 (89/11)	94
6	**1** + **29**	**22** (1)	166	-40	**30**	trace	--
7	**1** + **29**	**22-II** (1)	166	-40	**30**	84 (95/5)	95
8	**1** + **32**	**22** (1)	154	-50	**33**	trace	--
9	**1** + **32**	**22-II** (1)	154	-50	**33**	76 (94/6)	96

[a] LLB-II : LLB + H$_2$O (1 mol equiv) + BuLi (0.9 mol equiv); **22-II** : **22** + H$_2$O (1 mol equiv) + BuLi (0.9 mol equiv)

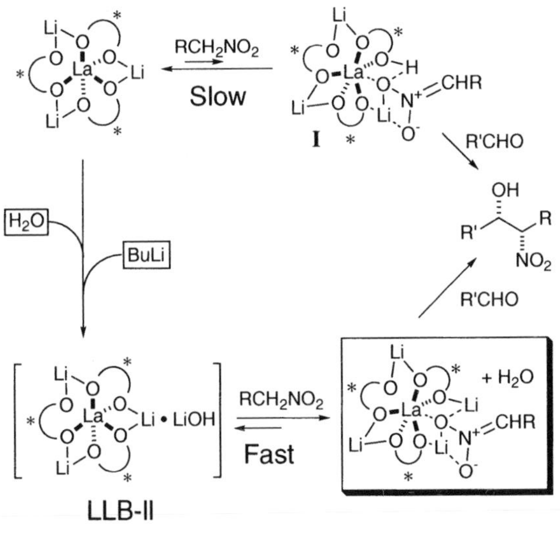

Scheme 4. Possible mechanism of catalytic asymmetric nitroaldol reaction.

while the former functions as a Lewis acid to activate the other carbonyl component.

We speculated that it might be possible to develop a direct catalytic asymmetric aldol reaction of aldehydes and unmodified ketones by employing heterobimetallic catalysts. However, our initial concerns were dominated by the possibility that our heterobimetallic asymmetric catalysts would be ineffective at promoting aldol reactions due to their rather low Brønsted basicity. We were thus pleased to find that, first of all, aldol reactions of the desired type using tertiary aldehydes proceeded smoothly in the presence of LLB as catalyst. It is noteworthy that **56** can be obtained in 94 % ee. The achievement of developing an efficient catalytic asymmetric aldol reaction using aldehydes with α-hydrogens clearly represents a much greater challenge than for cases such as those above, since self-aldol products can easily be formed. However, we found that, for example, the reaction of cyclohexanecarboxaldehyde **5** with acetophenone **43** proceeded smoothly without significant formation of the self-aldol product of **5**, giving **51** in 44 % ee and in 72 % yield. On the other hand, the reaction between hydrocinnamaldehyde **1**, which possesses two α-hydrogens, and **43** proved more difficult. Although **53** was obtained in 52 % ee, the yield was low (28 %) due to the formation of self-condensation byproducts (–20 °C). The results are summarized in Table 3.[11] Thus, we have achieved success in carrying out direct catalytic asymmetric aldol reactions of aldehydes with unmodified ketones for the first time. However, in order to attain a synthetically useful level for this methodology, the challenge remains to reduce the amounts of ketones and catalysts used, shorten reaction times and increase enantioselectivities. As mentioned above, we observed for an asymmetric nitroaldol reaction that the LLB·LiOH tight complex enhanced the catalytic activity of LLB. Encouraged by this result, development of a new strategy to activate LLB for the direct catalytic asymmetric aldol reaction was attempted. As a result the catalyst generated from

Having developed an efficient catalytic asymmetric nitroaldol reaction, we next paid our attention to a direct catalytic asymmetric aldol reaction. The aldol reaction is generally regarded as one of the most powerful of the carbon–carbon bond-forming reactions. The development of a range of catalytic asymmetric aldol-type reactions has proven to be a valuable contribution to asymmetric synthesis.[10] In all of these catalytic asymmetric aldol-type reactions, however, preconversion of the ketone moiety to a more reactive species such as an enol silyl ether, enol methyl ether or ketene silyl acetal is an unavoidable necessity (Scheme 5). Development of a direct catalytic asymmetric aldol reaction, starting from aldehydes and unmodified ketones, is thus a noteworthy endeavor. Such reactions are known in enzyme chemistry, with the fructose-1,6-bisphosphate and/or DHAP aldolases being characteristic examples. The mechanism of these enzyme catalyzed aldol reactions is thought to involve co-catalysis by a Zn^{2+} cation and a basic functional group in the enzyme's active site, with the latter abstracting a proton from a carbonyl compound

Scheme 5. Catalytic asymmetric aldol reaction.

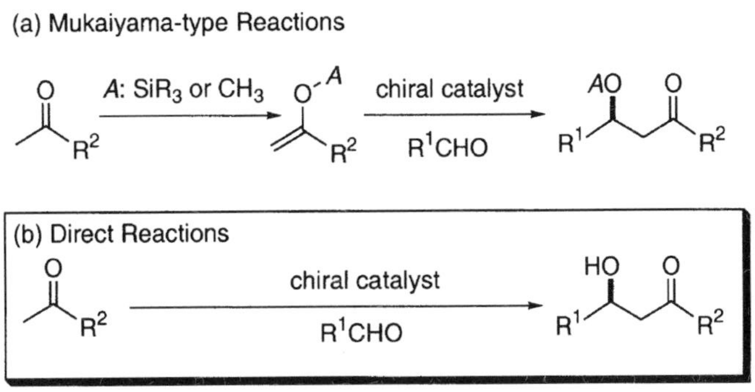

$$R^1CHO \ + \ \underset{R^2}{\overset{O}{\|}} \xrightarrow[\text{THF, -20 °C}]{\substack{(R)\text{-LLB} \\ (20 \text{ mol \%})}} \ \underset{R^1}{\overset{OH \quad O}{\|}} R^2$$

42: R^1 = *t*-Bu

45:
R^1 = PhCH$_2$C(CH$_3$)$_2$

5: R^1 = cyclohexyl

3: R^1 = *i*-Pr

1: R^1 = Ph(CH$_2$)$_2$

43: R^2 = Ph

46:
R^2 = 1-naphthyl

48: R^2 = CH$_3$

50: R^2 = Et

44: R^1 = *t*-Bu, R^2 = Ph

47: R^1 = *t*-Bu, R^2 = 1-naphthyl

49: R^1 = PhCH$_2$C(CH$_3$)$_2$, R^2 = Ph

51: R^1 = cyclohexyl, R^2 = Ph

52: R^1 = *i*-Pr, R^2 = Ph

53: R^1 = Ph(CH$_2$)$_2$, R^2 = Ph

54: R^1 = PhCH$_2$C(CH$_3$)$_2$, R^2 = CH$_3$

55: R^1 = *t*-Bu, R^2 = CH$_3$

56: R^1 = PhCH$_2$C(CH$_3$)$_2$, R^2 = Et

Table 3. Direct catalytic asymmetric aldol reactions promoted by (*R*)-LLB (20 mol%).

entry	aldehyde	ketone (equiv)	product	time (h)	yield (%)	ee (%)
1[a]	42	43 (5)	44	88	43	89
2	42	43 (5)	44	88	76	88
3	42	43 (1.5)	44	135	43	87
4	42	43 (10)	44	91	81	91
5	42	46 (8)	47	253	55	76
6	45	43 (7.4)	49	87	90	69
7	5	43 (8)	51	169	72	44
8[b]	3	43 (8)	52	277	59	54
9	1	43 (10)	53	72	28	52
10	45	48 (10)	54	185	82	74
11	42	48 (10)	55	100	53	73
12	45	50 (50)	56	185	71	94

[a] (*R*)-LLB and addition of 1 equiv of H$_2$O to LLB.

[b] The reaction was carried out at -30 °C.

LLB, KHMDS (0.9 equiv to LLB) and H$_2$O (1 equiv to LLB), which presumably forms a heteropolymetallic complex (LLB-II'), was found to be a superior catalyst for the direct catalytic asymmetric aldol reaction giving **49** in 89 % yield and 79 % ee (using 8 mol% of LLB). We employed this method to generate KOH *in situ* because of its insolubility in THF. The use of KO-*t*-Bu instead of KHMDS gave a similar result, indicating that HMDS dose not play a key role. Interestingly, further addition of H$_2$O (1 equiv with respect to LLB) resulted in the formation of **49** in 83 % yield and higher ee. The powder obtained from the cata-

lyst solution by evaporation of the solvent showed a similar result. This powder can be easily handled without the need of an inert atmosphere. In addition, we were pleased to find that as little as 3 mol% of the catalyst promoted the reaction efficiently to give **49** in 71% yield and 85% ee. Moreover, in contrast to catalytic asymmetric nitroaldol reactions, the generation of LiOH or other bases was found to give less satisfactory results. The results are summarized in Table 4.

This newly developed heteropolymetallic catalyst system (LLB-II') was applied to a variety of direct cat-

$$\underset{\substack{Ph \\ \mathbf{45}}}{\overset{O}{\|}}H \ + \ \underset{\mathbf{43} \ (5 \ eq)}{\overset{O}{\|}}Ph \xrightarrow[\substack{\text{base (7.2 mol \%)} \\ \text{THF, H}_2\text{O, -20 °C}}]{(R)\text{-LLB (8 mol \%)}} \ \underset{\substack{Ph \\ \mathbf{49}}}{\overset{OH \quad O}{\|}}Ph$$

Table 4. Direct catalytic asymmetric aldol reactions of **45** with **43** under various conditions.

entry	base	H$_2$O (mol %)	time (h)	yield (%)	ee (%)
1	- (LLB itself)	-	18	trace	-
2	KHMDS	0	18	83	58
3	KHMDS	8	18	89	79
4	KHMDS	16	18	83	85
5[a]	KHMDS	16	33	71	85
6	KHMDS	32	18	67	89
7	LHMDS	16	5	22	80
8	NHMDS	16	5	28	86
9	KHMDS	16	5	74	84

[a] 3 mol % of catalyst was used.

Table 5. Direct catalytic asymmetric aldol reactions promoted by heteropolymetallic asymmetric catalyst and following Baeyer-Villiger oxidations.

42: R^1 = *t*-Bu
45: R^1 = PhCH$_2$C(CH$_3$)$_2$
3: R^1 = *i*-Pr
1: R^1 = PhCH$_2$CH$_2$
57: R^1 = BnOCH$_2$C(CH$_3$)$_2$
58: R^1 = Et$_2$CH
59: R^1 = *n*-C$_5$H$_{11}$

43: R^2 = Ph
48: R^2 = CH$_3$ [O]
50: R^2 = Et
60: R^2 = 3-NO$_2$-C$_6$H$_4$
61: R^2 = -(CH$_2$)$_3$-

entry	aldehyde (R^1)	ketone[a] (R^2) (eq)	aldol	time (h)	yield (%)	ee (%)	yield of ester[b]
1	**42**	**43** (5)	**44**	15	75	88	
2	**45**	**43** (5)	**49**	28	85	89	**68**: 80%[c]
3	**45**	**48** (10)	**54**	20	62	76	
4[d]	**45**	**50** (15)	**56**	95	72	88	
5	**57**	**43** (5)	**62**	36	91	90	
6[e]	**57**	**43** (5)	**62**	24	70	93	**69**: 73%[f]
7[g]	**3**	**43** (5)	**52**	15	90	33	
8[h]	**3**	**60** (3)	**63**	70	68	70	**70**: 80%[i]
9[j]	**58**	**60** (3)	**64**	96	60	80	
10[h,k]	**59**	**60** (5)	**65**	96	55	42	
11[l]	**1**	**60** (3)	**66**	31	50	30	
12	**45**	**61** (5)	**67**	99	95	76/88 (syn/anti)	**71**: 85%[c]
						(syn/anti = 93/7)	

[a] Excess of ketone was recovered after reaction. [b] The yield from aldol product. [c] Conditions: SnCl$_4$ (cat.), (TMSO)$_2$, *trans-N,N'*-bis(*p*-toluenesulfonyl)-cyclohexane-1,2-diamine (cat.), MS 4A, CH$_2$Cl$_2$. [d] 8 mol % of H$_2$O was used. [e] The reaction was carried out in 5.7 mmol (**57**) scale. [f] Conditions: *m*CPBA, NaH$_2$PO$_4$, DCE. [g] The reaction was carried out at -30 °C. [h] The reaction was carried out at -50 °C. [i] Conditions: i) PtO$_2$, H$_2$, MeOH; ii) ZCl, Na$_2$CO$_3$, MeOH-H$_2$O; iii) SnCl$_4$ (cat.), (TMSO)$_2$, *trans-N,N'*-bis(*p*-toluenesulfonyl)cyclohexane-1,2-diamine (cat.), MS 4A, CH$_2$Cl$_2$. R^2 (**70**) = 3-ZNH-C$_6$H$_4$. [j] Conditions: (*R*)-LLB (15 mol %), KHMDS (13.5 mol %), H$_2$O (30 mol %), -45 °C. [k] Conditions: (*R*)-LLB (30 mol %), KHMDS (27 mol %), H$_2$O (60 mol %). [l] The reaction was carried out at -40 °C.

72

73

alytic asymmetric aldol reactions, giving aldol products **44–67** in modest to good ees as shown in Table 5. It is noteworthy that even **65** can be produced from hexanal **59** in 55 % yield and 42 % ee without the formation of the corresponding self-aldol product (–50 °C). This result can be understood by considering that in general aldehyde enolates are not generated by the catalyst at low temperature. In fact, this assumption was confirmed by several experimental results. It is also noteworthy that the direct catalytic asymmetric aldol reaction between **45** and cyclopentanone **61** also proceeded smoothly to afford **67** in 95 % yield (syn/anti = 93/7, syn = 76 % ee, anti = 88 % ee). Several of the aldol

products obtained were readily converted to their corresponding esters by Baeyer–Villiger oxidation. The results are summarized in Table 5. Ester **69** was further transformed into key epothilone A intermediate **72** and also a key synthetic intermediate **73** for bryostatin 7. What is the mechanism of the present direct catalytic asymmetric aldol reactions using LLB-II? It is obvious that the self-assembly of LLB and KOH takes place, because of the formation of a variety of aldol products in high ees and yields. In addition, the ^{13}C NMR spectrum of LLB·KOH and also the LDI-TOF(+)MS spectrum show that there is a rapid exchange between Li$^+$ and K$^+$. We have already found that LPB[LaK$_3$-tris(binaphthoxide)]

Scheme 6. Possible mechanism of direct catalytic asymmetric aldol reaction.

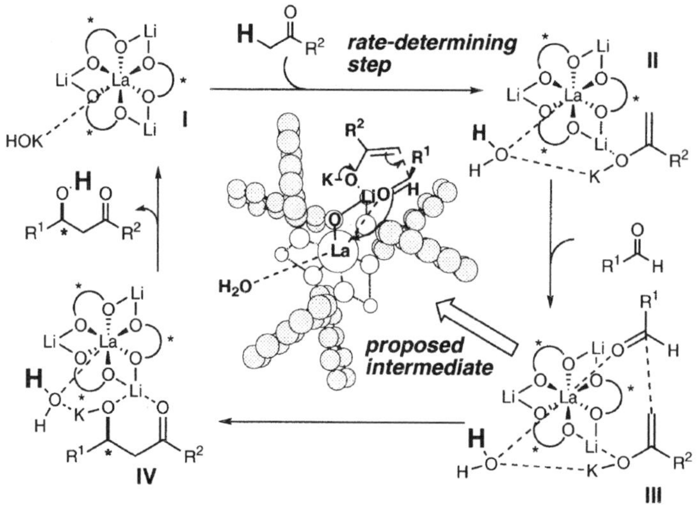

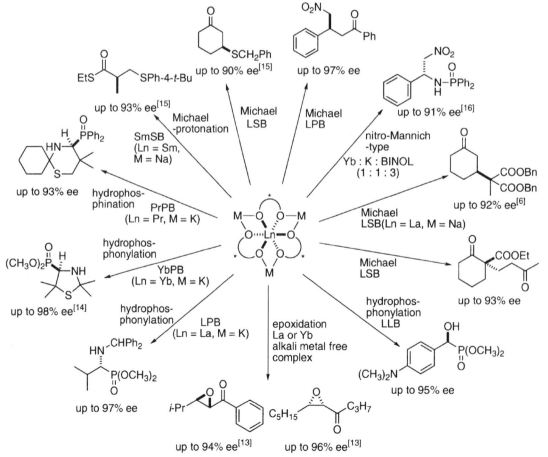

itself is not a useful catalyst for aldol reactions, and that the complexes LPB·KOH or LPB·LiOH give rise to much less satisfactory results. Consequently we believe that the BINOL core of the active complex is essentially LLB. Therefore, the heteropolymetallic complex of LLB and KOH, with KOH axially coordinated to La, among other possible complexes, would be the most effective catalyst for the present reaction. To clarify the reaction mechanism, we carried out kinetic studies. As a result, significant isotope effects ($k_H/k_D \sim 5$) were observed, and the reaction rate has been found to be independent of the concentration of the aldehyde. Both of these results indicate that the rate-determining step is the deprotonation of the ketone, and they also suggest that the catalyst readily forms a relatively tight complex with the aldehyde, thus activating it. This coordination of an aldehyde was supported by the ^{1}H NMR spectrum. Although the precise role of H$_2$O is not clear at present, we have suggested a working

Scheme 7. Catalytic asymmetric reactions promoted by heterobimetallic complexes.

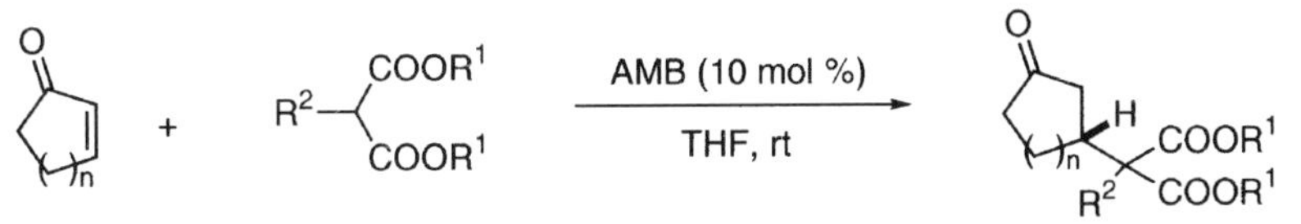

Table 6. Catalytic asymmetric Michael reactions promoted by the AlMbis(R)-binaphthoxide) complex (AMB).

74: n = 1
75: n = 2

76: R^1 = Et, R^2 = CH_3
77: R^1 = Bn, R^2 = H
78: R^1 = CH_3, R^2 = H
79: R^1 = Et, R^2 = H

80: n = 1, R^1 = Et, R^2 = CH_3
81: n = 1, R^1 = Bn, R^2 = H
82: n = 2, R^1 = Bn, R^2 = H
83: n = 2, R^1 = CH_3, R^2 = H
84: n = 2, R^1 = Et, R^2 = H

entry	enone	Michael donor	product	M	time (h)	yield (%)	ee (%)
1	74	76	80	Li	72	84	91
2	74	77	81	Li	60	93	91
3	75	77	82	Li	72	88	99
4	75	77	82	Na	72	50	98
5	75	77	82	K	72	43	87
6	75	77	82	Ba	6	100	84
7	75	78	83	Li	72	90	93
8	75	79	84	Li	72	87	95

model of the catalytic cycle and a possible mechanism which allows us to explain the observed absolute configurations of the products (Scheme 6). The stereoselectivities appear to be kinetically controlled. In fact, the ee of the aldol product was constant during the course of the reaction. Thus, we have succeeded in carrying out the first catalytic asymmetric aldol reaction between aldehydes and unmodified ketones by using LLB or LLB-II'. Several reactions are already synthetically useful especially in the case of tertiary aldehyde, leading to the catalytic asymmetric synthesis of key intermediates *en route* to natural products.[12] Further studies are currently underway.

Moreover, these rare earth heterobimetallic complexes can be utilized for a variety of efficient catalytic asymmetric reactions as shown in Scheme 7.

Next we began with the development of an amphoteric asymmetric catalyst assembled from aluminum and an alkali metal.[17] The new asymmetric catalyst could be prepared efficiently from $LiAlH_4$ and 2 mol equiv of (R)-BINOL, and the structure was unequivocally determined by X-ray crystallographic analysis (Scheme 8). This aluminum-lithium-BINOL complex (ALB) was highly effective in the Michael reaction of cyclohexenone 75 with dibenzyl malonate 77, giving 82 with 99% ee and 88% yield at room temperature. Although LLB and

Scheme 8. Preparation of ALB.

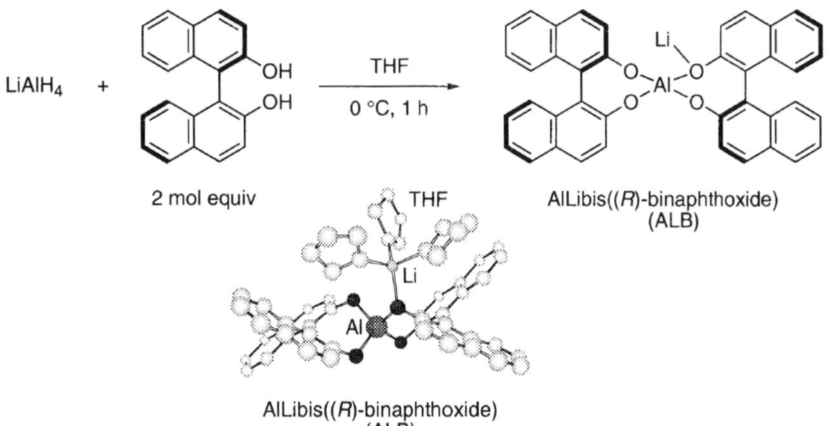

$LiAlH_4$ + 2 mol equiv

THF
0 °C, 1 h

AlLibis((R)-binaphthoxide)
(ALB)

AlLibis((R)-binaphthoxide)
(ALB)

LSB complement each other in their ability to catalyze asymmetric nitroaldol and Michael reactions, aluminum-M-(R)-BINOL complexes (M = Li, Na, K, and Ba) are useful for the catalytic asymmetric Michael reactions. Moreover, we have developed a strategy for the activation of ALB: the addition of nearly 1 equiv of base, such as BuLi or KO-t-Bu, to ALB can accelerate a catalytic asymmetric Michael addition without lowering the high enantiomeric excess. However, 3–5 mol% of the catalyst is still required to obtain the product in excellent yield and high enantiomeric excess. We intended to improve the catalytic asymmetric Michael addition to a practically useful level. We were pleased to find that adddition of MS 4A to the reaction medium greatly improved the catalytic asymmetric Michael addition. Actually, as shown in Table 7, the use of ALB (0.3 mol%), KO-t-Bu (0.27 mol%) and MS 4A gave 83 in 99% ee and 94% yield even at room temperature. Furthermore, we successfully carried out this reaction on a 100 g scale, and the product was purified without column chromatography. Using the Michael adduct 83, catalytic asymmetric syntheses of tubifolidine 85,[18] 19,20-dihydroakuammicine 86 and coronafacic acid 87 have been achieved.[19]

The mechanistic considerations of a catalytic asymmetric Michael reaction suggest that the reaction of a alkali metal enolate derived from a malonate derivative with an enone should lead to an intermediary aluminum enolate. Is it possible that such an Al enolate could then be trapped by an electrophile such as an aldehyde? As was anticipated, the reaction of 74, diethyl methylmalonate 76, and hydrocinnamaldehyde 1 in the presence of 10 mol% of ALB gave the three-component coupling product 88 as a single isomer in 91% ee (64% yield) (Table 8). This cascade Michael-aldol reaction has been successfully applied to a catalytic asymmetric synthesis of 11-deoxy-PGF$_1\alpha$ 89.[20]

Moreover, ALB was found to be also useful for the hydrophosphonylation of aldehydes. ALB and LLB can thus be used in a complementary manner for the hydrophosphonylation of aldehydes.

The enantioselective ring opening of epoxides is an attractive and quite powerful method in asymmetric synthesis. Although various types of stoichiometric or catalytic asymmetric epoxide ring openings have been reported, only a few practical methods have been reported so far.[21] We became very interested in the development of catalytic asymmetric epoxide ring openings using nucleophiles such as RSH, ROH, HCN, and HN$_3$. First of all, we examined a catalytic asymmetric ring opening of symmetrical epoxides with thiols using the heterobimetallic complexes. We envisioned that these complexes would prove to be useful for the catalytic asymmetric ring openings of 91 with a nucleophile such as PhCH$_2$SH. However, LaM$_3$tris(binaphthoxide) (M = Li or Na) or ALB showed only low catalytic activity, giving 2-(benzylthio)cyclohexanol in 1–10% yields, although modest to high ees (27–86% ee) were observed. We then examined the new hetero-

Table 7. A greatly improved catalytic asymmetric Michael addition of 78 to 75.

entry	ALB (x mol %)	KO-t-Bu	MS 4A	time (h)	yield (%)	ee (%)
1[b]	10	-	-	72	90	93
2[c]	5	+	-	48	97	98
3[c]	0.3	+	-	120	74	88
4[c]	0.3	+	+[e]	120	94	99
5[d]	1.0	+	+[f]	72	96	99

[a] (R)-AlLibis(binaphthoxide). [b] 200 mg scale reaction. [c] 400 mg scale reaction.
[d] 100 g scale reaction. [e] MS 4A (8.3 g) was used for ALB (1 mmol).
[f] MS 4A (2.0 g) was used for ALB (1 mmol).

85

86

87

Table 8. Tandem Michael-aldol reactions.

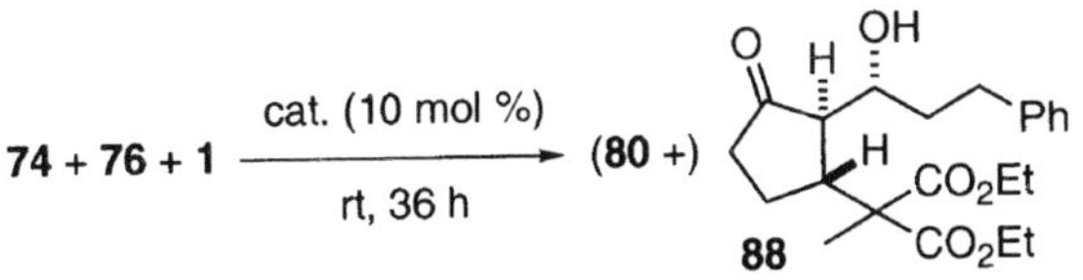

$$74 + 76 + 1 \xrightarrow[\text{rt, 36 h}]{\text{cat. (10 mol \%)}} (80 +) \quad 88$$

catalyst	80		88	
	yield (%)	ee (%)	yield (%)	ee (%)
(R)-ALB	7	90	64	91
(R)-LLB	46	3	30[a]	–
(R)-LSB	73	86	trace	–
Li-free-La-(R)-BINOL	57	83	trace	–

[a] Inseparable mixture.

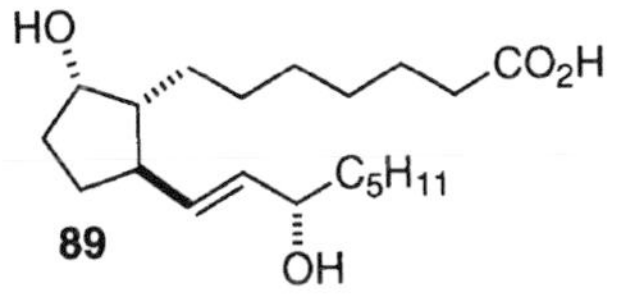

89

bimetallic asymmetric complexes with group 13 elements (B, Ga, In) other than Al. Of these, the GaLibis[(R)-binaphthoxide] complex [(R)-GaLB], which was readily prepared from GaCl₃, (R)-binaphthol (2 mol equiv to GaCl₃) and BuLi (4 mol equiv to GaCl₃) in THF, showed a high catalytic activity for the present reaction (Scheme 9).[22] After many attempts, GaLB was found to be quite useful for the asymmetric ring opening of symmetrical epoxides with t-BuSH as shown in Table 9. The almost optically pure **98** has been utilized for the

preparation of the attractive chiral ligand by Evans.[23]

Moreover, as shown in Table 10, GaLB was found to be suitable to the asymmetric ring opening of symmetrical epoxides with 4-methoxyphenol, giving products in good to excellent ees. However, chemical yield was only modest, despite the use of more than 20 mol% GaLB. This was due to the undesired ligand exchange between BINOL and 4-methoxyphenol, which resulted in the decomplexation of GaLB. Application of various known chiral

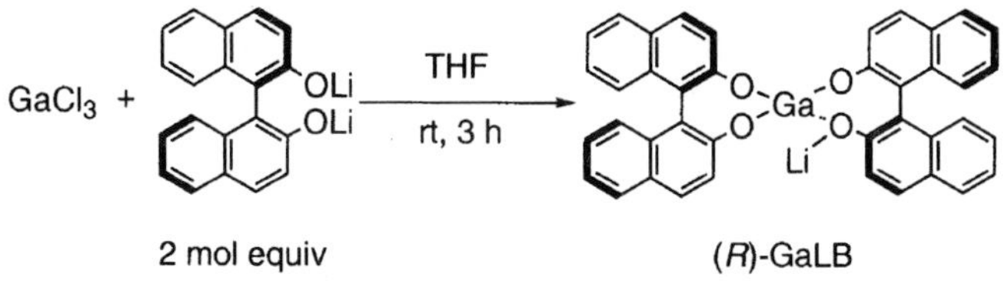

$$\text{GaCl}_3 + \quad \xrightarrow[\text{rt, 3 h}]{\text{THF}} \quad (R)\text{-GaLB}$$

2 mol equiv

Scheme 9. Preparation of GaLB.

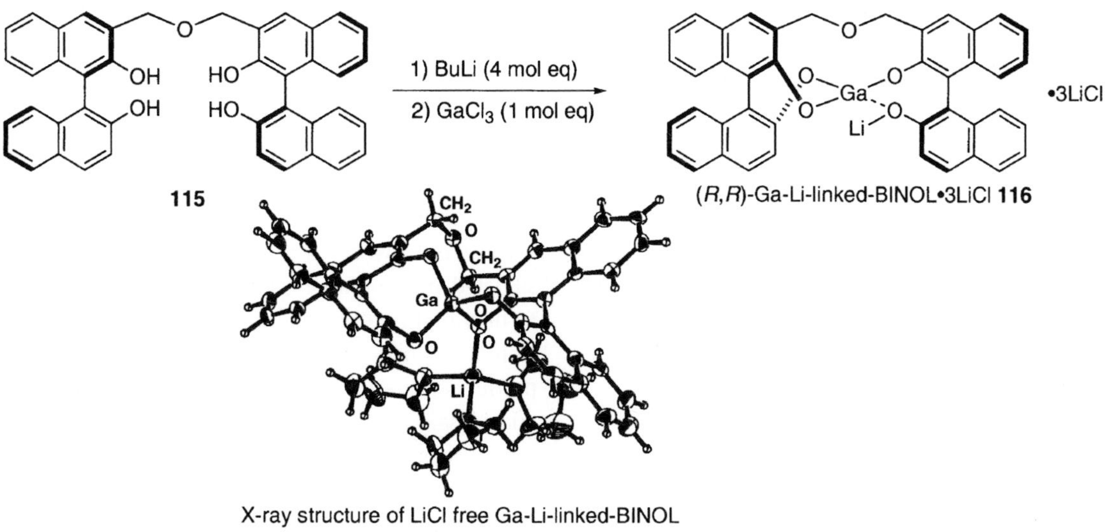

$$\mathbf{115} \xrightarrow[\text{2) GaCl}_3 \text{ (1 mol eq)}]{\text{1) BuLi (4 mol eq)}} (R,R)\text{-Ga-Li-linked-BINOL•3LiCl } \mathbf{116}$$

•3LiCl

X-ray structure of LiCl free Ga-Li-linked-BINOL

Scheme 10. Preparation of Ga-Li-linked-BINOL.

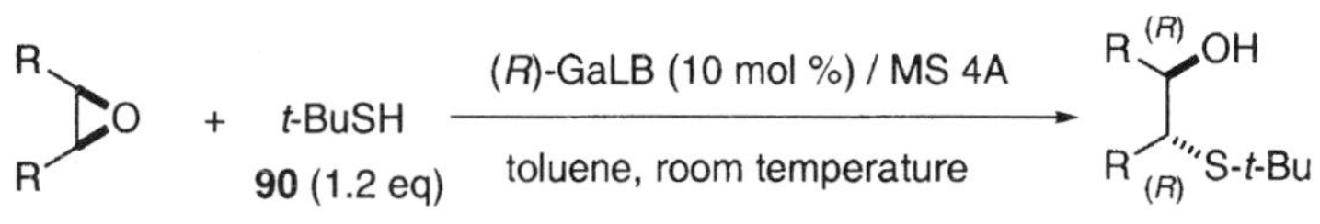

Table 9. Catalytic asymmetric ring openings of symmetrical epoxides with t-BuSH (**90**) catalyzed by (R)-GaLB with MS 4A.

entry	epoxide		MS 4A[a] (g)	time (h)	product	yield (%)	ee (%)
1		**91**	none	65	**98**	35	98
2		**91**	0.2	9	**98**	80	97
3		**92**	0.2	36	**99**	74	95
4[b]	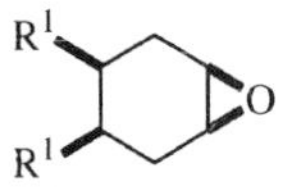	**93**	0.2	12	**100**	83	86
5[b]	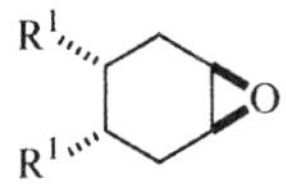	**94**	0.2	137	**101**	64	91
6	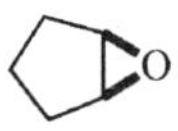	**95**	0.2	24	**102**	89	91
7[c]	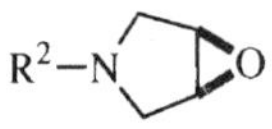	**96**	0.2	72	**103**	89	89
8	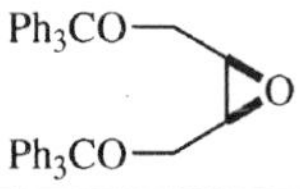	**97**	2.0	48	**104**	89	82

[a] Weight per 0.1 mmol of GaLB. [b] $R^1 = CH_2OSiPh_2t$-Bu. [c] Carried out at 50 °C in the presence of 30 mol % of GaLB. $R^2 = $ 2,4,6-Trimethylbenzenesulfonyl.

ligand such as 6,6'-bis((triethylsilyl)ethynyl)-BINOL and H$_8$-BINOL were examined, but satisfactory results were not obtained.[24] To overcome this problem a novel linked-BINOL **115** containing coordinative oxygen atom in the linker has been developed. By linking two BINOL units in GaLB, the stability of the Ga-complex was greatly improved. Using 3–10 mol% (R,R)-Ga-Li-linked-BINOL complex **116**, a variety of epoxide opening reactions were found to proceed smoothly, affording products in analogous ee (66–96% ee) and in much higher yield (y. 67–94%) compared to (R)-GaLB (Table 10). The structure of the LiCl free Ga-Li-linked-BINOL complex was elucidated by X-ray analysis.[25]

In addition, quite recently a direct catalytic asymmetric Mannich-type reaction has been achieved by the cooperative catalysis of ALB and La(OTf)$_3$·nH$_2$O.

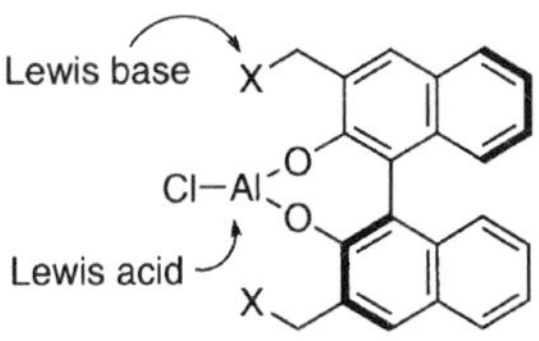

123: X = P(O)Ph$_2$
124: X = CH$_2$P(O)Ph$_2$
125: X = CHPh$_2$
126: X = P(O)(PhN(CH$_3$)$_2$-p)$_2$

Figure 3. Chiral Lewis acid–Lewis base catalysts.

Bifunctional Asymmetric Catalysis Promoted by Chiral Lewis Acid – Lewis Base Complexes

Based on the achievements desribed above, it seemed rational to design a new bifunctional asymmetric catalyst consisting of Lewis acid and Lewis base moieties, which activate both electophiles and nucleophiles at defined positions simultaneously. This type of asymmetric catalysis is seen only in a few examples.[26] We designed the chiral Lewis acid – Lewis base catalyst **123**. We assumed that the aluminum would work as a Lewis acid to activate the carbonyl group, and the oxygen atom of the phosphine oxide would work as a Lewis base to activate the silylated nucleophiles. Catalyst **123** has been found to be a highly efficient catalyst for the cyanosilylation of aldehydes[27] with broad generality, affording products in excellent chemical yields and excellent enantioselectivities. One of the key issues for designing a Lewis acid – Lewis base catalyst is how to prevent the internal complexation of these moieties.

Molecular modeling studies suggested that **123** would avoid such a problem, because the coordina-

Table 10. Catalytic enantioselective epoxide ring opening with 4-methoxyphenol 105 promoted by gallium hetero-bimetallic complexes in the presence of MS 4A.

entry	epoxide	product		GaLB time (h)	yield (%)	ee (%)	GaSO time (h)	yield (%)	ee (%)
1		95	107	72 $(72)^a$	75 $(73)^a$	86 $(89)^a$	4	77	54
2		91	108	72 $(72)^a$	48 $(60)^a$	93 $(94)^a$	4 $(4)^b$	73 $(61)^b$	56 $(51)^b$
3		106	109	72	31	67	4	67	58
4		92	110	72 $(72)^a$	70 $(69)^a$	87 $(92)^a$	24	90	55
5^c		93	111	96	34	80	48	83	43
6^d		96	112	160	51	90	19	44	34
7		97	113	72	e	e	7	75	50
8^f		91	114	–	–	–	4	75	61

^{a} Values in parentheses show the results of the GaLB* catalyzed reaction (B* = 6,6'-bis((triethylsilylethynyl)binaphthol)). ^{b} 5 mol % GaSO was used. ^{c} R^1 = CH$_2$OSiPh$_2$t-Bu. ^{d} R^2 = 2,4,6-Trimethylbenzenesulfonyl; 30 mol % GaLB was used. ^{e} No reaction. ^{f} 4-Methoxy-1-naphthol was used instead of **105**.

(R)-GaSO

tion of the Lewis base to the internal aluminum seemed to be torsionally unfavorable. When considering **124**, however, which has an ethylene linker, the internal coordination seemed to be quite stable without strain. In accordance with this expectation, the reaction of TMSCN with benzaldehyde **131**, catalyzed by **124** (9 mol%), proceeded slowly at –40 °C (37 h) and gave the cyanohydrin **141** in only 4% yield after hydrolysis. However, a solution of **123** (9 mol %), **131** and TMSCN, at –40 °C (37 h), afforded **141** in 91% yield and in 87% ee. Encouraged by the result of benzaldehyde, we next investigated the reaction of aliphatic aldehydes. Surprisingly, aliphatic aldehydes afforded very low ee values. We anticipated that there would be competition between two reaction pathways in the case of the more reactive aliphatic aldehydes. The desired pathway involves the dual activation between the Lewis acid and the aldehyde and between the Lewis base and TMSCN, whereas the undesired pathway involves mono-activation by the Lewis acid. We assumed that these two pathways could differ more significantly if the Lewis acidity of the catalyst was

decreased, and so we investigated the effect of additives which coordinate to the aluminum to reduce its Lewis acidity. Moreover, the additive could change the geometry of aluminum from tetrahedral to trigonal bipyramidal, which should allow the phosphine oxide to exist in a more favorable position relative to the aldehyde. We found that electron donating phosphine oxides had a beneficial effect on ee. In the case of **1**, the ee values of **134** significantly increased from 9% to 41% and 56% by the addition of 36 mol% of CH$_3$P(O)Ph$_2$ and Bu$_3$P(O), respectively. Further improvement of ee (up to 97%) was achieved by the slow addition of TMSCN (10 h), via syringe pump, in the presence of Bu$_3$P(O). In the case of **131**, however, addition of Bu$_3$P(O) resulted in a very sluggish reaction, affording only a trace amount of the product. However, the reaction proceeded in 98% yield and in 96% ee in the presence of CH$_3$P(O)Ph$_2$. Therefore, we used Bu$_3$P(O) as the additive for aliphatic and α,β-unsaturated aldehydes, and CH$_3$P(O)Ph$_2$ as the additive for aromatic aldehydes. This catalyst is practical and has a broad generality with respect to the variety of aldehydes that can be used (Table 12).

Table 11. Enantioselective ring opening of various *meso*-epoxides with 4-methoxyphenol (**105**) promoted by Ga-Li-linked-BINOL complex (**116**).

entry	epoxide	product	AroH (eq)	temp (°C)	time (h)	yield[a] (%)	ee[b] (%)
1	91	108	3.0	75	96	72	91
2	95	107	3.0	60	63	88	85
3	107	109	3.0	75	108	82	66
4	92	110	3.0 (1.2)[c]	75 (75)[c]	36 (117)[c]	94 (80)[c]	85 (91)[c]
5[d]	93	111	3.0	60	96	72	79
6[e]	96	112	3.0	60	160	77	78
7	117	120	3.0	60	48	67	87
8	118	121	2.0	60	70	85	96
9	119	122	3.0	60	140	72	91

[a] Isolated yield. [b] Determined by HPLC analysis. [c] 3 mol % catalyst was used. [d] R^1 = CH_2OSiPh_2t-Bu. [e] 30 mol % catalyst was used. Mts = 2,4,6-trimethylbenzenesulfonyl.

To the best of our knowledge, this is the most efficient and the most general catalytic asymmetric cyanosilylation of aldehydes. Preliminary kinetic studies, using catalyst **126** which contains a more electron-rich phosphine oxide, seem to support the dual Lewis acid – Lewis base activation pathway. The initial reaction rate with **126** (10 mol%) is 1.2 times faster than that with **123** (10 mol%) (k**126**/k**123** = 1.2), reflecting the higher Lewis basicity of the phosphine oxide in the reaction of **1** in the presence of $Bu_3P(O)$. Thus, the enantioselectivity of the reaction catalyzed by **123** may be explained by the working model **144**, with the external phosphine oxide coordinating to the aluminum, thus giving a pentavalent aluminum.[28] This concept could provide a guide for designing new asymmetric catalysts for the reaction of a variety of nucleo-philes, including silylated ones, with carbonyl compounds.

Moreover, it seemed to be a rational extension to apply the catalyst **123** to the asymmetric Strecker-type reaction.[29] Actually, as shown in Table 13, an efficient and general catalytic asymmetric Strecker-type reaction has been realized. Products were successfully converted to the corresponding amino acid derivatives in high yields without loss of enantiomeric purity.[30]

Epilogue

We believe that the successful development of the multifunctional concept has opened up a new field in asymmetric catalysis. There are still many possibilities for the design of asymmetric two-

Table 12. Asymmetric cyanosilylation of aldehydes catalyzed by **123**[a].

Entry	R	Aldehyde	Product	Additive	Time (h)	Yield (%)[b]	ee (%)[c]	S/R
1	Ph(CH₂)₂	**1**	**134**	Bu₃P(O)	37	97	97	S
2	CH₃(CH₂)₅	**127**	**135**	Bu₃P(O)	58	100	98	S
3	(CH₃)₂CH	**3**	**136**	Bu₃P(O)	45	96	90	S
4	(CH₃CH₂)₂CH	**58**	**137**	Bu₃P(O)	60	98	83	S
5	trans-CH₃(CH₂)₃CH=CH₂	**128**	**138**	Bu₃P(O)	58	94	97	–[g]
6	PhCH=CH	**129**	**139**	Bu₃P(O)	40	99	98	S
7[d]	(thiazolyl–CH=C(CH₃)–)	**130**	**140**	Bu₃P(O)	74	97	99	–[g]
8[e]	Ph	**131**	**141**	CH₃P(O)Ph₂	96	98	96	S
9	p-CH₃C₆H₄	**132**	**142**	CH₃P(O)Ph₂	79	87	90	S
10[f]	(furyl)	**133**	**143**	CH₃P(O)Ph₂	70	86	95	S

[a] TMSCN (1.8 equiv) was added over 10 h via syringe pump unless otherwise mentioned. [b] Isolated yield. [c] Determined by HPLC analysis. Configuration assigned by comparison to literature values of optical rotation. [d] 20 mol % of **123** and 80 mol % of the addtive were used. 1.2 equiv of TMSCN was used. [e] TMSCN (1.2 equiv) was added dropwise over 1 min. [f] 18 mol % of **123** and 72 mol % of the additive waer used. [g] The absolute configuration was not determined.

144

Table 13. Asymmetric Strecker-type reaction of imines catalyzed by **123**.

entry	R	145a-m	time (h)	yield (%)[b]	ee (%)[c]
1	Ph	a	44	92	95
2	p-ClPh	b	44	92	95
3	p-MeOPh	c	44	93	93
4	1-Naphthyl	d	68	95	88
5	2-furyl	e	44	93	79
6	3-furyl	f	44	92	90
7	(thienyl)	g	58	90	89
8	trans-PhCH=CH	h	41	80	96
9[d]	trans-CH₃(CH₂)₃CH=CH₂	i	24	66	86
10[e]	CH₃(CH₂)₅	j	24	80	80
11	CH₃CH₂	k	44	84	70
12	i-Pr	l	44	89	72
13	t-Bu	m	44	97	78

[a] PhOH (20 mol %) was added over 17 h via syringe pump unless otherwise mintioned. [b] Isolated yield. [c] Determined by HPLC analysis. [d] 50 mol % of PhOH was used. The aminonitrile was isolated as the corresponding trifluoroacetamide. [e] Without PhOH.

center catalysts. These include heterobimetallic transition metal catalysts, homobimetallic transition metal catalysts, early transition metal-late transition metal catalysts, and so on. Moreover, development of reusable asymmetric two-center catalysts are quite challenging research projects from now on.

References and Notes

1. (a) W. A. Herrmann, B. Cornils, Applied Homogeneous Catalysis with Organometallic Compounds, VCH: Weinheim, 1996. (b) Noyori, R., Asymmetric Catalysis in Organic Synthesis, John Wiley & Sons: New York, 1994. (c) I. Ojima, Catalytic Asymmetric Synthesis, VCH, New York, 1994. (d) B. Bosnich,Asymmetric Catalysis, Martinus Nijhoff Publishers: Dordrecht, 1986. (e) J. D. Morrison, Asymmetric Synthesis, Academic Press: Orland, 1985, Vol. 5.

2. M. Shibasaki, H. Sasai, T. Arai, Asymmetric Catalysis with Heterobimetallic Compounds, *Angew. Chem. Int. Ed. Engl.* **1997**, 36, 1236–1256.

3. H. Sasai, T. Suzuki, S. Arai, T. Arai, M. Shibasaki, Basic Character of Rare Earth Metal Alkoxides. Utilization in Catalytic C-C Bond-Forming Reactions and Catalytic Asymmetric Nitroaldol Reactions, *J. Am. Chem. Soc.* **1992**, 114, 4418–4420.

4. H. Sasai, S. Watanabe, M. Shibasaki, A New Practical Preparation Method for Lanthanum-Lithium-Binaphtol Catalysts (LLBs) for Use in Asymmetric Nitroaldol Reactions, *Enantiomer*, **1997**, 2, 267–271.

5. H. Sasai, T. Suzuki, N. Itoh, K. Tanaka, T. Date, K. Okamura, M. Shibasaki, Catalytic Asymmetric Nitroaldol Reaction Using Optically Active Rare Earth BINOL Complex: Investigation of the Catalyst Structure, *J. Am. Chem. Soc.* **1993**, 115, 10372–10373.

6. H. Sasai, T. Arai, Y. Satow, K. N. Houk, M. Shibasaki, The First Heterobimetallic Multifunctional Asymmetric Catalyst, *J. Am. Chem. Soc.* **1995**, 117, 6194–6198.

7. Purchased from Kojundo Chemical Laboratory Co. Saitama, Japan.

8. H. Sasai, T. Suzuki, N. Itoh, M. Shibasaki, Catalytic Asymmetric Synthesis of Propranolol and Metoprolol Using La-Li-BINOL Complex, *Appl. Organomet. Chem.* **1995**, 9, 421–426.

9. H. Sasai, T. Tokunaga, S. Watanabe, T. Suzuki, N. Itoh, M. Shibasaki, Efficient Diastereoselective and Enantioselective Nitroaldol Reactions from Prochiral Starting MaterialsStUtilization of La-Li-6,6'-Disubstituted BINOL Complexes as Asymmetric CatalystsUtJ. *Org. Chem.* **1995**, 60, 7388–7389.

10. For recent examples of catyltic asymmetric Mukayama-aldol reactions, see: (a) D. A. Evans, C. S. Burgey, N. A. Paras, T Vojkovsky, S. W. Tregay, D. A. Evans, C. S. Burgey, N. A. Paras, T. Vojkovsky, S. W. Tregay, C2-Symmetric Copper(II) Complexes as Chiral Lewis Acids. Enantioselective Catalysis of the Glyoxylate-Ene Reaction, *J. Am. Chem. Soc.* **1998**, 120, 5824–5825. (b) J. Krüger, E. M. Carreira, Apparent Catalytic Generation of Chiral Metal Enolates: Enantioselective Dienolate Additions to Aldehydes Mediated by Tol-BINAP·Cu(II) Fluoride Complexes, *J. Am. Chem. Soc.* **1998**, 120, 837–838. (c) A. Yanagisawa, Y. Matsumoto, H. Nakashima, K. Asakawa, H. Yamamoto, Enantioselective Aldol Reaction of Tin

Enolates with Aldehydes Catalyzed by BINAP·Silver(I) Complex, *J. Am. Chem. Soc.* 1997, 119, 9319–9320. (d) S. E. Denmark, K.-T. Wong, R. A. Stavenger, The Chirality of Trichlorosilyl Enolates. 2. Highly-Selective Asymmetric Aldol Additions of Ketone Enolates, *J. Am. Chem. Soc.* **1997**, 119, 2333–2334, and references cited therein.

11. Y. M. A. Yamada, N. Yoshikawa, H. Sasai, M. Shibasaki, Direct Catalytic Asymmetric Aldol Reactions of Aldehydes and Unmodified Ketones, *Angew. Chem. Int. Ed. Engl.* **1997**, 36, 1871–1873.

12. N. Yoshikawa, Y. M. A. Yamada, J. Das, H. Sasai, M. Shibasaki, Direct Catalytic Asymmetric Aldol Reaction, *J. Am. Chem. Soc.* **1999**, 121, 4168–4178.

13. (a) M. Bougauchi, S. Watanabe, T. Arai, H. Sasai, M. Shibasaki, Catalytic Asymmetric Epoxidation of α,β-Unsaturated Ketones Promoted by Lanthanoid Complexes, *J. Am. Chem. Soc.* **1997**, 119, 2329–2330. (b) S. Watanabe, Y. Kobayashi, T. Arai, H. Sasai, M. Bougauchi, M. Shibasaki, Water vs. Desiccant. Improvement of Yb-BINOL Complex Catalyzed Enantioselective Epoxidation of Enones, *Tetrahedron Lett.* **1998**, 39, 7353–7356. (c) S. Watanabe, T. Arai, H. Sasai, M. Bougauchi, M. Shibasaki, The First Catalytic Enatioselective Synthesis of cis-Epoxyketones from cis-Enones, *J. Org. Chem.* **1998**, 63, 8090–8091.

14. (a) H. Gröger, Y. Saida, S. Arai, J. Martens, H. Sasai, M. Shibasaki, First Catalytic Asymmetric Hydrophosphonylation of Cyclic Imines: Highly Efficient Enantioselective Approach to a 4-Thiazolidinylphosphonate via Chiral Titanium and Lanthanoid Catalysis, *Tetrahedron Lett.* **1996**, 37, 9291–9292. (b) H. Gröger, Y. Saida, H. Sasai, K. Yamaguchi, J. Martens, M. Shibasaki, A New and Highly Efficient Asymmetric Route to Cyclic α-Amino Phosphonates: The First Catalytic Enantioselective Hydrophosphonylation of Cyclic Imines Catalyzed by Chiral Heterobimetallic Lanthanoid Complexes, *J. Am. Chem. Soc.* **1998**, 120, 3089–3103.

15. E. Emori, T. Arai, H. Sasai, M. Shibasaki, A Catalytic Michael Addition of Thiols to α,β-Unsaturated Carbonyl Compounds: Asymmetric Protonations, *J. Am. Chem. Soc.* **1998**, 120, 4043–4044.

16. K. Yamada, S. J. Harwood, H. Gröger, M. Shibasaki, The First Catalytic Asymmetric Nitro-Mannich-Type Reaction Promoted by a New Heterobimetallic Complex, *Angew. Chem.* ,**1999**, 38, 3504–3506.

17. T. Arai, H. Sasai, K. Aoe, K. Okamura, T. Date, M. Shibasaki, A New Multifunctional Heterobimetallic Asymmetric Catalyst for Michael Additions and Tandem Michael-Aldol Reactions, *Angew. Chem. Int. Ed. Engl.* **1996**, 35, 104–106.

18. S. Shimizu, K. Ohori, T. Arai, H. Sasai, M. Shibasaki, A Catalytic Asymmetric Synthesis of Tubifolidine, *J. Org. Chem.* **1998**, 63, 7547–7551.

19. S. Nara, H. Toshima, A. Ichihara, Asymmetric Total Syntheses of (+)–Coronafacic Acid and (+)–Coronatine, Phytotoxins Isolated from Pseudomonas Syringae Pathovars, *Tetrahedron*, **1997**, 53, 9509–9524.

20. K. Yamada, T. Arai, H. Sasai, M. Shibasaki, A Catalytic Asymmetric Synthesis of 11-Deoxy-PGF1α Using ALB, a Heterobimetallic Multifunctional Asymmetric Complex, *J. Org. Chem.* **1998**, 63, 3666–3672.

21. M. Tokunaga, J. F. Larrow, F. Kakiuchi, E. N. Jacobsen, Asymmetric Catalysis with Water: Efficient Kinetic Resolution of Terminal Epoxides by Means of Catalytic Hydrolysis, *Science* **1997**, 277, 936–938, and references cited therein.

22. T. Iida, N. Yamamoto, H. Sasai, M. Shibasaki, New Asymmetric Reactions Using a Gallium Complex: A Highly Enantioselective Ring Opening of Epoxides with Thiols Catalyzed by a Gallium · Lithium · Bis(binaphthoxide) Complex, *J. Am. Chem. Soc.* **1997**, *119*, 4783–4784.

23. D. A. Evans, K. R. Campos, J. S. Tedrow, F. E. Michael, M. R. Gagné, Chiral Mixed Phosphorus/Surfur Ligands for Palladium-Catalyzed Allylic Alkylations and Aminations, *J. Org. Chem.* **1999**, *64*, 2994–2995.

24. T. Iida, N. Yamamoto, N. Matsunaga, H.-G. Woo, M. Shibasaki, Enantioselective Ring Opening of Epoxides with 4-Methoxyphenol Catalyzed by Gallium Heterobimetallic Complexes: An Efficient Method for the Synthesis of Optically Active 1,2-Diol Monoethers, *Angew. Chem. Int. Ed. Engl.* **1998**, *37*, 2223–2226.

25. S. Matsunaga, J. Das, J. Roels, E. M. Vogl, N. Yamamoto, T. Iida, K. Yamaguchi, M. Shibasaki, Catalytic Enantioselective meso-Epoxide Ring Opening Reaction with Phenolic Oxygen Nucleophile Promoted by Gallium Heterobimetallic Multifunctional Complexes, *J. Am. Chem. Soc.* in press.

26. (a) E. J. Corey, R. K. Bakshi, S. Shibata, Highly Enantioselective Borane Reduction of Ketones Catalyzed by Chiral Oxazaborolidines. Mechanism and Synthetic Implications, *J. Am. Chem. Soc.* **1987**, *109*, 5551–5553. (b) R. Noyori, M. Kitamura, Enantioselective Addition of Organometallic Reagents to Carbonyl Compounds: Chirality Transfer, Multiplication, and Amplification, *Angew. Chem., Int. Ed. Engl.* **1991**, *30*, 49–69. (c) S. Kobayashi, Y. Tsuchiya, T. Mukaiyama, Enantioselective Addition Reaction of Trimethylsilyl Cyanide with Aldehydes Using a Chiral Tin(II) Lewis Acid, *Chem. Lett.* **1991**, 541–544.

27. For other catalytic asymmetric cyanosilylation of aldehydes, see: C.-D. Hwang, D.-R. Hwang, B.-J. Uang, Enantioselective Addition of Trimethylsilyl Cyanide to Aldehydes Induced by a New Chiral Ti(IV) Complex, *J. Org. Chem.* **1998**, *63*, 6762–6763, and references cited therein.

28. Y. Hamashima, D. Sawada, M. Kanai, M. Shibasaki, A New Bifunctional Asymmetric Catalysis: An Efficient Catalytic Asymmetric Cyanosilylation of Aldehydes, *J. Am. Chem. Soc.* **1999**, *121*, 2641–2642.

29. For catalytic asymmetric Strecker-type reactions, see: (a) M. S. Iyer, K. M. Gigstad, N. D. Namdev, M. Lipton, Asymmetric Catalysis of the Strecker Amino Acid Synthesis by a Cyclic Dipeptide, *J. Am. Chem. Soc.* **1996**, *118*, 4910–4911. (b) M. S. Sigman, E. N. Jacobsen, Schiff Base Catalysts for the Asymmetric Strecker Reaction Identified and Optimized from Parallel Synthetic Libraries, *J. Am. Chem. Soc.* **1998**, *120*, 4901–4902. (c) M. S. Sigman, E. N. Jacobsen, Enantioselective Addition of Hydrogen Cyanide to Imines Catalyzed by a Chiral (Salen)Al(III) Complex, *J. Am. Chem. Soc.* **1998**, *120*, 5315–5316. (d) H. Ishitani, S. Komiyama, S. Kobayashi, Catalytic Enantioselective Synthesis of α-Aminonitriles with a Novel Zirconium Catalyst, *Angew. Chem., Int. Ed. Engl.* **1998**, *37*, 3186–3188. (e) C. A. Krueger, K. W. Kuntz, C. D. Dzierba, W. G. Wirschun, J. D. Gleason, M. L. Snapper, A. H. Hoveyda, Ti-Catalyzed Enantioselective Addition of Cyanide to Imines. A Practical Synthesis of Optically Pure α-Amino Acids, *J. Am. Chem. Soc.* **1999**, *121*, 4284–4285. (f) E. J. Corey, M. J. Grogan, Enantioselective Synthesis of α-Amino Nitriles from N-Benzhydryl Imines and HCN with a Chiral Bicyclic Guanidine as Catalyst, *Org. Lett.* **1999**, *1*, 157–160.

30. M. Takamura, Y. Hamashima, H. Usuda, M. Kanai, M. Shibasaki, A Catalytic Asymmetric Strecker-type Reaction: Interesting Reactivity Difference between TMSCN and HCN, *Angew. Chem. Int. Ed. Engl.* **2000**, *39*, 1650–1652.

Asymmetric Phase Transfer Catalysis

Takayuki Shioiri and Shigeru Arai

Faculty of Pharmaceutical Sciences, Nagoya City University, Nagoya 467-8603, Japan

Phone: +81 52 836 3439, Fax: +81 52 834 4172, e-mail: shioiri@phar.nagoya-cu.ac.jp

Keywords: ■ Phase Transfer ■ Quaternary Ammonium Salts ■ Crown Ethers ■ Enantiomeric Excess (ee) ■ Enantioselective Synthesis ■ Catalytic Reactions

Concept: Phase transfer catalysis (PTC)[1] is now a convenient and useful tool in chemistry, especially in preparative organic chemistry. In general, compounds (reactants) located in different phases of a reaction mixture such as water and benzene sluggishly react each other even by harsh stirring the mixture because the reactants can not easily contact together. Phase transfer catalysts transfer between different phases, become highly active species, and catalytically mediate desired reactions. The common catalysts include the salts of onium (ammonium, phosphonium, and perhaps arsonium) cations or neutral complexes for inorganic cations (crown ethers, polyethers, polyols, etc.), and quaternary ammonium salts are used in many cases. When chiral non-racemic phase transfer catalysts are employed in reactions producing new stereogenic centers, reactions may proceed stereoselectively to give chiral non-racemic products.

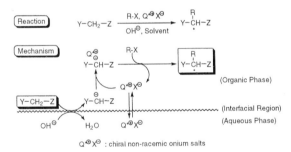

$Q^{\oplus}X^{\ominus}$: chiral non-racemic onium salts

Abstract: Phase transfer catalysts including onium salts or crown ethers transfer between heterogeneous different phases and catalytically mediate desired reactions. Chiral non-racemic phase transfer catalysts are useful for reactions producing new stereogenic centers, giving chiral non-racemic products. Recent developments in this rapid expanding area will be presented.

Prologue

C. M. Starks at Continental Oil Company in U. S. A. reported as followed almost thirty years ago[2] : "1-Chlorooctane **1** did not react with sodium cyanide in a mixture of water and decane even by heating for 3 h. However, addition of hexadecyltributylphosphonium bromide **2** caused the desired displacement reaction to give 1-cyanooctane **3** in 99 % yield after 1.8 h heating." He reported further: "Reaction between 1-octene **4** and aqueous neutral potassium permanganate was not observed at room temperature even after vigorous stirring for several hours. Addition of tricaprylmethylammonium chloride **5**, however, caused immediate oxidation of 1-octene **4** by evolution of so much heat that the reaction mixture could not contain in the flask. When the temperature of the reaction mixture was controlled to 35 °C by slow addition of 1-octene **4** to a mixture of potassium permanganate, water and a 5 % solution of the quaternary ammonium salt **5** in benzene, hexanoic acid **6** was obtained in essentially quantitative yield."

These two phase reactions summarized in Scheme 1 are strongly promoted by catalytic amounts of the quaternary onium salts which will transfer between aqueous and organic phases, as shown in Scheme 2 for the substitution of halides with cyanides. The effect was termed by Starks

$$C_8H_{17}Cl \xrightarrow[\text{decane} / H_2O]{\overset{\text{NaCN}}{C_{16}H_{33}P^+Bu_3Br^- \text{ (2)}}} C_8H_{17}CN$$

1   **3**

$$C_6H_{13}CH{=}CH_2 \xrightarrow[\text{benzene} / H_2O]{\overset{\text{KMnO}_4}{MeN^+(C_8H_{17})_3Cl^- \text{ (5)}}} C_6H_{13}CO_2H$$

4   **6**

Scheme 1. Representative phase transfer catalysis.

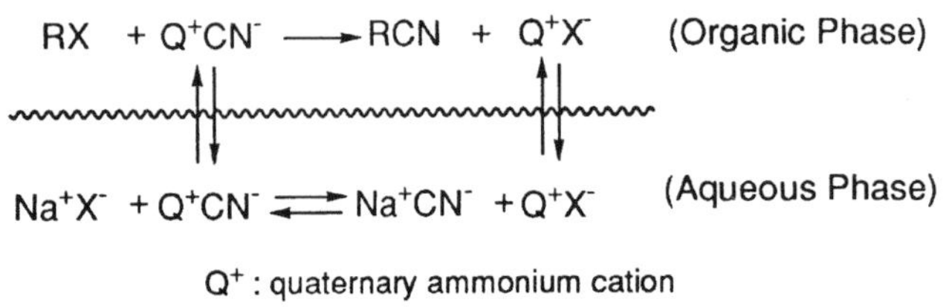

$$RX + Q^+CN^- \longrightarrow RCN + Q^+X^- \quad \text{(Organic Phase)}$$

$$Na^+X^- + Q^+CN^- \rightleftharpoons Na^+CN^- + Q^+X^- \quad \text{(Aqueous Phase)}$$

Q^+ : quaternary ammonium cation

Scheme 2. Starks' extraction mechanism.

"Phase Transfer Catalysis." Before Starks' work, there have been some reports about analogous phenomena. The foundations of phase transfer catalysis, however, were laid by Starks together with M. Makosza and A. Brändström in the mid to late 1960s. Since then, phase transfer catalysis has been very quickly developed and now becomes an indispensable tool in chemistry.[3–6]

Characteristics and Present Status of Phase Transfer Catalysis

Phase transfer catalysis has been recognized as a practical methodology for organic synthesis due to its operational simplicity, high yield processes, mild reaction conditions, use of safe and inexpensive reagents and solvents, safety considerations, environmental concerns, and possibility to conduct reactions on large scale. Thus, phase transfer catalysis is now a basic and useful tool in various organic chemistry such as heterocyclic chemistry, polymer chemistry, organometallic chemistry and so on. Furthermore, since phase transfer catalysis will increase the reaction rate and replace, reduce or eliminate solvents, it has a wide industrial application especially in pharmaceutical and agrochemical manufacturing as well as perfumes, flavors, and dyes industry. In addition, phase transfer catalysis has a vast spectrum of analytical applications in medicinal, forensic, and other laboratories. If the desired reactions sluggishly proceed, addition of a small amount of phase transfer catalysts to the reaction mixture will be recommended since the phase transfer catalysts might accelerate the reactions. Thus, phase transfer catalysis in general has now become a standard method in synthesis and it might be considered to be approaching maturity.[3–6]

In contrast, the progress of asymmetric synthesis by use of chiral non-racemic phase transfer catalysts had been slow compared to the ordinary phase transfer catalysis. However, recent achievements in this particular area are noteworthy and efficient asymmetric phase transfer catalysis has been increasingly explored.[7–10]

General Mechanism of Phase Transfer Catalysis

The mechanism of phase transfer catalysis is still a matter of discussion and remains a subject of some controversy. However, it will be roughly classified into two: the extraction mechanism proposed by Starks[2] and the interfacial mechanism by Makosza.[11]

Normal phase transfer catalysis conducted under neutral conditions such as the displacement of halides with cyanides (Scheme 1) generally follows the extraction mechanism, as shown in Scheme 2. In the extraction mechanism, the reaction takes place in the organic phase and the inorganic anion is transferred to the organic phase. The catalyst is necessary for both the reaction step and the transfer step. On the other hand, phase transfer catalysis conducted in the presence of strong bases will follow the interfacial mechanism in which the deprotonation of the CH organic acids with hydroxide anions takes place at the aqueous-organic interfacial region without phase transfer catalysts, as shown in Scheme 3. However, some reactions can be better explained by the assistance of the phase transfer catalysts in the deprotonation at the interfacial region, as shown in Scheme 4. This is the interfacial mechanism modified by Liotta.[11]

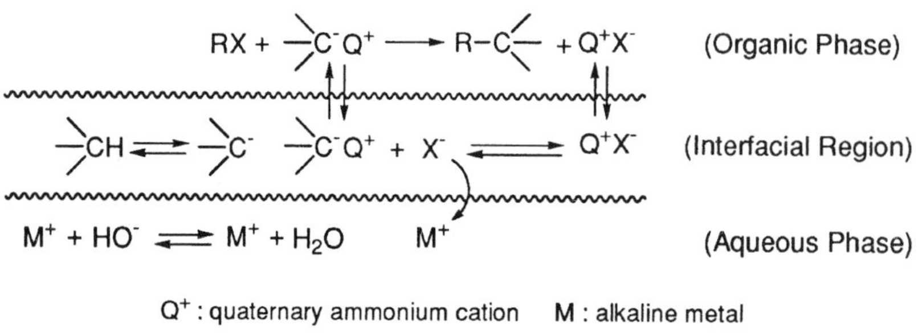

$$RX + {>}C^-Q^+ \longrightarrow R-C{<} + Q^+X^- \quad \text{(Organic Phase)}$$

$${>}CH \rightleftharpoons {>}C^- \quad {>}C^-Q^+ + X^- \rightleftharpoons Q^+X^- \quad \text{(Interfacial Region)}$$

$$M^+ + HO^- \rightleftharpoons M^+ + H_2O \quad M^+ \quad \text{(Aqueous Phase)}$$

Q^+ : quaternary ammonium cation   M : alkaline metal

Scheme 3. Makosza's interfacial mechanism.

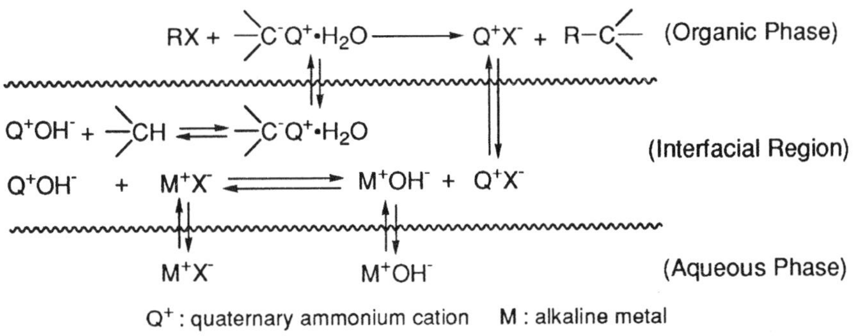

Scheme 4. Interfacial mechanism modified by Liotta.

Q^+ : quaternary ammonium cation   M : alkaline metal

Phase transfer catalysis can be conducted under liquid-liquid conditions, liquid-solid conditions, or liquid-liquid-solid triphasic conditions.

Chiral Non-racemic Phase Transfer Catalysts

Cinchona alkaloids now occupy the central position in designing the chiral non-racemic phase transfer catalysts because they have various functional groups easily derivatized and are commercially available with cheap price. The quaternary ammonium salts derived from cinchona alkaloids as well as some other phase transfer catalysts are

shown in Figure 1. The exact relationship between the onium salts derived from cinchonine (or quinidine) and cinchonidine (or quinine) is diastereomeric, but their amino alcohol parts, the key parts for asymmetric induction, are enantiomeric. Thus, the relationship between **7** and **9** (**8** and **10**; **11** and **12**) is called "pseudoenantiomeric". Typically, if one family of catalysts gives (S)-enantiomer as the major product, the other family yields the antipodal (R)-enantiomer in excess. The ammonium salts derived from the other chiral alkaloids and amines also serve as chiral phase transfer catalysts. These ammonium salts are prone to undergo the Hofmann degradation under strong alkaline conditions, hence it should be careful to evaluate the

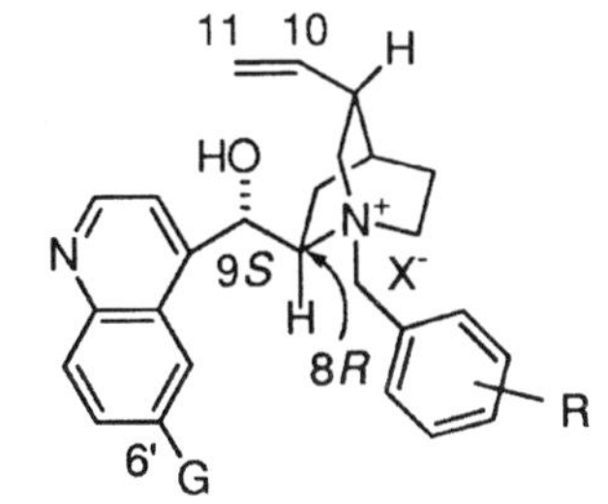

Figure 1. Representative chiral non-racemic phase transfer catalysts.

7 (G=H, derived from cinchonine)
8 (G=MeO, derived from quinidine)

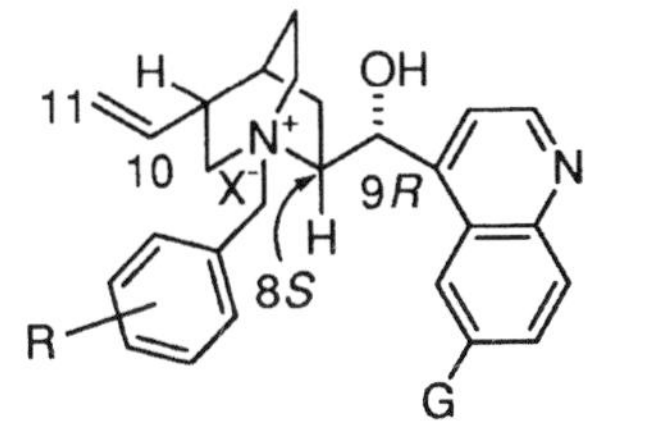

9 (G=H, derived from cinchonidine)
10 (G=MeO, derived from quinine)

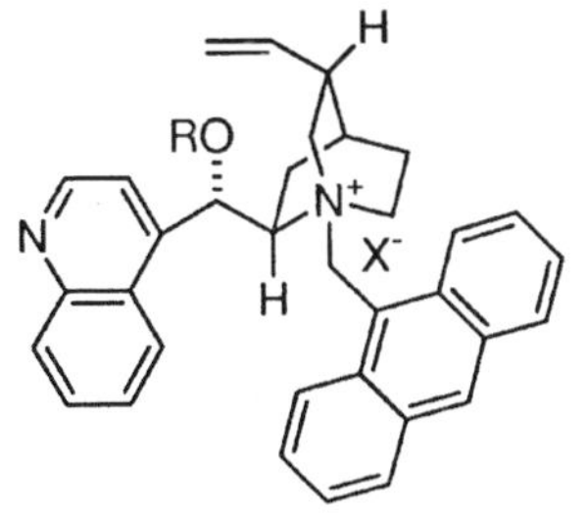

11 (derived from cinchonine)

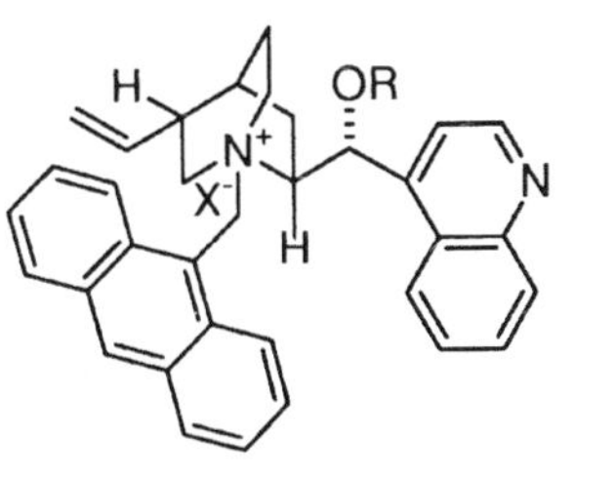

12 (derived from cinchonidine)

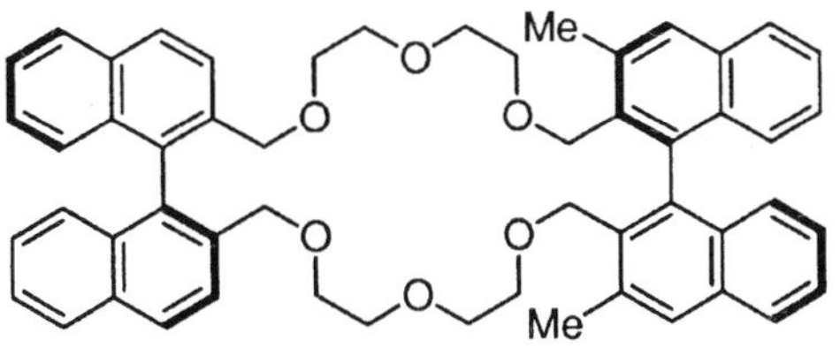

13

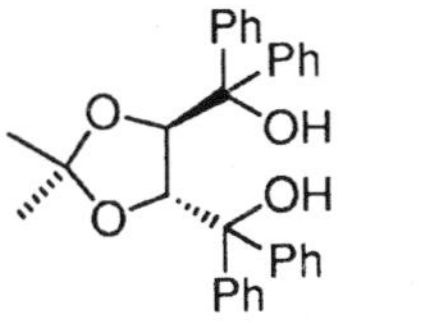

14 [(R,R)-TADDOL]

enantiomeric excess (ee) whether the products are contaminated with these degradation products from the onium salts.[12]

Chiral crown ethers such as **13** are suitable alternatives to the ammonium salts and not decomposed under alkaline conditions. They usually have higher catalyst turnover than the chiral ammonium salts, and the design of catalysts will be much easier. However, they are, in general, costly and difficult to prepare on large scale. Polyols (e.g., (*R,R*)-TADDOL **14**) also serve as phase transfer catalysts.

Effective stirring is generally essential to obtain rapid reactions in heterogeneous reactions while levels of asymmetric induction are not affected. Furthermore, it should be noted that the enantiomeric excess of the reactions should be determined by direct analysis of product mixtures or their derivatives by use of chiral high performance liquid chromatography, chiral gas chromatography or NMR spectroscopy with chiral shift reagents. Determination of optical purity by measuring optical rotation sometimes affords erroneous results and lacks accuracy.[13]

Phase Transfer Reactions

There are many examples on the asymmetric phase transfer catalysis, but highly efficient ones are not so many though they are increasing in recent years.[7-10] This review will highlight the notable examples with emphasis on recent reports.

Carbon–Carbon Bond Formation

Alkylation of Cyclic Ketones

The first practical and efficient asymmetric alkylation by use of chiral phase-transfer catalysts was the alkylation of the phenylindanone **15** (R[1]=Ph), reported by the Merck research group in 1984.[14-16] By use of the quaternary ammonium salt **7** (R=4-CF₃, X=Br) derived from cinchonine, the alkylated products **16** were obtained in excellent yield with high enantiomeric excess, as shown in

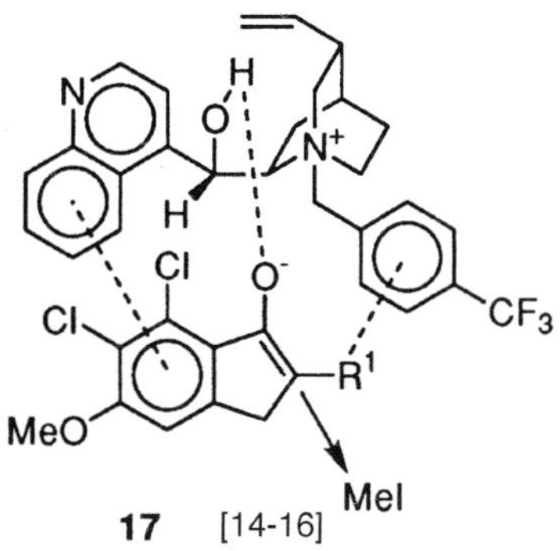

17   [14-16]

Figure 2. The tight ion pair **17** in the asymmetric alkylation of the cyclic ketones.

Scheme 5. An intermediacy of a tight ion pair **17** fixed by an electrostatic effect and hydrogen bonding as well as π–π stacking interactions was proposed to account for the result (Figure 2).[16]

α-Fluorotetralone **18** also undergoes the asymmetric alkylation under phase transfer catalyzed conditions using the quaternary ammonium salts **7** derived from cinchonine.[17] Variation of the arylmethyl groups attached at the quaternary ammonium center revealed that the electron donating functions in the benzene nucleus afforded much better results than the electron withdrawing functions. Thus, the 2,3,4,5,6-pentamethylphenylmethyl derivative **7** (R=2,3,4,5,6-Me₅, X=Br) with potassium hydroxide in toluene gave the corresponding arylmethyl derivatives **19** with 70–91% ee, shown in Scheme 6. Interestingly, α-methyltetralone was found to be an ineffective substrate for the analogous phase catalyzed asymmetric alkylation (up to 55% ee). Requirement of the higher planarity of the active enolate was suggested for the effective alkylation.

The phosphonium salt **21** having a multiple hydrogen-bonding site which would interact with the substrate anion was applied to the phase transfer catalyzed asymmetric benzylation of the β-keto ester **20**,[18, 19] giving the benzylated β-keto ester **22** in 44% yield with 50% ee, shown in Scheme 7. Although the chemical yield and enantiomeric excess remain to be improved, the method will suggest a new approach to the design of chiral non-racemic phase transfer catalysts.

R[1]=Ph, R[2]=Me : 98%, 94% ee[14, 16]
R[1]=Pr[n], R[2]=MeC(Cl)=CHCH₂ : 99%, 92% ee[15]

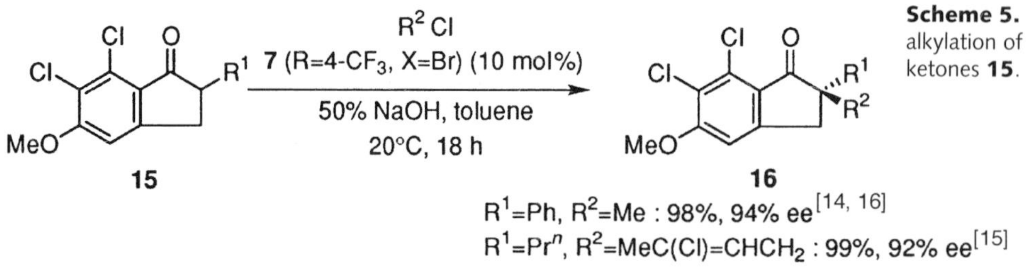

Scheme 5. Asymmetric alkylation of the cyclic ketones **15**.

Scheme 6. Asymmetric alkylation of α-fluorotetralone **18**.

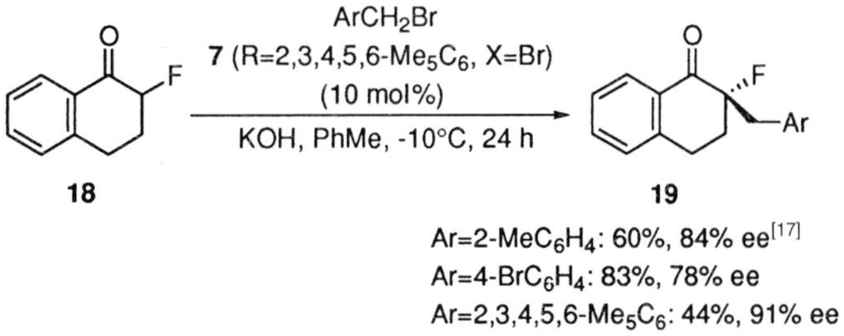

ArCH₂Br
7 (R=2,3,4,5,6-Me₅C₆, X=Br)
(10 mol%)
KOH, PhMe, -10°C, 24 h

Ar=2-MeC₆H₄: 60%, 84% ee[17]
Ar=4-BrC₆H₄: 83%, 78% ee
Ar=2,3,4,5,6-Me₅C₆: 44%, 91% ee

Scheme 7. Asymmetric benzylation of the β-keto ester **20**.

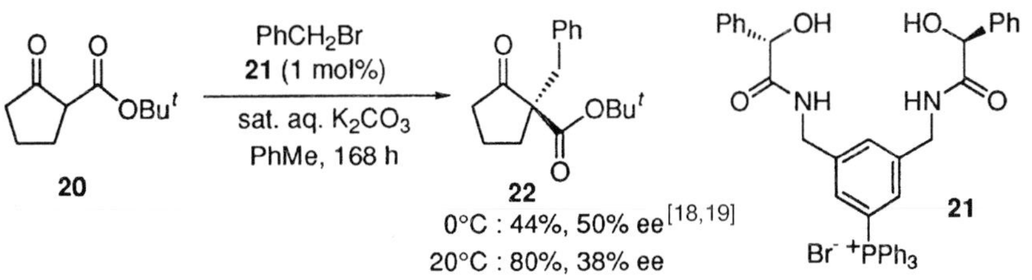

PhCH₂Br
21 (1 mol%)
sat. aq. K₂CO₃
PhMe, 168 h

0°C : 44%, 50% ee[18,19]
20°C : 80%, 38% ee

Alkylation of Imines of Glycine Esters

One of the most successful use of asymmetric phase transfer catalysis to date is the preparation of optically active amino acid derivatives **24** (and **25**) by asymmetric alkylation of the benzophenone imine of glycine ester **23**, the O'Donnell imine. The first promising result in this area was reported by O'Donnell and co-workers in 1989,[20] who obtained the alkylated products **24** by alkylation of the O'Donnell imine **23** using cinchonine and cinchonidine catalysts with up to 66% ee. Detailed studies of the alkylation revealed that an active catalyst will be the N,O-dialkyl-cinchona salt formed *in situ* during the reaction.[21] Thus, the asymmetric benzylation of **23** by use of the O-allyl cinchoninium catalyst proceeded at 5°C with rapid stirring (e.g., 2000 rpm) to give **24** in 87% yield with 81% ee.[22] Introduction of the bulkier N-anthracenylmethyl group at the nitrogen substituent in the quaternary ammonium salts derived from cinchona alkaloids has now opened a new era of asymmetric phase transfer catalysis.[23, 24] The Lygo group [23] used liquid-liquid phase transfer conditions (50% aq. KOH in toluene) with the catalysts **11** (R=H, X=Cl) and **12** (R=H, X=Cl) at room temperature while the Corey group[24] employed solid-liquid phase transfer conditions (CsOH·H₂O in CH₂Cl₂) at –78°C with the catalyst **12** (R=allyl, X=Br). The both groups achieved the benzylation of the O'Donnell imine **23** in good yield with efficient enantioselectivity, as shown in Table 1. The N-anthracenylmethyl derivatives **11** (R=H, X=Cl) and **12** (R=H, X=Cl) prepared from cinchonine and cinchonidine, respectively, show a comparable efficiency to give amino acid derivatives **24** antipodal to each other.[23] It should be noted that the phase transfer catalyst **12** (R=allyl,

X=Br) can be efficiently recovered from aqueous solution of the work-up for reuse.[24] Utilizing this N-anthracenylmethyl technology, various amino acid derivatives have been synthesized by Lygo and co-workers in 40–86% yields with 67–94% ee as the N-free *tert*-butyl esters **25** and by Corey and co-workers in 67–91% yields with 92–99.5% ee as the imines **24**, respectively. Some representative results are shown in Table 2.

One of the interesting application of **12** (R=allyl, X=Br) will be the synthesis of cyclic amino acid, (S)-pipecolic acid, as its *tert*-butyl ester **27**.[25] Monoalkylation of the O'Donnell imine **23** with 1-chloro-4-iodobutane afforded the alkylated product **26** with 99% ee. The conversion of **26** to the *tert*-butyl ester of pipecolic acid **27** was achieved in high yield by the sequence imine reduction, cyclization, and hydrogenolytic removal, as shown in Scheme 8.

The bulky N-anthracenylmethyl group clearly plays a key role in the efficient stereoselective alkylation. The steric screening would be provided by the N-anthracenylmethyl group and the tight ion pair **28** from the ammonium cation (N⁺) and the enolate (O⁻) would be stereoselectively alkylated as shown in Figure 3.[24] Furthermore, removal of the quinoline ring proved to be not useful to attain the good enantioselectivity.[26]

O'Donnell and co-workers devised a method of the homogeneous catalytic asymmetric alkylation of the imine **23** by use of the organic soluble, nonionic phosphazene (Schwesinger) bases, BEMP and BTPP, in conjunction with the Corey's ammonium catalyst **12** (R=allyl, X=Br) or its pseudoenantiomer **11** (R=allyl, X=Br).[27] This novel homogeneous phase transfer catalytic process avoids the efficient stirring crucial for accelerating the heterogeneous reactions using the organic insoluble, alkaline hydrox-

Table 1. Asymmetric benzylation of the O'Donnell imine **23**.

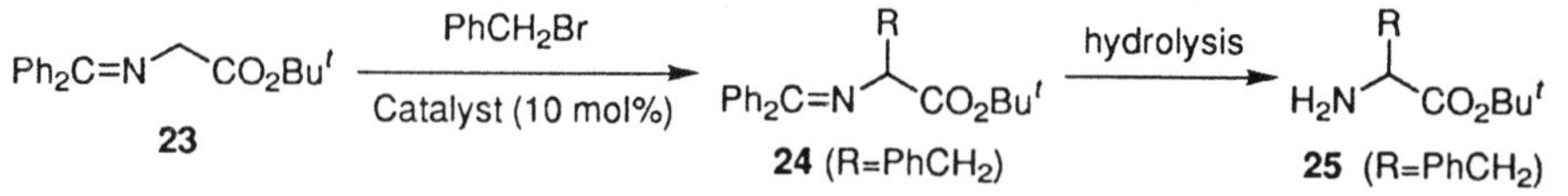

Catalyst	base	solvent	Temp. (°C)	Time (h)	Yield (%) [a]	%ee	Major isomer	Ref
7 (R=H, X=Cl)	50% aq. NaOH	CH$_2$Cl$_2$	rt	9	75	66	*R*	20
9 (R=H, X=Br, 10,11-dihydro)	50% aq. NaOH	CH$_2$Cl$_2$ toluene	5	0.5	87 [b]	81	*S*	22
11 (R=H, X=Cl)	50% aq. KOH	toluene	rt	18	63 [c] [d]	89	*R*	23
12 (R=H, X=Cl 10,11-dihydro)	50% aq. KOH	toluene	rt	18	85 [c] [d]	94	*S*	23
12 (R=allyl, X=Br)	CsOH·H$_2$O	CH$_2$Cl$_2$	−78	23	87 [e]	94	*S*	24
11 (R=allyl, X=Br)	BEMP [f]	CH$_2$Cl$_2$	−78	7	89	83	*R*	27
12 (R=allyl, X=Br)	BEMP [f]	CH$_2$Cl$_2$	−78	7	88	91	*S*	27
29	50% aq. KOH	toluene	0	0.5	95 [g]	96	*R*	28

a) Yield of **24**.
b) Rapidly stirred (2000 rpm).
c) Yield of **25**.
d) Rapidly stirred (1000 rpm).
e) An excess (5 equiv) of benzyl bromide was used.
f) The structure of BEMP is shown in Table 2.
g) One mol% of the catalyst was used.

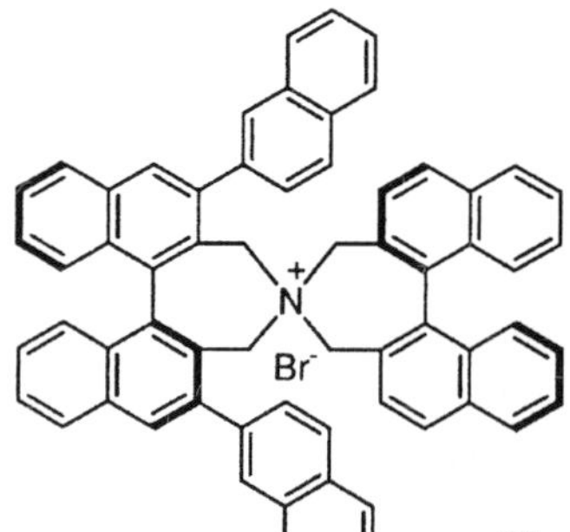

29

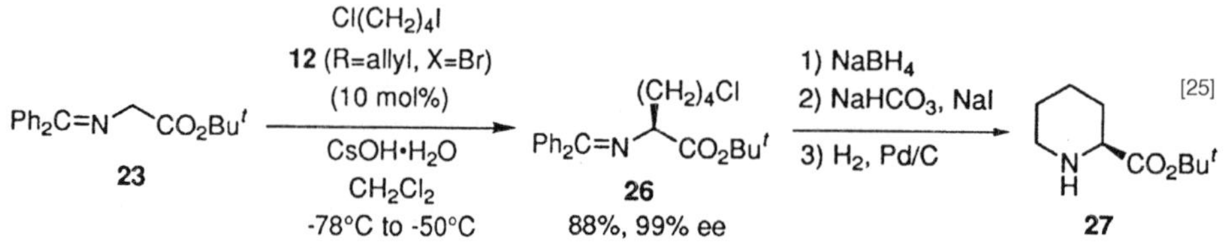

Scheme 8. Asymmetric synthesis of the pipecolic acid derivative **27**.

ides such as CsOH. As shown in Table 1, the benzylation of **23** in the presence of BEMP with either **11** (R=allyl, X=Br) or **12** (R=allyl, X=Br) proceeded smoothly and promptly to give **24** (R=benzyl) with good enantioselectivity. The other alkylation also efficiently proceeded using BEMP for active halides and BTPP for non-active halides, respectively, to give chiral α-amino acid precursors **24**, as summarized in Table 2.

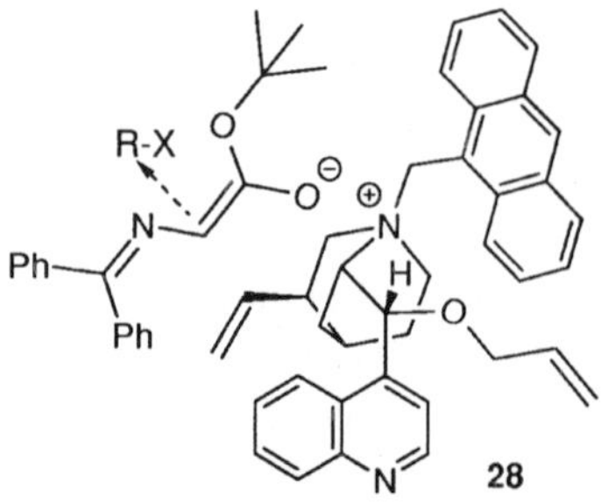

The ingeniously designed chiral non-racemic phase transfer catalyst **29** derived from binaphthol

Figure 3. The tight ion pair **28** in the asymmetric alkylation of the O'Donnell imine **23**.

Table 2. Asymmetric alkylation of the O'Donnell imine **23**.

$$Ph_2C=N\diagup CO_2Bu^t \xrightarrow[\text{Catalyst (10 mol\%)}]{RX} Ph_2C=N\diagdown CO_2Bu^t \xrightarrow{\text{hydrolysis}} H_2N\diagdown CO_2Bu^t$$

23   **24**   **25**

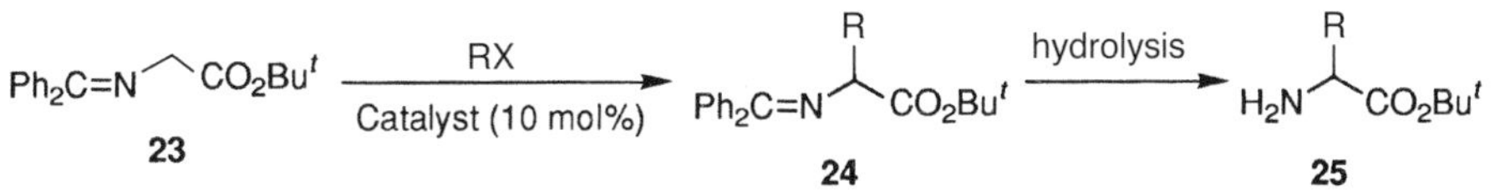

RX	Catalyst	Base / Solvent	Temp. (°C)	Time (h)	Yield (%)[a]	%ee	Major isomer	Ref
MeI	**12** (R=H, X=Cl)	50% aq. KOH / toluene	rt	3	41 [b]	89	*S*	23
	12 (R=allyl, X=Br)	CsOH·H₂O / CH₂Cl₂	−60	28	71 [c]	97	*S*	24
	12 (R=allyl, X=Br)	BEMP / CH₂Cl₂	−78	4	92	94	*S*	27
	29	50% aq. KOH / toluene	0	8	64 [d]	90	*R*	28
Allyl Br	**12** (R=H, X=Cl)	50% aq. KOH / toluene	rt	18	76 [b]	88	*S*	23
	12 (R=allyl, X=Br)	CsOH·H₂O / CH₂Cl₂	−78	22	89 [c]	97	*S*	24
	12 (R=allyl, X=Br)	BEMP / CH₂Cl₂	−78	6	96	90	*S*	27
	29	50% aq. KOH / toluene	0	1	84 [d]	94	*R*	28
BuI	**11** (R=H, X=Cl)	50% aq. KOH / toluene	rt	18	56 [b]	87	*R*	23
Hexyl I	**12** (R=allyl, X=Br)	CsOH·H₂O / CH₂Cl₂	−60	32	79 [c]	99.5	*S*	24
Octyl I	**12** (R=allyl, X=Br)	BTPP / CH₂Cl₂	−50	3.5	83	93	*S*	27

a) Yield of **24**. b) Yield of **25**. c) An excess (5 equiv) of alkyl halide was used. d) One mol% of the catalyst was used.

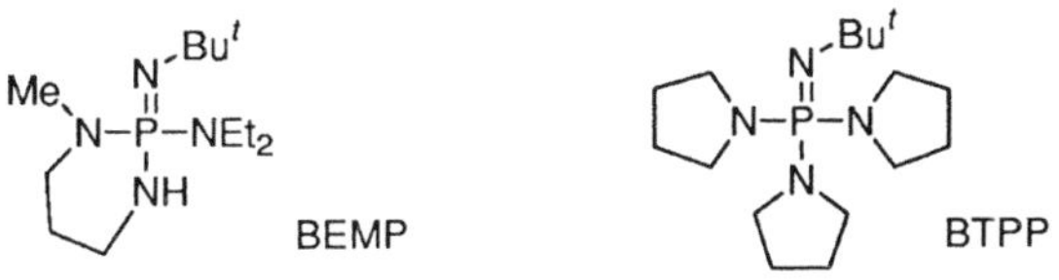

BEMP

BTPP

proved also effective in the asymmetric alkylation of the O'Donnell imine **23**.[28] It is noteworthy that the use of 1 mol% catalyst is sufficient to conduct the benzylation with good chemical yield and enantiomeric excess even at 0 °C for only 0.5 h. On the basis of the experimental findings as well as X-ray structure of the related catalyst, the authors proposed a reaction model in which the binaphthyl and the β-naphthyl moieties of the spiro ammonium salt **29** would effectively shield the *si*-face of the *E*-enolate of the O'Donnell imine **23** suitably situated in the molecular pocket of the catalyst **29**. Thus, alkyl halides could only approach the *re*-face of the enolate to give the *R*-isomer. The method is

also applicable to the preparation of various amino acid derivatives in 41–95 % yield with 90–96 % ee, some examples of which are shown in Table 2.

The method of the above homogeneous catalytic asymmetric alkylation utilizing Schwesinger base was applied to the solid phase synthesis of α-amino acid derivatives.[29] The Wang resin-bound derivative **30** of glycine Schiff base ester was alkylated at −78°C in the presence of the quaternary ammonium salt **11** (R=allyl, X=Br) or **12** (R=allyl X=Br) using the phosphazene bases, BEMP or BTTP, to give either enantiomer of the products α-amino acids **32** with 51–89 % ee, as shown in Scheme 9. Although the optimal conditions involve a full equivalent of

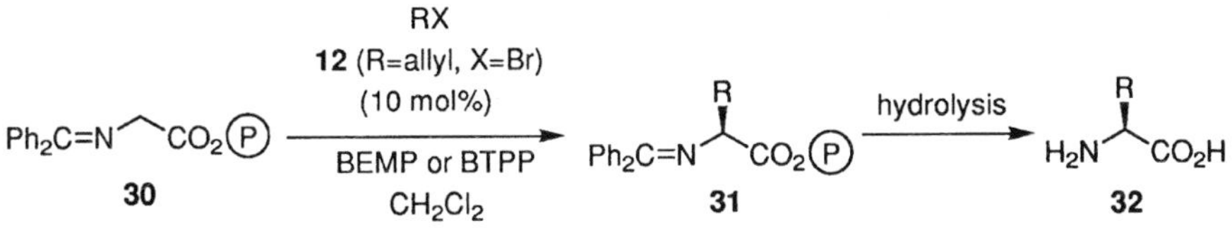

$$Ph_2C=N\diagup CO_2\text{(P)} \xrightarrow[\substack{\text{12 (R=allyl, X=Br)} \\ \text{(10 mol\%)} \\ \text{BEMP or BTPP} \\ \text{CH}_2\text{Cl}_2}]{RX} Ph_2C=N\diagdown CO_2\text{(P)} \xrightarrow{\text{hydrolysis}} H_2N\diagdown CO_2H$$

30   **31**   **32**

RX	Base	Temp. (°C)	Yield (%)	% ee
MeI	BEMP	-78	99	86
Octyl I	BTPP	-40	99	80
CH₂=C(Me)CH₂Br	BEMP	-78	94	89

[29] (P) : Wang resin

Scheme 9. Asymmetric solid phase synthesis of the α-amino acid **32**.

Table 3. Asymmetric synthesis of the *bis*-α-amino acid esters **35**.

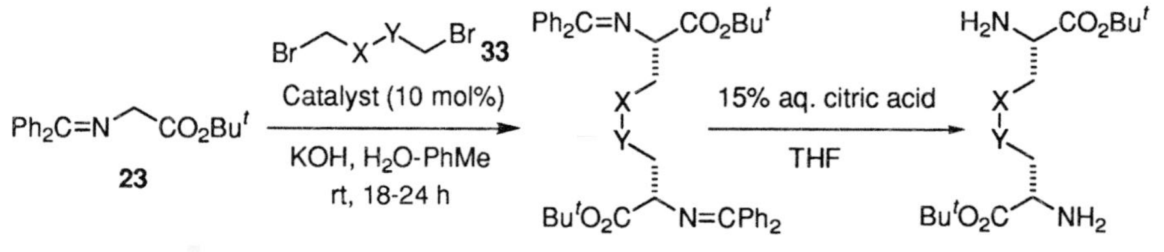

Catalyst	Br⌢X–Y⌣Br **33**	Amino Acid Ester **35**	Yield (%) of **35**	% de	% ee	Ref
12 (R=H, X=Br 10,11-dihydro)	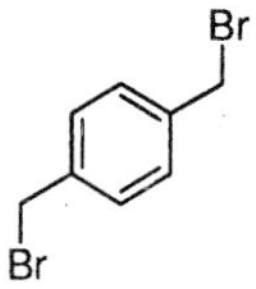	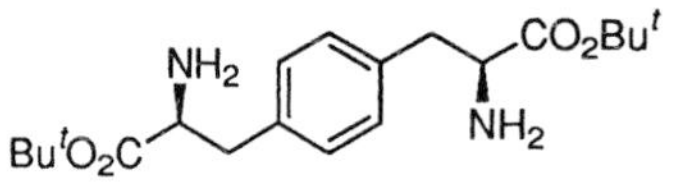	55	72	≧95	30
12 (R=H, X=Br 10,11-dihydro)	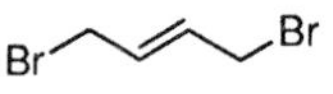	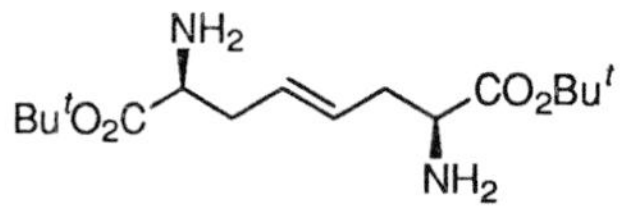	49	82	≧95	30
12 (R=H, X=Br)	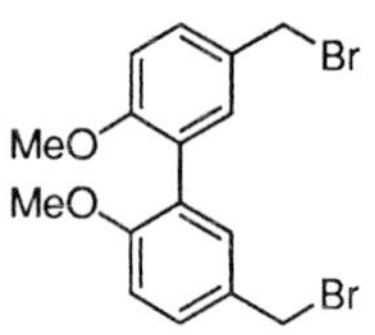	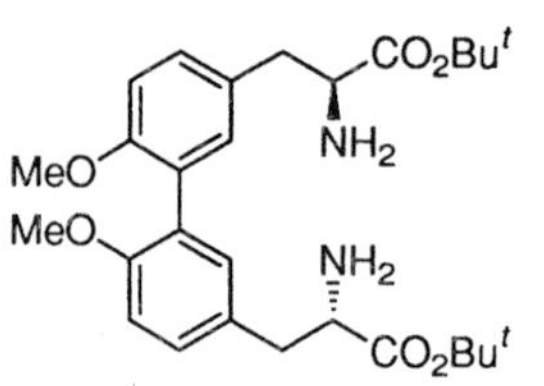	63	80	≧95	31
12 (R=H, X=Br)	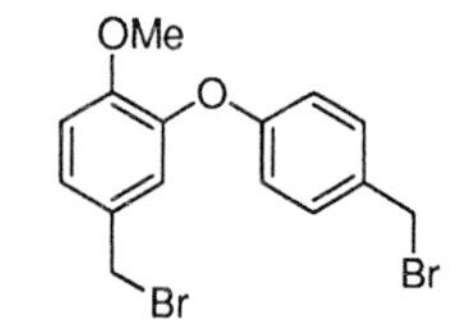	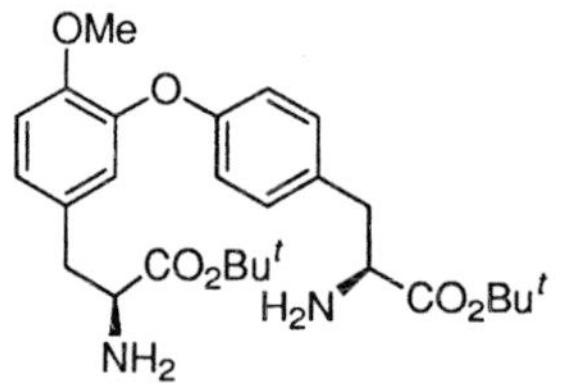	65	80	≧95	31

the quaternary catalyst but not a catalytic amount, the method will provide a new valuable utilization of the quaternary ammonium salts.

Bis-α-amino acid esters **35** were prepared by use of the *N*-anthracenylmethyl catalysts **12** (R=H, X=Cl) or its dihydro derivative **12** (R=H, X=Br, 10,11-dihydro) with high diastereo and enantioselectivities via alkylation of two equivalents of the O'Donnell imine **23** with appropriate dibromides **33**, as summarized in Table 3.[30,31]

N-anthracenylmethyl catalyst **12** (R=H, X=Br, 10,11-dihydro) is also useful for the effective preparation of chiral α,α-dialkyl-α-amino acids **37** by asym-

metric alkylation of the alanine-derived imines **36**, as shown in Scheme 10.[32] The *tert*-butyl ester in **36** was better than isopropyl and methyl esters while arylmethyl bromides showed a higher level of enantioselectivity.

(*R,R*)-TADDOL **14** was also found to catalyze the C-alkylation of the alanine-derived imine *rac*-**36** under phase transfer catalyzed conditions to give α,α-dialkyl amino acids **37** (Scheme 11).[33] The proposed mechanism is shown in Scheme 12. (*S*)-2-Amino-2'-hydroxy-1,1'-binaphthyl **38** [(*S*)-NOBIN] and its derivatives were also used for the same reaction in place of (*R,R*)-TADDOL with less enantiomeric efficiency.[34]

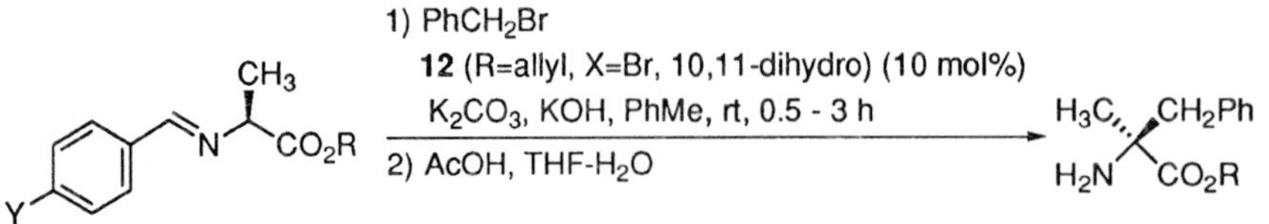

Scheme 10. Asymmetric synthesis of the α,α-dialkyl-α-amino acids **37** by use of the cinchona alkaloid derivative **12**.

Y=Cl, R=Bu^t: 95%, 87% ee[32]
Y=H, R=Bu^t: 87%, 84% ee
Y=Cl, R=Pr^i: 80%, 60% ee

1) PhCH2Br
14 or 38 (10 mol%)
solid NaOH, PhMe
2) 6N HCl aq.
rac-36
37
(R,R)-TADDOL 14 : 81%, 82% ee[33]
(S)-NOBIN 38 : 60%, 68% ee[34]
38

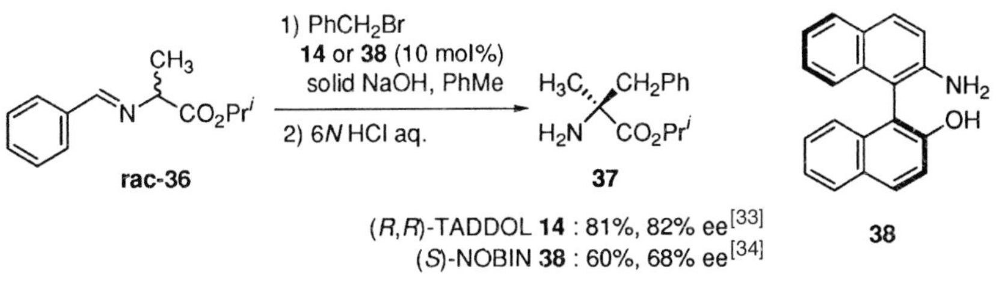

Scheme 11. Asymmetric synthesis of the α, α-dialkyl-α-amino acid **37** by use of (*R*,*R*)-TADDOL **14** and (*S*)-NOBIN **38**.

PhCH2Br
NaBr
TADDOL
Sodium enolate aggregates
solid NaOH
H2O

Scheme 12. The proposed mechanism by use of (*R*,*R*)-TADDOL.

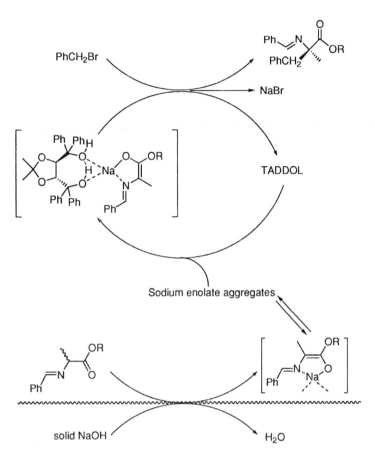

Other Alkylation

The efficient enantioselective alkylation of the β,γ-unsaturated ester **39** was achieved[35] by use of the *N*-anthracenylmethyl catalyst **12** (R=benzyl, X=Br) together with $CsOH \cdot H_2O$ under phase transfer conditions analogous to those in the alkylation of the O'Donnell imine **23**, as shown in Scheme 13. The enantioselectivity of the alkylation correlates with Hammett σ constants, and the *N*,*N*-dimethylamino substituents in **39** showed the most effective enantioselectivity. The tight ion pair in which the enolate oxygen of **39** closely approaches to the N^+ center in the catalyst **12** was proposed to explain the results.

Alkylation of diphenylmethylene benzylimine, $Ph_2C{=}NCH_2Ph$, by use of a proline-derived quaternary ammonium catalyst was reported to give up to 94% ee,[36] but the result has been disputed[13] because the enantioselectivity was determined by optical rotation and the high enantiomeric excess could not be reproduced.

RX
12 (R=benzyl, X=Br) (10 mol%)
CsOH • H2O (10 eq.)
CH2Cl2 - Et2O (1 : 1)
-45 ~ -65 ℃
62 - 83%, 94 - 98% ee[35]
39
40

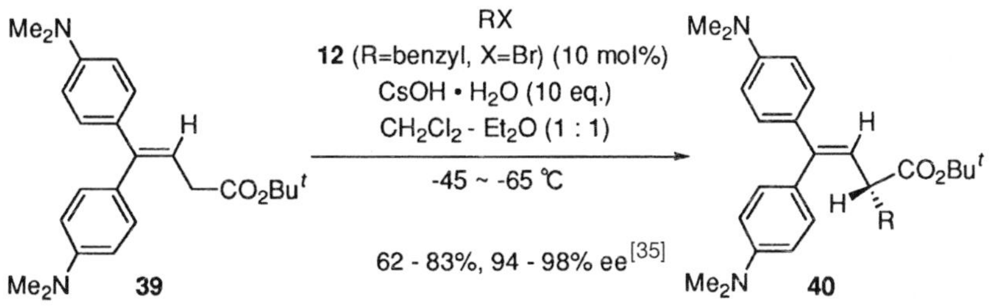

Scheme 13. Asymmetric alkylation of the β,γ-unsaturated ester **39**.

Aldol Reactions

The first promising asymmetric aldol reactions through phase transfer mode will be the coupling of silyl enol ethers with aldehydes utilizing chiral non-racemic quaternary ammonium fluorides,[37] a chiral version of tetra-*n*-butylammonium fluoride (TBAF). Various ammonium and phosphonium catalysts were tried[38,39] in the reaction of the silyl enol ether 41 of 2-methyl-1-tetralone with benzaldehyde, and the best result was obtained by use of the ammonium fluoride 7 (R=H, X=F) derived from cinchonine,[37] as shown in Scheme 14.

The *N*-anthracenylmethyl cinchonidinium catalyst 12 (R=PhCH₂, X=HF₂) was applied to the aldol reaction of the silyl ether 43 derived from the O'Donnell imine 23 with various aldehydes, giving β-hydroxy-α-amino acid esters 44 with high enantiomeric excess,[40] as shown in Scheme 15.

The enantioselective nitroaldol reaction of phenylalaninals 45 with nitromethane was also promoted with the *N*-anthracenylmethyl ammonium fluorides in the presence of potassium fluoride.[41] Interestingly, as shown in Scheme 16, the major product was the (2R,3S)-isomer 46a when N,N-dibenzyl-(S)-phenylalaninal and 12 (R=benzyl, X=F) were used while the (2S,3S)-isomer 46b was major when *N*-*tert*-butoxycarbonyl derivative 45b and 12 (R=allyl, X=Br) together with potassium fluoride were used. The nitroalcohols 46a and 46b were respectively converted to amprenavir 47a, a HIV protease inhibitor, and its diastereomer 47b. The

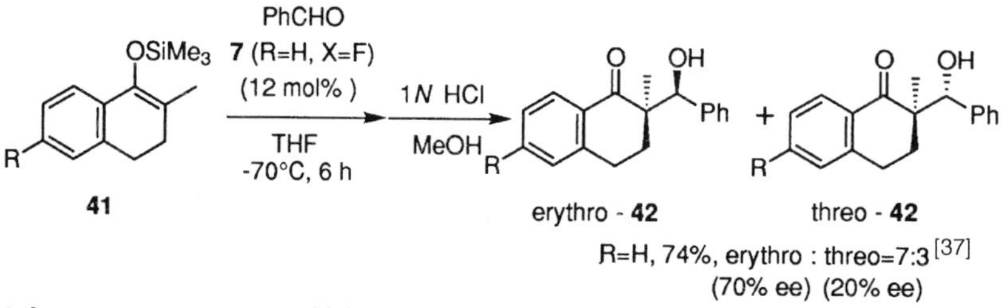

Scheme 14. Asymmetric aldol reaction of the silyl enol ethers **41**.

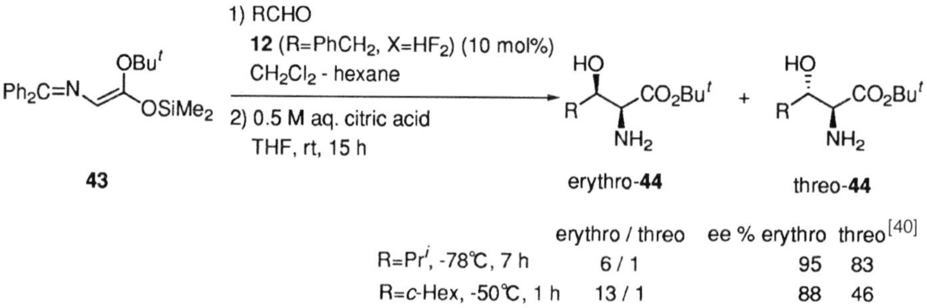

Scheme 15. Asymmetric aldol reaction of the silyl enol ether **43**.

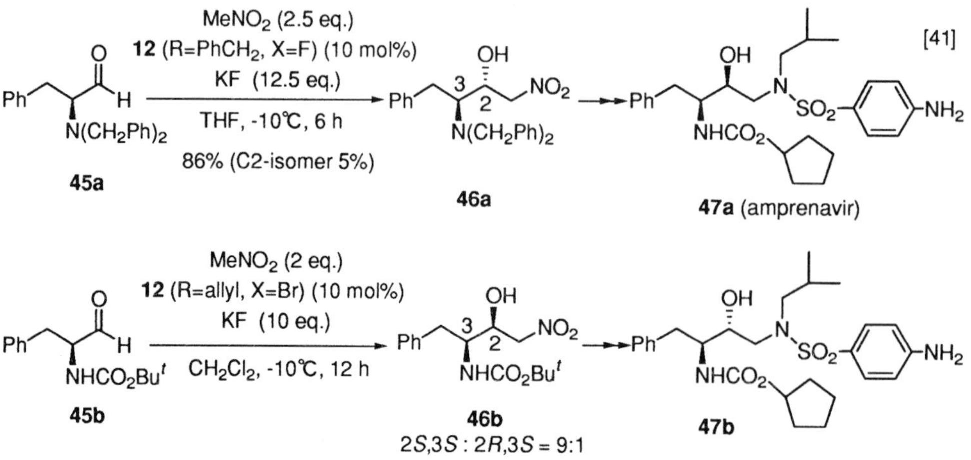

Scheme 16. Asymmetric nitroaldol reaction.

reversal of the nitroaldol products depending on the *N*-protecting group suggests a new strategy for stereocontrol of such reactions.

Other 1,2-Carbonyl Additions

The quaternary ammonium salt **48** derived from (+)-ephedrine was utilized for the enantioselective addition of diethyl zinc to aldehydes.[42] The chiral ammonium fluorides **7** (R=4-CF$_3$ or 2,4-(CF$_3$)$_2$, X=F) were also useful for the enantioselective trifluoromethylation of aldehydes and ketones with moderate enantioselectivity,[43] shown in Scheme 17.

Michael Addition

The first successful results of the asymmetric Michael addition under phase transfer catalyzed conditions were achieved by use of ingeniously designed chiral crown ethers **13** and **52**.[44] The β-keto ester **49** reacted with methyl vinyl ketone by use of **13** to give the Michael product **50** with excellent enantioselectivity but in moderate yield, as shown in Scheme 18. The Michael addition of methyl 2-phenylpropionate **51** to methyl acrylate afforded the diester **53** by use of another crown ether **52** in good yield with good enantioselectivity.[44] Various chiral crown ethers were studied to

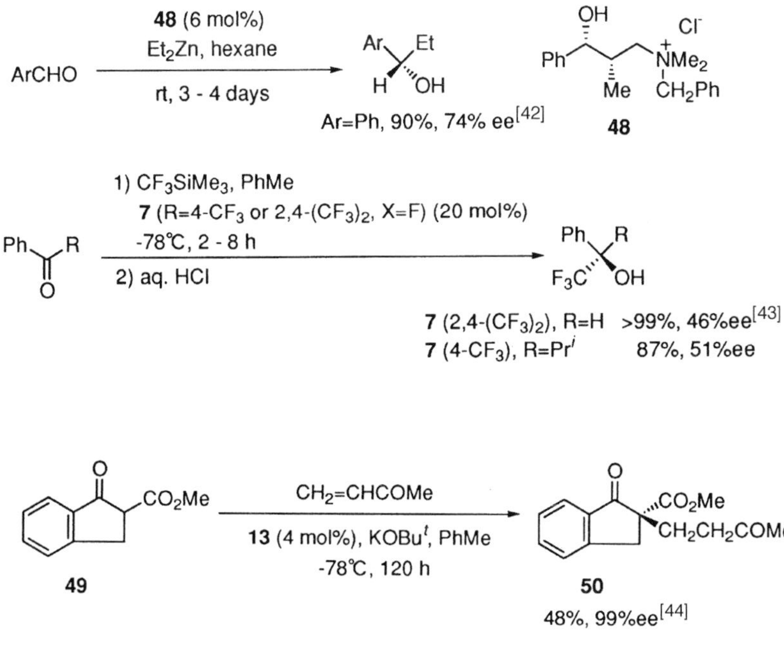

Scheme 17. Asymmetric addition and trifluoromethylation.

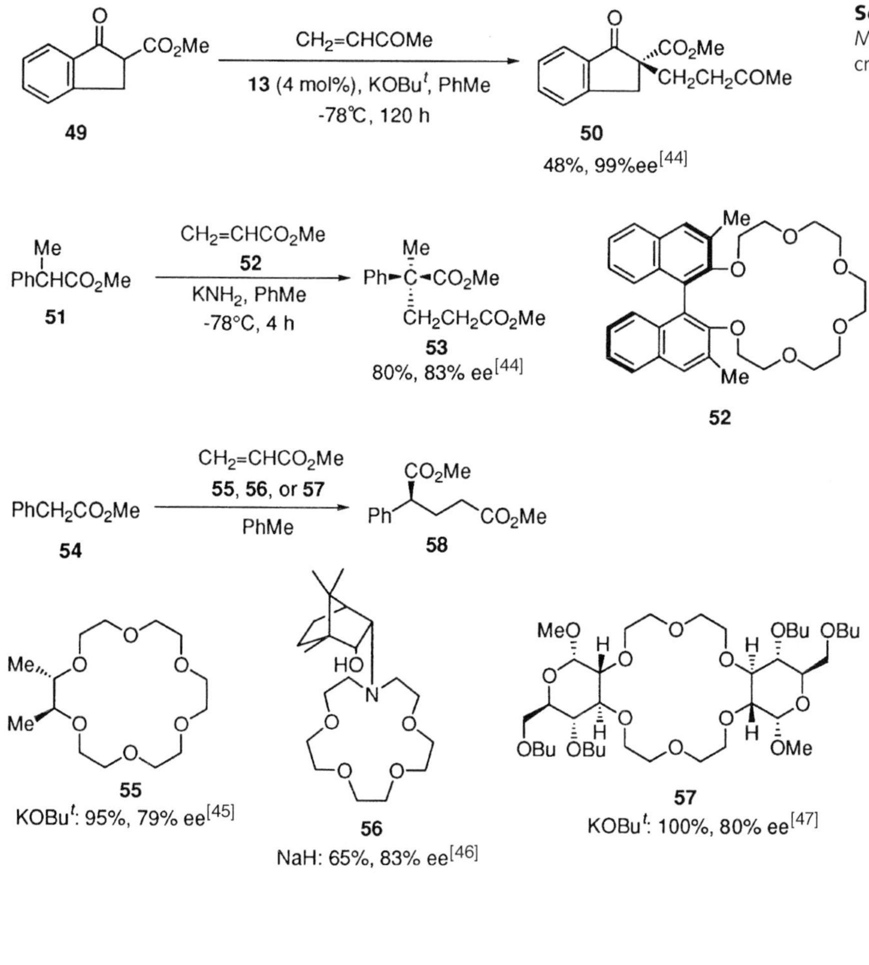

Scheme 18. Asymmetric Michael reaction by use of crown ethers.

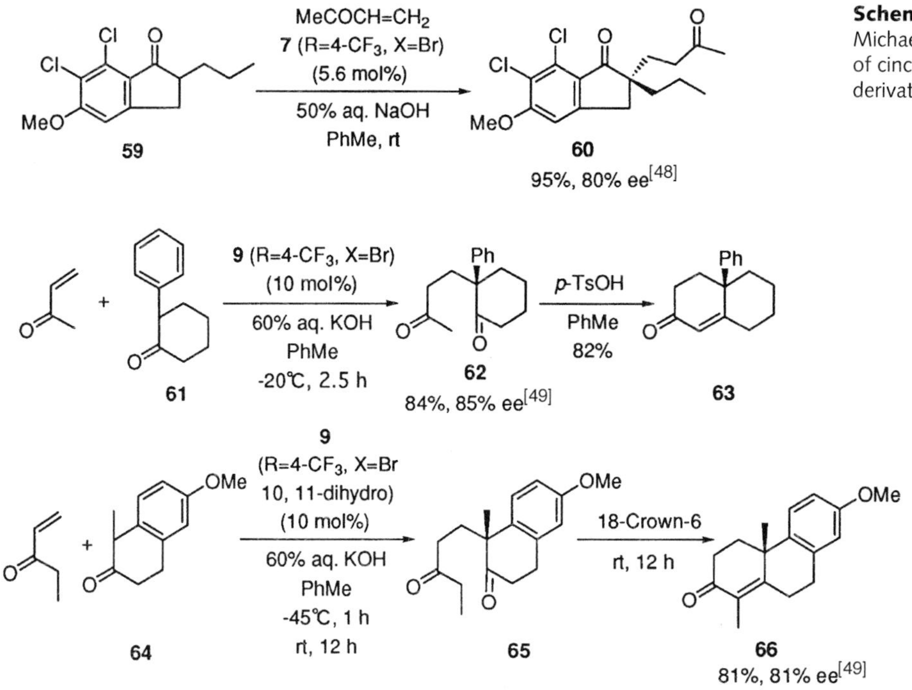

Scheme 19. Asymmetric Michael reaction by use of cinchona alkaloid derivatives.

achieve the enantioselective Michael addition of methyl phenylacetate **54** to methyl acrylate, some of which using **55-57** were shown in Scheme 18.[45–47] The catalyst **57** was used also for the deracemization of the Michael adduct **58** with moderate efficiency.[47]

Some chiral quaternary ammonium salts are also effective in Michael addition reactions. The Merck catalysts **7** (R=4-CF₃, X=Br) and **9** (R=4-CF₃, X=Br, 10,11-dihydro) were used for the Michael additions of **59**, **61**, and **64** to vinyl ketones to give the adducts **60**, **62**, and **65** (isolated as **66**), respectively,[48,49] with excellent enantioselectivity, as shown in Scheme 19. The Michael addition of the O'Donnell imine **23** to the α,β-unsaturated carbonyl compounds also efficiently proceeded by use of the N-anthracenylmethyl catalyst **12** (R=allyl, X=Br), giving the Michael adducts **67** (Scheme 20).[25]

Chalcone **68a** also efficiently underwent the asymmetric Michael addition with diethyl acetami-

domalonate **69** and 2-nitropropane **72**, as shown in Scheme 21. The quaternary ammonium salts **48b** and **70** derived from (+)-ephedrine were effective for the former[50] while the chiral azacrown ethers **73** and **74** were useful for the latter.[51–55] Interestingly, the enantioselectivity of the reaction was better when the reaction with the ephedrinium salts **48b** and **70** was carried out without solvent, and the importance of the π–π attractive interactions between the aryl groups of the benzyl moiety of the catalyst and one of the two phenyl groups of the electrophile **68a** was suggested to be responsible for the enantioselectivity.[50]

Cyclopropanation

Nitromethane, cyanomethyl phenyl sulfone **78a**, and benzyl cyanoacetate **78b** underwent the asymmetric cyclopropanation with 2-bromo-2-cyclo-

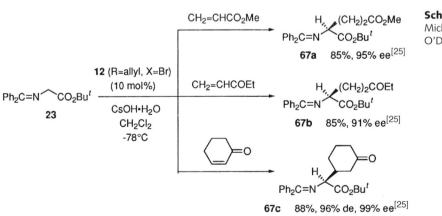

Scheme 20. Asymmetric Michael reaction of the O'Donnell imine **23**.

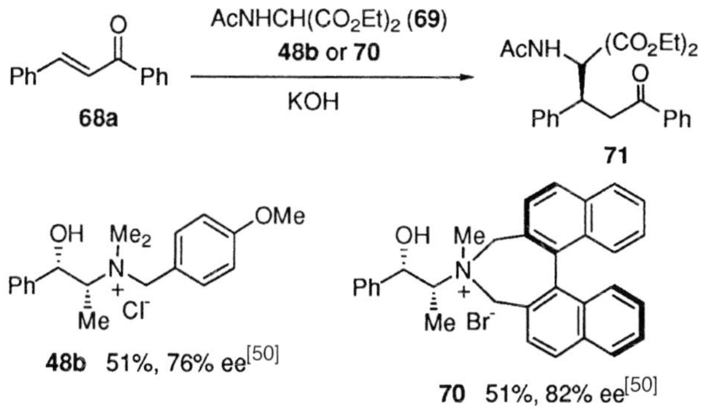
AcNHCH(CO2Et)2 (69)
48b or 70
KOH
68a
71
48b 51%, 76% ee[50]
70 51%, 82% ee[50]

Scheme 21. Asymmetric Michael reaction of chalcone **68a**.

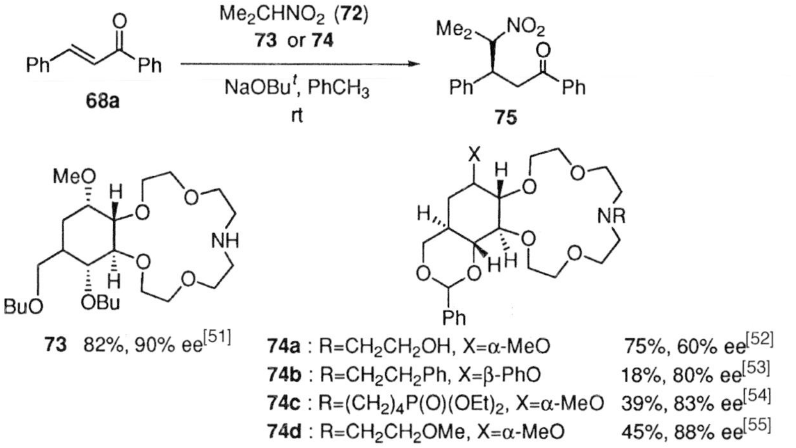
Me2CHNO2 (72)
73 or 74
NaOBu^t, PhCH3
rt
68a
75
73 82%, 90% ee[51]
74a : R=CH2CH2OH, X=α-MeO 75%, 60% ee[52]
74b : R=CH2CH2Ph, X=β-PhO 18%, 80% ee[53]
74c : R=(CH2)4P(O)(OEt)2, X=α-MeO 39%, 83% ee[54]
74d : R=CH2CH2OMe, X=α-MeO 45%, 88% ee[55]

alkanones **76** by use of the ammonium bromides **8** derived from quinidine to give the bicyclic compounds **77**,[56] as shown in Scheme 22. Although the efficiency should be more improved, this will be the first satisfactory results of the catalytic asymmetric cyclopropanation via the successive Michael addition, proton transfer, and intramolecular alkylation under phase transfer conditions, as demonstrated in Scheme 23. The reaction of nitromethane with 2-bromo-2-cyclopentenone **76a** required a catalytic amount (1 mol%) of tetrahexylammonium bromide[57,58] as a co-catalyst to promote the reaction rapidly. The effective catalysts in each reaction possess the substituent at the ortho position on the benzene ring. Presumably, the substituents at the 2-position of the arylmethyl moiety in the phase transfer catalyst prevent the free rotation of C–N$^+$ bond due to its steric hindrance and arrange the nucleophile in the favored direction.[56]

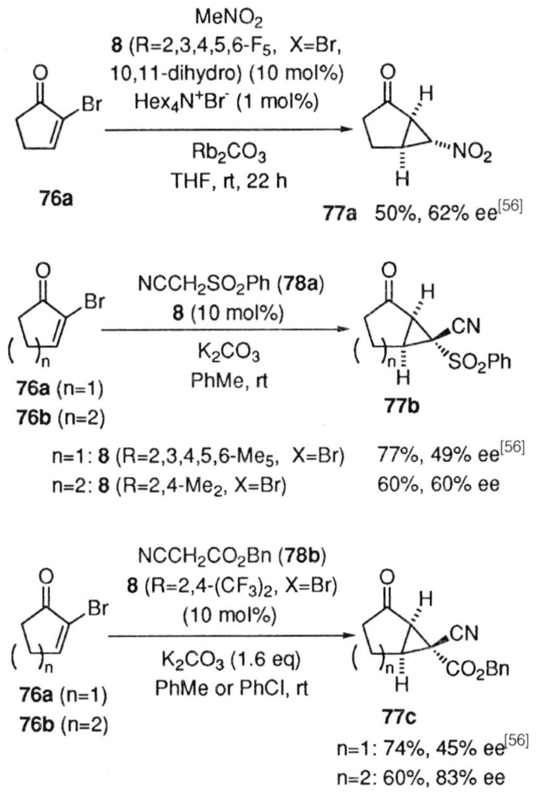
MeNO2
8 (R=2,3,4,5,6-F5, X=Br, 10,11-dihydro) (10 mol%)
Hex4N+Br- (1 mol%)
Rb2CO3
THF, rt, 22 h
76a
77a 50%, 62% ee[56]
NCCH2SO2Ph (78a)
8 (10 mol%)
K2CO3
PhMe, rt
76a (n=1)
76b (n=2)
77b
n=1: 8 (R=2,3,4,5,6-Me5, X=Br) 77%, 49% ee[56]
n=2: 8 (R=2,4-Me2, X=Br) 60%, 60% ee
NCCH2CO2Bn (78b)
8 (R=2,4-(CF3)2, X=Br)
(10 mol%)
K2CO3 (1.6 eq)
PhMe or PhCl, rt
76a (n=1)
76b (n=2)
77c
n=1: 74%, 45% ee[56]
n=2: 60%, 83% ee

Scheme 22. Asymmetric cyclopropanation.

Darzens Condensation

The asymmetric Darzens condensation, which involves both carbon-carbon and carbon-oxygen bond constructions, was realized by use of the chiral azacrown ether **75a**[52,53,55] and the quaternary ammonium salts derived from cinchona alkaloids[59–62] under phase transfer catalyzed conditions. The α,β-epoxy ketone **80** (R=Ph) was obtained with reasonable enantioselectivity by the reaction of

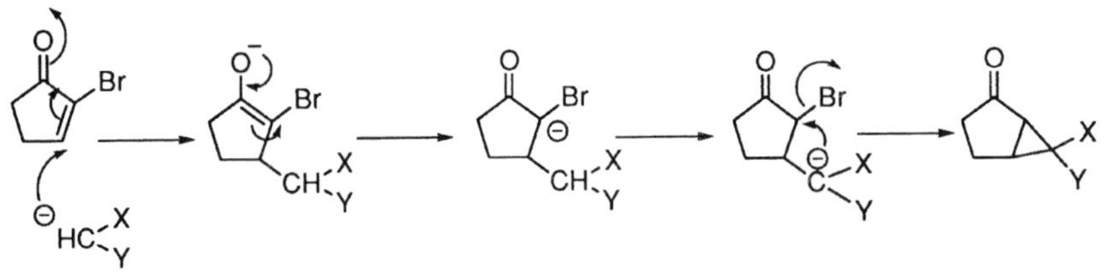

Scheme 23. The proposed mechanism of asymmetric cyclopropanation.

benzaldehyde with phenacyl chloride **79** using the chiral azacrown ethers **74**,[52,53,55] for which the Merck catalyst **7** (R=4-CF$_3$, X=Br) proved to be not effective.[59,60] However, the Darzens condensation with aliphatic aldehydes by use of the same catalyst **7** (R=4-CF$_3$, X=Br) gave the better results, as shown in Table 4.[59,60] The diastereoselectivity of these Darzens condensation proved to be perfect because no *cis* isomer could be detected. The detailed investigation of the mechanism of the Darzens condensation catalyzed by **7** (R=4-CF$_3$, X=Br) revealed that the reaction did not involve a chiral enolate process but a dynamic kinetic resolution which led to optically active products because of a slow cyclization step to produce the epoxyketone **80**, shown in Scheme 24.

The cyclic α-chloro ketone **81** which forms the (Z)-enolate only also underwent the asymmetric Darzens condensation with various aldehydes by use of the Merck catalyst **7** (R=4-CF$_3$, X=Br) under analogous conditions to furnish the α,β-epoxy ketones **82** with up to 86 % ee,[60,61] as shown in Scheme 25. It should be noted that this high enantioselectivity was attained by the reaction at room temperature.

α,β-Epoxysulfones **84** were prepared with good enantioselectivity by the analogous Darzens condensation utilizing the ammonium bromide **10** (R=4-CF$_3$, X=Br) derived from quinine, shown in Scheme 26.[62] Various ammonium catalysts including **7** (R=4-CF$_3$, X=Br) were examined, but the catalyst **10** (R=4-CF$_3$, X=Br) proved to be superior.

Table 4. Asymmetric Darzens reaction.

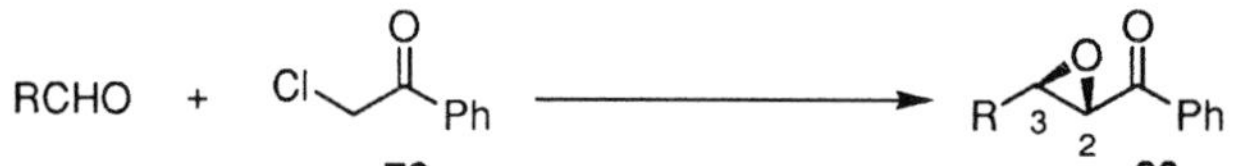

R	Catalyst	Base	Solvent	Temp. (°C)	Time (h)	Yield (%)	%ee[a]	Ref
Ph	**74a**	30% aq. NaOH	PhMe	−20	4	76.5	64 [b]	55
Ph	**74b**	30% aq. NaOH	PhMe	c)	2	68	74 [b]	53
Ph	**7** (R=4-CF$_3$, X=Br)	LiOH·H$_2$O	Bu$_2$O	4	69	43	42	59, 60
Et	**7** (R=4-CF$_3$, X=Br)	LiOH·H$_2$O	Bu$_2$O	4	117	32	79	59, 60
Pr^i	**7** (R=4-CF$_3$, X=Br)	LiOH·H$_2$O	Bu$_2$O	4	60	80	53	59, 60
Bu^i	**7** (R=4-CF$_3$, X=Br)	LiOH·H$_2$O	Bu$_2$O	4	134	73	69	59, 60

a) (2S,3R)-Isomer.   b) (2R,3S)-Isomer.   c) Not described.

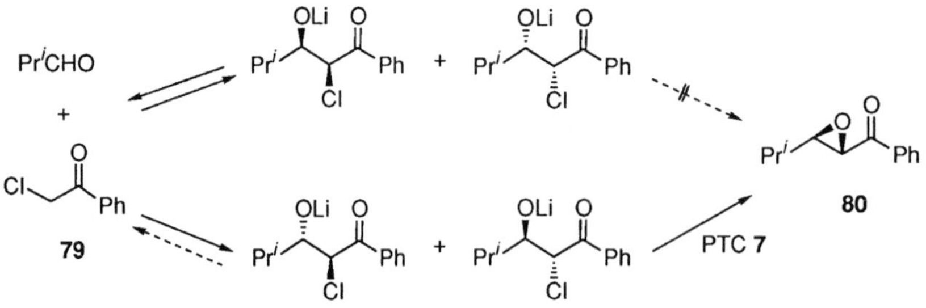

Scheme 24. The proposed mechanism of asymmetric Darzens reaction.

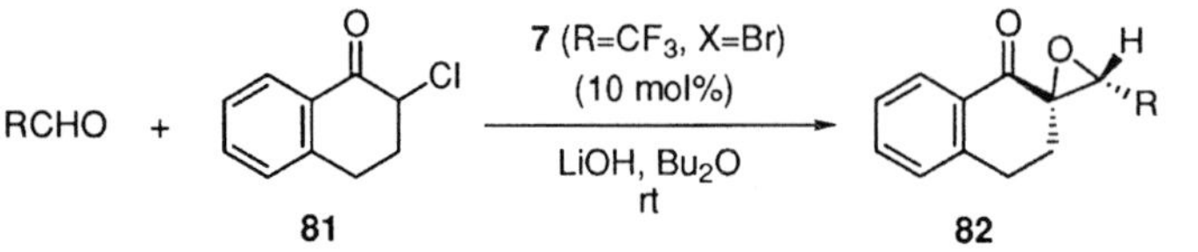

R=Bu^{t}CH$_2$ 86%, 86% ee[60, 61]
R=EtCH$_2$ 67%, 84% ee

Scheme 25. Asymmetric Darzens reaction of α-chlorotetralone **81**.

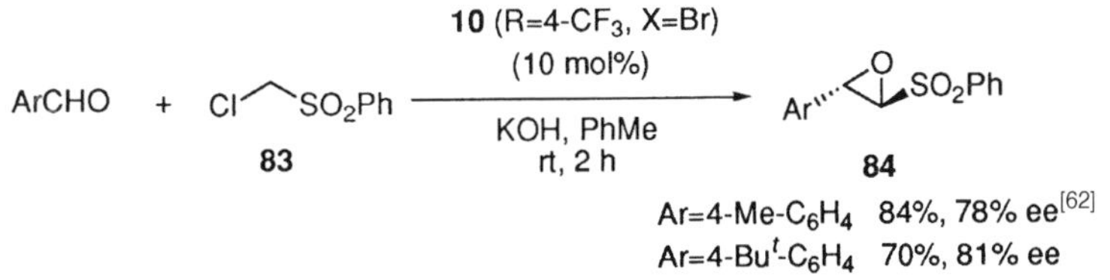

Scheme 26. Asymmetric synthesis of the α,β-epoxysulfones **84**.

$$ArCHO + Cl\diagup\diagdown SO_2Ph \xrightarrow[\substack{KOH, PhMe \\ rt, 2 h}]{\substack{\textbf{10} (R=4\text{-}CF_3, X=Br) \\ (10\ mol\%)}} Ar\cdots\triangleleft\diagdown SO_2Ph$$

83   **84**

Ar=4-Me-C₆H₄   84%, 78% ee[62]
Ar=4-Buᵗ-C₆H₄   70%, 81% ee

Horner–Wadsworth–Emmons Reaction

The catalytic asymmetric Horner–Wadsworth–Emmons reaction was realized by use of the quaternary ammonium salts **7** derived from cinchonine as a phase transfer catalyst.[63] Thus, *tert*-butylcyclohexanone **85** reacted with triethyl phosphonoacetate **86** together with RbOH·H₂O in the presence of the ammonium salts **7**, and then the product **87** was isolated after reesterification by treatment with acidic ethanol, as shown in Scheme 27. Among the

ammonium salt was quite ineffective in this case. Furthermore, importance of the *N*-benzyl and quinoline units was suggested.

The *N*-anthracenylmethyl ammonium catalysts **11** and **12** also proved to be very effective in the asymmetric epoxidation of α,β-unsaturated ketones utilizing aqueous sodium[65,66] or potassium[67] hypochlorite solution as an oxidant. Protection of the hydroxyl group in the *N*-anthracenylmethyl ammonium salts may be essential to attain high enantioselectivities in the epoxidation[65–67], and use of

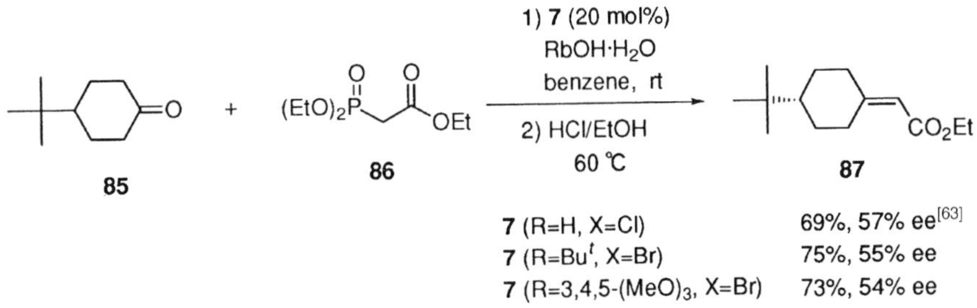

Scheme 27. Asymmetric Horner–Wadsworth–Emmons reaction.

1) **7** (20 mol%)
   RbOH·H₂O
   benzene, rt
2) HCl/EtOH
   60 °C

85   **86**   **87**

7 (R=H, X=Cl)   69%, 57% ee[63]
7 (R=Buᵗ, X=Br)   75%, 55% ee
7 (R=3,4,5-(MeO)₃, X=Br)   73%, 54% ee

catalysts **7** examined, the ammonium salts having the electron-donating groups at the benzene ring of the arylmethyl functions seem to give the better results. Although the reaction requires ca. 1 week and the enantioselectivity is moderate, the above reaction will be the first notable example of the catalytic asymmetric Horner-Wadsworth-Emmons reaction.

Carbon–Oxygen Bond Formation

Epoxidation

Epoxidation is another important area which has been actively investigated on asymmetric phase transfer catalysis. Especially, the epoxidation of various (E)-α,β-unsaturated ketones **68** has been investigated in detail utilizing the ammonium salts derived from cinchonine and cinchonidine, and highly enantioselective and diastereoselective epoxidation has now been attained. When 30 % aqueous H₂O₂ was utilized in the epoxidation of various α,β-unsaturated ketones **68**, use of the 4-iodobenzyl cinchoninium bromide **7** (R=I, X=Br) together with LiOH in Bu₂O afforded the α,β-epoxy ketones **88** up to 92 % ee,[64] as shown in Table 5. The *O*-substituted

aqueous H₂O₂ showed a remarkable decrease of the enantioselectivity.[65,66] Interestingly, the *N*-anthracenylmethyl salt **11** (R=benzyl, X=Br) in conjunction with aqueous hypochlorite afforded the opposite enantiomer **88** from that obtained with the catalyst **7** (R=I, X=Br) and aqueous H₂O₂. Thus, it has proved that the ammonium catalysts from the same cinchonine can give high selectivity for either enantiomer of the epoxide **88**. Although the mechanistic detail of these two reaction systems remains to be clearly defined, these results show that both the *O*- and *N*-substituents in the catalyst and the choice of oxidation conditions have a profound effect on the enantioselectivity of the epoxidation.[66] The aromatic functions at either position of R¹ or R² in the α,β-unsaturated ketones **68** may be essential to conduct asymmetric phase transfer catalysis with high stereoselectivity. The mechanism of the asymmetric epoxidation has been discussed in detail, and the specific arrangement of the reactants has been proposed.[67]

Excellent enantioselectivities have not been attained in the asymmetric phase transfer catalyzed epoxidation of (Z)-enones in contrast to that of (E)-enones. However, a few promising results[68,69] have been reported on the epoxidation of 2-substituted

Table 5. Asymmetric epoxidation of the (*E*)-enones **68**.

R¹	R²	Catalyst	Oxidant Solvent	Temp. (°C)	Time (h)	Yield (%)	%ee	Conf. α β	Ref
Ph	Ph	**7** (R=I, X=Br) (5 mol%)	30% H₂O₂ a) Bu₂O	4	37	97	84	S R	64
Ph	Ph	**11** (R=benzyl, X=Br) (10 mol%)	11% NaOCl PhMe	rt	48	81	90	R S	65, 66
Ph	Ph	**12** (R=benzyl, X=Br 10,11-dihydro) (10 mol%)	11% NaOCl PhMe	rt	48	90	86	S R	65, 66
Ph	Ph	**12** (R=benzyl, X=Br 10,11-dihydro) (10 mol%)	8 M KOCl PhMe	−40	12	96	93	S R	67
3-MeC₆H₄	Ph	**7** (R=I, X=Br) (5 mol%)	30% H₂O₂ a) Bu₂O	4	64	100	92	S R	64
Ph	4-MeC₆H₄	**7** (R=I, X=Br) (5 mol%)	30% H₂O₂ a) Bu₂O	4	36	95	89	S R	64
n-C₆H₁₃	4-NO₂C₆H₄	**12** (R=benzyl, X=Br 1011-dihydro) (10 mol%)	11% NaOCl PhMe	rt	4–48	79	90	S R	66
4-Cl-C₆H₄	4-F-C₆H₄	**12** (R=benzyl, X=Br 10,11-dihydro) (10 mol%)	8 M KOCl PhMe	4	12	94	98.5	S R	67

a) LiOH was used as a base.

naphthoquinones **89**, as summarized in Table 6. Interestingly, the absolute configurations of the products **91** were the same in the epoxidation by use of the quinidinium salt **90** together with 30 % aqueous H₂O₂[68] and the pseudoenantiomeric quininium salt **10** (R=H, X=Cl) in conjunction with *tert*-butyl hydroperoxide.[69]

The epoxidation of the dienone **92** proceeded to give the epoxide **93** using the cinchonidinium catalyst **9** (R=H, X=Cl) with *tert*-butyl hydroperoxide.[70]

Although the enantioselectivity was superior and the yield based on recovered starting material was good (82 %), the catalyst **9** should be used in a large quantity (100 mol%) and the reaction requires 7 days at room temperature (Scheme 28).

Interesting results were obtained in the asymmetric epoxidation of the (*Z*)-alkenes **94** using the (salen)Mn catalyst **95** in conjunction with sodium hypochlorite as an oxidant, giving the optically active (*E*)-epoxides **96** as the major products,[71] as

Table 6. Asymmetric epoxidation of the naphthoquinones **89**.

R	Catalyst	Oxidant Base / Solvent	Temp. (°C)	Time (h)	Yield (%)	% ee	Ref
Pr^i	**90**	30% aq. H₂O₂ LiOH / CHCl₃	−10	5	93	70	68
Ph	**90**	30% aq. H₂O₂ LiOH / CHCl₃	−10	23	47	76	68
2-MeO₂CC₆H₄	**10** (R=H, X=Cl)	*tert*-BuOOH NaOH / PhMe	0 rt	0.5 1	95	78	69

Scheme 28. Asymmetric epoxidation of the dienone **92**.

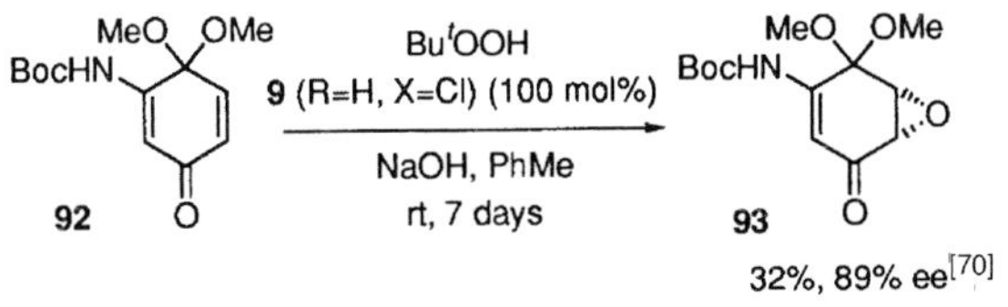

32%, 89% ee[70]

shown in Scheme 29. The co-use of chiral quaternary ammonium salts derived from cinchona and ephedra alkaloids (e. g., **10** (R=H, X=Cl)) greatly influenced diastereoselectivity of the epoxidation but not enantioselectivity. The role of the chiral ammonium salts are not clear, but this work will suggest a new role of chiral ammonium salts.

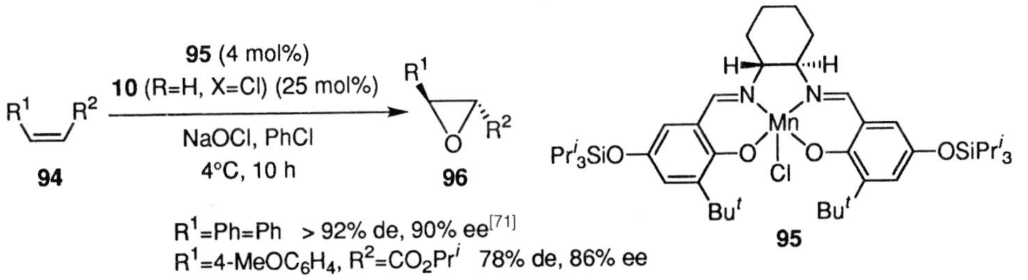

lyst **7** (R=4-CF$_3$, X=Br).[72] It is interesting to note that the *N*-anthracenylmethyl catalyst **12** (R=H, X=Cl) was suggested to be less efficient than the *N*-*p*-bromobenzyl catalyst **9** (R=4-Br, X=Br) in the α-hydroxylation of 2-ethyltetralone **97** (R=Et, n=6).[26]

R^1=Ph=Ph   > 92% de, 90% ee[71]
R^1=4-MeOC$_6$H$_4$, R^2=CO$_2$Pr^i 78% de, 86% ee

Scheme 29. Asymmetric epoxidation of the (*Z*)-alkenes **94**.

α-Hydroxylation of Ketones

Asymmetric α-hydroxylation of ketones **97** through phase transfer catalysis under alkaline conditions was realized by use of the Merck catalyst **7** (R=4-CF$_3$, X=Br)[72] as well as the chiral azacrown ether **98**[73] in conjunction with molecular oxygen, as shown in Scheme 30. The highest enantioselectivity of 79 % ee was attained in the α-hydroxylation of the tetralone **100** by use of the Merck cata-

Reduction

Enantioselectivities in the asymmetric reduction under phase transfer catalyzed conditions reported till now are still moderate, and should be improved further in future.

Most of the attempted asymmetric reductions have used sodium borohydride in conjunction with quaternary ammonium catalysts. Recently, the solution structures of ion pairs formed by quaternary ammonium ions derived from quinine with borohydride ion have been characterized by nuclear magnetic resonance methods in CDCl$_3$.[74]

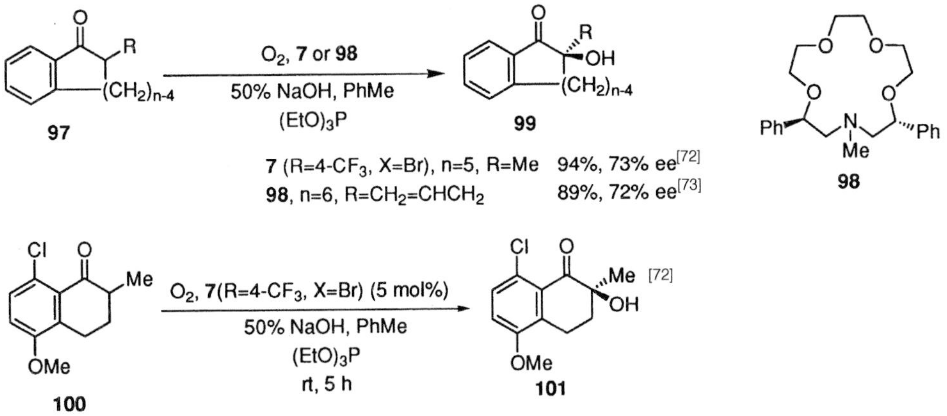

7 (R=4-CF$_3$, X=Br), n=5, R=Me   94%, 73% ee[72]
98, n=6, R=CH$_2$=CHCH$_2$ 89%, 72% ee[73]

Scheme 30. Asymmetric α-hydroxylation of the cyclic ketones **97**.

The quininium and quinidinium fluoride catalysts, **10** (R=H etc., X=F) and **8** (R=H, X=F), were used for the asymmetric reduction of alkyl aryl ketones in conjunction with silanes.[75] One of the most efficient silanes proved to be tris(trimethylsiloxy)silane, which together with **8** (R=H, X=F) reduced acetophenone to give the alcohol **102** in almost quantitative yield with 78 % ee, as shown in Scheme 31. The

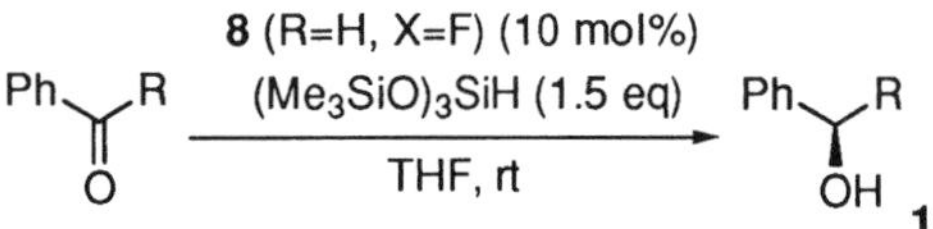

$$Ph \overset{O}{\underset{}{\wedge}} R \xrightarrow[\text{THF, rt}]{\substack{\textbf{8} \text{ (R=H, X=F) (10 mol\%)} \\ (Me_3SiO)_3SiH \text{ (1.5 eq)}}} Ph \overset{OH}{\underset{}{\wedge}} R$$

102

R=Me   99%, 78% ee

Scheme 31. Asymmetric reduction of alkyl aryl ketones.

quininium hydroxide catalyst **10** (R=H, X=OH) was also effective as the quininium fluoride **10** (R=H, X=F), and the fluoride catalyst **7** (R=H, X=F) derived from cinchonine showed no enantioselectivity.

Epilogue

Numbers of asymmetric phase transfer catalysis can now be accomplished efficiently to give a variety of chiral non-racemic products with high enantiomeric excesses. Thus, asymmetric phase transfer catalysis has grown up into practical level in numbers of reactions and some optically pure compounds can be effectively produced on large scale by use of chiral phase transfer catalysts.

The first breakthrough on asymmetric phase transfer catalysis has been achieved by the Merck research group[14–16] utilizing 4-trifluoromethyl-benzyl substituted cinchona alkaloids in enantioselective carbon-carbon bond formation. The efforts by the Cram's group[44] on the use of ingeniously designed new chiral crown ethers are also noteworthy. Recent significant developments have been attained independently by the research groups of Lygo[23,30,31,32,65,66] and Corey [24,25,35,40,41,67] through the introduction of the bulky *N*-anthracenylmethyl function to cinchona alkaloids. Cinchona alkaloids offer an powerful and convenient tool for the design of new quaternary ammonium salts because of their various functionalities, rigid quinuclidine skeleton, and pseudoenantiomeric nature in addition to easy availability with cheap price. Obviously, more syntheses and utilization of quaternary ammonium salts derived from other alkaloids will be carried out in future. Design of synthetic quaternary ammonium salts of non-natural products origin as well as chiral crown ethers will be conducted more and more.

Effective mixing of phases during reactions is generally essential in heterogeneous phase transfer catalysis to conduct the reactions rapidly, and notable will be the efforts of O'Donell's group[28,29] changing heterogeneous reactions to homogeneous ones by co-use of organic bases (Schwesinger bases) with the *N*-anthracenylmethyl catalysts. Furthermore, easy recovery of phase transfer catalysts will be important, and recovery of the *N*-anthracenylmethyl catalysts by the Corey group[24,25,41,67] is worth noticing. Polymer- or silica gel-supported catalysts will be attractive because they are insoluble in both organic and aqueous media and easily removed after the reaction for future reuse.

Numbers of reactions have been developed by utilizing the strong affinity of the fluoride anion to the silicon atom. In this context, the use of chiral non-racemic ammonium fluorides[37–41,43,75] for asymmetric silyl mediated reactions will be further investigated in future.

In contrast to the maturity of asymmetric synthesis utilizing chiral transition metal catalysts, asymmetric phase transfer catalysis is still behind it and covers organic reactions to lesser extent. Thus, it is further necessary in wide range to explore efficient asymmetric phase transfer catalysis keeping its superiority of easy operation, mild reaction conditions, and environmental binignancy.

The recent rapid progress in this area will definitely promise that asymmetric phase transfer catalysis is the reaction for the 21st century.

References and Notes

1. The term phase transfer catalysis is also spelled as phase-transfer catalysis. In this review, we will adopt the former without hyphen. PTC means both phase transfer *catalysis* and phase transfer *catalyst*. To avoid the confusion, full spellings will be adopted.
2. C. M. Starks, "Phase-Transfer Catalysis. I. Heterogeneous Reactions Involving Anion Transfer by Quaternary Ammonium and Phosphonium Salts", *J. Am. Chem. Soc.* **1971**, *93*, 195–199.
3. E. V. Dehmlow, S. S. Dehmlow, Phase Transfer Catalysis, 3rd ed., VCH, Weinheim, **1993**.
4. C. M. Starks, C. L. Liotta, M. Halpern, Phase-Transfer Catalysis, Chapman & Hall, New York, **1994**.
5. Y. Sasson, R. Neumann (ed.), Handbook of Phase Transfer Catalysis, Blackie Academic & Professional, London, **1997**.
6. M. E. Halpern (ed.), Phase-Transfer Catalysis: Mechanism and Syntheses (ACS Symposium Series 659), American Chemical Society, Washington, DC, **1997**.
7. M. J. O'Donnell, "Asymmetric Phase Transfer Reactions", In *Catalytic Asymmetric Synthesis* (Ed. I. Ojima), VCH, New York, **1993**, Chapter 8 (pp. 390–411).
8. T. Shioiri, "Chiral Phase Transfer Catalysis", In ref. 4, **1997**, Chapter 14 (pp. 462–479)

9. M. J. O'Donnell, "Asymmetric PTC Reactions. Part 1: Amino Acids", *Phases – The Sachem Phase Transfer Catalysis Review* **1998**, Issue 4, pp. 5–8.

10. M. J. O'Donnell, "Asymmetric PTC Reactions. Part 2: Further Reactions", *Phases – The Sachem Phase Transfer Catalysis Review* **1999**, Issue 5, pp. 5–8.

11. M. Makosza, C. L. Liotta, S. I. McCoy, "A Look at Kinetics and Mechanism", *Phases – The Sachem Phase Transfer Catalysis Review* **1997**, Issue 2, pp. 13–15.

12. E. V. Dehmlow, P. Singh, J. Heider, "A Cautionary Note on Optical Inductions by Chiral β-Hydroxy-Ammonium Catalysts", *J. Chem. Res. (S)* **1981**, 292–293.

13. E. V. Dehmlow, R. Klauck, S. Düttmann, B. Neumann, H.-G. Stammler, "Revisiting Optically Active Quaternary Derivatives made from Prolinol as Phase Transfer Catalysts", *Tetrahedron: Asymmetry* **1998**, *9*, 2235–2244.

14. U.-H. Dolling, P. Davis, E. J. J. Grabowski, "Enantioselective Synthesis of (+)-Indacrinone via Chiral Phase-Transfer Catalysis", *J. Am. Chem. Soc.* **1984**, *106*, 446–447.

15. A. Bhattacharya, U.-H. Dolling, E. J. J. Grabowski, S. Karady, K. M. Ryan, L. M. Weinstock, "Enantioselective Robinson Annelations via Phase-Transfer Catalysis", *Angew. Chem. Int. Ed. Engl.* **1986**, *25*, 476–477.

16. D. L. Hughes, U.-H. Dolling, K. M. Ryan, E. F. Schoenewaldt, E. I. J. Grabowski, "A Kinetic and Mechanistic Study of the Enantioselective Phase-Transfer Methylation of 6,7-Dichloro-5-methoxy-2-phenyl-1-indanone", *J. Org. Chem.* **1987**, *52*, 4745–4752.

17. S. Arai, M. Oku, T. Ishida, T. Shioiri, "Asymmetric Alkylation Reaction of α-Fluorotetralone under Phase-Transfer Catalyzed Conditions", *Tetrahedron Lett.* **1999**, *40*, 6785–6789.

18. K. Manabe, "Asymmetric Phase-Transfer Alkylation Catalyzed by a Chiral Quaternary Phosphonium Salt with a Multiple Hydrogen-Bonding Site", *Tetrahedron Lett.* **1998**, *39*, 5807–5810.

19. K. Manabe, "Synthesis of Nobel Chiral Quaternary Phosphonium Salts with a Multiple Hydrogen-Bonding Site, and Their Application to Asymmetric Phase-Transfer Alkylation", *Tetrahedron* **1998**, *54*, 14465–14476.

20. M. J. O'Donnell, W. D. Bennett, S. Wu, "The Stereoselective Synthesis of α-Amino Acids by Phase-Transfer Catalysis", *J. Am. Chem. Soc.* **1989**, *111*, 2353–2355.

21. M. J. O'Donnell, S. Wu, J. C. Huffman, "A New Active Catalyst Species for Enantioselective Alkylation by Phase-Transfer Catalysis", *Tetrahedron* **1994**, *50* 4507–4518.

22. M. J. O'Donnell, S. Wu, I. Esikova, A. Mi, *U. S. Patent* **1996**, 5,554,753 (*CA* **1995**, *123*, 9924v).

23. B. Lygo, P. G. Wainwright, "A New Class of Asymmetric Phase-Transfer Catalysts Derived from *Cinchona* Alkaloids – Application in the Enantioselective Synthesis of α-Amino Acids", *Tetrahedron Lett.*, **1997**, *38*, 8595–8598.

24. E. J. Corey, F. Xu, M. C. Noe, "A Rational Approach to Catalytic Enantioselective Enolate Alkylation Using a Structurally Rigidified and Defined Chiral Quaternary Ammonium Salt under Phase Transfer Conditions", *J. Am. Chem. Soc.*, **1997**, *119*, 12414–12415.

25. E. J. Corey, M. C. Noe, F. Xu, "Highly Enantioselective Synthesis of Cyclic Functionalized α-Amino Acids by Means of a Chiral Phase Transfer Catalyst", *Tetrahedron Lett.* **1998**, *39*, 5347–5350.

26. E. V. Dehmlow, S. Wagner, A. Müller, "Enantioselective PTC : Varying the Cinchona Alkaloid Motive", *Tetrahedron* **1999**, *55*, 6335–6346.

27. M. J. O'Donnell, F. Delgado, C. Hostettler, R. Schwesinger, "An Efficient Homogeneous Catalytic Enantioselective Synthesis of α-Amino Acid Derivatives", *Tetrahedron Lett.* **1998**, *39*, 8775–8778.

28. T. Ooi, M. Kameda, K. Maruoka, "Molecular Design of a C_2-Symmetric Chiral Phase-Transfer Catalyst for Practical Asymmetric Synthesis of α-Amino Acids", *J. Am. Chem. Soc.*, **1999**, *121*, 6519–6520.

29. M. J. O'Donnell, F. Delgado, R. S. Pottorf, "Enantioselective Solid-Phase Synthesis of α-Amino Acid Derivatives", *Tetrahedron* **1999**, *55*, 6347–6362.

30.. B. Lygo, J. Crosby, J. A. Peterson, "Enantioselective Synthesis of Bis-α-Amino Acid Esters via Asymmetric Phase-Transfer Catalysis", *Tetrahedron Lett.* **1999**, *40*, 1385–1388.

31. B. Lygo, "Enantioselective Synthesis of Dityrosine and Isodityrosine via Asymmetric Phase-Transfer Catalysis", *Tetrahedron Lett.* **1999**, *40*, 1389–1392.

32. B. Lygo, J. Crosby, J. A. Peterson, "Enantioselective Alkylation of Alanine-Derived Imines Using Quaternary Ammonium Catalysts", *Tetrahedron Lett.* **1999**, *40*, 8671–8674.

33. Y. N. Belokon', K. Kochetkov, T. D. Churkina, N. S. Ikonnikov, A. A. Chesnokov, O. V. Larionov, V. S. Parmár, R. Kumar, H. B. Kagan, "Asymmetric PTC C-Alkylation Mediated by TADDOL-Novel Route to Enantiomerically Enriched α-Alkyl-α-Amino Acids", *Tetrahedron: Asymmetry* **1998**, *9*, 851–857.

34. Y. N. Belokon', K. A. Kahetkov, T. D. Churkina, N. S. Ikonnikov, S. Vyskocil, H. B. Kagan, "Enantiomerically enriched (R)- and (S)-α-Methylphenylalanine via Asymmetric PTC C-Alkylation catalyzed by NOBIN", *Tetrahedron: Asymmetry* **1999**, *10*, 1723–1728.

35. E. J. Corey, Y. Bo, J. Busch-Peterson, "Highly Enantioselective Phase Transfer Catalyzed Alkylation of a 3-Oxygenated Propionic Ester Equivalent; Application and Mechanism", *J. Am. Chem. Soc.* **1998**, *120*, 13000–13001.

36. J. J. Eddine, M. Cherqaoui, "Chiral Quaternary Benzophenone Hydrazonium Salt Derivatives ; Efficient Chiral Catalysts for the Enantioselective Phase-Transfer Alkylation of Imines. Application to Synthesis of Chiral Primary Amines", *Tetrahedron: Asymmetry* **1995**, *6*, 1225–1228.

37. A. Ando, T. Miura, T. Tatematsu, T. Shioiri, "Chiral Quaternary Ammonium Fluoride. A New Reagent for Catalytic Asymmetric Aldol Reactions", *Tetrahedron Lett.* **1993**, *34*, 1507–1510.

38. T. Shioiri, A. Bohsako, A. Ando, "Importance of the Hydroxymethyl-quinuclidine Fragment in the Catalytic Asymmetric Aldol Reactions Utilizing Quaternary Ammonium Fluorides derived from Cinchona Alkaloids", *Heterocycles* **1996**, *42*, 93–97.

39. T. Shioiri, A. Ando, M. Masui, T. Miura, T. Tatematsu, A. Bohsako, M. Higashiyama, C. Asakura, "Use of Chiral Quaternary Salts in Asymmetric Synthesis", In ref. 5, **1997**, Chapter 11 (pp. 136–147).

40. M. Horikawa, J. Bush-Petersen, E. J. Corey, "Enantioselective Synthesis of β-Hydroxy-α-amino Acid Esters by Aldol Coupling Using a Chiral Quaternary Ammonium Salt as Catalyst", *Tetrahedron Lett.* **1999**, *40*, 3843–3846.

41. E. J. Corey, F.-Y. Zhang, "*re* and *si*-Face-Selective Nitroaldol Reactions Catalyzed by a Rigid Chiral Quaternary Ammonium Salt : A Highly Stereoselective Synthesis of the HIV Protease Inhibitor Amprenavir (Vertex 478)", *Angew. Chem. Int. Ed.* **1999**, *38*, 1931–1934.

42. K. Soai, M. Watanabe, "Chiral Quaternary Ammonium Salts as Solid State Catalysts for the Enantioselective Addition of Diethylzinc to Aldehydes", *J. Chem. Soc., Chem. Commun.* **1990**, 43–44.

43. K. Iseki, T. Nagai, Y. Kobayashi, "Asymmetric Trifluoromethylation of Aldehydes and Ketones with Trifluoromethyltrimethylsilane Catalyzed by Chiral Quaternary Ammonium Fluorides", *Tetrahedron Lett.* **1994**, *35*, 3137–3138.

44. D. J. Cram, G. D. Y. Sogah, "Chiral Crown Complexes catalyse Michael Addition Reactions to give Adducts in High Optical Yields", *J. Chem. Soc., Chem. Commun.* **1981**, 625–628.

45. S. Aoki, S. Sasaki, K. Koga, "Simple Chiral Crown Ethers Complexed with Potassium *tert*-Butoxide as Efficient Catalysts for Asymmetric Michael Additions", *Tetrahedron Lett.* **1989**, *30*, 7229–7230.

46. E. Brunet, A. M. Poveda, D. Rabasco, E. Oreja, L. M. Font, M. S. Batra, J. C. Rodrígues-Ubis, "New Chiral Crown Ethers derived from Camphor and Their Application to Asymmetric Michael Addition. First Attempts to Rationalize Enantioselection by AM1 and AMBER Calculations", *Tetrahedron: Asymmetry* **1994**, *5*, 935–948.

47. L. Tõke, P. Bakó, G. M. Keserü, M. Albert, L. Fenichel, "Asymmetric Michael Addition and Deracemization of Enolate by Chiral Crown Ether", *Tetrahedron* **1998**, *54*, 213–222.

48. R. S. E. Conn, A. V. Lovell, S. Karady, L. M. Weinstock, "Chiral Michael Addition: Methyl Vinyl Ketone Addition Catalyzed by Cinchona Alkaloid Derivatives", *J. Org. Chem.* **1986**, *51*, 4710–4711.

49. W. Nerinckx, M. Vandewalle, "Asymmetric Alkylation of α-Aryl Substituted Carbonyl Compounds by Means of Chiral Phase Transfer Catalysts. Applications for the Synthesis of (+)-Podocarp-8(14)-en-13-one and of (–)-Wy-16,225, A Potent Analgesic Agent", *Tetrahedron: Asymmetry* **1990**, *1*, 265–276.

50. A. Loupy, A. Zaparucha, "Asymmetric Michael Reaction under PTC Conditions without Solvent. Importance of π Interactions for the Enantioselectivity", *Tetrahedron Lett.* **1993**, *34*, 473–476.

51. P. Bakó, T. Kiss, L. Tõke, "Chiral Azacrown Ethers derived from D-Glucose as Catalysts for Enantioselective Michael Addition", *Tetrahedron Lett.* **1997**, *38*, 7259–7262.

52. P. Bakó, Á. Szöllõy, P. Bombicz, L. Tõke, "Asymmetric C–C Bond Forming Reactions by Chiral Crown Catalysts; Darzens Condensation and Nitroalkane Addition to the Double Bond", *Synlett* 1997, 291–292.

53. P. Bakó, K. Vízvárdi, S. Toppet, E. V. der Eycken, G. J. Hoonaert, L. Tõke, "Synthesis, Extraction Ability and Application in Asymmetric Synthesis of Azacrown Ethers derived from D-Glucose", *Tetrahedron* **1998**, *54*, 14975–14988.

54. P. Bakó, T. Novák, K. Ludányi, B. Pete, L. Tõke, G. Keglevich, "D-Glucose-based Azacrown Ethers with a Phosphonoalkyl Side Chain: Application as Enantioselective Phase Transfer Catalysis", *Tetrahedron: Asymmetry* **1999**, *10*, 2373–2380.

55. P. Bakó, E. Czinege, T. Bakó, M. Czugler, L. Tõke, "Asymmetric C-C Bond Forming Reactions with Chiral Crown Catalysts derived from D-Glucose and D-Galactose", *Tetrahedron: Asymmetry* **1999**, *10*, 4539–4551 and references cited therein.

56. S. Arai, K. Nakayama, T. Ishida, T. Shioiri, "Asymmetric Cyclopropanation Reaction under Phase-Transfer Catalyzed Conditions", *Tetrahedron Lett.* **1999**, *40*, 4215–4218.

57. S. Arai, K. Nakayama, K. Hatano, T. Shioiri, "Stereoselective Synthesis of Cyclopropane Rings under Phase-Transfer-Catalyzed Conditions", *J. Org. Chem.* **1998**, *63*, 9572–9575.

58. S. Arai, K. Nakayama, Y. Suzuki, K. Hatano, T. Shioiri, "Stereoselective Synthesis of Dihydrofurans under Phase-Transfer-Catalyzed Conditions", *Tetrahedron Lett.* **1998**, *39*, 9739–9742.

59. S. Arai, T. Shioiri, "Catalytic Asymmetric Darzens Condensation under Phase-Transfer-Catalyzed Conditions", *Tetrahedron Lett.* **1998**, *39*, 2145–2148.

60. S. Arai, Y. Shirai, T. Ishida, T. Shioiri, "Phase-Transfer-Catalyzed Asymmetric Darzens Reactions", *Tetrahedron* **1999**, *55*, 6375–6386.

61. S. Arai, Y. Shirai, T. Ishida, T. Shioiri, "Phase-Transfer Catalyzed Asymmetric Darzens Reactions of Cyclic α-Chloro Ketones", *J. Chem. Soc., Chem. Commun.* **1999**, 49–50.

62. S. Arai, T. Ishida, T. Shioiri, "Asymmetric Synthesis of α,β-Epoxysulfones under Phase-Transfer Catalyzed Darzens Reaction", *Tetrahedron Lett.* **1998**, 8299–8302.

63. S. Arai, S. Hamaguchi, T. Shioiri, "Catalytic Asymmetric Horner–Wadsworth–Emmons Reaction under Phase-Transfer-Catalyzed Conditions", *Tetrahedron Lett.* **1998**, *39*, 2997–3000.

64. S. Arai, H. Tsuge, T. Shioiri, "Asymmetric Epoxidation of α,β-Unsaturated Ketones under Phase-Transfer Catalyzed Conditions", *Tetrahedron Lett.* **1998**, *39*, 7563–7566.

65. B. Lygo, P. G. Wainwright, "Asymmetric Phase-Transfer Mediated Epoxidation of α,β-Unsaturated Ketones using Catalysts Derived from Cinchona Alkaloids", *Tetrahedron Lett.* **1998**, *39*, 1599–1602.

66. B. Lygo, P. G. Wainwright, "Phase-Transfer Catalyzed Asymmetric Epoxidation of Enones using N-Anthracenylmethyl-Substituted Cinchona Alkaloids", *Tetrahedron* **1999**, *55*, 6289–6300.

67. E. J. Corey, F.-Y. Zhang, "Mechanism and Conditions for Highly Enantioselective Epoxidation of α,β-Enones Using Charge-Accelerated Catalysis by a Rigid Quaternary Ammonium Salt", *Org. Lett.* **1999**, *1*, 1287–1290.

68. S. Arai, M. Oku, M. Miura, T. Shioiri, Catalytic Asymmetric Epoxidation of Naphthoquinone Derivatives under Phase-Transfer Catalyzed Conditions", *Synlett*, **1998**, 1201–1202.

69. Y. Harigaya, H. Yamaguchi, M. Onda, "Syntheses and Absolute Configurations of a Chiral Naphthoquinone Epoxide and Chiral Naphtho[1,2-c]isocoumarins", *Chem. Pharm. Bull.* **1981**, *29*, 1321–1327, and references cited therein.

70. G. Macdonald, L. Alcaraz, N. J. Lewis, R. J. K. Taylor, "Asymmetric Synthesis of the mC$_7$N Core of the Manumycin Family : Preparation of (+)-MT 35214 and a Formal Total Synthesis of (–)-Alsamycin", *Tetrahedron Lett.* **1998**, *39*, 5433–5436 and references therein.

71. S. Chang, J. M. Galvin, E. N. Jacobsen, "Effect of Chiral Quaternary Ammonium Salts on (salen)Mn-Catalyzed Epoxidation of *cis*-Olefins. A Highly Enantioselective,

Catalytic Route to Trans-Epoxides", *J. Am. Chem. Soc.* **1994**, *116*, 6937–6938.

72. M. Masui, A. Ando, T. Shioiri, "Asymmetric Synthesis of α-Hydroxy Ketones Using Chiral Phase Transfer Catalysts", *Tetrahedron Lett.* **1988**, *29*, 2835–2838.

73. E. F. J. de. Vries, L. Ploeg, M. Colao, J. Brussee, A. van der Gen, "Enantioselective Oxidation of Aromatic Ketones by Molecular Oxygen, Catalyzed by Chiral Monoaza-Crown Ethers", *Tetrahedron: Asymmetry* **1995**, *6*, 1123–1132.

74. C. Hofstetter, P. S. Wilkinson, T. C. Pochapsky, "NMR Structure Determination of Ion Pairs Derived from Quinine: A Model for Templating in Asymmetric Phase-Transfer Reductions by BH_4^- with Implications for Rational Design of Phase-Transfer Catalysts", *J. Org. Chem.* **1999**, *64*, 8794–8800.

75. M. D. Drew, N. J. Lawrence, W. Watson, S. A. Bowles, "The Asymmetric Reduction of Ketones Using Chiral Ammonium Fluoride Salts and Silanes", *Tetrahedron Lett.* **1997**, *38*, 5857–5860.

Asymmetric Catalysis in Target-Oriented Synthesis

Amir H. Hoveyda   *Department of Chemistry, Merkert Chemistry Center, Boston College, Chestnut Hill, MA 02467, USA,*

Phone: +617 552 3618, Fax: +617 552 1442, e-mail: Amir.Hoveyda@BC.edu

Concept: A variety of metal-based chiral catalysts effect enantioselective chemical transformations that allow for concise and efficient synthesis of optically pure organic molecules. These transformations, particularly when used in tandem with other catalytic reactions, enantioselective or not, serve a pivotal role in the total synthesis of a number of medicinally important target molecules.

Abstract: Diels-Alder cycloaddition, aldol addition, olefin epoxidation, olefin alkylation, olefin dihydroxylation, olefin metathesis and cyclopropanation are among the myriad important transformations that now have catalytic enantioselective variants. A number of total syntheses have emerged in the past decade that utilize these catalytic enantioselective reactions at the strategic level. These asymmetric transformations are reliable, highly enantioselective and require chiral catalysts that are often readily accessible or even commercially available. In certain instances, catalytic enantioselective reactions are used to prepare optically pure starting materials that serve to induce absolute asymmetry in the remaining stereogenic centers, the installment of which are required for the completion of the synthesis. In other cases, such transformations are employed to construct the main skeleton of the target molecule. In yet another type of total synthesis, referred to as "catalysis-based total syntheses", various catalytic enantioselective reactions are employed in combination with a number of other catalytic transformations.

Prologue

Synthesis of organic molecules in a selective and efficient manner stands as a central objective in modern chemistry. Within this context, in the past fifty years, significant advances in the field of natural product, or target-oriented, total synthesis have been made. Molecules of remarkable complexity, such as palytoxin or vancomycin, have been prepared in the laboratory by chemists. However, what still remains out of reach is the existence of a field – of a collection of transformations and strategies – that allows most chemists to prepare within a reasonable amount of time and in a cost-effective fashion *any* molecule in the *optically pure form* and in *high overall yield*. The day has not arrived, and may be far in the future, when the enantioselective synthesis of many complex molecules can be attained without traversing a large number of unsuccessful (and often undisclosed) routes, without the need to employ large groups of exceptionally dedicated team members, without repeated and numerous separation of regio-, diastereo- and enantiomers, and without generation of waste side products and solvents that by far outweigh the amount of the desired target molecule produced. That is, the science of stereo- and regioselective organic synthesis is far from mature. Even if we recapitulate that we can make any molecule we want, it would be impossible to claim that we can make any molecule with a reasonable degree of efficiency (> 50 % overall yield) and in sufficient quantities (> 10 g) for subsequent use or biological studies. Our weaknesses in synthesis, in spite of all of the spectacular successes of the past, become even more glaring when issues of environmental safety (minimum waste), enantioselectivity and cost are underscored.

Thus, asymmetric catalysis has and will continue to play an increasingly critical role in the future of

Keywords ■ *Asymmetric Catalysis* ■ *Natural Product Synthesis* ■ *Chiral Metal-Based Complexes* ■ *Catalysis-Based Total Syntheses* ■ *Enantioselective C–C Bond Formation* ■ *Enantioselective C–O Bond Formation* ■ *Enantioselective C–N Bond Formation* ■ *Enantioselective C–H Bond Formation*

organic synthesis. Asymmetric catalytic reactions are capable of affording optically pure materials with only the use of small amounts of chiral catalysts (favoring cost efficiency and waste minimization). In many instances, the reactions effected by catalysts are simply not feasible by any other method; asymmetric catalysis can influence a synthesis plan by offering an otherwise unviable route that is notably shorter and cost-effective.

This article provides a brief overview of several recent total syntheses of natural and unnatural products that have benefited from the use of catalytic asymmetric processes. The article is divided by the type of bond formation that the catalytic enantioselective reaction accomplishes (e.g., C–C or C–O bond formation). Emphasis is made on instances where a catalytic asymmetric reaction is utilized at a critical step (or steps) within a total synthesis; however, cases where catalytic enantioselective transformations are used to prepare the requisite chiral non-racemic starting materials are also discussed. At the close of the article, two recent total syntheses are examined, where asymmetric catalytic reactions along with a number of other catalyzed processes are the significant driving force behind the successful completion of these efforts (Catalysis-Based Total Syntheses).

Catalytic Asymmetric C–C Bond Forming Reactions in Total Synthesis

Catalytic Asymmetric Addition of Alkylzincs to Aldehydes

The inelegance of a total synthesis is often forgiven, rightfully or not, when the target is regarded as a "highly complex molecule". To synthesize a molecule that is viewed as relatively simple, the chemist cannot afford such a reprieve; the target synthesis must accomplish the task with a clear flair for creativity. The total synthesis of muscone (**6**), reported by Oppolzer,[1] is one such example. Following the seminal developmental works of Noyori in the area of Zn-catalyzed asymmetric alkylation of aldehydes, macrocyclic allylic alcohol **4** is prepared in a single vessel from alkynal **1** in 75% yield and 92% ee (enantiomeric excess). The requisite vinylzinc reagent is accessed through the hydroboration/metal exchange; in the presence of only 1 mol% of the chiral ligand (–)-DAIB (**2**), the macrocycle is obtained from **3** enantioselectively. The total synthesis is completed by a diastereoselective directed cyclopropanation, followed by oxidation of the secondary carbinol and regioselective reductive opening of the cyclopropyl ring (**5**→**6**).

Catalytic Asymmetric Alkylation of π-Allyl Metal Complexes

The synthesis of lycorane (**13**) by Mori and Shibasaki[2] is breathtaking for its use of three consecutive Pd catalyzed C–C bond forming reactions. Thus, Pd-catalyzed asymmetric allylic substitution of a benzoate in *meso* **7** in the presence of the chiral bisphosphine **8** leads to the regioselective formation of **10** in 40% ee. It is easy to overlook this low level of enantioselectivity when we are faced with the subsequent elegant Pd-catalyzed reactions: Pd-catalyzed intramolecular amination is followed by a Pd-catalyzed Heck coupling to afford **12**, which is then readily converted to the target molecule.

Pd-catalyzed asymmetric alkylation is also used in the synthesis of chanoclavine I by Genet

Scheme 1. Catalytic enantioselective aldehyde alkylation affords the chiral macrocyclic alcohol **3** in Oppolzer's total synthesis of muscone (1993).

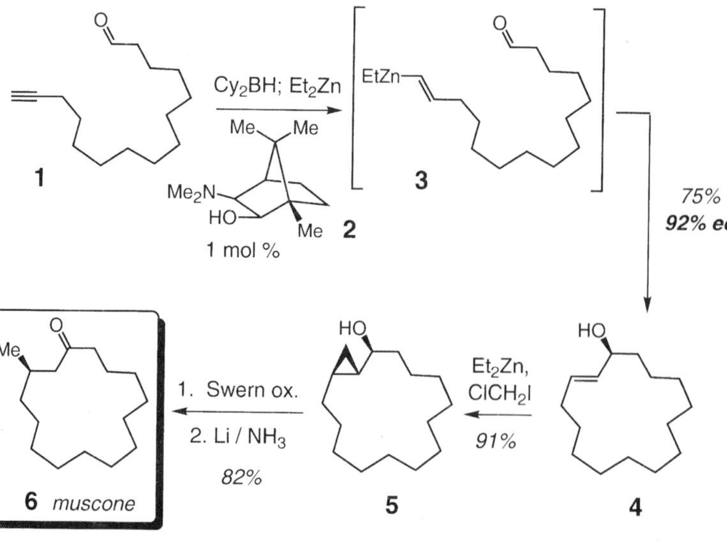

Scheme 2. Pd-catalyzed asymmetric allylic alkylation of **7** is followed by a Pd-catalyzed intramolecular C–N bond formation and a Pd-catalyzed intramolecular Heck-type alkylation in Mori and Shibasaki's total synthesis of lycorane (1995).

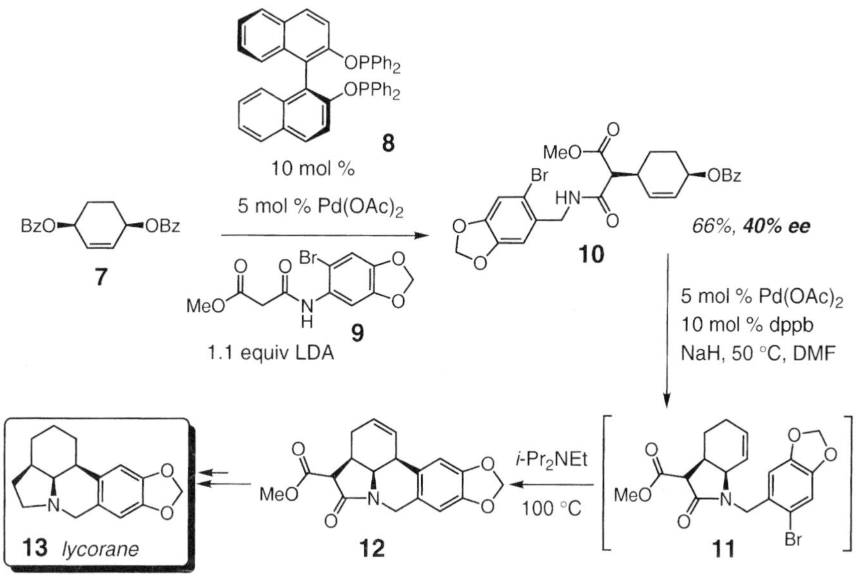

(Scheme 3).[3] In the presence of 6 mol% **15** and 3 mol% Pd(OAc)₂, intermediate **16** is formed in 60 % yield and 95 % ee; the resulting tricycle is then readily transformed to the desired target molecule.

Catalytic Asymmetric Heck Reactions

Pd-catalyzed Heck reactions are among the most effective methods for the formation of quaternary carbon centers. Considering the significance and the strategic difficulties associated with the synthesis of quaternary carbons, particularly in the optically enriched or pure form, it is not a surprise that the development of catalytic asymmetric Heck reactions has held center stage for the past few years. One of the leading labs in this area is that of Shibasaki, who in 1993 reported a concise total synthesis of eptazocine **23** (Scheme 4).[4] Thus, treatment of silyl ether **18** with 10 mol% Pd(OAc)₂ and 25 mol% (S)-**19** leads to the formation of **20** in 90 % yield and 90 % ee. As illustrated in Scheme 4, once the quaternary carbon center is synthesized efficiently and selectively, the target molecule is accessed in a few steps.

Significant developments in this area have been reported by Overman as well. As illustrated in Scheme 5, a Pd-catalyzed asymmetric Heck reaction leads to the formation of cyclic amide **24**; subsequent treatment with aqueous acid delivers **29** in 84 % yield and 93 % ee.[5] Optically pure **25** is obtined after recrystallization (80 % recovery). Follow-up functionalization, shown in Scheme 5, affords either physostigmine **26** or physovenine **27**. It is difficult to imagine an alternative, and nearly efficient or selective, approach to the construction of these target molcules.

Catalytic Asymmetric Diels-Alder Reactions

There are few transformations that can, in a single stroke, provide the structural and stereochemical complexities that are attained through a Diels-Alder cycloaddition. It is little surprise then that numerous studies have appeared in the past decade that aim to develop a catalytic asymmetric Diels-Alder process. It is only in a few instances, however, that these catalytic asymmetric protocols have been applied to target-oriented synthesis. Several impres-

Scheme 3. Pd-catalyzed intramolecular asymmetric allylic alkylation in Genet's synthesis of chanoclavine I (1994).

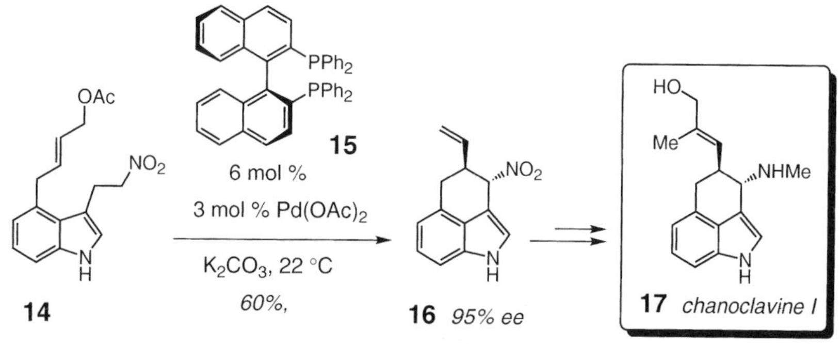

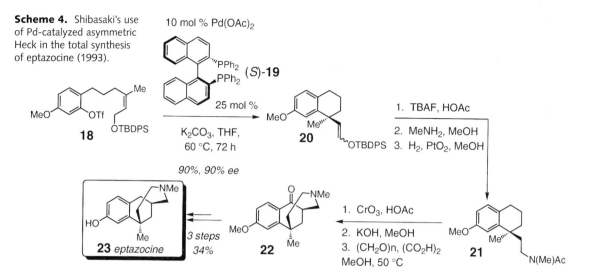

Scheme 4. Shibasaki's use of Pd-catalyzed asymmetric Heck in the total synthesis of eptazocine (1993).

sive examples have been reported by Corey, two of which are depicted in Scheme 6.[6a]

In the presence of 25 mol% **29**, diene **28** reacts with methacrolein to give **30** in high yield (83 %), excellent enantiofacial selectivity (97 % ee) and near perfect diastereocontrol (> 99 % diastereoselection). Four transformations later, cassiol is attained in 61 % yield and in high optical purity. In an enantioselective approach to gibberellic acid,[6b] Corey illustrates how functionality incorporated within the diene and dienophile units can be later exploited for further modification. The bromide in the dienophile segment not only enhances reactivity by its inductive effect, the resulting α-bromo-aldehyde is readily manipulated for subsequent stereoselective functionalization of the cycloadduct (**34**→**35**).

Another impressive catalytic asymmetric Diels-Alder process is that designed and developed by Evans, whose research group, as shown in Scheme 7, has applied it to an impressively concise total synthesis of isopulo'upone.[7] In an efficient and highly diastero- and enantioselective intramolecular catalytic cycloaddition, triene **39** is converted to **41** (96 % ee). Thus, the formation of two ring structures and four stereogenic centers is promoted in a single step by 5 mol% of Cu(II) catalyst **40**. The resident oxazolidone is converted to the requisite ketone and the pyridine tether is installed in six steps en route to the target molecule.

Scheme 5. The utility of the Pd-catalyzed asymmetric Heck in Overman's total syntheses of physostigmine and physovenine (1993, 1998).

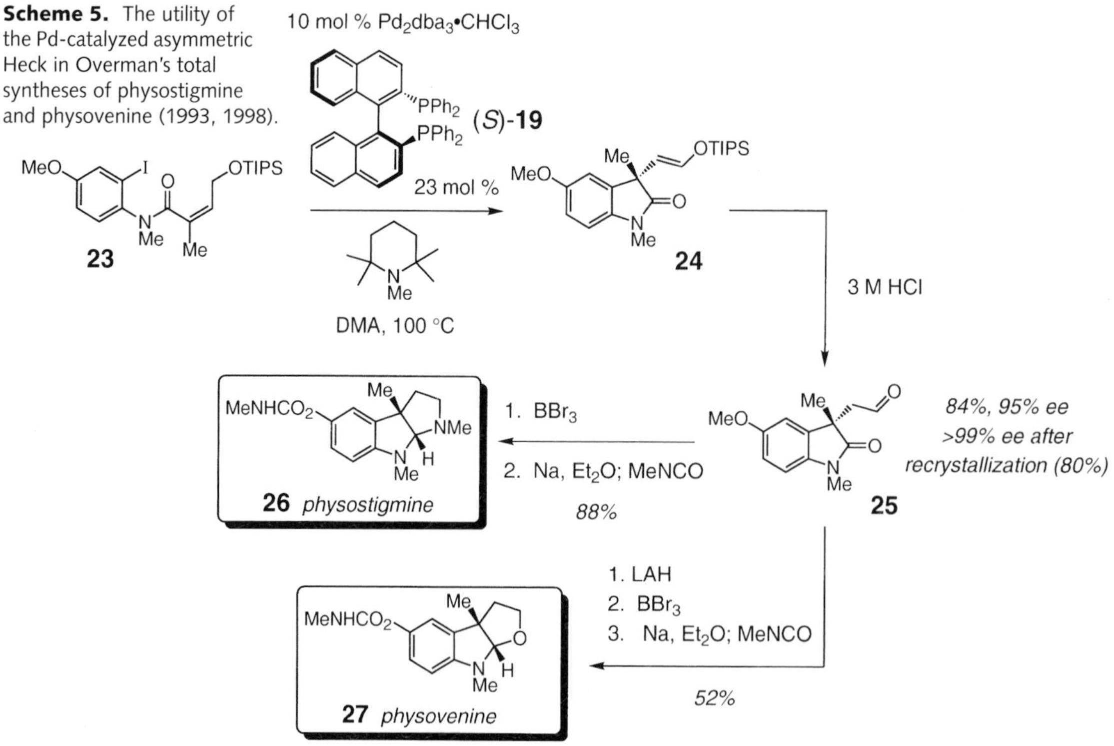

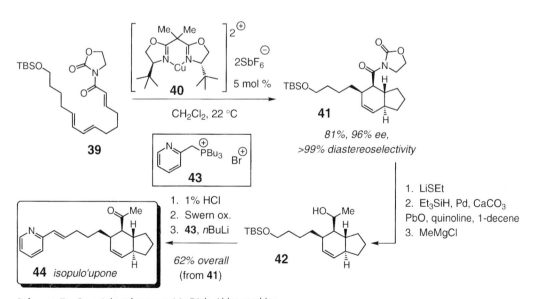

Scheme 6. Corey's approaches to cassiol and gibberellic acid via chiral intermediates built by catalytic asymmetric Diels-Alder reactions (1994).

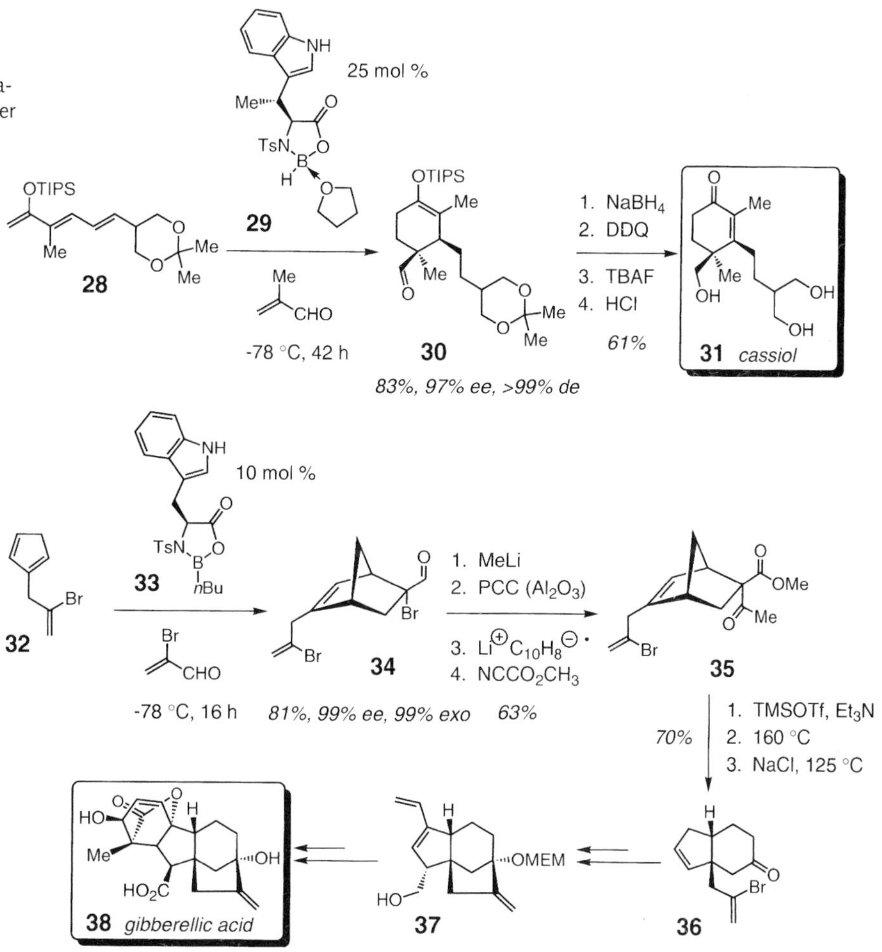

Catalytic Asymmetric Cyclopropanations

A catalyst (46) that is structurally closely related to that used by Evans in his intramolecular [4+2] cycloaddition (cf. Scheme 7) is utilized by Overman in an intramolecular asymmetric Cu(I)-catalyzed cyclopropanation (45→47, Scheme 8).[8] The resulting enantiopurity (88% ee) is improved by a single recrystallization (>99% ee, 62% recovery). As depicted in Scheme 8, the optically pure 47 is then converted to vinyl iodide 51; the cyclopropane unit is used for the construction of the required seven-

Scheme 7. Cu-catalyzed asymmetric Diels-Alder used by Evans in the total synthesis of isopulo'upone (1997).

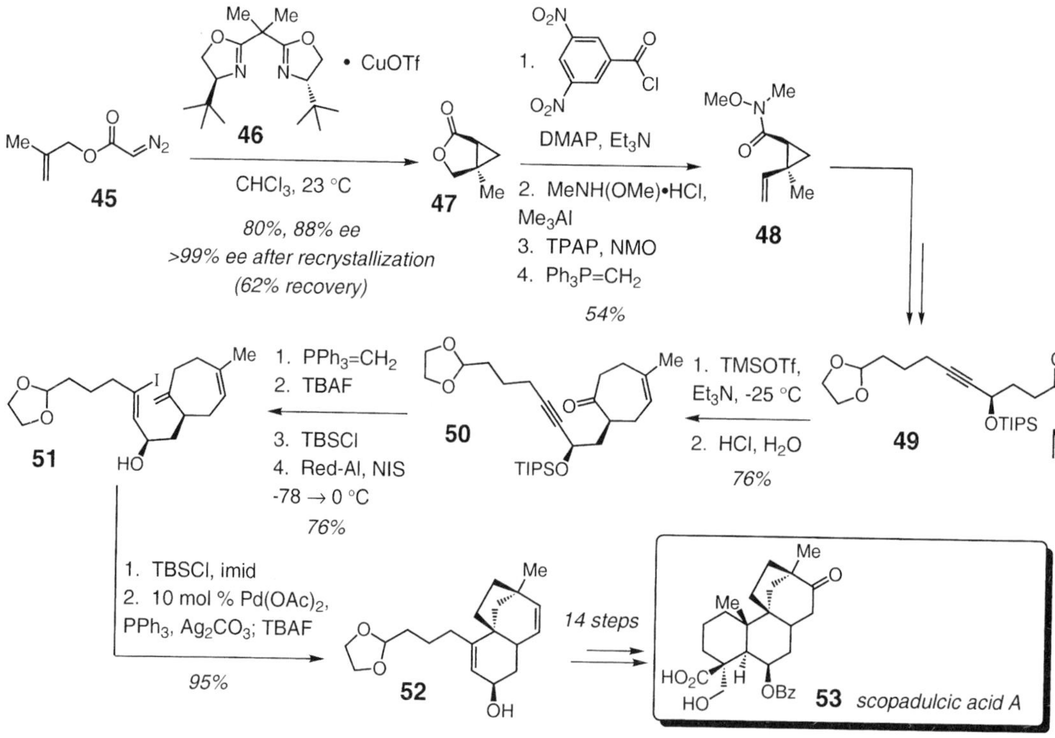

Scheme 8. Cu-Catalyzed asymmetric cyclopropanation used by Overman in the enantioselective total synthesis of scopadulcic acid A (1999).

membered ring through an Ireland Claisen rearrangement (**49 → 50**). Silylation of the carbinol center in **51**, followed by a spectacular tandem double-Heck alkylation leads to the formation of tricycle **52** in the optically pure form. The target molecule, Chinese folk medicine scopadulcic acid, is obtained

after fourteen additional steps. The Overman synthesis is particularly instructive, as it demonstrates a creative use of catalytic asymmetric cyclopropanation for the formation of medium-ring carbocycles in the optically pure form.

Scheme 9. The asymmetric Rh-catalyzed C–H insertion routes to ritalin by Davies and Winkler (1999).

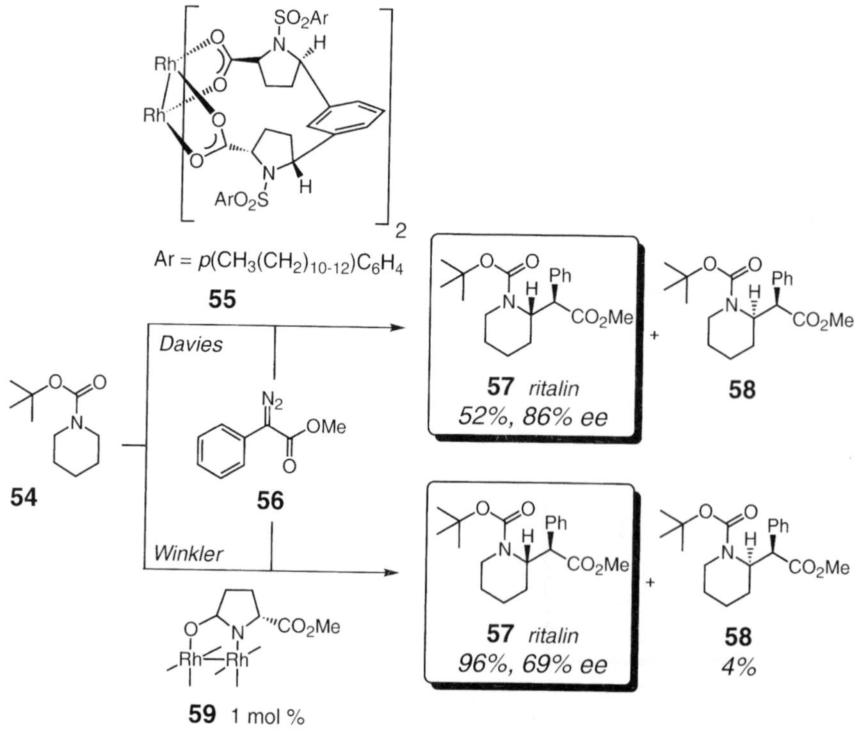

Catalytic Asymmetric Carbon-Hydrogen Insertions

This is another powerful catalytic C–C bond forming process. The eye-catching potential of this class of transformations is best illustrated by two efficient syntheses of ritalin **57** by Davies[9] and Winkler.[10a] In a single step, site- and enantioselective Rh-catalyzed insertion of **56** into a C–H bond of **54** leads to the formation of **57** with appreciable asymmetric induction. The Davies approach is more enantioselective but less diastereocontrolled; the Winkler plan is less enantioselective but only generates 4% of the unwanted diastereomer **58**. Once again, it is difficult to fathom a more direct route for the asymmetric synthesis of this most prescribed psychotropic medication for children. The strong impact of the Rh-catalyzed asymmetric protocol is brought into full light when we recall a nine-step recent synthesis of this target,[10b] where a well-known chiral auxiliary is employed.

Catalytic Asymmetric Aldol Additions

Another classic transformation, a catalytic asymmetric version of which has been the focus of serveral studies is the aldol reaction. Evans, Carriera and Shibasaki are three of the active researchers in this area.

Evans uses his Cu(II)-based Lewis acid **61** (Scheme 10) to effect an efficient and enantioselective aldol

addition to prepare **62** in the optically pure form.[11] Subsequent convesrion of **62** to pyran **65** in eight steps, affords a key intermediate in the enantioselective total synthesis of bryostatin 2.

Carreira's catalytic asymmetric aldol method has been used by his research group in an elegant total synthesis of macrolactin A.[12] However, it is a testimony to the utility of this Ti-catalyzed C–C bond forming reaction that other research teams, such as that of Simon, have also employed the technology. Simon's total synthesis of antitumor depsipeptide FR-901228 is summarized in Scheme 11.[13] Hydroxy ketone (S)-**69**, formed in > 98% ee and in quantitative yield through an asymmetric aldol addition promoted by (S)-**63**, is used to prepare macrocyclization precursor **70**. Acyclic polypeptide **70**, however, cannot be coaxed to undergo cyclization in an efficient manner. At this point, Simon exploits one of the most attractive features of a catalytic asymmetric transformation: he uses the alternative catalyst antipode (R)-**63**, to prepare **72** via (R)-**69**, which is then readily converted to macrocyclic **73** through a Mitsunobu reaction. The target molecule is accessed by formation of the disulfide bond upon treatment with I_2 (Scheme 11).

In addition to Evans' Cu(II)-catalyzed and Carreira's Ti-catalyzed asymmetric aldol reactions, there is omit Shibasaki's La-catalyzed protocol.[14] A recent total synthesis of one of the more celebrated targets of the nineties, epothilone A, utilizes both an enantioselective Al-catalyzed cyanide addition to an aldehyde (**75** → **77**) and a La-catalyzed enantioselect-

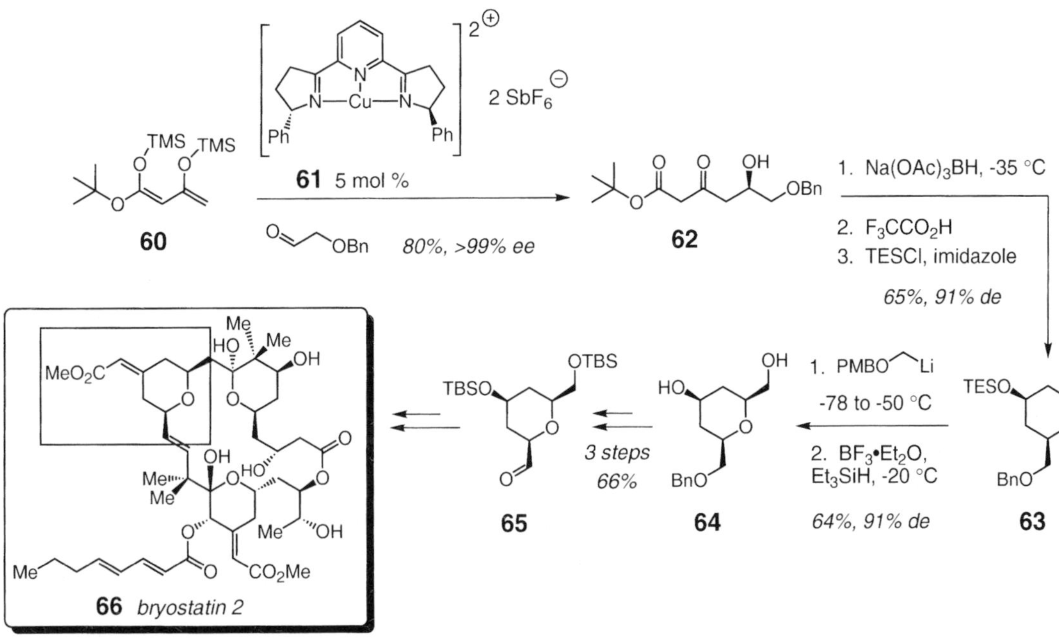

Scheme 10. Cu-catalyzed asymmetric aldol addition used in the enantioselective total synthesis of bryostatin **2** by Evans (1998).

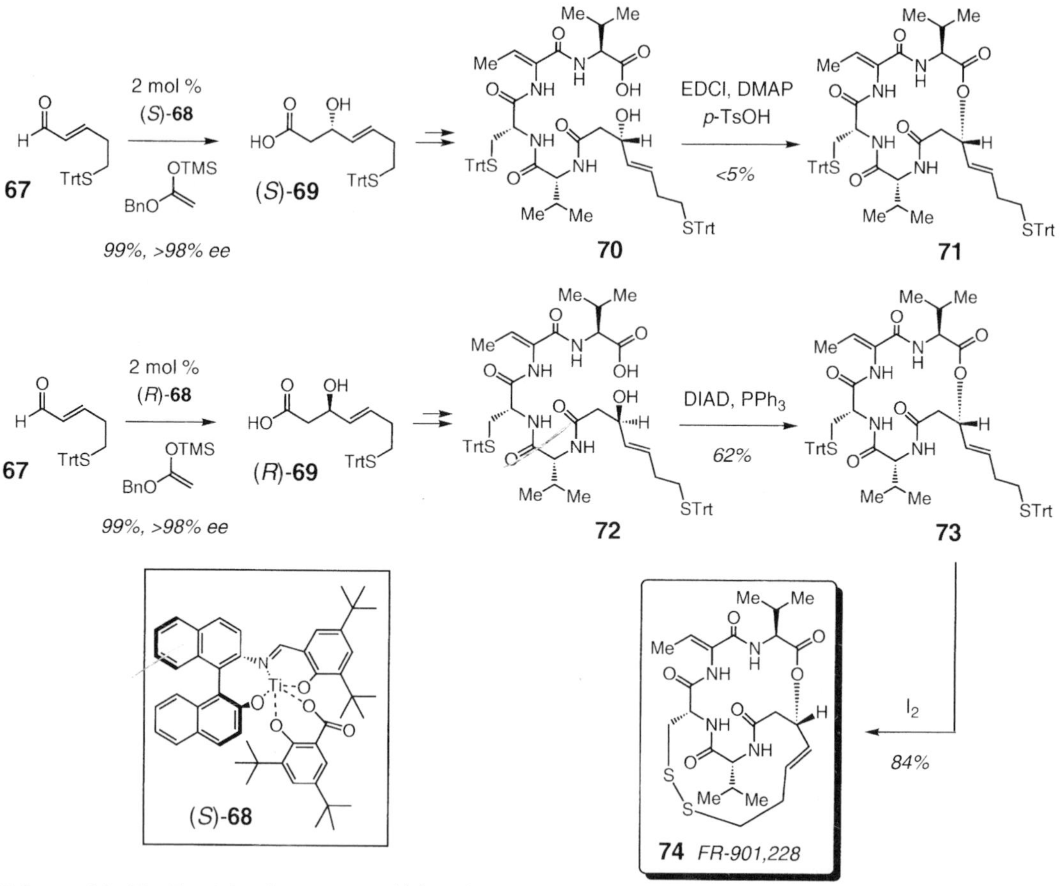

Scheme 11. The Ti-catalyzed asymmetric aldol renders both aldol adduct antipodes readily accessible, allowing Simon to complete the total synthesis of FR-901, 228 (1996).

ive aldol (*rac*-**82** → **84**, Scheme 12). Both transformations are effected by binol-based catalysts **76** and **83**, where the catalytic asymmetric aldol reaction is used to effect a kinetic resolution process.

Catalytic Asymmetric Ring-Closing Metathesis

Catalytic olefin metathesis, in only a few years, has risen to be one of the most important and reliable processes in organic synthesis. Recently, several reports by Schrock and Hoveyda[15b-c] have indicated that certain chiral Mo catalysts can effect these C–C bond forming transformations efficiently and enantioselectively. A recent concise and enantioselective synthesis of *exo*-brevicomin by Burke utilizes chiral catalyst **91** (Scheme 13) to effect the desymmetrization of **90** through a ring-closing metathesis.[15a]

Catalytic Asymmetric C–O Bond Forming Reactions in Total Synthesis

Catalytic Asymmetric Epoxidation

One of the earliest and most important discoveries in metal-catalyzed asymmetric synthesis is Sharpless's Ti-catalyzed epoxidation of allylic alcohols. A mere mention of all the total syntheses that have used this technology would require a separate review article. Here, we select Trost's masterful total synthesis of solamin (**100**, Scheme 14), for its beautiful and multiple use of Sharpless's asymmetric epoxidation.[16] Optically pure epoxy alcohol **95** is converted to both epoxy iodide **96** and diol **97**. The latter two intermediates are then united to give **98**, which is oxidized and converted to dihydrofuran **99** by a Ramberg-Backlund transformation. The Re-catalyzed butenolide annulation that is used to afford the requisite unsaturated lactone only adds to the efficiency of this beautiful total synthesis.

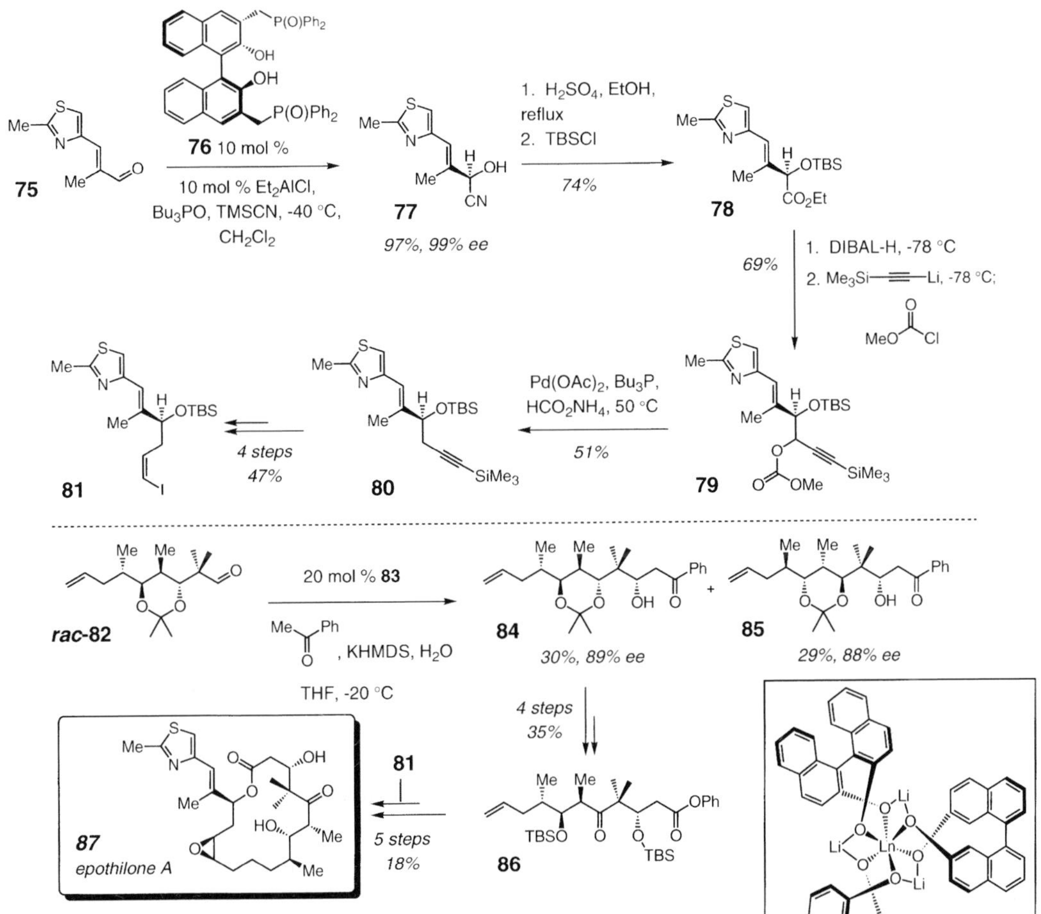

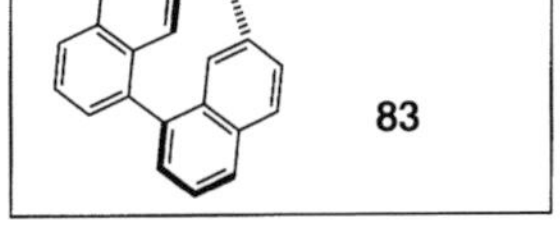

Scheme 12. Shibasaki utilizes a Al-catalyzed asymmetric CN addition to aldehydes and a La-catalyzed asymmetric aldol in the total synthesis of epothilone A (2000).

Scheme 13. Asymmetric Mo-catalyzed ring-closing metathesis used by Burke in a concise total synthesis of exo-brevicomin (1999).

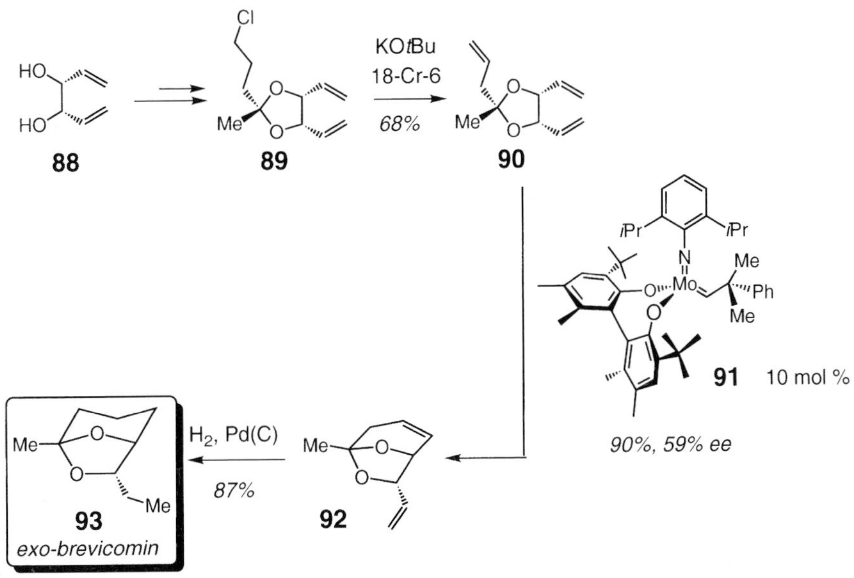

Scheme 14. Trost's use of directed asymmetric epoxidation in the enantioselective total synthesis of solamin (1994).

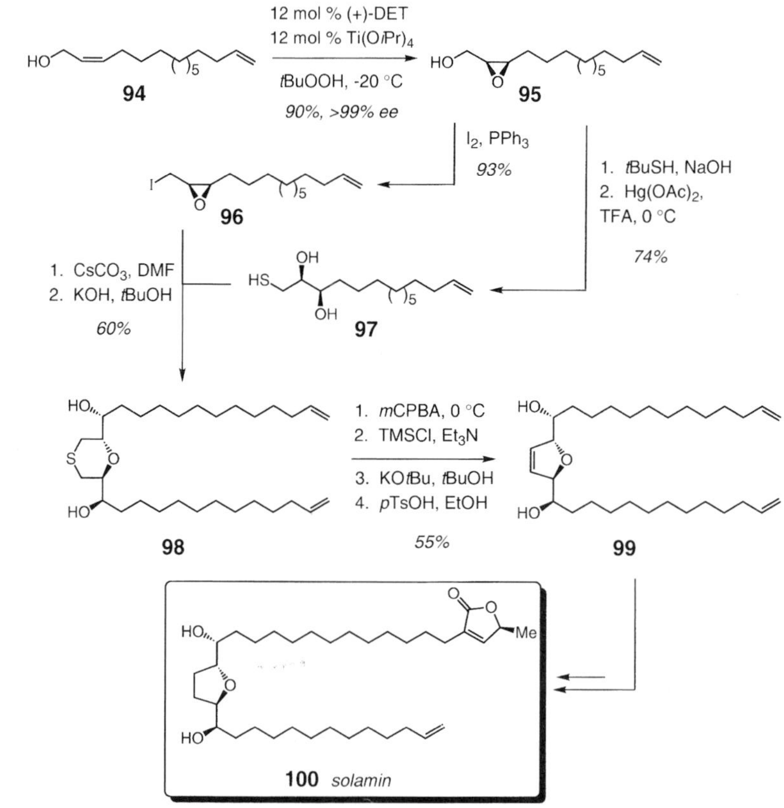

Catalytic Asymmetric Allylic Substitutions

There is little doubt that the discovery of a class of chiral catalysts by Trost, represented by **102** in Scheme 15, constitutes one of the most important recent developments in asymmetric catalysis. The total synthesis of aflatoxin B, depicted in Scheme 15,

involves the catalytic asymmetric Pd-catalyzed addition of **101** to **103**.[17] Trost's total synthesis introduces a cunning synthesis strategy: due to the presence of the carbonyl unit, the two enantiomeric Pd-π-allyl complexes can be rapidly interconverted, such that chiral racemic **103** is converted to **104** in *89% yield* and *>95% ee* (dynamic kinetic asymmetric reaction).

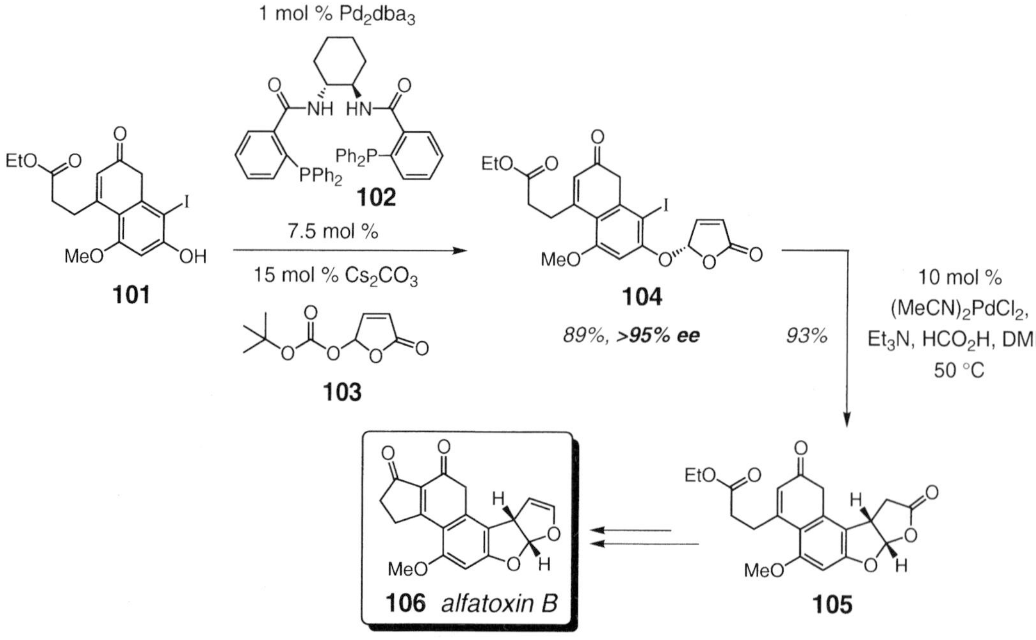

Scheme 15. Trost's approach to aflatoxin B involves the Pd-catalyzed dynamic asymmetric transformation of chiral racemic **103** to optically pure **104** (1999).

Scheme 16. Catalytic asymmetric dihydroxylation used by Corey to prepare 109, used in his 1985 total synthesis of ovalicin, in the optically pure form (1994).

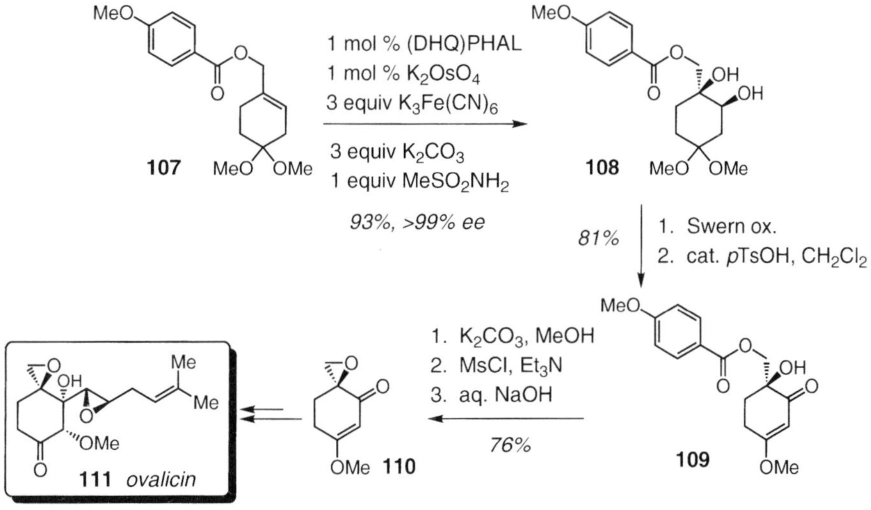

Catalytic Asymmetric Dihydroxylation

Along with catalytic asymmetric epoxidation, the related dihydroxylation of olefins is another venerable catalytic enantioselective process that is widely used by the modern organic chemist. An application of this important transformation may be found in Corey's 1994 preparation of optically pure 109 (Scheme 16), an intermediate in Corey's 1985 total synthesis of ovalicin.[18] The catalytic asymmetric dihydroxylation that affords 108 solves one of the most challenging problems in the total synthesis: installment of the tertiary alcohol center with the appropriate relative and absolute stereochemistry.

Catalytic Asymmetric C–N Bond Forming Reactions in Total Synthesis

Catalytic Asymmetric Allylic Substitution

In addition to its utility in the enantioselective formation of C–O bonds (cf. Scheme 15), Trost's chiral ligand 102 has been used in the catalytic asymmetric synthesis of C–N bonds. An impressive application of this protocol is in the enantioselective total synthesis of pancrastatin by Trost (Scheme 17).[19] Thus, Pd-catalyzed desymmetrization of 112 leads to the formation of 113 efficiently and in > 95 % ee. The follow-up use of the N₃ group to fabricate the requisite cyclic amide via isocyanate 117 demonstrates the impressive versatility of this asymmetric technology.

Catalytic Asymmetric C–H Bond Forming Reactions in Total Synthesis

Catalytic Asymmetric Reduction of Ketones

This important class of catalytic enantioselective reactions has been used for the preparation of a large variety of optically pure intermediates in a number of natural product total syntheses. Two examples, involving total syntheses of immunosuppressive agents FK506 (123)[20a] and rapamycin (127) by Schreiber, shown in Scheme 18, are illustrative.[20b] In these impressive efforts, Noyori's Ru-catalyzed asymmetric hydrogenation of β-ketoesters is used to establish secondary alcohol absolute stereochemistries that later serve to install other neigboring centers of asymmetry. In certain cases, the secondary alcohol center, after being used to incorporate other stereogenic ceners, is oxidized to the derived carbonyl unit (122 → 123).

Another important class of catalysts, introduced by Corey, that readily and enantioselectively effect ketone reductions are represented by 129. An excellent recent review article by Corey provides an in-depth coverage of this topic.[21a] In Scheme 19, two representative examples are illustrated. In the examples shown, these versatile catalysts are employed, in conjunction with a borohydride reagent, in the syntheses of important target molecules. One is a recent effort by Corey, leading to the enantioselective contruction of (–)-wodeshiol. Optically enriched 130, obtained from asymmetric reduction of 128, is used to prepare vinylstannane 131, which is transformed to 132 through a Pd-catalyzed dimerization. A subsequent Ti-catalyzed diastereoselective double-epoxidation and acid hydrolysis yields the target molecule.[21b] Another

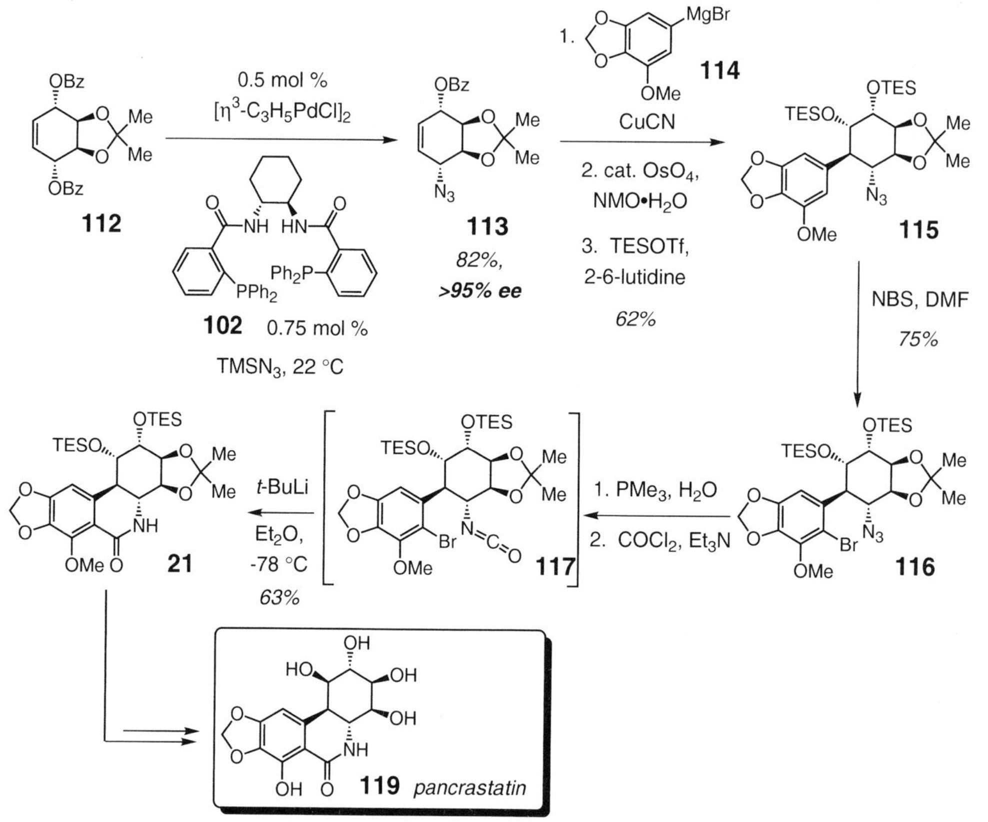

Scheme 17. Trost's total synthesis of pancrastatin; chiral ligand 102 is utilized in a Pd-catalyzed asymmetric allylic alkylation (1995).

Scheme 18. Ru-catalyzed asymmetric ketone reduction used by Schreiber in the total syntheses of immunosuppressive agents FK506 and rapamycin (1990, 1993).

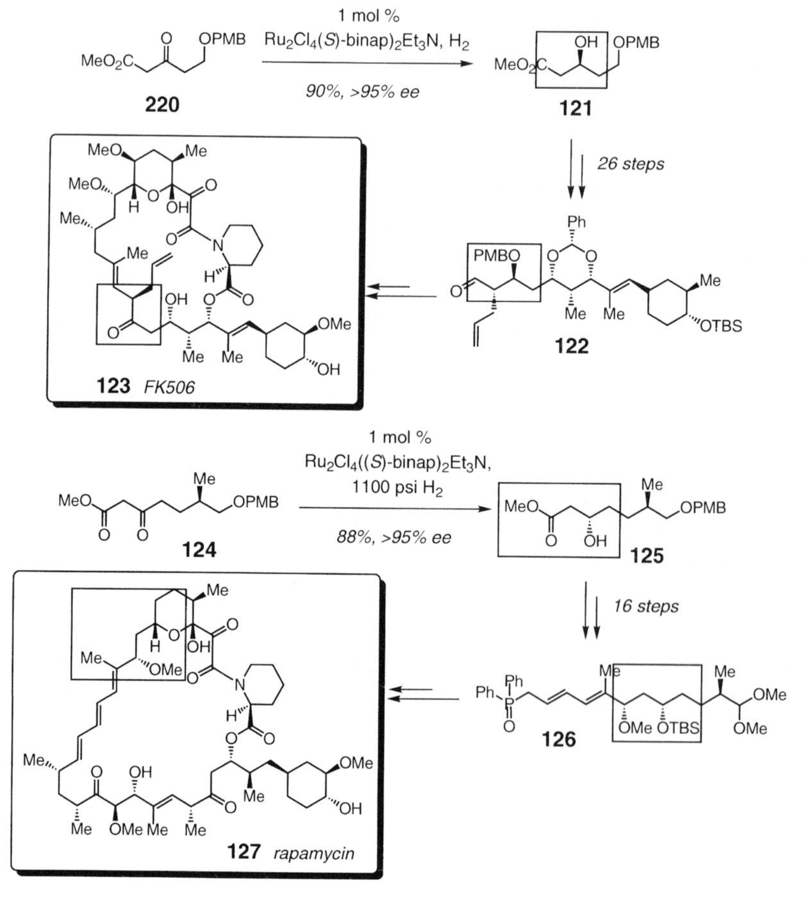

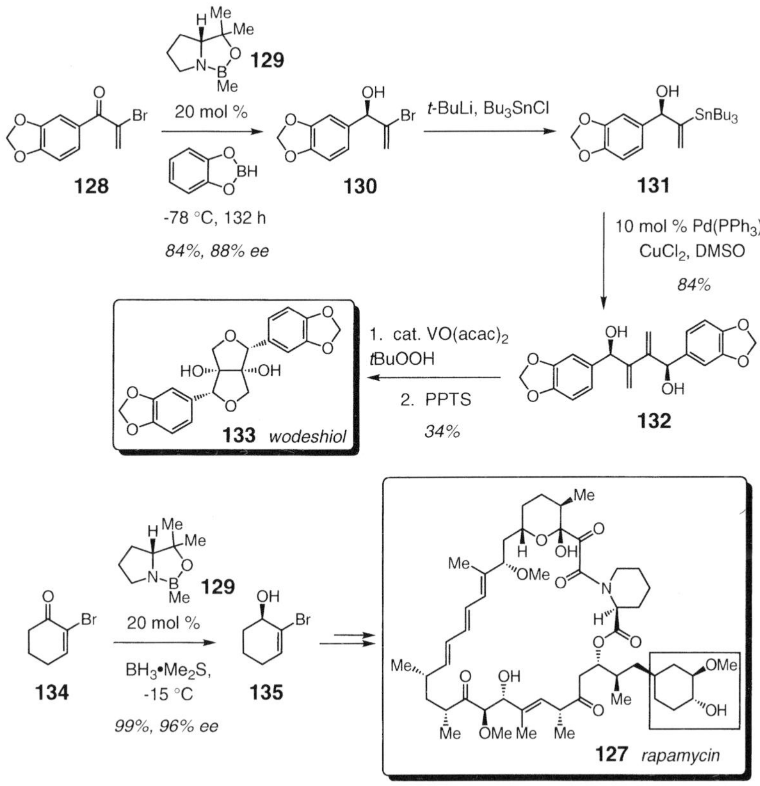

Scheme 19. Corey's utilization of B-methyl CBS catalyst in the asymmetric reduction of a ketone en route to wodeshiol (1999), and Nicolaou's application of the Corey technology to a total synthesis of rapamycin (1993).

example, involves the asymmetric catalytic synthesis of **135** by Nicolaou, in his total synthesis of rapamycin (**127**, Scheme 19).[21c]

Catalysis-Based Total Syntheses

Due to many impressive advances in metal-catalyzed transformations, both asymmetric and non-asymmetric, several efforts have been directed towards designing total synthesis routes that very heavily depend on various catalytic methods. These total syntheses benefit from the economic efficiency and environmental consciousness that are two of the inherent attributes of catalytic reactions. The total synthesis of wodeshiol **133** by Corey, discussed above (Scheme 19) is one such example. Two additional catalysis-based enantioselective total syntheses are briefly discussed below. In both efforts, all centers of asymmetry are attained by a catalytic enantioselective method, and the synthesis is completed through the use of several other catalytic reactions.

Catalytic Enantioselective Total Synthesis of Taurospongin A

The first example involves the total synthesis of taurospongin A **144**, reported by Jacobsen in 1998 (Scheme 20).[22] Jacobsen uses his celebrated salen ligand to accomplish a Cr-catalyzed kinetic resolu-

tion of terminal epoxide **136**. The resulting (S)-**136** is alkylated and then coupled with optically pure **140**, prepared by the above mentioned Ru-catalyzed reduction of β-ketoester **139**. With **141** in hand, Jacobsen effects a Ru-catalyzed and highly diastereoselective ketone reduction to prepare the key intermediate **143**, which is then transformed into the target molecule in six steps and 34 % overall yield. Jacobsen's total synthesis is an impressive demonstration of the notable potential that can be tapped when various metal-catalyzed reactions are used in tandem.

Catalytic Enantioselective Total Synthesis of Nebivolol

The second example is the enantioselective total synthesis of the antihypertensive agent nebivolol by Hoveyda, also reported in 1998.[23] Here, *rac*-**146** is prepared by a highly diastereo- and stereoselective Pd-catalyzed addition of an alkoxystanane; this Pd-catalyzed reaction constitutes a net S_N2 reaction with retention of stereochemistry. The resulting styrenyl ether is then resolved catalytically through a Zr-catalyzed enantioselective alkylation reaction to afford optically pure (S,R)-**145** in 40 % isolated yield (50 % maximum). A subsequent catalytic tandem ring-opening/ring-closing metathesis, effected by 4 mol% of Schrock's Mo catalyst **149**, leads to the formation of chromene **150**; optically pure **151** is sub-

Scheme 20. Jacobsen's
sequential use of catalytic
asymmetric reactions,
including his Cr-catalyzed
kinetic resolution of epoxides in the total synthesis of
taurospongin A (1998).

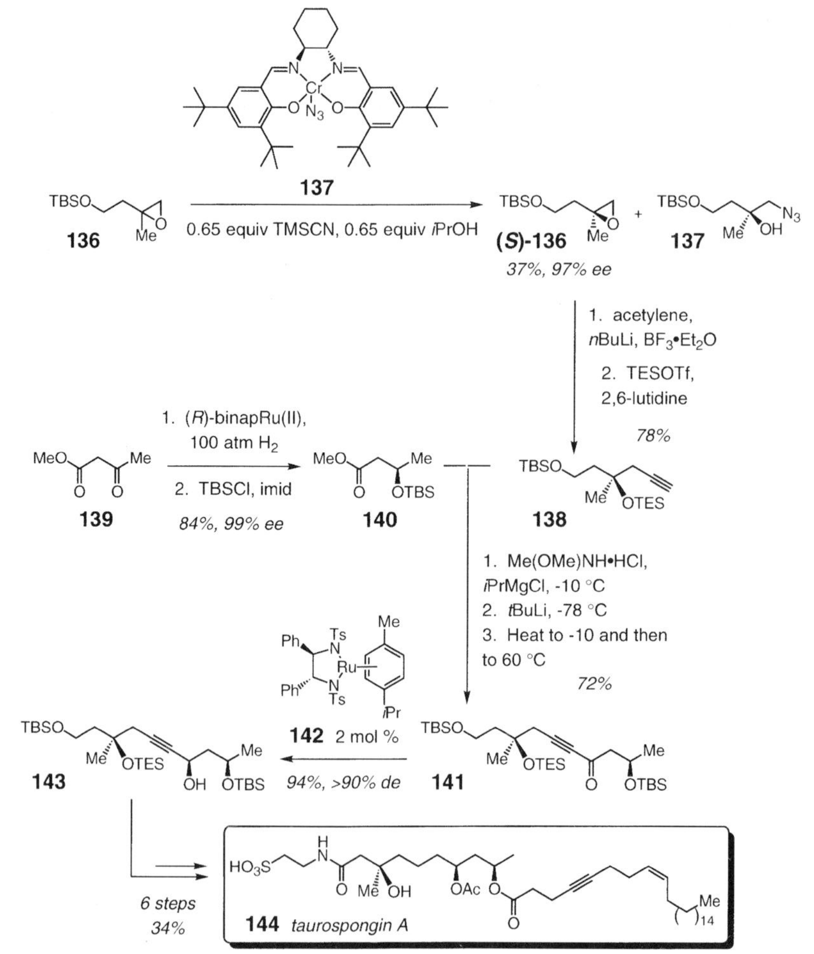

sequently accessed in four steps and 70 % overall yield. Optically pure amine **154** is obtained by a similar procedure, where *rac*-**152** serves as the starting material. It is noteworthy that whereas catalytic resolution of *rac*-**146** requires (*S*)-**147**, that of **152** necessitates the use of the (*R*) – antipode of the same chiral metallocene. Hoveyda uses the resolution procedure of Buchwald[24] to attain both (*R*)- and (*S*)-**147** in a single resolution. The total synthesis is completed by the reductive amination of the optically pure amine and aldehyde, followed by a deprotection step.

Epilogue

Only twenty years ago, effecting a chemical reaction catalytically and in an enantioselective manner appeared to be a difficult and relatively unattainable goal to organic chemists. Clearly, many significant advances have since been made in this area, as is evident by the total syntheses summarized in this article. These accomplishments indicate that catalytic enantioselective methods can, and should be, utilized in the planning and execution stages of multistep syntheses of target molecules. In this

manner, absolute stereochemistry is established efficiently, without recourse to installment and removal of chiral auxiliaries and without circuitous and lengthy schemes involving modification of carbohydrate starting materials. As shown in a number of total syntheses, such as those in Scheme 19 and 20 by Corey and Jacobsen, or in Scheme 8 by Overman, when such processes are combined with other catalytic reactions, remarkably economical, selective and efficient total syntheses emerge.

The field of asymmetric catalysis in synthesis is far from mature however. There are a considerable number of important chemical transformations that do not yet have a catalytic enantioselective variant, and few of the existing catalytic asymmetric reactions – perhaps none of them – are truly general. Some argue that there are no general methods, asymmetric or not; several asymmetric variations of the same general transformation may therefore be needed. Furthermore, many of the applications that were discussed here were employed by researchers that also developed the asymmetric methods; we have not reached a stage yet, when scientists not involved in the discovery and development of catalytic asymmetric reactions regularly utilize such protocols. It is hoped that this article

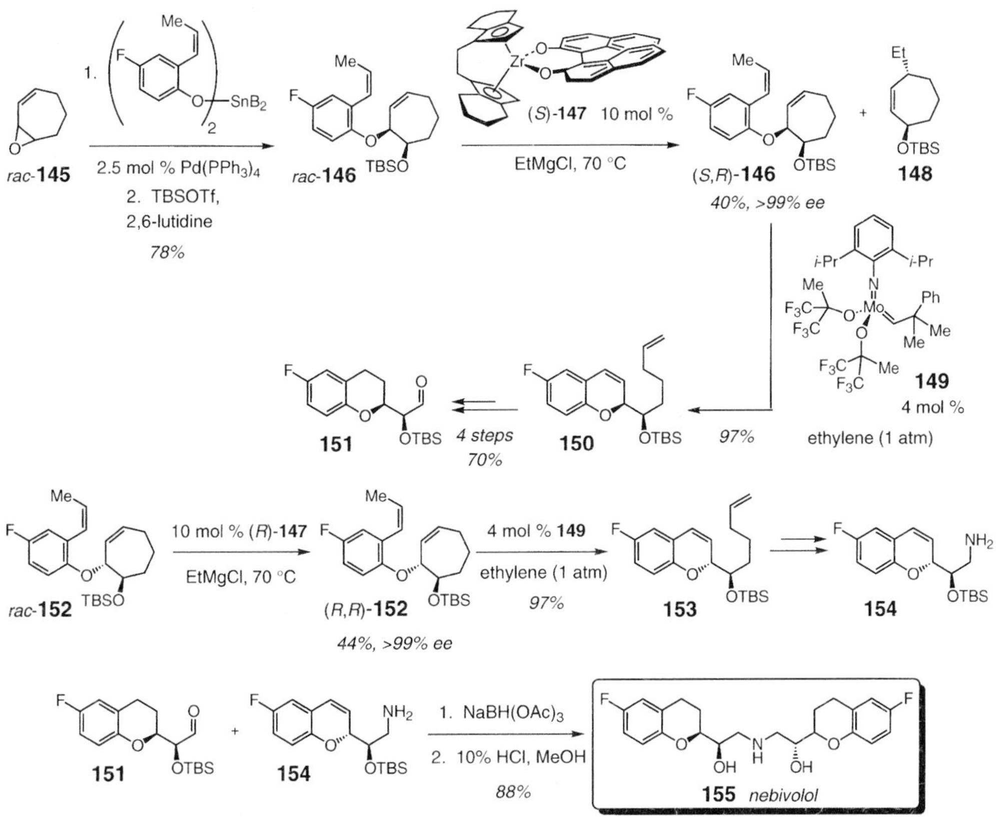

Scheme 21. Zr-catalyzed asymmetric olefin alkylation is used in conjunction with Pd-catalyzed addition of an aryloxide to an epoxide and Mo-catalyzed olefin metatheses in Hoveyda's total synthesis of nebivolol (1998).

demonstrates to the practitioners of the art of organic synthesis that catalytic asymmetric technologies can offer an exciting and more efficient alternative for the preparation of organic molecules in the optically pure form.

References and Notes

1. W. Oppolzer, R. N. Radinov, "Synthesis of (*R*)-(–)-Muscone by an Asymmetrically Catalyzed Macrocyclization of an ς-Alkynal", *J. Am. Chem. Soc.* **1993**, *115*, 1593–1594.

2. H. Yoshizake, H. Satoh, Y., Satoh, S. Nukui, M. Shibasaki, M. Mori, "Palladium-Mediated Asymmetric Synthesis of *Cis*-3,6-Disubstituted Cyclohexenes. A Short Total Synthesis of Optically Active (+)-γ-Lycorane", *J. Org. Chem.* **1995**, *60*, 2016–2021.

3. N. Kardos, J-P. Genet, "Synthesis of (–)-Chanoclavine I", *Tetrahedron Asymmetry.* **1994**, *5*, 1525–1533.

4. T. Takemoto, M. Sodeoka, H. Sasai, M. Shibasaki "Catalytic Asymmetric Synthesis of Benzylic Quaternary Carbon Centers. An Efficient Synthesis of (–)-Eptazocine", *J. Am. Chem. Soc.* **1993**, *115*, 8477–8478.

5. (a) A. Ashimori, T. Matsuura, L. E. Overman, D. J. Poon "Catalytic Asymmetric Synthesis of Either Enantiomer of Physostigmine. Formation of Quaternary Carbon Centers with High Enantioselection by Intramolecular Heck Reactions of (Z)-2-Butenanolides", *J. Org. Chem.* **1993**, *58*, 6949–6951. (b) T. Matsuura, L. E. Overman, D. J. Poon "Catalytic Asymmetric Synthesis of Either Enantiomer of the Calabar Alkaloids Physostigmine and Physovenine" *J. Am. Chem. Soc.* **19987**, *120*, 6500–6503.

6. (a) E. J. Corey, A. Guzman-Perez, T-P. Loh "Demonstration of the Synthetic Power of Oxazaborolidine-Catalyzed Enantioselective Diels-Alder Reactions by Very Efficient Routes to Cassiol and Gibberellic Acid", *J. Am. Chem. Soc.* **1994**, *116*, 3611–3612. (b) E. J. Corey, T-P. Loh "First Application of Attractive Intramolecular Interactions to the Design of Chiral Catalysts for Highly Enantioselective Diels-Alder Reactions", *J. Am. Chem. Soc.* **1991**, *113*, 8966–8967.

7. D. A. Evans, J. S. Johnson "Chiral C2-Symmetric Cu(II) Complexes as Catalysts for Enantioselective Intramolecular Diels-Alder Reactions. Asymmetric Synthesis of (–)-Isopulo'upone", *J. Org. Chem.* **1997**, *62*, 786–787.

8. M. E. Fox, C. Li, J. P. Marino, Jr, L. E. Overman "Enantiodivergent Total Synthesis of (+)- and (–)-Scopadulcic Acid A", *J. Am. Chem. Soc.* **1999**, *121*, 5467–5480.

9. (a) H. M. L. Davies, T. Hansen, D. W Hopper, S. A. Panaro "Highly Regio-, Diastereo-, and Enantioselective C–H Insertions of Methyl Aryldiazoacetates into Cyclic N-Boc-Protected Amines. Asymmetric Synthesis of Novel C2-Symmetric Amines and *threo*-Methylphenidate", *J. Am. Chem. Soc.* **1999**, *121*, 6509–6510. (b) For the study of the asymmetric method, see: H. M. L. Davies, T. Hansen "Asymmetric Intermolecular Carbenoid C-H Insertions Catalyzed by Rhodium(II) (*S*-N-(*p*-Dodecylphenyl)sulfonylprolinate", *J. Am. Chem. Soc.* **1997**, *119*, 9075–9076.

10. J. M. Axten, R. Ivy, L. Krim, J. D. Winkler "Enantioselective Synthesis of D-*threo*-Methylphenidate", *J. Am. Chem.*

Soc. **1999**, 6511–6512. (b) M. Prashad, H-Y. Kim, Y. Lu, Y. Liu, D. Har, O. Repic, T. J. Blacklock, P. Giannousis "The First Enantioselective Synthesis of (2*R*,2'*R*)-*threo*-(+)-Methylphenidate Hydrochloride", *J. Org. Chem.* **1999**, *64*, 1750–1753.

11. D. A. Evans, P. H. Carter, E. M. Carreira, J. A. Prunet, A. B. Charette, M. Lautens "Asymmetric Synthesis of Bryostatin 2", *Angew. Chem., Int. Ed. Engl.* **1998**, *37*, 2354–2359. For methodological studies on asymmetric Cu-catalyzed aldol addition, see: D. A. Evans, J. Murry, M. C. Kozlowski "*C*2-Symmetric Cu(II) Complexes as Chiral Lewis Acids. Catalytic Enantioselective Aldol Additions of Silylketene Acetals to (Benzyloxy)acetaldehyde", *J. Am. Chem. Soc.* **1996**, *118*, 5814–5815.

12. Y. Kim, R. A. Singer, E. M. Carreira "Total Synthesis of Macrolactin A with Versatile Catalytic, Enantioselective Dienolate Aldol Addition Reactions", *Angew. Chem., Int. Ed. Engl.* **1998**, *37*, 1261–1263. For initial methodological studies, see: (a) E. M. Carreira, R. A. Singer, W. Lee "Catalytic Enantioselective Aldol Additions with Methyl and Ethyl Acetate *O*-Silyl Enolates: A Chiral Tridentate Chelate as a Ligand for Titanium(IV)", *J. Am. Chem. Soc.* **1994**, *116*, 8837–8838. (b) R. A. Singer, E. M. Carreira "Catalytic, Enantioselective Dienolate Additions to Aldehydes: Preparation of Optically Active Acetoacetate Aldol Adducts", *J. Am. Chem. Soc.* **1995**, *117*, 12360–12361. (c) J. Kruger, E. M. Carreira "Apparent Catalytic Generation of Chiral Metal Enolates: Enantioselective Dienolate Additions to Aldehydes Mediated by Tol-BINAP-Cu(II) Fluoride Complexes", *J. Am. Chem. Soc.* **1998**, *120*, 837–838.

13. K. W. Li, J. Wu, W. Xing, J. A. Simon "Total Synthesis of the Antitumor Depsipeptide FR-901, 228", *J. Am. Chem. Soc.* **1996**, *118*, 7237–7238.

14. D. Sawada, M. Shibasaki "Enantioselective Total Synthesis of Epothilone A Using Multifunctional Asymmetric Catalyses", *Angew. Chem., Int. Ed. Engl.* **2000**, *39*, 209–213.

15. (a) S. D. Burke, N. Muller, C. M. Beudry "Desymmetrization by Ring-Closing Metathesis Leading to 6,8-Dioxabicyclo[3.2.1]octones: A New Route for the Synthesis of (+)-*exo*- and *endo*-Brevicomin", *Org. Lett.* **1999**, *1*, 1827–1829. For initial methodological work, see: (b) D. S. La, J. B. Alexander, D. R. Cefalo, D. D. Graf, A. H. Hoveyda, R. R. Schrock "Mo-Catalyzed Asymmetric Synthesis of Dihydrofurans. Catalytic Kinetic Resolution and Enantioselective Desymmetrization through Ring-Closing Metathesis", *J. Am. Chem. Soc.* **1998**, *120*, 9720–9721. (c) S. S. Zhu, D. R. Cefalo, D. S. La, J. Y. Jamieson, W. M. Davis, A. H. Hoveyda, R. R. Schrock "Chiral Mo-Binol Complexes: Activity, Synthesis, and Structure. Efficient Enantioselective Six-Membered Ring Synthesis through Catalytic Meathesis", *J. Am. Chem. Soc.* **1999**, *121*, 8251–8259.

16. B. M. Trost, Z. Shi "A Concise Convergent Strategy to Acetogenins. (+)-Solamin and Analogues", *J. Am. Chem. Soc.* **1994**, *116*, 7459–7460.

17. B. M. Trost, L. S. Chupak, T. Lubbers "Total Synthesis of (±)-and (+)-Valienamine via a Strategy Derived from New Palladium-Catalyzed Reactions", *J. Am. Chem. Soc.* **1998**, *120*, 1732–1740.

18. (a) E. J. Corey, A. Guzman-Perez, M. C. Noe "Short Enantioselective Synthesis of (–)-Ovalicin, a Potent Inhibitor of Angiogenesis, Using Substrate Enhanced Catalytic Asymmetric Dihydroxylation", *J. Am. Chem. Soc.* **1994**, *116*, 12109–12110. For the 1985 total synthesis of ovalicin,

see: E. J. Corey, J. P. Dittami "Total Synthesis of (±)-Ovalicin", *J. Am. Chem. Soc.* **1985**, *107*, 256–257.

19. B. M. Trost, S. R. Pulley, "Asymmetric Total Synthesis of (+)-Pancratistatin", *J. Am. Chem. Soc.* **1995**, *117*, 10143–10144.

20. (a) M. Nakatsuka, J. A. Ragan, T. Sammakia, D. B. Smith, D. E. Uehling, S. L. Schreiber "Total Synthesis of FK506 and an FKBP Probe Agent, (C8,C9–13C2)-FK506", *J. Am. Chem. Soc.* **1990**, *112*, 5583–5601. (b) D. Romo, S. D. Meyer, D. D. Johnson, S. L. Schreiber "Total Synthesis of (–)-Rapamycin Using an Evans-Tishchenko Fragment Coupling", *J. Am. Chem. Soc.* **1993**, *115*, 7906–7907. For the original study of the asymmetric reduction method, see: R. Noyori, T. Ohkuma, M. Kitamura, H. Takaya, N. Sayo, H. Kumobayashi, S. Akutagawa, "Asymmetric Hydrogenation of β-Keto Carboxylic Esters. A Practical, Purely Chemical Access to β-Hydroxy Esters in High Enantiomeric Purity", *J. Am. Chem. Soc.* **1987**, *109*, 5856–5858.

21. (a) E. J. Corey, C. J. Helal "Reduction of Carbonyl Compounds with Chiral Oxazaborolidine Catalysts: A New Paradigm for Enantioselective Catalysis and a Powerful New Synthetic Method", *Angew. Chem., Int. Ed. Engl.* **1998**, *37*, 1986–2012. (b) X. Han, E. J. Corey "A Catalytic Enantioselective Total Synthesis of (–)-Wodeshiol", *Org. Lett.* **1999**, *1*, 1871–1872. (c) K. C. Nicolaou, P. Bertinato, A. D. Piscopio, T. K. Chakraborty, N. Minowa "Stereoselective Construction of C21-C42 Fragment of Rapamycin", *J. Chem. Soc. Chem. Commun.* **1993**, 619–622.

22. (a) H. Lebel, E. N. Jacobsen "Enantioselective Total Synthesis of Taurospongin A", *J. Org. Chem.* **1998**, *63*, 9624–9625. For the development of the Cr-catalyzed kinetic resolution of epoxides, see: M. Tokunaga, J. F. Larrow, F. Kakiuchi, E. N. Jacobsen "Asymmetric Catalysis with Water: Efficient Kinetic Resolution of Terminal Epoxides by Means of Catalytic Hydrolysis", *Science*, **1997**, *277*, 936–938.

23. C. W. Johannes, M. S. Visser, G. S. Weatherhead, A. H. Hoveyda "Zr-Catalyzed Kinetic Resolution of Allylic Ethers and Mo-Catalyzed Chromene Formation in Synthesis. Enantioselective Total Synthesis of the Antihypertensive Agent (S,R,R,R)-Nebivolol" *J. Am. Chem. Soc.* **1998**, *120*, 8340–8347. For development of the enantioselective methodology, see: M. S. Visser, J. P. A. Harrity, A. H. Hoveyda "Zirconium-Catalyzed Kinetic Resolution of Cyclic Allylic Ethers. An Enantioselective Route to Unsaturated Medium Ring Systems", *J. Am. Chem. Soc.* **1996**, *118*, 3779–3780.

24. B. Chin, S. L. Buchwald "An Improve Procedure for the Preparation of Enantiomerically Pure Ethylenebis(tetrahydroindenyl) Zirconoim Derivatives", *J. Org. Chem.* **1996**, *61*, 5650–5651.

II Architecture, Organization and Assembly

From Fullerenes to Novel Carbon Allotropes: Exciting Prospects for Organic Synthesis

Yves Rubin[a] and François Diederich[b]

[a] Department of Chemistry and Biochemistry, University of California, Los Angeles, CA 90095–1569, USA
Phone: +310 206 2338
Fax: +310 206 7649
e-mail: rubin@chem.ucla.edu,
Web:
http://siggy.chem.ucla.edu

[b] Laboratorium für Organische Chemie, ETH Zürich, Universitätstrasse 16, 8092 Zürich, Switzerland
Phone: +41 1 632 2992,
Fax: +41 1 632 1109
e-mail:
diederich@org.chem.ethz.ch.

Keywords ■ Carbon Allotropes ■ Fullerenes ■ Cyclocarbons ■ Nanomaterials ■ Advanced Materials ■ Optical Activity

Concept: With the discovery of the fullerenes, it has become evident that elemental *carbon* can exist in almost an infinite number of stable allotropes that are either molecular or polymeric in nature. Whereas achiral and chiral fullerenes can now be prepared in bulk quantities and methods for their regio- and stereoselective multiple functionalization are being developed in increasing numbers, the search for stable molecular and polymeric acetylenic carbon allotropes is still ongoing. The two areas encompassing fullerene chemistry and the construction of acetylenic all-carbon and carbon-rich networks have generated fascinating perspectives in organic synthesis and promise to provide new classes of advanced functional materials for technological applications.

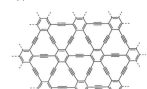

Prologue

The preparation of novel carbon allotropes and the exploration of their chemistry has driven a good part of targeted organic synthesis efforts over the past two decades. While several early proposals for new synthetic allotropes of carbon were published, it was the group of Orville L. Chapman at UCLA that first became involved in the early nineteen eighties with a rational total synthesis of a carbon allotrope, namely "soccerene" C_{60}.[1] In 1987, we initiated a program for the synthesis of the cyclo[n]carbons (cyclo-C_n) that are n-membered monocyclic rings constituted purely of sp-hybridized carbon atoms. These novel molecules were expected to have unique electronic structures stemming from

Abstract: The past two decades have profoundly changed the view that we have of elemental carbon. The discovery of the fullerenes, spherically-shaped carbon molecules, has permanently altered the dogma that carbon can only exist in its two stable natural allotropes, graphite and diamond. The preparation of molecular and polymeric acetylenic carbon allotropes, as well as carbon-rich nanometer-sized structures, has opened up new avenues in fundamental and technological research at the interface of chemistry and the materials sciences. This article outlines some fascinating perspectives for the organic synthesis of carbon allotropes and their chemistry. Cyclo[n]carbons are the first rationally designed molecular carbon allotropes, and

their formation in the gas-phase from defined precursors has been demonstrated. Evidence has also been provided that these monocyclic rings made-up of sp-hybridized carbon atoms are intermediates in the gas phase formation of fullerenes. In approaches to several two-dimensional all-carbon networks, a variety of perethynylated annulenes, radialenes, and other carbon-rich conjugated perimeters of nanometer dimensions have been prepared, and their unusual optoelectronic properties subsequently investigated. As a spin-off from this work, molecularly disperse π-conjugated rods incorporating a poly(triacetylene) backbone of remarkable stability have been synthesized. With a length of up to 12 nm, they represent the lon-

gest known linearly π-conjugated systems ("molecular wires") without aromatic rings in the conjugated backbone.

Following the discovery of a bulk fullerene preparation process in 1990, the covalent chemistry of these carbon allotropes has developed at a phenomenal pace. Frontier orbital (LUMO) and tether-directed functionalization concepts have been successfully applied to the regio- and stereoselective preparation of multiple covalent adducts of C_{60}. These have found increasing applications in the construction of functional supramolecules. More recently, the sequence of *Bingel* reaction – *retro-Bingel* reaction has provided an elegant access to isomerically pure higher fullerenes and, in particular, to pure carbon enantiomers.

Fullerene networks themselves have become fascinating targets. The preparation of a dimeric form of C_{60} (C_{120}) is already documented. A scheme for the rational opening, filling, and closing of a fullerene is close to becoming reality, while a total synthesis of C_{60} appears within reach.

their two perpendicular sets of conjugated π-orbitals, both in-plane and out-of-plane.[2] Results from this work were initially summarized in our *Angewandte Chemie* review.[1]

Five years after the first experimental evidence for the existence of C_{60}[3], the discovery of a bulk preparation method for fullerenes by Krätschmer, Huffman, and coworkers[4] ignited synthetic work on carbon allotropes in a truly explosive fashion. In addition to C_{60} and C_{70}, a variety of higher fullerenes were quickly discovered in the soot formed by resistive or arc heating of graphite under inert atmosphere.[5] Quite excitingly, some of these carbon molecules, e.g. D_2-C_{76}, were found to be chiral,[6] implying that carbon is one of the rare elements that can exist in stable enantiomerically pure modifications.[7] The functionalization chemistry of fullerenes developed thereafter with exponential growth,[8] providing a unique array of three-dimensional building blocks for molecular and supramolecular construction.[9] New frontiers arising from that body of work have been typified by rational synthetic approaches to endohedral fullerene complexes,[10] the formation of all-carbon fullerene-based networks,[11] and ultimately, a total synthesis of C_{60}. These fascinating topics will be discussed in greater detail in this article.

Despite formidable challenges, the construction of two- and three-dimensional all-carbon and carbon-rich networks has been pursued with great vigor in the past ten years.[12] Fueled by the availability of new synthetic methods, in particular Pd(0)-catalyzed cross-coupling reactions,[13] versatile "molecular construction kits" for perethynylated

molecules have been developed for the synthesis of acetylenic molecular scaffoldings.[14] The formation of exotic functional acetylenic molecular architectures with dimensions on the *multi*-nanometer scale, some displaying fascinating optoelectronic properties, has reached new heights. Progress in this field is surveyed in this article.

Cyclocarbons: All-Carbon Molecules through Rational Synthesis. Likely Intermediates in Fullerene Formation

One of the most debated questions since the discovery of the fullerenes has been that of the mechanism of their formation, especially considering the remarkable yields that can be achieved from the highly energetic plasma from which they originate. This is, in fact, a very complex issue because a number of intermediates of different structures are involved (Scheme 1).[15,16] Much of this debate has been fueled by the lack of solid experimental data on medium to large-sized carbon clusters. With the advent of the "ion chromatography" technique developed by Bowers, Jarrold, and coworkers, critical information has been gathered.[17,18] This technique correlates flight times of ions passing through a diffusion cell (helium as the collision gas) with their calculated diffusivities, which correlate themselves very well with the average cross-section of a molecule. This technique can be performed to such a high resolution that even "subtle" changes in molecular shape, hence of structure, can be

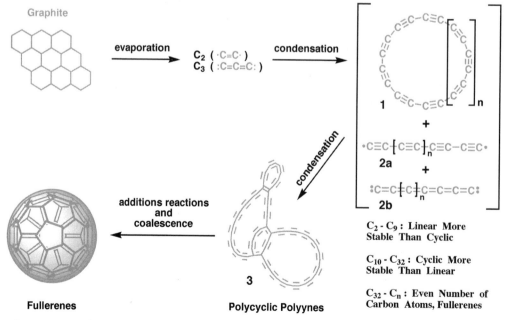

Scheme 1. Gas-phase chemistry of evaporated carbon.

Figure 1. The Stone-Wales rearrangement.

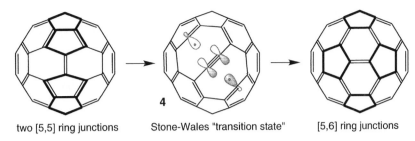

two [5,5] ring junctions   Stone-Wales "transition state"   [5,6] ring junctions

inferred.[19] From the results in a number of these studies, correlated further with high level computational work in Scuseria's group,[20] a general picture for the formation of fullerenes from carbon sources has emerged: Evaporation of carbon (graphite) at around 2500 °C first gives very reactive small clusters (mainly C_2 and C_3), which condense quickly in cooler regions to either linear or cyclic carbon molecules. The structures of these clusters depend on their size as indicated in Scheme 1, and can be a combination of both linear and cyclic forms for the same C_n.[16] Further condensations by cycloaddition reactions ([4+2], [6+4]) are then invoked to explain the formation of larger, fullerene-sized clusters (3), which coalesce ultimately to the most stable fullerenes by intramolecular rearrangements favored by the large energy released upon sp-to-sp^2 carbon rehybridization.[16,20] These multicyclic intermediates are indeed detected in the ion mobility experiments.[19]

Another important issue regarding the formation mechanism of fullerenes in the gas phase lies in the fact that a number of isomers, each with the molecular formula C_{2n}, must be formed during carbon coalescence, yet only one or very few structures are isolated, for example the unique isomer of C_{60} with its beautiful icosahedral symmetry. These lowest strain structures all follow the empirical isolated-pentagon rule (IPR),[16] which states that fullerenes with two-adjacent 5-membered rings are energetically unfavorable. The formation of the final fullerene structures can be explained by a bond-rearrangement process advanced by Stone and Wales.[21] This rearrangement is a general mechanism for the conversion of energetically unfavor-

able [5,5] ring junctions in fullerenes to more stable [5,6] ring junctions through a permutation involving the high-energy "transition state" 4 having an alkynyl unit separating two radical centers.[21] Since the calculated activation barrier for this process is very high (6–7 eV, 140–160 kcal·mol⁻¹), different proposals have been advanced.[22] One of these involves initial C_1-addition to the fullerene, lowering the barrier of this process by 2 eV through saturation of one of the radical centers of 4 by cyclopropanation. A similar argument can be made to explain a fullerene-growth process where formation of lower (or higher) fullerenes from condensing carbon vapor is achieved by sequential fragmentations (or additions) of C_2-units that lead ultimately to the most stable fullerenes *via* the Stone-Wales mechanism (Figure 1). Accordingly, a highly dynamic picture for fullerene formation seems to operate, in which cyclocarbons play a critical role, but where fullerene-to-fullerene conversions also participate in a phenomenal "symphonic orchestration".

Another important key experiment has helped shed light on this mechanism: 1) Carbon evaporation in a fullerene generation apparatus was carried out in the presence of dicyanogen (N≡C–C≡N) in an attempt to prepare heterofullerenes ($C_{59}N$, $C_{58}N_2$, etc.).[23] Instead, the linear dicyanopolyynes 5a–e and higher were formed with remarkable efficiency (Scheme 2). The interception of the singlet or triplet species 2a or 2b (Scheme 1) by cyano radicals can be invoked from this experiment and constitutes a good indication for the intermediacy of these carbon clusters. 2) A similar experiment in the presence of chlorine affords, among other chlorinated PAHs, decachlorocorannulene (6).[23]

Scheme 2. Trapping of carbon condensates with dicyanogen and chlorine.

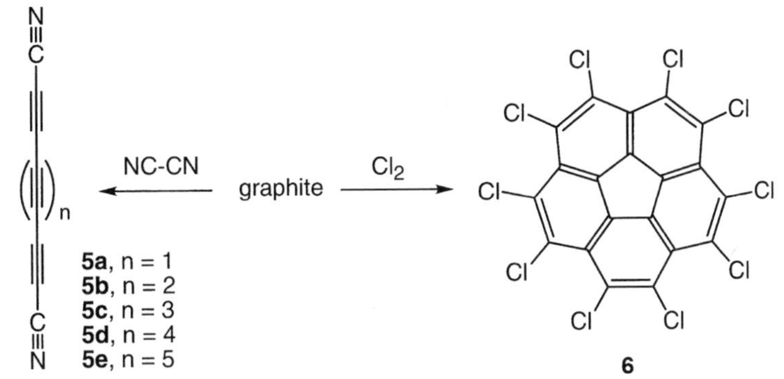

5a, n = 1
5b, n = 2
5c, n = 3
5d, n = 4
5e, n = 5

6

The question of the intermediacy of the cyclic polyynes **1** in fullerene formation is at the origin of an ambitious research program initiated by Diederich in the mid-eighties. The target molecule chosen for this project was C_{18}, in part because it was of an ideal size to accommodate ring strain based on its linearly *sp*-hybridized carbons, and because it was not so large that potential aromatic stabilization would become too small, as is the case with the larger [n]annulenes. More evidently, the availability of a suitable synthetic sequence was needed. This work has been reviewed previously[1] and will only be summarized here.

There is an important and theoretically intriguing question entirely proper to this class of molecules: Because these compounds consist only of *sp*-hybridized atoms, there are actually two sets of circularly conjugated pathways for the π-electrons, one *out-of-plane* (as usual), and one *in-plane* (Figure 2). Both can, in principle, contribute to some degree of aromatic stabilization. Interestingly, the HOMO/LUMO frontier orbitals for C_{18} are made up of two degenerate sets with *in-plane* and *out-of-plane* *p*-orbital configurations (Figure 2). This orthogonality implies that additions of nucleophilic or electrophilic reagents to these molecules may occur by in-plane or out-of-plane attacks, i.e. these molecules are rather "naked" and may be quite reactive depending on the amount of aromatic stabilization they experience (see below).

Several structural predictions for *cyclo*-C_{18} have been proposed and hotly debated.[24] Initially, self-consistent-field (SCF) calculations with a 3–21G or larger basis set by Houk and coworkers[24] predicted that the cyclic bond-localized acetylenic D_{9h}-structure **1a** (Figure 3) represents the ground-state geometry. However, electron correlation included in Møller-Plesset second-order perturbation theory (MP2), as well as in density functional theory (DFT), favors the fully delocalized cumulenic D_{18h}-structure **1b**. These levels of theory, however, exaggerate the contribution of diffuse orbitals and favor delocalized structures in general. The most recent calculations with the corrected Becke-LYP DFT func-

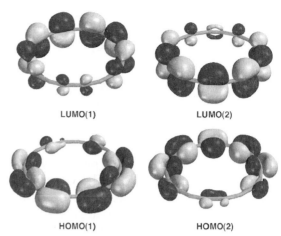

Figure 2. Frontier orbitals of D_{9h}-C_{18} at the RHF/6–31G* level (Spartan 5.0). Both sets of HOMO and LUMO orbitals are degenerate.

tional and RHF/6–31G* basis sets give strong evidence that *cyclo*-C_{18} is a polyyne of C_{9h}-symmetry, similar to **1a** but with two different, alternating bond angles at each –C≡C– unit.[24,25] These calculations also reveal that in the (4n+2) π-electron cyclocarbons, there exists a stronger tendency toward bond localization with increasing ring size than in the sp^2-hybridized annulene series. This trend can be compared to that seen in the two series of linear –CH=CH– and –C≡C– chains, where there is for example a marked difference between the UV-Vis spectra of Ph–(CH=CH)$_n$–Ph and Ph–(C≡C)$_n$–Ph as seen from the larger increase in bathochromic shifts in the polyene series (more color for the same *n*) resulting from a larger degree of delocalization.[26] It is clear that a synthesis and characterization of *cyclo*-C_{18} is required to provide a definitive answer to these questions.

Syntheses of *cyclo*-C_{18}, and subsequently of lower and higher members of this series, have been approached through the placement of good leaving groups at the three corners of a dehydroannulene (Scheme 3).[1,27] The choice of the leaving groups is critical because it is necessary to avoid nucleophilic reagents, toward which polyynes are very sensitive. Accordingly, cyclization reactions of alkene-functionalized *cis*-enediynes under oxidative condi-

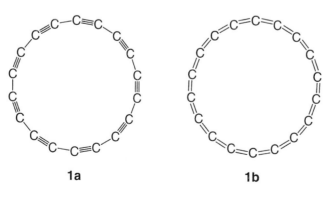

1a   **1b**

Figure 3. Possible structures of C_{18}.

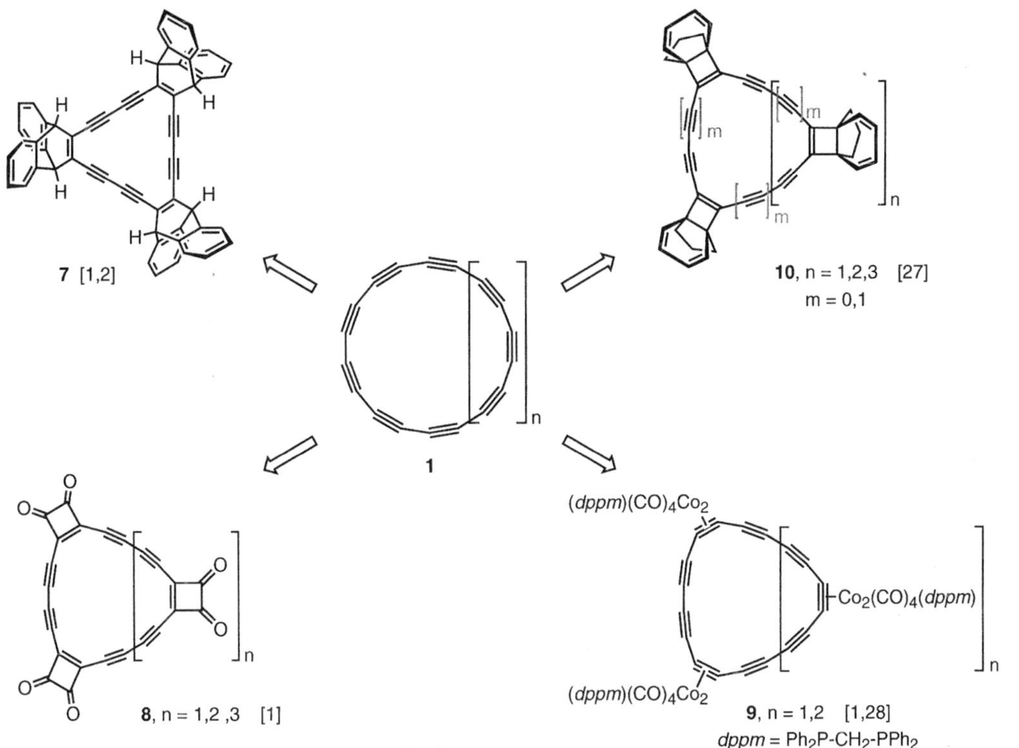

Scheme 3. Retrosynthetic routes to C_{18} and other cyclo[n]carbons.

tions give the 18-, and 24-, and 30-membered macrocycles **7–10** in good yields.[1,27] The formation of C_{18}, C_{24}, and C_{30}, and subsequently of C_{12}, C_{16}, and C_{20}, from these compounds has been demonstrated by mass spectrometric analysis of the flash-heated or flash-photolyzed precursors (LDMS).[1,27] One of the most interesting observations, unexpected at the time, was that the cyclocarbons coalesce in the gas phase to give rise to fullerenes.[1]

Although the preparation of bulk quantities of free *cyclo*-C_{18} remains elusive, one of its transition metal complexes has been prepared and characterized by X-ray crystallography.[1,28] Its synthesis benefited from the fact that alkynes can be protected with $Co_2(CO)_8$ to give (μ-acetylene)dicobalt hexacarbonyl complexes having dramatically reduced

C≡C–C angles of values close to 140C°, permitting unrestricted coupling cyclization.[1,28]

The starting material, 1,6-bis(triisopropylsilyl)-1,3,5-hexatriyne, was subjected to reaction with $Co_2(CO)_8$, followed by ligand exchange with the bridging *bis*(diphenyphosphino)methane (*dppm*) ligand to stabilize the dinuclear cobalt cluster. Deprotection of this intermediate to afford the stable dark-red dicobalt complex **11**, and ensuing oxidative coupling at high dilution, afforded the very stable cobalt complexes of *cyclo*-C_{18} (**12a**) and *cyclo*-C_{24} (**12b**) (Scheme 4). The X-ray crystal structure of **12a** revealed significant angle distortion of the three butadiyne moieties within the nearly planar C_{18} ring.[28] With C≡C–C angle values as low as 161C°, these distortions approach the bending expected for

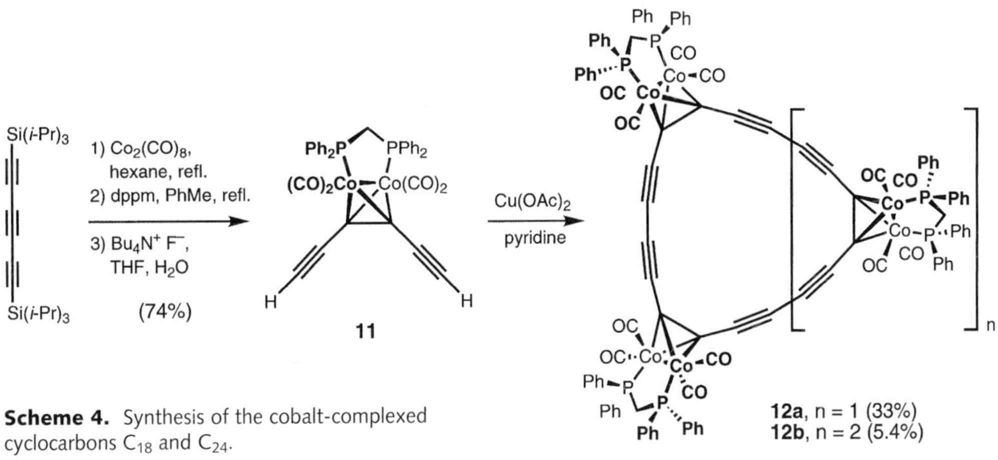

Scheme 4. Synthesis of the cobalt-complexed cyclocarbons C_{18} and C_{24}.

free *cyclo*-C_{18}. As anticipated from the steric shielding and intrinsic stability of the dicobalt moiety provided by the *dppm* ligand, oxidative or ligand-exchange reactions did not free *cyclo*-C_{18} from complex **12a**. Attempts to do this chemistry with the "unstabilized" $Co_2(CO)_6$-complexed alkynes unfortunately were thwarted by their unusual lability.

The availability of *cyclo*-C_{18} or higher members of this class of molecules, if stable to a point that they can be isolated and manipulated, would not only permit the study of their chemical and physical properties, but would also perhaps lead to oligomerizations that provide either planar graphdiyne sheets[1,29] or narrow buckytubes (Figure 4). Such oligomerizations would be favored by the proper arrangement of individual moieties; for example, self-assembled monolayers of *cyclo*-C_{18} on a metal or other surface, or through "doughnut" stacks of these molecules templated by a crystal lattice or a pseudorotaxane-like thread.

Acetylenic Molecular Scaffoldings: From Perethynylated Building Blocks to All-Carbon Networks and Carbon-Rich Nanostructures

Although a novel polymeric carbon network different from diamond or graphite still awaits preparation, efforts directed towards its synthesis have generated a great variety of theoretically interesting molecules.[29] Accordingly, a variety of two- and three-dimensional macrocyclic oligo(phenylacetylene)s and oligo(phenyldiacetylene)s have been prepared as models for such allotropes.[30] Large planar organometallic carbon-rich molecular objects have also been prepared starting from multiply ethynylated π-complexes of iron, manganese, and cobalt.[31,32]

The tetrabenzohexadehydro[20]annulene **13** was found by Vollhardt and coworkers[33] to undergo explosive decomposition under formation of graphitized carbon particles in low yield aside from

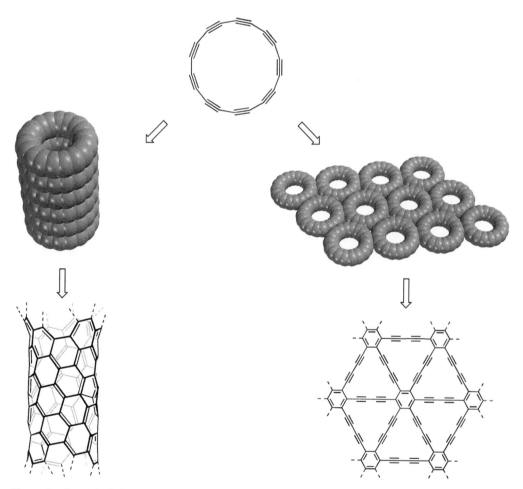

Figure 4. Conceptual oligomerization schemes for cyclo-carbons to planar sheets (graphdiynes)[1,29] and buckytubes.

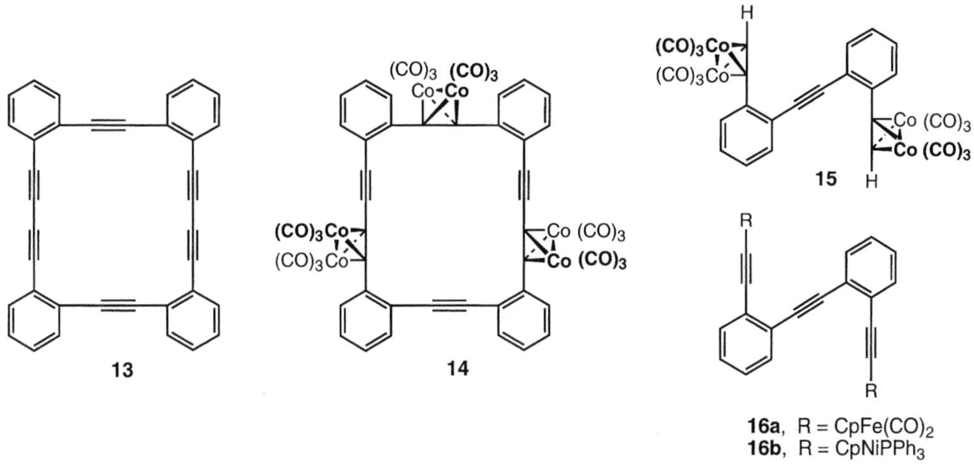

Figure 5. Tetrabenzodehydroannulene **13** was discovered to undergo explosive decomposition under formation of graphitized carbon particles in low yield. High yields of car-bon onions and multiwalled nanotubes, partially containing crystalline metal deposits in their interior were obtained by pyrolysis of the transition metal complexes **14–16**.

amorphous carbon and graphite as the main components (Figure 5). A much improved result was obtained when the dicobalt hexacarbonyl complexes **14** or **15** were pyrolyzed.[34] Thermal treatment of **14** at 800 °C for 6 h produced large quanti-

ties of well-formed carbon onions and multiwalled nanotubes. Some of the tubes and onions were shown by transmission electron microscopy (TEM) to contain crystalline metal deposits in their interior. Similar results were obtained by thermolysis of

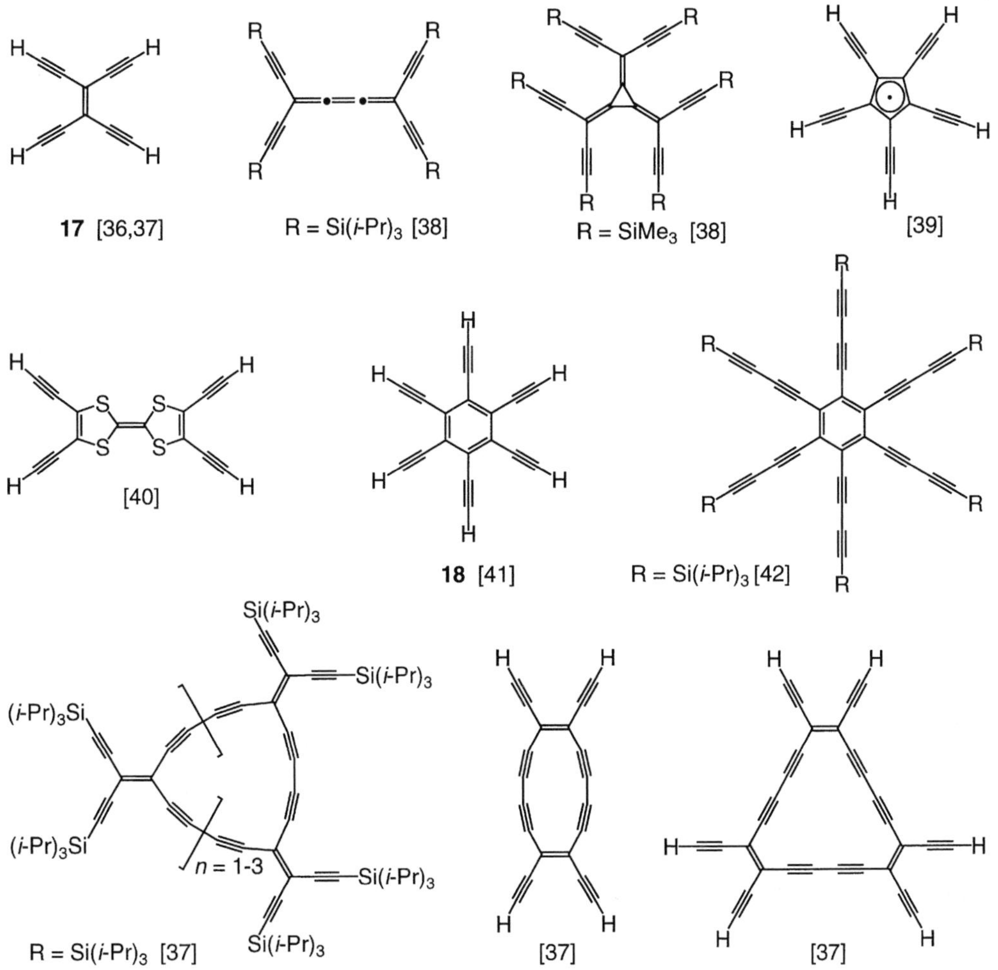

Figure 6. Examples for known perethynylated building blocks for two-dimensional acetylenic scaffoldings.

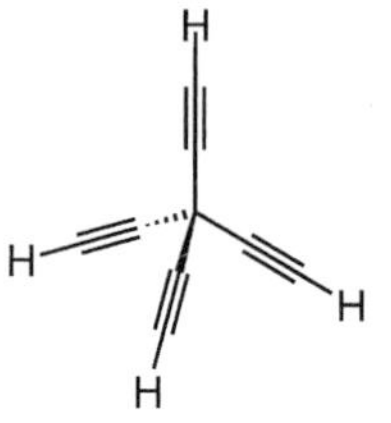

[43]

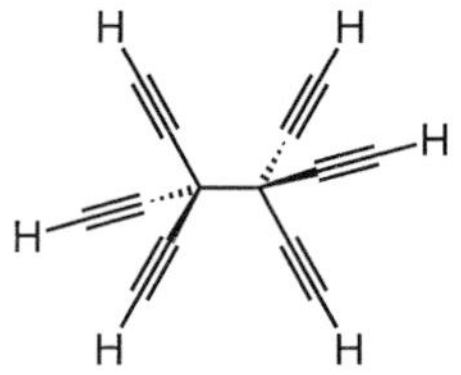

[43]

Figure 7. Known perethynylated modules for three-dimensional all-carbon scaffolding.

the transition metal-terminated iron and nickel complexes **16a** and **16b**. It is conceivable that further experimentation with complexes formed by acyclic and macrocyclic oligo(phenylacetylenes) will provide valuable, unprecedented insight into the mechanisms behind transition-metal mediated carbon nanoparticle assembly.[35]

Several two-dimensional all-carbon networks comprise the core of tetraethynylethene (TEE, **17**) as a monomeric repeat unit.[1,12] In 1991, we synthesized the hitherto elusive "perethynylated ethene" **17** on the way to these novel materials.[36] In the

mean-time, a rich diversity of perethynylated derivatives have been prepared as modules for two-dimensional acetylenic molecular scaffoldings. They extend from simple benzene derivatives to heterocycles, radialenes and expanded radialenes, and to dehydroannulenes. Examples are shown in Figure 6. Several perethynylated modules for three-dimensional scaffoldings, including tetraethynylmethane, have been prepared (Figure 7). Several perethynylated modules for acetylenic nanoconstruction that are attractive targets for future research are depicted in Figure 8.

Molecular scaffoldings with tetraethynylethenes (TEEs, 3,4-diethynylhex-3-ene-1,5-diynes) and *trans*-1,2-diethynylethenes [DEEs, (*E*)-hex-3-en-1,5-diynes] are at a particularly advanced stage.[14,37,38,44] A collection of close to one hundred partially protected and functionalized derivatives have been prepared in the meantime, providing starting materials for the perethynylated dehydroannulenes and expanded radialenes shown in Figure 6.[36–44] TEEs and DEEs, as well as dimeric derivatives substituted at the terminal alkynes with donor (D, *p*-(dimethyl-

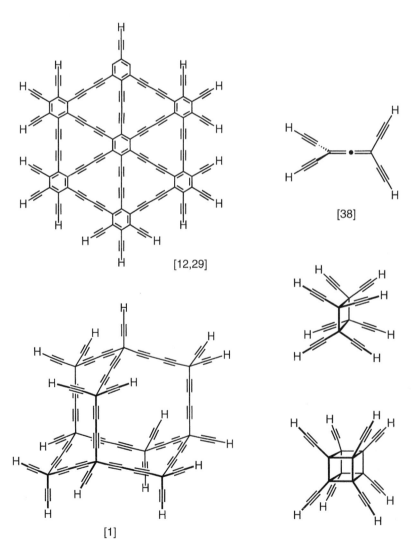

[12,29]

[38]

[1]

Figure 8. Some hitherto elusive modules for two- and three-dimensional all-carbon scaffolding: challenges for future targeted synthesis.

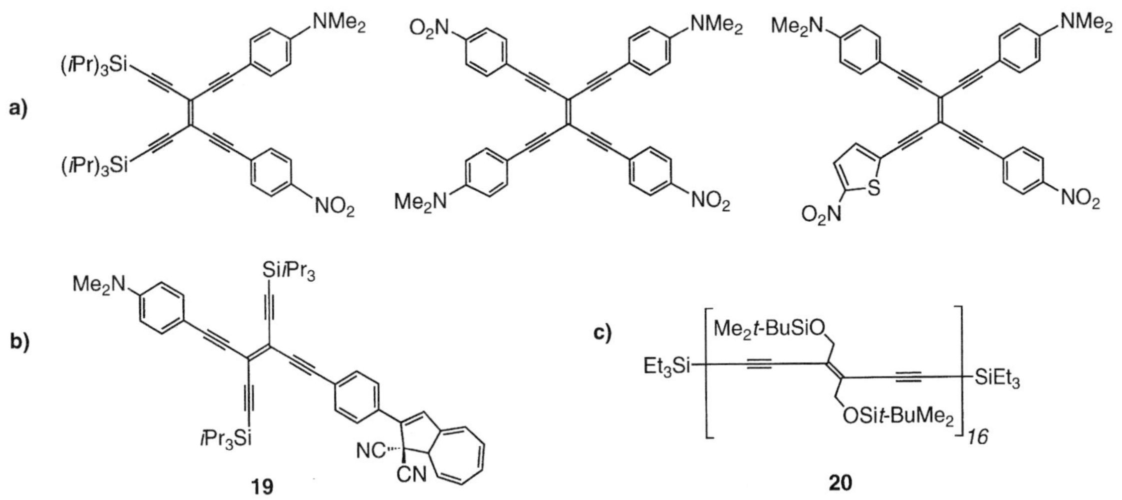

Figure 9. Tetraethynylethene (TEE) and diethynylethene (DEE) molecular scaffoldings. **a)** Donor-acceptor substituted chromophores for nonlinear optical studies. **b)** A three-way chromo-phoric molecular switch (**19**). **c)** A poly(triacetylene) molecular wire (**20**) of 11.9 nm length.

amino)phenyl and/or acceptor (A, *p*-nitrophenyl) groups, have been found to display very large second hyperpolarizabilities (γ) as determined by THG measurements (Figure 9).[45] Comprehensive investigations on a large series of such chromophores has provided useful structure-property relationships for third-order nonlinear optical effects. These compounds have also very appealing second-order nonlinear optical properties.[46] Additionally, these compounds belong to the rare class of molecular switches that undergo *cis* → *trans* and *trans* → *cis* isomerization exclusively under the stimulus of light. This property has been exploited in the construction of a three-way chromophoric molecular switch (**19**) that displays three independently addressable pH and light controllable switching cycles leading to eight different states.[47] The three switching functions operating in **19** are: 1) reversible protonation of the dimethylanilino group, 2) photochemical *cis/trans* isomerization of the central TEE chromophore, and 3) photochemical ring-opening of the dihydroazulene moiety into a vinylheptafulvene moiety and thermal back cyclization.

A rapid and versatile covalent assembly technique starting from DEE oligomers has provided the 11.9 nm long hexadecameric poly(triacetylene) rod **20**.[48] With its linearly conjugated 16 double and 32 triple bonds spanning in-between the terminal silicon atoms, compound **20** is currently the longest linear, fully π-conjugated molecular wire without aromatic repeat units in the backbone.

Another fascinating all-carbon molecule, the molecular belt **21a**, is an isomer of C_{60} and can be constructed from six non-planar TEE subunits (Figure 10).[49] Semiempirical calculations (AM1) indicate

that **21a** should have an electron-rich interior circular cavity with an open space of ca. 8.3 Å in diameter and a relatively electrophilic external surface. Although this electron distribution is indicative of increased reactivity, formation of porous crystals for substrate inclusion or separation can be envisaged in principle. Additional interest in **21a** arises from theoretical calculations on similar conjugated oligoacene molecular belts that are predicted to display superconductivity. Furthermore, this strained all-carbon belt could undergo thermal or photochemical rearrangement to the thermodynamically more stable, spherical C_{60} isomer, providing a stepwise, controlled synthesis of buckminsterfullerene. Likewise, polymerization by [2+2] cycloadditions of the reactive C=C units would release strain and lead to the stereoregular network **21b**. Although several approaches have been pursued to construct this fascinating molecule, its synthesis remains elusive at this time.

Progress towards the formation of *planar* fragments of two-dimensional carbon allotropes in which the graphitic texture is extended by acetylene or butadiyne units (graphynes) has advanced tremendously, thanks to novel methodologies for the construction of differentially protected hexaethynylbenzenes (HEBs, e.g. **22**).[29] The availability of these compounds has allowed the synthesis of interesting and stable graphyne fragments. For example, dimerization of the protected HEB **22** gave the "first generation" graphdiyne fragment **23b** (Figure 11).[29]

The principle of building these planar carbon networks can also be extended to the third dimension in a manner similar to the construction of fullerenes from planar graphite sheets: Thus, fullere-

Figure 10. Molecular belt **21a**, an isomer of buckminsterfullerene C_{60}, and its hypothetical [2+2] polymer **21b**.

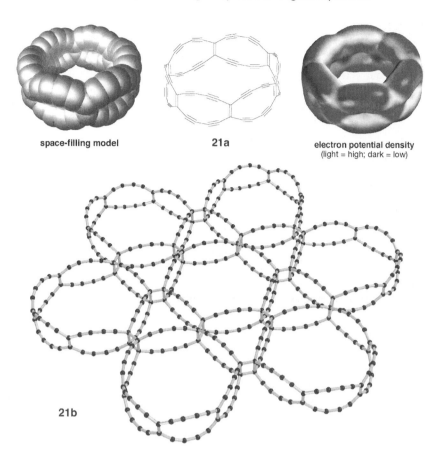

space-filling model   21a   electron potential density
(light = high; dark = low)

21b

Figure 11. a) Synthesis of a "first generation" fragment **23b** from the protected hexaethynylbenzene **22**. **b)** A large graphdiyne fragment.

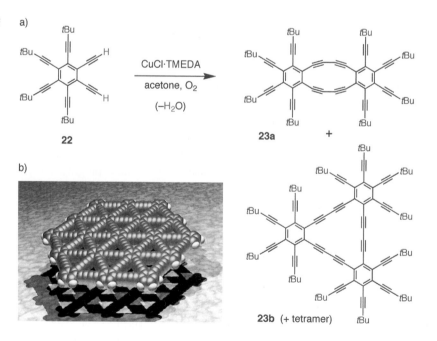

a)

22

CuCl·TMEDA
acetone, O_2
($-H_2O$)

23a   +

23b (+ tetramer)

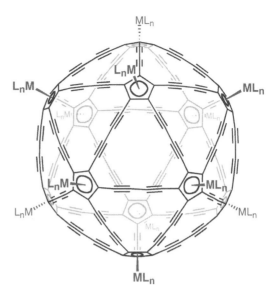

Figure 12. Hypothetical fullereneyne C_{180}, which can be stabilized in principle by complexation at the five-membered rings with metal-ligand moieties.

neyne C_{180} can be envisaged (Figure 12).[12,29,31,32] Advances towards the synthesis of these three-dimensional carbon allotropes have been centered so far around the synthesis of the basic building blocks.[29,31,32]

Aspects of Covalent Fullerene Chemistry: Regioselective Multiple Functionalization, Optically Active Carbon Allotropes, and Electroluminescent Devices (LEDs)

The covalent functionalization of buckminsterfullerene, C_{60}, has developed dramatically over the past decade. A great diversity of synthetic protocols for the formation of monoadducts has become available. One area of intense ongoing interest deals with the regio- and stereoselective multiple functionalization of the carbon sphere.[53-63] Sequential reactions with fullerenes rarely are regioselective, often making product separation tedious. Furthermore, all the addition patterns that are possible cannot be distinguished readily from spectroscopic data. A very useful approach to resolving these issues lies in the tether-directed remote functionalization strategy.[55]

One of the most convenient of these tether-directed functionalization strategies exploits the macrocyclization of C_{60} with bismalonate derivatives in a double "*Bingel*" reaction.[56,57] With the exception of the *cis-1* bisaddition product, all possible bisaddition patterns have been obtained by this macrocyclization reaction (Figure 13). As a general

synthetic protocol, diols are transformed into *bis* (ethyl malonyl) derivatives and the one-pot reaction of C_{60}, *bis*(ethyl malonate), I_2 (to prepare the corresponding halomalonate *in situ*) and 1,8-diazabicyclo [5.4.0]undec-7-ene (DBU) in toluene at 20 °C generates the macrocyclic bisadducts in yields of usually between 20 % and 40 % with good-to-high regio- and diastereoselectivities.

Importantly, the tethers in these bisadducts can be readily removed, making them true templating units. This point is illustrated by the cleavage of the crown ether tether in conjugate **24** to give untethered *trans-1* functionalized fullerenes that are of interest for use as further molecular scaffoldings (Scheme 5).[56,60,61]

A sequence of a highly diastereoselective *Bingel* macrocyclization using a non-racemic tether, followed by removal of the tether via transesterification provides an enantioselective synthesis of optically active *cis-3* bis-adducts in which the chirality results exclusively from the C_2-symmetric addition pattern.[53,54,57] Starting from (*R,R*)-**25** and (*S,S*)-**25**, prepared from the corresponding optically pure diols, the two enantiomeric *cis-3* bisadducts (*R,R*,^{f}*A*)-**26** and (*S,S*,^{f}*C*)-**26**[57] were obtained with high selectivity (diastereomeric excess d.e. > 97 %, Scheme 6). In each macrocyclization, two diastereomeric *out-out/cis-3* bisadducts are possible on account of the chiral addition pattern; however, the high asymmetric induction in the second intramolecular *Bingel* addition by the optically active tether leads to the formation of (*R,R*,^{f}*A*)-**26** and (*S,S*,^{f}*C*)-**26** exclusively. Transesterification of (*R,R*,^{f}*A*)-**26** and (*S,S*,^{f}*C*)-**26** yielded the *cis-3* tetraethyl esters (^{f}*A*)-**27** and (^{f}*C*)-**27** with an enantiomeric excess (e.e.) higher

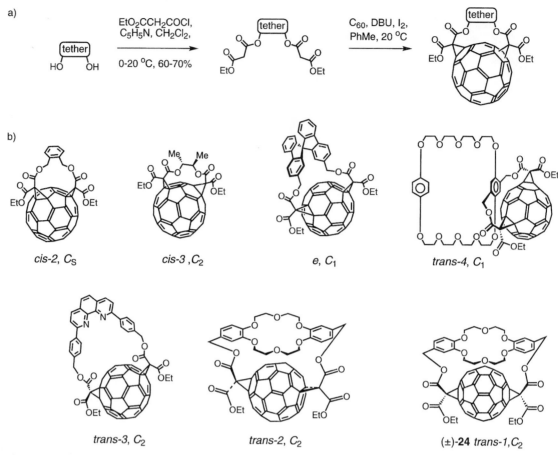

Figure 13. **a)** General synthetic protocol for the regio- and stereoselective preparation of *bis*(cyclopropanated) C_{60} derivatives by macrocyclization via double *Bingel* addition. **b)** Examples of bisadducts produced.

than 99 % [(^fA)-**27**] and 97 % [(^fC)-**27**] (HPLC), reflecting the e.e.s of the corresponding commercial starting diols. The absolute configurations of these optically active fullerene derivatives could be assigned from their calculated circular dichroism (CD) spectra.

The *Bingel* macrocyclization has found successful application in supramolecular fullerene chemistry.[9] Using this method, organic chromophores or receptor sites can be precisely positioned in close proximity to the fullerene surface, thus offering the potential for inducing changes in the physical properties of the carbon allotropes. Examples are the fullerene-crown ether conjugates depicted in Figure 13: When a potassium cation binds to their crown ether binding site, the first reduction potential encounters a marked anodic shift, i.e. it becomes facilitated.[56]

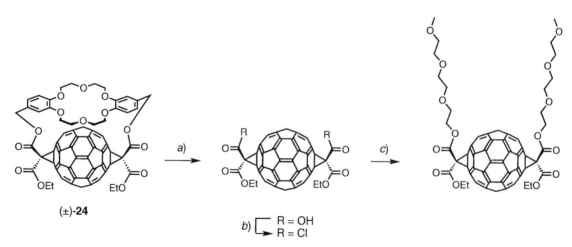

Scheme 5. Selective removal of the dibenzo[18]crown-6 tether from conjugate (±)-**24**. **a)** TsOH, toluene, Δ, 3 h. **b)** (COCl)$_2$, CH$_2$Cl$_2$, 40 °C, 2 h. **c)** Triethyleneglycol mono- methyl ether, pyridine, CH$_2$Cl$_2$, 20 °C, 14 H; 68 % from (±)-**24**.

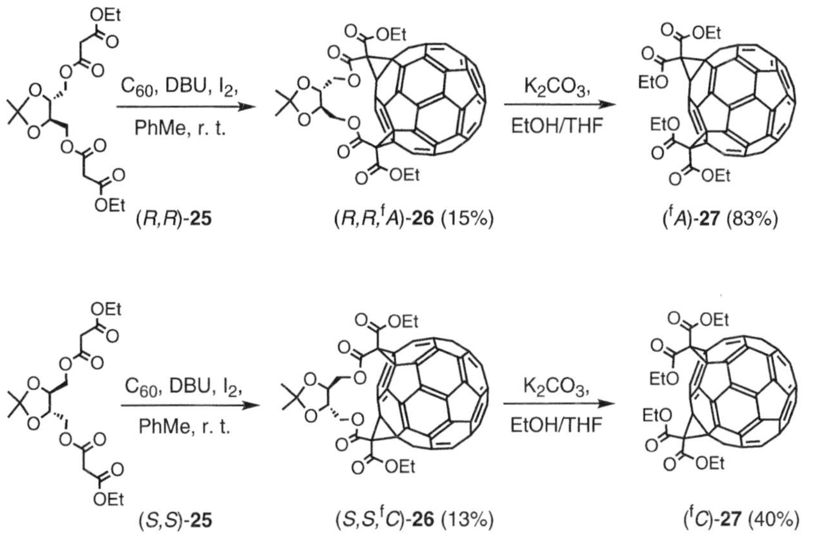

Scheme 6. Enantioselective synthesis of both enantiomers (^{f}A)-**27** and (^{f}C)-**27** by diastereoselective tether-directed biscyclopropanation of C$_{60}$, followed by transesterification.

(R,R)-**25** $\xrightarrow{\text{C}_{60}, \text{DBU}, \text{I}_2, \text{PhMe, r. t.}}$ $(R,R,^fA)$-**26** (15%) $\xrightarrow{\text{K}_2\text{CO}_3, \text{EtOH/THF}}$ (^{f}A)-**27** (83%)

(S,S)-**25** $\xrightarrow{\text{C}_{60}, \text{DBU}, \text{I}_2, \text{PhMe, r. t.}}$ $(S,S,^fC)$-**26** (13%) $\xrightarrow{\text{K}_2\text{CO}_3, \text{EtOH/THF}}$ (^{f}C)-**27** (40%)

Bis(alkoxycarbonyl)methano addends (*Bingel* addends) introduced at 6,6-bonds (bonds at the intersect between two hexagons) on the carbon sphere are of high thermal and chemical stability. However, these addends can also be removed in high yield by the recently discovered electrochemical retro-*Bingel* reaction.[58] Exhaustive electrolytic reduction at constant potential transforms the bis(alkoxycarbonyl)-methano adducts of C$_{60}$ and C$_{70}$ in high yield into the parent fullerenes. The *Bingel*/retro-*Bingel* reaction sequence has been successfully applied to the isolation of enantiomerically pure higher fullerenes.[58,59] The chromatographic separation of constitutional isomers of the higher fullerenes beyond C$_{70}$ is extremely tedious. Even more difficult is the chromatographic enantiomer separation of chiral fullerenes on chiral stationary phases. On the other hand, mono- and higher adducts of higher fullerenes are quite readily separable by HPLC.[57] Racemic D_2-C$_{76}$ was therefore first transformed by *Bingel* addition with a non-racemic malonate into the mixture of diastereomeric C$_{76}$ monoadducts $(S,S,^fA)$-**28** and $(S,S,^fC)$-**28** (Scheme 7). HPLC on silica gel provided the two optically active, pure diastereomers. When a pure diastereomer was subjected to controlled potential electrolysis, the pure enantiomer of D_2-C$_{76}$ was obtained. The two separated enantiomers of the higher fullerene displayed perfectly mirror-image CD spectra with very large Cotton effects ($\Delta\varepsilon$ values up to 350 M^{-1} cm^{-1}), extending from the UV into the near infrared region. This versatile *Bingel*/retro-*Bingel* strategy was also successfully applied to the enantiomer separation of one of the D_2-symmetrical constitutional isomers of C$_{84}$.[59] These examples demonstrate, in an impressive way, how advances in the chemistry of carbon allotropes, such as the separation of pure element enantiomers, have benefited from new synthetic methodology developments.

The tether approach has been exploited also to obtain icosahedral (*pseudo*-octahedral) addition patterns that have any desired malonate permutation on the generic tris- or hexakis-*Bingel* adducts **30** and **31** (A–F = C(CO$_2$R)$_2$ with any R group), or any of the adducts with an intermediate degree of addition (Figure 14).[60–62] The tethered *trans*-1 bisadduct **29** can be formed by a Diels-Alder cycloaddition in relatively good yield.[10,60] It is designed to protect three of the *pseudo*-octahedral positions on the fullerene moiety with two covalently-attached and one sterically shielding units, which force subsequent *Bingel*

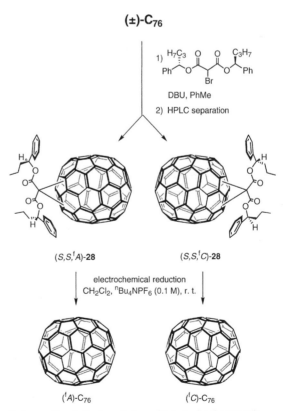

Scheme 7. Optical resolution of (±)-C$_{76}$ by the *Bingel*/retro-*Bingel* reaction sequence.

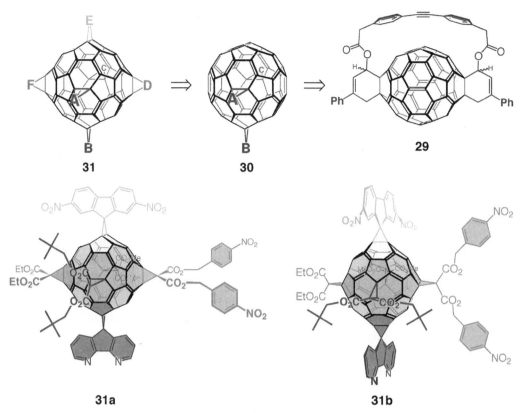

Figure 14. Formation of fully differentiated hexa-*Bingel* adducts by a "3+3" addition scheme.

additions to the remaining three *pseudo*-octahedral positions with very high regioselectivity in the order A → B → C. Removal of the tether and the covalently attached "protecting" groups affords a trisadduct displaying further high selectivity in the next three additions toward the hexakisadducts **31**.[61] In this manner, the two totally differentiated hexakisadducts **31a** and **31b** have been prepared in high overall yields.[62] These two interesting diastereomers differ only by the *permutation of all six malonates* between *edge* and *facial* equatorial *pseudo*-octahedral locations. This example illustrates the remarkable geometric diversity that can be achieved with the three-dimensional frameworks

of fullerenes and invites an exploration of molecular diversity based on these frameworks.

A surprising functionalization result was obtained with a quite different reaction: exhaustive addition to C_{60} of the relatively bulky azomethine ylide $Me_2C^--HN^+=CMe_2$ resulted in the remarkable formation of only two hexakisadducts (**32a** and **32b**) in good overall yield in a single operation (up to 80% combined, Figure 15).[63] X-ray structures confirmed the averaged T_h-symmetry of **32a** and the D_3-symmetry (chiral) of **32b**. Both adducts are surprisingly fluorescent and appear to result from the preferred formation of a bisadduct (*trans*-3) that channels all subsequent additions to the final products through LUMO and steric control.

Figure 15. Structures of the T_h and D_3 hexapyrrolidines **32a** and **32b**.

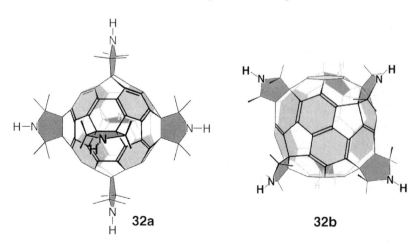

The unusual yellow-green fluorescence of compounds **32a** and **32b** was also studied in detail.[63] The results are surprising: at 25 °C, fluorescence is quite strong (Φ = 0.024), by far the strongest reported for any fullerene. Upon cooling, intense orange-red phosphorescence is observed (MTHF glass, 77 K) with a half-life of 4.4 s. To put these observations in perspective, fullerenes are universally known as highly efficient luminescence quenchers because they easily intersystem cross to a triplet excited state (i.e., singlet excited state lifetime is too short to fluoresce). The triplet generally looses its energy by non-radiative processes. With the hexapyrrolidines **32a** and **32b**, the respective singlet excited states live long enough to decay by fluorescence (Φ ≈ 2–3%), and the remaining energy is transferred to the triplet (Φ = 97%) which, in this case, decays by phosphorescence. Remarkably, the steady-state population of the triplet can be maintained as high as 30% by a simple photon source such as that in a fluorimeter.

The exceptional photophysical properties of compound **32a** suggested that it may also function as a light-emitting device (LED).[64] Within the context of LEDs, the fluorescence of **32a** is not overwhelming; however, electroluminescence generated with a fullerene derivative could bring a possible use for this class of compounds with future improved emitters. The device was based on an unconventional but easy to fabricate *single*-layer arrangement that combines the conducting polymer (poly(vinylcarbazole, PVK), the hole transporter (2,5-*bis*-1-naphthyloxadiazole, BND), and the charge transporter (**32a**, THP; Figure 16) in one blend. It emitted bright white light at relatively low turn-on voltages (~13 V) as a result of blue and yellow-green emission overlap of the two fluorescence emitters (THP and BND). Although higher electroluminescence efficiencies need to be achieved first of all, white light generation in a similar manner could find use as a general source of ambient lighting in closed areas.

Novel Fullerene-Based Carbon Allotropes and Total Synthesis of C₆₀

The availability of fullerenes allows one to imagine novel carbon allotrope structures based on hybrid frameworks. A number of such structures can be envisaged (e.g., **34a/b**, **38**, **41**, and **42**, Schemes 8, 10, and 11). Initial reports on this possibility came out of our two respective groups (Schemes 8 and 9).[65–68] The diethynylmethanofullerene **33** was synthesized as a precursor of the carbon allotropes C₁₉₅ (**34a**) and C₂₆₀ (**34b**), but it was quickly found that the products of its oxidative coupling are highly insoluble. This frustrating experimental difficulty has been solved in two ways: 1) Formation of a solubilized analog of **33** in the form of a penta-*Bingel* hexaadduct gave unequivocal and clean formation of the trimeric and tetrameric macrocycles **35a** and **35b** which constitute derivatized forms of the parent carbon molecules **34a** and **34b** (Scheme 9).[68] Interestingly, laser-desorption mass spectrometry experiments provided evidence for the easy fragmentation of **35a** and **35b** to the monofullerene adducts **36a** and **36b** of the cyclocarbons C₁₅ and C₂₀ proceeding by loss of two or three penta-*Bingel*-C₆₀ units. 2) The other approach to solving the problem of solubility in a more direct way was to exploit the reversible Diels-Alder/*retro*-Diels-Alder reaction of 9,10-dimethylanthracene (DMA) with fullerene double bonds to keep the intermediate linear oligomers and final cycles **34a** and **34b** in solution (Scheme 8).[66] LDMS of the product mixture with DMA as the matrix gave a clear indication that **34a** and **34b** are formed in this process. A more direct experimental characterization by STM is necessary and will permit the study of self-assembled monolayers of these interesting materials. These macrocycles may ultimately lead to single-sized *giant* fullerenes by catalyzed coalescence reactions, since fullerene coalescence has been observed in electron diffraction images of crystallized C₆₀.

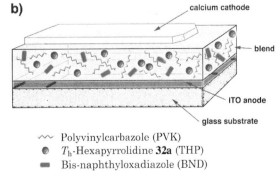

Figure 16. a) Electroluminescence of the reference system PVK/BND (left) and the white-light emitter PVK/THP/BND (right). **b)** Device configuration.

Scheme 8. Formation of
the parent fullerene-
cyclocarbon hybrids C_{195}
and C_{260}.

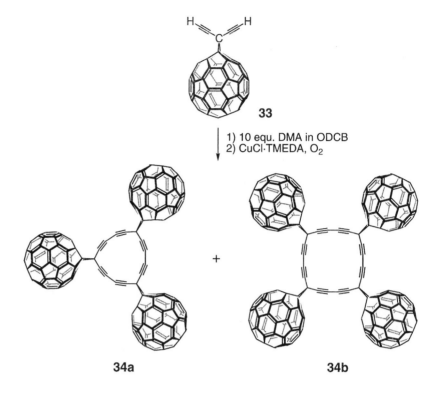

33

1) 10 equ. DMA in ODCB
2) CuCl·TMEDA, O_2

+

34a   **34b**

A spectacular solid-state [2 + 2] dimerization reaction of C_{60} was discovered by Komatsu et al.[11] When a mixture of C_{60} and potassium cyanide is intimately ground by high-speed vibration milling, an unusual addition-elimination reaction involving the cyanide adduct **37** ensues which joins two fullerene moieties to afford the C_{120} dimer **38** (Scheme 10). Presumably, the cyclodimerization involves

sequential single-electron transfer steps. Unequivocal proof of this structure was obtained by X-ray diffraction. The crystallographic data show that the 4-membered ring in **38** is closed as drawn rather than open with two C=C bonds tying up the fullerene units. This work helped bring a definitive answer to a long-debated issue about the structure of photo-polymerized C_{60}. The isolation of higher oligomers

Scheme 9. Fragmentation
of the fullerene-cyclocarbon
hybrids.

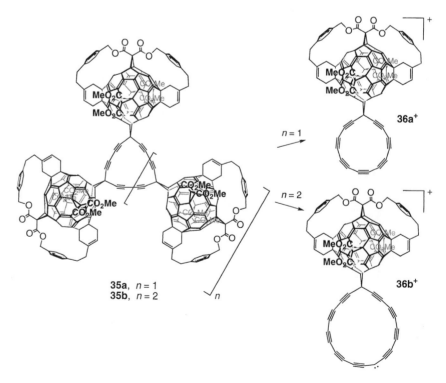

35a, $n = 1$
35b, $n = 2$

$n = 1$

$n = 2$

36a$^+$

36b$^+$

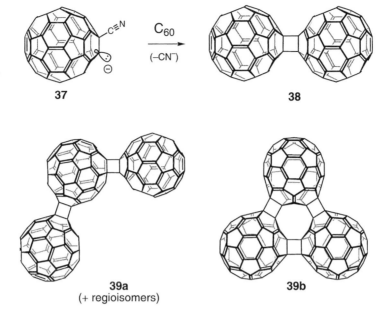

Scheme 10. Oligomerization of C_{60} catalyzed by cyanide to the unusual dimer **38** (C_{120}). Trimers that are also formed have been difficult to characterize but are likely "bent" and "cyclized" isomers **39a** and **39b**, respectively.

39a
(+ regioisomers)

39b

was also made possible by this discovery using more active catalysts, e.g., 4-aminopyridine. The structures of two types of trimers (**39a** and **39b**) have been advanced, but considerable experimental difficulty stemming from the poor solubility of these compounds has so far hampered their full structural characterization.[69]

Two other types of fullerene-based carbon allotropes have been synthesized (Scheme 11).[70–72] Both C_{121} (**40**) and C_{122} (**41**) are methanofullerenes with the bridging unit being either a spirocarbon (C_1) or a bisethylidene (C_2) linker. These compounds have been approached by carbene formation methods involving the generation of either atomic carbon from the highly explosive diazotetrazole (**42**), or by α-elimination of two bromine atoms from dibromomethanofullerene **43** by pyrolysis in the presence of excess C_{60}. The extreme difficulty of this work in terms of characterization was addressed by carbon-13 labeling (^{13}C NMR) and by scanning tunneling microscopy (STM) in addition to mass spectrometry.[71,72]

The total synthesis of fullerenes was pursued early on in a classical, carbon framework-building approach by the groups of Scott, Rabideau, and Siegel.[73–75] Scott sparked a "revival" of this field by developing a highly efficient, straightforward approach to corannulene. This curved aromatic hydrocarbon was first synthesized in 1965 by Barth and Lawton in 17 steps and 0.40 % overall yield from acenaphthene.[1] Considerable progress has been made by these groups to obtain ever larger bowl-shaped structures mapping onto the framework of C_{60}. This approach may ultimately lead to the synthesis of fullerenes, especially since a low-yielding flash vacuum pyrolysis step can now be performed in solution through reductive coupling.[76]

A highly convergent approach to fullerenes has been pursued by the groups of Rubin and Tobe.[10,29] This approach disconnects fullerenes into highly energetic perethynylated cyclophanes whose structures are inspired from the cyclocarbon coalescence mechanism presented in Scheme 1. One of the possible key building blocks is hexaethynylbenzene (**18**,

Scheme 11. Formation of the C_1- and C_2-linked dumbbell structures **40** and **41**.

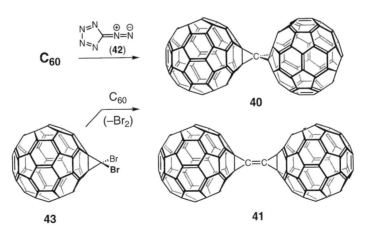

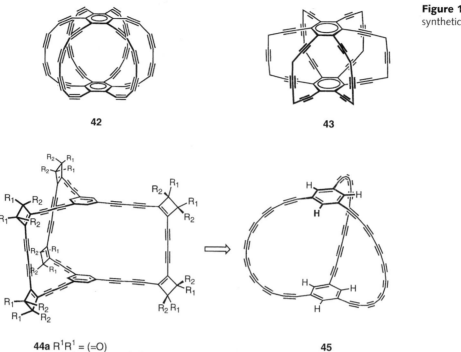

Figure 17. Carbon-rich synthetic precursors of C_{60}.

44a $R^1R^1 = (=O)$
44b $R^1R^1 = (CH_2)_3$, $R^2R^2 = (CH=CH)_2$

Fig. 6), which can be used as a bridgehead scaffold for the all-carbon cage molecule C_i-C_{60} (**42**), an isomer of I_h-buckminsterfullerene that is 767 kcal·mol^{-1} higher in energy by the calculated heats of formation (RHF/3–21G).[10,29] A similar macrocycle (**45**, $C_{60}H_6$) of easier synthetic accessibility is based on the 1,3,5-triethynylbenzene moiety.

The laser desorption mass spectra of **44a** and **44b** generate ions corresponding to $C_{60}H_6$·$^-$ by decarbonylation or elimination of indane fragments.[29] Importantly, both sets of experiments find that C_{60}·$^-$ is the base peak. Fragment ions resulting from C_2-loss (C_{58}·$^-$, C_{56}·$^-$) are also observed, indicating that the C_{60}·$^-$ ions have a fullerene structure. The involvement of **45** in their formation shows that the polyyne cyclization mechanism is valid, although this hypothesis will have to be demonstrated unequivocally when isolable quantities of C_{60} can be prepared by this method using ^{13}C-labeled precursors.[29] These results constitute an important step toward understanding the formation of C_{60}, and signal a fast approaching total synthesis of this molecule and metal-incarcerated or shell-substituted analogs.

Filled, Opened, and Shell-Substituted Fullerenes

The hollowness of C_{60} and other fullerenes has tempted many to encapsulate one or more elements to change the physical and chemical properties of

these carbon shells.[10,77] Elements that have been incorporated are still few, but the rare gases (He, Ne, Ar, Kr, Xe), atomic nitrogen (N), and the lanthanide and alkaline-earth metals have been included.[77] The endohedral metallofullerenes (endohedral ≡ inside the cage) were prepared for the first time in 1991 by arc vaporization of graphite impregnated with metal oxides. Representative lanthanide metallofullerenes that have been characterized include $Ln@C_{76}$, $Ln@C_{78}$, $Ln@C_{82}$ (Ln = La, Y, Sc, Gd, Tm), $La_2@C_{72}$, $Sc_2@C_{74}$, $La_2@C_{80}$, $Ce_2@C_{80}$, $Sc_2@C_{82}$, $Sc_2@C_{84}$, $Sc_3@C_{82}$, and $Sc_3N@C_{80}$, the symbol @ denoting that the metal is encapsulated within the cage. In the alkaline-earth series, $M@C_{72}$, $M@C_{74}$, $M@C_{80}$, $M@C_{82}$, $M@C_{84}$ where M = Ca, Sr, Ba, have been prepared and characterized more recently.

One can expect from the spectacular physical properties of C_{60} (superconductivity, photophysics, non-linear optical activity, etc.) that its endohedral complexes could display a number of important properties as well, especially in the area of materials science and magnetic resonance imaging (MRI; Gd complexes). Most endohedral metallofullerenes are still very difficult to obtain in pure form in quantities greater than the milligram. To allow further developments in the study of these attractive materials, a general synthetic-organic approach to endohedral metallofullerenes has been pursued by one of us to bring a practical answer to the limitations of the graphite evaporation method.[10,77] The following section present some of the first results toward this goal.

Figure 18. a) Addition of six saturating groups on a fullerene. **b)** [2+2+2] ring-opening reaction.

The reversible formation of an opening within a fullerene framework could possibly constitute one of the most powerful methods to introduce practically any atom inside these hollow structures. The task of opening a cavity on the surface of fullerenes has proven very challenging, since mechanisms that realize this goal were recognized only relatively late in the fullerene functionalization "game". More importantly, one has to form an opening wide enough to allow a atom to pass through before closing it back. A breakthrough in this endeavor was reached when Wudl's group observed that the azafulleroid **46** spontaneously reacts in a [2+2] cycloaddition reaction with singlet oxygen ($^1O_2$, Scheme 12). This produces a dioxetane (**47**) which rearranges to a ketolactam (**48**) by ring-fragmentation. The 11-membered opening in **48** – although sizable – was found to be too small to allow even the escape of helium, one of the smallest neutral atoms, from its fullerene "prison".[77]

This work, however, provided guidance for the next generation of ring fragmentations. A strategy based on a [2+2+2] sigmatropic rearrangement was developed in the Rubin group whereby a fully saturated six-membered ring (Figure 18) opens to three edge C=C bonds, a process that is calculated favorable when addends are bulky (e.g. $C_{60}Me_6$).[10,77] Interestingly, this energetic advantage is largely due to aromatic resonance stabilization of the edge six-membered rings.

For a first practical look at this concept, the rigid "preorganized" trifunctional reagent **49** was prepared with the prediction that the likely outcome for its reaction with C_{60} would be the wide-open bistriazole **51** from reactivity precedents and AM1 calculations.[78] However, in an interesting twist, the unexpected bislactam **53** was obtained instead as a result of the lability of the triazoline moieties in intermediate **50**. Nevertheless, bislactam **53** has the largest orifice formed thus far on a fullerene. It is generated by the rearrangement of the putative intermediate **50** to the diamine **52** with loss of two N_2 moieties, followed by a [4+2]-addition of $^1O_2$ and [2+2+2] ring opening.[78] The involvement of the unusually reactive diaminobutadiene **52** was established unequivocally by trapping with *p*-benzoquinone and by characterization using 2D ^{1}H-NMR (T-ROESY) of the corresponding Diels-Alder adducts (*syn/anti*).

Compound **53** has an orifice just wide enough to allow the potential entry of small ions, atoms (He), or molecules (H_2, D_2, T_2) inside its cavity as judged from high-level transition state calculations at the density functionally theory level (B3LYP/6–31G**, Figure 19).[79] The low barrier of insertion for helium makes it a specially attractive target, but other spe-

Scheme 12. Formation of the ketolactam **48**.

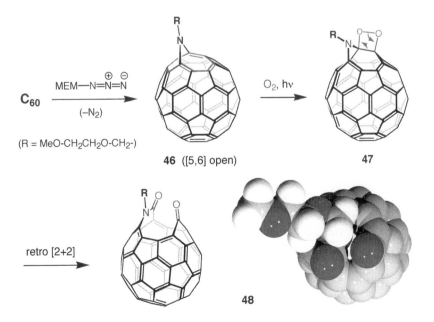

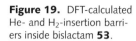

Figure 19. DFT-calculated He- and H₂-insertion barriers inside bislactam **53**.

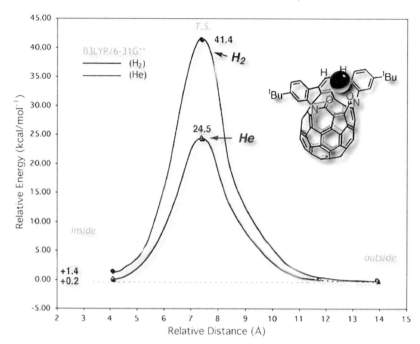

cies will be examined as well. Further work toward the formation of larger openings is also necessary, constituting a particularly exciting prospect in the fullerene area.

One particularly striking aspect of endohedral fullerene complexes is the surprising ability of the carbon cage to contain highly reactive species such as atomic nitrogen.[80] N@C₆₀ (**54**) has been prepared

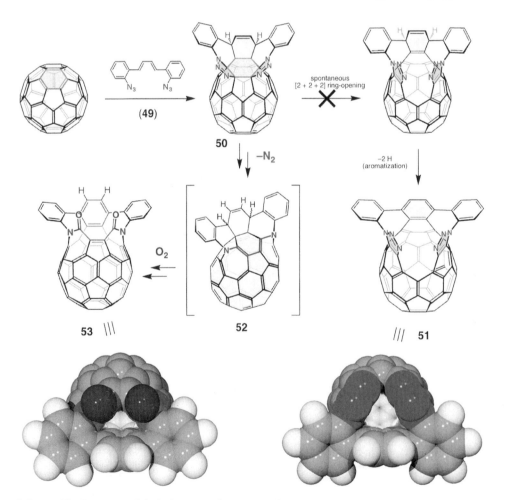

Scheme 13. Formation of the bislactam **53** by reaction of bisazide **49** with C₆₀.

Scheme 14. Probable mechanism of insertion of atomic nitrogen inside C_{60}.

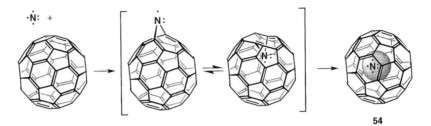

54

in a single step by sublimation of C_{60} into a nitrogen plasma produced in a discharge tube similar to regular fluorescent light bulbs.[80] The incorporation ratio is very small (~0.01%) but the ground state of atomic N in this molecule is a quartet revealed by an amazingly sharp and strong ESR signal. A probable mechanism of formation has been advanced in light of the rather low-activation barrier for this insertion (~40 kcal·mol^{-1}, Scheme 14).[77] The insertion may occur by intermediate bonding of highly energetic nitrogen atoms to [5,6] or [6,6] bonds to give bridged, open intermediates allowing easier insertion and bond release once forced inside the cage. Bonding of nitrogen to carbons on the inside of the cage is energetically very unfavorable, shown and erideneed by the lack of hyperfine interactions in the sharpness of the ESR lines. Although the preparation of this compound is a feat in itself, it is clear that incorporation ratios near 100% would make it even more interesting for the study of spin correlations in the bulk state, where ferromagnetic materials may result from either pristine N@C_{60} or functionalized derivatives, or for its use as a highly sensitive biological spin label.

Another early suggestion was that some of the fullerene carbon atoms could be replaced by another element such as nitrogen or boron.[81] Such *hetero*fullerenes are envisaged to have strongly altered chemical properties in comparison to the all-carbon structures.[81] For example, as the substitution number of an electronegative element such

as nitrogen increases in a heterofullerene ($C_{60-n}N_n$, n = 1–12), its electron-accepting ability greatly increases. Salts of these heterofullerenes may have higher superconducting temperatures than the current records for organic materials with K_3C_{60} or Rb_3C_{60} (T_c = 18 and 30 K, respectively) and thus constitute attractive targets.

However, replacement of a carbon within the shell of C_{60} by a heteroatom has constituted a daunting task. Until recently, there was no mechanism available to do so, and only syntheses from "scratch" appeared to be up to the challenge.[10,77] The first and so far only reported syntheses are for hydroazafullerene $C_{59}N^+$ (**55**) and its α-substituted derivatives (Scheme 15). The formation of this heterofullerene has been achieved in a remarkable acid-catalyzed rearrangement of the opened ketolactam **48** described earlier (Schemes 12 and 15).

A proposed mechanism for this rearrangement is supported by fragmentation patterns observed in the mass spectrum of **48**.[81] In the preparative scale reaction, hydroazafullerene $C_{59}NH$ (**56a**) is formed predominantly in the presence of a reducing agent such as hydroquinone. Otherwise, an intermediate radical formed by spontaneous reduction of **55** couples to give the stable dimer $C_{118}N_2$ (**56c**). Other α-substitution products, e.g. **56b**, can be obtained by electrophilic aromatic substitution of anisole or other activated aromatic compounds. A similar rearrangement reaction is observed with the bisazafulleroid **57**.

Scheme 15. Mechanism for the formation of heterofullerene derivatives $C_{59}NH$ (**56a**), $C_{59}N(C_6H_4OMe)$ (**56b**), and $C_{118}N_2$ (**56c**).

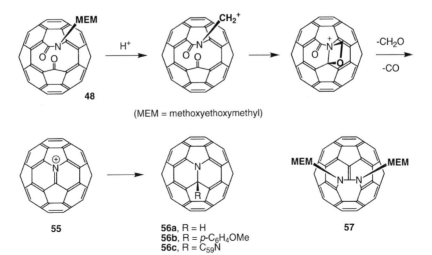

(MEM = methoxyethoxymethyl)

48

55

56a, R = H
56b, R = p-C_6H_4OMe
56c, R = $C_{59}N$

57

Epilogue

This article describes highlights from an area that has dominated a large part of chemistry over the past 15 years and, with exceptionally bright fundamental and technological perspectives, will likely continue to do so in the 21st century. Organic synthesis has taken a pivotal role in the investigation and exploitation of carbon allotropes. The use of advanced synthetic methodologies is essential to reach desirable targets, such as new all-carbon networks, modified fullerenes with different shapes, shells and their interior content, as well as functional fullerene derivatives for applications in materials science or in biology and medicine. At the same time, light will be increasingly shed on the fundamental properties of the element that is used for the construction of complex biological edifices. Carbon allotropes will continue to provide a rich playground for discovery in the decades to come.

Acknowledgements

We thank the U.S. National Science Foundation and the Office of Naval Research (Y.R.), as well as the Swiss National Science foundation and the ETH Research Council (F.D.), for their generous support of this research.

References and Notes

1. For a review, see: F. Diederich, Y. Rubin, "Synthetic Approaches Towards Molecular and Polymeric Carbon Allotropes", *Angew. Chem. Int. Ed. Engl.* **1992**, *31*, 1101–1123.

2. F. Diederich, Y. Rubin, C. B. Knobler, R. L. Whetten, K. E. Schriver, K. N. Houk, Y. Li, "All-Carbon Molecules: Evidence for the Generation of Cyclo[18]carbon from a Stable Organic Precursor", *Science* **1989**, *245*, 1088–1090.

3. H. W. Kroto, J. R. Heath, S. C. O'Brien, S. C. Curl, R. F. Smalley, "C$_{60}$: Buckminsterfullerene", *Nature* **1985**, *318*, 162–163.

4. W. Krätschmer, L. D. Lamb, K. Fostiropoulos, D. R. Huffman, "Solid C$_{60}$: a New Form of Carbon", *Nature* **1990**, *347*, 354–358.

5. F. Diederich, R. Ettl, Y. Rubin, R. L. Whetten, R. Beck, M. Alvarez, S. Anz, D. Sensharma, F. Wudl, K. C. Khemani, A. Koch, "The Higher Fullerenes: Isolation and Characterization of C$_{76}$, C$_{84}$, C$_{90}$, C$_{94}$, and C$_{70}$O, an Oxide of *D$_{5h}$*-C$_{70}$" *Science* **1991**, *252*, 548–551.

6. R. Ettl, I. Chao, F. Diederich, R. L. Whetten, "Isolation of *D$_2$*-C$_{76}$: A Chiral Allotrope of Carbon", *Nature* **1991**, *352*, 149–153.

7. J. M. Hawkins, A. Meyer, "Optically Active Carbon – Kinetic Resolution of C$_{76}$ by Asymmetric Osmylation", *Science* **1993**, *260*, 1918–1920.

8. A. Hirsch, "The Chemistry of the Fullerenes", Thieme, Stuttgart, 1994.

9. F. Diederich, M. Gomez-Lopez, "Supramolecular Fullerene Chemistry", *Chem. Soc. Rev.* **1999**, *28*, 263–277.

10. Y. Rubin, "Organic Approaches to Endohedral Metallofullerenes: Cracking Open or Zipping Up Carbon Shells?", *Chem. Eur. J.* **1997**, *3*, 1009–1016.

11. G.-W. Wang, K. Komatsu, Y. Murata, M. Shiro, "Synthesis and X-ray Structure of Dumb-Bell-Shaped C$_{120}$", *Nature* **1997**, *387*, 583–586.

12. F. Diederich, "Carbon Scaffolding: Building Acetylenic All-Carbon and Carbon-Rich Compounds", *Nature* **1994**, *369*, 199–207.

13. F. Diederich, P. J. Stang, "Metal-Catalyzed Cross-Coupling Reactions", Wiley-VCH, Weinheim, 1997.

14. F. Diederich, L. Gobbi, "Cyclic and Linear Acetylenic Molecular Scaffolding", *Top. Curr. Chem.* **1999**, *201*, 43–79.

15. H. Schwarz, "The Mechanism of Fullerene Formation", *Angew. Chem. Int. Ed. Engl.* **1993**, *32*, 1412–1415.

16. N. S. Goroff, "Mechanism of Fullerene Formation", *Acc. Chem. Res.* **1996**, *29*, 77–83.

17. G. von Helden, N. G. Gotts, M. T. Bowers, "Experimental Evidence for the Formation of Fullerenes by Collisional Heating of Carbon Rings in the Gas Phase", *Nature* **1993**, *363*, 60–63.

18. J. Hunter, J. Fye, M. F. Jarrold, "Annealing C$_{60}^{+}$: Synthesis of Fullerenes and Large Carbon Rings", *Science* **1993**, *260*, 784–786.

19. P. Dugourd, R. R. Hudgins, D. E. Clemmer, M. F. Jarrold, "High-Resolution Ion Mobility Measurements", *Rev. Sci. Instr.* **1997**, *68*, 1122–1129.

20. D. L. Strout, G. E. Scuseria, "A Cycloaddition Model for Fullerene Formation", *J. Phys. Chem.* **1996**, *100*, 6492–6498.

21. A. J. Stone, D. J. Wales, "Theoretical Studies of Icosahedral C$_{60}$ and some Related Species", *Chem. Phys. Lett.* **1986**, *128*, 501–503.

22. B. R. Eggen, M. I. Heggie, G. Jungnickel, C. D. Latham, R. Jones, P. R. Briddon, "Autocatalysis During Fullerene Growth", *Science* **1996**, *272*, 87–89.

23. T. Grösser, A. Hirsch, "Dicyanopolyynes: Formation of New Rod-Shaped Molecules in a Carbon Plasma", *Angew. Chem. Int. Ed. Engl.* **1993**, *32*, 1340–1342.

24. D. A. Plattner, Y. Li, K. N. Houk, "Modern Computational and Theoretical Aspects of Acetylene Chemistry"; in *Modern Acetylene Chemistry*; Stang, P. J.; Diederich, F., Ed.; VCH: New York, 1995; pp 1–32.

25. D. A. Plattner, K. N. Houk, "C$_{18}$ is a Polyyne", *J. Am. Chem. Soc.* **1995**, *117*, 4405–4406.

26. L. T. Cheng, W. Tam, S. R. Marder, A. E. Stiegman, G. Rikken, C. W. Spangler, "Experimental Investigations of Organic Molecular Nonlinear Optical Polarizabilities. 2. A Study of Conjugation Dependences", *J. Phys. Chem.* **1991**, *95*, 10643–10652.

27. Y. Tobe, T. Fujii, H. Matsumoto, K. Naemura, "Towards the Synthesis of Monocyclic Carbon Clusters – [2+2] Cycloreversion of Propellane-Annelated Dehydroannulenes", *Pure Appl. Chem.* **1996**, *68*, 239–242.

28. Y. Rubin, C. B. Knobler, F. Diederich, "Synthesis and Crystal Structure of a Stable Hexacobalt Complex of Cyclo[18]carbon", *J. Am. Chem. Soc.* **1990**, *112*, 4966–4968.

29. U. H. F. Bunz, Y. Rubin, Y. Tobe, "Polyethynylated Cyclic π-Systems: Scaffoldings for Novel Two- and Three-dimensional Carbon Networks", *Chem. Soc. Rev.* **1999**, *28*, 107–119.

30. M. M. Haley, J. J. Pak, S. C. Brand, "Macrocyclic Oligo(phenylacetylenes) and Oligo(phenyldiacetylenes)", *Top. Curr. Chem.* **1999**, *201*, 81–130.

31. U. H. F. Bunz, "Carbon-Rich Molecular Objects from Multiply Ethynylated *p*-Complexes", *Top. Curr. Chem.* **1999**, *201*, 131–161.

32. U. H. F. Bunz, G. Roidl, M. Altmann, V. Enkelmann, K. D. Shimizu, "Synthesis and Structural Characterization of Novel Organometallic Dehydroannulenes with Fused CpCo-Cyclobutadiene and Ferrocene Units Including a Cyclic Fullerenyne Segment", *J. Am. Chem. Soc.* **1999**, *121*, 10719–10726.

33. R. Boese, A. J. Matzger, K. P. C. Vollhardt, "Synthesis, Crystal Structure, and Explosive Decomposition of 1,2:5,6:11,12:15,16-Tetrabenzo-3,7,9,13,17,19-hexadehydro[20]annulene: Formation of Onion- and Tube-Like Closed-Shell Carbon Particles", *J. Am. Chem. Soc.* **1997**, *119*, 2052–2053.

34. P. I. Dosa, C. Erben, V. S. Iyer, K. P. C. Vollhardt, I. M. Wasser, "Metal Encapsulating Carbon Nanostructures from Oligoalkyne Metal Complexes", *J. Am. Chem. Soc.* **1999**, *121*, 10430–10431.

35. See ref. [34] for a detailed listing of reviews on carbon nanotubes and their transition-metal derivatives.

36. Y. Rubin, C. B. Knobler, F. Diederich, "Tetraethynylethene", *Angew. Chem. Int. Ed. Engl.* **1991**, *30*, 698–700.

37. R. R. Tykwinski, F. Diederich, "Tetraethynylethene Molecular Scaffolding", *Liebigs Ann./Recueil* **1997**, 649–661.

38. F. Diederich, "Functional Acetylenic Molecular Architecture", *Pure Appl. Chem.* **1999**, *71*, 265.

39. N. Jux, K. Holczer, Y. Rubin, "An Unusually Stable Pentaethynylcyclopentadienyl Radical", *Angew. Chem. Int. Ed. Engl.* **1996**, *35*, 1986–1990.

40. D. Solooki, T. C. Parker, S. I. Khan, Y. Rubin, "Synthesis and Redox Properties of Tetraethynyl Tetrathiafulvalenes", *Tetrahedron Lett.* **1998**, *39*, 1327–1330.

41. R. Diercks, J. C. Armstrong, R. Boese, K. P. C. Vollhardt, "Hexaethynylbenzene", *Angew. Chem. Int. Ed. Engl.* **1986**, *25*, 268–269.

42. R. Boese, J. R. Green, J. Mittendorf, D. L. Mohler, K. P. C. Vollhardt, "The First Hexabutadiynylbenzenes – Synthesis and Structures", *Angew. Chem. Int. Ed. Engl.* **1992**, *31*, 1643–1645.

43. K. S. Feldman, C. K. Weinreb, W. J. Youngs, J. D. Bradshaw, "Preparation and Some Subsequent Transformations of Tetraethynylmethane", *J. Am. Chem. Soc.* **1994**, *116*, 9019–9026.

44. R. E. Martin, F. Diederich, "Linear Monodisperse π-Conjugated Oligomers: Model Compounds for Polymers and More", *Angew. Chem. Int. Ed. Engl.* **1999**, *38*, 1350–1377.

45. R. R. Tykwinski, U. Gubler, R. E. Martin, F. Diederich, C. Bosshard, P. Günter, "Structure-Property Relationships in Third-Order Nonlinear Optical Chromophores", *J. Phys. Chem. B* **1998**, *102*, 4451–4465.

46. R. Spreiter, C. Bosshard, G. Knöpfle, P. Günter, R. R. Tykwinski, M. Schreiber, F. Diederich, "One- and Two-Dimensionally Conjugated Tetraethynylethenes: Structure versus Second-Order Optical Polarizabilities", *J. Phys. Chem. B* **1998**, *102*, 29–32.

47. L. Gobbi, P. Seiler, F. Diederich, "A Novel Three-Way Chromophoric Molecular Switch: pH and Light Controllale Switching Cycles", *Angew. Chem. Int. Ed. Engl.* **1999**, *38*, 674–678.

48. R. E. Martin, T. Mäder, F. Diederich, "Monodisperse Poly(triacetylene) Rods: Synthesis of a 11.9 nm Long Molecular Wire and Direct Determination of the Effective Conjugation Length by UV/Vis and Raman Spectroscopies", *Angew. Chem. Int. Ed. Engl.* **1999**, *38*, 817–821.

49. R. R. Tykwinski, F. Diederich, V. Gramlich, P. Seiler, "1,1,2,2-Tetraethynylethanes: Synthons for Tetryethynylethenes and Modules for Acetylenic Molecular Scaffolding", *Helv. Chim. Acta* **1996**, *79*, 634–645.

50. F. Diederich, C. Thilgen, "Covalent Fullerene Chemistry", *Science* **1996**, *271*, 317–323.

51. S. Samal, S. K. Sahoo, "An Overview of Fullerene Chemistry", *Bull. Mater. Sci.* **1997**, *20*, 141–230.

52. A. Hirsch, " Principles of Fullerene Reactivity", *Top. Curr. Chem.* **1999**, *199*, 1–65.

53. F. Diederich, R. Kessinger, "Templated Regioselective and Stereoselective Synthesis in Fullerene Chemistry", *Acc. Chem. Res.* **1999**, *32*, 537–545.

54. F. Diederich, R. Kessinger, "Regio- and Stereoselective Multiple Functionalization of Fullerenes" in "Templated Organic Synthesis", Eds. P. J. Stang, F. Diederich, Wiley-VCH, Weinheim, 1999, pp. 189–218.

55. L. Isaacs, R. F. Haldimann, F. Diederich, "Tether-directed Remote Functionalization of Buckminsterfullerene: Regiospecific Hexaadduct Formation", *Angew. Chem. Int. Ed. Engl.* **1994**, *33*, 2339–2342.

56. J.-P. Bourgeois, P. Seiler, M. Fibbioli, E. Pretsch, F. Diederich, L. Echegoyen, "Cyclophane-Type Fullerene-dibenzo[18]crown-6 Conjugates with *trans-1*, *trans-2*, and *trans-3* Addition Patterns: Regioselective Templated Synthesis, X-ray Crystal Structure, Ionophoric Properties, and Cation-Complexation-Dependent Redox Behavior", *Helv. Chim. Acta* **1999**, *82*, 1572–1595.

57. For origins of fullerene chirality and definitions of configurational description factors ^fC and ^fA (f = fullerene, C = clockwise, A = anticlockwise), see: C. Thilgen, A. Herrmann, F. Diederich, "The Covalent Chemistry of Higher Fullerenes: C_{70} and Beyond", *Angew. Chem. Int. Ed. Engl.* **1997**, *36*, 2269–2280.

58. R. Kessinger, J. Creassous, A. Herrmann, M. Rüttimann, L. Echegoyen, F. Diederich, "Preparation of Enantiomerically Pure C_{76} via a General Electrochemical Method for the Removal of Di(alkoxycarbonyl)methano Bridges from Methanofullerenes: The Retro-Bingel Reaction", *Angew. Chem. Int. Ed. Engl.* **1998**, *37*, 1919–1922.

59. J. Crassous, J. Rivera, N. S. Fender, L. Shu, L. Echegoyen, C. Thilgen, A. Herrmann, F. Diederich, "Chemistry of C_{84}: Separation of Three Constitutional Isomers and Optical Resolution of D_2-C_{84} Using the "*Bingel*-Retro-*Bingel*" Strategy, *Angew. Chem. Int. Ed. Engl.* **1999**, *38*, 1613.

60. W. Y. Qian, Y. Rubin, "Towards Sixfold Functionalization of Buckminsterfullerene (C_{60}) at Fully Addressable Octahedral Sites" *Angew. Chem. Int. Ed. Engl.* **1999**, *38*, 2356–2360.

61. W. Qian, Y. Rubin, "A Parallel Library of all Icosahedral Stereoisomeric 2+2+2-Hexakisadducts of C_{60}: Inspiration from Werner's Octahedral Stereoisomerism"; Qian, W.; Rubin, Y. *Angew. Chem. Int. Ed. Engl.* **2000**, *39*, in press.

62. N. Qian, Y. Rubin, "Complete Control over Addend Permutation at all Six *pseudo*-Octahedral Positions of Fullerene C_{60}, *J. Am. Chem. Soc.* **2000**, *122*, in press.

63. G. Schick, M. Levitus, L. D. Kvetko, B. A. Johnson, I. Lamparth, R. Lunkwitz, B. Ma, S. I. Khan, M. A. Garcia-Garibay, Y. Rubin, "Unusual Luminescence of C_{60}-

Hexapyrrolidines with T_h and a Novel D_3-Symmetry", *J. Am. Chem. Soc.* **1999**, *121*, 3246–3247.

64. K. Hutchison, J. Gao, G. Schick, Y. Rubin, F. Wudl, "Bucky Light Bulbs: White Light Electroluminescence from a Fluorescent C_{60} Adduct-Single Layer Organic LED", *J. Am. Chem. Soc.* **1999**, *121*, 5611–5612.

65. Y.-Z. An, Y. Rubin, C. Schaller, S. W. McElvany, "Synthesis and Characterization of Diethynylmethanobuckminsterfullerene, a Building Block for Macrocyclic and Polymeric Carbon Allotropes" *J. Org. Chem.* **1994**, *59*, 2927–2929.

66. G. A. Ellis, Ph.D. thesis, UCLA, 1996.

67. H. L. Anderson, R. Faust, Y. Rubin, F. Diederich, "Fullerene-Acetylene Hybrids: On the Way to Synthetic Molecular Carbon Allotropes", *Angew. Chem. Int. Ed. Engl.* **1994**, *33*, 1366–1368.

68. L. Isaacs, F. Diederich, R. F. Haldimann, "Multiple Adducts of C_{60} by Tether-Directed Remote Functionalization and Synthesis of Soluble Derivatives of New Carbon Allotropes $C_{n(60+5)}$", *Helv. Chim. Acta* **1997**, *80*, 317–342.

69. K. Komatsu, personal communication.

70. J. Osterodt, F. Vögtle, "$C_{61}Br_2$: A New Synthesis of Dibromomethanofullerene and Mass Spectrometric Evidence of the Carbon Allotropes C_{121} and C_{122}", *J. Chem. Soc., Chem. Commun.* **1996**, 547–548.

71. T. S. Fabre, W. D. Treleaven, T. D. McCarley, C. L. Newton, R. M. Landry, M. C. Saraiva, R. M. Strongin, "The Reaction of Buckminsterfullerene with Diazotetrazole. Synthesis, Isolation, and Characterization of $(C60)_2C_2$", *J. Org. Chem.* **1998**, *63*, 3522–3523.

72. N. Dragoe, S. Tanibayashi, K. Nakahara, S. Nakao, H. Shimotani, L. Xiao, K. Kitazawa, Y. Achiba, K. Kikuchi, K. Nojima, "Carbon Allotropes of Dumbbell Structure: C_{121} and C_{122}", *Chem. Commun.* **1999**, 85–86.

73. L. T. Scott, "Fragments of Fullerenes: Novel Syntheses, Structures and Reactions", *Pure Appl. Chem.* **1996**, *68*, 291–300.

74. P. W. Rabideau, A. Sygula, "Buckybowls: Polynuclear Aromatic Hydrocarbons Related to the Buckminsterfullerene Surface", *Acc. Chem. Res.* **1996**, *29*, 235–242.

75. J. S. Siegel, T. J. Seiders, "From Bowls to Saddles", *Chem. Brit.* **1995**, *31*, 313–316.

76. T. J. Seiders, E. L. Elliott, G. H. Grube, J. S. Siegel, "Synthesis of Corannulene and Alkyl Derivatives of Corannulene", *J. Am. Chem. Soc.* **1999**, *121*, 7804–7813.

77. Y. Rubin, "Ring Opening Reactions of Fullerenes: Designed Approaches to Endohedral Metal Complexes", *Top. Curr. Chem.* **1999**, *199*, 67–91.

78. G. Schick, T. Jarrosson, Y. Rubin, "Formation of an Effective Opening within the Fullerene Core of C_{60} by an Unusual Reaction Sequence", *Angew. Chem. Int. Ed. Engl.* **1999**, *38*, 2360–2363.

79. T. Jarrosson, G.-W. Wang, M. D. Bartberger, G. Schick, M. Saunders, R. J. Cross, K. N. Hock, Y. Rubin, "Insertion of Helium and Molecular Hydrogen Inside an Open Fullerene", manuscript in preparation.

80. A. Weidinger, M. Waiblinger, B. Pietzak, T. A. Murphy, "Atomic Nitrogen in C_{60}: N@C_{60}", *Appl. Phys. A* **1998**, *66*, 287–292.

81. J. C. Hummelen, C. Bellavia-Lund, F. Wudl, "Heterofullerenes", *Top. Curr. Chem.* **1999**, *199*, 93–134.

Dendritic Architectures

Sven Gestermann, Richard Hesse, Björna Windisch, and Fritz Vögtle

Kekulé-Institut für Organische Chemie und Biochemie der Universität Bonn, Gerhard-Domagk-Str. 1, 53121 Bonn, Germany,

Phone: +49 228 73 3495, Fax: +49 228 73 56 62, e-mail: voegtle@uni-bonn.de Web: www.chemie.uni-bonn.de

Keywords ■ Macromolecules ■ Dendrimers ■ Fractal Structure ■ Hyperbranched Polymers ■ Nanostructure ■ Divergent Synthesis ■ Convergent Synthesis ■ Dendrylation

Concept: The structural motif of a defined branching of branches leads to dendritic molecules reaching the nanometer scale. The development of this concept has given rise to a great variety of macromolecules which cover a broad range of chemical functionalities. These so-called dendrimers afford access to new properties, resulting from the topological peculiarities of their dendritic architectures.

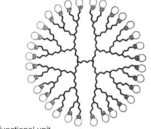

○▯ = functional unit

Abstract: The concept of a defined branching of branches leads to dendrimers, highly branched macromolecules with well-defined structures. The topologies of dendritic molecules result in them having some structural features, which give access to new, unique properties. These features are a result of both the porous inner region with its own micro-environment as well as of the great number of end groups. Besides dendrimers with different skeletons, functional dendrimers extend the concept even further. Whereas most dendrimers are prepared by *divergent* or *convergent* synthesis, the majority of functional dendrimers are obtained, either by the dendrylation of a functionality using preformed dendrons, or by the functionalization of already synthesized dendrimers on their peripheries. By following these pathways, nearly every chemist can make his or her chemistry dendritic and thus take advantage of the various dendritic ways to change or amplify the properties of the functionality in question. Since the concept is a structural one and hence not limited to any particular chemical functionality, dendrimers cover the whole range of organic and, to some extent, also inorganic chemistry.

Prologue

Dendrimers are nanometer-sized, highly branched molecules with fascinatingly symmetrical architectures. They always consist of a core unit, branching units, and end groups located on their peripheries (Figure 1).[1] Dendrimers are based on the concept of "branching branches" in their structural design, as well as in their synthesis. It yields a fractal arrangement. In the field of mathematics, a fractal is a geometrical formation which is of a complex and detailed structure at every level of extension. Fractals possess the property of being *self-resembling*, i.e., each small section of the fractal is of the same structure as the entire object.

Synthesis of Dendrimers

Dendrimer architectures are largely determined by the types of branching units they display. To be realised in a dendrimer synthesis, these branching units have to fulfil certain conditions.

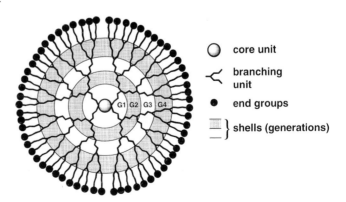

Figure 1. General subdivision of dendrimers into core unit, branching units, and end groups.

They need to have three functionalities (coupling sites), with one functionality being different from the two others. The two different functionalities must not be able to react directly with each other, but only after one of them has been modified.[2] If they were the same, all functional groups of the branching units would react in one construction step resulting in highly branched polydisperse polymers (molecules with different molecular masses). Therefore a branching unit consists of at least a group A and two groups B (Figure 2). A must not be able to react directly with B, for once again, in this case, highly branched polymers would arise. If one wants to obtain monodisperse dendrimers (only consisting of identical molecules), it must be possible to convert A or B into an activated form C which

can then subsequently react with **A** or **B**, as the case might be. For reasons of better understanding we will distinguish later on different groups of branching units, depending on the use in a *divergent* (Figure 2) or *convergent* (Figure 3) dendrimer synthesis.

A *divergent synthesis* (Figure 2) is carried out starting from the dendrimer core towards the dendrimer periphery – that is, from the inside out. The **C**-groups of the core are reacted with the **A**-groups of the branching units. Then, the **B**-groups, also introduced in this step, are converted into reactive **C**-groups to enable their subsequent reaction with **A**-groups. The synthesis of a structurally perfect, monodisperse dendrimer requires a quantitative conversion in each of these reaction steps, since it is difficult to separate fully converted products

Figure 2. Divergent synthesis: a general scheme.

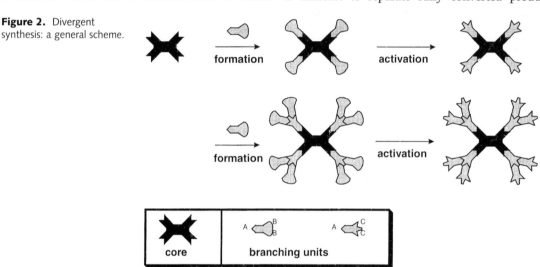

Table 1. Examples for branching units employed in dendrimer synthesis.

from not completely converted ones. One or two end groups missing do not change, for example, the column-chromatographic behavior enough to allow convenient isolation of a pure dendrimer.

A *convergent synthesis* (Figure 3) is performed in the opposite direction – that is, from the outside in. In the first step, the end groups are reacted with the **B**-groups of the branching units. Following deprotection or modification, the **A**-group is then converted into a **C**-group in the second step, so that it can react with a **B**-group. In the course of the synthesis, preprepared dendrons (piece-of-cake-like dendritic substituents) are bound to a branching unit, yielding dendrons, constantly increasing in size. In this case, defective molecules differ significantly from the structurally perfect ones (by one dendron unit) and can hence be separated easily. The covalent coupling of the sterically demanding dendrons can, however, be difficult, particularly for higher generations, because of a lack of space.

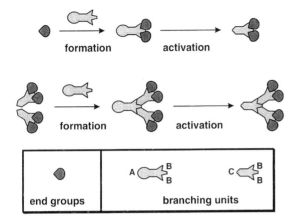

end groups | **branching units**

Figure 3. Convergent synthesis: a general scheme.

Thus, the synthesis of dendrimers consists of two constantly repeating reaction steps. The first step deals with the linkage of a branching unit to two other units – the *construction step*. In the second reaction, groups are transformed, so that they can react further – the *activation step*. This procedure is also referred to as an iterative (repetitive) strategy.[3]

In these two methods (*divergent/convergent*), the core unit and the end groups are implicated either at the beginning or at the end of the dendrimer synthesis. The *divergent method* starts with the core unit carrying – if possible – several C-groups which can be reacted with the A-groups of the branching units. The end groups, which are added at the end of the dendrimer synthesis, only have to be able to react with the B- or C-groups. In the *convergent synthesis*, exactly the opposite sequence is followed. Hence, the end groups must be able to react with

the B-groups and ought to be inert to the other reactions. In the last reaction step, piece-of-cake-like dendrons are connected to a core unit. Therefore, the core unit must be capable of reacting with the A- or C-groups. In special cases, linking of the single dendrons to form the dendrimer can also occur as a result of the core unit being synthesized in the final reaction step. For this purpose, the dendrons must carry appropriate precursors of the core unit at their focal points.

The size of a dendrimer is also characterized by its *generation*. The term generation relates to how many times the construction step is repeated. In other words, the generations correlate to the number of "shells" of branching units around the core unit (Figure 1). On account of the fractal structure of dendrimers, the number of end groups in any generation can be calculated exactly according to the following equation:

$$n_g = F_k \cdot (F_v-1)^g$$

n_g: number of end groups in generation g.
F_k: functionality in the core (e.g., 4 in Figures 1 and 2)
F_v: functionality of the branching unit (e.g., $AB_2 = 3$)
g: generation of the dendrimer

In dendrimers, a high number of end groups is obtained after only a few generations. That is why they are so interesting for applications. One can accumulate numerous building blocks, e.g., fluorescent or complexing ones, particularly on the peripheries and in this way intensify certain properties, e.g., fluorescence, solubility, metal-ion complexation, and so on.

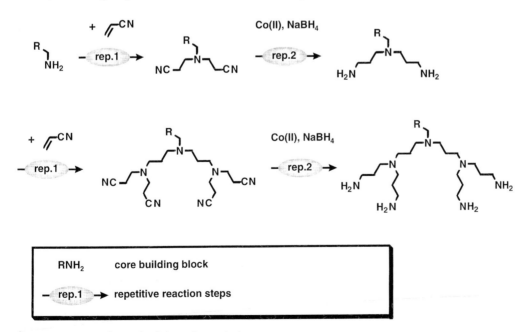

RNH$_2$ core building block

— rep.1 →   repetitive reaction steps

Figure 4. First synthesis of poly(propyleneamine) dendrimers (abbreviated POPAM) by Vögtle et al.

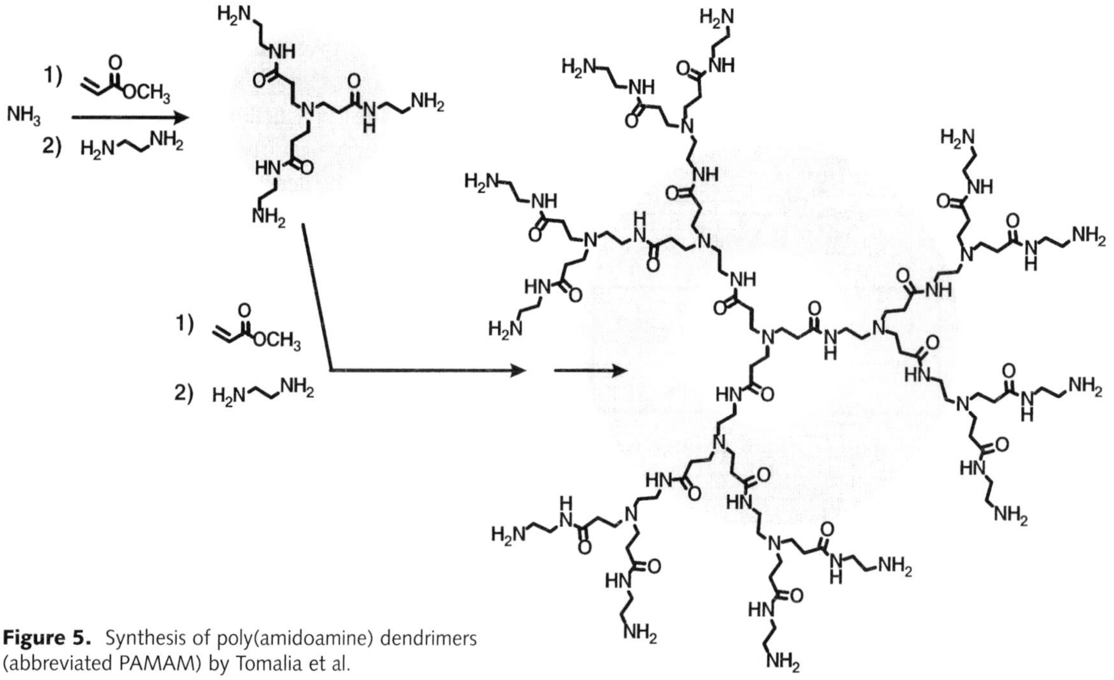

Figure 5. Synthesis of poly(amidoamine) dendrimers (abbreviated PAMAM) by Tomalia et al.

In recent years, such a variety of dendrimer syntheses have been described in the literature, that they cannot all be mentioned in this essay. Many reviews have been published, reflecting the enormous attention which has been given to dendrimers.[4] The general architectural and synthetical concepts of dendrimer chemistry can, however, be explained by some important examples.

In 1978, Vögtle et al.[5] reported the first synthesis of a poly(propyleneamine) dendron (Figure 4) based on the combination of a *Michael* condensation of acrylonitrile to amines and the subsequent reduction by hydrogenation of nitrile groups to yield amine functions. Whereas the *Michael* condensation represents the constructing step, the reduction is used to activate dendrons (regenerate the reactive NH_2 groups). In this synthesis, the branching is based on the treatment of primary amines with two molecules of acrylonitrile at a time.

The *Michael* condensation has also been employed as the construction step in the synthesis of another dendrimer type (Figure 5). In these poly(amido-amine) dendrimers, the branching is achieved by reacting methyl acrylate with amines.[6] In the second step, the carboxylic ester functions are converted to amides by using excessive ethylenediamine, resulting once again in primary amino groups on the peripheries of the dendrimers. The use of ethylenediamine – being bifunctional – actually violates the rules for a consistently *divergent* dendrimer synthesis, since the activation step can lead to inter- as well as intramolecular cross-linking. However, by using a large excess of ethylenediamine, side reactions can be suppressed. Nowadays,

poly(propyleneamine) (POPAM) dendrimers, as well as poly(amidoamine) (PAMAM) dendrimers, are commercially available and therefore often employed as starting materials in dendrimer chemistry.[7]

Fréchet et al.[8] designed a frequently used and subsequently adopted *convergent* dendrimer synthesis, applying this principle to chemistry for the first time. In these polyarylether dendrimers, 3,5-dihydroxybenzyl alcohol is employed as the branching unit which is transformed to higher generations using two reactions (Figure 6). First a benzylic bromide (**C** group) reacts with the phenolic hydroxyl groups (**B** groups) of the 3,5-dihydroxybenzyl alcohol. In the activating step, the benzylic alcohol (**A** group) is converted to the bromide (**C** group), whereupon this group is activated for further branching to be promoted convergently.

Besides the two traditional methods of dendrimer synthesis already described, a combination of both synthetic approaches also yields new dendrimers (Figure 7). This method makes use of the benefits of both the *convergent* and the *divergent* strategies. Employing a *divergent* synthesis, a core unit bearing several functionalities on its periphery is built up. In one reaction step, several dendrons are then attached to this core unit. Altogether, this method represents a simple way to rapidly enlarge dendrimers. If, for example, different branching units are employed, a dendrimer is obtained which contains onion-like shells with varying microenvironments.[9]

Some of the more recent dendrimer syntheses are based on the simultaneous growth in *divergent*

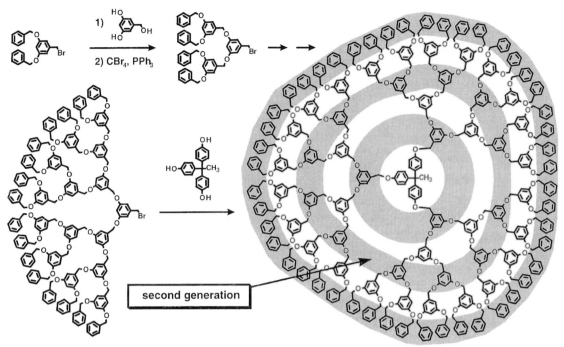

Figure 6. Convergent synthesis of poly(arylether) dendrimers by Fréchet et al.

and *convergent* directions. They rely on branching units (AB₂) which are protected at both the A and B groups. These branching units are deprotected separately at their A and B groups. After that, the two products are reacted with each other, resulting in a second generation dendron with protecting groups at the focal point and on its periphery. If the procedure is repeated with the second generation dendron, a fourth generation dendron is obtained. Because of the rapid growth, this synthetic method is also referred to as being *"doubly exponential"* (Figure 8).

It seems that dendrimer chemists are particularly creative, for they have always invented new types of

synthetic methods, when temporarily it seemed to be impossible to overcome certain difficulties – such as solubility problems, structural perfection, yields, steric hindrance, etc. *Orthogonal* dendrimer syntheses have also been reported in the literature. These syntheses proceed without an activating step by employing two different branching units with coupling functions which are complementary to each other. The units are built up as **AB₂** and **CD₂** systems, **A** and **D** reacting under altered conditions from **B** and **C**. This means that the functionalities **A** and **D** have to be chosen in such way that they, as well as their coupled products, are inert under the reaction conditions of the second pair of function-

Figure 7. A combination of *divergent* and *convergent* synthesis, yielding a dendrimer with pockets in different micro-environments.

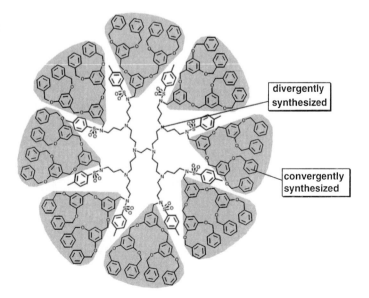

Figure 8. Doubly exponential dendrimer synthesis involving simultaneous growth in divergent and convergent directions.

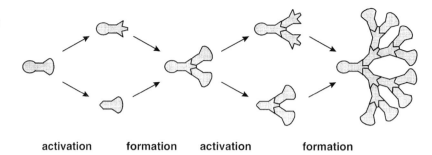

activation   formation   activation   formation

alities **B** and **C**. In principle, the construction can be performed *divergently* or *convergently* this way. The reactions employed must, however, meet the stringent requirements just outlined. Therefore, this path has not been followed very frequently.

Especially, the *divergent* and *convergent* strategies have been applied to many dendrimer syntheses. Nowadays, there are dendrimers with branching units which are based on silicon, phosphorus and other heteroatoms,[10] dendrimers which are built up supramolecularly (rather than covalently)[11] and even pure hydrocarbon dendrimers.[12,13] A dendrimer of the tenth generation with 3072 gold ions at its periphery, which is 15 nanometers in diameter, has been prepared successfully by Majoral et al.[10]

By the development of these concepts leading to the efficient synthesis of dendritic architectures,[5] a new impetus has been given in organic and inorganic chemistry which is leading to the production of new materials. Possible applications will be dis-

cussed in the following section, since they strongly depend on their functionalization.

Functionalization of Dendrimers

After a variety of very different dendrimers had become available up to high generations, many groups directed their attention more and more to designing and synthesizing functionalized dendrimers, i.e., dendrimers adorned with functionalities that are intended to give them characteristic properties favourable for application purposes.[4,14]

Generally, several different methods can be envisaged to obtain functionalized dendrimers. They include:

(1) *Dendrylation* of a central molecule which carries functional groups by attaching (grafting) suitable dendrons (dendrylic substituents; Figure 9);[15]

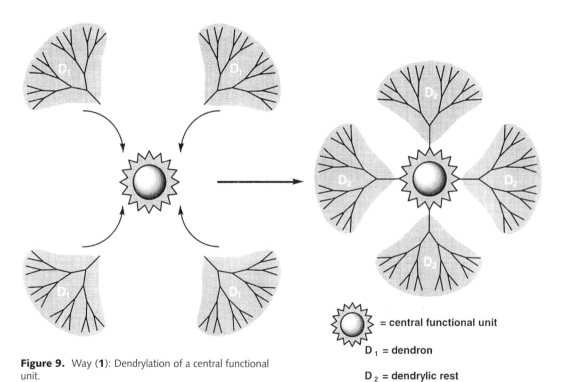

= central functional unit

D₁ = dendron

D₂ = dendrylic rest

Figure 9. Way (**1**): Dendrylation of a central functional unit.

(2) Attaching the functionality to already constructed dendrimer skeletons by suitably "refunctionalizing" the end groups or other moieties in the dendrimers;[4,14]

(3) Divergent or *convergent* synthesis of dendrimers which already carry the desired functionality in the branching unit, i.e., within the dendrimer skeleton.[1,10,11,19,21,22]

The dendrylation of a central functional unit according to the approach illustrated in Figure 9 results in core-functionalized dendrimers. Receptors, self-assembling components, chromophores, luminophores, photo- and redox-responsive groups, redox-active moieties, catalytically active components, chiral centers, polymerizable monomers and polymers themselves have been employed as central functional units.[4,10,20] In the majority of the described syntheses, the *convergent* approach has been followed, that is, the linkage of a central molecule to one or more pre-synthesized dendrons. The *divergent* approach in which the construction of the dendrimer skeleton generation by generation starting from the central unit and leading to topologically similar dendrimers has been used infrequently.

The potential influence of the dendrylation on the functional core unit includes sometimes a drastically increased molecule size as well as a steric shielding (encapsulation) and a micro-environment different and isolated from its external surroundings, e.g., unimolecular micellar structures, electron-rich shells, solubilization. It is even possible to activate the core unit by both energy and electron transfer processes. In the following subsections, these design possibilities will be dealt with in more detail.

Size: The molecular enlargement achieved by dendrylation can be variously utilized. For example, the separation of the dendrylated component from smaller molecular species by means of ultrafiltration makes use of a size difference. This procedure is particularly advantageous with respect to catalytically active focal functionalities, because it simplifies the recovery of the catalyst and allows even continuous diaphragm processes.[4,10,14,17–22]

Size and shape: Certain steric effects can be achieved using characteristically wedge-shaped dendrons. Thus, self-assembling dendrons have been connected to supramolecular aggregates with defined dimeric or hexameric structures. Such aggregates can form columnar superstructures which reveal liquid-crystalline properties. Spherical superstructures arise from the self-assembly process when conical dendrons are used.[4,14,16] Similar

effects have been obtained from the dendrylation of polymers. Lateral covalent attaching of dendrons forces these polymers to take up a stretched cylindrical shape. Also, when polymerizing dendrons with a polymerizable functionality at the focal point, more spherical or cylindrical polymer shapes can be achieved depending on the degree of polymerization.[16]

Size, shape, and density: The shielding effects of dendritic shells can likewise be caused by steric factors. Thus, the access of foreign molecules to the central functional unit can be hindered or prevented according to size and density of the dendritic shell. Sometimes, even a certain size selectivity is observed. These effects are especially interesting for electrochemically, catalytically active, redox- and photo-active functional units, since interactions with foreign molecules, such as oxygen quenching of the luminescence (photo-active units) or the access of substrates (catalytically active units) can be influenced.[4,11,17,22]

Polarity, electron density: As a result of the dendritic shell sheathing and shielding the core, a micro-environment is created which can differ from the outer environment and which can change the properties of the central functional unit depending on polarity and electron density. Distinct changes in the solvatochromism and the redox behavior of appropriate central functionalities, for example, have been ascribed to a changed micro-environment.[17,22] Moreover, a change in the micro-environment can enable the solubilization of functionalities which are insoluble in certain solvents. The solubility of the whole molecule results from dendrylating such units with dendrons that carry solvatophilic groups. Such dendrimers represent unimolecular micelles.[11,14] Photo-active central units could even be actively influenced by electron and energy transfer, respectively, through the dendritic skeleton in some cases. For this purpose, appropriate active components have been either electrostatically docked to the dendron periphery or covalently incorporated.[4,11,14]

Thanks to the simplicity of some dendron syntheses, way (1) (Figure 9) is a much followed path, along which a multitude of research groups have extended their chemistry or developed it in a dendritic manner without having a lot of previous experience in the fields of dendrimer or polymer chemistry.

By functionalizing the end groups according to way (2) (Figure 10) polyfunctionalized dendrimers can be achieved in which the end groups of the dendrimer are connected to functional moieties in one synthetic step. Not least of all because of the

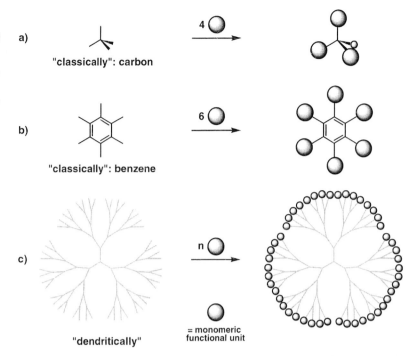

Figure 10. Way (**2**): Possible ways to "multiply" functional groups: **a)** and **b)** "classically" with carbon (four times) and benzene (six times), respectively, and **c)** "dendritically" (*n* times) by functionalizing the dendrimer periphery.

fact that poly(propyleneamine) (POPAM) and poly(amidoamine) (PAMAM) dendrimers are commercially available now,[7] numerous functionalizations of this type have been carried out on them.[4,9,11,14,17,18,20,22]

In order to functionalize dendrimers on their peripheries, reactions of all the functional groups of the same kind in the dendrimer with a reactive moiety of the functionality to be introduced require new strategic considerations and make particular demands on the reaction in question. On account of the large number of groups, to be converted per dendrimer, only those reagents can be used which react very selectively with these groups and at the same time reveal high reactivity. These requirements are explained by the following considerations:

Assuming that, in the course of the functionalization, each reaction between the reagent and a still free peripheral group occurs without being influenced by the number of already converted moieties,[23] the total yield A of perfectly reacted dendrimer can be calculated by means of the *product formula for a stochastically independent event*, if the yield of the individual reaction a is known. The yield A is given by a^n where n is the number of peripheral functional groups of the dendrimer.

One can clearly appreciate that only highly selective reactions with almost quantitative conversion are suitable for functionalizing higher generation dendrimers (Figure 11). Even reactions leading to 90 % yield in each reaction step result in a total yield of only 43 % of the desired product for a fifth generation dendrimer (64 peripheral groups). Thus, there are 57 % of different dendritic by-products formed,

which very much resemble the actual target molecule, since a major portion of the amino groups react as expected, while some groups enter into a different or no reaction. Due to their similarity, it is nearly impossible to remove these by-products. A further requirement on potential functionalizations is the separability from excessive coupling reagent and from eventual by-products, which are produced by the reagent itself. Furthermore, the stability of the newly-linked bonds is an important criterion.

Whereas in dendrylations using way (**1**) (Figure 9) the formation of a dendritic environment around a central unit was in the limelight, in way (**2**) (Figure 10), the multiplication of functional moieties is the principal aim. In this way, also chromophoric, luminophoric, photo-active switches, redox-active units, catalytically active moieties, polymerizable groups, complexing components, mesogenic groups, solubilizing units, chiral molecules, and saccharides have also been introduced using the functionalization strategy.[4,14,17,22] In general, by multiplying these identical functional units, the effect caused by these groups is drastically increased. In some cases, however, a decrease in the effect per functional unit has been noted, a fact that has been ascribed to the steric crush of these groups, growing with the increasing dendrimer generation and the accompanying mutual influence. Hitherto, a "*dendritic effect*" in the sense that the effect increases more than additively with a growing number of functional groups has been found in few cases.[24] A further important effect of peripheral functionalization is the dominating influence of the end groups on dendrimer solubilities. Thus, non-polar dendrimer skeletons

Figure 11. The implications for the total yield, assuming an idealized dendrimer functionalization for the example of a dendrimer with **AB**$_2$-branching and bifunctional core unit (G = generation).

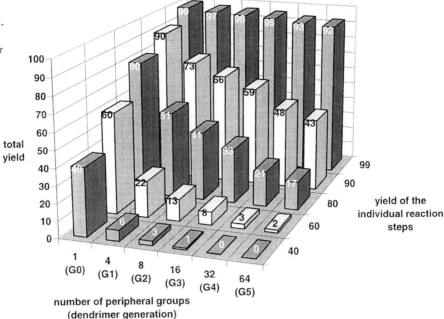

are successfully solubilized in polar solvents by means of polar end groups and vice versa.

Way (3) (Figure 12) also leads to polyfunctionalized molecules. In most cases, it involves a considerable synthetic effort, since it is necessary to synthesize the whole dendrimer skeleton *de novo*. Moreover, the functionalities which are introduced have to tolerate the often quite drastic reaction conditions of the dendrimer synthesis. The advantage of this synthesis is, however, that the complete dendritic skeleton can be tailor-made for the desired purpose. Along with numerous other examples, dendrimers with chiral skeleton building blocks,[18] with coupling sites for subsequent functionalizations located in the skeleton,[4,10,22] with an energy gradient from the periphery to the redox- and photo-active core, have been prepared success-

fully.[4,17,19,21] Also supramolecularly constructed dendrimers can be synthesized this way, the branching units themselves being supramolecular moieties.[11,19,20,21] Likewise various dendrimers based on heteroatoms have been realized.[10]

Employing ways (1), (2), and (3) a broad range of dendritic architectures can be envisaged. It has enabled many research groups to expand their special interests into this particular macromolecular

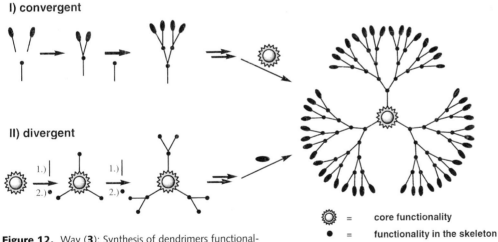

Figure 12. Way (3): Synthesis of dendrimers functionalized in the core, in the skeleton, and on the periphery.

field without having to acquire previous expertise in polymer or dendrimer chemistry. Thanks to the simple availability of the dendrons, these three general ways allow straightforward access to dendrimers and their comparatively easier functionalizability. This is not least of all important because of the fact that dendrimers represent a structural principle, which is not limited to a certain class of substances.

Characteristics of the Dendritic Structure and Some Resulting Potential Applications

Dendrimers reveal a variety of properties, which clearly distinguish them from usual polymers of similar size.[1,22] Firstly, one must mention the defined molecular size, which – despite the structural defects occurring at higher generations – clearly shelters them from the molecular-mass distribution of conventional polymers which is broader by several orders of magnitude. This property becomes especially important with respect to the separability of dendritic functional units, both for chemical applications (ultrafiltration, catalysis) and for medicinal applications *in vivo* (therapeutics, diagnostics) owing to the filtration characteristics of blood vessels and the kidney. Additionally, for higher dendrimer generations, there is their globular structure with an easily accessible molecular surface. This allows ready access to large (space-filling) substituents which are needed, e.g., for stoppers in rotaxanes with large wheels.[11] Classically, it would be more difficult to prepare large space-demanding and highly soluble moieties. Moreover, the surface carries a large, defined number of functional or functionalizable groups. In this way, identical functions can be multiplied and so in many cases the effect is enhanced. These considerations are interesting for applications in the field of catalysis, particularly in the field of the *in vivo* diagnostics (peak amplification MRI) and therapeutics (concentration of active agent, boron neutron capture ther-

apy). One particular application is just about to be introduced onto the market (MRI-Schering).[4,22] By multiplying identical functions, as possible in dendrimers, polyvalent interactions can occur. Polyvalency is ubiquitous, particularly in the biological field (cell-cell recognition, peak transduction, attachment of viruses, etc.). Therefore, it turns out to be useful to apply dendrimers in these fields. The functionalization of dendrimers, for example, with appropriate ligands, such as saccharides, has yielded sugar balls, which competitively inhibit the attachment of certain viruses to complementary cell receptors.[11,17,22,23]

Furthermore, a characteristical feature of the dendritic structure principle is the occurrence of cavities, i.e., of alcoves in the inside of the skeleton which have their own micro-environments. More recent investigations have shown that these cavities are not of a permanent nature, but that – for flexible arms – they are filled up in time-averaged manner because of peripheral groups backfolding and the dendritic skeleton collapsing. Nevertheless, dendrimers are capable of accepting guests into these latent cavities. Besides statistical inclusions, mainly electrostatic interactions and the hydrophobic effect serve as driving forces. Hence, dendrimers of a micellar structure (solvatophilic peripheries and solvatophobic interiors) have been employed successfully in liquid-liquid extractions [Figure 13 (II)]. Also, irreversible encapsulation has been achieved by functionalization with sterically demanding substituents and the consolidation of the dendrimer periphery as a consequence thereof.

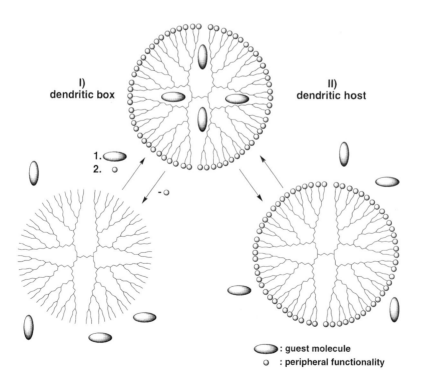

I)
dendritic box

II)
dendritic host

1.
2.

Figure 13. **I)** Inclusion of guest molecules inside a dendritic box; **II)** insertion of guest molecules into the dendritic interior.

: guest molecule

o : peripheral functionality

In this case, the guest was successfully released only after these peripheral groups had been cleaved off [Figure 13 (I), dendritic box]. Likewise, metal nanoclusters have been stabilized in the dendritic interior.[4,11,17,22] This property could allow future applications such as molecular containers which show a directed release of the active agent (depot effect). A further characteristic feature of the dendritic structure principle is the number of branching units in the current "shell" increasing exponentially with the distance from the core. With this property, dendritic systems are predestined to form systems with the property of collecting energy and focusing it towards the core, provided the core, skeleton and peripheral units are appropriately tailored (*light harvesting*).[17]

Epilogue

Since they were first mentioned in 1978, the *concept* of "cascade" or "dendritic structures" has witnessed a meteoric rise. On the one hand, thanks to their aesthetic beauty, and on the other hand, because of their broad applicability, dendrimers are in considerable demand. As the concept is not limited to one class of substances and, furthermore, allows a simple access on several routes, there is – besides the researchers who started dendrimer chemistry – a growing number of research groups which are dendritically expanding their special areas of interest in different ways and describing new or varying already existing properties.

Since this topic is still booming, this essay can only give a general idea about dendrimers, the basic principles surrounding them amd some promising developments involving them, without paying the many pioneers of this research field the respect they deserve. Therefore, more detailed and comprehensive reviews have been featured and hardly any original publications have been cited.

Acknowledgement

We are indebted to Prof. Dr. J. F. Stoddart for help with this contribution.

References and Notes

1. For comprehensive review of the synthesis of different dendrimers, see: G. R. Newkome, C. N. Moorefield, F. Vögtle, *"Dendritic Molecules: Concepts, Synthesis, Perspectives"*, WILEY-VCH, Weinheim **1996**.

2. Subsequently an AB_2-system is assumed for reasons of simplification, even if higher branches are known to the literature.

3. N. Feuerbacher, F. Vögtle, "Iterative Synthesis in Organic Chemistry", *Top. Curr. Chem.*, Vol. 197: *"Dendrimers"* (Ed.: F. Vögtle), Springer-Verlag, Berlin, Heidelberg **1998**, 1–18.

4. M. Fischer, F. Vögtle, "Dendrimers: From Design to Application – A Progress Report", *Angew. Chem.* **1999**, *111*, 934–955; *Angew. Chem. Int. Ed.* **1999**, *38*, 884–905.

5. E. Buhleier, W. Wehner, F. Vögtle, "'Cascade'- and 'Nonskid-Chain-Like' Syntheses of Molecular Cavity Topologies", *Synthesis* **1978**, 155–158.

6. D. A. Tomalia, H. Baker, J. R. Dewald, M. Hall, G. Kallos, S. Martin, J. Roeck, J. Ryder, P. Smith, *Polym. J.* **1985**, *17*, 117–132.

7. Poly(propyleneimine) dendrimers are sold by the company DSM, Netherlands, under the name of Astramol®. Poly(amidoamine) dendrimers are available as PAMAM® dendrimers at Aldrich.

8. C. J. Hawker, J. M. J. Fréchet, "Preparation of Polymers with Controlled Molecular Architecture: A New Convergent Approach to Dendritic Macromolecules", *J. Am. Chem. Soc.* **1990**, *112*, 7638–7647.

9. A. Archut, S. Gestermann, R. Hesse, C. Kauffmann, F. Vögtle, "Selective Activation, Different Branching and Grafting of Poly(propyleneamine)-Dendrimers", *Synlett* **1998**, 546–548.

10. J.-P. Majoral, A.-M. Caminade, "Dendrimers Containing Heteroatoms (Si, P, B, Ge or Bi)", *Chem. Rev.* **1999**, *99*, 845–880.

11. F. Zeng, S. C. Zimmerman, "Dendrimers in Supramolecular Chemistry: From Molecular Recognition to Self-Assembly", *Chem. Rev.* **1997**, *97*, 1681–1712.

12. J. S. Moore, Z. Xu, "Rapid Construction of Large-Size Dendrimers up to 12.5 Nanometers in Molecular Diameter", *Angew. Chem.* **1993**, *105*, 1394–1396; *Angew. Chem. Int. Ed. Engl.* **1993**, *32*, 1354–1357.

13. F. Morgenroth, E. Reuther, K. Müllen, "Polyphenylene Dendrimers: From Three-Dimensional to Two-Dimensional Structures", *Angew. Chem.* **1997**, *109*, 647–649; *Angew. Chem. Int. Ed. Engl.* **1997**, *36*, 631–634.

14. A. Archut, F. Vögtle, "Functional Cascade Molecules", *Chem. Soc. Rev.* **1998**, *27*, 233–240.

15. F. Vögtle, M. Plevoets, G. Nachtsheim, U. Wörsdörfer, "Monofunktionalisierte Dendrons verschiedener Generationen – als Reagenzien zur Einführung dendritischer Reste", *J. prakt. Chem.* **1998**, *340*, 112–120.

16. H. Frey, "From Random Coil to Extended Nanocylinder, Dendrimer Fragments Shape Polymer Chains", *Angew. Chem.* **1998**, *110*, 2313–2318; *Angew. Chem. Int. Ed.* **1998**, *37*, 2193–2197.

17. D. K. Smith, F. Diederich, "Functional Dendrimers: Unique Biological Mimics", *Chem. Eur. J.* **1998**, *4*, 1353–1360.

18. D. Seebach, P. B. Rheiner, G. Greiveldinger, T. Butz, H. Sellner, "Chiral Dendrimers", *Top. Curr. Chem.*, Vol. 197: *"Dendrimers"* (Ed.: F. Vögtle), Springer-Verlag, Berlin, Heidelberg **1998**, 125–164.

19. M. Venturi, S. Serronni, A. Juris, S. Campagna, V. Balzani, "Electrochemical and Photochemical Properties of Metal-Containing Dendrimers", *Top. Curr. Chem.*, Vol. 197: *"Dendrimers"* (Ed.: F. Vögtle), Springer-Verlag, Berlin, Heidelberg **1998**, 193–228.

20. M. A. Hearshaw, J. R. Moss, "Organometallic and Related Metal-Containing Dendrimers", *Chem. Commun.* **1999**, 1–8.
21. C. Gorman, "Metallodendrimers", *Adv. Mater.* **1998**, *10*, 295–309.
22. O. A. Matthews, A. N. Shipway, J. F. Stoddart, "Dendrimers – Branching out from Curiosities into New Technologies", *Progr. Polymer Science* **1998**, *23*, 1–56.
23. M. Mammen, S.-K. Choi, G. M. Whitesides, "Polyvalent Interactions in Biological Systems: Implication for Design and Use of Multivalent Ligands and Inhibitors", *Angew. Chem.* **1998**, *110*, 2908–2953; *Angew. Chem. Int. Ed.* **1998**, *37*, 2754–2794.

Chemical Encapsulation in Self-Assembling Capsules

Christoph A. Schalley[a] and Julius Rebek, Jr.[b]

[a] Kekulé-Institut für Organische Chemie und Biochemie der Universität Bonn, Gerhard-Domagk-Str. 1, 53121 Bonn, Germany, Phone: +49 228 73 57 84, Fax: +49 228 73 56 62, e-mail: c.schalley@uni-bonn.de

[b] The Skaggs Institute for Chemical Biology and Department of Chemistry, The Scripps Research Institute, 10550 North Torrey Pines Road, La Jolla, CA 92037, USA, Phone: + 1858 7842250 Fax: + 1858 7842876 e-mail: Jrebek@scripps.edu

Keywords ■ Supramolecular Chemistry ■ Host-Guest Chemistry ■ Self-Assembly ■ Encapsulation ■ Chiral Recognition ■ Catalysis ■ Liquid Crystals ■ Hydrogen-Bonded Polymers ■ Non-covalent Interactions

Concept: Chemical encapsulation involves the inclusion of a guest molecule in the inner cavity of a closed-shell host. Encapsulation is a case of mechanical bonding since the guest experiences steric barriers imposed upon it by the host. Viewed in this manner, the situation is one of host and hostage, but since many weak attractive forces also exist between the two components, the friendler connotations of the familiar host/guest terminology are preferred. Either way, new phenomena can emerge within the host-guest complex. For example, the stabilization of reactive intermediates and new types of stereoisomerism can be witnessed in *Cram*-type carcerands – molecule-within-molecule complexes held together by covalent bonds. When, instead, the container host is assembled from non-covalently connected, concave subunits, the resulting capsules exhibit a dynamic quality. The steric barriers are temporary: the guest molecule binds reversibly and the subunits of the host are also exchanging: the system is kinetically labile. This self-assembly of the host allows new, more complex properties and *functions* to emerge that are characteristic of the complete assembly, rather than of its individual subunits. Among them are chiral recognition, catalysis, the formation of liquid crystalline phases, hydrogen-bonded polymers, and fibers. shape to be recognized. The capsules can be made chiral and show enantioselective recognition. They can also encapsulate more than one guest in which case they serve as reaction vessels: *Diels-Alder* reactions are accelerated, even catalyzed by encapsulation and spacial preorganization of the reactants. Linking two capsule monomers by a covalent spacer provides the basis for "polycaps" – hydrogen-bonded polymeric capsules. The polycaps form lyotropic, nematic liquid crystals at higher concentrations. Polymer fibers can be pulled out of these samples with a surprising load carrying capacity. In particular, these material properties indicate that molecular encapsulation is leaving the playground of fundamental science and is coming closer to applications in, for example, sensor technology.

Prologue

Encapsulation of guest molecules by closed-shell, covalent container molecules includes phenomena that range from endohedral complexes of buckminster fullerene[1] to the capture of guest molecules by *Cram*-type carcerands.[2] Because these containers are held together by strong, covalent bonds, their cavities provide inert matrix-like conditions for trapping and characterizing reactive intermediates such as *o*-benzyne[3] or cyclobutadiene.[4] The capsules isolate these species and protract their lifetimes; they prevent bimolecular reactions such as, for example, the dimerization of cyclobutadiene through a *Diels-Alder* reaction or of *o*-benzyne by a [2+2]-type cycloaddition. If the container is narrow enough, it restricts the rotation of the guest inside and new types of stereoisomerism have been observed. The detection of, for example, "up" and "down" diastereoisomers becomes possible with the guest in different, non-interconverting orientations with respect to the host.[5] It can be argued that this kind of isomerism is conceptually similar to translational isomerism found in appropriate catenanes and rotaxanes.

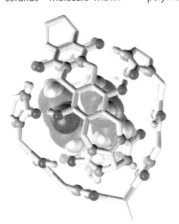

Abstract: Concave molecules with a periphery of complementary hydrogen-bond donors and acceptors dimerize to form non-covalently bound capsules. These host structures feature the ability to bind appropriate guest molecules inside their cavities. The host-guest complexes self-assemble under thermodynamic control, and a fine balance of entropic and enthalpic effects determines whether the capsules are formed and what guests are of appropriate size and

In marked contrast, the self-assembly[6,7] of appropriate concave subunits to capsules held together by non-covalent forces provides containers that are kinetically labile and are prone to the exchange of guests and parts of their walls. Nature makes extensive use of this principle. The tobacco-mosaic virus is just one of the most prominent examples. Its genetic information is inscribed on an RNA string which templates the self-assembly of 2130 identical cage proteins for its protection. Similarly, natural virus cages have been wrapped around artificial material such as anionic polymers and might be useful for drug delivery.[8] For a detailed investigation of the factors that govern the assembly of such hosts and the encapsulation of guests, artificial, minimalistic models were designed.[9,10] These capsules consist of two (or four) identical, self-complementary mono-

mers held together by hydrogen bonds. The first examples **1·1** – **3·3** resembled a tennis ball (Figure 1). The information for assembly is written into the monomers by chemical synthesis that defines the concave shape and the positions of hydrogen bond donors and acceptors appropriately for self-complementarity. These bonds form a seam that holds the dimers temporarily together.

Typically, the characterization of supramolecular systems requires the application of a whole set of analytical techniques. Without going into details, the capsular structure of the tennis ball **1·1** has been confirmed by X-ray crystallography, vapor phase osmometry (VPO), and NMR experiments.[11] The encapsulation of guests can be routinely analyzed by NMR experiments. Recently, a mass spectrometric method was developed that not only

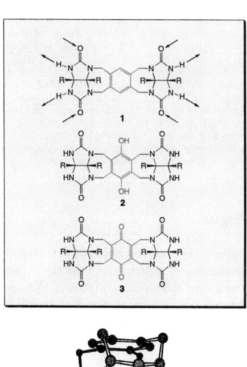

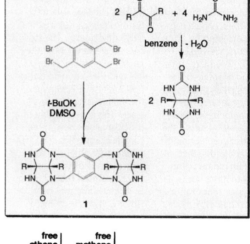

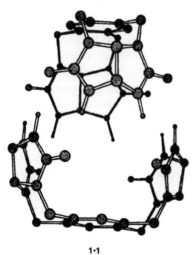

1·1

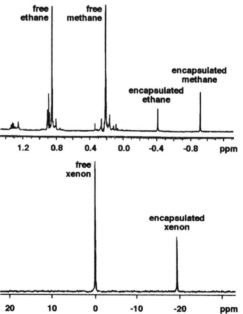

Figure 1. Top: Original "tennis ball" monomer **1**, its synthesis, and analogs **2** and **3**. Bottom left: X-ray structure of **1·1**. Bottom right: Upfield region of the ¹H-NMR spectrum of **1·1** in CDCl₃ after addition of ethane and methane as the guests and ¹²⁹Xe-NMR spectrum of a solution of **1·1** with Xe added. Due to the aromatic centerpieces of the capsule monomers, signals for the encapsulated guests are generally shifted upfield with respect to the free species.

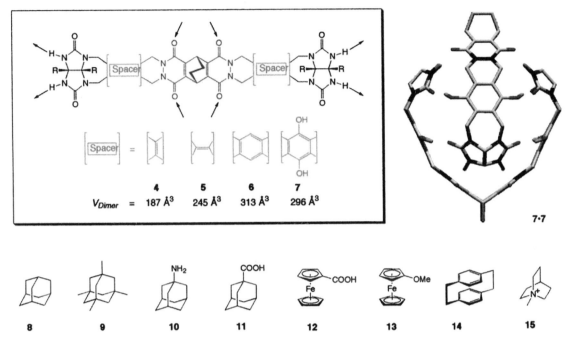

Figure 2. Left: Family of "softball" monomers **4** – **7**. Different spacers between the centerpiece and the glycoluril binding sites permit assembly of dimers with cavity volumes between 186 Å^3 and 313 Å^3. Right: Computer-minimized structure of the hydroxysoftball dimer **7**·**7** (solubilizing side chains and carbon-centered hydrogen atoms omitted). Bottom: Selection of guests **8** – **15** for the softballs **6**·**6** and **7**·**7**.

allows the generation of ionic encapsulation complexes in the gas phase, but also their structural characterization.[12]

Capsules with a variety of hydrogen bonding patterns and different sizes and shapes have been synthesized and studied using these methods, among them the "softballs" **4**·**4** – **7**·**7** (Figure 2), tetraurea calixarene dimer **16**·**16** (Figure 3), resorcinarene-based **20**·**20**, and the tetrameric "football" (**25**)$_4$. Each of these capsules contributes new aspects to our knowledge of self-assembling hosts and the rules for the encapsulation of guest molecules. From the assembly of two, four, or even multiple copies of the same molecule, new properties and functions emerge, which are not found for the monomeric subunits. This chapter intends to illustrate these aspects from different points of view and mostly covers recent results which have not yet been reviewed in detail.

Encapsulation in Self-Assembling Hosts: Complex Balance of Enthalpy and Entropy

Self-assembly is a thermodynamically controlled process. The formation of the capsules and the exchange of guest molecules proceed within seconds to hours, sometimes days, but finally, an equilibrium is reached which is governed by a finely balanced interplay of enthalpy and entropy. A detailed statistical analysis of the most successful host-guest pairs leads to the conclusion that – provided a congruent shape of guest and cavity – a guest is ideal if it occupies roughly 55 % of the cavity volume.[13] This value is close to typical packing coefficients of liquid organic solvents. With increasing guest size, first its motions, in particular rotations within the host, become more and more restricted due to close contacts with the capsule walls. The entropic costs of this effect must be paid by the binding enthalpy. Even larger guest molecules would stretch the hydrogen bonds or even break part of them quickly, causing energetic problems that most likely prevent the molecule from being a guest at all. Guests with a packing coefficient of significantly less than 55 % solvate only part of the inner surface of the capsule. The importance of these forces is illustrated by the encapsulation of methane in the tennis balls **1**·**1** – **3**·**3**. These capsules are very similar in the size of their cavities, but differ significantly with respect to the electronic properties of the center pieces. While electron rich **2**·**2** encapsulates methane with a larger binding constant of $K = 70 \ l \ mol^{-1}$ than **1**·**1** ($K = 33 \ l \ mol^{-1}$), methane is less strongly bound by the electron poor dimer **3**·**3** ($K = 10 \ l \ mol^{-1}$).[9] This result suggests that binding of the guest is not purely mechanical, but also depends on a favorable complementarity of the electronic properties of guest and capsule walls.

How selective encapsulation can be is clear from the examination of capsule **20**·**20**. In benzene,

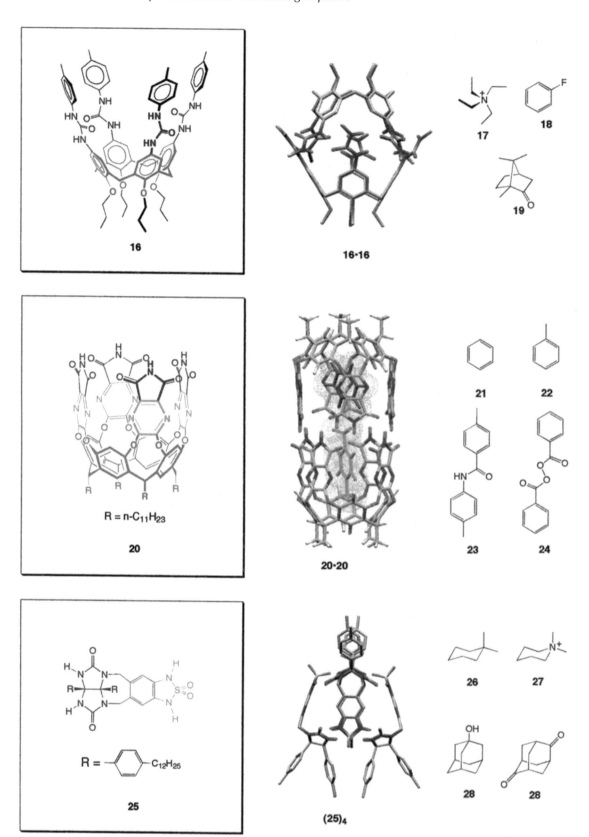

Figure 3. Capsules based on other hydrogen-bonding patterns, each with a selection of guests (right column) and a computer-generated picture of the dimer or tetramer (solubilizing side chains and carbon-centered hydrogen atoms omitted): Tetraurea calixarene dimer **16 · 16**, resorcinarene-based **20 · 20** (shown with the benzene/xylene pair as the guests), and the tetrameric "american football" **(25)₄**.

20·20 binds two benzene molecules inside its cavity. In a 1:1 mixture of benzene **21** and xylene **22**, however, the pairwise selection of one benzene and one xylene molecule is observed exclusively (Figure 3).[14] Neither capsules containing two benzene molecules are detected, nor those with two xylene molecules inside the cavity. Only with a large excess of benzene over xylene (> 70:1) can the container filled with two benzene molecules be detected in minor amounts.

Particular effects are found if the capsule is large enough to encapsulate more than one solvent molecule in the absence of appropriate guests.[15] Addition of such a guest liberates all the solvent molecules upon encapsulation. The total number of particles increases in this reaction and with it, entropy favors encapsulation of the guest over that of the solvent. If, for example, adamantane carboxylic acid **11**, ferrocene carboxylic acid **12**, or other similar guests are added to a chloroform solution of **7·7**, the temperature dependence of the guests' binding constants yield positive entropies of formation ($\Delta S > 0$) which overcompensate the enthalpically unfavorable ($\Delta H > 0$) inclusion of the guests.

The packing coefficient can be significantly higher than 55%, if additional forces favor binding of a certain guest over that of solvent molecules or other guests. One example is the encapsulation of quaternary ammonium cations inside the football **(25)₄**.[16,17] This capsule does not form at all, if no guest is present. The monomers are insoluble in dichloromethane, but addition of a guest, e.g. the ammonium ion **27**, adamantane **28**, or adamantanedione **29**, nucleates capsule formation and dissolves all of the material in form of its tetramer. Addition of even a large excess (up to 200-fold) of 1,1-dimethyl cyclohexane **26** does not result in capsule formation. In marked contrast, addition of two equivalents of **27**, a molecule with the same size and shape as **26**, to **(25)₄** leads to complete capsule formation. ¹H-NMR experiments yield a lower limit for the energy difference of these two complexes of more than 4 kcal mol⁻¹. This energy difference cannot be due to differences in size or shape of the two guests, but must be attributed to the positive charge of the ammonium ion. Consequently, cation-π interactions[18] of the charged guest with the aromatic capsule walls stabilize the charged over the neutral capsule. In general, most of the atoms that line the capsules are sp² hybridized with the π electrons directed into the cavity. The guests experience a thin layer of negative charge around them. This effect is even more pronounced with the tetraurea calixarene dimer **16·16**. The best charged guest found so far, the tetraethylammonium ion **17**, fills

ca. 78% of the available space inside the cavity which represents the record of all packing coefficients calculated for capsule-guest pairs.[19]

Chiral Molecular Recognition

A more challenging task is the synthesis of chiral self-assembling capsules capable of distinguishing the two enantiomers of a chiral guest. The first, modestly successful approach is based on the modified softballs **30** and **31** (Figure 4).[20] The two "arms" of the softball monomers bear spacers of different length, and one of the two mirror planes of the symmetric softball monomers **4–7** is broken. The other one does still exist and, consequently, the monomers are achiral. For the dimer, this situation does not hold true. Since, in the dimeric capsule, one monomer is rotated relative to the other, the dimers **30·30** and **31·31** do not bear any planar symmetry. In achiral solvents, these capsules exist as a pair of enantiomers. If, however, a chiral, enantiopure guest is added, two diastereoisomers are formed by encapsulation of the guest inside the racemic capsules. Accordingly, the ¹H-NMR spectra of **30·30** and **31·31** with **38** as the guest (Figure 4) show two sets of signals, one for each of the diastereoisomers. From the integrations, the diastereoisomeric excess (de) for each of the guests can be determined, as given in Figure 4.

Another approach utilizes chiral monomers which bear the chiral information on the outer surface of the capsule. Tetraurea calixarenes **37** and **38** synthesized from the tetraisocyanates and enantiopure β-branched amino acids such as isoleucine and valine preferentially heterodimerize with arylurea functionalized calixarene **16** to yield **16·37** and **16·38** (Figures 3 and 5).[21] The formation of homodimers **37·37** and **38·38** is not observed. The seam of hydrogen bonds connecting the eight urea functions is organized in a head-to-tail arrangement, so that two different directionalities – a clockwise and a counterclockwise array – might be realized. However, the point chirality of the amino acid residues is transferred to the hydrogen-bonding pattern. Only one of the two directionalities is realized, while the other is suppressed completely. Addition of a racemic mixture of a chiral guest, e.g. norcamphor **39**, to the **16·37** heterodimer results in a 1.3:1 ratio of both diastereomers. This diastereomeric excess is not large, but in view of the fact that the chiral information is located on the outside of the capsule, it seems striking that the guest inside is sensing the chirality of the capsule at all.

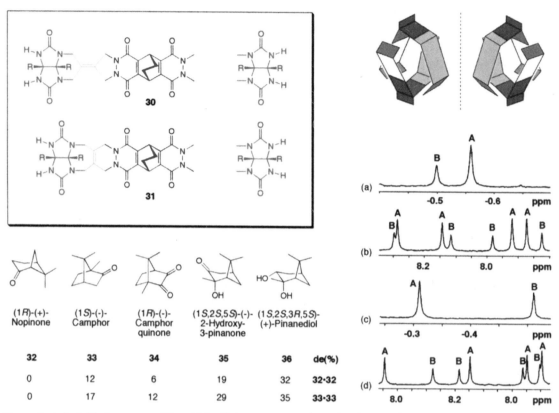

Figure 4. Left: Softball monomers **30** and **31** with two different glycoluril "arms" and de values (%) obtained for encapsulation of guests **32 – 36** in **30·30** and **31·31**. Right: Schematic representation of the two enantiomeric capsules in the absence of a chiral guest. Upfield (a and c) and aromatic (b and d) regions of the ^{1}H-NMR spectra of **30** and **31** in CDCl$_3$ with **36** as the guest (**A** and **B** denote signals from each of the two diastereomeric capsules).

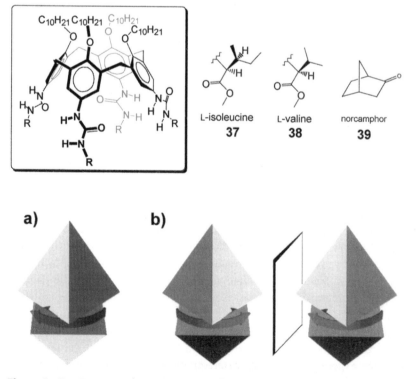

Figure 5. Top: Tetraurea calixarene monomers **37** and **38** bearing chiral amino acid ester residues (isoleucine and valine methyl esters, respectively) attached to the urea functions. Norcamphor **39** was the chiral guest used to detect the chirality transfer from the outside to the inner cavity. Bottom: Pictorial representation of (**a**) the achiral, S$_8$ symmetric homodimer **16·16** and of (**b**) the two enantiomeric forms of a calixarene capsule heterodimer. The head-to-tail directionality of the urea groups (arrow) defines the chirality.

Figure 6. Top: The two enantiomeric forms of the "american football" monomer **40** and views of the tetrameric capsule formed from four identical monomers along the mutually perpendicular C_2-axes. Bottom: Chiral guests **41–46** and the corresponding de values obtained from ^{1}H-NMR integrations.

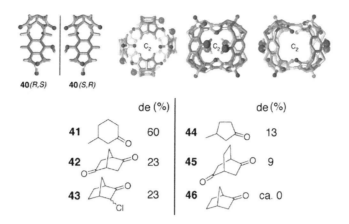

40 *(R,S)*   40 *(S,R)*

	de (%)			de (%)
41	60	**44**	13
42	23	**45**	9
43	23	**46**	ca. 0

Finally, the tetrameric football **25** has been substituted with a hydroxy group at the aromatic center piece to yield chiral monomers **40**.[22] The hydroxy groups take part in the hydrogen bonding pattern resulting in chiral tetramers **(40)₄** that do not bear any symmetry elements except three mutually perpendicular C_2-axes (Figure 6). Addition of chiral, racemic guests again leads to the preferential formation of one of the two possible diastereomers with de values up to 60 % for 3-methyl cyclohexanone **46**.

The Formation of Heterodimers and the Mechanism of Guest Exchange

The formation of heterodimers of two different capsules from the same series, e.g. the softball family, has been regarded as another criterion for the formation of capsules and also confirms the reversible dimerization of two subunits by non-covalent forces. If a solution of softball **6·6** is mixed with an equimolar solution of hydroxy softball **7·7**, the heterodimer **6·7** is observed in almost the statistically expected amount.[12] Both monomers have spacers of the same length and the geometric fit in the heterodimer is good. Overall, the number of hydrogen bonds does not change and thus heterodimerization is a more or less thermoneutral process. This picture changes significantly, if two softballs are combined that bear spacers of different length, e.g. **5·5** and **7·7**. The two monomers do not fit as well as **6** and **7** and the heterodimer **5·6** is underrepresented in the mixture.

Two different dynamic processes take place and can be followed by kinetic measurements: (i) the exchange of guests and (ii) the exchange of the capsule monomers. In a first experiment, [2.2]paracyclophane **14** is added to softball **7·7** loaded with adamantane **8**. The cyclophane replaces adamantane within several hours. Then, in a second experiment, the heterodimerization of **6·6** and **7·7** is studied.

Interestingly, the exchange of the monomers to yield the heterodimer **6·7** proceeds within days and is ca. 2–3 orders of magnitude slower than guest exchange.[23] Consequently, there must be a way to replace one guest by another one without completely separating the two monomers of the capsule.

In order to account for these results, "window" mechanisms have been proposed.[23,24] By a simple ring inversion of one of the six-membered rings in its arms, the softball **7·7** is able to open a flap large enough to enable the guest inside to leave the capsule. In a second step, the new guest could enter the capsule. This scenario resembles a classical S_N1 reaction. Of course, the empty capsule that is generated as an intermediate is an energetically disfavored species, because solvation of the inner surface is lost. Accordingly, two windows are more likely involved in guest exchange; the process then proceeds in analogy to a S_N2 reaction (Figure 7). Two different possibilities exist: If two vicinal windows open (Figure 7a) – one in each monomer – ten of the 16 hydrogen bonds of **7·7** must be cleaved and the guest exchange occurs in an approximate 90° angle between the incoming and leaving guests. In contrast, a "backside attack" mechanism reminds one even more of a classical S_N2 transition structure with its 180° trajectory (Figure 7b). This pathway requires 12 hydrogen bonds of **7·7** to be broken, two more than the former one. Consequently, the mechanism depicted in Figure 7a is energetically favored over the "backside attack" pathway. In addition, it is also favored by entropy, because statistically there exist four possibilities to select two vicinal windows and only two ways to open opposite arms in the same softball monomer.

Catalysis Through Encapsulation

The capsules bind guest molecules with quite a high selectivity. Why should they not also be able to bind transition structures more strongly than reac-

Figure 7. Top: The formation of heterodimer **6·7** from softballs **6·6** and **7·7**. Bottom: Two S_N2-like "window" mechanisms for guest exchange in the softballs. Attack in an approximate 90°C angle between incoming and leaving guest is favored over backside attack energetically – only ten instead of twelve hydrogen bonds are broken – and entropically – statistically, there are four ways to open two vicinal windows, but only two ways to open windows opposite to each other.

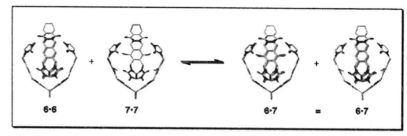

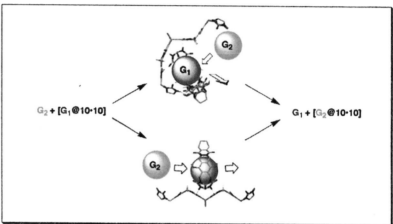

tants and products? It should be possible to use the capsules as reaction vessels and catalyze reactions by encapsulation of the corresponding reactants. The capsules would then combine two important features of enzyme catalysts: substrate selectivity and catalytic activity.

Indeed, it has been found that the *Diels-Alder* reaction of benzoquinone **47** and cyclohexadiene **48** is accelerated by encapsulation in the hydroxysoftball **7·7**.[25] It is known that the capsule in its resting state is occupied with two benzoquinone molecules,

one of which is exchanged by the diene prior to the cycloaddition step. Control experiments with naphthoquinone **49** did not result in rate enhancement. Consequently, catalysis mediated by mere hydrogen bonding of the quinone to the amide protons of the capsule monomer can be excluded, because it should affect **47** as well as **49**. Only the pairs of diene and dienophile that fit in the capsule together react at a higher rate and encapsulation can be identified as the reason for catalysis. Unfortunately, the catalyst is inhibited by the *Diels-Alder*

Figure 8. Catalytic cycles for the *Diels-Alder* reactions of benzoquinone **47** with cyclohexadiene **48** (top) and tetramethyl thiophene dioxide **51**. While the reaction of **47** with **48** suffers from product inhibition, adduct **52** is easily replaced by new reactants inside the capsule.

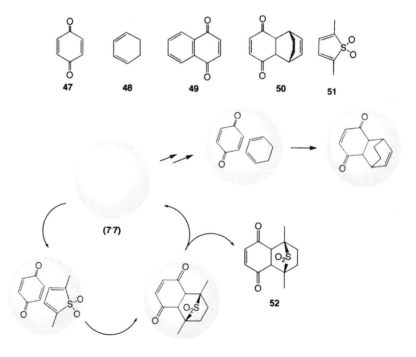

adduct **50** which is a better guest for the capsule than the reactants. This situation, however, changes, if tetramethyl thiophene dioxide **51** is used as the diene (Figure 8).[26] Real catalysis is observed, because the adduct **52** is a poor guest and is easily replaced by the reactants; product inhibition does not occur. The catalytic cycloaddition of **47** and **51** can be suppressed by a competitive inhibitor like [2.2]para-cyclophane **14** which is such a good guest for the hydroxy softball **7·7** that it completely replaces the reactants inside the capsule's cavity.

There are two possible explanations for the catalytic activity. The first one implies that the *effective* concentration of diene and dienophile in the capsule is much higher than their concentration in the surrounding solution. The experiments were conducted at a concentration of ca. 4 mM. From volume data of the capsule and the two reactants, the concentration inside the capsule can be calculated to be roughly three orders of magnitude higher (ca. 5 M), resulting in the observed acceleration of the *Diels-Alder* reaction. The second explanation takes into account that the *Diels-Alder* reaction has a negative volume of activation, i.e. the transition structure is smaller than the reactants. If the diene and the dienophile together occupy a little more space inside the capsule than required for the ideal guest, the transition structure would fit better. Consequently, the binding energy of the reactants is somewhat lower than that of the transition structure. The barrier for the cycloaddition is lowered by this difference and the reaction is accelerated accordingly. In other words, catalysis can be expected if the transition structure is a better guest than the reactants. Without excluding the former one, the latter rationalization is supported by the surprising finding that capsule **20·20** (Figure 3) is capable of stabilizing dibenzoyl peroxide **24** against decomposition upon warming the mesitylene solution to 70°C for several days. Outside the capsule the peroxide decomposes within three hours.[27] The transition structure for the cleavage of the peroxide O–O bond is surely larger than the peroxide itself; the decomposition reaction thus has a positive activation volume. The peroxide likely is a better guest than the transition structure and – in contrast to the *Diels-Alder* reaction – the barrier for the reaction inside the capsule is higher than that for the same reaction in solution resulting in the observed stabilization of the peroxide inside the cavity of **20·20**. The alternative explanation – that the dissociation occurs inside the capsule but the cage effect results in recombination of the radicals – must await the outcome of appropriate labeling experiments.

New Materials: Liquid Crystals and Fibers from "Polycaps"

So far, we have discussed the properties of single capsule dimers. This section will be devoted to capsule polymers and the hierarchy of order that can emerge from them. One way to produce polymeric capsules is to connect two or more monomers with covalent linkers so that the hydrogen-bonding sites of each monomer diverge. The calixarene dimer **53** (Figure 9), for example, bears a rigid spacer connecting the two tetraurea calixarene halves at their bottom rim. In chloroform solution, **53** forms hydrogen-bonded polymers[28] with capsules like beads on a string, when appropriate guest molecules are present.[29,30] The polymerization process can thus be fine-tuned by subtle changes in solvent conditions or guest properties. It can even be completely reversed by addition of a protic solvent that destroys the hydrogen bonds.

A more concentrated solution of **53** in aprotic organic solvents such as $CHCl_3$ (20–30%) is viscous and at even higher concentrations (> 36%), it becomes turbid. When viewed between crossed polarizers in a light microscope, Schlieren textures (Figure 9a,b) typical for nematic, lyotropic liquid crystals are visible and the material is birefringent.[31] The material can be deformed with pressure and is self-healing – two other characteristic properties of liquid crystals. Competitive solvents diffusing into the sample slowly erode the textures and finally the isotropic state is reached. These results imply that polymerization generates a material which is able to self-organize within small domains and is evidence for order on a higher degree of complexity.

Transmission electron microscopy (TEM) of replicas from freeze-fractured samples (Figure 9c) generates micrographs that show rounded structures throughout the whole sample. Their dimensions (1.6 nm × 2.2 nm) match those obtained for a single capsule from diffraction measurements and molecular modeling. In some areas (white little arrow in the center of the picture) the rounded structures appear to be linked together along strings, which in these micrographs are ordered randomly. The application of external forces to these samples, however, provides a non-random order of these strings. Shearing a sample leads to oriented fibers (Figure 9d). They can also be generated up to a length of several centimeters by pulling them out of the solution with a needle (Figure 9e). The fibers generated in both experiments have an approximately uniform width of 6 μm, a number that translates to several thousands of mutually well-aligned poly-

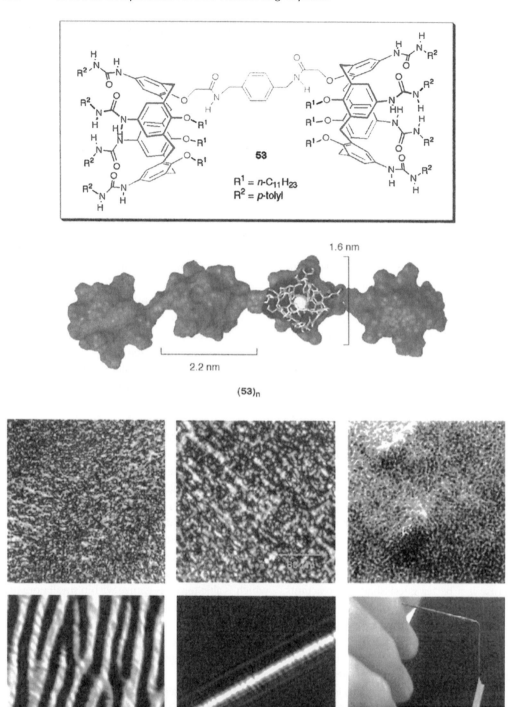

Figure 9. Top: Two tetraurea calixarene monomers connected by a rigid spacer at their bottom rim display diverging hydrogen-bonding sites ideally suited for polymerization. The polymer **(53)**$_n$ bears capsules of ca. 1.6 nm × 2.2 nm dimensions like beads on a string. Bottom: Photomicrographs of typical Schlieren textures of **53** in chloroform (top row, left) and *p*-difluorobenzene (top row, middle) as viewed between crossed polarizers. TEM image of **53** prepared by freeze-fracture from a CHCl$_3$ solution (top row, right). Laser confocal microscopic images of fibers formed from **53** in CHCl$_3$ by shearing the sample (bottom row, left) and pulling a fiber from the solution (bottom row, middle). The fiber can carry the weight of an NMR tube cap and tape (bottom row, right).

mer strings. Such a fiber can carry the weight of an NMR tube cap and tape (Figure 9f).

These higher-order phenomena clearly demonstrate how far self-assembly of simple subunits may lead. The problem of controlling self-assembly and designing tailored subunits for a certain purpose is not yet completely solved, but the more we understand from examples like those presented here and many others, the closer we come to real-world applications.

Epilogue

Encapsulation research is now at a state where it is useful to take a longer view and draw a few conclusions concerning it as a concept. The parallels to phenomena in biology and biochemistry are quite obvious. These cavities open and close reversibly in solution, are able to select a narrow range of guests, can discriminate between stereoisomers, and can catalyze chemical transformations. These features also apply to the pockets at the active centers of enzymes. The capsules probably select guests for the same reasons as do enzymes but with imperfect substrate specifity and modest rate enhancements. The finely balanced interplay of enthalpy and entropy decides which guests are good and which are not. The aspect of space filling might also be applicable to pharmacology. If an ideal guest occupies ca. 55 % of the available space inside the capsule in the absence of specific interactions, this is likely to be true for enzyme pockets as well. This figure may be an advantageous starting point for the rational design and choice of new lead structures for potential drugs.

The future prospects for the capsule project emerge from these considerations. Further increasing the size of the capsule and building chemical functionalities into the inner cavity would allow a closer emulation of the functions of enzymes, especially those that require cofactors in order to catalyze chemical transformations. Another important aspect is to design capsules that can combine stereospecificity and catalysis – that is accelerate stereoselective transformations. Capsules that reversibly dimerize in water would probably contribute a lot more to our understanding of non-covalent forces and solvent effects in this most biorelevant medium. So far, water solubility and assembly have not been achieved with hydrogen-bonded capsules.

Above all stand the principles of self-organization and self-assembly. Creating new functional entities by assembling simple building blocks into a more complex structure of higher order not only simplifies the synthetic efforts necessary, but it is also a fascinating challenge for the chemist. Frequently, the synthesis of building blocks leads to self-organization of higher order structures with unexpected properties, properties that the building blocks alone cannot exhibit.

In closing, encapsulation research is another example of the slow, but sometimes successful shift from basic to applied sciences and underlines the importance of understanding the fundamentals before translating them into applications. The original idea for the tennis ball came from the world of sports, and then developed into a project intended to understand the nature of non-covalent forces and to uncover the rules which govern the self-organization of self-complementary molecules. Now, it impinges upon into the field of materials science. What problems capsules can solve depends on our ability to identify needs in smart materials, sensor technology and other applications where an extreme form of molecular recognition is required.

Acknowledgements

We are indebted to the experimentally gifted and imaginative coworkers who were involved in this project; their names are given in the list of references. For generous, continuous financial support, we wish to thank the NIH and the Skaggs Research Foundation. C.A.S. thanks the Deutsche Akademie der Naturforscher Leopoldina/BMBF and the Stipendien-Fonds des Verbandes der Chemischen Industrie for fellowship support.

References and Notes

1. T. Weiske, T. Wong, W. Krätschmer, J. K. Terlouw, H. Schwarz, "The Neutralization of He@C$_{60}^{+\cdot}$ in the Gas Phase: Compelling Evidence for the Existence of an Endohedral Structure for He@C$_{60}$", *Angew. Chem. Int. Ed. Engl.* **1992**, *31*, 183–185.

2. D. J. Cram, J. M. Cram, *Container Molecules and Their Guests*, Royal Society of Chemistry, Cambridge, 1994.

3. R. Warmuth, "*o*-Benzyne: Strained Alkyne or Cumulene? – NMR Characterization in a Molecular Container", *Angew. Chem. Int. Ed. Engl.* **1997**, *36*, 1347–1350.

4. D. J. Cram, M. E. Tanner, R. Thomas, "The Taming of Cyclobutadiene", *Angew. Chem. Int. Ed. Engl.* **1991**, *30*, 1024–1027.

5. P. Timmerman, W. Verboom, F. C. J. M. van Veggel, J. P. M. van Duynhoven, D. N. Reinhoudt, "A Novel Type of Stereoisomerism in Calixarene-based Carceplexes", *Angew. Chem. Int. Ed. Engl.* **1994**, *33*, 2345–2348.

6. G. M. Whitesides, J. P. Mathias, C. T. Seto, "Molecular Self-Assembly and Nanochemistry: A Chemical Strat-

egy for the Synthesis of Nanostructures", *Science* **1991**, *254*, 1312–1319.

7. D. Philp, J. F. Stoddart, "Self-Assembly in Natural and Unnatural Systems", *Chem. Rev.*, **1996**, *35*, 1154–1196.

8. T. Douglas, M. Young, "Host-guest encapsulation of materials by assembled virus protein cages", *Nature* **1998**, *393*, 152–155.

9. M. M. Conn, J. Rebek, Jr., "Self-Assembling Capsules", *Chem. Rev.* **1997**, *97*, 1647–1668.

10. J. de Mendoza, "Self-Assembling Cavities: Present and Future", *Chem. Eur. J.* **1998**, *4*, 1373–1377.

11. R. Wyler, J. de Mendoza, J. Rebek, Jr., "A Synthetic Cavity Assembles Through Self-Complementary Hydrogen Bonds," *Angew. Chem. Int. Ed. Engl.*, 1993, *32*, 1699–1701.

12. C. A. Schalley, J. M. Rivera, T. Martín, J. Santamaría, G. Siuzdak, J. Rebek, Jr., "Structure Determination of Supramolecular Architectures by Electrospray Ionization Mass Spectrometry", *Eur. J. Org. Chem.*, **1999**, 1325–1331.

13. S. Mecozzi, J. Rebek, Jr., "The 55 % Solution: A Formula for Molecular Recognition in the Liquid State", *Chem. Eur. J.* **1998**, *4*, 1016–1022.

14. T. Heinz, D.M. Rudkevich, J. Rebek, Jr., "Pairwise selection of guests in a cylindrical molecular capsule of nanometre dimensions", *Nature* **1998**, *394*, 764–766.

15. J. Kang, J. Rebek, Jr., "Entropically driven binding in a self-assembling molecular capsule", *Nature* **1996**, *382*, 239–241.

16. T. Martín, U. Obst, J. Rebek, Jr., "Molecular Assembly and Encapsulation Directed by Hydrogen-Bonding Prefernces and the Filling of Space", *Science* **1998**, *281*, 1842–1845.

17. C. A. Schalley, T. Martín, U. Obst, J. Rebek, Jr., "Characterization of Self-Assembling Encapsulation Complexes in the Gas Phase and Solution", *J. Am. Chem. Soc.* **1999**, *121*, 2133–2138.

18. J. C. Ma, D. A. Dougherty, "The Cation-π Interaction", *Chem. Rev.* **1997**, *97*, 1303–1324.

19. C. A. Schalley, R. K. Castellano, M. S. Brody, D. M. Rudkevich, G. Siuzdak, J. Rebek, Jr., "Investigating Molecular Recognition by Mass Spectrometry: Characterization of Calixarene-Based Self-Assembling Capsule Hosts with Charged Guests", *J. Am. Chem Soc.*, **1999**, *121*, 4568–4579.

20. J. M. Rivera, T. Martín, J. Rebek, Jr., "Chiral Spaces: Dissymmetric Capsules Through Self-Assembly", *Science* **1998**, *279*, 1021–1023.

21. R. K. Castellano, C. Nuckolls, J. Rebek, Jr., "Transfer of Chiral Information through Molecular Assembly", *J. Am. Chem. Soc.* **1999**, *121*, 11156–11163.

22. C. Nuckolls, F. Hof, T. Martín, J. Rebek, Jr., "Chiral Microenvironments in Self-Assembled Capsules", *J. Am. Chem. Soc.* **1999**, *121*, 10281–10285.

23. J. Santamaría, T. Martín, G. Hilmersson, S. L. Craig, J. Rebek, Jr., "Guest exchange in an encapsulation complex: A supramolecular substitution reaction", *Proc. Natl. Acad. Sci. USA* **1999**, *96*, 8344–8347.

24. K. N. Houk, K. Nakamura, C. Sheu, A. E. Keating, "Gating as a Control Element in Constrictive Binding and Guest Release by Hemicarcerands", *Science* **1996**, *273*, 627–629.

25. J. Kang, J. Rebek, Jr., "Acceleration of a Diels-Alder reaction by a self-assembled molecular capsule", *Nature* **1997**, *385*, 50–52.

26. J. Kang, J. Santamaría, G. Hilmersson, J. Rebek, Jr., "Self-Assembled Molecular Capsule Catalyzes a Diels-Alder Reaction", *J. Am. Chem. Soc.* **1998**, *120*, 7389–7390.

27. S. K. Körner, F. C. Tucci, D. M. Rudkevich, T. Heinz, J. Rebek, Jr., "A Self-Assembled Cylindrical Capsule: New Supramolecular Phenomena through Encapsulation", *Chem. Eur. J.*, **2000**, *6*, 187–195.

28. J. H. K. K. Hirschberg, F. H. Beijer, H. A. van Aert, P. C. M. M. Magusin, R. P. Sijbesma, E. W. Meijer, "Supramolecular polymers from linear telechelic siloxanes with quadruple hydrogen bonded units", *Macromolecules* **1999**, *32*, 2696–2705.

29. R. K. Castellano, D. M. Rudkevich, J. Rebek, Jr., "Polycaps: reversibly formed polymeric capsules", *Proc. Natl. Acad. Sci. USA* **1997**, *94*, 7132–7137.

30. R. K. Castellano, J. Rebek, Jr., "Formation of Discrete, Functional Assemblies and Informational Polymers through the Hydrogen-Bonding Preferences of Calixarene Aryl and Sulfonyl Tetraureas", *J. Am. Chem. Soc.* **1998**, *120*, 3657–3663.

31. R. K. Castellano, C. Nuckolls, S. H. Eichhorn, M. R. Wood, A. J. Lovinger, J. Rebek, Jr., "Hierarchy of Order in Liquid Crystalline Polycaps", *Angew. Chem. Int. Ed.* **1999**, *38*, 2603–2606.

Slippage and Constrictive Binding

Matthew C. T. Fyfe, Françisco M. Raymo, and J. Fraser Stoddart

Department of Chemistry and Biochemistry, University of California, Los Angeles, CA 90095, USA,

Phone: +310 206 7078, Fax: +310 206 1843, e-mail: stoddart@chem.ucla.edu

Keywords ■ Constrictive Binding ■ Crown Ethers ■ Hemicarcerands ■ Self-Assembly ■ Host-Guest Chemistry ■ Noncovalent Interactions ■ Rotaxanes ■ Slippage ■ Supramolecular Chemistry

Concept: Rotaxane-like entities are created by the slippage protocol in which the cavities of macrocyclic polyethers "dilate", at elevated temperatures, so as to allow their passage over the relatively bulky stoppers of chemical dumbbells (see Cartoon below). Similarly, hemicarceplexes are produced when the portals of hemicarcerands expand, on heating, to allow guests to enter their internal voids. The rotaxane-like and hemicarceplex products obtained are stabilized both by non-covalent bonds and by mechanical coercion.

Slippage

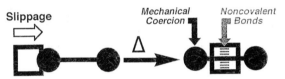

Abstract: Kinetically stable superarchitectures can be constructed efficiently by relying upon the assistance that mechanical constraints can offer to noncovalent bonding interactions. Thus, the free energies of binding and activation associated with complexation/dissociation processes are determined by the combined strengths of the noncovalent bonds formed by, and by the size complementarity between, the constituents of a complex, as well as by the differential solvation between the complexed and dissociated states. Using this information, kinetically stable complexes can be constructed through the careful design of stereoelectronically-matching components and the judicious selection of the experimental conditions under which they are brought together. Rotaxane-like complexes, prepared via the slippage of appropriately sized macrocycles over the stoppers of chemical dumb-bells, and hemicarceplexes, created via the ingression of a guest into a hemicarcerand's cavity, are examples of kinetically stable species that can be synthesized noncovalently through the combined action of noncovalent bonding and mechanical coercion. The synthetic protocol brought to light by the syntheses of these complexes holds considerable promise for the future construction of nanosized devices, with specific shapes, sizes and functions, the fabrication of which is impracticable by classical synthetic routes.

Prologue

At a fundamental level, the creation of a supramolecular[1–3] complex may be analyzed from both thermodynamic and kinetic standpoints (Figure 1). In solution, a complex's thermodynamic stability (i.e., the free energy of binding $\Delta G°$) is determined by: (1) the strength of the noncovalent bonds formed between the recognition sites embedded in its host and guest components, and (2) the difference between the solvation energies of the complexed and dissociated species. When the free host and guest are solvated poorly relative to their aggregated sum, differential solvation and noncovalent bonds cooperate to favor complexation, but if the complex is solvated inadequately relative to its individual constituents, $\Delta G°$ becomes finely balanced between the conflicting factors of differential solvation and noncovalent bonding. On the kinetic side, the free energy barrier accompanying the complexation process ($\Delta G_c^{\ddagger}$) is determined by the complementarity between the size of the entrance on the host, which allows the guest access to its internal recognition site, and the steric bulk associated with either all, or some, of the guest. A monotonic correlation exists between $\Delta G_c^{\ddagger}$ and entrance/guest sizes; $\Delta G_c^{\ddagger}$ increases when the guest becomes bulkier and/or the entrance becomes smaller. The complex's kinetic stability (i.e., the free energy barrier for dissociation $\Delta G_d^{\ddagger}$) is associated with the rate of decomplexation. When a complex is more stable than its separate constituents, $\Delta G_d^{\ddagger}$ is related to both $\Delta G°$ and $\Delta G_c^{\ddagger}$ through the relationship $\Delta G_d^{\ddagger} = \Delta G° + \Delta G_c^{\ddagger}$. As a result of the thermodynamic and kinetic issues discussed above, three factors must be considered if the supramolecular synthesis[4] of a kinetically stable complex (i.e., a complex that does not dissociate) is to be realized:[5,6] these are, specifically: (1) noncovalent bond strengths, and (2) differ-

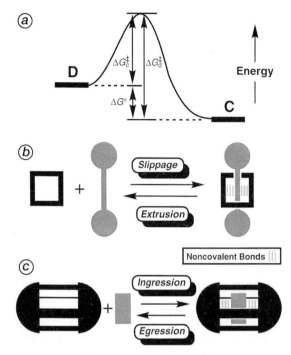

Figure 1. (a) Free energy profile for complexation/decomplexation processes, illustrating the transition between dissociated (**D**) and complexed (**C**) states. (b) The slippage process generates kinetically stable rotaxane-like species when one of the stoppers of a chemical dumbbell slips through the cavity of a macrocycle. (c) Forcing a guest through the portals of an empty hemicarcerand leads to hemicarceplexes that are stable under standard practical conditions.

ential solvation, both of which are reflected by $\Delta G°$, as well as (3) host–guest size complementarity, which is encompassed by $\Delta G_c^{\ddagger}$. When all three criteria are well met, the self-assembly[7] of the separate components into a kinetically stable species occurs efficiently upon simple solution-phase mixing at an appropriate temperature. This article discusses how the kinetic and thermodynamic factors mentioned above may be exploited for the syntheses of such kinetically stable entities, with particular reference being made to the preparation of: (1) rotaxane-like[8] species, through a process that has been termed[5] "slippage" and (2) hemicarceplexes, which have been popularized by the seminal research[6] of Cram and his associates.

Rotaxane-Like Complexes via Slippage

Several years ago in our research laboratories, a novel route to the synthesis of kinetically stable rotaxane-like[9–12] species was discovered,[13] which relies upon the aforementioned balance between kinetics and thermodynamics. This route is termed "slippage" inasmuch as the rotaxane-like species are obtained when a preformed macrocycle bears a cav-

ity that is just of the right size to allow it to slip (Figure 1b) over the bulky stopper units of a chemical "dumbbell" at elevated temperatures. The slippage process is facilitated when both macrocycle and dumbbell possess complementary recognition elements[2] within their covalent skeletons that make $\Delta G°$ more negative as a result of noncovalent association,[14] and increase, ultimately, the rotaxane-like product's kinetic stability (i.e., $\Delta G_d^{\ddagger}$). By virtue of its simplicity, the slippage protocol is an attractive technique that several researchers,[15,16] including ourselves,[5,17–19] are now utilizing routinely for the production of a diverse range of rotaxane-like entities.

The macrocyclic polyether bis-*p*-phenylene[34]crown-10 (BPP34C10, Scheme 1) binds[16] 4,4′-bipyridinium-based guests with pseudorotaxane co-conformations,[20] both in solution and in the solid state. The resulting complexes are stabilized by means of [C–H⋯O] hydrogen bonds, between the α-4,4′-bipyridinium hydrogen and polyether oxygen atoms, along with π–π stacking interactions, between the 4,4′-bipyridinium and 1,4-dioxybenzene units. The kinetics of the complexation and decomplexation processes are governed by the size of the substituents attached to the 4,4′-bipyridinium N atoms. Thus, when equimolar amounts of BPP34C10 and 1,1′-dimethyl-4,4′-bipyridinium bis(hexafluorophosphate) were mixed, in MeCN at 25 °C, a red–orange color developed[21] immediately as a result of complex formation. In contrast, there was no observable color change when equimolar amounts of BPP34C10 and any of the dumbbell-shaped compounds $\mathbf{1}\cdot 2PF_6$–$\mathbf{4}\cdot 2PF_6$ (Scheme 1) were mixed,[17] under otherwise identical conditions, suggesting that the tetraarylmethane-based stoppers were too bulky for BPP34C10 to slip over them under these circumstances. However, the cavity of this rather flexible macrocyclic polyether "expands" when the temperature is increased. When equimolar MeCN solutions of BPP34C10 and any of $\mathbf{1}\cdot 2PF_6$–$\mathbf{3}\cdot 2PF_6$ were heated at 50 °C, the red–orange coloration appeared gradually. The equilibrium between the complexed and uncomplexed species was achieved after approximately ten days, from which time onward there was no further increase in the intensity of the color. After cooling the solution back to ambient temperature, the cavity of macrocyclic component "shrunk" so that it could no longer slip back over the tetraarylmethane-based stoppers of the dumbbell-shaped components.

As a consequence, $[BPP34C10\cdot\mathbf{1}][PF_6]_2$–$[BPP34C10\cdot\mathbf{3}][PF_6]_2$ became kinetically stable and were isolated from their "unreacted" components by conventional chromatographic techniques. Interestingly,

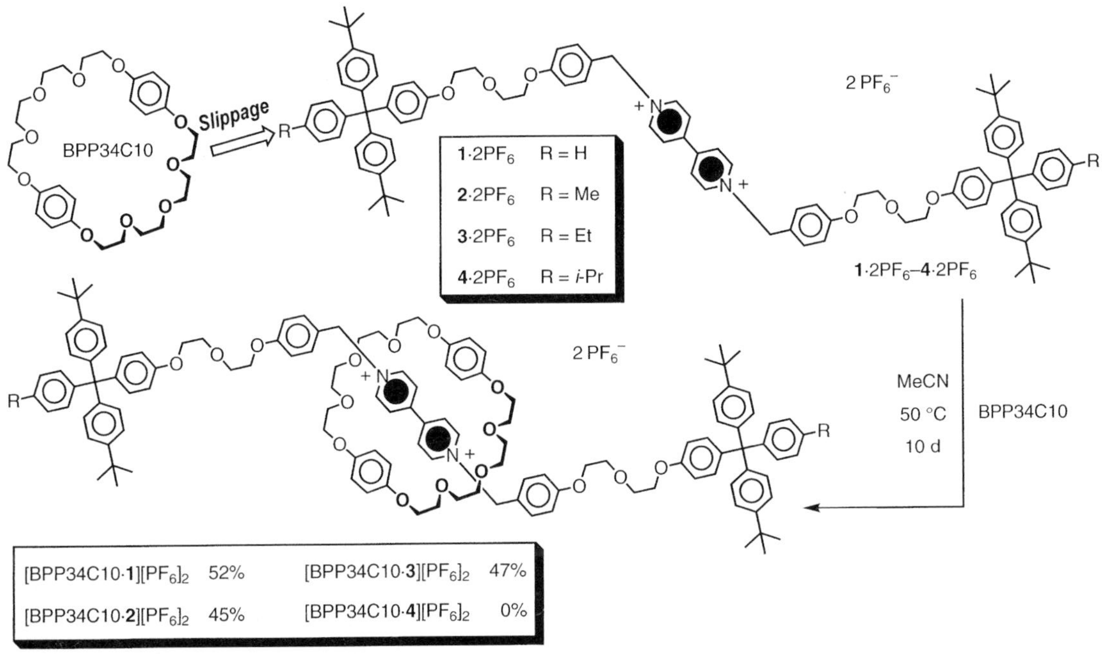

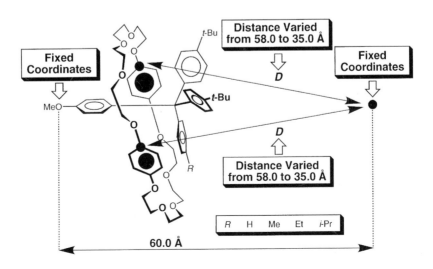

Scheme 1. BPP34C10 slips over the stoppers of $1\cdot2PF_6$–$3\cdot2PF_6$ to generate rotaxane-like entities that are stabilized by mechanical and noncovalent bonding interactions. However, the "*i*-Pr-substituted" stoppers of $4\cdot2PF_6$ are too bulky to be transcended by BPP34C10.

no color change was detected, even after several weeks, when an MeCN solution of BPP34C10 and the dumbbell-shaped compound $4\cdot2PF_6$ was heated at elevated temperatures, i.e., the passage of BPP34C10 over the stoppers of 4^{2+} cannot take place because of the increased bulk rendered by the *i*-Pr groups. Clearly, slippage can occur only when the right balance between the size of the macrocycle's cavity and stoppers is achieved, in this instance, when R (Scheme 1) is smaller than *i*-Pr. In order to gain further insight into this "all-or-nothing" substituent effect, the mechanism associated with the slippage of BPP34C10 over model tetraarylmethane-based stoppers was analyzed[22] by computational techniques. One of the aryl rings of the stopper was inserted (Figure 2) through the cavity of BPP34C10.

The coordinates of the stopper's oxygen atom were fixed, as were those of a reference point located at a distance of 60.0 Å away. The distances D were varied stepwise from 58.0 to 35.0 Å so that the BPP34C10 macroring was forced to slip over the remaining three aryl rings of the stopper. At each step, the superstructure was subjected to molecular dynamics at a simulated temperature of 500 K, then energies of 200 randomly selected co-conformers were minimized. The energies calculated for the lowest energy point obtained for each step were plotted against D.

This simulation protocol was employed to study the slippage processes associated with R equal to H, Me, Et and *i*-Pr. In all cases, the resulting energy profiles displayed two energy barriers. The first of

Figure 2. Initial geometry and constraints employed for simulating the passage of BPP34C10 over the model tetraarylmethane-based stoppers.

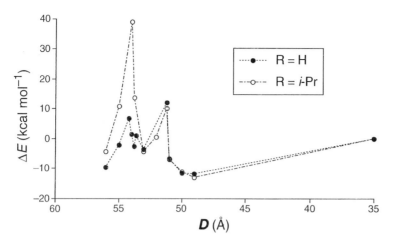

Figure 3. Energy profiles associated with the slippage of BPP34C10 over the model stoppers with R equal to H and *i*-Pr.

these is associated with the slippage of one of the polyether chains over the R group, while the second corresponds to the slippage of the other polyether chain over one of the two *t*-Bu groups. When the size of the R group is increased, the first energy barrier rises whereas the second remains constant. When R is equal to either H or Me, the second energy barrier is rate determining. On the other hand, when R is equivalent to either Et or *i*-Pr, the first energy barrier is rate determining. However, the energy barrier observed for slippage of BPP34C10 over the *i*-Pr-bearing stopper is more than 20 kcal mol⁻¹ higher than for any of its smaller congeners.

Figure 3 plots the energy profiles associated with R equal to H and *i*-Pr. It is interesting to note that the first energy barrier increases significantly and becomes rate determining when R is *i*-Pr. Comparison of the transition states calculated for R equal to Et or *i*-Pr shows (Figure 4) that one of the polyether chains of BPP34C10 and the *i*-Pr-C₆H₄ ring of the stopper are distorted significantly. As a result of this steric effect, the energy barrier associated with the slippage process increases by ca. 20 kcal mol⁻¹ on going from Et to *i*-Pr, a fact that prevents rotaxane formation when R is *i*-Pr. However, the cavity of the macrocyclic polyether enlarges sufficiently to permit slippage when the 1,4-dioxybenzene rings of BPP34C10 are replaced with 1,5-dioxynaphthalene units; passage of 1,5-dinaphtho[38]crown-10 (1/5DN38C10, Scheme 2) over the stoppers occurs even when R is *i*-Pr. Accordingly, the corresponding rotaxane-like species was isolated[18] (57% yield) when an MeCN solution of 4·2PF₆ and 1/5DN38C10 was heated, for only two days, at 50 °C. Whereas BPP34C10 does not slip when R is *i*-Pr, 1/5DN38C10 does!

Reaction (Scheme 2) of salt 5·2PF₆ with an excess of 6, in the presence of the macrocyclic polyether BPP34C10, provided[18] [BPP34C10·7][PF₆]₄ via a "threading-followed-by-stoppering" approach.[23] This rotaxane incorporates a free 4,4'-bipyridinium recognition site within its dumbbell-shaped compo-

nent and its "*i*-Pr-substituted" stoppers provide a steric barrier large enough to prevent its macrocyclic component (BPP34C10) from slipping off, while, at the same time, allowing the larger macrocycle 1/5DN38C10 to slip on. Indeed, [BPP34C10·1/5DN38C10·7][PF₆]₄, which incorporates two constitutionally-different macrocyclic components, was isolated when an MeCN solution of 1/5DN38C10 and [BPP34C10·7][PF₆]₄ was heated at 50 °C for two days.

The macrocyclic polyether dibenzo[24]crown-8 (DB24C8, Scheme 3) binds[24,25] dialkylammonium-containing guests in solution, as well as in the solid state, to generate complexes with pseudorotaxane co-conformations that are stabilized primarily by [O···H–N⁺] and [O ···H–C] hydrogen bonds. The complexation and decomplexation kinetics of these recognition events are influenced by the size of the

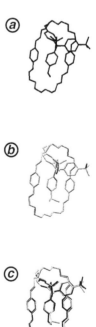

Figure 4. Transition states calculated for the slippage of BPP34C10 over the model tetraarylmethane stoppers when R is equal to (**a**) Et and (**b**) *i*-Pr. (**c**) Superposition of both transition states.

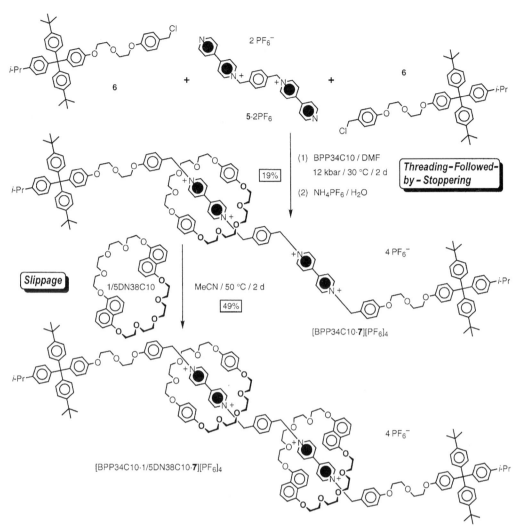

Scheme 2. Rotaxane synthesis combining different methodologies. The synthesis of the three-component rotaxane-like species [BPP34C10 ·1/5DN38C10 ·**7**][PF$_6$]$_4$ employs (1) threading-followed-by-stoppering then (2) slippage approaches.

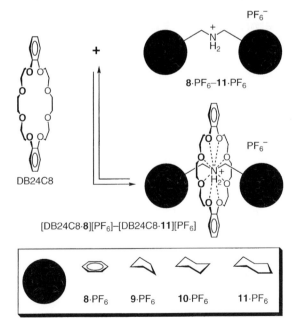

Scheme 3. Some of the compounds utilized for the discovery of a slippage synthesis involving the binding of dialkylammonium ions by DB24C8.

substituents located on the dialkylammonium guest. By way of illustration, ^{1}H NMR spectroscopy revealed that [DB24C8·**8**][PF$_6$] equilibrated immediately with its components, viz, DB24C8 and **8**·PF$_6$ (equimolar ratio), in CDCl$_3$–CD$_3$CN (3:1) at 20 °C. On the other hand, the equilibrium between complex [DB24C8·**9**][PF$_6$] and its separate components DB24C8 and **9**·PF$_6$ (1:1 mixture) took 90 minutes to be established after mixing in CDCl$_3$–CD$_3$CN (3:1) at 20 °C.[17] Obviously, the planar, unsaturated Ph stoppers of **8**$^+$ can penetrate the cavity of the DB24C8 macroring much more easily than the nonplanar, fully saturated c-C$_5$H$_9$ stoppers of **9**$^+$. In addition to these results, we discovered that an equimolar solution of DB24C8 and **10**·PF$_6$ had to be heated, at 40 °C for ca. 16 weeks, in order to encourage the macrocyclic polyether to slip over the bulky c-C$_6$H$_{11}$ stoppers of the cation **10**$^+$. The rotaxane-like complex [DB24C8·**10**][PF$_6$] is kinetically stable at ambient temperature and was isolable (90 % yield after crystallization) after an equimolar CH$_2$Cl$_2$ solution

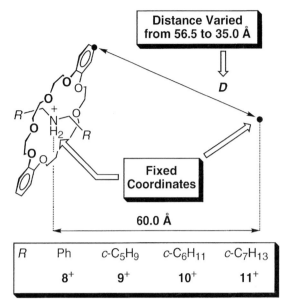

Figure 5. Initial co-conformation[20] and constraints used for simulating the passage of DB24C8 over the R groups of the dialkylammonium-containing cations **8⁺–11⁺**.

of DB24C8 and **10·**PF₆ had been heated at 40 °C for 36 days. No dissociation of [DB24C8·**10**][PF₆] was detected, even after several weeks at 20 °C, in nonpolar solvents (e.g., CD₂Cl₂). Nonetheless, when the [**2**]rotaxane was dissolved in (CD₃)₂SO, complete dissociation into its constituents occurred after only 18 hours.[26] In other words, [DB24C8·**10**][PF₆] is kinetically stable (high $\Delta G_d^\ddagger$) in CD₂Cl₂, while it is kinetically unstable (low $\Delta G_d^\ddagger$) in (CD₃)₂SO. When another CH₂ unit was inserted into **10·**PF₆'s cycloalkyl stoppers, no complex could be obtained under any circumstances; DB24C8 did not slip over the (relatively) bulky c-C₇H₁₃ stoppers of **11·**PF₆, even when the two components were heated together (1:1 solution) at 50 °C. As we had done beforehand with systems based on larger crown ethers and tetraarylmethane-based stoppers, the extrusion of DB24C8 from the dialkylammonium salts by passing over the stoppers of **8⁺–11⁺** was simulated (Figure 5) computationally. The coordinates of the acy-

clic component's nitrogen atom, together with those of a reference point located at a distance of 60.0 Å away, were fixed. The distance D was varied stepwise from 56.5 to 35.0 Å, forcing the DB24C8 macroring to slip over one of the two R groups. At each step, the energy of the superstructure was minimized and the resulting values were plotted (Figure 6) against D. The energy profiles associated with **8⁺** and **9⁺** are remarkably "flat", suggesting that, as observed experimentally, the passage of DB24C8 over either Ph or c-C₅H₉ groups is relatively easy. However, the energy profiles for the systems involving **10⁺** and **11⁺** show pronounced maxima, indicating that the passage of DB24C8 over the R groups becomes increasingly difficult as the size of these groups is enlarged.

Recently, a variant of the slippage approach that does not require a solvent has been reported by Vögtle and his coworkers.[16] In this variant, rotaxane-like structures were obtained when preformed macrocycles and dumbbells were melted together momentarily. For instance, the rotaxane-like species **12·13** was isolated (Scheme 4) when its components were heated together, in the melt at 350 °C, for a short time. It is believed that hydrogen bonding interactions between the amide groups of the acyclic and cyclic components are responsible for the stabilization that leads to rotaxane formation. The slippage[27] protocol that Vögtle and his group have developed further simplifies the general method and may well be extended for the preparation of a diverse range of rotaxane-like architectures in the future.

Hemicarcerands and Constrictive Binding

In 1991, Cram's research group reported a new class of container molecules, named *hemicarcerands*,[6] that are involved (Figure 1c) in several processes related to slippage. Guests can be imprisoned

Figure 6. Energy profiles associated with the slippage of DB24C8 over the R groups of **8⁺–11⁺**.

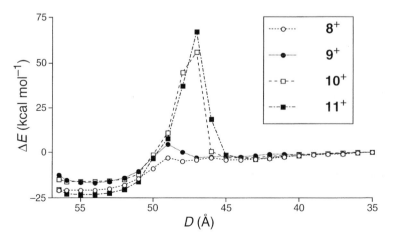

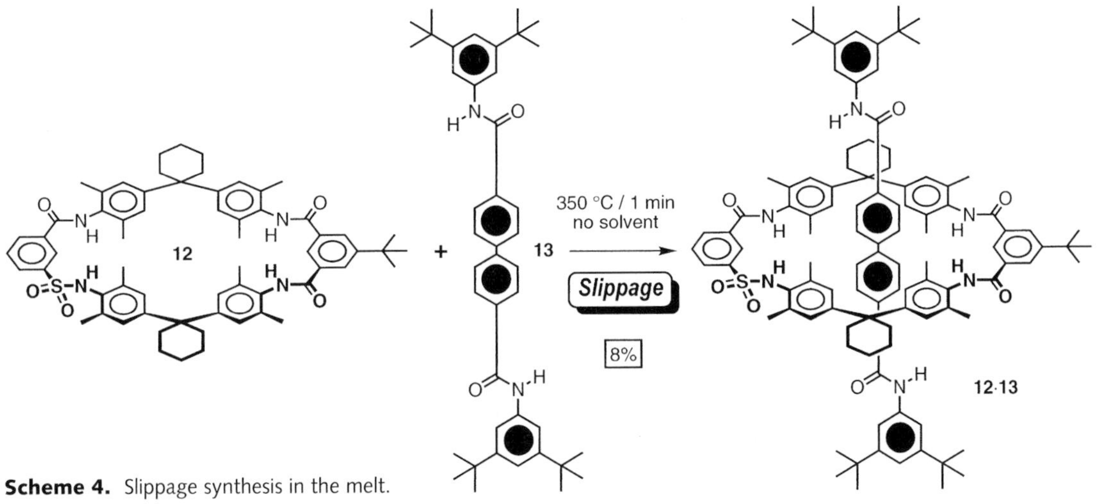

Scheme 4. Slippage synthesis in the melt.

within the internal cavities of these cage-like molecules to generate complexes, termed *hemicarceplexes*, that are isolable under standard laboratory conditions. However, at high temperatures, the portals of the hemicarcerand hosts open sufficiently to allow their imprisoned guests to escape[28] from the complexes in a process reminiscent of a "molecular jailbreak". A case in point[29] is hemicarceplex **16·DMA** (DMA = *N,N*-dimethylacetamide), which was prepared (Scheme 5) by reaction of tetrol **14** with dibromide **15** in the presence of DMA–Cs$_2$CO$_3$.[30] It can be purified easily by typical laboratory procedures, has unlimited stability at ambient temperature and is fully characterizable by conventional techniques.

When **16·DMA** was heated for 24 hours at 160 °C in *p-i*-Pr$_2$C$_6$H$_4$ – a solvent that is too large to fit through **16**'s portals – the DMA guest was liberated to furnish the free hemicarcerand **16**. New complexes of **16** can be generated either from this "free cage" or from **16·DMA**. When **16·DMA** was heated, at 80–200 °C, in the presence of a fresh guest solvent, the new hemicarceplex **16·solvent** was produced, by virtue of mass action, through a two-step mechanism involving the empty hemicarcerand intermediate **16**. Cram identified two free energy terms that govern the stabilities of the complexes formed between hemicarcerands and their guests. *Intrinsic binding* is equivalent to the free energy of

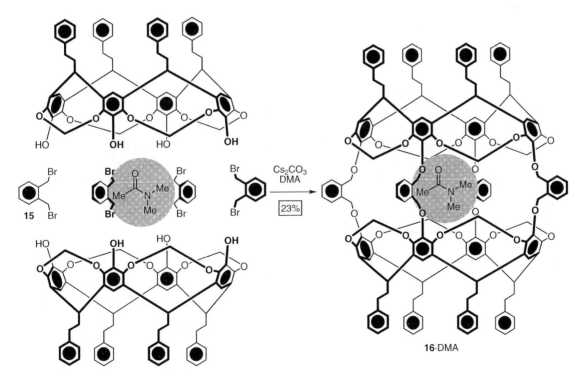

Scheme 5. The synthesis of hemicarcerand **16** is templated around a molecule of DMA, leading to the production of hemicarceplex **16·DMA**.

binding, $\Delta G°$; it is favored both enthalpically and entropically[31] for molecules such as DMA or EtOAc. *Constrictive binding* is commensurate with the free energy of activation for complex formation, i.e., $\Delta G_c^{\ddagger}$, and arises as a result of the physical barriers to decomplexation. The sum of the constrictive and intrinsic binding terms is equal to the free energy that must be fed in to the system in order to arrive at the complexed–uncomplexed transition state from the associated state, i.e., $\Delta G_d^{\ddagger}$.

Houk's research group has described[32] how the magnitude of constrictive binding is controlled by gating processes. These processes are coupled with conformational motions in the host molecule that modify the size of the entrances allowing guest ingression and egression. Gating permits hemicarceplexes to be prepared with guests of varying sizes. Gate-bearing molecules can form guest complexes that are stable at low temperatures, but which dissociate/associate reversibly at higher temperatures. The gates open by two conformational processes, viz, through (Figure 7) the French door and sliding door mechanisms. By employing computational

A Fuzzy Branch of Contemporary Chemistry!

The species that we have described so far provide a bit of a quandary for those who wish to have chemical science defined strictly with explicit boundaries between different classes of substances. For instance, when[19] does a non-interlocked pseudorotaxane complex become an interlocked rotaxane molecule? By the same token, where do we draw the line between hemicarceplexes and carceplexes? These are difficult questions to answer, for complexes that are completely stable to dissociation under one set of conditions may dissociate under others, i.e., a species that seems to be a rotaxane or a carceplex may, in fact, be a pseudorotaxane or a hemicarceplex in a different environment. Take [DB24C8·**10**][PF$_6$] for instance. It appears to be a rotaxane in nonpolar solvents, but it dissociates in more polar environments. Thus, this two-component assemblage must be a pseudorotaxane, as its c-C$_6$H$_{11}$ stoppers do not maintain its integrity under all conditions. However, it must have more

Figure 7. Guest molecules egress from hemicarcerands (Host) through conformational motions known as gating. Gate-opening occurs by (**a**) French door and (**b**) Sliding door mechanisms.

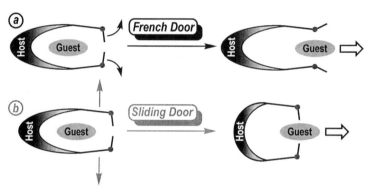

methods, Houk's team established the mechanistic route through which guests can escape from hemicarcerand **16**. Their simulations suggest that the guest does not vacate **16**'s cavity directly, but rather that its extrusion involves its initial progression into an antechamber created between the xylylene connector units. The guest completes its escape when it leaves the antechamber by freeing itself from noncovalent interactions with the xylylene rings. It was found that, in all cases, sliding door gating influences the rate of decomplexation and is not dependent on the nature of the guest. On the other hand, French door gating is highly dependent on the stereoelectronic properties of the guest and becomes the rate-determining factor only occasionally.

rotaxane-like traits than [DB24C8·**8**][PF$_6$], a species that is formed much more rapidly without the need for heating, as its planar unsaturated Ph stoppers can pass relatively easily through the cavity of the DB24C8 macroring. This example shows that the concept of a pseudorotaxane is inherently fuzzy;[33] rather than impose an arbitrary and complicated cut-off parameter that distinguishes pseudorotaxanes from rotaxanes, it is best to say that the transition between the two species is a gradual (Figure 8) one, in which pseudorotaxanes progressively acquire more rotaxane-like characteristics until they are completely interlocked rotaxanes.

Along the same lines, there is a fuzzy boundary between hemicarceplexes and carceplexes, some hemicarceplexes having more carceplex-like traits than others. The schematic diagram illustrated in Figure 8 shows how the concepts of a pseudorotaxane and of a hemicarceplex may be explained using fuzzy sets. The sets of these species belong simulta-

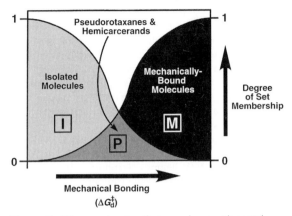

Figure 8. Diagram showing that complexes with partial mechanical bonding (**P**) character, i.e., pseudorotaxanes and hemicarceplexes, are represented by the intersection set [**M** ∩ **I**] of the set of (wholly) mechanically-bound molecules (**M**) and the set of isolated molecules (**I**) – in other words, the fuzzy region in between these two sets. Thus, the complexes in set **P** are endowed simultaneously with characteristics associated with species belonging to both **M** and **I**. The numbers 1 and 0 have been assigned arbitrarily to the species that belong either entirely or not at all to the sets **M** and **I**.

neously to both the set of their mechanically bound congeners and to the set of two isolated chemical entities. Indeed, pseudorotaxanes and hemicarcerands may be considered to be complexes that are endowed with partial mechanical bonds, i.e., in addition to noncovalent bonds, these species are held together, to some extent, by non-valence forces. The fuzzy explanation that Figure 8 provides imparts much more meaning than one in which a definite cut-off is made between either pseudorotaxanes and rotaxanes or hemicarceplexes and carceplexes.

Epilogue

For decades, chemical synthesis has relied upon multistep reaction sequences, involving the (occasionally monotonous) making and breaking of kinetically stable covalent bonds through the application of appropriate reagents, catalysts and/or protecting groups. Supramolecular chemistry is now providing the synthetic chemist with alternative routes to complicated chemical entities. In this article, we have described how it has allowed the preparation of an intriguing set of species that are kinetically stable under ambient conditions. This set lies at the interface between the molecular and supramolecular worlds, the syntheses of its "fuzzy" chemical members relying on the use of thermodynamically stable noncovalent bonds, in conjunction with mechanical coercion. The secrets of these syntheses are rooted in designing the complexes' stereoelectronically matching components prudently and

selecting the experimental conditions, through which the components are united, with care. The synthetic protocol that has been used to prepare the rotaxane-like entities and hemicarceplexes described herein can be extended, in principle, to the construction of very large, but yet well defined, kinetically stable species that, quite simply, cannot be made by traditional synthetic approaches. Toward this goal, Meijer's group has recently accomplished[34] the multiple encapsulation of guests by large dendritic hosts through a process reminiscent of Cram's hemicarceplex syntheses, while molecular shuttles, incorporating moving parts that are not bound covalently to one another, have been prepared[17,18] by slippage. In systems such as these, the mechanically-constrained entities' co-conformations can be controlled by external stimuli,[35] an outcome that has led to the preparation, via synthetic supramolecular approaches, of molecular devices[12,36] that operate like certain everyday macroscopic devices (e.g., bearings, motors, logic gates, sensors, switches and transistors). The demand for the further miniaturization of such devices means that, in all certainty, approaches based on synthetic supramolecular chemistry will become the methods of choice for the synthetic chemist in the 21st century.

References and Notes

1. F. Vögtle, *Supramolecular Chemistry*, Wiley, New York, **1991**.
2. J.-M. Lehn, *Supramolecular Chemistry*, VCH, Weinheim, **1995**.
3. *Comprehensive Supramolecular Chemistry, 11 Vols.* (Eds. J. L. Atwood, J. E. D. Davies, D. D. MacNicol, F. Vögtle), Pergamon, Oxford, **1996**.
4. M. C. T. Fyfe, J. F. Stoddart, "Synthetic Supramolecular Chemistry", *Acc. Chem. Res.* **1997**, *30*, 393–401.
5. F. M. Raymo, J. F. Stoddart, "Slippage–A Simple and Efficient Way to Self-Assemble [*n*]Rotaxanes", *Pure Appl. Chem.* **1997**, *69*, 1987–1997, and references cited therein.
6. D. J. Cram, J. M. Cram, "Hemicarcerands and Constrictive Binding", In *Container Molecules and Their Guests* (Ed. J. F. Stoddart), The Royal Society of Chemistry, Cambridge, **1994**, pp. 149–169, and references cited therein.
7. R. E. Gillard, F. M. Raymo, J. F. Stoddart, "Controlling Self-Assembly", *Chem. Eur. J.* **1997**, *3*, 1933–1940.
8. The term *rotaxane* derives (G. Schill, "History", In *Catenanes, Rotaxanes, and Knots*, Academic, New York, **1971**, pp. 1–4) from the Latin words *rota*, meaning wheel, and *axis*, meaning axle. In these molecular compounds, several wheel and axle components are constrained to be bound to one another mechanically, i.e., without the aid of any valence forces, since the axle(s) is/are endowed with bulky stopper groups that prevent the extrusion of the wheel(s).
9. D. B. Amabilino, J. F. Stoddart, "Interlocked and Intertwined Structures and Superstructures", *Chem. Rev.* **1995**, *95*, 2725–2828.

10. H. W. Gibson, "Rotaxanes", In *Large Ring Molecules* (Ed. J. A. Semlyen), Wiley, Chichester, **1996**, pp. 191–262.

11. R. Jäger, F. Vögtle, "A New Synthetic Strategy towards Molecules with Mechanical Bonds: Nonionic Template Synthesis of Amide-Linked Catenanes and Rotaxanes", *Angew. Chem. Int. Ed. Engl.* **1997**, *36*, 930–944.

12. J.-P. Sauvage, "Transition Metal-Containing Rotaxanes and Catenanes in Motion: Toward Molecular Machines and Motors", *Acc. Chem. Res.* **1998**, *31*, 611–619.

13. P. R. Ashton, M. Belohradsky, D. Philp, J. F. Stoddart, "Slippage–An Alternative Method for Assembling [2] Rotaxanes", *J. Chem. Soc. Chem. Commun.* **1993**, 1269–1274.

14. Kinetically stable rotaxane-like species have been synthesized without the aid of strong noncovalent bonds previously (I. T. Harrison, "Preparation of Rotaxanes by the Statistical Method", *J. Chem. Soc. Perkin Trans. 1* **1974**, 301–304). However, the yields reported were very low.

15. D. H. Macartney, "The Self-Assembly of a [2]Pseudorotaxane of α-Cyclodextrin by the Slippage Mechanism", *J. Chem. Soc. Perkin Trans. 2* **1996**, 2775–2778.

16. M. Händel, M. Plevoets, S. Gestermann, F. Vögtle, "Synthesis of Rotaxanes by Brief Melting of Wheel and Axle Components", *Angew. Chem. Int. Ed. Engl.* **1997**, *36*, 1199–1201.

17. P. R. Ashton, R. Ballardini, V. Balzani, M. Belohradsky, M. T. Gandolfi, D. Philp, L. Prodi, F. M. Raymo, M. V. Reddington, N. Spencer, J. F. Stoddart, M. Venturi, D. J. Williams, "Self-Assembly, Spectroscopic, and Electrochemical Properties of [*n*]Rotaxanes", *J. Am. Chem. Soc.* **1996**, *118*, 4931–4951, and references cited therein.

18. M. Asakawa, P. R. Ashton, R. Ballardini, V. Balzani, M. Belohradsky, M. T. Gandolfi, O. Kocian, L. Prodi, F. M. Raymo, J. F. Stoddart, M. Venturi, "The Slipping Approach to Self-Assembling [*n*]Rotaxanes", *J. Am. Chem. Soc.* **1997**, *119*, 302–310.

19. P. R. Ashton, I. Baxter, M. C. T. Fyfe, F. M. Raymo, N. Spencer, J. F. Stoddart, A. J. P. White, D. J. Williams, "Rotaxane or Pseudorotaxane? That Is the Question!", *J. Am. Chem. Soc.* **1998**, *120*, 2297–2307.

20. The term *co-conformation* has been employed hitherto (M. C. T. Fyfe, P. T. Glink, S. Menzer, J. F. Stoddart, A. J. P. White, D. J. Williams, "Anion-Assisted Self-Assembly", *Angew. Chem. Int. Ed. Engl.* **1997**, *36*, 2068–2070) to characterize the three-dimensional spatial arrangement of the atoms in molecular and supramolecular entities that are comprised of two or more distinct components. The co-conformation of a *pseudorotaxane* (P. R. Ashton, D. Philp, N. Spencer, J. F. Stoddart, "The Self-Assembly of [*n*]Pseudorotaxanes", *J. Chem. Soc. Chem. Commun.* **1991**, 1677–1679) resembles that of a rotaxane, with the exception that the dumbbell's extremities – in other words, its stopper units – are not large enough to prevent its extrusion from the macrocycle.

21. The macrocyclic polyether BPP34C10 and 1,1'-dimethyl-4,4'-bipyridinium bis(hexafluorophosphate) give rise to colorless solutions when they are dissolved by themselves in MeCN. The sudden appearance of a color, upon mixing the two compounds in MeCN, is indicative of complex formation and is a result of charge transfer interactions between the complementary aromatic units of the host and guest.

22. F. M. Raymo, K. N. Houk, J. F. Stoddart, "The Mechanism of the Slippage Approach to Rotaxanes. Origin of the All-or-Nothing Substituent Effect", *J. Am. Chem. Soc.* **1998**, *120*, 9318–9322.

23. The *threading-followed-by-stoppering* approach is another route (P. R. Ashton, P. T. Glink, J. F. Stoddart, P. A. Tasker, A. J. P. White, D. J. Williams, "Self-Assembling [2]- and [3]Rotaxanes from Secondary Dialkylammonium Salts and Crown Ethers", *Chem. Eur. J.* **1996**, *2*, 729–736) to the synthesis of rotaxanes. This approach involves the threading of a linear molecule through a macroring to generate a pseudorotaxane that is then stoppered by reaction with a bulky reagent.

24. P. R. Ashton, E. J. T. Chrystal, P. T. Glink, S. Menzer, C. Schiavo, N. Spencer, J. F. Stoddart, P. A. Tasker, A. J. P. White, D. J. Williams, "Pseudorotaxanes Formed between Secondary Dialkylammonium Salts and Crown Ethers", *Chem. Eur. J.* **1996**, *2*, 709–728.

25. M. C. T. Fyfe, J. F. Stoddart, "(Supra)molecular Systems Based on Crown Ethers and Secondary Dialkylammonium Ions", *Adv. Supramol. Chem.* **1998**, *5*, 1–53.

26. In polar solvents with high donor numbers, the complex is solvated poorly relative to its separate components. This difference in solvation energies nullifies the hydrogen bonding interactions present in the complex, which, as a result, dissociates into its host and guest components.

27. Although an alternative mechanism, proceeding through covalent bond cleavage, threading and covalent bond reformation, can be envisaged for the formation of species like **12·13**, it is almost certain that these reactions in the melt occur via a slippage mechanism, as no catenanes were isolated from the reaction mixtures.

28. Thus, the *hemicarcerands* differ[6] from the related *carcerands*, which are similar in shape and constitution, but from which, guests cannot escape without the breaking of covalent bonds. The mechanically bound complexes formed between carcerands and their guests are termed *carceplexes*.

29. D. J. Cram, M. T. Blanda, K. Paek, C. B. Knobler, "Constrictive and Intrinsic Binding in a Hemicarcerand Containing Four Portals", *J. Am. Chem. Soc.* **1992**, *114*, 7765–7773.

30. One molecule of DMA acts as a template for the shell-closure reaction, leading to its encapsulation within **16** and the creation of hemicarceplex **16·DMA**.

31. The entropic driving force for intrinsic binding may be considered to originate as a result of two factors. Entropy is increased because, during incarceration, the solvated guest (1) liberates fairly organized solvent molecules and (2) converts the large empty space within the cavity into numerous smaller spaces throughout the bulk solvent.

32. K. N. Houk, K. Nakamura, C. Sheu, A. E. Keating, "Gating as a Control Element in Constrictive Binding and Guest Release by Hemicarcerands", *Science* **1996**, *273*, 627–629.

33. D. H. Rouvray, "That Fuzzy Feeling in Chemistry", *Chem. Br.* **1995**, *31*, 544–546.

34. J. F. G. A. Jansen, E. M. M. de Brabander-van den Berg, E. W. Meijer, "Encapsulation of Guest Molecules into a Dendritic Box", *Science* **1994**, *266*, 1226–1229.

35. Chemical, electrochemical and/or photochemical stimuli can be used, for instance, to effect co-conformational changes in molecular devices.

36. V. Balzani, M. Gomez-Lopez, J. F. Stoddart, "Molecular Machines", *Acc. Chem. Res.* **1998**, *31*, 405–414.

Crystal Engineering with Soft and Topologically Adaptable Molecular Host Frameworks

K. Travis Holman and **Michael D. Ward**

Department of Chemical Engineering and Materials Science, University of Minnesota, Minneapolis, MN 55455, USA

Phone: +612 625 3062, Fax: +612 625 7805, e-mail: wardx004@tc.umn.edu

Keywords ■ Architectural Isomerism ■ Clathrates ■ Crystal Engineering ■ Host-Guest Chemistry ■ Hydrogen-Bonding ■ Inclusion Compounds ■ Supramolecular Chemistry ■ Template Synthesis

Concept: Crystal engineering – a field devoted to guiding the assembly of molecules into desirable solid state structures through skillful contrivance – is commonly thwarted by the inability to maintain architectural control when making even the slightest changes to the molecular components. This barrier can be circumvented by incorporating "soft" and topologically adaptable supramolecular modules that allow the lattice to achieve dense packing through low-energy deformations while retaining their inherent dimensionality and supramolecular connectivity. These properties can be exemplified by crystalline inclusion compounds based on lamellar hydrogen-bonded host frameworks constructed from guanidinium cations and anionic organodisulfonates.

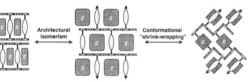

Abstract: The two-dimensional (2D) hydrogen-bonded sheet comprising guanidinium (**G**) cations and the sulfonate (**S**) moieties of organosulfonate anions displays a remarkable persistence that can be exploited for the synthesis of crystalline low-density host frameworks capable of including a variety of guest molecules. The organic portions of organo*di*sulfonates function as "pillars" that connect opposing **GS** sheets to create porous galleries, occupied by guest molecules, between the sheets. The use of different pillars permits adjustment of pore size and character. The guest molecules, however, also play an important role in directing the assembly of the framework into two possible *architectural isomers* – a discrete pillared *bilayer* isomer and a lower density, more open, continuous *brick* isomer, the latter promoted by larger guest molecules. The **GS** host frameworks possess an intrinsic conformational "softness" associated with accordion-like flexibility of the **GS** sheet, and rotational and conformational freedom of the organic pillars. This conformational softness enables the host to "shrink-wrap" around included guests so that dense packing, required for achieving sufficient cohesive energy for crystallization, can be achieved. This structural adaptability bestows the **GS** hosts with the unique ability to include a variety of differently sized and functionalized guest molecules. Host systems with this property significantly facilitiate the design of new inclusion compounds with molecular scale cavities that can serve as miniature reaction chambers, catalytic environments and chemical storage compartments, while providing reliable architectures that enable control of the spatial organization of guest molecules, so that solid state properties can be systematically manipulated.

Prologue

The design and synthesis of "nanoporous" materials, or low-density host frameworks with well-defined pores that may or may not be occupied by guest molecules, has recently emerged as a prolific sub-discipline of solid state chemistry. The intense interest in these materials stems from both fundamental issues, including the mechanism of pore formation during assembly of the host framework, control of pore dimensionality, structure and stability, as well as their potential application in various technologies, including catalysis, separations, chemical storage, and optoelectronics. Considerable attention has been given to porous inorganic materials, such as zeolites and pillared clays,[1,2] which have frameworks supported by strong covalent bonds capable of maintaining structural integrity in the absence of guest molecules. Although the technological importance of such materials cannot be argued, they remain rather limited in some respects. The void shapes and sizes supplied by inorganic porous materials typically are dictated by rather rigid frameworks (*e.g.*, metal oxides) that are not readily amenable to the precise chemical modification that is required for many applications. Consequently, the past several years have witnessed considerable efforts directed toward the design and synthesis of organic[3–6] or metal-organic[7–10] analogues of inorganic host frameworks. These efforts have relied largely on modular strategies based on molecular building blocks that assemble into supramolecular motifs by directional non-covalent bonding such as metal coordination or hydrogen bonding. In principle, libraries of molecular building blocks can be rationally designed and synthesized, using the principles of organic synthesis, to create host frameworks with void shapes, sizes, and chemical attributes that can be systematically and precisely adjusted. These features are important for fine chemical separations that rely on precise host-guest molecular recognition events; molecular hosts may have a clear advantage over their inorganic relatives in that they can be reversibly disassembled by simple dissolution of the non-covalent host framework under mild conditions so that the guests can be easily liberated. With regard to the synthesis of functional materials, the modular approach allow the separation of crystal architecture, which is provided by the host framework, from function, which can be introduced by the included guests.

Historically, organic solid state host frameworks have been discovered by chance, for example the well known (thio)urea, tri-*o*-thymotide, phenolic, perhydrotriphenylene, choleic acid, or cyclotriveratrylene hosts.[11,12] The numerous investigations of inclusion compounds based on these frameworks has advanced the understanding of inclusion phenomena considerably. However, these hosts are considerably limited in that their molecular components cannot be modified without destroying the basic crystal architecture required for guest inclusion, thereby preventing significant modification of the inclusion cavities. This inability to maintain architectural control when making even the slightest changes to the molecular components is actually a general problem in crystal engineering. The challenge, therefore, is to design molecular hosts that are structurally robust toward modification of a generic framework. Numerous attempts have been made to introduce structural robustness by modules that generate *rigid* frameworks. However, inclusion of a specific guest molecule in a rigid framework would require that the shape of the inclusion cavity framework be precisely engineered to match that of the guest or, as is often the case, that inclusion involves multiple guest molecules that, as an ensemble, can conform to the shape of the rigid cavity. This argues that crystal engineering of inclusion compounds is likely to be more successful when design strategies are based on soft frameworks that can adapt to the steric landscape of guest molecules while retaining their *general* architectural features, particularly their inherent dimensionality and supramolecular connectivity. Such systems can facilitate the development of the guiding principles required for the construction of host frameworks with adjustable pore characteristics. Furthermore, the ability to make systematic adjustments to structurally robust host frameworks would provide insight into the underlying factors that govern the creation of low-density structures and their stability, polymorphism, interpenetration, and molecular recognition.

Designer Inclusion Compounds

Crystal engineering[13,14] studies in our laboratory have recently demonstrated the persistence of a two-dimensional hydrogen-bonded network of topologically complementary guanidinium (**G**, $[C(NH_2)_3]^+$) cations and sulfonate (**S**) moieties of organosulfonate anions (Figure 1).[15–17] The threefold symmetry of the **G** ions and **S** moities, and the equivalent numbers of guanidinium hydrogen-bond donors and sulfonate oxygen acceptor sites, guide molecular assembly into a quasihexagonal hydrogen-bonded sheet that can be viewed as edge-connected GS "ribbons". The hydrogen bonding in this supramolecular network is further fortified by its ionicity. The robustness of the **GS** network is evidenced by over 30 different crystalline phases based on a variety of substituted organo-monosulfonate salts.

The crystal architecture of guanidinium monosulfonate salts can be classified according to two unique lamellar stacking motifs, one a discrete "bilayer" form and the other a "continuously interdigitated layer" structure. In both cases,[18] all the organic substituents of each **GS** ribbon project from the same side of the ribbon, but the two stacking motifs differ with respect to the relative projections of the organic substituents on adjacent connected ribbons. Bilayer organization ensues when all the organic groups of each **GS** ribbon project to the

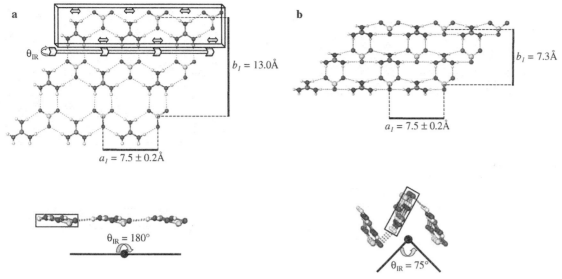

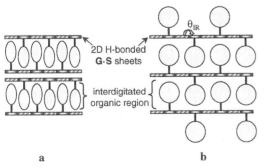

Figure 1. The flexible quasihexagonal 2D H-bonded **GS** sheet as viewed normal to (upper) and along (lower) the ribbon direction (boxed). The translational repeat distances

(*a₁* and *b₁*) of (**a**) a flat (ϑ_{IR} = 180°) and (**b**) highly puckered (ϑ_{IR} = 75°) **GS** sheet are depicted.

same side of the **GS** sheet, whereas the continuously interdigitated layer motif is formed when the projection of the organic groups on adjacent ribbons alternates up/down (Figure 2). Crystal assembly along the stacking direction in both forms results from interdigitation of the organic groups on opposing **GS** sheets, creating lamellar structures consisting of alternating non-polar organic regions and polar H-bonded **GS** sheets.

The solid state structures of these materials suggest that the 2D **GS** network is more accurately described as an assembly of 1D **GS** "ribbons" connected to each other via lateral (G)N–H···O(S) H-bonds that serve as flexible hinges. These hinges allow the **GS** sheet to pucker, like an accordion, without an appreciable change in the near-linear geometries, which are considered to be optimal, of the (G)N–H···O(S) H-bonds. With respect to crystal metrics, the range of repeat distances within the **GS**

ribbons is quite narrow (7.5 ± 0.2 Å, hereafter referred to as a_1), reflecting stiffness (although by the standards of covalent chemical bonds these H-bonded ribbons are "soft"). In contrast, repeat distances, referred to as b_1, normal to the ribbon direction, within the plane of the **GS** sheet vary considerably due to changes in the *inter-ribbon puckering angle* (ϑ_{IR}),[2] ranging from as small as b_1 = 7.3 Å for highly puckered sheets (ϑ_{IR} = 75°; [G][4-nitrobenzenesulfonate]) to b_1 = 13.0 Å for perfectly flat sheets (ϑ_{IR} = 180°; [G][triflate]).

The observation of bilayer or single layer motifs appears to depend primarily on the steric requirements of organic portion of the anion. Generally, if the bilayer arrangement does not allow interdigitation of the organic groups projecting from opposing sheets (*e.g.* for sterically bulky organic groups), or if the organic substituent sterically hinders coplanar arrangement of ribbons, the continuously

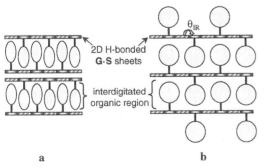

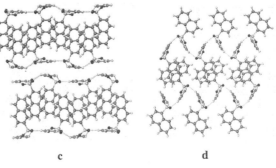

Figure 2. The two layering motifs observed for guanidinium organomonosulfonate salts. (**a**) The " discrete bilayer" motif in which all of the organic groups project from the same side of a given **GS** sheet. (**b**) The "continuously interdigitated layer" motif, in which the organic groups on adjacent **GS** ribbons project to opposite sides of the sheet.

(**c**) The discrete bilayer motif observed in the crystal structure of [**G**][1-naphthalenesulfonate] reveals only a slight puckering between adjacent ribbons (ϑ_{IR} = 170°). (**d**) The significantly puckered continuously interdigitated layer motif observed in the sterically hindered [**G**][1-naphthalenesulfonate] (ϑ_{IR} = 84°).

interdigitated layer motif is manifested. Figure 2 depicts the bilayer and single layer structures of [G][2-naphthalenesulfonate] and [G][1-naphthalenesulfonate], respectively. Despite essentially identical molecular volumes of these anions, 1-substitution of the naphthalene residue sterically prohibits the coplanar approach of adjacent GS ribbons required for the planar, or near-planar, bilayer structure. Consequently, a *highly puckered* continuously interdigitated layer motif results.

Although bilayer structures can exhibit GS sheets which pucker to a slight extent, typically the continuously interdigitated layer structures are more highly puckered. The ability to pucker confers a resilience to the GS sheet structure in that it provides a mechanism for achieving dense-packing of the organic groups while retaining the dimensionality and the intermolecular connectivity for this module. The puckering, and the free rotation of the organic groups about the C–S bond, allow for optimization of the intermolecular interactions so that the cohesive energy required for crystallization can be achieved.

In the case of the continuously interdigitated layer motif, the GS sheet repeat distance normal to the ribbon direction (b_l) becomes a simple function of the puckering angle (ϑ_{IR}). Highly puckered sheets ($\vartheta_{IR} < 110°$) tend to direct their organic substituents normal to the GS *ribbons* and b_l satisfies equation (1)

(within 4%). Less puckered sheets ($\vartheta_{IR} > 110°$) tend to direct their organic substituents normal to the mean plane of the GS *sheet* and b_l satisfies equation (2). For example, equation (1) predicts $b_l = 7.9$ Å for the highly puckered [G][1-naphthalenesulfonate] based on the experimentally observed $\vartheta_{IR,exp} = 84°$, whereas equation (2) predicts $b_l = 11.1$ Å for the less puckered [G][ferrocenesulfonate] based on $\vartheta_{IR,exp} = 118°$. These values compare favorably to the actual values of $b_l = 8.07$ Å and 10.74 Å, respectively, as determined from single crystal diffraction analysis. The ability to use rather simple geometric models to describe the metrics of the GS sheet structure stems from its resilience and well-defined dimensionality (2D). The advantages of this approach in structure prediction and its relevance to crystal engineering will become more apparent below.

$$b_l = 11.8 \cdot \sin (\vartheta_{IR}/2) \ \text{Å} \qquad (1)$$
$$b_l = 13.0 \cdot \sin (\vartheta_{IR}/2) \ \text{Å} \qquad (2)$$

The persistent and distinctive structural features of the GS network effectively reduce crystal engineering to the last remaining (third) dimension. These features prompted our laboratory to synthesize related materials in which the monosulfonate components were replaced with disulfonates. We surmised that this would lead to a two-fold reduction in the amount of space occupied by the organic residues between the GS sheets, creating frameworks

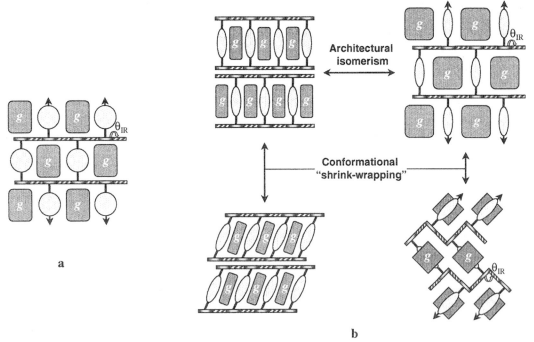

Figure 3. Schematic representations of feasible pillared **GS** host frameworks. (**a**) Sterically demanding pillars force a continuous "brick" architecture characterized by **GS** sheets that are topologically homologous to the continuously interdigitated layer motif observed in some guanidinium organomonosulfonate salts. (**b**) Sterically undemanding pillars can from either "discrete bilayer" or continuous brick architec-

tural isomers, depending on the size and shape of the included guests. Both the brick and bilayer host frameworks are conformationally flexible and can "shrink-wrap" around included guest molecules, providing a mechanism for achieving optimal host-guest packing. The guests are denoted here as **g**.

with porous structures in which organodisulfonate "pillars" connect opposing **GS** sheets in a manner reminiscent of pillared metal organophosphonates.[19] Importantly, the close packing of atoms within the **GS** sheet would preclude self-interpenetration of equivalent networks, a problem that plagues the design of many open framework structures.[20] In principle, the sizes, heights, shapes, and chemical environment of the resulting voids, formed in 2D gallery regions between opposing sheets, could be manipulated by the choice of molecular pillar. We also expected a topological homologism between the **GS** sheet structures of the disulfonate host frameworks and the simple monosulfonate salts. Sterically *demanding* pillars would lead exclusively to a fully interconnected "continuous brick" architecture, in which the up/down arrangement of the pillars on each **GS** sheet would be topologically homologous with the continuously interdigitated layer of their corresponding monosulfonates. Sterically *undemanding* pillars would lead to either "discrete bilayer" *or* continuous brick host frameworks. Smaller guests would favor the bilayer motif by filling the voids while allowing favorable coulombic interactions between the ionic **GS** sheets of stacked bilayers. In contrast, guests too large to fit within the voids of a bilayer framework of a specific pillar were expected to promote *architectural isomerism*, in which the larger guests would template the more open brick architecture.

We have successfully implemented this design strategy, synthesizing pillared host frameworks based upon a series of organodisulfonate pillars [Scheme 1: **I** = dithionate; **II** = 1,2-ethanedisulfonate; **III** = 1,4-butanedisulfonate; **IV** = 1,5-naphthalenedisulfonate; **V** = 2,6-naphthalenedisulfonate; **VI** = 4,4'-biphenyldisulfonate; **VII** = 2,6-anthracenedisulfonate; **VIII** = 4,4'-azobenzenedisulfonate; **IX** = 1,2-bis(*p*-sulfophenoxy)ethane; **X** = 1,3-bis(*p*-sulfophenoxy)propane] of varying lengths ranging from l = 2.1 Å to 15.6 Å (where l is the S⋯S separation).[22] The ease with which disulfonates can be synthesized has

enabled us to create a substantial library of pillars that will afford a diverse set of inclusion compounds in which pore characteristics can be systematically adjusted and crystal engineering principles tested and developed. In most cases, the inclusion compounds are prepared by treatment of the acid form of the pillars with [G][BF$_4$] in acetone, which results in the immediate precipitation of **GS** salts, generally as acetone clathrates. These materials readily lose solvent upon standing in air to yield pure apohosts, which can then be used, in dissolved form, for the crystallization of desired inclusion compounds. Single crystals of the inclusion compounds can be obtained easily by standard crystallization techniques, using methanol or water solutions containing the apohost and the appropriate guest molecule. With the exception of pillars **I** and **II**, which are too short to provide voids of sufficient size for guest inclusion, **G** salts of pillars **III**–**X** form crystalline inclusion compounds in the presence of appropriate guests. These compounds crystallize as the anticipated bilayer or continuous brick motifs, with inclusion cavity volumes that depend on the architectural isomer, pillar length and included guests. The guests range from small alcohols and nitriles to large aromatic molecules.

Illustrative examples of **GS** bilayer inclusion compounds are depicted in Figure 4. The gallery heights, as defined by the shortest distance between the mean planes of the nitrogen and oxygen atoms of the **GS** sheets, and thus the available volume for guests increase systematically with increasing pillar length. This is evident from a comparison of **G$_2$III**, which includes only two equivalents of acetonitrile, and **G$_2$X**, which includes two equivalents of the larger nitrobenzene guest. This trend notwithstanding, the bilayer host frameworks are not rigid. A systematic investigation of over thirty bilayer inclusion compounds with the composition **G$_2$VI**·*n*guest has revealed a remarkable adaptability of the bilayer host framework, creating pores with sizes and shapes that accommodate the steric or elec-

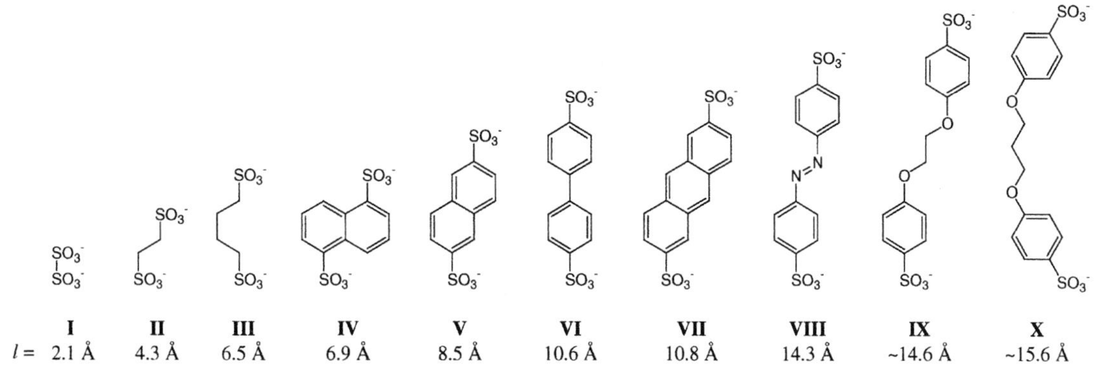

	I	**II**	**III**	**IV**	**V**	**VI**	**VII**	**VIII**	**IX**	**X**
l =	2.1 Å	4.3 Å	6.5 Å	6.9 Å	8.5 Å	10.6 Å	10.8 Å	14.3 Å	~14.6 Å	~15.6 Å

Scheme 1.

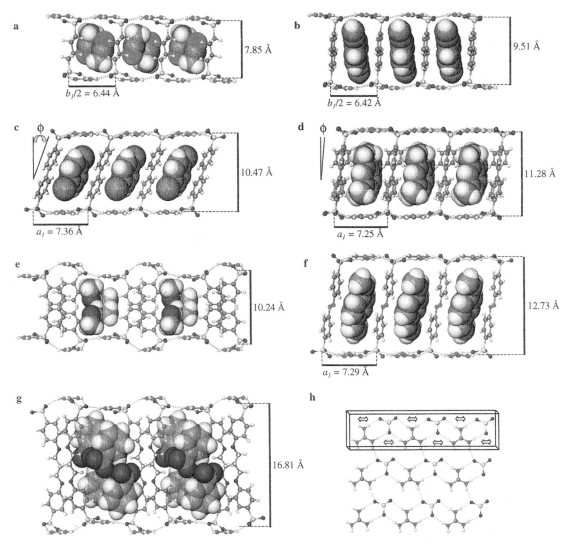

Figure 4. The various mechanisms for structural "shrink-wrapping" in bilayer **GS** inclusion compounds can be gleaned from these illustrative examples, here viewed down the channel direction. (**a**) quasihexagonal $G_2III\cdot$ 2(CH_3CN), (**b**) quasihexagonal $G_2V\cdot$ 1,4-diethynylbenzene, (**c**) shifted ribbon $G_2VI\cdot$ 1,4-dichlorobenzene, (**d**) shifted ribbon $G_2VI\cdot o$-xylene, (**e**) quasihexagonal $G_2VI\cdot$ 2(MeOH), (**f**) shifted ribbon $G_2VIII\cdot$ 1,4-divinylbenzene, and (**g**) quasihexagonal $G_2X\cdot$ 2(nitrobenzene) (the guest molecules are disordered). (**h**) The "shifted ribbon" **GS** sheet motif.

tronic requirements of a diverse collection of included guests.[23] The volume of the host framework, calculated from the host-only structures, is 335 ± 3 $Å^3$ per G_2VI unit.[24] However, the bilayer framework can conform to the shape of the guests, "shrink-wrapping" about the guests to achieve dense packing and optimization of intermolecular interactions. The inclusion cavity volume in this system varies by as much as 34% for guest molecules, independent of temperature.

The structural adaptation exhibited by **GS** bilayer frameworks is achieved through a variety of mechanisms associated with the intrinsic *conformational softness* of the host framework, including: (1) formation of a "shifted-ribbon" **GS** sheet motif (Figure 4h) in which adjacent ribbons are shifted from the quasihexagonal arrangement, by as much as $a_1/2$, such that they are connected by one strong (**G**)N–H···O(**S**)

H-bond ($d_{O···H} \approx 2.0$ Å) and one very weak one ($d_{O···H} \approx 2.5$ Å), (2) slight puckering of the **GS** sheet, (3) turnstile rotation of the pillars about the C-S bonds, (4) tilting of the pillars with respect to a normal to the **GS** sheets, and (5) twisting and flexing of the pillar, if possible.

Although the shifted ribbon motif might be considered to be energetically less favorable than the quasihexagonal motif due to the loss of one strong inter-ribbon H-bond, it is fairly common in the bilayer architectures. Quasihexagonal motifs are generally observed, however, when the molecular volume of the guests are near the limit of available inclusion cavity volume in the bilayer architecture, suggesting that the shifted ribbon motif facilitates the achievement of dense-packing about smaller guests. Interestingly, we have yet to observe non-stoichiometric, incommensurate inclusion compounds in these systems. This clearly argues that the observed **GS** host deformations, including the shifted ribbon configuration, are energetically preferable to forming incommensurate host-guest structures. We note that urea inclusion compounds exhibit phase transitions, described as "spring-loaded,"[25] associated with slightly deformed commensurate structures that are energetically more favorable than incommensurate forms with an ideal host structure.

In cases where guests are too small to adequately fill the inclusion cavities of the bilayer framework, such as $G_2VI\cdot2(MeOH)$ and $G_2X\cdot2(nitrobenzene)$, the host can adapt by puckering of the quasihexagonal **GS** sheet, which reduces the gallery heights. This is reminiscent of the bilayer *monosulfonate* salts (*e.g.,* [G][2-naphthalenesulfonate]), which pucker in order to achieve close packing (Figure 2).

In either the quasihexagonal or shifted ribbon arrangements, the pillars can rotate freely about their C–S bonds like turnstiles such that one-dimensional channels, flanked by the pillars, are created within the galleries between the **GS** sheets of the bilayer host framework. The pillars in $G_2V\cdot$ 1,4-diethynylbenzene, which exhibits the quasihexagonal motif, are rotated with their arene planes nominally *parallel* to the **GS** ribbons. Consequently, the guest filled channels run parallel to the ribbon direction. The widths of these channels are defined by $b_1/2$ (*ca.* 6.5 Å for these nearly flat **GS** sheets). In contrast, the pillars in $G_2VI\cdot$1,4-dichlorobenzene, $G_2VI\cdot p$-xylene, and $G_2VIII\cdot$1,4-divinylbenzene, which adopt the shifted ribbon motif, are rotated with the arene planes, and the guest-occupied channels, nearly *orthogonal* to the **GS** ribbons. The widths of the channels in these compounds are roughly defined by a_1 (*ca.* 7.3 Å), but can be described more accurately as $a_1\cdot cos\ \phi$, where ϕ is the tilt angle of the pillars with respect to the normal to the **GS** sheet. For a given host, larger values of ϕ are synonymous with shorter gallery heights and, consequently, smaller pore volumes. However, the gallery heights in the bilayer architectures of G_2VI, which range from 10.24 Å for $G_2VI\cdot2(MeOH)$ to 11.75 Å for $G_2VI\cdot$2,6-dimethylnaphthalene, do not correlate exactly with the guest volumes (V_g). For example, despite the similar volumes of methyl and chloro substituents, $G_2VI\cdot$1,4-dichlorobenzene and $G_2VI\cdot p$-xylene (not shown) have substantially different gallery heights (10.47 Å and 11.23 Å respectively) and pillar tilt angles (ϕ = 25.85° and 15.24°, respectively). Similarly, the gallery height for the 1:1 inclusion of toluene (V_g = 95 Å^3) in the G_2VI bilayer framework is 10.9 Å, exceeding the value of 10.8 Å observed when the larger bromobenzene guest (V_g = 99 Å^3) is

included. The unit cell volume of $G_2VI\cdot$toluene also is greater than that of $G_2VI\cdot$bromobenzene. These examples demonstrate that simple *sterics, based on molecular volume, are not the sole structure directing influence* in these inclusion compounds. The absence of a monotonic relationship between pore and guest volumes is evident in many inclusion compounds in this series, particularly when comparing isostructural methylated with halogenated guests. Inspection of these compounds indicates that attractive ion-dipole interactions between the guanidinium ions, which are exposed at the floor and ceiling of the channels, and the C–X dipoles of the guest promote shrinking of the gallery height.

The role of the pillars as structural supports for the bilayer framework suggests that axial rigidity would be important for stabilizing their inclusion compounds. However, flexible pillars such as **IX** and **X** also form bilayer inclusion compounds. Torsional twisting about the central C–C bond of the axially rigid biphenyl pillar in bilayer G_2VI inclusion compounds enables the pillar to conform to the shape of the guests while behaving as synchronous molecular gears that relay instructions for guest ordering from one pore to another. For example, the two arenes of the biphenyl pillar in $G_2X\cdot m$-xylene subtend a dihedral angle of 27.4°. Though rare, the biphenyl pillar can also bend significantly about the central C–C bond, as we have observed in $G_2VI\cdot$2,6-dimethylnaphthalene (not shown).[26] While this may seem energetically unfavorable, the observed deformation (each arene ring bends 6.5° out-of-plane relative to an ideal planar biphenyl) has precedent in cyclophanes, bowl-shaped polyaromatics, and fullerenes. The bent pillars in $G_2VI\cdot$2,6-dimethylnaphthalene are a consequence of the guest having a molecular volume that is at the upper limit of available space in the inclusion cavities of the G_2VI bilayer framework. Clearly, the cohesive energy of the pillar-guest ensemble in the gallery must override the energetic penalty associated with the strain of the biphenyl pillars.

We have observed *topological homology,* with respect to the up/down arrangement of the organic residues projecting from the **GS** sheet, when comparing the [G][organomonosulfonate] salts with their [G][organodisulfonate] counterparts. For example, the bilayer host framework based on the 2,6-naphthalenedisulfonate pillar in G_2V inclusion compounds is topologically homologous to [G][2-naphthalenesulfonate], with all the organic residues attached to a given sheet projecting from the same side of the sheet. Topological homology is also observed for [G][1-naphthalenesulfonate] and the G_2IV host framework, the former described by a continuously interdigitated layer motif and the latter by a "continuous brick" host architecture. The organic residues in both [G][1-naphthalenesulfonate] and the **G₂IV** host framework alternate their orientation up/down on adjacent **GS** ribbons. This is attributed to the 1,5-naphthalenedisulfonate pillar sterically prohibiting the co-planar arrangement of **GS** ribbons that is required for the formation of a bilayer host framework, instead promoting the formation of a highly puckered **G₂IV** host with the brick architecture (Figure 5).

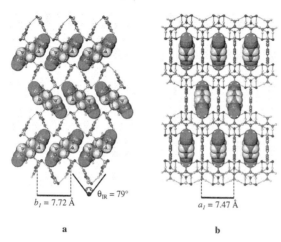

$b_1 = 7.72$ Å $\vartheta_{IR} = 79°$ $a_1 = 7.47$ Å

a   b

Figure 5. The sterically promoted "continuous brick" host architecture for **G₂IV**·*n*guest, as viewed along (**a**) the ribbon direction and (**b**) the channel direction in **G₂IV**·1-hexanenitrile. Note that the 1-hexanenitrile guests are disordered and appear as 1,6-hexanedinitrile molecules.

The **G₂IV** host framework includes methanol or linear substituted alkane guests such as 1-alkanenitriles, α,ω-alkanedinitriles, 1-alkanols, or triglyme. These guest molecules are incorporated into 1D channels, of width a_1, that run orthogonal to the ribbon direction and are flanked by the naphthalene pillars. Notably, the 1,5-naphthalenedisulfonate pillar of the **G₂IV** host lacks the conformation flexibility of the biphenyl pillar described above, leaving puckering and the formation of shifted ribbon motifs as the only mechanisms for structural adaptation. The ϑ_{IR} values vary from 79° (for **G₂IV**·1-hexanenitrile; $b_1 = 7.72$ Å) to 99° (**G₂IV**·1-octanenitrile; $b_1 = 8.93$ Å), reflecting the limits imposed by steric hindrance between the pillars and the **GS** sheets and between neighboring pillars in the gallery. As with the highly puckered **GS** sheets of monosulfonate salts, equation (1) adequately describes the dependence of b_1 on ϑ_{IR}, predicting 7.5 Å and 9.0 Å, respectively).

Although certain sterically demanding pillars, such as **IV**, are only capable of forming brick host architectures, it is conceivable that pillars observed in bilayer frameworks (*i.e.* **III** and **V-X**) can also, under the appropriate conditions, form continuous

brick frameworks. We refer to this as *architectural isomerism*, wherein the bilayer and brick frameworks are true isomers, having identical chemical compositions. The ideal brick framework, with flat **GS** sheets, would possess up to twice the void space of its corresponding bilayer isomer.

We recently demonstrated that the bilayer-to-brick isomerism in **G₂VI** can be promoted by guests that were too large to fit into the inclusion cavities of the bilayer frameworks.[27] *In this respect, the guests serve as templates that direct the assembly of the molecular framework.* The templating role of guest molecules in this system is reminiscent of that ascribed to surfactant or organic microstructures in the templation of open framework zeolites.[28,29] Figure 6 depicts a comparison of the molecular volumes of included guests (V_g) with the observed framework isomer in the **G₂VI**·*n*guest system. Though there seems to be no definitive guest volume threshold that induces isomerism, the preference for brick architectures with increasing V_g is quite clear. However, the observation of the bilayer isomer for some rather large guests indicates that V_g is not the only factor governing the isomer selectivity. This is not particularly surprising considering the marked structural differences in bilayer frameworks that contain guests of similar volumes (*vide supra*). In general, the role played by guest molecules in templating solid-state frameworks is not completely understood, but certainly sterics, in *combination* with guest-guest and host-guest interactions, plays an important role. Guest aggregates can template the more open brick architecture even though the individual molecular volumes are small. For example, the apparent optimization of guest-guest interactions through the formation of 1D stacked chains of guest molecules in **G₂VI**·4(nitrobenzene) ($V_g = 103$ Å³) allows these comparatively small guests to template the brick architecture of the **G₂VI** host (*vide infra*). Similarly, host-guest interactions can promote the formation of polar crystals in some cases.[30] We have also established architectural isomerism for **G₂V** and **G₂VII** with similar dependencies on guest volume, although to date the number of examples in these systems is limited compared to the **G₂VI**·*n*-guest series.

Brick host architectures adapt to the steric demands of different guests through mechanisms similar to the bilayer frameworks, including turnstile rotation and tilting of the pillars. For the brick frameworks, however, the formation of shifted ribbon **GS** sheet motifs is extremely rare and the most important mechanism for structural adaptation is the puckering of the quasihexagonal **GS** sheets. Puckering, which can occur to a much greater

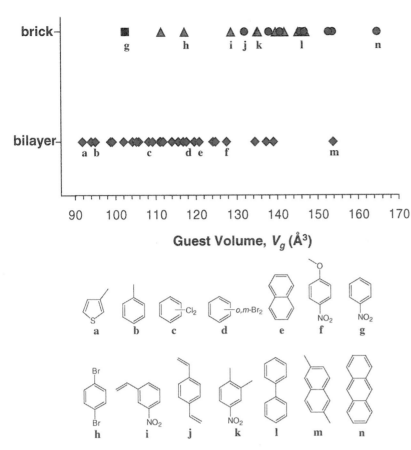

Figure 6. A comparison of the molecular volumes of fifty guests (V_g) with the observed architectural isomer in the **G₂VI·*n*guest** system of inclusion compounds reveals that smaller guests tend to promote the formation of the bilayer framework whereas larger guests template the brick isomer. Some representative guests are labeled (**a–n**). The three symbols used here for the brick inclusion compounds denote the different channel structures observed in the galleries of these frameworks (■ = 1D channels of width b_1; ▲ = 1D channels of width a_1; ● = 2D channels). The dimensionality and the direction of the channel structures depends upon the degree of **GS** sheet puckering and the rotation of the pillars about their long axes (*vide infra*).

extent in the brick architectures because of steric considerations, provides the brick framework with a substantially greater range of conformational flexibility than the bilayer framework. Consequently, the available volume for included guests (*i.e.* the volume of the crystal not occupied by the host), in these brick frameworks is extremely variable. In the brick **G₂VI·*n*guest** system, this value varies from 346 Å³/pillar in **G₂VI·3-nitrostyrene** to 859 Å³/pillar in **G₂VI·3(anthracene)**! The available volume for included guests is maximized when the **GS** sheets are flat, whereas highly puckered sheets can result in brick architectures with void volumes that are only slightly greater than the corresponding bilayer architecture.

The combined conformational mechanisms for structural adaptation – puckering, turnstile rotation, tilting, twisting and flexing of the pillars – permit the identification of three distinct sub-architectures of the brick framework (Figure 7). The first is characterized by a substantially puckered **GS** sheet and highly titled pillars whose arene planes are aligned roughly orthogonal to the **GS** ribbon direction, enforcing gallery regions with 1D channels of width a_1. This sub-architecture represents the most shrunken form of the brick framework. This is evident on comparing the brick structure of **G₂VI·1,4-dibromobenzene** to **G₂VI·1,2-dibromobenzene** and **G₂VI·1,2-dibromobenzene**, which

both form bilayer inclusion compounds. The guest volumes in these compounds are nearly identical, yet the *para* isomer templates a brick architecture which must shrink considerably in order to pack densely about the guest. In compounds where the **GS** sheets are much less puckered, however, the distance between the pillars is increased and they cannot form continuous walls flanking 1D channels. This results in a brick sub-architecture with a 2D pore structure in the galleries [*e.g.* **G₂VI·3(1,4-divinylbenzene)**] and large void volumes capable of including a significant amount of guest, typically with a 1:3 host:guest stoichiometry. A third sub-architecture arises when the arene planes of the pillars are aligned *parallel* to the ribbon direction. Similar to the highly puckered one, this sub-architecture possesses gallery regions with 1D channels, but the width of the channels is now defined by b_1, which can vary (up to 13.0 Å) with the degree of puckering as defined by ϑ_{IR} [*e.g.* **G₂VI·4(nitrobenzene)**].

The *architectural isomerism* described above, in which a framework based on a specific pillar exhibits two different forms depending upon the guest size, suggests that **GS** hosts with longer pillars would revert to the bilayer framework for the larger guests. Indeed, we demonstrated that the bilayer-to-brick isomerism observed for **G₂VI** with 1,4-dibromobenzene, 1,4-divinylbenzene, 1-nitro-

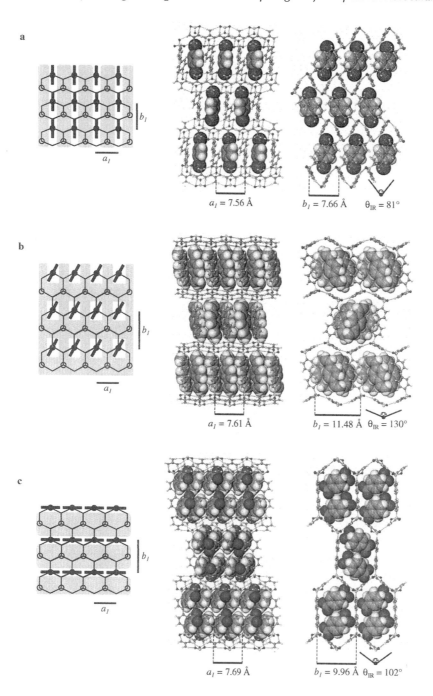

Figure 7. Conformational flexibility of the brick host architecture leads to three distinct sub-architectures with respect to the rotation of the pillars and the dimensionality of the pore structure in the galleries. (**a**) Highly puckered **GS** sheets, with pillars aligned orthogonal to the ribbon direction, yield gallery regions with 1D channels, flanked by the pillars, of width a_1. Such a structure is observed for $G_2VI\cdot$1,4-dibromobenzene. (**b**) Less puckered sheets in $G_2VI\cdot$3(1,4-divinylbenzene) yield a 2D continuous guest network in which the guests surround the pillars. (**c**) Alignment of the pillars parallel to the ribbon direction yields gallery regions with 1D channels of width b_1 in $G_2VI\cdot$4(nitrobenzene). The schematic illustrations at the left depict the gallery regions within their respective inclusion compounds as viewed normal to a **GS** sheet. The **GS** sheets are represented as hexagons. Filled and open circles represent pillars that project above and below the **GS** sheet, respectively. The guest-filled channels are highlighted in grey.

a) $a_1 = 7.56$ Å $b_1 = 7.66$ Å $\theta_{IR} = 81°$

b) $a_1 = 7.61$ Å $b_1 = 11.48$ Å $\theta_{IR} = 130°$

c) $a_1 = 7.69$ Å $b_1 = 9.96$ Å $\theta_{IR} = 102°$

naphthalene, and nitrobenzene guests can be *reversed* by employing the longer pillar **VIII**.[31] That is, G_2VIII inclusion compounds of these guests consist of shifted-ribbon bilayer frameworks rather than brick frameworks. For example, the 1,4-divinylbenzene guests in the $G_2VI\cdot$3(1,4-divinylbenzene) brick inclusion compound are organized as 2D continuous arrays in the pillared galleries whereas the same guests in the $G_2VIII\cdot$1,4-divinylbenzene bilayer form are confined to 1D channels flanked by the azobenzene pillars. The ability to achieve different guest organizations, through architectural control of the host framework, suggests opportunities for directing reaction pathways for reactive guests, in this case the polymerizable 1,4-divinylbenzene.

Though by considering the steric demands of the guest it is often possible to predict the general architecture of **GS** inclusion compounds (*i.e.* bilayer *vs.* brick), the reliable prediction of the more intricate features (*e.g.* channel and unit cell dimensions, sub-architectures, host-guest stoichiometry) remains a challenge. Recently, however, we recognized a near match between the dimensions of the native 2D herringbone motifs observed in some crystalline aromatic hydrocarbons (*ca.* 8 × 6 Å) such as naphthalene, biphenyl and anthracene (Figure 8), and the $a_1 \times b_1$ lattice parameters (*ca.* 7.5 × 12 Å) that can be achieved by the flexible **GS** sheet. This prompted us to examine **GS** inclusion compounds for which the organic portion of the pillar and these guests were

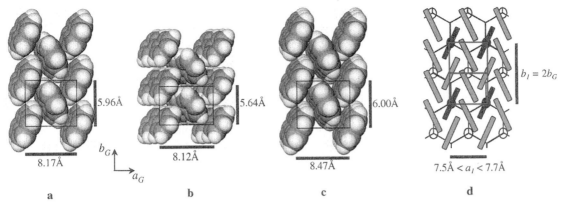

Figure 8. The 2D herringbone packing motif in (**a**) naphthalene, (**b**) biphenyl, and (**c**) anthracene as viewed normal to the *ab* plane. (**d**) Anticipated 1:3 pillar (black)-guest (grey) herringbone packing within the gallery regions of a templated brick architecture, as viewed normal to the **GS** sheets. The lattice match between the **GS** structure and that of the pure guest crystals is depicted by the rectangles in the middle, which represent unit cells of the herringbone packed guests.

identical.[32] We anticipated that the flexibility of the **GS** sheet, particularly along b_1, would permit the formation of inclusion compounds with pillar-guest ensembles that mimicked the herringbone packing motif of the crystalline guest alone.

Simple steric and geometric considerations dictate that herringbone pillar-guest packing is not sterically possible within the bilayer **GS** framework. Consequently, structural mimicry of the herringbone packing motif requires the **GS** host framework to adopt the more flexible brick architecture with 2D channels, as illustrated in Figure 8d. The organodisulfonate pillars of the **GS** host would effectively replace every fourth molecule in the 2D herringbone structure of the pure guest, resulting in inclusion compounds of 1:3 host:guest stoichiometry. Furthermore, the a_1 and b_1 lattice parameters of the anticipated inclusion compounds should be predictable. To achieve ideal herringbone packing, a_1 must match a_G, the *a*-axis of the pure guest. The stiffness of the **GS** ribbons along the ribbon direc-

tion, however, may prevent a_1 from matching a_G exactly. Therefore, a_1 can be expected to fall within the upper limit of possible values, between 7.5 and 7.7 Å. However, variable puckering of the conformationally flexible **GS** sheets would allow b_1 to match $2b_G$, where b_G is the *b*-axis of the pure guest. The value of ϑ_{IR}, reflecting the degree of host puckering required to achieve the native herringbone packing observed in the crystal structures of the respective pure guests be predicted by equation (3) whereby $b_1 = 2b_G$. Note that equation (3) is simply a modification of equation (2) and only requires the value of b_G as input.

$$\vartheta_{IR,calc} = 2\sin^{-1}(2b_G/13.0) \qquad (3)$$

Indeed, the conformational softness of the **GS** brick framework, and its importance to structure prediction based on the native 2D packing of aromatic guests, is evident in the structures of $G_2V\cdot3$(naphthalene), $G_2VI\cdot3$(biphenyl), and $G_2VII\cdot3$(anthracene). These compounds are isostructural, their host

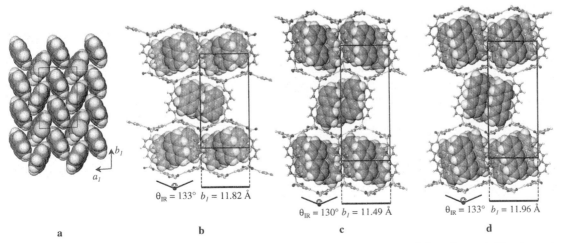

Figure 9. (**a**) Herringbone pillar-guest packing in the *ab* plane of **G₂V·3**naphthalene. **G** ions and sulfonate oxygen atoms have been removed for clarity. (**b – d**) The crystal structures of **G₂V·3**naphthalene, **G₂VI·3**biphenyl, and **G₂VII·3**anthracene as viewed along the *a* axis. The observed inter-ribbon puckering angles (ϑ_{IR}) are provided below each structure.

frameworks adopting the continuous brick architecture. In each compound, the guests and their isostructural pillars pack between the GS sheets in a manner nearly identical to that observed in the crystal structures of the respective pure guests, with the organodisulfonate pillars replacing every fourth molecule in the herringbone motif of the pure guests to afford a 1:3 pillar:guest stoichiometry. The values of b_1, the variable lattice parameter, *are nearly identical* to the $2b_G$ values of the respective guests (Table 1). Additionally, the average dihedral angles (χ) between the mean planes of the arenes (pillars and guests) in the inclusion compounds agrees with the values expected from the native herringbone packing of the pure guests. This reflects the ability of the pillars to rotate about their long axes, which is crucial to achieving the herringbone motif. Using the b_G lattice parameters of the pure naphthalene, biphenyl and anthracene guests, equation (3) affords $\vartheta_{IR,calc}$ values that compare favorably to $\vartheta_{IR,obs}$, measured from the crystal structures of the inclusion compounds. The observed lamellar spacings, which are affected by host puckering, also agree with the values expected from b_G and the length of each pillar.

Epilogue

Three decades of crystal engineering, since its inception at the Wiezmann Institute with the seminal studies of solid state reactions by Schmidt,[13] have provided substantial insights into the molecular recognition events that govern the assembly of molecules into solid state motifs. However, our ability to predict and control crystal packing precisely is still in a rather embryonic stage. Further advances rely on the identification of the key fundamental factors that govern molecular organization and the development of sound engineering principles for the design and synthesis of predictable crystal architectures. The examples described above illustrate some principles which we believe to be universal.

First, the use of higher dimensional robust networks (such as the two-dimensional GS network) simplifies crystal engineering because is reduces crystal design to the last remaining dimension. The use of two-dimensional supramolecular modules, in particular, provides an easily conceptualized mechanism for topological adaptation (in the case of GS networks the arrangement of the pillars).

Table 1. Comparison of structural parameters of GS inclusion compounds and pure guest structures.

Compound	a_1 or a_G	b_1 or $2b_G$	$\vartheta_{IR,obs}$	$\vartheta_{IR,calc}$	χ (avg.)
G₂V·3(naphthalene)	$a_1 = 7.64$ Å	$b_1 = 11.82$ Å	133°	133°	54°
naphthalene	$a_G = 8.17$ Å	$2b_G = 11.92$ Å			52°
G₂VI·3(biphenyl)	$a_1 = 7.66$ Å	$b_1 = 11.49$ Å	130°	120°	62°
biphenyl	$a_G = 8.12$ Å	$2b_G = 11.28$ Å			67°
G₂VII·3(anthracene)	$a_1 = 7.56$ Å	$b_1 = 11.96$ Å	133°	135°	51°
anthracene	$a_G = 8.47$ Å	$2b_G = 12.00$ Å			51°

The agreement of b_1 and χ with the corresponding values in the pure guest structure, and $\vartheta_{IR,obs}$ and the observed lamellar spacings with calculated values based on the pure guest lattice constant b_G, strongly argues that the *major driving force for pillar-guest organization in the inclusion compounds is the achievement of an innate herringbone motif of the pure guests.* More important, these observations illustrate the importance of conformational softness in the host lattice – puckering of the GS sheets and freely rotating pillars allow the formation of pillar – guest ensembles that mimic, *almost exactly,* the molecular packing in the crystal structures of the guests alone. In the absence of such softness the achievement of optimal host-guest packing, and the attendant cohesive energy necessary for assembly and crystallization, would be unlikely. *The conformational softness of the brick framework in these inclusion compounds also enables prediction of specific metric parameters in a series of crystalline materials based on the crystal structures of another, one of the principle goals of crystal engineering.*

Second, the robustness of such networks, and therefore the ensuing crystal architectures, improves with conformational softness, not rigidity. Soft architectures can *adapt* to different steric requirements of substituents appended to the framework, or in the case of inclusion compounds to differently sized guests, while maintaining their overall dimensionality and supramolecular connectivity. This softness compensates for our inability to engineer precisely the optimum molecular packing required for crystallization. In general, inclusion compounds based on rigid frameworks will only achieve the cohesive energy needed for crystallization by very precise size and shape matching between the rigid host and the guest or, alternatively, the inclusion of multiple guest molecules that individually are much smaller than the inclusion cavity. Inefficient packing in such molecular materials is tantamount to the introduction of stress and unless the host framework has an unusually high yield strength, such networks will not be stable and the system may form alternative and unanticipated structures. We note, however, that robust rigid architectures are now emerging.[8]

Third, the accelerating discovery of inclusion compounds reveals the versatility of these materials. This versatility can be attributed to the inherent modular character of inclusion compouds – the host modules providing the structural framework while the guest modules, which are readily interchanged if the host is accommodating, providing function. A simple analogy can be made to a modern skyscraper – its framework (the host) is provided by the floors and ceilings (*e.g.*, the **GS** sheets), supported by walls and a skeleton of metal beams (*e.g.*, the organodisulfonate pillars) while the function is provided by the office furniture, or anthropomorphically, the office workers (the guests). The rapidly growing advances in this area will likely lead to a new generation of designer materials based on supramolecular assembly that will impact numerous technologies, including separations, catalysis, optoelectronics, magnetics, and chemical storage.

References and Notes

1. G. Alberti, R. Vivani, F. Marmottini, P. Zappelli, P., "Microporous solids based on pillared metal(IV) phosphates and phosphonates", *J. Porous Mater.* **1998**, 5, 205–220.
2. A. Clearfield, "Organically pillared micro- and mesoporous materials", *Chem. Mater.* **1998**, 10, 2801–2810.
3. "Molecular inclusion and molecular recognition. Clathrates I & II", *Topics in Current Chemistry, Vol. 140* and *149* (Ed. E. Weber), Springer-Verlag, New York, **1987** and **1988**.
4. O. Ermer, L. Lindenberg, Lorenz, "Double-diamond inclusion compounds of 2,6-dimethylideneadamantane-1,3,5,7-tetracarboxylic acid", *Helv. Chim. Acta* **1991**, 74, 825–8277.
5. P. Brunet, M. Simard, J. D. Wuest, "Molecular tectonics. Porous hydrogen-bonded networks with unprecedented structural integrity", *J. Am. Chem. Soc.* **1997**, 119, 2737–2738.
6. K. Endo, T. Koike, T. Sawaki, O. Hayashida, H. Masuda, Y. Aoyama, "Catalysis by organic solids. Stereoselective Diels-Alder reactions promoted by microporous molecular crystals having an extensive hydrogen-bonded network", *J. Am. Chem. Soc.* **1997**, 119, 4117–4122.
7. S. Subramanian, M. J. Zaworotko, "Porous solids by design: [Zn(4,4'-bpy)₂(SiF₆)]ₙ·xDMF, a single framework octahedral coordination polymer with large square channels", *Angew. Chem., Int. Ed. Engl.* **1995**, 34, 2127–2129.
8. H. Li, M. Eddaoudi, M. O'Keeffe, O. M. Yaghi, "Design and synthesis of an exceptionally stable and highly porous metal-organic framework", *Nature* **1999**, 402, 276–279.
9. S. S.-Y. Chui, S. M.-F. Lo, J. P. H. Charmant, A. G. Orpen, I. D. Williams, "A chemically functionalizable nanoporous material [Cu₃(TMA)₂(H₂O)₃]ₙ." *Science*, **1999**, 283, 1148–1150.
10. P. J. Hagrman, D. Hagrman, J. Zubieta, "Organic-inorganic hybrid materials: from "simple" coordination polymers to organodiamine-templated molybdenum oxides" *Angew. Chem., Int. Ed.* **1999**, 38, 2639–2684, and references therein.
11. "Structural aspects of inclusion compounds formed by organic host lattices", *Inclusion Compounds, Vol. 2*, (Eds. J. L. Atwood, J. E. D. Davies, D. D. MacNicol), Academic, London, **1984**.
12. "Solid-State Supramolecular Chemistry: Crystal Engineering", *Comprehensive Supramolecular Chemistry, Vol. 6* (Eds.: J. L. Atwood, J. E. D. Davies, D. D. MacNicol, F. Vögtle, K. S. Suslick), Pergamon, Oxford, **1996**.
13. G. M. J. Schmidt, "Photodimerization in the solid state", *Pure Appl. Chem.* **1971**, 27, 647–678.
14. G. R. Desiraju, *Crystal engineering: The design of organic solids*, Elsevier, New York, **1989**.
15. V. A. Russell, M. C. Etter, M. D. Ward, "Layered materials by molecular design: structural enforcement by hydrogen bonding in guanidinium alkane- and arenesulfonates", *J. Am. Chem. Soc.* **1994**, 116, 1941–1952.
16. V. A. Russell, M. C. Etter, M. D. Ward, "Guanidinium *para*-substituted benzenesulfonates: competitive hydrogen bonding in layered structures and the design of nonlinear optical materials", *Chem. Mater.* **1994**, 6, 1206–1217.
17. V. A. Russell, M. D. Ward, "Two-dimensional hydrogen-bonded assemblies: the influence of sterics and competitive hydrogen bonding on the structures of guanidinium arenesulfonate networks", *J. Mater. Chem.* **1997**, 7, 1123–1133.
18. To date, we have only observed one exception to these topologies. [G][1-butanesulfonate] exhibits a structure resembling the "continuously interdigitated layer", but with organic substituents alternating their up/down arrangement within the **GS** ribbons, as well as between adjacent ribbons of the **GS** sheet.
19. C. Cao, H.-G. Hong, T. E. Mallouk, "Layered metal phosphates and phosphonates: from crystals to monolayers", *Acc. Chem. Res.* **1992**, 26, 420–427.
20. S. Batten, R. Robson, "Interpenetrating nets: ordered, periodic entanglement", *Angew. Chem., Int. Ed.* **1998**, 37, 1461–1494.
21. The values of ϑ_{IR} are determined from the angle subtended by the centroid of two sulfur atoms on a selected **GS** ribbon and the nearest sulfur atom on the two adjacent ribbons.
22. V. A. Russell, C. C., Evans, W. Li, M. D. Ward, "Nanoporous molecular sandwiches: pillared two-dimensional hydrogen bonded networks with adjustable porosity", *Science* **1997**, 276, 575–579.
23. J. A. Swift, A. M. Reynolds, M. D. Ward, "Cooperative host-guest recognition in crystalline clathrates: steric guest ordering by molecular gears", *Chem. Mater.* **1998**, 10, 4159–4168.
24. The volume occupied by the host framework was calculated by subtracting the "available volume", after removal of the guest molecules, from the volume of the unit cell. "Available volumes" were calculated with Molecular Simulations Inc. Cerius2 (v. 3.5) software using a probe radius of 0.5Å and "fine" grid spacing.
25. M. D. Hollingsworth, U. Werber-Zwanziger, M. E. Brown, J. D. Chaney, J. C. Huffman, K. D. M. Harris, S. P. Smart, "Spring-loaded at the molecular level: relaxation of guest-induced strain in channel inclusion compounds", *J. Am. Chem. Soc.* **1999**, 121, 9732–9733.

26. A. M. Pivovar, M. D. Ward, unpublished results.
27. J. A. Swift, A. M. Pivovar, A. M. Reynolds, M. D. Ward, "Template-directed architectural isomerism of open molecular frameworks: engineering of crystalline clathrates", *J. Am. Chem. Soc.* **1998**, *120*, 5887–5894.
28. M. E. Davis, A. Katz, W. R. Ahmad, "Rational catalyst design via imprinted nanostructured materials", *Chem. Mater.* **1996**, *8*, 1820–1839.
29. Q. Huo, D. I. Margolese, G. D. Stucky, "Surfactant control of phases in the synthesis of mesoporous silica-based materials", *Chem. Mater.* **1996**, *8*, 1147–1160.
30. J. A. Swift, M. D. Ward, "Cooperative Polar Ordering of Acentric Guest Molecules in Topologically Controlled Host Frameworks", *Chem. Mater.* **2000**, *12*, 1501–1504.
31. C. C. Evans, L. Sukarto, M. D. Ward, "Sterically controlled architectural reversion in H-bonded crystalline clathrates", *J. Am. Chem. Soc.* **1998**, *121*, 320–325.
32. K. T. Holman, M. D. Ward, "Metric Engineering of Crystalline Inclusion Compounds by Structural Mimicry", *Angew. Chem. Int. Ed.*, **2000**, *39*, 1653–1656.

III Molecular Devices and Material Properties

Molecular Wires and Devices

James M. Tour

Rice University, Department
of Chemistry and Center for
Nanoscale Science and
Technology, MS 222,
6100 Main Street, Houston,
Texas 77005,

Phone: +713-737-6246,
Fax: +713-737-6250,
e-mail: tour@rice.edu,
Web: www.jmtour.com.

Keywords ■ *Molecular Wires* ■ *Molecular Devices*
■ *Alligator Clips* ■ *Conjugated Materials*

Concept: Standard electronic componentry is scaled to the nanometer level by using individual or small packets of molecules in lieu of traditional solid-state wires and devices.

Microelectronic Componentry

Nanoscale Molecular Electronic Analogues

Scale

Tunnel Diode

Molecular Tunnel Diode

Abstract: Described are the synthetic routes to precisely defined molecular wires which are of discrete length and constitution. They are fully conjugated systems and are expected to have nearly linear current-voltage response curves. Their ends are functionalized with molecular alligator clips, based on chalconides and isonitriles, for adhesion between proximal probes. Both solution and solid-phase approaches have been used to prepare these molecular wires that are based on oligo(thiophene ethynylene)s and oligo(phenylene ethynylene)s. Molecular device syntheses are also described that would be expected to have nonlinear current-voltage responses. In some cases, they have more than two termini. The testing of the wires or two-terminal device packets is described using proximal probes that have been litho-graphically patterned or constructed in nanopore configurations. Single molecules have been addressed using either scanning probe microscopy or mechanically controllable break junctions. While the testing of two terminal devices was carried out experimentally in the nanopore, we have only been able to evaluate the multi-terminal systems using computational models.

Prologue

The term "molecular wire" has been used in several contexts in the recent literature. Just like the term "nano" is being used to describe a plethora of systems that were traditionally embodied within the context of chemistry, the term "molecular wire" is being applied to any molecular system that may have traditionally been characterized by a conducting polymer, or more generally, a conjugated molecule. For the purposes of the discussions here, we will reserve the use of the term "molecular wires" for those systems that (1) are fully conjugated, (2) are of precise length, (3) are of precise constitution, and (4) bear one or two terminal functionalities, namely molecular alligator clips, for adhesion to microscopic metallic leads. Similarly, "molecular devices" could be two- or more termini systems with electrical current/voltage responses that would be expected to be nonlinear due to intermediate barriers or heterofunctionalities in their molecular framework. The molecular devices are also of precise length and constitution and they bear molecular alligator clips at, at least, one terminus. These molecular systems possess numerous interesting physical properties; [1] however, here we will focus exclusively on their electronic properties. Finally, although carbon nanotubes, single-walled or multi-walled, will undoubtedly be useful for electronic conduction, they are not being considered here because they have yet to be prepared with precise lengths. Additionally, the separation between the carbon nanotubes' metallic, semiconducting and chiral species remains unrealized, and electrical contacts made to them have, up to this point, only been formed by contacting micron-sized regions along their sidewalls rather than via discrete nanometer-sized terminal functionalities.[2]

Why even consider molecular wires and devices? The semiconductor industry has been surviving quite well without the use of such entities. The big question is: Can they continue to make these stellar achievements in size reduction with concomitant increases in computational performance? Due to the inevitable roadblocks ensuing through traditional downsizing of silicon devices, alternate technologies must be considered. Contrary to the view of some, this is not simply a synthetic chemist with a hidden agenda speaking. The limits that will be realized in the next 5–10 years have recently been outlined by a solid-state silicon industry insider, Paul Packan of Intel Corporation.[3] The roadblocks arise from both fundamental physical constraints and monetary hills that are too steep to climb. With regard to the physical limitations of solid-state size reduction, issues are raised such as: (1) oxide barriers at the 3-atom thick level are inadequately insulating resulting in charge leakage, and (2) silicon no longer possesses its fundamental band structure when it is restricted to very small sizes. Molecules, on the other hand, have large energy level separations at room temperature and at the nanometer scale level due to their discrete orbital structures, which are independent of broad band properties. From the monetary standpoint, a current fabrication line costs 2.5 billion US dollars to construct, and that cost is projected to rise to 15 billion US dollars by the year 2010; a sum that will likely be beyond the reach of even large industrial consortia. Molecular systems, on the contrary, can easily be constructed on the nanometer size scale at costs that are relatively insignificant. And considering devices as individual entities, in a state-of-the-art desktop computer such as a Pentium-based system for scientific work or large-scale graphics manipulations, there are approximately 10^7–10^8 transistors on the processor chip plus 10^9 transistors in the electronic memory. Sound like a lot? Let us compare these numbers to numbers of molecules: there are far more water molecules in one drop of water than there are transistors in all the world's computers combined. Period! Hence, if we can simply tap into the enormous numbers of individual entities afforded us by molecular synthesis, it could be one of the methods used to tackle the electronics needs of the new millennium. To be fair to present electronic systems, each one of those 10^9 transistors in the Pentium is addressable and is connected to a power supply. Nonetheless, the richness of chemistry to produce exorbitant numbers of entities with precise control of constitution, regiochemistry, and stereochemistry, with a vast number of functional groups, provides us with a tremendous impetus to explore chemical routes to molecular wires and devices. Are all the details of the path outlined? Of course not. But for each obstacle, there is a light at the end of the tunnel. Therefore, we embark.

Synthesis of Molecular Wires

Solution-Phase Syntheses of Molecular Wires

There has been considerable recent effort to prepare large conjugated molecules of precise length and constitution.[1] Our initial approach to these compounds maintained several key features that made it well suited for the requisite large molecular architectures for molecular scale electronics studies.[4] Specifically, the route involved: (1) a rapid construction method that permitted a doubling of the molecular length at each coupling stage to afford unbranched 100+ Å oligomers, the approximate size of present nanopatterned probe gaps, (2) an iterative approach so that the same high yielding reactions can be used throughout the sequence, (3) the syntheses of conjugated compounds that are semiconducting in the bulk, (4) products that are stable to light and air so that subsequent engineering manipulations will not be impeded, (5) products that would permit facile and independent functionalization of the ends to serve as molecular alligator clips, which would allow for surface contacts to be made to metal probes, (6) products that are rigid in their frameworks so as to minimize conformational flexibility, yet containing substituents for maintaining solubility and processability, (7) alkynyl units (cylindrically symmetric) separating the aryl units so that ground state contiguous π-overlap will be minimally affected by rotational variations, and (8) molecular systems that do not have degenerate ground state resonance forms and are thus not subject to Peierls distortions.[4,5]

The iterative divergent/convergent approach is outlined in Figure 1.[4] A batch of monomer material M, with inactive end groups X and Y, is divided into two portions. In one portion, the end group X is activated by conversion to X'. In the second portion, Y is activated by conversion to Y'. The two portions are then brought back together to form the dimer XMMY with loss of X'Y'. Since the same end groups that were present in the monomer are now present in the dimer, the procedure can be repeated with a doubling of molecular length at each iteration. The advantages of this approach are that the molecular length grows rapidly, at a rate of 2^n where n = the number of iterations, and incomplete

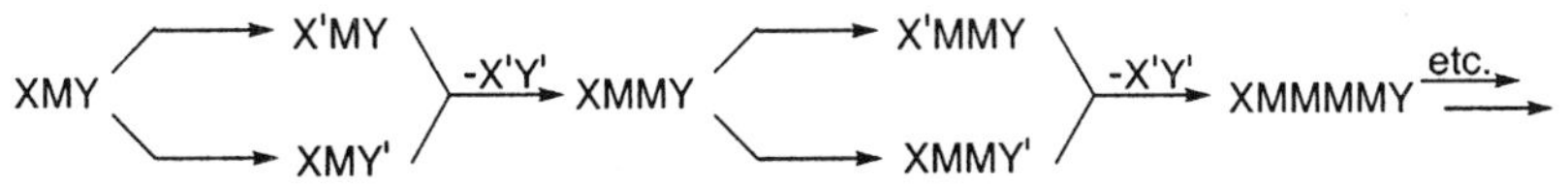

Figure 1. Schematic outline of the iterative divergent/convergent approach to molecular length doubling.

reactions yield unreacted material that is half the size of the desired compound. Thus, purification at each step is far simpler since separation involves, for example, an octamer from a 16-mer. This iterative divergent/convergent approach is therefore a particularly attractive one.

The specific iterative divergent/convergent synthetic approach is outlined in Figure 2.[4] The sequence involves partitioning the starting monomer into two portions; iodinating the 5-position in one of the portions and desilylating the alkynyl end in the other portion. Bringing the two portions

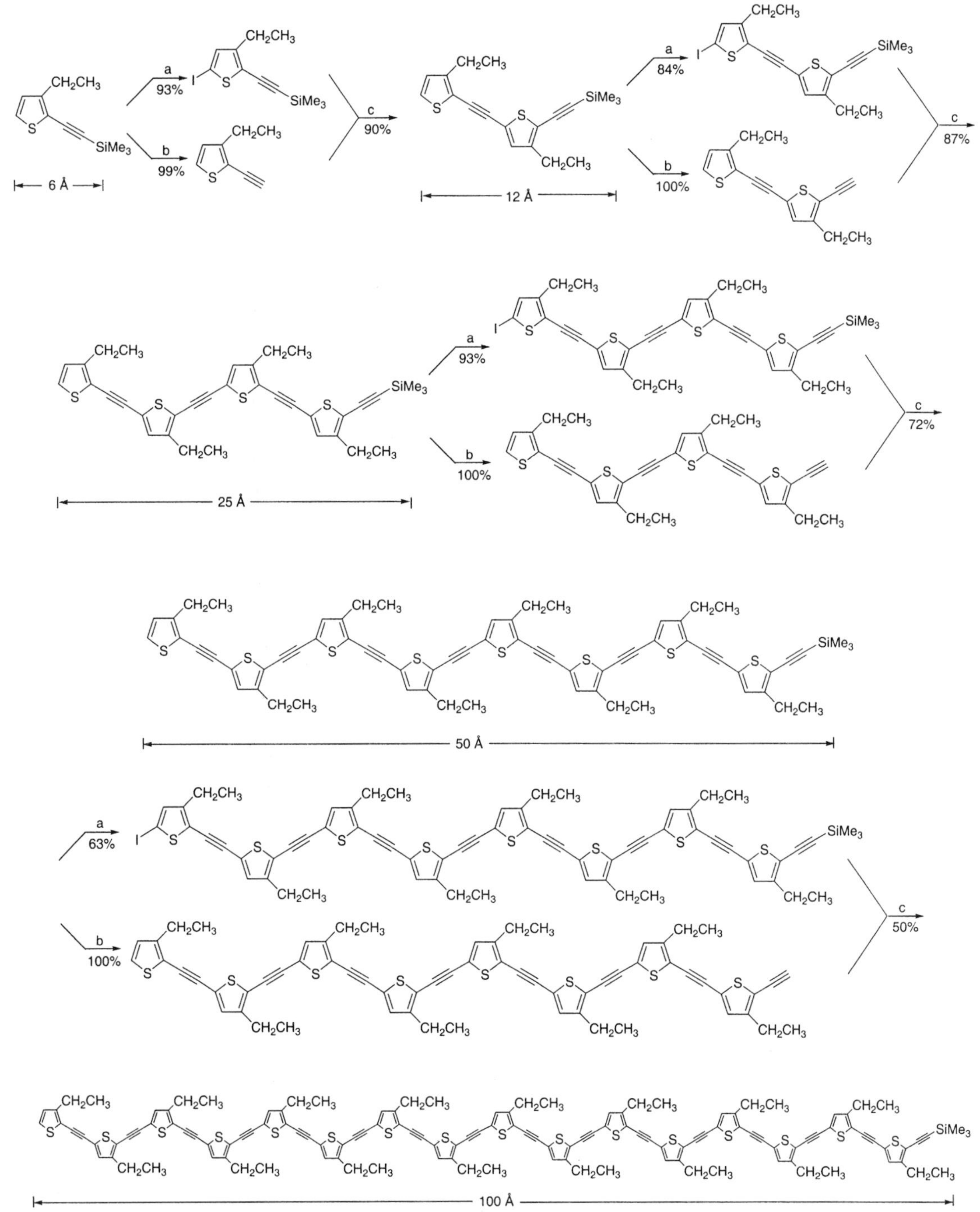

Reagents: (a) LDA, Et$_2$O, −78° to 0°C then I$_2$, −78°. (b) K$_2$CO$_3$, MeOH, 23°C. (c) Cl$_2$Pd(PPh$_3$)$_2$ (2 mol %), CuI (1.5 mol %), THF, *i*-Pr$_2$NH, 23°C.

Figure 2. Solution phase synthesis of the oligo(thiophene ethynylene)s by the divergent/convergent doubling approach.

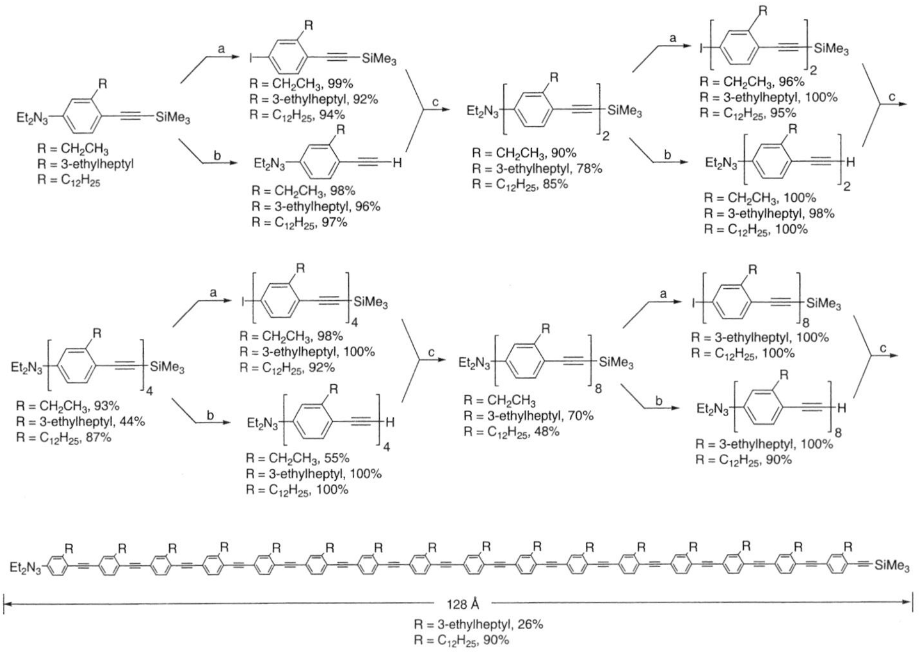

Reagents: a. MeI as solvent, 120°C in a screw cap tube. b. K$_2$CO$_3$, MeOH, 23°C or *n*-Bu$_4$NF, THF, 23°C c. Pd(dba)$_2$ (5 mol %), CuI (10 mol %), PPh$_3$ (20 mol %), *i*-Pr$_2$NH/THF (1:5), 23°C.

Figure 3. Solution phase synthesis of the oligo(phenylene ethynylene)s by the divergent/convergent doubling approach.

back together in the presence of a soluble Pd/Cu catalyst mixture couples the aryl iodide to the terminal alkyne, thus generating the dimer. Iteration of this reaction sequence doubles the length of the dimer to afford the tetramer, and so on to the octamer, and finally the 16-mer. Although the silylated alkynes showed good oxidative stability, upon protodesilylation, the tetramer and octamer became air sensitive and immediate work-up and further coupling was necessary to minimize oxidative decomposition of these terminal alkyne intermediates.

A similar molecular doubling approach has been used for preparing oligo(phenylene ethynylene)s (Figure 3).[5] We were able to achieve the synthesis of the 16-mer. More recently, we developed an approach in which the molecular systems are extended from both ends after each iteration (Figure 4).[6] This approach proved to be a very simple route for the rapid construction of these oligomers in solution using an in situ deprotection and coupling so that the oxidatively unstable α,ω-bis(terminal alkyne)s need not be isolated.

Further, attachment of alligator clips to one or both sides of related molecules has been demonstrated (Figure 5).[4,5] We have utilized several types of molecular alligator clips in a quest to minimize

the contact resistance with metallic probes.[7] These include sulfur, selenium, tellurium, and isonitrile end groups. We have studied these clips both experimentally and theoretically.[8–15] There remains a need to develop molecular alligator clips that minimize the impedance mismatches between molecular structures and metal surfaces, thereby affording a better energy match between the lowest unoccupied molecular orbital of the molecule and the Fermi level of the metallic contact.

Solid-Supported Syntheses of Molecular Wires

We have extended these synthetic molecular doubling approaches to solid phase methods in an effort to streamline the syntheses and make them suitable for automation. For example, the hexadecyl(phenylene ethynylene) has been prepared starting from Merrifield's resin (Figure 6).[5] More recently, we used the solid-support methodology to grow the oligomers from both ends, a process that can not be carried out in solution due to polymerization of the compounds (Figure 7). This methodology gave a route to molecular length tripling at

Figure 4. (Phenylene ethynylene)-based conjugated oligomers prepared in solution using a bi-direction growth approach with in situ deprotection and coupling.

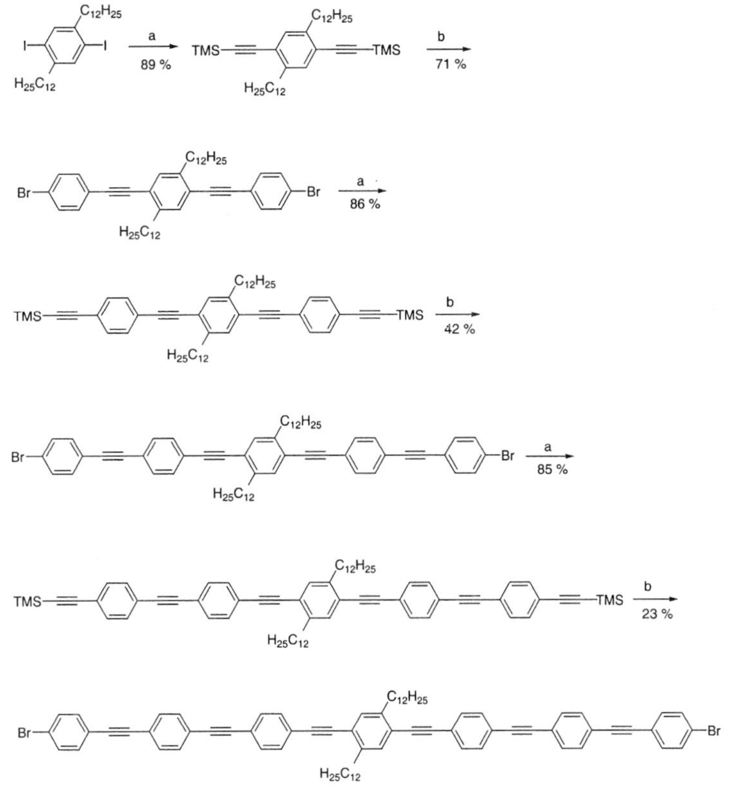

Conditions: a. Trimethylsilylacetylene, Pd(PPh₃)₂Cl₂, CuI, PPh₃, Et₃N, THF. b. 1-Bromo-4-iodobenzene, Pd(dba)₂, CuI, PPh₃, K₂CO₃, MeOH, THF.

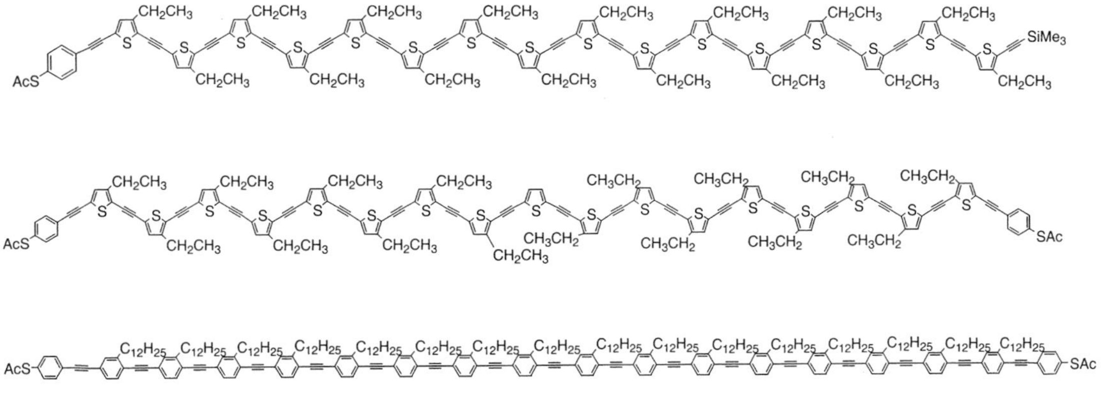

Figure 5. Precise molecular wires bearing acetyl-protected alligator clips (SAc) at one and two ends.

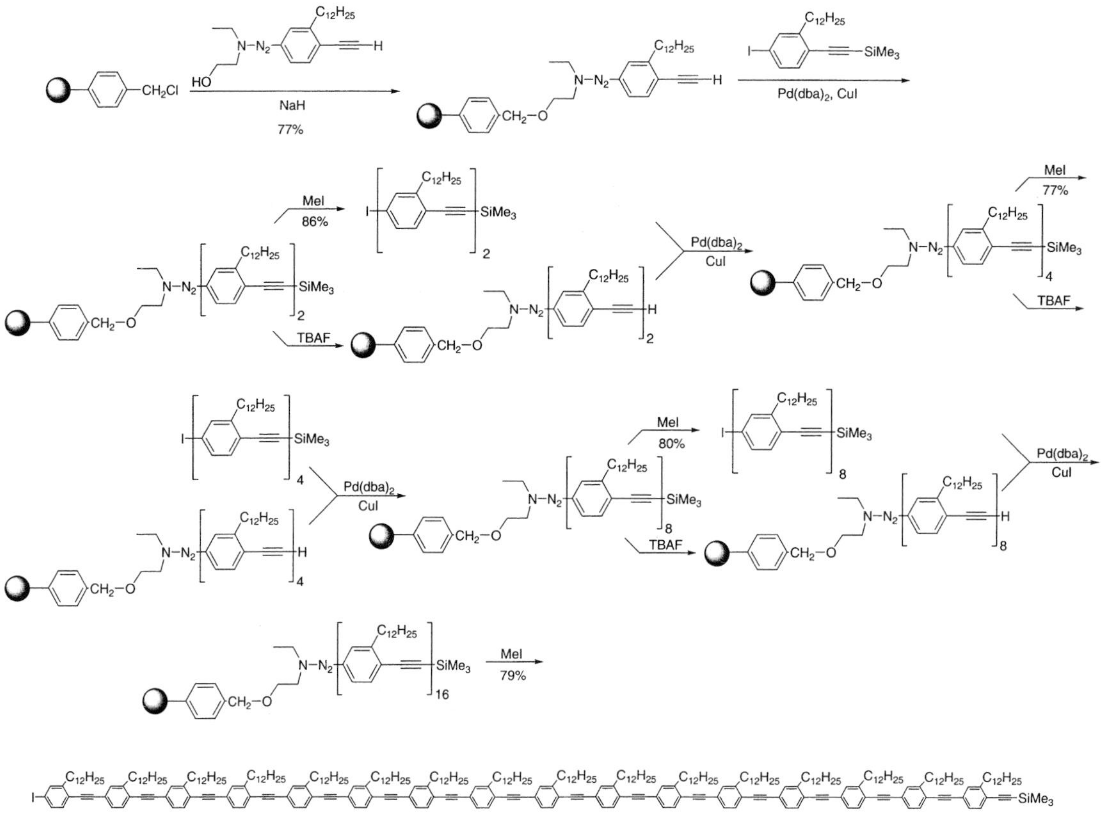

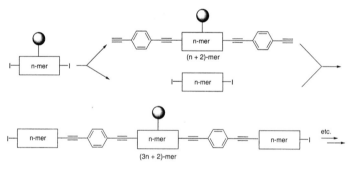

Figure 6. Iterative divergent/convergent molecular doubling growth on a polymer support.

Figure 7. General scheme for the iterative divergent/convergent molecular tripling growth approach on a polymer support.

each iteration and the homooligomers proved an easy target for this process (Figure 8).[16] Additionally, the alternating block co-oligomers could be prepared rapidly by this method (Figure 9).[17] This route

could be particularly attractive for preparing devices based upon linear systems since all device structures must have some element of heterogeneity for computational use.

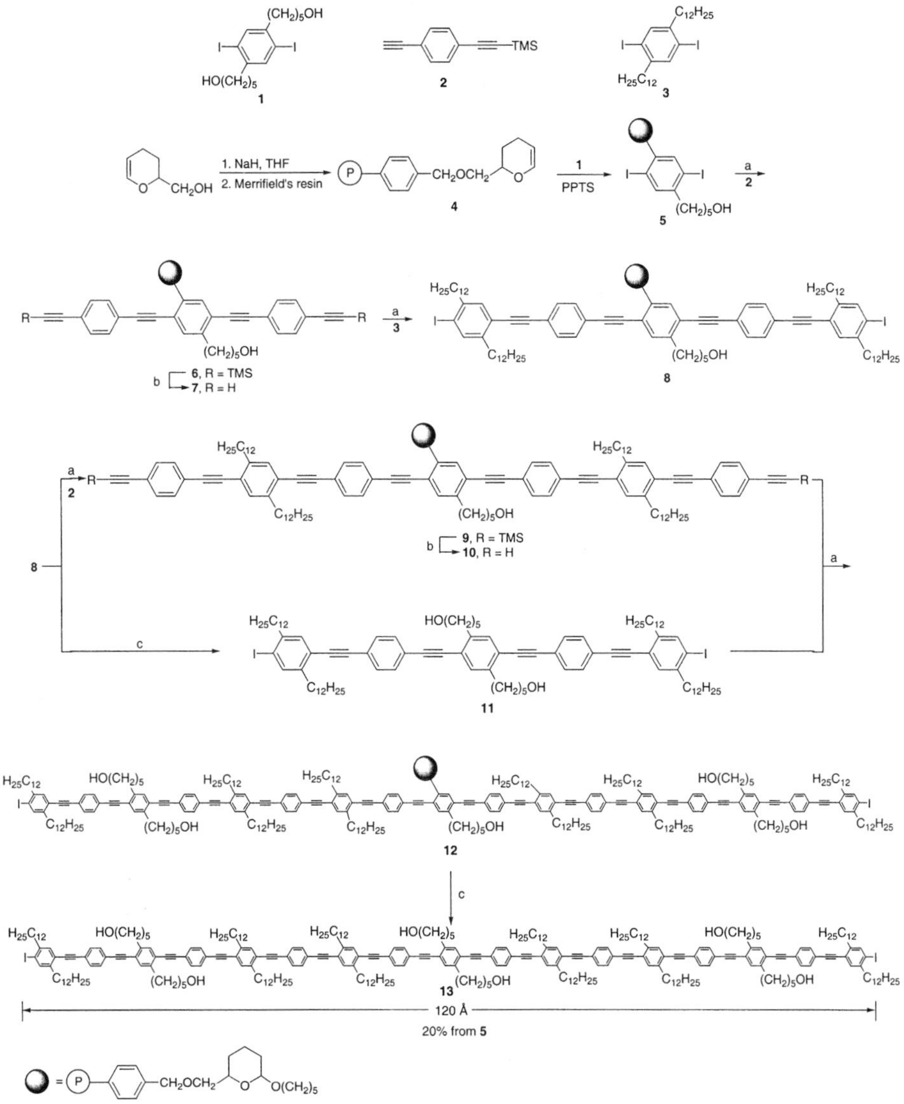

Reagents: a. Pd(dba)₂, PPh₃, and CuI (5 mol %, 10 mol %, and 5 mol %, respectively, per iodide atom), Et₂NH/THF (1:4). b. TBAF, THF. c. PPTS, *n*-C₄H₉OH, ClCH₂CH₂Cl

Figure 8. Tripling iterative divergent/convergent molecular growth approach on a polymer support to prepare homo-oligomers.

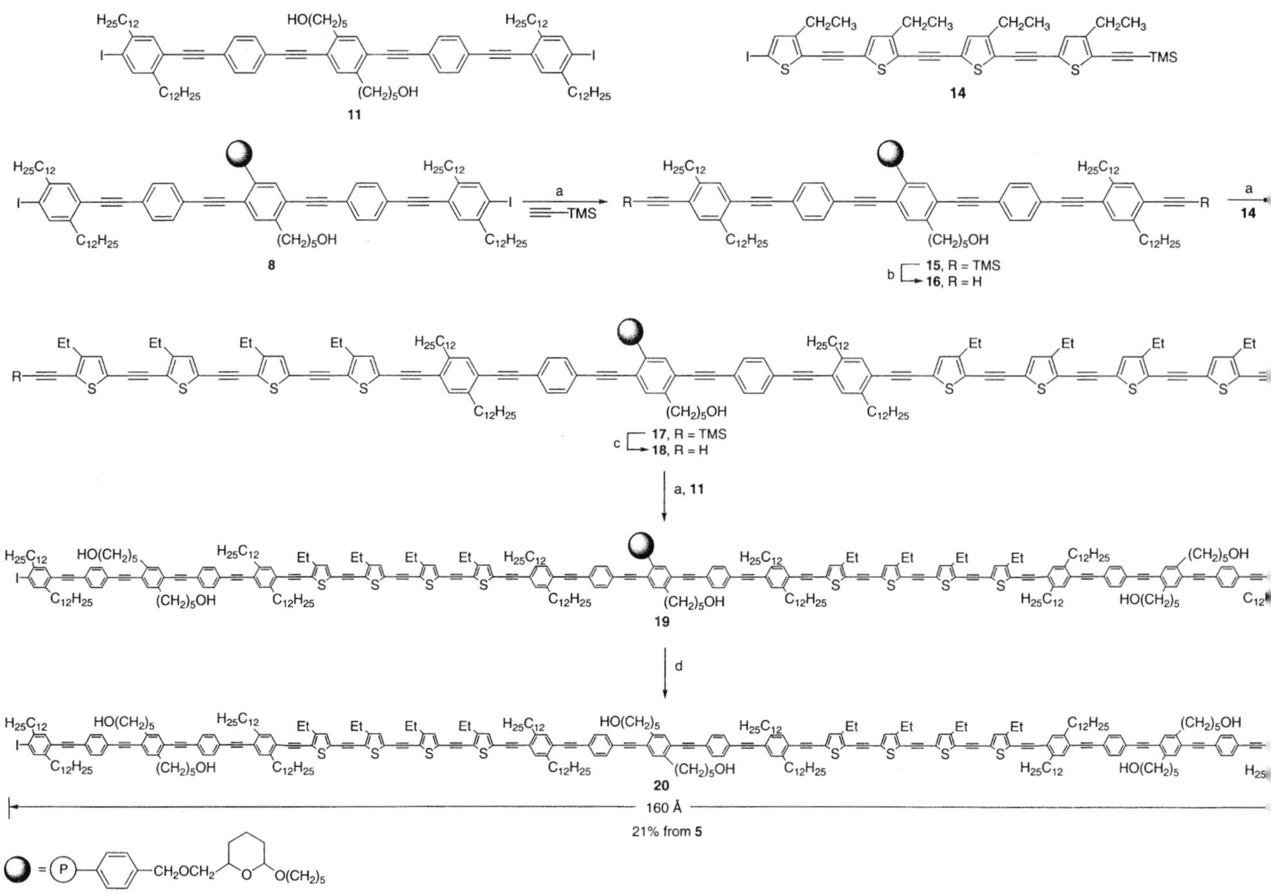

Figure 9. Tripling iterative divergent/convergent molecular growth approach on a polymer support to prepare [AB] block-co-oligomers.

Synthesis of Molecular Devices

Two-Terminal Molecular Devices

The molecular devices described here are of precise length and constitution and they bear molecular alligator clips on, at least, one terminus. Molecular devices could be two- or more termini systems with current-voltage responses that would be expected to be nonlinear due to intermediate barriers or heterofunctionalities in the molecular framework. For example, we began our molecular device syntheses by inserting tunnel-barrier methylene fragments into the conjugated systems as shown in Figure 10.[18] Notice that both one- and two-barrier systems have been prepared. A number of porphyrin-based molecular devices have also been prepared (Figure 11).[19] In addition, we have recently synthesized molecular systems that have heterofunctionality along the backbone of the molecule (Figure 12).[20] These systems have proven particularly useful for generating non-linear device properties from conjugated molecules.

Three- and Four-Terminal Molecular Devices

Although three-terminal systems have, to date, been unadressable as single molecules, several such systems have been prepared. Using Pd/Cu coupling methodologies, we have prepared molecular analogues of simple junctions, switches and logic gates (Figure 13).[18] Notice that in the two device cases, at least one tunnel barrier is needed for device functionality. Additionally, we have prepared orthogonally fused systems that have also been targeted for switch-like applications (Figure 14).[21] However, we have not yet affixed molecular alligator clips to these structures as these compounds will require a six-probe testing array – a structure that is unlikely to be constructed in the near future.

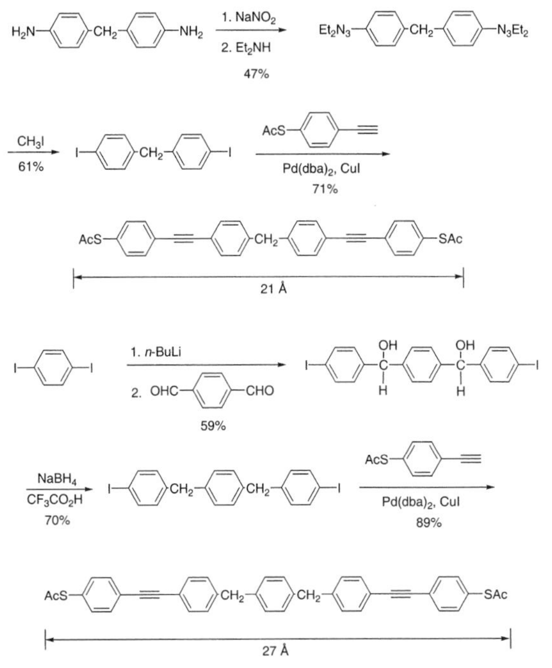

Figure 10. Two-ended molecular devices with methylene tunnel barriers.

Testing of Molecular Wires

Since any potential molecular wire must bridge two electrodes, the question arises as to how these rigid rod difunctional oligomers will order on metallic surfaces. For example, will the oligomers bridge the gold – gold gap as in Figure 15a, or will they reside nearly parallel to the surface of the gold by either dithiol or aromatic adsorption to the gold surfaces as in Figures 15b and 15c, respectively?[1] By making self-assembled monolayers (SAMs) on gold surfaces, we have demonstrated using ellipsometry, XPS, and grazing angle IR measurements, that the rigid rod systems stand nearly perpendicular to the surface; the thiol groups dominating the adsorption

sites on the gold. Even when the oligomers were α,ω-dithiol-substituted, the rigid molecules tended to stand on end as judged by the ellipsometric thickness of the adsorbate layer.[14,22–24] This trend holds true for molecules that are up to approximately 50 Å; however, beyond that length, it becomes difficult to obtain densely packed, well-ordered monolayers. Note that we utilized thioacetate end groups since they could be selectively deprotected in THF, to the free thiol, using ammonium hydroxide during the deposition process. Alkali metal salts were avoided since they tend to disrupt electronic measurements. Use of the free thiols, rather than the thioacetates, proved to be problematic since the free aromatic thiols are prone

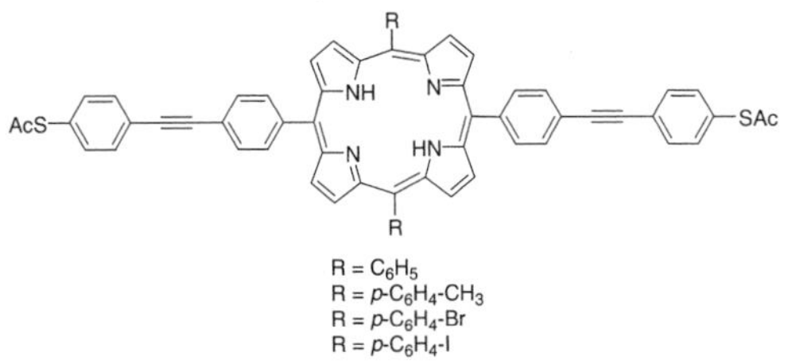

Figure 11. Porphyrin-based molecular devices with alligator clips.

R = C6H5
R = p-C6H4-CH3
R = p-C6H4-Br
R = p-C6H4-I

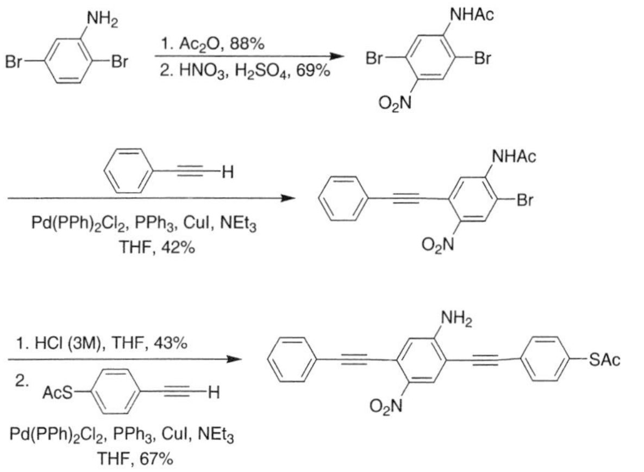

Figure 12. Synthetic route to a molecular device that bears functionality for electron capture. This nitroaniline exhibits negative differential resistance.

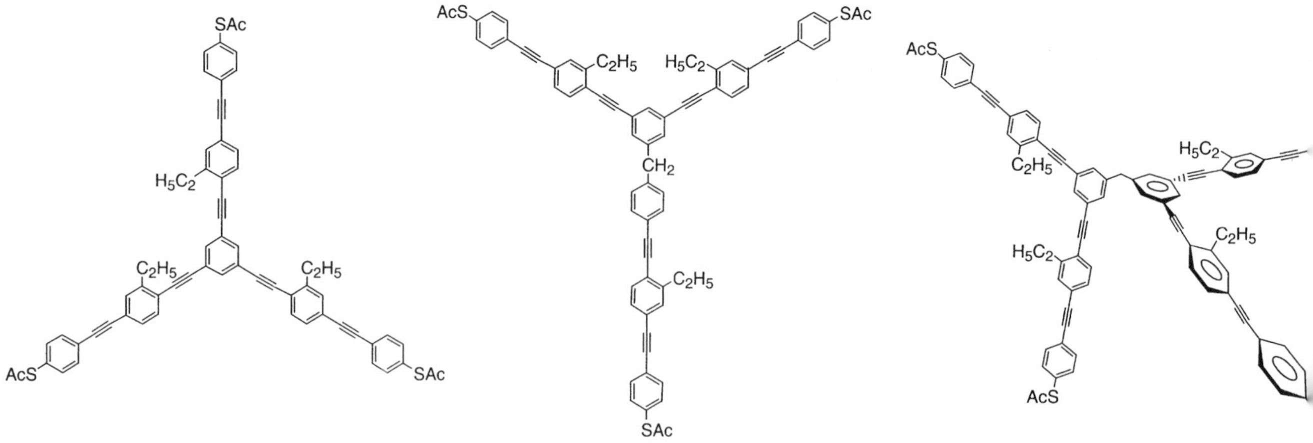

Figure 13. A three-terminal molecular junction (left), molecular switch (center) and four-terminal molecular logic gate (right).

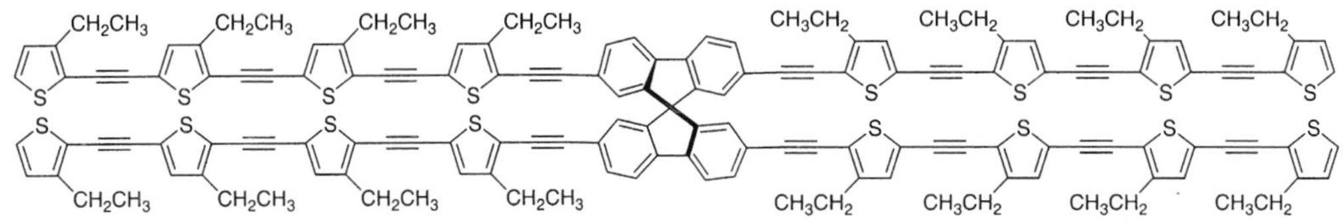

Figure 14. An orthogonally-fused system that could act as a molecular switching device.

to very rapid oxidative disulfide bond formation. Although disulfides can self-assemble on gold, the assembly is ~1000 times slower than with the thiols. Moreover, when using the α,ω-dithiols, oxidative polymerization ensues which rapidly results in insoluble material. Hence, in situ deprotection proved to be quite effective.

Our initial efforts were directed toward straddling longer molecular wires across lithographically patterned proximal gold-coated probes that were separated by approximately 100 Å. Although we tried many experiments over a period of two years, we were never able to record conductances

across these gap regions. The reasons for the failure could have been the inability to bridge these patterned gaps with molecular systems of this length, or that the molecules did assemble across those gaps, but they were too resistive over such length scales.[24] Efforts to make smaller lithographically patterned gaps were hampered by technical patterning limitations and by our inability to perform scanning probe microscopy within the crevice formed by the proximal probe gap. Hence, definitive gap separation determinations were not achievable.

Our long-time collaborator, Mark Reed, developed a technique for determining conductances of

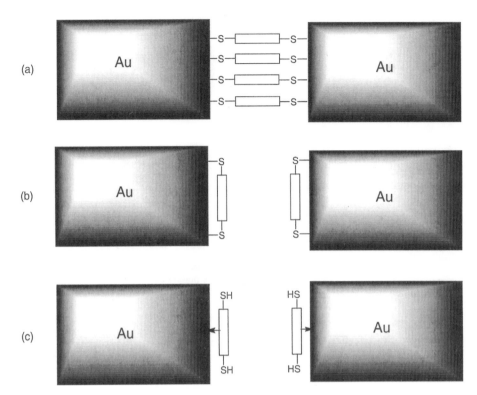

Figure 15. Modes for molecular wire assembly (**a**) as desired between proximal gold probes, (**b**) adsorbed via the thiols without bridging the probes, or (**c**) having the adsorption sites dominated by the aromatic units.

our molecular systems using a nanopore arrangement. In this embodiment, electronic measurements are performed in a nanostructure that has a metal top contact, a SAM active region, and a metal bottom contact.[25] The essential feature of the fabrication process is the use of a nanoscale device, with a diameter of 30 nm, that gives rise to a small number of self-assembled molecules (≈1000) sandwiched between two metal contacts. Using this small area, we eliminated pinhole and other defect mechanisms that hamper through-monolayer elec-

tronic transport measurements. Excellent control over the device area was achieved while also maintaining good intrinsic contact stability. Using this procedure, current-voltage characteristics, I(V) curves, could be recorded on a series of molecular wire systems[25] as well as on two-terminal devices (vide infra). Additionally, the contact barrier between molecule and the metallic probes, via the alligator clips, could be determined.[25-27] Recently Metzger[28] and the team of Heath, Stoddart and Williams,[29] were able to use micron-sized regions of

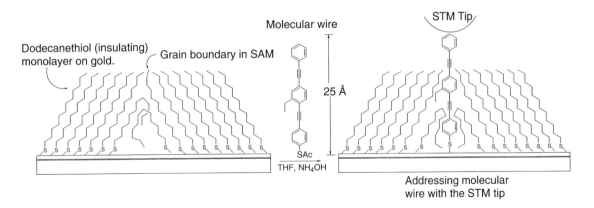

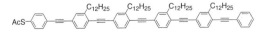

Figure 16. Protocol for inserting molecular wires into dodecanethiolate SAMs at grain boundaries. Relative conductance recording was done with a STM tip. The molecule at the bottom has also been used in this study.

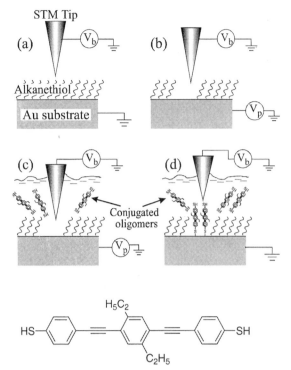

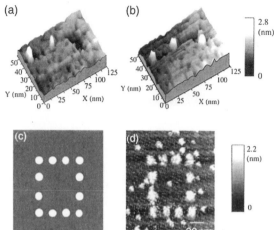

Figure 18. (**a**) A dodecane thiol SAM surface after three consecutive voltage pulsing events. The first two pulsed locations have molecular wires inserted while the third location remains to be filled. (**b**) The image taken a few minutes later shows that wire insertion at the third pulse location is now complete. (**c**) A programmed rectangular pattern for controlled voltage pulses. (**d**) The image of the patterned SAM after pulsing and molecular wire insertion. Some random insertions at grain boundaries or other defect sites are also evident.

Figure 17. Schematic of the lithographic patterning and replacement of conjugated molecules in an alkanethiol matrix. (**a**) Normal STM imaging of an alkanethiolate SAM with tip bias V_b. (**b**) SAM removal by applying a voltage pulse V_p to the substrate. (**c**) Carrying out the same voltage pulse as in (**b**), but under a solution of molecular wires (expanded structure at bottom) causes (**d**) insertion of the wires into the newly vacated site.

Langmuir-Blodgett (LB) films for evaporation of top contacts without forming shorts due to defects in the LB structure.

In collaboration with Paul Weiss and David Allara, we were able to address single molecular wires that had been inserted at grain boundaries within a self-assembled monolayer of dodecane-thiolate on gold (Figure 16). Using scanning tunneling microscopy (STM), the molecules could be individually imaged. Qualitative results of the conductance levels showed that the molecular wires, although topographically higher above the gold surface, were more highly conducting than the sur-

rounding alkanethiolate structures.[23,30] While such a result is intuitive, it had never before been demonstrated in a single conjugated molecule that was projected on end and isolated from all its neighbors. We further developed methods to insert the molecular wires at controlled locations rather than at random locations along grain boundaries. By applying controlled voltage pulses to an alkanethiolate SAM under a solution of molecular wires (Figure 17), we were able to achieve precise placements of molecular wire bundles (< 10 molecules/bundle) at preprogrammed positions (Figure 18).[31]

In order to quantify the amount of current that could be passed through a single molecule, we again worked with Reed's group who developed a mechanically controllable break junction.[32] Using this device, two gold tips could be generated and moved in picometer increments with respect to each other by use of a piezo element (Figure 19).[26,33]

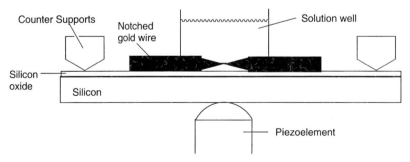

Figure 19. A schematic of the mechanically controllable break junction showing the bending beam formed from a silicon wafer, the counter supports, the notched gold wire which is glued to the surface, the pizeo element for control-

ling the tip-to-tip distance through bending of the silicon platform, and the glass capillary tube containing the solution of molecular wires.

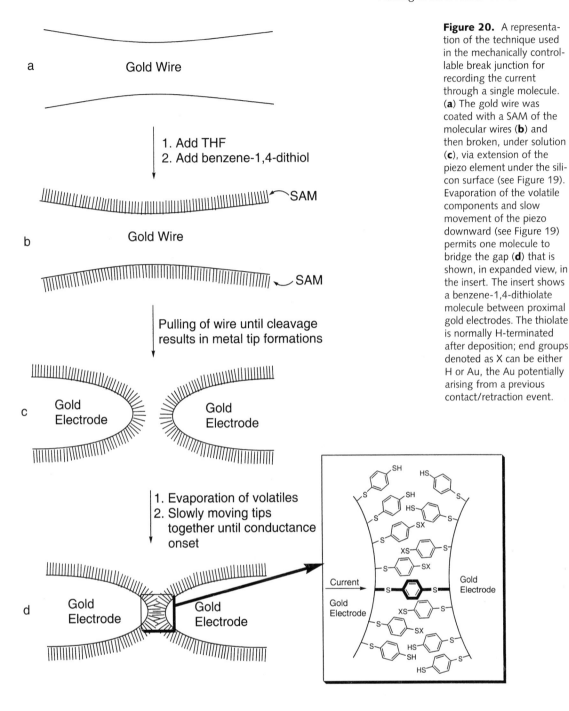

Figure 20. A representation of the technique used in the mechanically controllable break junction for recording the current through a single molecule. (**a**) The gold wire was coated with a SAM of the molecular wires (**b**) and then broken, under solution (**c**), via extension of the piezo element under the silicon surface (see Figure 19). Evaporation of the volatile components and slow movement of the piezo downward (see Figure 19) permits one molecule to bridge the gap (**d**) that is shown, in expanded view, in the insert. The insert shows a benzene-1,4-dithiolate molecule between proximal gold electrodes. The thiolate is normally H-terminated after deposition; end groups denoted as X can be either H or Au, the Au potentially arising from a previous contact/retraction event.

Conductance quantization (changes in the conductance in discrete steps of $2e^2/h$, where e is elementary charge and h is Planck's constant) observed in the probe system ensured that tip movement was controllable in subatomic-length increments. Benzene-1,4-dithiol (generated in situ from the dithioacetate) was permitted to self-assemble on the two tips that were, initially, widely separated. The two tips were then moved together, in picometer steps during the close contact point, until one molecule bridged the gap (Figure 20).[33] Current/voltage responses were recorded for a single molecule bridging the gap. Remarkably, 0.1 microamps of current could be recorded through a single molecule.

However, few or none of those 10^{12} electrons per second were colliding with the nuclei of the molecule, therefore all the heat was dissipated in the contact. Note that the mean free path of an electron in a metal is hundreds of angstroms. Hence, it is not surprising that collisions, within a small molecule, did not take place. Most importantly, since most computing instruments operate on microamps of current, the prospects for molecular scale electronics are quite intriguing.

Figure 21. I(V) characteristics of a Au-(2′-amino-4,4′-di(ethynylphenyl)-5′-nitro-1-benzenethiolate)-Au device at 60 K. The peak current density is ~50 A/cm², the NDR is ~ −400 μΩ-cm², and the PVR is 1030:1.

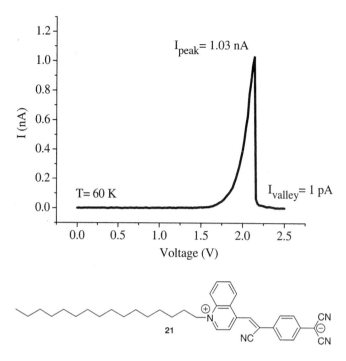

Testing of Molecular Devices

Two-terminal devices can be fashioned from many types of molecular systems. In 1997, Metzger and co-workers [28] used a multilayer structure of **21** to observe electrical rectification. Similarly, the groups of Heath, Williams and Stoddart[29] used a monolayer film of a rotaxane to generate a switchable device structure, albeit irreversible, when sandwiched between aluminum and titanium electrodes. We have recently shown that simple functionalizations of oligo(phenylene ethynylene)s permit the formation of several device structures. For example, using the nanopore set-up as described above, the nitroaniline (Figure 12) showed an intense negative differential resistance (NDR) effect with a peak-to-valley (PVR) ratio of > 1000:1 (Figure 21). NDR is an inflection in the current/voltage curve where there is a sudden spike in the current as the voltage is scanned. While silicon devices do not show this effect, gallium arsenide exhibits this property at levels typically on the order of < 100:1. Hence, the fact that the molecular system showed a response of > 1000:1 was remarkable indeed. Modifications of the structure have also permitted the storage of current in related systems, and this can be used as a molecular memory, or dynamic random access memory (DRAM), to be more precise.[34]

Three- and four-terminal systems have yet to be addressed either as single molecules or as nanometer-sized bundles. This situation stems from the fact that it is far easier to bring two probes into close proximity than it is to bring three probes into near contact. Macroscopic leads are akin to basket-

balls; touching two basketballs together at the same junction point is simple, however, trying to touch three together at a common contact point is impossible. It is the same with molecular addressing. Unless the three probes have very large aspect ratios (extreme sharpness at the molecular level), or unless one cleverly devises another method, testing of three- or four terminal systems will remain a challenge. However, in collaboration with Jorge Seminario, we have tested these systems computationally and have shown that they can respond as logic gates in numerous configurations such as OR, AND, NOR, and NAND gates (Figure 22).[18] Specifically, we considered the use of electrostatic potentials for the transfer of information. Using these scenarios, one could address several key issues that are sure to arise from the dense architectures. For example, the enormous heat dissipation constraints that may occur when using molecular systems. Therefore, the use of molecular devices is indeed attractive, and although there are obstacles, numerous methods are being considered to overcome these obstacles.

Epilogue

The use of molecular wires and devices for electronics applications is destined to occur. The ability to control molecular structures at the sub-nanometer scale is obvious throughout chemical synthesis. These are the same techniques that have been optimized over the last 50 years for the synthesis and modification of compounds for pharmaceutical, dye, petroleum, and fine chemical indus-

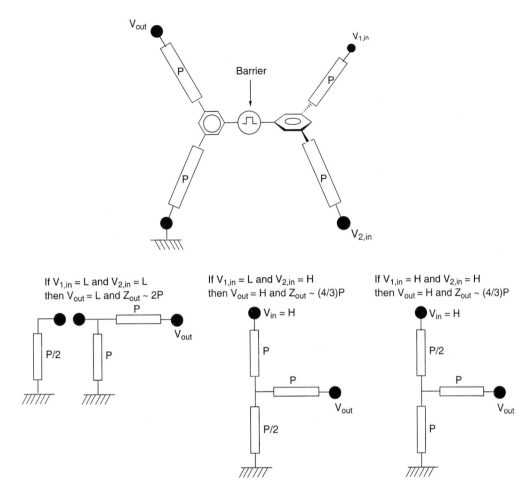

If $V_{1,in}$ = L and $V_{2,in}$ = L
then V_{out} = L and Z_{out} ~ 2P

If $V_{1,in}$ = L and $V_{2,in}$ = H
then V_{out} = H and Z_{out} ~ (4/3)P

If $V_{1,in}$ = H and $V_{2,in}$ = H
then V_{out} = H and Z_{out} ~ (4/3)P

If the output is active (as a potential)

$V_{1,in}$	$V_{2,in}$	V_{out}
L	L	L
L	H	H
H	L	H
H	H	H

OR (AND if negative logic is used)

If the output is passive (as an impedance)

$V_{1,in}$	$V_{2,in}$	Z_{out}
L	L	H
L	H	L
H	L	L
H	H	L

NOR (NAND if negative logic is used)

Figure 22. Shown is the schematic outline of the four-terminal device in Figure 13 with its reduced logic profiles and subsequent truth table outputs where the electrostatic potential is varied through inputs $V_{1,in}$ and $V_{2,in}$. The central methylene is depicted as an electrical barrier. Monitoring of the electrostatic potential output is observed at V_{out}, and H and L represent high and low values, respectively. While P refers to the molecular electrostatic potential impedance of a particular segment of the molecule, Z_{out} represents the total electrostatic potential impedance of the molecule, measured as an output. The electrostatic potentials were calculated using density functional theory.

tries. So why not exploit these same processes for the precise manipulations needed for electronics effects? And with the fundamental limits in solid-state structures being approached, molecules will become the devices of the future for ultradense computational systems. Not only will they be smaller and less expensive to produce than their solid state counterparts, but in some instances they will show superior behavior which is hitherto unattainable from solid-state devices. The key, however, is that molecular systems will likely not supplant solid-state systems, but that they will be complements to traditional electronic devices. Hence, the development of hybrid architectures, molecules working in concert with silicon, is among the most likely paths that this technology will take in the coming years as we develop ultradense and ultrafast computational systems.

Acknowledgements

Financial Support from the Defense Advanced Research Projects Agency (DARPA) via the Office of Naval Research (ONR) N00014-99-1-0406, and the Army Research Office (ARO) DAAD19-99-1-0085 is gratefully acknowledged.

References and Notes

1. J. M. Tour, "Conjugated Macromolecules of Precise Length and Constitution. Organic Synthesis for the Construction of Nanoarchitectures," *Chem. Rev.* **1996**, *96*, 537–553.

2. B. I. Yakobson, R. E. Smalley, "Fullerene Nanotubes: C1,000,000 and Beyond," *American Scientist* **1997**, *85*, 324–337.

3. P. Packan, "Pushing the Limits," *Science* **1999**, *285*, 2079–2081.

4. D. L. Pearson, J.M. Tour, "Rapid Syntheses of Oligo(2,5-thiophene-ethynylene)s with ThioesterTermini: Potential Molecular Scale Wires With Alligator Clips," *J. Org. Chem.* **1997**, *62*, 1376–1387.

5. L. Jones II, J. S. Schumm, J. M. Tour, "Rapid Solution and Solid Phase Syntheses of Oligo(1,4-phenylene-ethynylene)s With Thioester Termini: Molecular Scale Wires With Alligator Clips. Derivation of Iterative Reaction Efficiencies on a Polymer Support," *J. Org. Chem.* **1997**, *62*, 1388–1410.

6. S. Huang, J. M. Tour, "Rapid Bi-directional Synthesis of Oligo(1,4-phenylene ethynylene)s," *Tetrahedron Lett.* **1999**, *40*, 3347–3350.

7. A. G. Zacarias, M. Castro, J. M. Tour, J. M. Seminario, "Lowest Energy States of Small Pd Clusters Using Density Functional Theory and Standard ab Initio Methods. A Route to Understanding Metallic Nanoprobes," *J. Phys. Chem. A.* **1999**, *103*, 7692–7700.

8. J. S. Schumm, L. Jones II, D. L. Pearson, R. Hara, J. M. Tour, "Iterative Divergent/Convergent Doubling Approach to Linear Conjugated Oligomers. A Rapid Route to a 128 Å Long Potential Molecular Wire and Molecular Alligator Clips," *Polym. Prepr. (Am. Chem. Soc., Div. Polym. Chem.)* **1994**, *35*(2), 687–688.

9. C. J. Muller, B. J. Vleeming, M. A. Reed, J. S. Lamba, R. Hara, L. Jones II, J. M. Tour, "Atomic Probes: a Search for Conduction Through a Single Molecule," *Nanotechnology* **1996**, *7*, 409–411.

10. J. S. Schumm, D. L. Pearson, L. Jones II, R. Hara, J. M. Tour, "Potential Molecular Wires and Molecular Alligator Clips," *Nanotechnology* **1996**, *7*, 430–433.

11. W. A. Reinerth, T. P. Burgin, T. D. Dunbar, L. A. Bumm, J. J. Arnold, J. J. Jackiw, C.-w. Zhou, M. R. Deshpande, D. L. Allara, P. S. Weiss, M. A. Reed, J. M. Tour "Self-Assembled Monolayer Formation of Organoselenium Compounds on Gold and its Significance for Molecular Scale Electronics," *Polym. Mater., Sci. Engin. (Am. Chem. Soc., Div. Polym. Mater.)* **1998**, *78*, 178–179.

12. W. A. Reinerth, J. M. Tour, "Protecting Groups for Organoselenium Compounds," *J. Org. Chem.* **1998**, *63*, 2397–2400.

13. J. M. Tour, W. A. Reinerth, L. Jones II, T. P. Burgin, C.-w. Zhou, C. J. Muller, M. R. Deshpande, M. A. Reed, "Recent Advances in Molecular Scale Electronics," in *Ann. N.Y. Acad. Sci. Molecular Electronics: Science and Technology, Vol. 852*, Aviram, A., Ratner, M., Eds.; **1998**, Ann. N.Y. Acad. Sci., pp. 197–204.

14. D. L. Allara, T. D. Dunbar, P. S. Weiss, L. A. Bumm, M. T. Cygan, J. M. Tour, W. A. Reinerth, Y. Yao, M. Kozaki, L. Jones II, "Evolution of Strategies for Self Assembly and Hookup of Molecule-Based Devices," in *Ann. N.Y. Acad. Sci., Molecular Electronics: Science and Technology, Vol. 852*, Aviram, A., Ratner, M., Eds.; Ann. N.Y. Acad. Sci., **1998**, pp. 349–370.

15. J. M. Seminario, A. G. Zacarias, J. M. Tour, "Molecular Alligator Clips for Single Molecule Electronics. Studies of Group 16 and Isonitriles Interfaced with Au Contacts," *J. Am. Chem. Soc.* **1999**, *121*, 411–416.

16. S. Huang, J. M. Tour, "Rapid Solid Phase Synthesis of Oligo(1,4-phenyleneethynylene)s by a Divergent/Convergent Tripling Strategy," *J. Am. Chem. Soc.* **1999**, *121*, 4908–4909.

17. S. Huang, J. M. Tour. "Rapid solid-phase synthesis of conjugated homooligomers and [AB] alternating block cooligomers of precise length and constitution," *J. Org. Chem.* **1999**, *64*, 8898–8906.

18. J. M. Tour, M. Kozaki, J. M. Seminario, "Molecular Scale Electronics: A Synthetic/Computational Approach to Digital Computing," *J. Am. Chem. Soc.* **1998**, *120*, 8486–8493.

19. R. C. Jagessar, J. M. Tour, "Synthesis of Porphyrins Bearing trans-Thiols," *Org. Lett.* **2000**, *2*, 111–113.

20. A. Rawlett, J. Chen, M. A. Reed, J. M. Tour, "Advances in Molecular Scale Electronics: Synthesis and Testing of Molecular Scale Resonant Tunneling Diodes and Molecular Scale Controllers," *Polym. Mater., Sci. Engin. (Am. Chem. Soc., Div. Polym. Mater.)* **1999**, *81*, 140–141.

21. R. Wu, J. S. Schumm, D. L. Pearson, J. M. Tour, "Convergent Synthetic Routes to Orthogonally Fused Conjugated Oligomers Directed Toward Molecular Scale Electronic Device Applications," *J. Org. Chem.* **1996**, *61*, 6906–6921.

22. J. M. Tour, L. Jones II, D. L. Pearson, J. S. Lamba, T. P. Burgin, G. W. Whitesides, D. L. Allara, A. N. Parikh, S. Atre, "Self-Assembled Monolayers and Multilayers of Conjugated Thiols, α,ω-Dithiols, and Thioacetyl-Containing Adsorbates. Understanding Attachments Between Potential Molecular Wires and Gold Surfaces," *J. Am. Chem. Soc.* **1995**, *117*, 9529–9534.

23. M. T. Cygan, T. D. Dunbar, J. J. Arnold, L. A. Bumm, N. F. Shedlock, T. P. Burgin, L. Jones II, D. L. Allara, J. M. Tour, P. S. Weiss, "Insertion, Conductivity, and Structure of Conjugated Organic Oligomers in Self-Assembled Alkanethiol Monolayers on Au{111}," *J. Am. Chem. Soc.* **1998**, *120*, 2721–2732.

24. P. S. Weiss, L. A. Bumm, T. D. Dunbar, T. P. Burgin, J. M. Tour, D. L. Allara, "Probing Electronic Properties of Conjugated and Saturated Molecules in Self-Assembled Monolayers," in *Ann. N.Y. Acad. Sci., Molecular Electronics: Science and Technology, Vol. 852*, Aviram, A., Ratner, M., Eds.; Ann. N.Y. Acad. Sci., **1998**, pp. 145–168.

25. Reed, M. A.; Zhou, C.; Deshpande, M. R.; Muller, C. J.; Burgin, T. P.; Jones, L., II; Tour, J. M. "The Electrical Measurement of Molecular Junctions," in *Ann. N.Y. Acad. Sci., Molecular Electronics: Science and Technology, Vol. 852*, Aviram, A., Ratner, M., Eds.; Ann. N.Y. Acad. Sci., **1998**, pp. 133–144.

26. C. Zhou, C. J. Muller, M. A. Reed, T. P. Burgin, J. M. Tour, "Mesoscopic Phenomena Studied with Mechanically

Controllable Break Junctions at Room Temperature", in *Molecular Electronics*, Jortner J., Ratner, M., Eds., Blackwell Science: Oxford, **1997**, pp. 191–213.

27. C. Zhou, M. R. Deshpande, M. A. Reed, L. Jones II, J. M. Tour, "Nanoscale Metal/Self-Assembled Monolayer/Metal Heterostructures," *Appl. Phys. Lett.* **1997**, *71*, 611–613.

28. R. M. Metzger, B. Chen, U. Hopfner, M.V. Lakshmikantham, D. Vuillaume, T. Kawai, X. Wu, H. Tachibana, T. V. Hughes, H. Sakurai, J. W. Baldwin, C. Hosch, M. P. Cava, L. Brehmer, G. J. Ashwell, "Unimolecular Electrical Rectification in Hexadecylquinolinium Tricyanoquinodimethanide" *JACS* **1997**, *119*, 10455–10466.

29. C. P. Collier, E. W. Wong, M. Belohradsky, F. M. Raymo, J. F. Stoddart, P. J. Kuekes, R. S. Williams, J. R. Heath, "Electronically Configurable Molecular-Based Logic Gates," *Science* **1999**, *285*, 391–394.

30. L. A. Bumm, J. J. Arnold, M. T. Cygan, T. D. Dunbar, T. P. Burgin, L. Jones II, D. L. Allara, J. M. Tour, P. S. Weiss, "Are Single Molecular Wires Conducting?" *Science* **1996**, *271*, 1705–1706.

31. J. Chen, M. A. Reed, C. L. Asplund, A. M. Cassell, M. L. Myrick, A. M. Rawlett, J. M. Tour, P. G. Van Patten, "Placement of Conjugated Oligomers in an Alkanethiol Matrix by Scanned Probe Microscope Lithography," *App. Phys. Lett.* **1999**, *75*, 624–626.

32. C. J. Muller, J. M. van Ruitenbeck, L. J. de Jough, "Experimental Observation of the Transition from Weak Link to Tunnel Junction," *Physica C* **1992**, *191*, 485–504.

33. M. A. Reed, C. Zhou, C. J. Muller, T. P. Burgin, J. M. Tour, "Conductance of a Molecular Junction," *Science* **1997**, *278*, 252–254.

34. M. A. Reed, J. Chen, A. M. Rawlett, D. W. Price, J. M. Tour, "Molecular Random Access Memories" submitted.

Molecular-Level Devices and Machines

Vincenzo Balzani, Alberto Credi, Margherita Venturi

Dipartimento di Chimica "G. Ciamician", Università di Bologna, via Selmi 2, 40126 Bologna, Italy

Phone: +051 2099560, Fax: +051 2099456, e-mail: vbalzani@ciam.unibo.it, Web: www.ciam.unibo.it/ photochem.html

Keywords ■ Molecular Devices ■ Molecular Machines ■ Molecular Wires ■ Antenna Systems ■ Molecular Switches ■ Plug/socket Systems ■ Pseudorotaxanes ■ Rotaxanes ■ Catenanes ■ Supramolecular Chemistry ■ Photochemistry ■ Electrochemistry ■ Luminescence

Concept: The concept of (macroscopic) device can be extended to the molecular level. A *molecular-level device* can be defined as an assembly of a discrete number of molecular components designed to achieve a specific function. A *molecular-level machine* is a particular type of molecular-level device in which the component parts can display changes in their relative positions as a result of some external stimulus.

macroscopic components — macroscopic device

molecular components — supramolecular system

simple acts — complex function

Abstract: The concept of a macroscopic device can be extended to the molecular level by designing and synthesizing (supra)molecular species capable of performing specific functions. Molecular-level devices operate via electronic and/or nuclear rearrangements and, like macroscopic devices, need energy to operate and signals to communicate with the operator. The energy needed to make the device work can be supplied as chemical energy, electrical energy, or light. Among the most useful techniques to monitor the operation of molecular-level devices are spectroscopy (particularly luminescence) and electrochemistry. A *molecular-level electronic set* for energy and electron transfer (wires, switches, antennas, plug/ socket, and extension systems) and various kinds of *molecular-level machines* (tweezers, pyston/cylinder systems, shuttles, systems based on catenanes, rotary motors) have already been synthesized and studied. The extension of the concept of a device to the molecular level is of interest, not only for basic research, but also for the growth of nanoscience and the development of nanotechnology. Molecular-level devices should find applications in information storage, display, and processing; in the long run, they are expected to lead to the construction of molecular-based (chemical) computers.

Prologue

In everyday life we make extensive use of *macroscopic devices.* A macroscopic device is an assembly of components designed to achieve a specific function. Each component of the device performs a simple *act,* while the entire device performs a more complex *function,* characteristic of the assembly. For example, the function performed by a hairdryer (production of hot wind) is the result of acts performed by a switch, a heater, and a fan, suitably connected by electric wires and assembled in an appropriate framework.

The concept of a device can be extended to the molecular level.[1–4] A *molecular-level* device can be defined as an assembly of a discrete number of molecular components (that is, a *supramolecular* structure) designed to achieve a specific function. Each molecular component performs a single act, while the entire assembly performs a more complex function, which results from the cooperation of the various molecular components.

Molecular-level devices operate via electronic and/or nuclear rearrangements and, like macroscopic devices, are characterized by: *(i)* the kind of energy input supplied to make them work, *(ii)* the way in which their operation can be monitored, *(iii)* the possibility to repeat the operation at will (cyclic process), *(iv)* the time scale needed to complete a cycle, and *(v)* the performed function.

As far as point *(i)* is concerned, a chemical reaction can be used, at least in principle, as an energy input. In such a case, however, if the device is to work cyclically [point *(iii)*], it will need addition of reactants at any step of the working cycle, and the accumulation of by-products resulting from the repeated addition of matter can compromise the operation of the device. On the basis of this consideration, the best energy inputs to make a molecular device work are photons and electrons (or holes). With appropriately chosen photochemically and electrochemically driven reactions, it is indeed possible to design very interesting molecular devices.

In order to control and monitor the device operation [point *(ii)*], the electronic and/or nuclear rearrangements of the component parts should cause readable changes in some chemical or physical property of the system. In this regard, photochemical and electrochemical techniques are very useful since both photons[4] and electrons (or holes)[5] can play the dual role of "writing" (i.e., causing a change in the system) and "reading" (i.e., reporting the state of the system).

The operation time scale of molecular devices [point *(iv)*] can range from less than picoseconds to seconds, depending on the type of rearrangement (electronic or nuclear) and the nature of the components involved.

Finally, as far as point *(v)* is concerned, molecular-level devices performing various kinds of functions can be imagined; some specific examples will be discussed below.

The extension of the concept of a device to the molecular level is also of interest for the growth of nanoscience and the development of nanotechnology. Indeed, the miniaturization of components for the construction of useful devices, which is an essential feature of modern technology, is currently pursued by the large-downward (top-down) approach. This approach, however, which leads physicists and engineers to manipulate progressively smaller pieces of matter, has its intrinsic limitations. An alternative and promising strategy is offered by the small-upward (bottom-up) approach. Chemists, by the nature of their discipline, are already at the bottom, since they are able to manipulate molecules (i.e., the smallest entities with distinct shapes and properties) and are therefore in the ideal position to develop bottom-up strategies for the construction of nanoscale devices.

In this chapter we will illustrate examples of three families of molecular-level devices: *(i)* devices for the transfer of electrons or electronic energy, *(ii)* devices capable of performing extensive nuclear motions, often called *molecular-level machines*, and *(iii)* devices whose function implies the occurrence of both electronic and nuclear rearrangements. Most of the examples that will be illustrated refer to devices studied in our laboratories.

Devices Based on the Transfer of Electrons or Electronic Energy

Apart from futuristic applications related, e.g., to the construction of a chemical computer,[6] the design and realization of *a molecular-level electronic set* (i.e., a set of molecular-level systems capable of playing functions that mimick those performed by macroscopic components in electronic devices) is of great scientific interest since it introduces new concepts into the field of chemistry and stimulates the ingenuity of research workers engaged in the emerging field of nanotechnology. In the last few years, many systems that could prove useful for information processing at the molecular level (e.g., wires, antennas, on/off switches, plug/socket devices, memories, logic gates) have been investigated.

Wires and Related Systems

An important function at the molecular level is photoinduced energy and electron transfer over long distances and/or along predetermined directions. This function can be obtained by linking donor and acceptor components by a rigid spacer, as illustrated in Figure 1a. An example[7] is given by the $Ru(bpy)_3^{2+}$–$(ph)_n$–$Os(bpy)_3^{2+}$ compounds (bpy = 2,2'–bipyridine; ph = 1,4-phenylene; n = 3, 5, 7) in which excitation of the $Ru(bpy)_3^{2+}$ moiety is followed by electronic energy transfer from the excited $Ru(bpy)_3^{2+}$ unit to the $Os(bpy)_3^{2+}$ one, as shown by the sensitized emission of the latter. For the compound with n = 7 (Figure 1b), the rate constant for energy transfer over the 4.2 nm metal-to-metal distance is 1.3×10^6 s^{-1}. In the $Ru(bpy)_3^{2+}$–$(ph)_n$–$Os(bpy)_3^{3+}$ compounds, obtained by chemical oxidation of the Os–based moiety, photoexcitation of the $Ru(bpy)_3^{2+}$ unit causes the transfer of an electron to the Os-based one with a rate constant of 3.4×10^7 s^{-1} for n=7 (Figure 1c). Unless the electron added to the $Os(bpy)_3^{3+}$ unit is rapidly removed, a back electron transfer reaction (rate constant 2.7×10^5 s^{-1} for n=7) takes place from the $Os(bpy)_3^{2+}$ unit to the $Ru(bpy)_3^{3+}$ one.

Spacers with energy levels or redox states in between those of the donor and acceptor may help energy or electron transfer (hopping mechanism). Spacers whose energy or redox levels can be manipulated by an external stimulus can play the role of switches for the energy- or electron-transfer processes.[4]

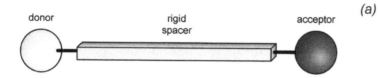

(a)

(b)

energy transfer

$R = n\text{-}C_6H_{13}$

hν   hν'

$Ru(bpy)_3{}^{2+}\text{--}(ph)_7\text{--}Os(bpy)_3{}^{2+}$

(c)

electron transfer

$R = n\text{-}C_6H_{13}$

hν

$Ru(bpy)_3{}^{2+}\text{--}(ph)_7\text{--}Os(bpy)_3{}^{3+}$

Figure 1. Schematic representation of a molecular-level wire (**a**) and examples of photoinduced energy (**b**) and electron (**c**) transfer processes.

Antenna Systems

In suitably designed dendrimers, electronic energy transfer can be channeled towards a specific position of the array. Compounds of this kind play the role of antennas for light harvesting.[4] A number of tree-like (dendritic) multicenter transition-metal complexes based on Ru and Os as metals, 2,3-bis(2-pyridyl)pyrazine (2,3-dpp) and 2,5-bis(2-pyridyl) pyrazine (2,5-dpp) as bridging ligands, and 2,2'-bipyridine (bpy) and 2,2'-biquinoline (biq) as terminal ligands have been prepared with this aim.[8] The largest compounds contain 22 metal atoms, 21 bridging ligands (2,3-dpp), and 24 terminal ligands (bpy). They comprise 1090 atoms, with a molecular weight of 10 890 Dalton (for the docosanuclear Ru complex), and an estimated size of 5 nm.

Since the properties of the modular components are known and different modules can be located in the desired positions of the dendrimer array, synthetic control of the various properties can be obtained. It is therefore possible, as schematically shown in Figure 2, to construct arrays where the electronic energy migration pattern can be predetermined, so as to channel the energy created by light absorption on the various components towards a selected module (antenna effect).

Devices Based on Nuclear Motions (Molecular-Level Machines)

A *molecular-level machine* is a particular type of molecular-level device in which the component parts can display changes in their relative positions as a result of some external stimulus.[3,9–13] Although there are many chemical compounds whose structure and/or shape can be modified by an external stimulus (see, e.g., the photoinduced *cis-trans* isomerization processes), the term molecular-level machines is only used for systems showing large amplitude movements of molecular components.

It is very important that such molecular-level motions are accompanied by changes of some chemical or physical property of the system, resulting in a "readout" signal that can be used to monitor the operation of the machine. The reversibility of the movement, i.e., the possibility to restore the initial situation by means of an opposite stimulus, is an essential feature of a molecular machine. Since such induced motions correspond to a binary logic, systems of this kind could also prove useful for information processing.

The concept of machine at the molecular level is not new. Our body can be viewed as a very complex ensemble of molecular-level machines that power our motions, repair damage, and orchestrate our

(a)

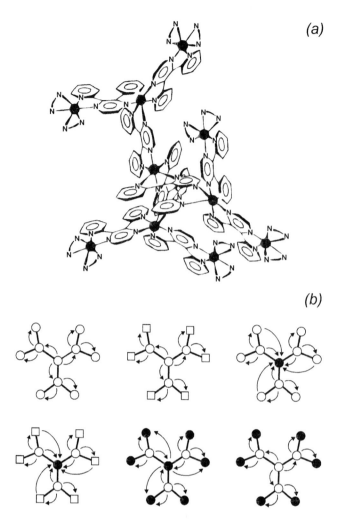

(b)

Figure 2. (*a*) Pictorial representation of a homodecanuclear dendrimer (N–N stands for bpy). (*b*) Schematic representation of the different energy-transfer patterns that can be obtained in such a structure on choosing different metals and ligands. The arrows indicate the exoergonic energy-transfer steps; empty and full circles indicate Ru(II) and Os(II), respectively; in the peripheral position, circles and squares indicate M(bpy)$_2$ and M(biq)$_2$ components, respectively. The compounds shown have +20 electric charge.

inner world of sense, emotion, and thought.[14] The problem of the construction of artificial molecular-level machines, however, is quite new. Although this possibility was proposed by Richard P. Feynman, Nobel Laureate in Physics, in his famous address to the American Physical Society in 1959,[15] and clever examples of molecular-level machines based on the photoisomerization of azobenzene derivatives were reported in the early 1980s,[16] research on artificial molecular machines has only begun to develop in the last few years.[9–13] Since most of the recently designed molecular-level machines are based on pseudorotaxanes, rotaxanes, and catenanes, it is worthwhile recalling some relevant features of such compounds.

Pseudorotaxanes, Rotaxanes, and Catenanes

Rationale and efficient synthetic approaches for the preparation of complicated (supra)molecular systems like pseudorotaxanes, rotaxanes and catenanes have been devised only recently.[17] The strategies chosen by Stoddart and coworkers[18] are based

on: (*i*) charge-transfer and C–H···O hydrogen-bonding interactions between an electron acceptor (e.g., 1,1′-dibenzyl-4,4′-bipyridinium dication) and an electron donor (e.g., 1,5-dinaphtho[38]crown-10, 1/5DN38C10), and/or (*ii*) hydrogen-bonding interactions between secondary ammonium functions (e.g., dibenzylammonium ion) and a suitable crown ether (dibenzo[24]crown-8, DB24C8) (Figure 3).

Systems Featuring Charge-Transfer Interactions

The charge-transfer interaction between electron-donor and electron-acceptor units has several important consequences from the spectroscopic and electrochemical viewpoints[19] and plays a fundamental role as far as the machine-like behaviour of these (supra)molecular species is concerned. The donor/acceptor interaction introduces low energy charge-transfer (CT) excited states which are responsible not only for the color of these (supra)molecular species (because of the presence of broad and weak absorption bands in the visible region), but also for the quenching of the potentially luminescent excited states localized on the molecular components (Figure 4). As far as the electrochemical behaviour is concerned, it should be

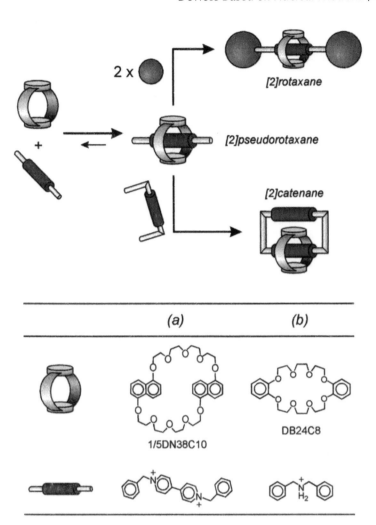

Figure 3. Pictorial representation of the self-assembly of pseudorotaxanes based on (a) charge-transfer and C–H···O hydrogen–bonding interactions between 1,1'-dibenzyl-4,4'–bipyridinium dication and 1,5-dinaphtho[38]crown-10 (1/5DN38C10), and (b) hydrogen-bonding interactions between dibenzyl ammonium ion and dibenzo[24]crown-8 (DB24C8). A possible route towards the synthesis of rotaxanes and catenanes is also schematized.

noted that, when engaged in CT interactions, the electron-donor and electron-acceptor units become more difficult to oxidize and to reduce, respectively. Furthermore, units which are topologically equivalent in an isolated component may not be so when the component is engaged in non-symmetric interactions with another component.

Consider, for example, the macrocyclic component **1** shown in Figure 5.[20] Such a species exhibits a two-electron reduction process, which corresponds to the simultaneous first reduction of the two equivalent bipyridinium units and, at a more negative potential, another two-electron process, which corresponds to the second reduction of such

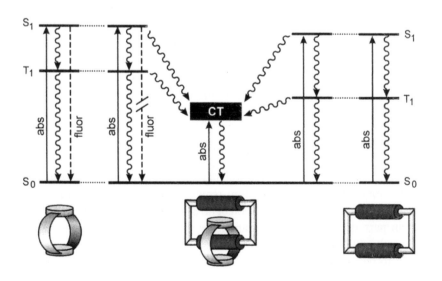

Figure 4. Schematic energy level diagram for a catenane based on charge-transfer (CT) interactions and for its separated components. The wavy lines indicate nonradiative decay paths of the electronic excited states.

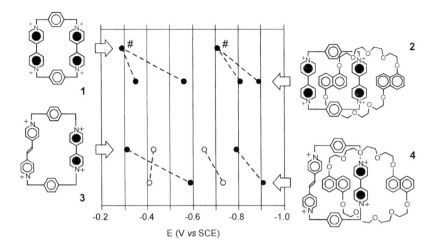

Figure 5. Correlations between the reduction potentials of two electron acceptor macrocycles and their catenanes with an electron donor crown ether. Black and white circles refer to reduction of bipyridinium and bis(pyridinium)ethylene units, respectively; processes marked with # involve two electrons.

units. When macrocycle **1** is interlocked in the catenane **2** with 1/5DN38C10, which contains two dioxynaphthalene electron-donor units, its electrochemical behaviour changes drastically: *(i)* all the reduction processes take place at more negative potentials, as expected because the bipyridinium units are engaged in CT interactions with the electron-donor units of the crown ether, and *(ii)* the two electron-acceptor units are no longer equivalent because the unit that resides inside the crown experiences a stronger CT interaction than the unit which resides alongside. Therefore, four distinct one-electron reduction processes are observed (Figure 5) which are assigned, starting from less negative potential values, to the first reduction of the alongside and inside bipyridinium units (first and second process), and to the second reduction of the alongside and inside units (third and fourth process).

In macrocycle **3**,[20] the two electron-acceptor units, a bipyridinium and a bis(pyridinium)ethylene, are different and therefore are reduced at different potentials, as expected on the basis of their electron-acceptor ability (Figure 5). When this macrocycle is interlocked with 1/5DN38C10 in the catenane **4**, the bipyridinium unit occupies the inside position and therefore it becomes more difficult to reduce compared with the bis(pyridinium)ethylene one since it experiences a stronger CT interaction. As a consequence, the first reduction of the bis(pyridinium)ethylene unit becomes the first reduction process of the whole system and therefore is displaced toward less negative potentials with respect to same process in the free macrocycle **3**, in which such a process follows the first reduction of the other unit.

In this kind of pseudorotaxanes, rotaxanes, and catenanes, the stability of a specific (supra)molecular structure is a result, at least in part, of the CT interaction. In order to cause mechanical movements, such a CT interaction has to be destroyed.

This requirement can be fulfilled by reduction of the electron-acceptor unit(s) or by oxidation of the electron-donor unit(s) by chemical, electrochemical, or photochemical redox processes. In most cases, the CT interaction can be restored by an opposite redox process, which thus promotes a reverse mechanical movement leading to the original structure.

Systems Based on Hydrogen-Bonding Interactions

Contrary to what happens in the case of CT interactions, hydrogen-bonding interactions between secondary ammonium centers and suitable crown ethers (Figure 3b) do not introduce low lying energy levels.[21] Therefore, even if the absorption bands of the molecular components of pseudorotaxanes, rotaxanes and catenanes based on this kind of interaction are often perturbed compared with the corresponding absorption bands of the isolated molecular components, no new band is present in the visible region. As far as luminescence is concerned, in the supramolecular architecture each component maintains its potentially luminescent levels, but intercomponent energy-transfer processes may often cause quenching and sensitization processes, as will be demonstrated better in the case of the plug/socket systems described in the next section. In principle, intercomponent electron transfer can also occur. The electrochemical properties of the separated components are more or less modified when the components are assembled.[21] In these compounds, mechanical movements can be caused by destroying the hydrogen bonding interaction which is responsible for assembly and spatial organization. This process can be easily caused by addition of a suitable base that is able to deprotonate the ammonium center. The movement can be reversed by addition of an acid that is able to reprotonate the amine function.

Photochemically Driven Piston/Cylinder Systems

Dethreading/rethreading of the wire and ring components of a pseudorotaxane reminds the movement of a piston in a cylinder. We have shown that, in suitably designed systems, the movement of such a rudimentary molecular machine can be driven by chemical energy or electrical energy and, most importantly, by light.

The first attempt at designing[22] a photochemically driven molecular-level machine of a pseudorotaxane type was based (Figure 6a) on the use of an external electron-transfer photosensitizer. As a result of CT interactions, the electron-acceptor ring **1** and a dioxynaphthalene-based electron-donor wire self-assemble in aqueous solution. Irradiation with visible light of an external electron-transfer photosensitiser [e.g., Ru(bpy)$_3^{2+}$] causes reduction of one of the bipyridinium units of the ring (the back electron-transfer reaction is prevented by the presence of a sacrificial reductant like triethanolamine). Once the ring has received an electron, the interaction responsible for self-assembly is partly destroyed and therefore the wire dethreads from the reduced ring. If oxygen is allowed to enter the solution, oxidation of the reduced bipyridinium unit restores the interaction and causes

rethreading. The threading, dethreading, and rethreading processes can be easily monitored by absorption and fluorescence spectroscopy.

Second generation photochemically driven machines were subsequently designed where the piston/cylinder pseudorotaxane superstructure incorporates the "light-fueled" motor (i.e., the photosensitizer) in the wire (Figure 6b)[23] or in the macrocyclic ring (Figure 6c).[24] In both cases, excitation of the photosensitiser with visible light in the presence of a sacrificial donor causes reduction of the electron-acceptor unit and, as a consequence, dethreading. Rethreading can be obtained by allowing oxygen to enter the solution.

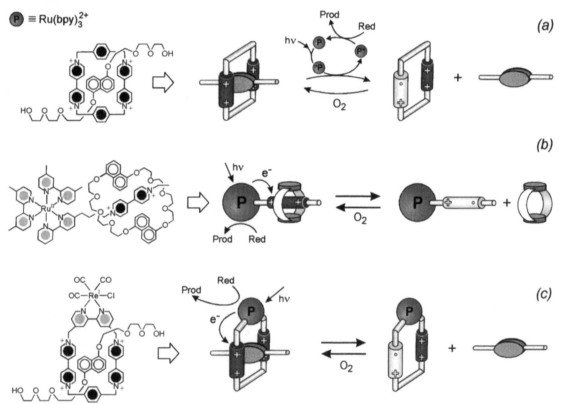

Figure 6. Light-driven dethreading of: (**a**) a pseudorotaxane by excitation of an external photosensitizer, (**b**) a pseudorotaxane incorporating a photosensitizer as a stopper in the wire-type component, (**c**) a pseudorotaxane incorporating a photosensitizer in the macrocyclic component.

Chemically induced dethreading/rethreading in similar pseudorotaxane systems has been shown to behave according to the XOR logic function.[25]

A Molecular Abacus

Rotaxanes are made of dumbbell-shaped and ring components which exhibit some kind of interaction originating from complementary chemical properties. In rotaxanes containing two different recognition sites in the dumbbell-shaped component, it is possible to switch the position of the ring between the two "stations" by an external stimu-

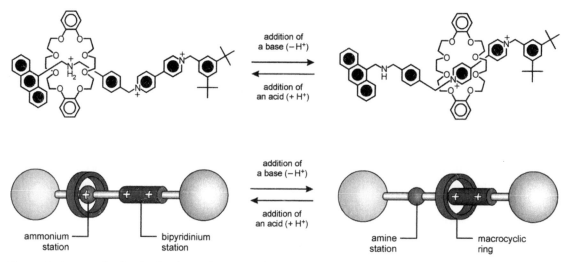

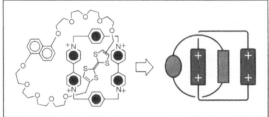

Figure 7. An example of molecular abacus: the ring can be switched between the two "stations" of the dumbbell-shaped component by base/acid inputs.

lus.[21] A system which behaves as a molecular abacus is shown in Figure 7. It is made of a DB24C8 ring and a dumbbell-shaped component containing a dialkylammonium center and a 4,4'-bipyridinium unit. An anthracene moiety is used as a stopper because its absorption, luminescence, and redox properties are useful to monitor the state of the system. The DB24C8 ring exhibits 100 % selectivity for the ammonium recognition site and therefore the rotaxane exists as only one of the two possible translational isomers, as evidenced by X-ray crystallography. Deprotonation of the ammonium center, however, causes 100 % displacement of the ring component to the bipyridinium unit. Reprotonation directs the crown ring back onto the ammonium center. Such a switching process has been investigated by ^{1}H NMR spectroscopy and by electrochemical and photophysical measurements. Futhermore, in the deprotonated rotaxane, it is possible to displace the crown ring from the bipyridinium station by destroying the charge-transfer interaction through: (i) electrochemical reduction of the bipyridinium station or (ii) electrochemical oxidation of the crown ring.

Electrochemically Driven Motion of a Ring in Catenanes

In a catenane, structural changes caused by circumrotation of one ring with respect to the other can be clearly evidenced when one of the two rings contains two non-equivalent units. In the catenane shown in Figure 8,[26] the ring containing the electron-acceptor units is "symmetric", whereas the other ring is non-symmetric since it contains two different electron-donor units, namely, a tetrathia-

fulvalene (TTF) and a 1,5-dioxynaphthalene (DMN) unit. In a catenane structure, the inside electron donor experiences the effect of two electron-acceptor units, whereas the alongside electron donor experiences the effect of only one electron acceptor. Therefore, in the catenane shown in Figure 8, the better electron donor (i.e., TTF) enters the ring and the less good one (i.e., DMN) remains alongside, as shown by a variety of techniques, including X-ray crystallography. On electrochemical oxidation, the first unit that undergoes oxidation is TTF, which thus loses its electron-donating properties. The disruption of the CT interaction and the electrostatic repulsion between TTF+ and the tetracat-

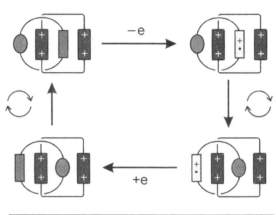

Figure 8. Electrochemically controlled movements of the ring components upon one-electron oxidation/reduction in a catenane containing a non-symmetric ring.

ionic macrocycle cause circumrotation of one ring to yield the translational isomer with the DMN moiety positioned inside the acceptor (Figure 8). Upon reduction of TTF$^+$, the switching is fully reversible. The oxidation/reduction cycle, which is accompanied by a clearly detectable color change, can be monitored by ^{1}H NMR spectroscopy, UV/Vis spectroscopy, and cyclic voltammetry.[26]

Devices Based on Electronic and Nuclear Motions

In the two preceding sections, we have illustrated examples of molecular-level devices working on the basis of either electron or nuclear movements. In other devices, which will be described in this section, the function they perform is based on both electronic and nuclear rearrangements, that take place in distinct steps.

Plug/Socket and Related Systems

A macroscopic plug/socket system is characterized by the following two features: *(i)* the possibility to connect/disconnect the two components in a reversible way, and *(ii)* the occurrence of electron flow from the socket to the plug when the two components are connected. Supramolecular systems have recently been designed that may be considered as molecular-level plug/socket devices. Plug in/plug out is reversibly controlled by acid/base reactions, and the photoinduced flow of electronic energy (or electrons) takes place in the plug-in state. The plug-in function can be based (see Figure 9)[27] on the threading of a (±)-binaphthocrown ether by

a (9-anthracenyl)benzylammonium ion. The association process can be reversed quantitatively (plug out) by addition of a suitable base like tributylamine. In the plug-in state (pseudorotaxane), the quenching of the binaphthyl-type fluorescence is accompanied by the sensitization of the fluorescence of the anthracenyl unit of the ammonium ion. The rate constant for electronic energy transfer from the binaphthyl unit of the crown to the anthracenyl unit of the wire is higher than 4×10^9 s^{-1}. Addition of a stoichiometric amount of base to the pseudorotaxane structure causes the revival of the binaphthyl fluorescence and the disappearance of the anthracenyl fluorescence upon excitation in the binaphthyl bands, demonstrating that plug out has happened.

The plug/socket molecular-level concept can be extended straightforwardly to the construction of molecular-scale extensions and to the design of systems where: *(i)* light excitation induces an electron flow instead of an energy flow, and *(ii)* the plug in/plug out function is stereoselective (the enantiomeric recognition of chiral ammonium ions by chiral crown ethers is well known). Thus, the plug/socket molecular-level systems are devices whose function is based on both electronic and nuclear movements, caused by two distinct external input(s).

Electrochemically Controlled Switches

In host-guest systems based on electron donor/acceptor interactions, association/dissociation can be driven by redox processes so that it is possible to design electrochemical switches than can be used to control energy- and electron-transfer processes.

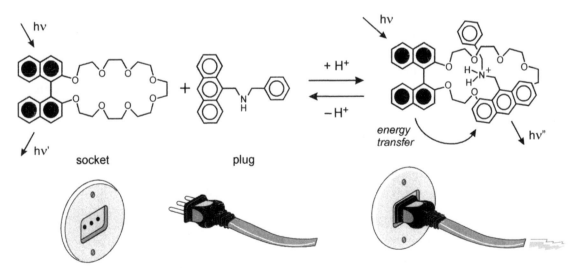

Figure 9. Acid/base controlled "plug in/plug out" of (9-anthracenyl)benzylammonium ion with a (±)-binaphthocrown ether. The occurrence of photoinduced energy transfer in the "plug in" state is schematized.

An example of a system that can be switched reversibly in three different states through electrochemical control of the guest properties of one component is illustrated in Figure 10.[28] Tetrathiafulvalene is stable in three different oxidation states, TTF(0), TTF$^+$, and TTF^{2+}. On oxidation, the electron-donor power of tetrathiafulvalene decreases with a

The mechanical movements taking place in this supramolecular system, in which a "free" molecule can be driven electrochemically to associate with either of two different receptors, opens the way to futuristic applications in the field of molecular-level signal processors.[6] One can conceive second generation systems wherein the electrochemically driven

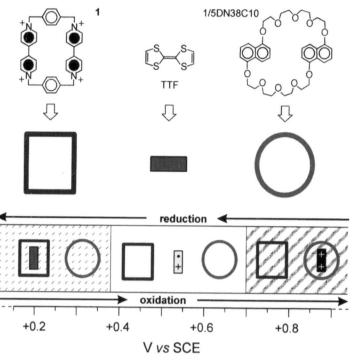

Figure 10. Components of a three-state system and schematic representation of the ranges of electrochemical stability of the three states available to the system.

concomitant increase in the electron-acceptor properties. Whereas TTF(0) plays the role of electron donor and gives a 1:1 charge-transfer complex with the cyclobis(paraquat-*p*-phenylene) electron-acceptor macrocycle **1**, TTF^{2+} plays the role of an electron acceptor and gives rise to a charge-transfer complex with the 1/5DN38C10 electron-donor macrocycle. TTF$^+$ does not show any electron donor/acceptor character. The system illustrated in Figure 10 consists of acetonitrile solutions of TTF and the two macrocycles **1** and 1/5DN38C10. Electrochemical experiments carried out on such a system show that, depending on the potential range, TTF can be: *(i)* free in the TTF$^+$ state, *(ii)* complexed with the electron-acceptor host in the TTF(0) state, or *(iii)* complexed with the electron-donor host in the TTF^{2+} state. The reversibility of the electrochemical processes shows that complexation/decomplexation and, as a consequence, the exchange of the guest between the two hosts, are fast processes compared to the time scale of the electrochemical experiments. In such a three-state system, switching ("writing") can be performed electrochemically and the state of the system can be monitored ("reading") by absorption, emission, and NMR spectroscopies, in addition to electrochemical techniques.

movements can control the selection of the partner in energy- or electron-transfer processes. Consider, e.g., a system (Figure 11) where a chromophoric group **A** is appended to the potential guest and chromophoric groups **B** and **C**, whose lowest excited state is lower than that of **A**, are appended to the potential hosts **1** and 1/5DN38C10, respectively. In such a system, light excitation of **A** will lead to no energy transfer (OFF), energy transfer to **B** (ON 1), or energy transfer to **C** (ON 2), depending on the potential value selected by the operator. Admittedly, the donor/acceptor-based molecular-level connection poses severe limitations as to the choice of the chromophoric groups to be used if energy transfer has to proceed through the charge-transfer connections. However, it does not seem unlikely that systems can be designed where molecular association: *(i)* relies on a different kind of interaction (e.g., hydrogen-bonding), or *(ii)* simply plays the role of bringing the two chromophoric groups to a suitable distance for through-space energy transfer. Similar switching of electron-transfer processes could also be performed. More complex energy- and/or electron-transfer patterns are conceivable. The strategy described in this section could also be used, in principle, to catalyze chemical reactions.

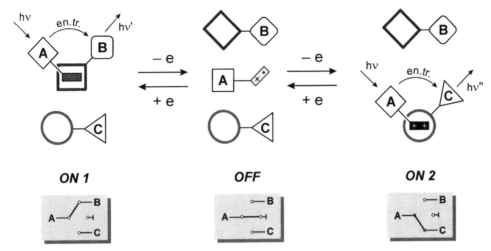

Figure 11. Schematic representation of the concept of a three-pole supramolecular switching of energy transfer. The cartoons correspond to the molecular components shown in Figure 10; **A**, **B**, and **C** are suitably chosen chromophoric groups. For more details, see the text.

Epilogue

In his 1959 address to the American Physical Society, when discussing the possibility of constructing molecular-level machines, R. P. Feynman said:[15] *"An internal combustion engine of molecular size is impossible. Other chemical reactions, liberating energy when cold, can be used instead"*. The described examples of molecular machines driven by redox or acid/base "cold" chemical reactions fulfil Feynman's prediction. Furthermore, the reported examples show that the primary energy source to drive such mechanical movements can be electricity or, even more interesting, light. In the same address, Feynman concluded his reflection on the construction of molecular-scale machines as follows: *"What would be the utility of such machines? Who knows? I cannot see exactly what would happen, but I can hardly doubt that when we have some control of the rearrangement of things on a molecular scale we will get an enormously greater range of possible properties that substances can have, and of different things we can do"*. We believe that these sentences are the most appropriate final comment to the work described in this chapter.

Acknowledgements

We would like to thank our colleagues and coworkers, whose names appear in the references quoted below, and particularly Prof. J. F. Stoddart and his group for a long lasting and most profitable collaboration. Financial support from EU (TMR grants FMRX-CT96-0031 and FMRX-CT96-0076), the University of Bologna (Funds for Selected Research Topics), and MURST (Supramolecular Devices Project) is gratefully acknowledged.

References and Notes

1. J.-M. Lehn, *Supramolecular Chemistry*, VCH, Weinheim, **1995**.
2. V. Balzani, F. Scandola, *Supramolecular Photochemistry*, Horwood, Chichester, **1991**.
3. V. Balzani, A. Credi, M. Venturi, "Molecular-Level Devices", in *Supramolecular Science: Where It Is and Where It Is Going* (Eds. R. Ungaro, E. Dalcanale), Kluwer, Dordrecht, **1999**, pp. 1–22.
4. V. Balzani, F. Scandola, "Photochemical and Photophysical Devices", in *Comprehensive Supramolecular Chemistry* (Eds. J. L. Atwood, J. E. D. Davies, D. D. MacNicol, F. Vögtle), Pergamon Press, Oxford, **1996**, Vol. 10, pp. 687–746.
5. P. L. Boulas, M. Gomez–Kaifer, L. Echegoyen, "Electrochemistry of Supramolecular Systems", *Angew. Chem. Int. Ed.* **1998**, *37*, 216–247.
6. D. Rouvray, "Reckoning on Chemical Computers", *Chem. Br.* **1998**, *34* (2), 26–29.
7. B. Schlicke, P. Belser, L. De Cola, E. Sabbioni, V. Balzani, "Photonic Wires of Nanometric Dimensions. Electronic Energy Transfer in Rigid Rod-like Ru(bpy)$_3^{2+}$–(ph)$_n$–Os(bpy)$_3^{2+}$ Compounds (ph=1,4–phenylene, *n*=3,5,7)", *J. Am. Chem. Soc.* **1999**, *121*, 4207–4212, and unpublished results.
8. V. Balzani, S. Campagna, G. Denti, A. Juris, S. Serroni, M. Venturi, "Designing Dendrimers Based on Transition-Metal Complexes. Light–harvesting Properties and Predetermined Redox Patterns", *Acc. Chem. Res.* **1998**, *31*, 26–34.
9. V. Balzani, M. Gomez–López, J. F. Stoddart, "Molecular Machines", *Acc. Chem. Res.* **1998**, *31*, 405–414.
10. J.-P. Sauvage, "Transition Metal-Containing Rotaxanes and Catenanes in Motions: Toward Molecular Machines and Motors", *Acc. Chem. Res.* **1998**, *31*, 611–619.
11. T. R. Kelly, H. De Silva, R. A. Silva, "Unidirectional Rotary Motion in a Molecular System", *Nature* **1999**, *401*, 150–152.
12. N. Koumura, R. W. J. Zijlstra, R. A. van Delden, N. Harada, B. L. Feringa, "Light-driven Monodirectional Molecular Rotor", *Nature* **1999**, *401*, 152–155.
13. V. Balzani, A. Credi, F. M. Raymo, J. F. Stoddart, "Artificial Molecular Machines", *Angew. Chem. Int. Ed.*, **2000**, *39*, 3348–3391.

14. D. S. Goodsell, *Our Molecular Nature: The body's Motors, Machines, and Messages*, Copernicus, New York, **1996**.

15. R. P. Feynman, "There's Plenty of Room at the Bottom", *Sat. Rev.* **1960**, *43*, 45–47.

16. S. Shinkai, "Switchable Guest-binding Receptor Molecules", in *Comprehensive Supramolecular Chemistry* (Eds. J. L. Atwood, J. E. D. Davies, D. D. MacNicol, F. Vögtle), Pergamon Press, Oxford, **1996**, Vol. 1, pp. 671–700, and references therein.

17. *Molecular Catenanes, Rotaxanes and Knots* (Eds. J.-P. Sauvage, C.O. Dietrich–Buckecher), VCH–Wiley, Weinheim, **1999**.

18. F. M. Raymo, J. F. Stoddart, " Interlocked Macromolecules", *Chem. Rev.* **1999**, *99*, 1643–1663.

19. M. Asakawa, P. R. Ashton, R. Ballardini, V. Balzani, M. Belohradsky, M. T. Gandolfi, O. Kocian, L. Prodi, F. M. Raymo, J. F. Stoddart, M. Venturi, "The Slipping Approach to Self–Assembling [n]Rotaxanes", *J. Am. Chem. Soc.* **1997**, *119*, 302–310.

20. P.R. Ashton, R. Ballardini, V. Balzani, A. Credi, M. T. Gandolfi, D. J.-F. Marquis, S. Menzer, L. Perez-García, L. Prodi, J. F. Stoddart, M. Venturi, A. J. P. White, D. J. Williams, "The Self-Assembly of [2]Catenanes Incorporating Photo-Active and Electro-Active π-Extended Systems", *J. Am. Chem. Soc.* **1995**, *117*, 11171–11197.

21. P.R. Ashton, R. Ballardini, V. Balzani, I. Baxter, A. Credi, M. C. T. Fyfe, M. T. Gandolfi, M. Gomez–López, M. V. Martínez–Díaz, A. Piersanti, N. Spencer, J.F. Stoddart, M. Venturi, A. J. P. White, D. J. Williams, "Acid–base Controllable Molecular Shuttles", *J. Am. Chem. Soc.* **1998**, *120*, 11932–11942.

22. R. Ballardini, V. Balzani, M. T. Gandolfi, L. Prodi, M. Venturi, D. Philp, H. G. Ricketts, J. F. Stoddart, "A Photochemically–Driven Molecular Machine", *Angew. Chem. Int. Ed. Engl.* **1993**, *32*, 1301–1303.

23. P. R. Ashton, R. Ballardini, V. Balzani, E. C. Constable, A. Credi, O. Kocian, S. J. Langford, J. A. Preece, L. Prodi, E. R. Schofield, N. Spencer, J. F. Stoddart, S. Wenger, "Ru(II)-Polypyridine Complexes Covalently Linked to Electron Acceptors as Wires for Light–Driven Pseudorotaxane-Type Molecular Machines", *Chem. Eur. J.* **1998**, *4*, 2411–2422.

24. P. R. Ashton, V. Balzani, O. Kocian, L. Prodi, N. Spencer, J. F. Stoddart, "A Light-Fueled "Piston-Cylinder" Molecular-Level Machine", *J. Am. Chem. Soc.* **1998**, *120*, 11190–11191.

25. A. Credi, V. Balzani, S. J. Langford, J. F. Stoddart, "Logic Operations at the Molecular Level. An XOR Gate Based on a Molecular Machine", *J. Am. Chem. Soc.* **1997**, *119*, 2679–2681.

26. M. Asakawa, P. R. Ashton, V. Balzani, A. Credi, C. Hamers, G. Mattersteig, M. Montalti, A. N. Shipway, N. Spencer, J. F. Stoddart, M. S. Tolley, M. Venturi, A. J. P. White, D. J. Williams, "A Chemically and Electrochemically Switchable [2]Catenane Incorporating a Tetrathiafulvalene Unit", *Angew. Chem. Int. Ed.* **1998**, *37*, 333–337.

27. E. Ishow, A. Credi, V. Balzani, F. Spadola, L. Mandolini, "A Molecular-level Plug/socket System. Electronic Energy Transfer from a Binaphthyl Unit Incorporated into a Crown Ether to an Anthracenyl Unit Linked to an Ammonium Ion", *Chem. Eur. J.* **1999**, *5*, 984–989.

28. P. R. Ashton, V. Balzani, J. Becher, A. Credi, M. C. T. Fyfe, G. Mattersteig, S. Menzer, M. B. Nielsen, F. M. Raymo, J. F. Stoddart, M. Venturi, D. J. Williams, "A Three-pole Supramolecular Switch", *J. Am. Chem. Soc.* **1999**, *121*, 3951–3957.

Electron and Energy Transfer

Michael N. Paddon-Row

School of Chemistry, University of New South Wales, Sydney, NSW, 2052, Australia

Phone: +2–9385-4724, Fax: +2–9385-6141, e-mail: m.paddonrow@unsw.edu.au

Keywords ■ Electron Transfer ■ Energy Transfer ■ Through-Bond Coupling ■ Superexchange ■ Molecular Wires ■ Solvent-Mediated Electron Transfer ■ Electron Transfer in DNA ■ Charge Separation ■ Electron Transfer Through H-Bonds

Concept: The rates of long-range electron transfer (ET) and excitation energy transfer (EET) processes between a pair of chromophores (redox couple) may be strongly facilitated by the presence of an intervening non-conjugated medium, such as saturated hydrocarbon bridges, solvent molecules and π-stacks, e.g.,

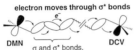

e^- ET over 13.5 Å

DMN · DCV · electron moves through σ* bonds · σ and σ* bonds,

base pairs in duplex DNA molecules. With respect to hydrocarbon bridges, the ET and EET processes are mediated by a through-bond or superexchange coupling mechanism, the dynamics of which can be controlled by changes in bridge geometry, bridge length and orbital symmetry.

Abstract: Evidence is presented in support of the concept that electron transfer (ET) and excitation energy transfer (EET) between a pair of chromophores may take place efficiently over large distances (> 10 Å) by the mediation of a *nonconjugated* intervening medium. For example, ET is found to take place on a subnanosecond timescale through saturated norbornylogous bridges greater than 12-bonds in length, by a through-bond coupling (superexchange) mechanism. The dependence of the ET dynamics on bridge constitution, configuration and orbital symmetry is consistent with the operation of a through-bond (TB) mechanism. ET mediated by hydrogen bonds (H-bonds) is also important, as is the modulation of ET dynamics by electric fields. The distinction between molecular wire behaviour and TB-mediated ET is made. TB-mediated EET processes have been investigated and analyzed in terms of double ET processes. Solvent-mediated ET is being investigated using novel rigid U-shaped systems and evidence is mounting that unsaturated solvent molecules may indeed mediate ET, although the interpretation of the rate data is clouded by predictions of large electrostatically-induced geometric distortions in the charge-separated states. Recent investigations of ET mechanisms in DNA are discussed and interpreted in terms of superexchange and hole hopping mechanisms, both mediated by the π-stack of base pairs.

Prologue

Electron transfer (ET) is the most fundamental and ubiquitous of all chemical reactions, playing a key role in many essential biological processes. A particularly important type of ET is photoinduced charge separation between two bichromophores (or redox centres), as illustrated in Figure 1. The first step involves local excitation of either the Donor (**D**) or Acceptor (**A**) chromophore to generate the locally excited state, *e.g.* ****D**–**A**. Exergonic[1] charge separation (k_{cs}) competes with decay modes (k_d), leading to the ground state, **D**–**A**, to generate the charge-separated (CS) state, $^+$**D**–**A**$^-$. The importance of the photoinduced charge separation process lies in its transduction of light energy into useful chemical potential, measured by the free energy change, ΔG_{cr}, for charge recombination. Such CS states may therefore be regarded as molecular photovoltaic devices. A classic example of the usefulness of photoinduced charge separation in biology is the vital role it plays in the primary events of photosynthesis.

The successful design of both natural and artificial molecular photovoltaic devices rests on meeting three fundamental requirements, namely:[2] (1) The quantum yield for the charge separation process should be as high as possible. That is, $k_{cs} \gg k_d$ (Figure 1). (2) The lifetime, τ_{cr} (= $1/k_{cr}$), of the CS state must be sufficiently long to enable it to carry out

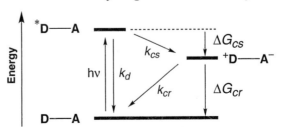

Figure 1. Energy diagram illustrating the possible pathways available in a photoinduced charge separation process between a donor (D) and acceptor (A).

"useful" chemical work. In practice, τ_{cr} should be substantially greater than 1 μs. (3) The energy content of the CS state should be as high as possible, thereby ensuring maximum conversion of photonic energy into chemical potential. Thus, $|\Delta G_{cs}|$ should be as small as practicable, while ensuring that requirement (1) is met.

Meeting these design criteria requires a sound knowledge of the fundamental mechanisms of electron transfer processes. Although ET is the most fundamental of all chemical reactions, it is by no means simple! This realization is particularly evident in the case of long-range ET processes which forms the basis of this essay.

One may easily visualise ET taking place between two chromophores that lie within van der Waals contact (Figure 2a). In this so-called through-space

While this pessimistic view indeed holds true for chromophores separated in a vacuum, such need not be the case if a medium is placed in between the chromophores. Indeed, rapid ET is expected, and is found, to take place if the chromophores are connected by a conjugated network of π bonds (*vide infra*).

Less obvious, and more intriguing, is whether an intervening non-conjugated medium could facilitate ET. Such a medium, which may comprise a saturated hydrocarbon bridge, a protein or oligopeptide, π-stacks, or even solvent molecules, could facilitate ET by what is known as a superexchange mechanism; to put it crudely, the medium provides "orbitals" $(\pi, \pi^*, \sigma, \sigma^*, \text{etc})$ which the migrating electron can use to tunnel between the chromophores (Figure 2b). In those situations where the medium is

Figure 2. (**a**) Short-range ET. The donor and acceptor orbitals overlap strongly and ET takes place by a direct, through-space mechanism. (**b**) Long-range ET. Direct overlap between the donor and acceptor orbitals is negligible and ET occurs by an indirect mechanism involving electron tunnelling through the orbitals of the intervening medium, *e.g.* solvent molecules (upper) or a covalently linked saturated bridge (lower).

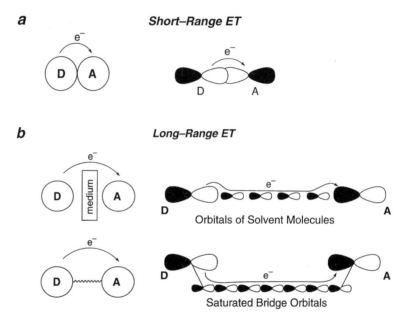

mechanism, there is strong overlap between the two active orbitals – that is, the orbital of the donor from which the electron is being transferred and the orbital of the acceptor into which the electron is being transferred. This orbital overlap provides a well-defined path for the electron to "move" from donor to acceptor.

Now, consider the case of long-range ET in which the donor and acceptor separation greatly exceeds the sum of their van der Waals radii. If the chromophores are in a vacuum, the direct, through-space overlap between the donor and acceptor orbitals is negligible; consequently, any electronic coupling between the donor and acceptor groups, resulting from this overlap, should be unimportant. This type of cursory analysis might lead one to conclude that *all* long-range ET processes are doomed to being very slow, inefficient, and of limited interest.

saturated and only σ and σ* orbitals are available for coupling with the chromophores, the superexchange mechanism is then often referred to as a through-bond (TB) coupling mechanism.

The notion that ET could take place efficiently through saturated bonds was, until the 1980s, regarded with scepticism, largely because it was felt that the energy gap between the π,π* manifolds of the chromophores and the σ,σ* manifolds of the bridge is too large to permit efficient electronic coupling. In spite of this scepticism, several examples of extremely rapid and efficient long-range ET processes were known which suggested that they were taking place by some sort of superexchange mechanism. One of the most spectacular of these long-range ET processes occurs in the photosynthetic reaction centres (PRCs) of certain photosynthetic bacteria (Figure 3). The primary photoinduced ET events in the PRC are observed to take place on a

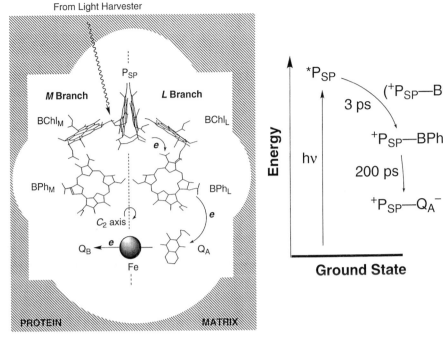

Figure 3. Schematics of the photosynthetic reaction centre and the energetics of the charge separation steps, together with the individual ET rates along the *L* branch.

picosecond timescale, with near unit efficiency, over interchromophore separations exceeding 10 Å! Moreover, long-range ET has been observed to be not only rapid, but also to take place in some instances with exquisitely controlled directionality. Again, using the PRC as an example, it is found that the primary ET cascade, emanating from the P_{SP} "special pair", and terminating at the quinone groups, Q_A and Q_B, takes place exclusively along the so-called *L* branch, rather than along the *M* branch, even though both branches are apparently related by near perfect C_2 symmetry (Figure 3). Since the chlorophyll-like and quinone cofactors (chromophores) are clearly not connected to each other by bonds, conjugated or otherwise, how does the ET process take place so rapidly and how does it "know" which path, the *L* or *M* branch, to take? Although the mechanism of ET in the PRC remains unknown, it is widely believed that its speed and directionality are mediated by the protein matrix surrounding the centre.[3]

The characteristics of the superexchange or though-bond mechanism for ET have, until recently, remained somewhat obscure. Characteristics primarily describe the distance and orientation dependence of the dynamics of long-range ET, and how this dependency is affected by the nature and composition of the intervening medium. It is this aspect of long-range ET processes that has captured the attention of a broad cross section of the chemical community, partly because a deeper under-

standing of medium effects on long-range ET is essential to the successful design of molecular photovoltaic systems and other intriguing molecular electronic devices.

The best approach to tackling the problem of the dependence of ET dynamics on the nature of the intervening medium is to design model "probe" systems possessing structurally well-defined and controllable characteristics. This goal is best achieved by fulfilling some or all of the following design criteria: (1) If possible, the chromophores should be linked covalently to the medium or bridge to form systems, henceforth denoted as **D–B–A** (**B** is a bridge). This attachment sharply defines the electronic coupling between the medium and the chromophores. (2) The bridge should be as rigid as possible, thereby ensuring that the interchromophore separation and orientation are well-defined. (3) The **D–B–A** systems should possess symmetry. This feature facilitates the analysis of the electronic coupling between the chromophores and the bridge as well as enabling one to explore the effect of orbital symmetry on ET dynamics. (4) The synthesis of the bridge should be sufficiently versatile to permit the length and configuration of the bridge to be altered systematically. It should be noted that the bridge serves the dual role of providing: (a) a molecular scaffolding to which the chromophores are affixed at well-defined separations and orientations, and (b) a medium which modulates the ET dynamics by the superexchange mechanism.

Figure 4. The norbornylogous bridge.

Several types of saturated hydrocarbon bridges have been employed in ET studies. They include cyclohexanes, decalin, and steroid-based systems, oligobicyclo[2.2.2]octanes, triptycene, and polyspiro-cyclobutanes.[4] None, however, has surpassed the versatility of the so-called norbornylogous bridges, comprising a mixture of linearly fused norbornane and bicyclo[2.2.0]hexane ring systems (Figure 4). This bridge, on account of its rigidity, symmetry, and the comparative synthetic ease by which its length and configuration can be altered and controlled, coupled with the ability to attach a wide range of different chromophores to its termini, has provided valuable insights into various aspects of long-range intramolecular ET and electronic excitation energy transfer (EET) processes.[5] Before discussing experimental results of ET and EET processes obtained using norbornylogous bridges and other systems, it is necessary to provide a simple theoretical underpinning which is used to analyze ET.

Basic Theoretical Background

The essentials of ET theory are shown in Figure 5. The energy surface for a charge separation process, conveniently represented by a one-dimensional reaction coordinate which is supposed to describe changes in both geometry of the D–A system and solvent orientation, may be regarded in terms of two diabatic surfaces, one representing the electronic configuration of the reactant, **D–A**, and the other representing the electronic configuration of the product, $^+$**D–A**$^-$. In the region where the diabatic surfaces intersect, the two configurations mix, symmetry permitting, resulting in an avoided crossing. The magnitude of the avoided crossing is given by $2V_{el}$, where V_{el} is the electronic coupling term and may be regarded as a rough measure of the strength of orbital interactions between **D** and **A**. In the case of long-range ET, V_{el} is generally very small (< 1.5 kJ mol^{-1}) and ET occurs nonadiabatically and, within the context of classical Marcus–Hush theory, the ET rate, k_{et} is given by the expression[6] shown in Figure 5. Three important quantities that affect the magnitude of k_{et} are V_{el}, ΔG_{cs} and λ, the reorganization energy.[7] Full understanding of medium-mediated ET requires knowledge of how the aforementioned three quantities are affected by the

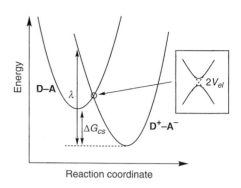

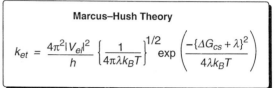

Marcus–Hush Theory

$$k_{et} = \frac{4\pi^2|V_{el}|^2}{h}\left\{\frac{1}{4\pi\lambda k_B T}\right\}^{1/2}\exp\left(\frac{-\{\Delta G_{cs}+\lambda\}^2}{4\lambda k_B T}\right)$$

Figure 5. Energy diagram for charge separation resolved into reactant-like and product-like diabatic surfaces. The two diabatic curves do not intersect, but interact, to give an avoided crossing, whose energy gap is twice the electronic coupling, V_{el}, for the interaction. Also depicted is the Marcus–Hush classical rate expression for nonadiabatic ET.

medium. In terms of electronic interactions and their distance and orientation dependence, V_{el}, is the most important parameter and we will focus on this quantity.

It is generally assumed that V_{el} and the associated ET rate constant, k_{et}, fall off approximately exponentially with increasing interchromophore separation, r, viz:

$$V_{el} \propto \exp(-0.5\beta_{el}r) \qquad (1)$$
$$k_{et} \propto \exp(-\beta r) \qquad (2)$$

where β_{el} and β are damping factors. It is often assumed that β_{el} and β have identical magnitudes, but this is not strictly correct because β, being a phenomenological quantity, incorporates distance dependence contributions, not only from V_{el}, but also from the reorganization energy, λ (see the Marcus–Hush expression, in Figure 5). Thus, β is expected to be slightly larger than β_{el} although, for the purpose of this review, we shall assume that β_{el} and β are the same.

Electron Transfer Mediated by Saturated Hydrocarbon Bridges

A crucial issue in ET concerns the dependence of the magnitude of β on the nature of the intervening medium. Consider ET occurring between two degenerate π-type orbitals separated in a vacuum (Figure 6a). The orbital degeneracy is lifted by through-space (TS) orbital overlap and gives rise to a

Figure 6. Distance dependence of the interactions between two equivalent π-type orbitals. The magnitude of the interaction, $\Delta E(\pi)$, at any given distance is approximately equal to twice the electronic coupling, V_{el}, for hole transfer between the two orbitals. (**a**) Through-space (TS) coupling. (**b**) Through-bond (TB) coupling. Using HF/3–21G theory, $\beta \approx 3.0$ Å^{-1}, for TS interactions, and 0.8 per bond for TB interactions in the diene series **1(n)**.

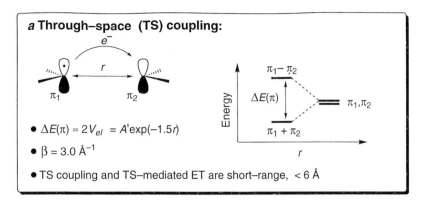

a **Through–space (TS) coupling:**

- $\Delta E(\pi) \approx 2V_{el} = A'\exp(-1.5r)$
- $\beta = 3.0$ Å^{-1}
- TS coupling and TS–mediated ET are short–range, < 6 Å

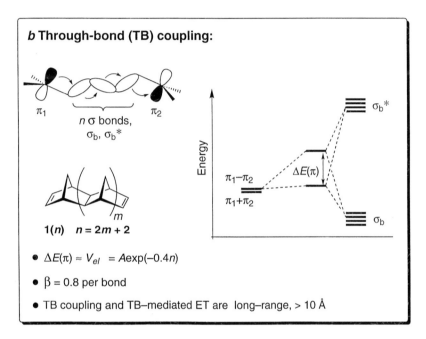

b **Through-bond (TB) coupling:**

1(n)   n = 2m + 2

- $\Delta E(\pi) \approx V_{el} = A\exp(-0.4n)$
- $\beta = 0.8$ per bond
- TB coupling and TB–mediated ET are long–range, > 10 Å

splitting energy, $\Delta E(\pi)$. It is readily shown that $\Delta E(\pi)$ is twice the electronic coupling V_{el} for ET in the system (Figure 6a) and a simple *ab initio* MO calculation predicts that $\beta \approx 3.0$ Å^{-1} for TS coupling.[8] In marked contrast, covalent attachment of the two π-type orbitals to a saturated hydrocarbon bridge, as in the polynorbornanediene series **1(n)**[9] (Figure 6b), leads to a significant through-bond splitting energy which is only weakly attenuated with distance; β = 0.8 per bond (or 0.67 Å^{-1}).[8,10] The difference in the TS and TB β values has a large impact on ET rates. Thus, for ET taking place over 15 Å, the TB-mediated rate is predicted to be 13 orders of magnitude faster than that proceeding by a TS mechanism.

The prediction of weak distance dependence for ET that is mediated through saturated bonds was based on indirect evidence obtained from photoelectron spectroscopic studies and Koopmans' theorem calculations on the dienes **1(n)** (Figure 6).[8] Experimental verification necessitates the measurement of actual ET rates as a function of interchromophore separation. This verification was forthcoming with the advent of the series of bichromo-

phoric systems (dyads) **2(n)**, in which the dimethoxynaphthalene (DMN) donor and dicyanovinyl (DCV) acceptor chromophores are fused to the norbornylogous bridge, ranging in length from four to 13 bonds (Figure 7).

Photoinduced charge separation rates for **2(n)** were found to be extremely rapid, even for the 13-bond system (Figure 7) and they were remarkably insensitive to solvent.[11] That charge separation had, indeed occurred, to form **⁺DMN–B–DCV⁻**, was unequivocally demonstrated using time-resolved microwave conductivity which provides a good estimate of the dipole moment, μ^*, of the product state. The enormous experimental values of μ^* for **2(n)** (Figure 7) are entirely consistent with the formation of the giant dipolar state **⁺DMN–B–DCV⁻**.[12]

Having thereby demonstrated that we were indeed measuring rates of charge separation, the distance dependence of the ET dynamics in **2(n)** could be determined and was found to have a β value of about 0.9 per bond. This value is quite close to that predicted from photoelectron spectroscopic measurements (*vide supra*) and nicely confirms the

Figure 7. Rate data for photoinduced charge separation in the dyads **2(n)**. Charge separation rates, k_{cs}, were measured in THF. Dipole moments, μ^*, were measured in benzene as were the charge recombination lifetimes, τ_{cr}, with the exception of **2(13)**, whose τ_{cr} was measured in 1,4-dioxane (that for **2(4)** was measured in cyclohexane).

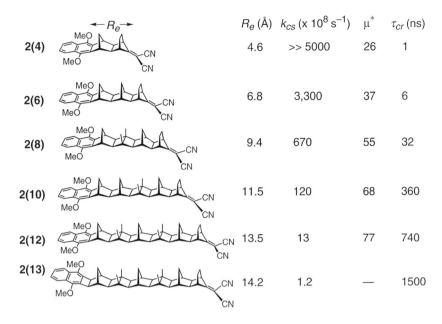

	R_e (Å)	k_{cs} ($\times 10^8$ s^{-1})	μ^*	τ_{cr} (ns)
2(4)	4.6	>> 5000	26	1
2(6)	6.8	3,300	37	6
2(8)	9.4	670	55	32
2(10)	11.5	120	68	360
2(12)	13.5	13	77	740
2(13)	14.2	1.2	—	1500

operation of the TB-mediated ET mechanism for charge separation in **2(n)**. An additional and elegant method of confirming the operation of TB-mediated ET in **2(n)** is to utilize one of the most important properties of TB interactions, namely the *all-trans* rule. This rule states that the magnitude of TB coupling depends on the configuration of the sigma-bridge and is maximized for an *all-trans* configuration and becomes progressively weaker with an increasing number of *cis* or *gauche* "kinks" in the bridge (Figure 8a).[13, 14]

If the TB coupling mechanism were mediating the photoinduced ET processes observed for **2(n)**, then the rates of these processes should be modulated by changes in the configuration of the norbornylogous bridge, in accordance with the *all-trans* rule. This hypothesis was tested using "kinked" systems, such as **3(8)** which possesses two *cisoid* arrangements of bridge bonds (Figure 8b). The photoinduced ET rate for the *all-trans* system **2(8)** was found to be faster than that for the "kinked" molecule **3(8)** by as much as an order of magnitude (Figure 8b), thereby confirming the operation of TB-mediated ET in the **2(n)**–**3(n)** series.[15]

It is amazing just how efficacious the TB mechanism can be in promoting ET. Thus, even for the 12-bond system, in which the DMN and DCV chromophores are 13.5 Å apart, photoinduced ET takes place on a nanosecond timescale with an efficiency approaching that found in the photosynthetic reaction center! Given the remarkable ability of the norbornylogous bridge and other saturated bridges[16] to promote ET over large distances, it is tempting to regard such bridges as molecular wires. However, it would be erroneous to do so, for reasons that will be given later.

The lifetimes, τ_{cr}, of the CS states, $^+$**DMN–B–DCV**$^-$ for direct charge recombination to the corresponding ground states are revealing (Figure 7). For the same bridge length, τ_{cr} is significantly larger than the lifetime, τ_{cs} (= $1/k_{cs}$) for charge separation. This outcome is mainly a consequence of the CS state lying well within the Marcus inverted region.[6,17] As for the charge separation process, charge recombination in $^+$**DMN–B–DCV**$^-$ also proceeds by a TB-

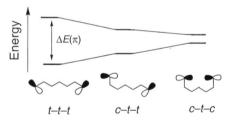

a   The *all-trans* rule of TB coupling:

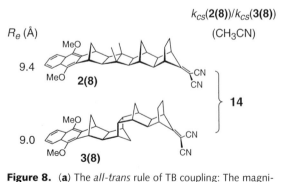

b   Experimental verification

Figure 8. (**a**) The *all-trans* rule of TB coupling: The magnitude of the interaction, $\Delta E(\pi)$, is maximized for an *all-trans* alignment of sigma bonds and becomes progressively smaller with increasing number of *cisoid* conformations. (**b**) Ratio of the photoinduced charge separation rate for the *all-trans* **2(8)** to that for **3(8)** in acetonitrile.

mediated mechanism, although the magnitude of V_{el} is different for the two processes since different orbitals are involved. The distance dependence for the charge recombination rate in $2(n)$ is about 1.0 per bond and is slightly larger than that for charge separation (*ca* 0.9 per bond). The charge recombination lifetimes become progressively longer with increasing bridge length. This trend is to be expected because V_{el} for charge recombination decays exponentially with distance (eq (1)). A fascinating finding is the τ_{cr} value of 1.1 μs for $2(13)$ (in 1,4-dioxane) which is unusually large for a simple dyad.

Orbital symmetry can influence ET dynamics through the electronic coupling term, V_{el}. The rigid, symmetric norbornylogous systems are well-suited for investigating orbital symmetry effects, two examples of which are shown in Figure 9. Charge recombination from the CS state of $2(8)$ is symmetry forbidden since the ground and CS states have opposite state symmetries. This situation arises because the DMN HOMO and the DCV LUMO in the CS state are each singly occupied but have different symmetries with respect to the molecular plane of symmetry, *i.e.*, *a*" and *a*', respectively (the molecular point group is C_s). Consequently, the overall symmetry of the CS state (which is the product of the HOMO and LUMO symmetries) is *A*", whereas that for the ground state is *A*' (the HOMO is now doubly occupied and the LUMO is empty).

That charge recombination in the CS states of the series $2(n)$ is symmetry forbidden is partly responsible for the unusually long lifetimes observed for the CS states (Figure 7). To what extent this orbital forbiddenness influences the CR process in $2(n)$ was gauged by comparing the magnitude of the electronic coupling element, V_{el} for CR in $2(8)$ with that for $4(8)$. In the latter system, the CR process is symmetry allowed because the maleonitrile LUMO has *a*" symmetry, thereby endowing both the CS and ground states of $4(8)$ with *A*' symmetry. From the analysis of charge transfer fluorescence bands that accompanied CR in both $2(8)$ and $4(8)$, the V_{el} values for CR in these molecules were calculated to be 36 and 129 cm^{-1}, respectively (in benzene).[18] This result implies that orbital symmetry influences electronic coupling by a factor of 3.6 and the charge recombination rates by a factor of 13 (*i.e.*, the CR rate varies with the square of the electronic coupling; see the Marcus–Hush expression, Figure 5).[19]

The second example (Figure 9) of orbital symmetry effects concerns photoinduced charge separation in the norbornylogous dyads, $5(7)$ and $6(7)$, which is symmetry forbidden in the former system but symmetry allowed in the latter. In these systems, orbital symmetry effects are more pronounced than in $2(8)$ and $4(8)$, amounting to a 28-fold modulation in the magnitude of V_{el}.[20]

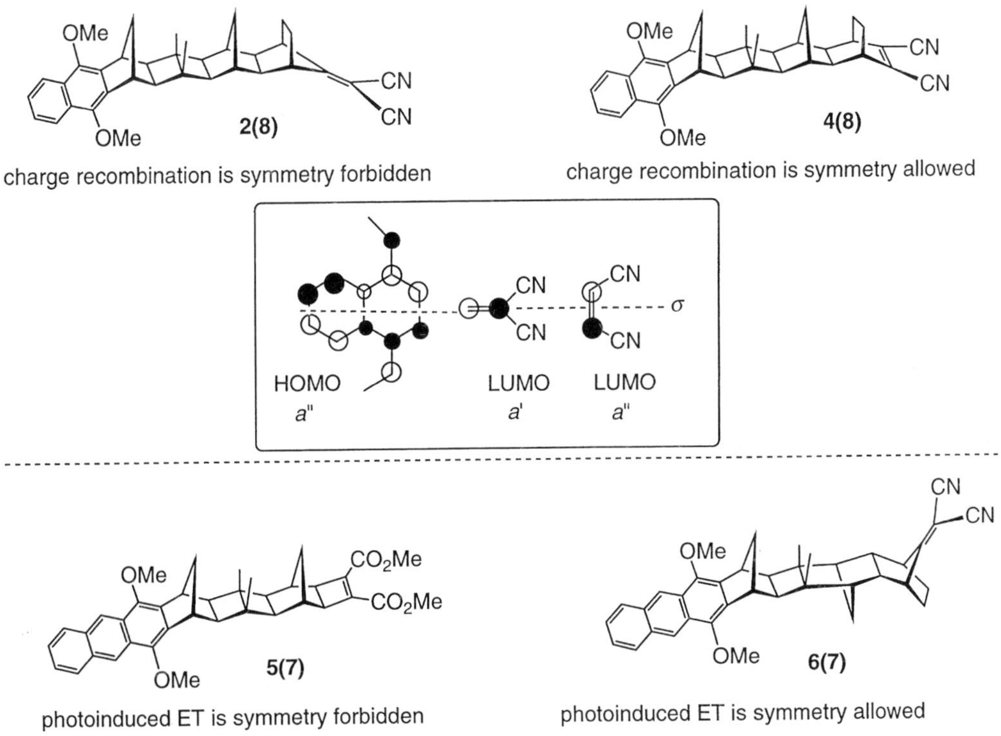

Figure 9. Dyads used for investigating orbital symmetry effects on ET rates. Centre: Orbital symmetries of the active MOs involved in the ET processes for **2(8)** and **4(8)**.

Other Chromophores

The major bulk of studies of ET processes in norbornylogous systems has used the DMN and DCV chromophores. The reason for this choice of donor and acceptor is that, in fairly polar solvents, the magnitude of the free energy change, ΔG_{CS}, for photoinduced charge separation in $2(n)$ happens to be nearly equal in magnitude but opposite in sign to the total reorganization energy, λ. The exponential term in the Marcus–Hush rate expression (Figure 5) is therefore unity, making the photoinduced charge separation process for $2(n)$ barrierless. This situation simplifies the analysis and interpretation of the ET rate data. Nevertheless, other chromophores – representative of which are **7** – **9** in Figure 10, have been

hydrocarbon bridges to mediate long-range ET processes. The CR rate from the CS state of **7(6)** is about one order of magnitude slower than the charge separation rate (Figure 11).

The two C_{60}-based dyads, **8(9)** and **9(11)**, display interesting photophysical features. Rapid (subnanosecond timescale) photoinduced charge separation, from the locally excited porphyrin to the C_{60} acceptor in **8(9)** takes place in benzonitrile (Figure 11). The amazing feature of this system is not so much the rapid rate of charge separation (which is quite unusual!), but the extremely long lifetime of 0.4 μs for the CS state.[22] It is instructive to compare the rate data for this system with those for **10** in which the same chromophores are connected by a single strand *flexible* tether.[23] Whereas the charge separa-

Figure 10. Some porphyrin- and C_{60}-based dyads.

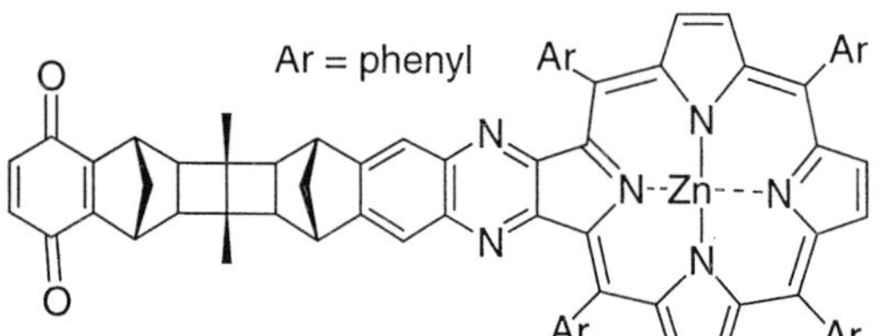

7(6)

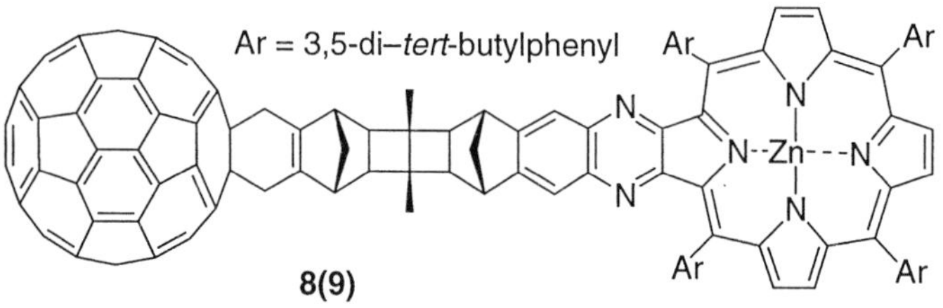

8(9)

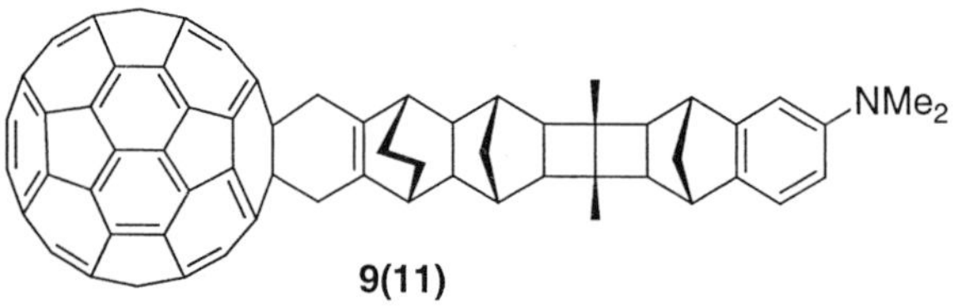

9(11)

investigated and they have provided interesting insights. The 6-bond porphyrin–quinone dyad serves as a model for aspects of ET in the photosynthetic reaction centre since congeners of these chromophores are present in the centre as cofactors (Figure 3). Photoinduced ET in **7(6)** takes place on a nanosecond timescale (Figure 11) with high efficiency, even though the charge separation process for this system is far from barrierless.[21] The charge separation rate is comparable to that found in the photosynthetic reaction centre. As with the $2(n)$ series, the rapid charge separation rate observed for **7(6)** testifies to the extraordinary ability of rigid

tion rates for the two systems are similar, the charge recombination rates are dramatically different, that for **8(9)** being *three* orders of magnitude smaller than that for **10**! This huge difference in the CR rates is attributed to symmetry effects. The rigidity of **8(9)** ensures that the molecule has C_s point group symmetry. Because the singlet CS state of **8(9)** has A'' symmetry, but the ground state has A' symmetry, the CR process is formally symmetry forbidden and this restriction slows down the CR rate. The flexible nature of the tether in **10** removes this symmetry restriction and the CR rate is, therefore, more rapid in this system.

Figure 11. Rates of photoinduced charge separation and subsequent charge recombination processes in dyads (in benzonitrile).

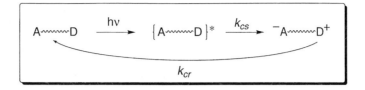

$A\!\!\sim\!\!\sim\!\!\sim\!\!D \xrightarrow{h\nu} \{A\!\!\sim\!\!\sim\!\!\sim\!\!D\}^* \xrightarrow{k_{cs}} {}^-\!A\!\!\sim\!\!\sim\!\!\sim\!\!D^+$

k_{cr}

	k_{cs} (s^{-1})	k_{cr} (s^{-1})
7(6) Ar = phenyl	2.7×10^9	2.0×10^8
8(9) Ar = 3,5-di-*tert*-butylphenyl	1.0×10^{10}	2.4×10^6
10 Ar = 3,5-di-*tert*-butylphenyl	9.0×10^9	2.0×10^9

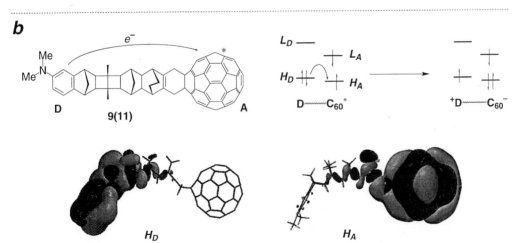

a

four σ–π interactions

twelve
σ–π interactions

b

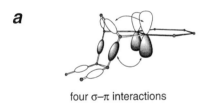

9(11)

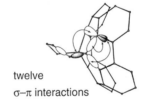

L_D —

H_D ⥮   ⥮ L_A
         ⥮ H_A

$D\!\!\sim\!\!\sim\!C_{60}{}^*$

→

—   ⥮
⥮   ⥮

$^+D\!\!\sim\!\!\sim\!C_{60}{}^-$

H_D

H_A

Figure 12. (**a**) Interactions between the orbitals of the bridge and those of a planar aromatic molecule and C$_{60}$. In the former, there are four interactions, two of which are shown, the other two involve the corresponding orbitals in the background. In the case of C$_{60}$, three interactions are shown and there are three more similar groups of three interactions involving symmetry-related orbitals, making a total of twelve interactions. (**b**) The active orbitals involved in photoinduced charge separation. H_D is the HOMO of the aniline donor and H_A is the HOMO of the C$_{60}$ acceptor; the orbital plots were obtained from AM1 calculations.

The aniline-C_{60} dyad **9(11)** was no less spectacular than **8(9)**, considering the greater interchromophore separation that exists in this system (11 bonds), compared to **8(9)** (9 bonds). The photoinduced charge separation rate for **9(11)** is about $6 \times 10^9 \ s^{-1}$ and the CR lifetime is 250 ns (both in benzonitrile).[24] Why are charge separation processes involving the C_{60} acceptor so rapid in our systems? Intuitively, one would have predicted the opposite behavior; because of its large size, the MOs of the C_{60} chromophore have small atomic coefficients compared to those of most other chromophores in use, such as DMN. Thus, the magnitude of the coupling of the C_{60} group to the bridge would be expected to be small. However, unlike the normal chromophores, which are generally planar, the curved nature of the C_{60} surface enables it to couple to the bridge using many more atoms than, say the DMN group, as illustrated in Figure 12a. There are twelve interactions between the orbitals of the C_{60} chromophore and the bridge, but only four interactions between the orbitals of a planar aromatic chromophore and the bridge. The result is strong coupling between the C_{60} group and the bridge and this is manifested by the extensive delocalization of the

C_{60} HOMO into the bridge (Figure 12b).[24,25] Another factor which generally facilitates photoinduced charge separation in *all* dyads containing the C_{60} acceptor is the smaller reorganization energy, λ, for C_{60}, compared to most acceptor groups.[26] From the Marcus–Hush expression (Figure 5), it may be seen that, for ET taking place within the normal region,[17] the activation barrier decreases and the corresponding ET rate *increases* with *decreasing* λ.[27]

The Dependence of the Magnitude of β on the Nature of the Bridge

Some experimentally determined β values for ET processes mediated by different bridges are shown in Figure 13. For the norbornylogous bridge, β varies only slightly with the nature of the ET process, *e.g.* hole transfer (HT)[28] in cation radicals of **1(n)**, determined by photoelectron spectroscopy (Figure 13a),[13] photoinduced charge separation (Figure 13b), charge recombination (Figure 13c) and optically induced charge-shift (Figure 13d). The β values for alicyclic and steroid bridges (Figure 13e)[16] and oligospiro bridges (Figure 13f)[29] are similar to those for the

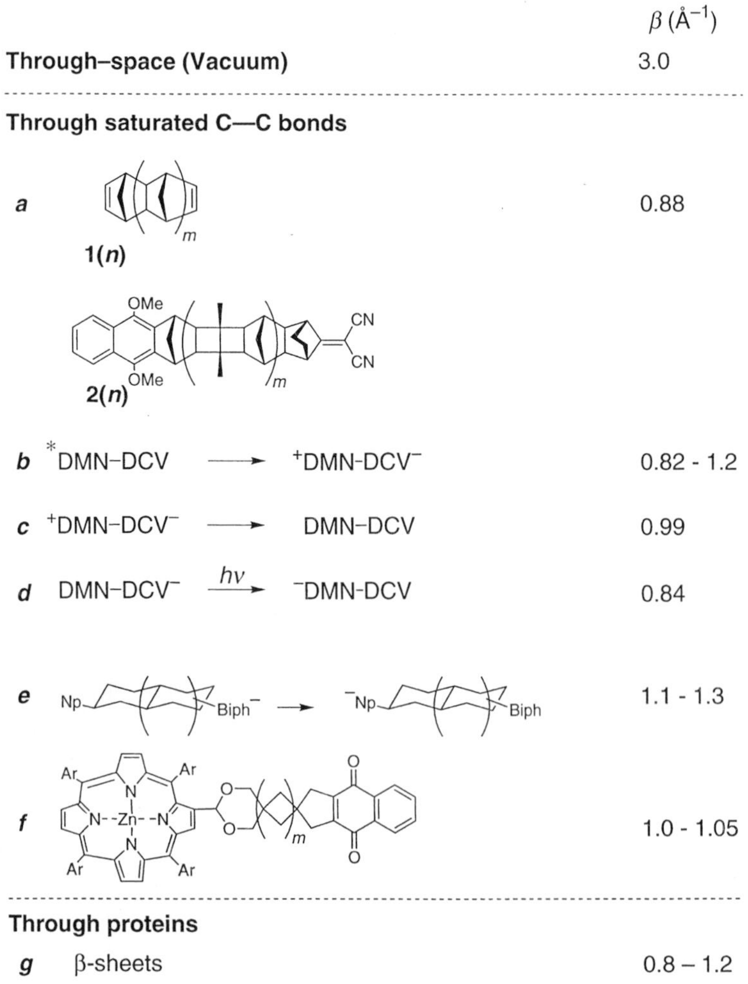

Figure 13. Distance dependence β values for various bridges.

		$\beta \ (\text{Å}^{-1})$
	Through–space (Vacuum)	3.0
	Through saturated C—C bonds	
a	**1(n)**	0.88
	2(n)	
b	*DMN–DCV $\longrightarrow$ +DMN–DCV−	0.82 - 1.2
c	+DMN–DCV− $\longrightarrow$ DMN–DCV	0.99
d	DMN–DCV− $\xrightarrow{hv}$ −DMN–DCV	0.84
e	Np...Biph− $\longrightarrow$ −Np...Biph	1.1 - 1.3
f		1.0 - 1.05
	Through proteins	
g	β-sheets	0.8 – 1.2

norbornylogous bridge. Thus, electronic coupling through saturated bridges appears to be somewhat insensitive to the structure of the bridge although it should be borne in mind that the majority of experimentally determined β values have not been corrected for contributions from the distance dependence of reorganization energy (*vide supra*). Interestingly, protein bridges display a range of β values similar to those displayed by the hydrocarbon bridges (Figure 13g).[30]

In summary, electron transfer dynamics, mediated through saturated hydrocarbon bridges and proteins, displays a surprisingly weak distance dependence behaviour (β = 0.8–1.2 Å^{-1}), compared to that predicted for a pure through-space mechanism ($\beta \approx 3.0$ Å^{-1}).

An important question remains to be answered: although the phenomenological β value appears to be fairly insensitive to the nature of the hydrocarbon bridge, does it necessarily follow that β_{el}, for the distance dependence of the electronic coupling, V_{el}, should likewise display a similar insensitivity? Extracting V_{el} values, and hence β_{el}, from experimental ET rate data is presently fraught with difficulties and uncertainties. An alternative approach to this problem is to calculate the couplings. A par-

ticularly simple, but approximate, way of doing this calculation is to obtain the π-orbital splitting energies in symmetrical dienes using *ab initio* MO theory. These splittings are equal to twice the electronic coupling for hypothetical HT in the cation radicals of the dienes. The results from these types of calculations revealed some nice surprises and led to the development a new stimulating concept, that of a superbridge! The relevant data, shown in Figure 14a, indicate that the magnitude of β_{el} actually does depend on the nature of the hydrocarbon bridge.

Interestingly, β_{el} is significantly larger for the hybrid polynorbornane-bicyclo[2.2.0]hexane bridge in the diene series 11(n), than for the polynorbornane diene system, 1(n), meaning that the distance dependence for the decay of the electronic coupling is much stronger in the former system. Thus, for the 10-bond systems, the TB-coupling is already three times weaker for 11(10) than for 1(10). The clue to solving this mystery came from noting that the β_{el} values for both doubly stranded bridges in 11(n) and 1(n) are larger than that for the singly stranded alkane bridge in 12(n).[31]

Naïvely, one might have expected that TB coupling in 11(n) and 1(n) should be stronger and display

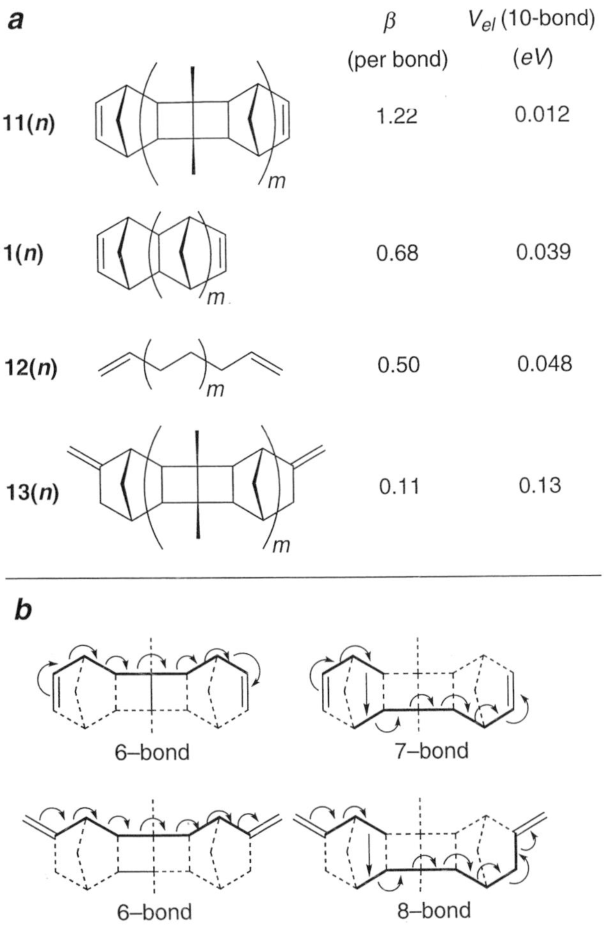

Figure 14. (a) HF/3–21G calculated β values for four different bridges. **(b)** Examples of the parities of some coupling pathways; *n*-bond means coupling through *n* bridge bonds. The bonds involved in each coupling path are highlighted.

		β (per bond)	V_{el} (10-bond) (eV)
11(n)		1.22	0.012
1(n)		0.68	0.039
12(n)		0.50	0.048
13(n)		0.11	0.13

6–bond   7–bond

6–bond   8–bond

a weaker distance dependence than that in **12(n)** because the former pair of systems each has two main TB-coupling relays, compared to only one in the latter. However, the two relays in **11(n)** and **1(n)** are spatially sufficiently close to each other to interact and this interaction establishes new coupling pathways between the two double bonds that "jump" from one main relay to the other (Figure 14b). Now, it is a rule of TB theory that the *sign* of the TB splitting energy through a single stranded relay depends on the parity of the number of bonds in the relay.[13,14] Thus, the net coupling through two relays having the same parity is given by the sum of the absolute magnitude of the coupling associated with each relay and is therefore strengthened by what may be called *constructive interference*. In contrast, the net coupling through relays having opposite parities is given by the difference in the absolute magnitude of the coupling associated with each relay and is therefore diminished by *destructive interference*.

From Figure 14b, it is easily seen that, for **11(n)** and also **1(n)** (not shown), coupling pathways that jump from one main relay to the other have opposite parity (*e.g.* 7-bonds) to that for the main relay (6-bonds). Consequently, the coupling in these dienes is degraded by destructive interference effects. In addition, the destructive interference is more severe in **11(n)** than in **1(n)**, the reason for this being that the main relays are held closer together in the former diene by the bicyclo[2.2.0]hexane units. This reasoning is fully consistent with the calculated trends in the β_{el} and V_{el} values along the series **11(n)**, **1(n)**, **12(n)** (Figure 14a).

This simple and intuitive analysis of interference effects on coupling through bridges, based on the parity rule of TB interactions, may be used advantageously to design systems in which the TB coupling is greatly enhanced by *constructive interference* effects.[31] This approach is exemplified by the series **13(n)**. In this system, all reasonable TB coupling pathways have the same (even) parity (Figure 14b); the prediction may therefore be made that the net TB coupling in **13(n)** should be superior to that in **11(n)**, even though both systems possess the same type of bridge. Gratifyingly, calculations confirmed this expectation. The calculated β_{el} value for **13(n)** is *ten times smaller* than that for **11(n)** and the strength of the coupling for a given bridge length is greater for the former. For example, V_{el} for **13(10)** is ten times larger than that for **11(10)** and this translates into a predicted two orders of magnitude rate enhancement for HT in the cation radical of **13(10)**, relative to **11(10)** (remember that rate depends on the *square* of V_{el}). Bridges such as **13(n)** deserve the epithet superbridges!

Before closing this section on TB-mediated ET, it is necessary to address the oft-asked question of whether the hydrocarbon bridges, on account of their ability to strongly mediate ET, may be considered to possess molecular wire (or electrical conducting) behaviour. The answer is an unequivocal no: they may not be regarded as wires. Electrical conduction through a bridge requires that the electron from the donor becomes thermally injected into the conduction band of the bridge and that it actually becomes localized within the bridge and is transported through the bridge, from donor to acceptor, by an incoherent scattering mechanism, such as a polaron. The distance dependence of the ET rate in such a molecular wire is determined by Ohmic scattering and therefore varies inversely with bridge length.[32] The molecular wire mechanism is summarized in Figure 15. It is expected to operate only when the energy gap, ΔE, between the donor level and the bridge conduction band is very small, of the order k_BT (k_B is the Boltzmann constant and T is the temperature). This condition is satisfied for long, conjugated bridges, such as graphite.

In contrast, in the superexchange (or TB) mechanism, ΔE is very large, at least 2 eV[18] and so the electron cannot be thermally injected into the bridge. Instead, the electron moves coherently, in one sudden "jump", from donor to acceptor and it never becomes localized within the bridge. The distance dependence of the ET rate for this mechanism is exponential decay (Figure 15). The conduction band for saturated hydrocarbon bridges corresponds roughly to the empty σ^* orbital levels. Since these levels are several eV higher in energy than the π and π^* levels of most unsaturated donor systems, ET mediated by such bridges can normally occur only by the superexchange mechanism.

It is useful to introduce a nomenclature for distinguishing between ET occurring by the conduction and superexchange mechanisms. The term electron *transport* is used in the context of molecular wire behavior, while electron *transfer* is used in the context of the superexchange mechanism.

There is a simple way of visualising the superexchange mechanism. Consider ET taking place from dimethylaniline donor to the locally excited singlet state of C_{60} in **9(11)** (Figure 12b). The active orbitals which are primarily involved in the ET process are the aniline HOMO (H_D) and the half-filled C_{60} HOMO (H_A). Because of TB (superexchange) coupling of the bridge to the chromophores, the HOMOs of the chromophores are no longer localized to their respective chromophores, but are extended into the bridge. This characteristic is evident from the AM1 plots for H_A and H_D (Figure 12b). These orbital exten-

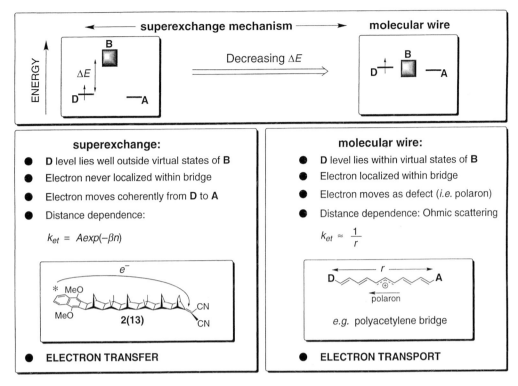

Figure 15. A schematic illustrating the difference between the superexchange mechanism and molecular wire behavior in a **D–B–A** dyad. Superexchange: the bridge states lie above the **D** level; consequently the electron is transferred in one coherent jump and is never localized within the bridge. The distance dependence behavior is exponential decay.

Molecular wire behavior: The bridge states are energetically comparable to the donor level; the electron may be thermally injected into the bridge, whereupon, it moves from donor to acceptor incoherently, as a defect such as a polaron. The distance dependence behavior is Ohmic (varies inversely with distance).

sions into the bridge result in favourable overlap between H_A and H_D and this enables ET to take place.

In principle, it should be possible to design a series of **D–B–A** dyads with a bridge, **B** possessing suitable electronic characteristics such that the superexchange mechanism prevails for short bridge lengths, but switches over to molecular wire behavior for bridge lengths exceeding a certain critical value. In order to achieve this possibility, the bridge must meet the requirement that its conduction band level falls sufficiently steeply with increasing bridge length so that it approaches the highest filled level of the donor, at which point electron transport could be the dominant mechanism (Figure 15).

A suitable conjugated bridge should meet this requirement and Wasielewski's group has recently used this stratagem to demonstrate that the mechanism does, indeed, switch from electron transfer to electron transport for a critical bridge length.[32] The systems, shown in Figure 16, comprise a tetracene (TET) donor and pyromellitimide (PI) acceptor, both covalently linked to *p*-phenylenevinylene oligomers of increasing length, ranging from one aromatic ring, in **14(1)**, to five, in **14(5)**. Rates of photoinduced

charge separation, from locally excited TET to PI, were measured for this series in methyltetrahydrofuran. Two distinct ET behaviours were observed (Figure 16, inset). For the two shortest bridges, the charge separation rate falls quite strongly with a β value of about 0.4 Å⁻¹. This distance dependence behaviour is consistent with the operation of the superexchange mechanism for charge separation in **14(1)** and **14(2)**. In contrast, there is an abrupt change in mechanism beginning with **14(3)** for which the charge separation rate is greater than that for **14(1)**, even though the bridge in the former system is 13 Å longer than in the latter! Moreover, extremely weak distance dependence of the charge separation rates is observed for the longer members of the series **14(3)** – **14(5)**, for which β is only 0.04 Å⁻¹. The distance decay characteristics for the charge separation rates for **14(3)**–**14(5)** were interpreted in terms of molecular wire behaviour: in these systems, the LUMO of the bridge approaches that for the locally excited TET chromophore and thermal injection of a TET electron into the bridge becomes energetically plausible.[33]

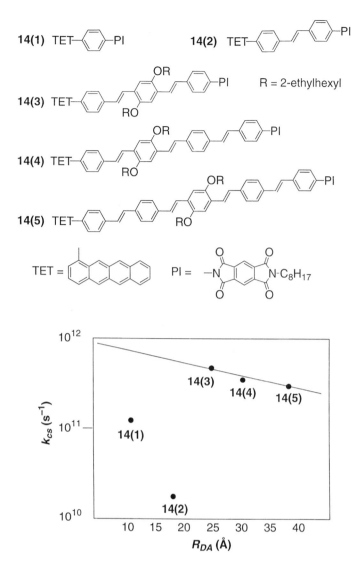

Figure 16. The series of dyads used for investigating the switchover from super-exchange characteristics to molecular wire behavior. Also shown is a schematic of the photoinduced charge separation rate *versus* donor-acceptor distance.[33]

Electron Transfer Mediated Through Hydrogen Bonds

Given that ET through proteins is fundamental to many biological processes and that H-bonds are prevalent in proteins, it is important to determine how efficiently ET can be mediated by H-bonds. Of the several studies which have been reported, that by Therien is of particular interest since it provides a comparative study of ET mediated by saturated bonds and by H-bonds.[34]

The Zn(II)porphyrin–bridge–Fe(III)porphyrin systems, together with their photoinduced ET rate constants are shown in Figure 17. The ET rate constant for the H-bonded system **15** is comparable to that for **16**, which possesses a partially unsaturated hydrocarbon bridge, whereas the rate for **17**, which contains a fully saturated bridge, is notably slower than the ET rate for **15**. These data suggest that electronic coupling mediated by H-bonds is significant and that, therefore, H-bonded networks may provide dominant pathways for ET in proteins.[35]

Modulation of Electron Transfer Dynamics by Electric Fields

We have discussed how bridge configuration and orbitals symmetry effects may modulate ET dynamics. We now turn to another important influence on ET dynamics, namely electric fields. It is reasonable that the presence of a strong electric field should influence the dynamics of charge separation processes because the dipole moment associated with the newly formed CS state will interact with the field. This interaction will modify the barrier height for the charge separation process. Indeed, it has been postulated that electric field effects might be the cause of the observed directionality of electron transfer in the photosynthetic reaction centre (Figure 3).

The dependence of ET rates on the strength and direction of the electric field has been investigated using the rigid synthetic helical peptides **18** and **19** (Figure 18), possessing covalently-linked dimethylaniline (An) donor and pyrene (Pyr) acceptor.[35] The

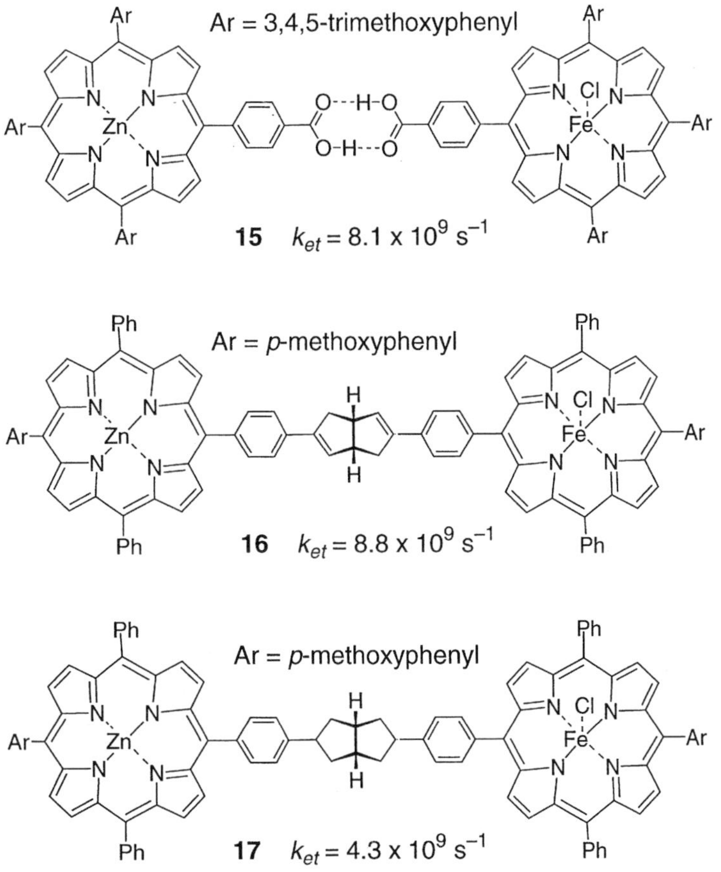

Figure 17. Three dyads possessing Zn(II) porphyrin donor and Fe(III) porphyrin acceptor linked by (from top to bottom) H-bonded bridge, a partially unsaturated bridge and a saturated bridge.[34]

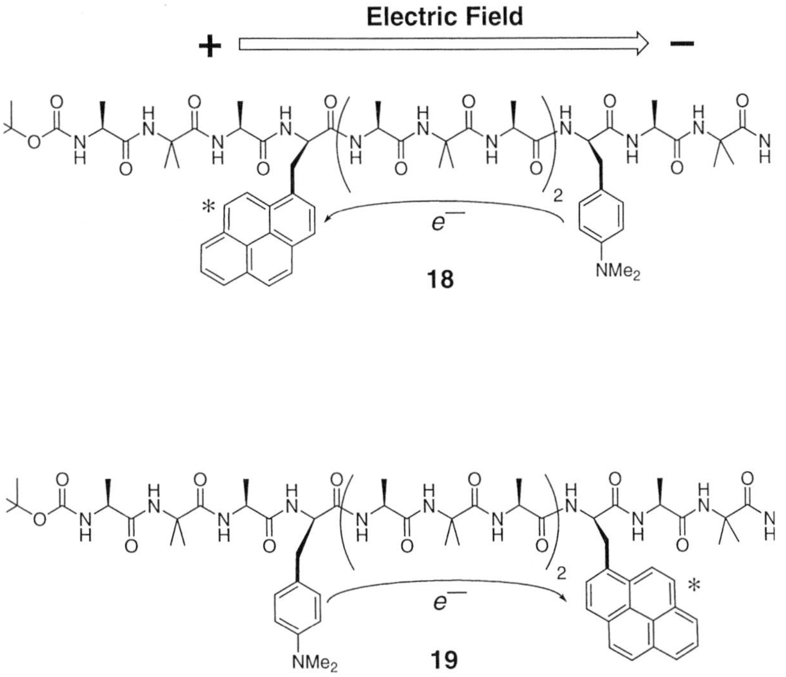

Ar = 3,4,5-trimethoxyphenyl

15 $k_{et} = 8.1 \times 10^9$ s^{-1}

Ar = p-methoxyphenyl

16 $k_{et} = 8.8 \times 10^9$ s^{-1}

Ar = p-methoxyphenyl

17 $k_{et} = 4.3 \times 10^9$ s^{-1}

Figure 18. The helical oligopeptide-based dyads used for investigating electric field effects on ET dynamics.

Electric Field

18

19

photoinduced charge separation process involves transfer of an electron from the aniline to the locally excited singlet state of the pyrene (Pyr*) to give the CS state, $^+$An–peptide–Pyr$^-$. In **18**, the dipole of the resulting CS state is aligned antiparallel to the electric field vector generated by the peptide backbone. The CS state of **18** is therefore stabi-

lized by the presence of the electric field (the positive and negative ends of which are located at the nitrogen and carboxylate termini, respectively). In **19**, the positions of the two chromophores are switched, everything else remaining unchanged; in this case, the CS state is now destabilized (relative to **18**) by interaction with the electric field.

Consequently, it is predicted that the charge separation rate should be greater for **18** than for **19**, since the former is associated with a more negative free energy change and therefore, from the Marcus–Hush expression (Figure 5), with a lower activation barrier. In agreement with qualitative prediction, it was found that the ratio of the photoinduced charge separation rates $k_{cs}(18):k_{cs}(19)$ is 27 in THF, a low polarity solvent, but fell to only five in the more polar solvent, methanol.[35] The attenuation of the ET rate ratio with increasing solvent polarity is also consistent with the electric field mechanism since the interaction between the electric field of the protein and the dipole of the CS state will be weakened in more polar solvents.

Electron Transfer Mediated by Solvent Molecules

That electron transfer may be facilitated by a TB or superexchange mechanism raises the possibility that solvent molecules may also mediate ET. Presumably, the electronic coupling through solvent will not be as strong as that through a covalently-linked bridge because the solvent molecules are further apart, both from each other and from the chromophores. In addition, the solvent molecules are tumbling about and this disorder will lead to an electronic coupling whose magnitude is less than that resulting from freezing the solvent molecules in the optimal configuration. An indication that ET could be effectively mediated by saturated hydrocarbon solvents came from early photoelectron spectroscopic and ESR studies on U-shaped dienes such as **20** (Figure 19).[8] These studies revealed a remarkably large π-splitting energy of 0.52 eV for **20** which could only be explained in terms of substantial interactions between the π MOs of the double bonds and the pseudo-π orbitals of the central

methano bridge (Figure 19, inset). Subsequent calculations predicted a large π-splitting of 0.32 eV for the molecular sandwich **21** which models electronic coupling between the double bonds and a methane "solvent" molecule. Given that a coupling of only 0.003 eV is sufficient to accelerate long-range ET, a predicted splitting of 0.32 eV for **21** is amazing and suggests that solvent-mediated ET should be an important process, although not as dramatic as that suggested by the idealized solvent-substrate configuration shown by **21**.

Experimental progress is being made in the investigation of solvent-mediated ET processes through the synthesis of rigid U-shaped systems based on the norbornylogous bridge, representative of which are **22–24** (Figures 20–22). In these systems, the terminal chromophores face each other across a "rigid" cavity created by a norbornylogous bridge. The length and configuration of the connecting bridge are designed to minimize TB-mediated ET through the bridge while encouraging solvent-mediated ET to occur by providing a cavity within which a certain number of solvent molecules are present (in a dynamic sense).

The dyad **22** possesses a cavity *ca.* 10 Å wide which is sufficiently large to accommodate a solvent molecule, such as benzonitrile (Figure 20).[36]

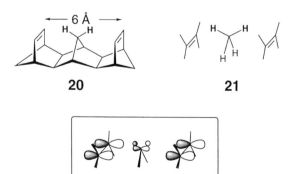

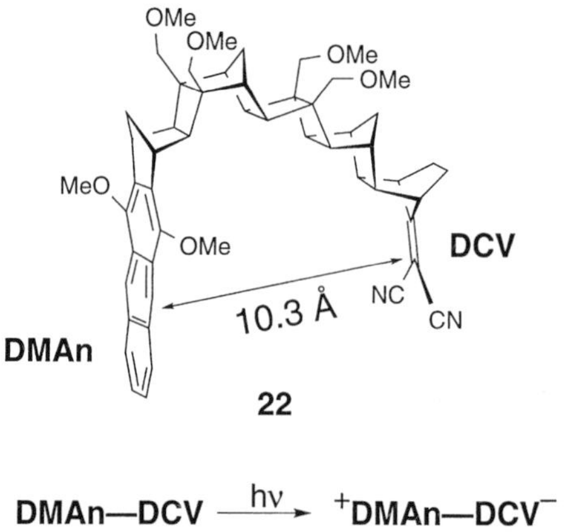

$$DMAn\text{—}DCV \xrightarrow{\ h\nu\ } {}^{+}DMAn\text{—}DCV^{-}$$

Figure 20. U-shaped dyad **22**.

Such a molecule would be able to provide a coupling pathway for mediating photoinduced charge separation between the locally excited singlet state of the dimethoxyanthracene (DMAn) donor and the DCV acceptor. Values of the electronic coupling, V_{el}, for photoinduced charge separation in **22** in various solvents were extracted from the rate data. For benzonitrile solvent, $V_{el} = 14$ cm^{-1}, whereas for acetonitrile, $V_{el} < 2$ cm^{-1}. These data were interpreted in terms of solvent-mediated ET using the following argument. Because photoinduced charge separation in **22** involves the LUMOs of the two chromophores, the strongest coupling with the solvent molecule

20   **21**

Figure 19. Interactions between the π MOs of the two double bonds and the pseudo-π orbitals of the methano group in **20** and the methane molecule in **21**.

will occur with the latter's LUMO rather than with its HOMO. Consequently, the magnitude of V_{el} for solvent-mediated charge separation in **22** should *increase* with *increasing* electron affinity of the solvent molecule. This prediction is in accord with observation since the electron affinity of benzonitrile is greater than that of acetonitrile.[36]

The dynamics of charge recombination from the giant CS states of ***syn*-23** and ***anti*-23**, $^+$**An–DMN–DCV**$^-$, were surprising (Figure 21).[37] The two molecules are diastereoisomers, the former possessing a rather large U-shaped cavity that is absent in the latter. It was found that charge recombination took place at least two orders of magnitude faster in the *syn* isomer compared to the *anti* isomer. Solvent-mediated charge recombination taking place in the *syn* isomer could explain this result but there is a problem with this explanation in that the CR rates are rapid, even in saturated solvents such as decalin, which have very high lying LUMOs. Electronic coupling of the donor and acceptor LUMOs to this type of solvent molecule will, therefore, be weak and it will be further weakened by the large cavity size in the *syn* isomer which would require coupling through at least two solvent molecules.

A plausible solution to this dilemma has been found. Gas phase *ab initio* MO calculations were carried out on the ground and CS states of an analogue

of ***syn*-23** in which 1,4-dimethoxybenzene (DMB) replaces the aniline chromophore (Figure 21b). Whereas the terminal chromophores in the optimized ground state structure of **DMB–DMN–DCV** are about 15 Å apart, this separation is predicted to be only about 4.3 Å in the giant CS state $^+$**DMB–DMN–DCV**$^-$![38] This large contraction in the inter-terminal separation, upon creation of the CS state is the result of electrostatic attractions between the oppositely charged terminal chromophores. The contraction is achieved by out-of-plane bendings of

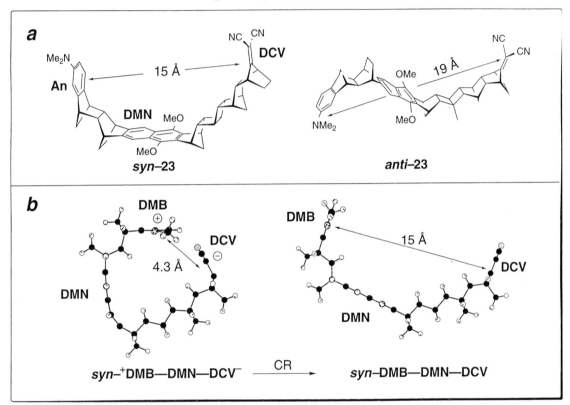

Figure 21. (a) Two trichromophoric systems used for investigating through-solvent-mediated charge recombination processes. (b) Profiles of the (U)HF/3–21G gas phase structures of the CS state (left) and ground state (right) of an analogue of ***syn*-23** in which the dimethylaniline donor is replaced by 1,4-dimethoxybenzene (DMB).

the aromatic rings and the DCV anion radical, both energetically not very costly processes. Although these calculations refer to the gas phase, such geometric distortions should also occur in solvents of low dielectric constant, such as decalin. It is noteworthy that the small inter-terminal separation in the CS state of ***syn*-23** might enable *direct* through-space mediated charge recombination to occur, thereby nicely explaining why CR is so fast in weakly polar solvents of low electron affinity.

The photophysical results[39] for the giant porphyrin-dimethoxynaphthalene-naphthoquinone-viologen tetrad, **P–DMN–NQ–MV^{2+}**, **24** (Figure 22), complement those obtained for **23**. Photoinduced charge separation, to form the giant CS state, $^+$**P–DMN–NQ–MV**$^+$, readily occurs in the *syn,syn* isomer, but not in the *anti,syn* isomer. This result

Figure 22. (**a**) Two tetra-chromophoric systems **24**. (**b**) Profiles of the (U)HF/3–21G gas phase structures for the "collapsed" ground state and giant CS state of **syn,syn-24**. The "normal" ground state structure was located using the AM1 semi-empirical method.

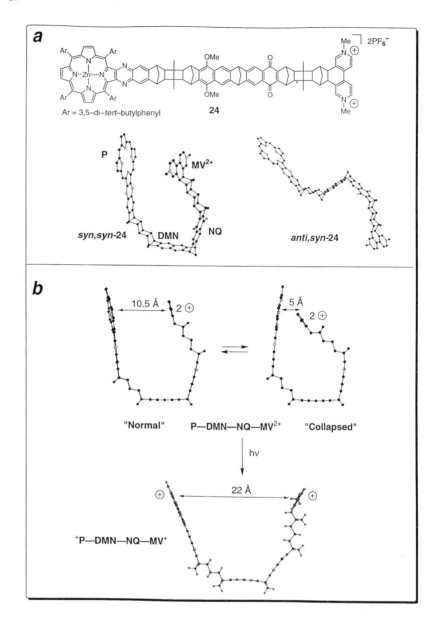

immediately informs us that the charge separation process is not occurring through the bridge in **syn,syn-24** because if it did, then photoinduced charge separation should also be observed in the *anti,syn* isomer, since the bridges in the two systems are identical in terms of length and composition. As observed for the charge recombination behavior for **syn-23**, the rate of photoinduced charge separation in **syn,syn-24** appears to be insensitive to solvent electron affinity. Another puzzle was the finding[39] that the giant CS state is quite long-lived, 500 ns, which stands in marked contrast to the very short lifetime of *ca* 0.1 ns for the CS state of **syn-23**.

The puzzle may be explained by the operation of electrostatically induced geometric distortion in *both* ground and giant CS states of **syn,syn-24**. Gas phase *ab initio* MO calculations indicate that there is an electrostatic attraction between the zinc porphyrin and the viologen dication in the ground state of

syn,syn-24. This attraction results in a "collapsed" conformer, in which the porphyrin and viologen groups are only about 5 Å apart (Figure 22b).[40] This "collapsed" structure is largely brought about by out-of-plane bending of the central pair of naphthalene rings. AM1 calculations suggest the presence of another, higher energy conformer in which the termini are at the "normal" separation of *ca* 10 Å, corresponding to a fully relaxed bridge. Perhaps these conformers are in rapid equilibrium. Photoinduced charge separation in the "collapsed" conformer could take place by a direct through-space mechanism without requiring electronic participation of the solvent. This mechanism would explain the solvent independence of the rate of photoinduced charge separation observed for **syn,syn-24**. Because of the large inter-terminal separation of 10.5 Å in the "normal" conformer of **syn,syn-24**, charge separation in this conformer may be too slow to compete with unproductive decay of the locally excited state of the porphyrin moiety to the ground state.

Gas phase geometry optimization of the giant CS state $^+$P–DMN–NQ–MV$^+$ offers an explanation for its very long lifetime. In this species, the terminal chromophores, both being positively charged, repel each other. As a result, the inter-terminal separation increases markedly, to about 22 Å (Figure 22b)! Both through-space and through-solvent-mediated charge recombination in $^+$P–DMN–NQ–MV$^+$ possessing this geometry would be unfavourable and this would explain its long lifetime.

Gas phase calculations on the CS state of **22** also predict a large change in the cavity size, from 10 Å in the ground state, to only about 6 Å in the CS state, and that this change is mainly caused by out-of-plane bending of both DCV and DMAn groups.[38] Similar geometric distortions are predicted for other U-shaped dyads in their CS states.

In summary, solvent-mediated ET studies are proving a challenge! The prediction that "rigid" U-shaped systems undergo large structural distortions in their CS states, particularly in low dielectric solvents, raises issues of interpretation of the data that are still being resolved.

Through-Bond-Mediated Electronic Excitation Energy Transfer

The transfer of excitation energy from a locally excited donor chromophore to an acceptor, resulting in the formation of the electronically excited acceptor state, is known as electronic excitation energy transfer (EET). There are two basic EET mechanisms, namely the Förster and the Dexter (exchange) mechanisms. In the Förster mechanism (Figure 23a), EET takes place through the interaction of the oscillating dipole of the excited donor with that of the acceptor. This dipole–dipole interaction is long-range and the rates of EET by the Förster mechanism decay inversely with the sixth power of the interchromophore separation. Note that because electrons one and two (Figure 23a) do not exchange their locations during the EET process, there is no dependence of EET dynamics on the donor–acceptor orbital overlap in the Förster mechanism. In the exchange mechanism (Figure 23b), the two electrons exchange orbitals and so the EET dynamics now depend on inter-orbital overlap and therefore should display an exponential decay with distance. In fact, the exchange mechanism may be regarded as a double ET process, one electron moving from the donor LUMO to the acceptor LUMO and the other from the acceptor HOMO to the

Figure 23. Two principal mechanisms of excitation energy transfer (EET). (**a**) The Förster dipole–dipole mechanism, in which the active electrons, one and two, remain, respectively, on **D** and **A** throughout the process. (**b**) In the (Dexter) exchange mechanism, electrons one and two exchange locations.

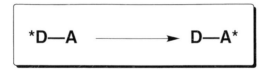

a Förster Mechanism

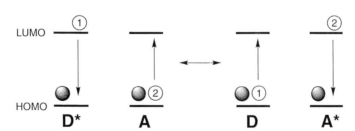

b Exchange Mechanism (Dexter)

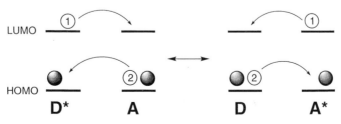

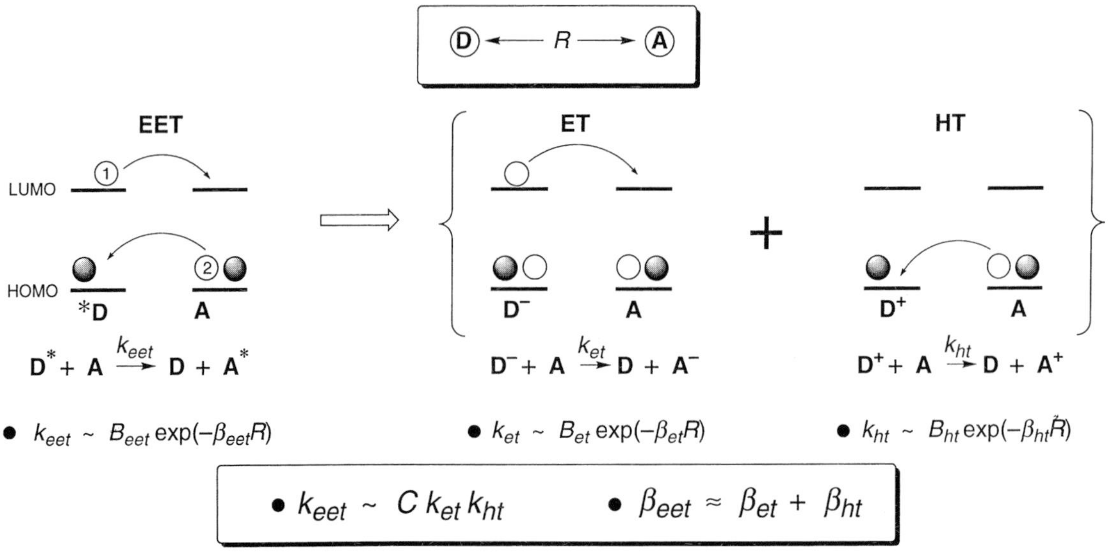

Figure 24. Analysis of the exchange mechanism for EET in terms of two electron transfer processes, one involving ET in the corresponding anion (centre) and the other involving hole transfer (HT) in the corresponding cation (right-hand side).

donor HOMO. This important insight is illustrated in Figure 24, from which it should be clear that the damping factor, β_{eet}, for EET occurring by an exchange mechanism is approximately given by the sum of the damping factors for two separate ET processes, β_{et} for electron transfer in the anion radical and β_{ht} for hole transfer in the cation radical. This prediction has been confirmed.[41]

Although the exchange mechanism was originally formulated in terms of direct orbital overlap between the donor and acceptor chromophores, it is clear that it can be extended to cover the case of through-bond mediated EET. The reason is because TB coupling provides a mechanism for spatially extending the active orbitals of the two chromo-

phores into the intervening medium, thereby facilitating interchromophore orbital overlap. TB-mediated EET has been explored using our rigid norbornylogous systems (Figure 25).

The experiments involved measuring the rates of singlet-singlet EET from the first excited singlet state of DMN to the ketone or dione chromophore. The distance dependence of the EET rates in **25(n)** follow an exponential decay, rather than an inverse sixth power law, thereby confirming that an exchange mechanism and not a Förster mechanism is operating in these systems.[42] The magnitude of β_{eet} was found to be about 2.1 per bond and is about double that found for photoinduced charge separation in the structurally related **DMN–B–DCV** sys-

Figure 25. Rate data for singlet-singlet EET between locally excited DMN donor and carbonyl or dione acceptor in various norbornylogous systems.

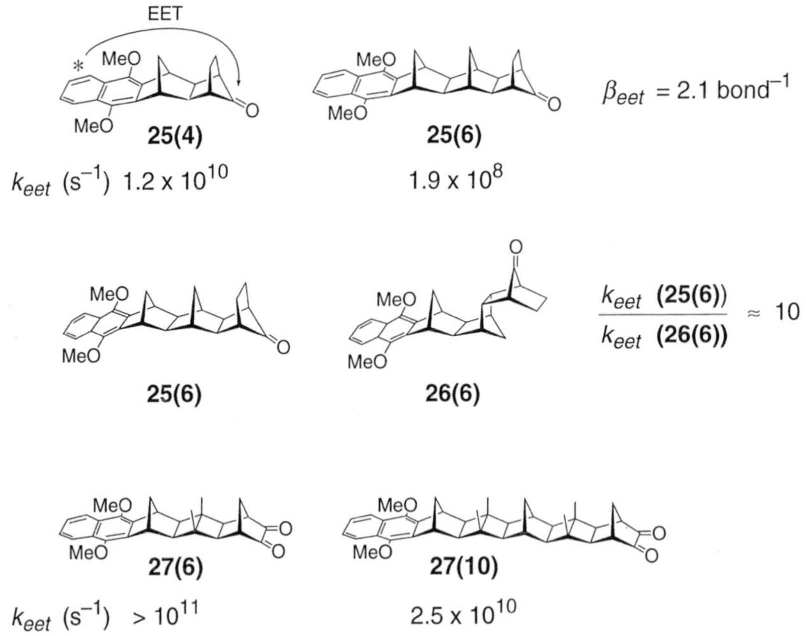

tems **2(n)** (*vide supra*). This relationship agrees with the postulate that β_{eet} is the sum of damping factors for two ET processes. Consistent with the TB-mediated EET mechanism was the finding that singlet-singlet EET occurs about 12 times more rapidly in the *all-trans* system **25(6)** than in the "kinked" system **26(6)**.

How rapidly can TB-mediated EET be propagated? Considering that the rate of TB-mediated EET falls off much more rapidly than that for a single ET process, it is not surprising that most TB-mediated EET processes are insignificant for interchromophore separations exceeding 10 Å. However, there are exceptions, the most dramatic of which are provided by the series **27(n)**. Thus, the rate of singlet-singlet EET for the 10-bond system **27(10)** is a massive 2.5×10^{10} s^{-1}.[43] This rate is two orders of magnitude larger than that measured for the 6-bond system, **25(6)**, and six orders of magnitude faster than that predicted for the 10-bond system **25(10)**.

Just why TB-mediated EET takes place so rapidly in **27(10)** is currently a mystery, but this example certainly underscores the importance of the through-bond mechanism in mediating not only ET but also EET processes.

Electron Transfer Mediated by π-Stacks in DNA

Imagine a columnar stack of aromatic rings placed about 3.4 Å apart in parallel planes (Figure 26). How well would such an array mediate ET and HT (hole transfer) processes, and by what mechanism would the charge transfer take place, superexchange or by a conduction-like mechanism with the electron or hole "hopping" between aromatic rings?

This problem is receiving intense scrutiny because of its relevance to ET and HT processes in duplex DNA in which the base pairs are approximately stacked in the manner described above. Although DNA's main biological function is not the transfer of electrons or holes, there is mounting evidence that it may suffer long-range oxidative damage by the migration of holes through the π-stack over extremely large distances (> 50 Å).[44]

Whether ET or HT in DNA occurs by charge hopping or superexchange will largely depend on the relative energy levels of the bridge states (i.e. the DNA bases) and the initial state (donor state). If the bridge states are energetically higher than the initial state, then the electron or hole cannot reside in the bridge and the charge transfer must occur by a superexchange mechanism, i.e. by a single coherent "jump", from donor to acceptor. The charge transfer dynamics will then follow an exponential decay with distance according to eq (2). If, on the other hand, the bridge and donor levels are similar, then thermal injection of an electron or hole into the bridge will take place, and the transport will occur by a hopping mechanism. If each single hopping step occurs over the same distance, then the hopping dynamics will display a weak distance dependence, varying inversely with respect to a small power of the number, N, of hopping steps, viz:[35]

$$k_{et} = N^{-\eta} \quad \eta \approx 1\text{--}2 \tag{3}$$

Thus, because of its exponential distance dependence, superexchange-mediated ET and HT in DNA will not be significant for donor-acceptor separations exceeding 15 Å, whereas ET and HT occurring by a hopping mechanism are expected to be much longer range processes, extending beyond 50 Å.

The composition and sequence of the base pairs in DNA should have an important effect on the ET and HT mechanisms. With regard to HT, guanine (*G*) is much more easily oxidized than adenine (*A*) or thymine (*T*); consequently, hole injection into the DNA helix will generate a $G^{+\cdot}$ radical cation. The hole may now randomly walk over large distances along the DNA helix by hopping between adjacent *G* bases, which have similar redox properties. The hole, however, will never reside on either *A* or *T* bases since their oxidation potentials are higher than that of *G*. The *A–T* base pairs serve, instead, as a superexchange medium through which the charge tunnels between two adjacent *G* bases. The

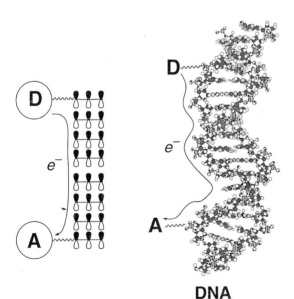

DNA

Figure 26. Schematic of the mediation of ET by a π-stack of aromatic rings, such as the base pairs of duplex DNA. The **D** and **A** groups associated with the DNA molecule may be covalently linked to the helix, they may be intercalators, or they may be DNA base pairs.

rate of hopping between a pair of adjacent *G* bases is therefore governed by the number of intervening *A–T* base pairs according to the superexchange mechanism. The overall HT hopping rate between donor and acceptor is therefore determined by the largest number of *A–T* base pairs between a pair of adjacent *G* bases in the charge transfer pathway. If each hopping step in an HT pathway involves the same number of intervening *A–T* base pairs, then the hopping mechanism obeys the distance dependence relationship given by eq (3).[45]

A number of distance dependence studies have been carried out using synthetic DNA duplexes. In most of these investigations, the data were found to be consistent with the superexchange mechanism since the ET and HT rates followed exponential decays with increasing distance, with β ranging from 0.1 Å^{-1} to 1.4 Å^{-1}.[46]

An example of the types of distance dependence experiments that are being carried out on DNA is shown in Figure 27.[47] The DNA strand is a synthetic

thetic duplex DNA leads to the formation of a sugar radical (Figure 28a) which then suffers heterolysis to give the radical cation. This sugar radical cation initiates the HT process by transferring the positive charge to a nearby *G* base (Figure 28a). The hole is then free to leave the initial site, $G_{23}^{+\cdot}$, and wander through the DNA molecule, hopping between adjacent *G* bases; this meandering is only terminated when the hole encounters a *GGG* unit and is irreversibly trapped (*GGG* has a lower oxidation potential than *G*). By varying the length, composition and sequence of the π-stack spanning $G_{23}^{+\cdot}$ and *GGG*, the distance dependence of the HT dynamics could be measured (Figure 28b). The results are revealing and exciting!

For **28(1)–28(4)**,[49] in which one to four *A–T* base pairs are inserted between $G_{23}^{+\cdot}$ and *GGG*, the relative HT rate from $G_{23}^{+\cdot}$ to *GGG* follows an exponential decay with increasing G_{23} –*GGG* separation, with an associated β value of 0.7 Å^{-1}. The operation of the superexchange mechanism for HT in this

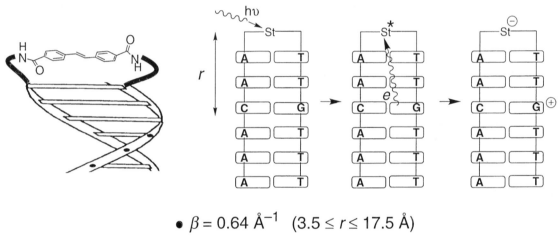

$$\bullet \ \beta = 0.64 \ \text{Å}^{-1} \quad (3.5 \leq r \leq 17.5 \ \text{Å})$$

Figure 27. Schematic of photoinduced ET process in a synthetic 6-mer comprising five *A–T* base pairs and one *G–C* base pair located at various distances from the stilbene acceptor.[47] The locally excited stilbene is a potent enough acceptor to effect an ET from *G*, but not from *A* or *T*.

6-mer duplex comprising five *A–T* base pairs and a single *G–C* base pair located at various positions along the helix. The duplex is capped by a stilbene hairpin which serves as a photo-oxidant. Photoexcitation of the stilbene group initiates electron transfer from the *G* donor to the locally excited stilbene acceptor (or equivalently, HT from stilbene to *G*). Importantly, neither *A* nor *T* is able to transfer an electron to the locally excited stilbene since their oxidation potentials are too high. By varying the stilbene – *G* separation it was possible to obtain a distance dependence for the HT rate, which turned out to be exponential with β = 0.64 Å^{-1}.[37]

Recent experiments have demonstrated the operation of the hole hopping mechanism between *G* bases in DNA.[45,48] In these experiments, Norrish I photocleavage of an acylated nucleoside in a syn-

series of DNA molecules is to be expected since the *A–T* radical cation states are significantly higher in energy than $G^{+\cdot}$.

Replacement of one of the four *A–T* base pairs in **28(4)** with a *G–C* base pair, to give **29(4)** and **30(4)**, led to a two orders of magnitude increase in the HT rate. This rate enhancement was attributed to reversible hole hopping between G_{23} and the inserted *G* unit.[48] The longest superexchange pathway in both **29(4)** and **30(4)** has been reduced to only two *A–T* base pairs, compared to four in **28(4)**. It is not surprising, therefore, that the HT rates for **29(4)** and **30(4)** are larger than that for **28(4)** and almost the same as that for **28(2)**, which also has a superexchange pathway comprising two *A–T* base pairs.

By using synthetic duplex DNA molecules in which every hopping step between adjacent *G–C*

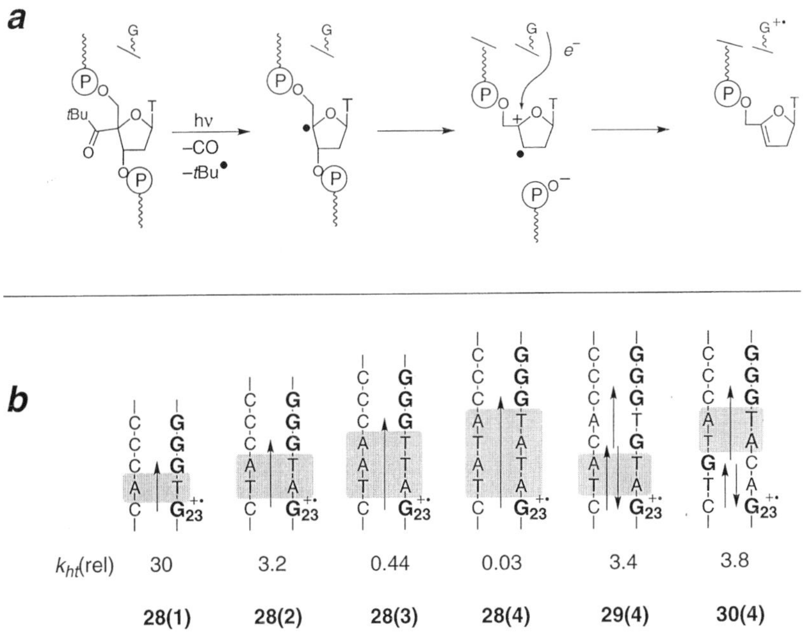

Figure 28. Exploring hole transfer propagation through DNA.[48] (**a**) The hole is generated by photocleavage, followed by heterolysis and subsequent hole generation at a nearby **G** base. (**b**) Relative rates of HT, from **G₂₃**⁺ to **GGG**, through various DNA duplexes. The shaded areas highlight the longest path between adjacent **G** bases.

pairs involves the same number (two) of intervening *A–T* pairs, the validity of eq (3), with $\eta = 1.7$ was demonstrated.[45]

In summary, these experiments suggest that the hole (or electron) is able to hop incoherently among the *G* bases in a random walk which gives rise to a weak algebraic decay of the HT rate with distance (eq (3), $\eta = 1.7$). Moving the electron from one *G* unit to the next requires the operation of the superexchange mechanism, because of the unfavorable redox properties of the intervening *A* and *T* bases. Because of the strong distance dependence of the superexchange mechanism, HT dynamics will be sensitive to the base sequence; the higher the density of *G* bases along the pathway, the faster will be the HT rate and the further the hole will be able to travel. Indeed, by judicious placement of *G* units charge transfer over 54 Å has been observed to occur as rapidly as over 17 Å.[45]

With regard to electron transfer in DNA radical anions (as opposed to HT in the radical cations), the hopping mechanism involves the intermediacy of radical anions of the *C* and *T* bases which have similar reduction potentials. Since one of these bases is present in each base pair, it is predicted that ET proceeding by the hopping mechanism would not be sequence dependent.[45] This important prediction has yet to be verified.

Epilogue

The past fifteen years have witnessed remarkable progress in the understanding of the character of long-range ET (electron transfer) processes. Through the elegant combination of synthesis, photophysical measurements and computational quantum chemical calculations, the scope and significance of superexchange-mediated ET have been delineated. For example, it is now known that ET mediated by saturated hydrocarbon bridges can take place rapidly over distances exceeding 12 Å and that the ET dynamics may be modulated in a predictable manner – for example, by changing the bridge configuration, by orbital symmetry constraints and by the presence of strong electric fields.

These significant findings form the basis of a set of design principles for the construction of molecular photovoltaic cells and other nanoscale electronic devices in which the control of both the rate and directionality of ET processes is an essential requirement. The successful construction of an artificial light-driven proton pump, based on principles of long-range ET processes illustrates the promise of this approach.[50]

Of course, it is impracticable to synthesise nanoscale devices based solely on covalently-linked bridged systems and it is reassuring to learn that ET is also strongly mediated by H-bonded networks and by solvent molecules, thereby opening the way for the construction of photovoltaic supramolecular assemblies.

The conditions which favour the incoherent *electron transport* mechanism, associated with molecular wire behaviour, over the coherent superexchange *electron transfer* mechanism are being elucidated, mainly through studies of systems possessing covalently-linked *conjugated* bridges and DNA molecules. This is an exciting, emerging area of research which lies at the heart of molecular electronics.

Recent HT (hole transfer) studies on DNA suggest that the double helix is able to act as a hole conductor and that its conducting ability may be fine-tuned by altering the location and density of guanine sites in the helix. These findings offer the possibility of using DNA as a charge transport device in nanoscale architecture. This is an important challenge to be met in the near future! Hopefully, this article will inspire chemists from various areas to pursue, not only applications of ET processes, but also those fundamental issues of ET theory that still need to be resolved.[51]

Acknowledgements

I am indebted to my collaborators whose names appear in the references. Support from the Australian Research Council is gratefully acknowledged, as is the award of an ARC Senior Research Fellowship.

References and Notes

1. An exergonic process is one for which the free energy change is negative.
2. K. A. Jolliffe, S. J. Langford, M. G. Ranasinghe, M. J. Shephard, M. N. Paddon-Row, "Design and Synthesis of Two (Pseudo)Symmetric Giant Trichromophoric Systems Containing the C_{60} Chromophore", *J. Org. Chem.* **1999**, *64*, 1238–1246.
3. B. A. Heller, D. Holten, C. Kirmaier, "Effects of *Asp* Residues near the L-Side Pigments in Bacterial Reaction Centers", *Biochemistry* **1996**, *35*, 15418–15427.
4. M. Bixon, J. Jortner, "Electron Transfer – From Isolated Molecules to Biomolecules", *Adv. Chem. Phys.* **1999**, *106*, 35–202.
5. M. N. Paddon-Row, "Investigating Long-Range Electron-Transfer Processes With Rigid, Covalently Linked Donor-(Norbornylogous Bridge)-Acceptor Systems", *Acc. Chem. Res.* **1994**, *27*, 18–25.
6. R. A. Marcus, "Electron Transfer Reactions in Chemistry – Theory and Experiment (Nobel Lecture)", *Angew. Chem., Int. Ed.* **1993**, *32*, 1111–1121.
7. Reorganization energy is an important concept in ET theory and may be defined as the energy of the product possessing the relaxed geometry of the reactant minus the energy of the product in its relaxed geometry (Figure 5). The solvent is also included in this definition.
8. M. N. Paddon-Row, K. D. Jordan, "Through-Bond and Through-Space Interactions in Unsaturated Hydrocarbons: Their Implications for Chemical Reactivity and Long-Range Electron Transfer", in J. F. Liebman, A. Greenberg (Eds.): *Modern Models of Bonding and Delocalization, Vol. 6*, VCH Publishers, New York 1988, pp. 115–194.
9. Unless indicated otherwise in this chapter, *n* is equal to the number of C–C sigma bonds that span the bridge termini. This definition should become perfectly clear upon inspection of the structures and their associated numbers given in Figure 7. Several structures in the

Figures also display an index *m* (see, for example, **1(*n*)** in Figure 6) and this serves to indicate the number of repeating units of the bridge. For **1(*n*)**, *n* = 2*m* + 2.
10. The distance used for calculating β may be measured either in Å or by the number of bonds which connect the two chromophores. For the norbornylogous bridge, one bond is equivalent to 1.2 Å.
11. H. Oevering, M. N. Paddon-Row, H. Heppener, A. M. Oliver, E. Cotsaris, J. W. Verhoeven, N. S. Hush, "Long-Range Photoinduced Through-Bond Electron Transfer and Radiative Recombination via Rigid Non-Conjugated Bridges: Distance and Solvent Dependence", *J. Am. Chem. Soc.* **1987**, *109*, 3258–3269.
12. M. N. Paddon-Row, A. M. Oliver, J. M. Warman, K. J. Smit, M. P. de Haas, H. Oevering, J. W. Verhoeven, "Factors Affecting Charge Separation and Recombination in Photo-Excited Rigid Donor-Insulator-Acceptor Molecules", *J. Phys. Chem.* **1988**, *92*, 6958–6962.
13. M. N. Paddon-Row, "Some Aspects of Orbital Interactions Through Bonds: Physical and Chemical Consequences", *Acc. Chem. Res.* **1982**, *15*, 245–251.
14. R. Hoffmann, "Interaction of Orbitals through Space and through Bonds", *Acc. Chem. Res.* **1971**, *4*, 1–9.
15. A. M. Oliver, D. C. Craig, M. N. Paddon-Row, J. Kroon, J. W. Verhoeven, "Strong Effects of the Bridge Configuration on Photoinduced Charge Separation in Rigidly Linked Donor-Acceptor Systems", *Chem. Phys. Lett.* **1988**, *150*, 366–373.
16. G. L. Closs, J. R. Miller, "Intramolecular Long-Distance Electron Transfer in Organic Molecules", *Science* **1988**, *240*, 440–447.
17. This origin of the inverted and normal regions may be understood using the Marcus-Hush expression (Figure 5). For those exergonic ET processes for which $-\Delta G_{cs} < \lambda$, the reactions are said to lie in the Marcus normal region. In this region, increasing the exergonicity leads to a reduction in the activation barrier, $\Delta G^{\neq} = (\Delta G_{cs} + \lambda)^2$, until it becomes zero when $-\Delta G_{cs} = \lambda$. At this point the ET rate is at its fastest. If the exergonicity is increased further still, the activation barrier again reappears and the ET rate now falls with progressively increasing exergonicity. This region is known as the Marcus inverted region.
18. Units: $1\ cm^{-1} = 0.012\ kJ\ mol^{-1}$; $1\ eV = 96.5\ kJ\ mol^{-1}$.
19. A. M. Oliver, M. N. Paddon-Row, J. Kroon, J. W. Verhoeven, "Orbital Symmetry Effects on Intramolecular Charge Recombination", *Chem. Phys. Lett.* **1992**, *191*, 371–377.
20. Y. Zeng, M. B. Zimmt, "Symmetry Effects in Photoinduced Electron Transfer Reactions", *J. Am. Chem. Soc.* **1991**, *113*, 5107–5109.
21. M. Antolovich, P. J. Keyte, A. M. Oliver, M. N. Paddon-Row, J. Kroon, J. W. Verhoeven, S. A. Jonker, J. M. Warman, "Modelling Long-Range Photosynthetic Electron Transfer in Rigidly Bridged Porphyrin-Quinone Systems", *J. Phys. Chem.* **1991**, *95*, 1933–1941.
22. T. D. M. Bell, T. A. Smith, K. P. Ghiggino, M. G. Ranasinghe, M. J. Shephard, M. N. Paddon-Row, "Long-lived Photoinduced Charge Separation in a Bridged C_{60}–Porphyrin Dyad", *Chem. Phys. Lett.* **1997**, *268*, 223–228.
23. H. Imahori, K. Hagiwara, M. Aoki, T. Akiyama, S. Taniguchi, T. Okada, M. Shirakawa, Y. Sakata, "Linkage and Solvent Dependence of Photoinduced Electron Transfer in Zinc Porphyrin–C_{60} Dyads", *J. Am. Chem. Soc.* **1996**, *118*, 11771–11782.

24. R. M. Williams, M. Koeberg, J. M. Lawson, Y. Z. An, Y. Rubin, M. N. Paddon-Row, J. W. Verhoeven, "Photoinduced Electron Transfer to C-60 Across Extended 3- and 11-Bond Hydrocarbon Bridges – Creation of a Long-Lived Charge-Separated State", *J. Org. Chem.* **1996**, *61*, 5055–5062.

25. Note that the active MO of the C_{60} unit associated with the charge separation process depicted in Figure 12 is the HOMO since an electron from the aniline is transferred into the half-filled HOMO of the locally excited C_{60} group.

26. H. Imahori, K. Hagiwara, T. Akiyama, M. Aoki, S. Taniguchi, T. Okada, M. Shirakawa, Y. Sakata, "The Small Reorganization Energy of C_{60} in Electron Transfer", *Chem. Phys. Lett.* **1996**, *263*, 545–550.

27. Most photoinduced CS processes take place within the Marcus normal region. However, charge recombination processes take place within the inverted region. Thus, the small reorganization energy of C_{60} will *slow* down CR processes.

28. Hole transfer (HT) is the movement of a positive charge, say in cation radicals. In reality, it is still an ET process but it is often more convenient to focus on the migration of the positive hole (vacated by an electron) rather than the electron (which moves in the opposite direction to the hole).

29. S. Knapp, T. G. M. Dhar, J. Albaneze, S. Gentemann, J. A. Potenza, D. Holten, H. J. Schugar, "Photoinduced Porphyrin-to-Quinone Electron Transfer across Oligospirocyclic Spacers", *J. Am. Chem. Soc.* **1991**, *113*, 4010–4013.

30. R. Langen, J. L. Colon, D. R. Casimiro, T. B. Karpishin, J. R. Winkler, H. B. Gray, "Electron Tunneling in Proteins – Role of the Intervening Medium", *J. Biol. Inorg. Chem.* **1996**, *1*, 221–225.

31. M. N. Paddon-Row, M. J. Shephard, "Through-Bond Orbital Coupling, the Parity Rule, and the Design of Superbridges Which Exhibit Greatly Enhanced Electronic Coupling – a Natural Bond Orbital Analysis", *J. Am. Chem. Soc.* **1997**, *119*, 5355–5365.

32. W. B. Davis, M. R. Wasielewski, M. A. Ratner, V. Mujica, A. Nitzan, "Electron Transfer Rates in Bridged Molecular Systems – a Phenomenological Approach to Relaxation", *J. Phys. Chem.* **1997**, *101*, 6158–6164.

33. W. B. Davis, W. A. Svec, M. A. Ratner, M. R. Wasielewski, "Molecular-Wire Behaviour in *p*-Phenylenevinylene Oligomers", *Nature* **1998**, *396*, 60–63.

34. P. J. F. de Rege, S. A. Williams, M. J. Therien, "Direct Evaluation of Electronic Coupling Mediated by Hydrogen Bonds – Implications for Biological Electron Transfer", *Science* **1995**, *269*, 1409–1413.

35. E. Galoppini, M. A. Fox, "Effect of the Electric Field Generated by the Helix Dipole on Photoinduced Intramolecular Electron Transfer in Dichromophoric Alpha-Helical Peptides", *J. Am. Chem. Soc.* **1996**, *118*, 2299–2300.

36. H. Han, M. B. Zimmt, "Solvent-Mediated Electron Transfer – Correlation Between Coupling Magnitude and Solvent Vertical Electron Affinity", *J. Am. Chem. Soc.* **1998**, *120*, 8001–8002.

37. These states were generated by local excitation of the DMN chromophore, to give **An–*DMN–DCV**, followed by charge separation, to give **An–⁺DMN–DCV⁻** and subsequent thermal HT from **⁺DMN** to **An**.

38. M. J. Shephard, M. N. Paddon-Row, "Large Predicted Changes in Geometry Accompanying Charge Separation in Various "Rigid" Multichromophoric Systems in the Gas Phase: An ab Initio MO Study", *J. Phys. Chem. A* **1999**, *103*, 3347–3350.

39. K. A. Jolliffe, T. D. M. Bell, K. P. Ghiggino, S. J. Langford, M. N. Paddon-Row, "Efficient Photoinduced Electron Transfer in a Rigid U-Shaped Tetrad Bearing Terminal Porphyrin and Viologen Units", *Angew. Chem., Int. Ed.* **1998**, *37*, 916–919.

40. K. A. Jolliffe, S. J. Langford, A. M. Oliver, M. J. Shephard, M. N. Paddon-Row, "A New Class of Giant Tetrads for Studying Aspects of Long–Range Intramolecular Electron Transfer Processes: Synthesis and Computational Studies", *Chem. Eur. J.* **1999**, *5*, 2518 – 2530.

41. G. L. Closs, M. D. Johnson, J. R. Miller, P. Piotrowiak, "A Connection between Intramolecular Long-Range Electron, Hole, and Triplet Energy Transfers", *J. Am. Chem. Soc.* **1989**, *111*, 3751–3753.

42. J. Kroon, A. M. Oliver, M. N. Paddon-Row, J. W. Verhoeven, "Observation of a Remarkable Dependence of the Rate of Singlet-Singlet Energy Transfer on the Configuration of the Hydrocarbon Bridge in Bichromophoric Systems", *J. Am. Chem. Soc.* **1990**, *112*, 4868–4873.

43. N. Lokan, M. N. Paddon-Row, T. A. Smith, M. La Rosa, K. P. Ghiggino, S. Speiser, "Highly Efficient Through-Bond-Mediated Electronic Excitation Energy Transfer Taking Place Over 12 Å", *J. Am. Chem. Soc.* **1999**, *121*, 2917–2918.

44. R. E. Holmlin, P. J. Dandliker, J. K. Barton, "Charge Transfer through the DNA Base Stack", *Angew. Chem., Int. Ed.* **1998**, *36*, 2715–2730.

45. B. Giese, S. Wessely, M. Spormann, U. Lindemann, E. Meggers, M. E. Michel-Beyerle, "On the Mechanism of Long-Range Electron Transfer through DNA", *Angew. Chem., Int. Ed.* **1999**, *38*, 996–998.

46. A. Harriman, "Electron Tunneling in DNA", *Angew. Chem., Int. Ed.* **1999**, *38*, 945–949.

47. F. D. Lewis, T. F. Wu, Y. F. Zhang, R. L. Letsinger, S. R. Greenfield, M. R. Wasielewski, "Distance-Dependent Electron Transfer in DNA Hairpins", *Science* **1997**, *277*, 673–676.

48. E. Meggers, M. E. Michel-Beyerle, B. Giese, "Sequence Dependent Long Range Hole Transport in DNA", *J. Am. Chem. Soc.* **1998**, *120*, 12950–12955.

49. For **28(*n*)**, **29(*n*)** and **30(*n*)**, *n* is equal to the number of intervening base pairs between the G_{23} and **GGG** groups.

50. G. Steinberg-Yfrach, P. A. Liddell, S. C. Hung, A. L. Moore, D. Gust, T. A. Moore, "Conversion of Light Energy to Proton Potential in Liposomes by Artificial Photosynthetic Reaction Centres", *Nature* **1997**, *385*, 239–241.

51. P. F. Barbara, T. J. Meyer, M. A. Ratner, "Contemporary Issues in Electron Transfer Research", *J. Phys. Chem.* **1996**, *100*, 13148–13168.

The Supramolecular Synthon in Crystal Engineering

Gautam R. Desiraju *School of Chemistry, University of Hyderabad, Hyderabad 500046, India*

Phone: +91 40 3010 221, Fax: +91 40 3010 567, e-mail: desiraju@uohyd.ernet.in

Keywords ■ Synthesis ■ Crystal engineering ■ Supramolecular synthon ■ Retrosynthesis ■ Robustness ■ Interaction insulation ■ Interaction interference ■ Strong and weak hydrogen bonds ■ Crystallization ■ Polymorphism

Concept: Synthetic approaches to crystal targets will not, in general, be reliable if the design strategy involves independent manipulations of molecular functionality. This condition arises because the recognition between molecules, which precedes crystallization, is necessarily complementary. Yet, correspondences between molecular structures and crystal structures must exist. In reality, these matchings amount to convolutions of molecular functionalities from adjacent molecules into small repetitive units that economically yet fully define the kernel of a crystal structure. These units encapsulate the recognition information during crystallization and are called *supramolecular synthons*.

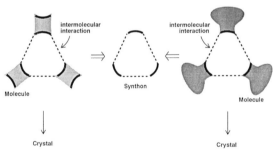

Abstract: A retrosynthetic dissection of a crystal structure into its constituent supramolecular synthons permits an identification of equivalent structures and, provided the levels of interaction insulation are high, crystal synthesis of these structures is possible. This approach to crystal engineering has been shown in a group of iodo nitro aromatics and in a family of inclusion complexes of biphenyl hosts. Interaction interference is detrimental to such crystal engineering efforts but, when understood, it may be incorporated into the design strategy. This principle has been demonstrated in a series of cubanecarboxylic acids. Interference is very high in *gem*-ethynyl alcohols because of the many types of strong and weak hydrogen bonds that are possible for these compounds, but here too some control over structural diversity is possible by identifying structures where the hydrogen-bonding and close-packing domains are orthogonal. Supramolecular synthons are important also to the crystallization process and may, in this context, be profitably invoked in studies of enthalpic driven polymorphism, multiple molecules in the asymmetric unit and solvation. These phenomena may be termed, in turn, alternative crystallization, incomplete crystallization and interrupted crystallization. The synthon is a representative structural unit that links molecules and crystals and is implicated in all the stages through which molecules progress as they form crystals.

Prologue

Crystal engineering, the design of solid state structures, is based upon the properties of intermolecular interactions and the ways in which they are manifested in crystal structures.[1] Crystal engineering is a form of supramolecular synthesis and like any other synthetic activity, it gains from a simplification of the target structure. The concept of a supramolecular synthon recognises the need to be able to simplify a three-dimensional structure into modular units. Supramolecular synthons are spatial arrangements of intermolecular interactions and play a focusing role in crystal engineering similar to that of the conventional synthon in molecular synthesis.[2] Thus, they have been defined as structural units within supermolecules which can be formed and/or assembled by known or conceivable intermolecular interactions.[3] The supramolecular synthons referred to in this essay are shown in Scheme 1.

As in molecular chemistry, an inherent advantage of the synthon concept in crystal engineering is that it can lead naturally to retrosynthetic thinking.[4] Analysis of the often complex interplay between close packing, hydrogen bonding and other interactions in crystal structures may be termed *supramolecular retrosynthesis* and, insofar as analysis and synthesis are carried out in opposite senses, the term 'retrosynthesis' aptly describes the procedure for logical analysis of a crystal structure.[5] Consider, for instance, the crystal structure of 4-iodonitrobenzene, **1**, which is based upon the $NO_2 \cdots I$ synthon, **I**, in itself constituted with two polarisation-based $O \cdots I$ interactions (Scheme 2).[6] In crystalline **1**, synthons **I** are connected with spacer phenyl rings to give a ribbon structure.[7] If the sense of the synthon alternates in the ribbon, the result is the crystal structure of the 1:1 complex, **2**, of 1,4-diiodobenzene and 1,4-dinitrobenzene. If the

Scheme 1.

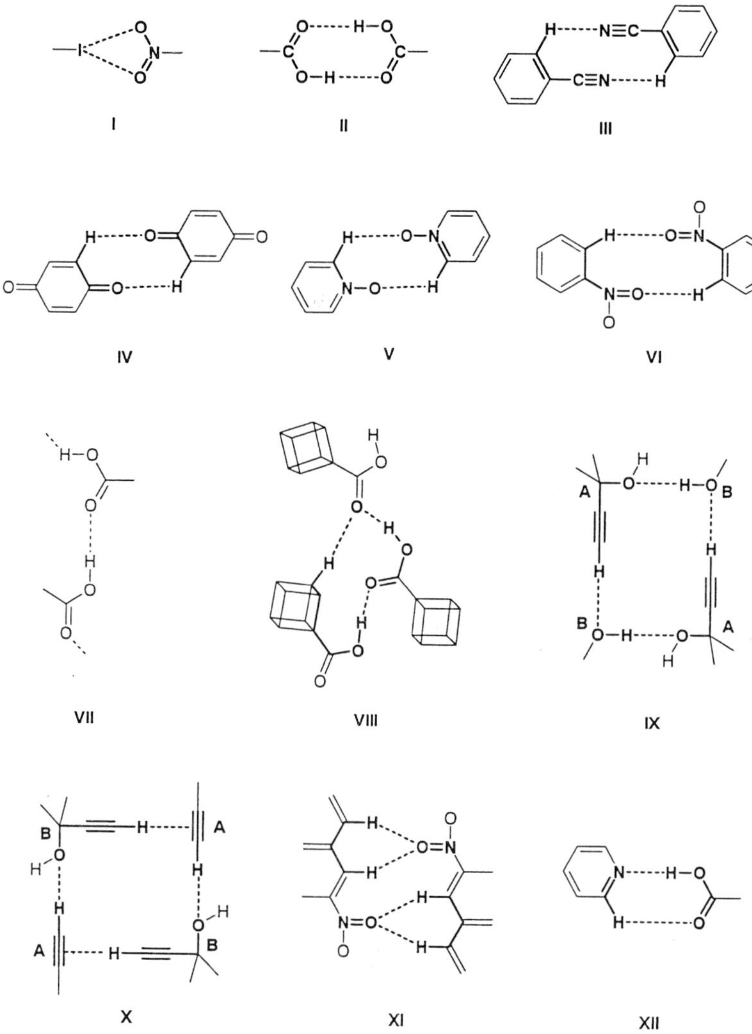

Scheme 2.

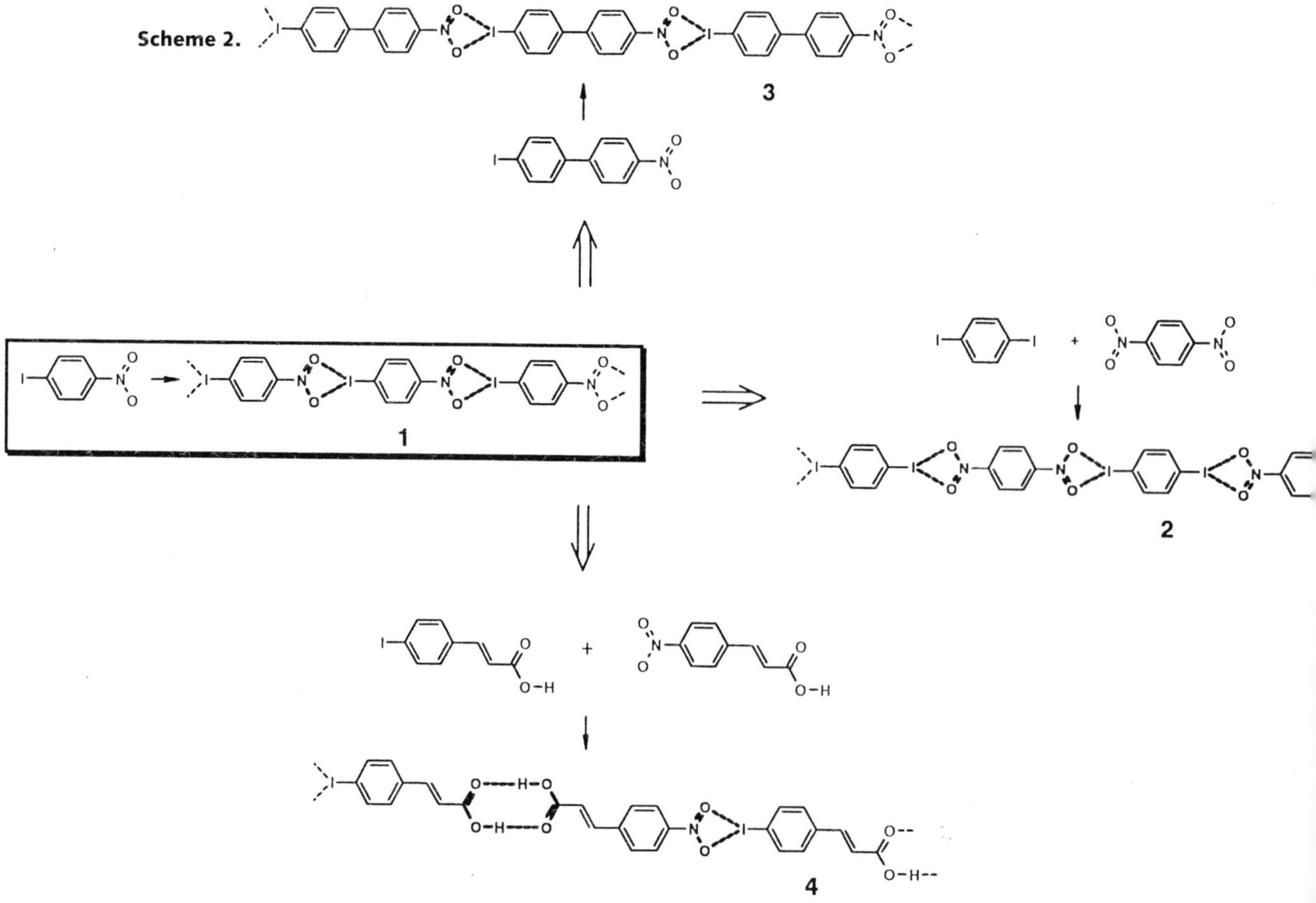

spacer is changed to a biphenyl moiety, the non-centrosymmetric structure of 4-iodo-4'-nitrobiphenyl, **3**, is obtained.[8–10] Finally, and more ambitiously, the crystal structure of the 1:1 complex, **4**, of 4-iodocinnamic acid and 4-nitrocinnamic acid shows that synthon **I** can act as a ribbon-former, even in the presence of the carboxyl dimer synthon **II**, that is constituted with the stronger O–H···O hydrogen bond.[6]

Though simple, this example reveals much that is noteworthy. Firstly, a comparison of the structures of **1** and **2** shows that there need be little distinction between structures of single-molecule and multi-molecular crystals. Indeed, structures **1** and **2** are nearly the same, at least in one dimension, and one might in a sense have expected that such parallels would exist in the world of supramolecular chemistry where the ruling paradigm is the supermolecule rather than the molecule.[11] To continue, a comparison of the structures of **1** and **3** shows that one may handle the synthon and the spacer independently in synthesis. Such a practice is advantageous but perhaps unsurprising in these compounds where the spacer is a more or less passive hydrocarbon residue. Further, one finds the same kind of *structural insulation* in complex **4**, but here it is more impressive. All the three constituents of the ribbon: synthon **I**, synthon **II** and the styryl hydrocarbon spacer may be treated in a building block fashion;[12] the structure is built in a modular fashion so that the element of predictability in its formation is high.

The NO_2···I synthon, **I**, is particularly *robust* and tends to occur often in crystal structures of molecules containing iodo and nitro functional groups. This tendency is not so pronounced for the NO_2···Br synthon and even less so for NO_2···Cl, indicating the importance of polarization in these NO_2···halogen interactions.[13] In this sense, one need not expect that replacement of an I atom by Br or Cl, in any of the NO_2···I based crystal structures above, would automatically preserve the structure. Insulation then is an attribute arising from chemical rather than geometrical factors, and the science and art of crystal engineering of molecular crystals lies in being able to maximize it in any given system.[14]

The Synthon in a Supramolecular Context

The functional group approach to molecular organic chemistry is almost axiomatic because functional groups are the correspondences or transforms that relate structure with properties. Each functional group is associated with its distinctive chemistry, even as functional groups communicate among themselves electronically and sterically in well understood and systematically derivable ways. As organic chemistry ascends to higher realms, communications between functional groups become more subtle and it becomes increasingly challenging for the chemist to comprehend and utilize these subtleties, especially in synthesis. Despite this, there is still much modularity in molecular chemistry and it is this characteristic more than any other which permits a classification of the oceanic body of facts pertaining to organic molecules and reactions within a rubric of a small number of basic concepts and principles.

In contrast, the grammar of supramolecular chemistry is still to be written. In any kind of molecular recognition, and this includes the recognition between identical molecules, it is the dissimilar rather than the similar functionalities that come into closest contact.[15, 16] Steric and electronic complementarity characterise the recognition events that are a prelude to crystallization and other forms of supramolecular assembly. In effect, the supramolecular behavior of any functional group depends not only on the group *but also on the nature and location of all other functional groups in the molecule*, including all hydrocarbon residues.[17] It is therefore a formidable venture, perhaps even a foolhardy and futile one, to try and predict crystal structures of molecular solids based on a functional group approach. The functional group, which is a molecular attribute, cannot be a descriptor of crystal-packing information, a property which is supramolecular in nature.

Since functional groups do not adequately reflect molecule-to-crystal transforms, how then are recognition and crystallization events reconciled in chemical terms? It is here that the concept of the supramolecular synthon is particularly valuable. Synthons are multimolecular units that take into account the complementary association of functional groups from different molecules. In that they are approximations of actual crystal structures, they are essentially supramolecular in nature. They are much closer to actual crystal structures than are molecules, functional groups or even single interactions. A given set of molecules will yield a given set of crystal structures if they are capable of leading to a given set of supramolecular synthons. Synthons therefore more properly reflect the relationships between molecular and supramolecular structures. The synthon is the supramolecular equivalent of the functional group and is as useful in supramolecular chemistry as the functional group is in molecular chemistry.

Synthons are representations of molecular recognition, and in a formalistic sense, the number of all possible synthons in a crystal structure is very large. However, in the most useful of them, the maximum amount of recognition information is held within a structural unit of minimum size. Very large structural units will naturally contain more information but these are almost always accompanied by unnecessary detail, while very small units may lack the critical information that is needed to define the essential elements of the supramolecular structure.[18] In practical terms then, crystal engineering of molecular solids consists in extracting the most robust synthons of intermediate size in a group of structures – in other words, in strategically maximizing structural insulation so that the design exercise towards other related structures becomes more modular. Robust synthons introduce this element of modularity in crystal design in the same way that functional groups do in molecular chemistry. Specific examples are now discussed.

Interaction Insulation and Interference

Figure 1 shows the crystal structure of the 1:1 complex, **5** formed by 4,4'-dicyanobiphenyl, **6** and urea. The novel feature here is a hexagonal channel composed of molecules of **6**. The channel is filled with urea molecules. The hexagonal host structure is mediated by a zig-zag networking of weak C–H⋯N hydrogen bonds (2.77–3.00 Å, 140–155°) from which may be dissected out synthon **III**. This structure formed the basis of further crystal engineering experiments directed at obtaining new host materials.[19, 20] Arguing that synthon **III** is equivalent to the well-known benzoquinone synthon, **IV** which, in turn, should be mimicked by synthon **V**, 4,4'-bipyridine *N,N'*-dioxide, **7**, was identi-

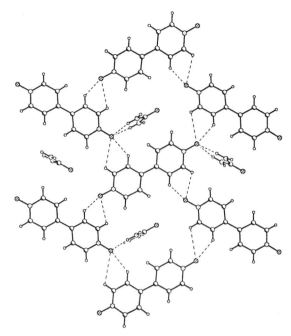

Figure 2. Skewed arrangement of 4,4'-bipyridine *N,N'*-dioxide molecules in the structure of its 1:1 complex with urea. O-atoms are shaded. Contrast this with Figure 1.

fied as a material for molecular complexation with urea. A 1:1 complex was formed but it was quite different from complex **5** (Figure 2). At this stage, it was felt that the desired structure was not obtained because the size of the urea guest is too large for the targetted cavity size. Noting that, if formed, a channel mediated by C–H⋯O based synthon **V** would be smaller than that found in complex **5**, complexation with a smaller guest, namely water, was attempted. This strategy was successful and the desired synthon **V** constituted structure was obtained with the water molecules enclosed within the hexagonal channels (Figure 3).

The next logical step was to try and obtain another structure similar to that of **5** and again with urea as the guest. For this purpose, a channel larger than what could be formed by **7** is required,

Figure 1. Hexagonal C–H⋯N based channels in the crystal structure of complex **5** formed by 4,4-dicyanobiphenyl and urea. Synthon **III** is highlighted. Note that there are two geometrical variations of this synthon.

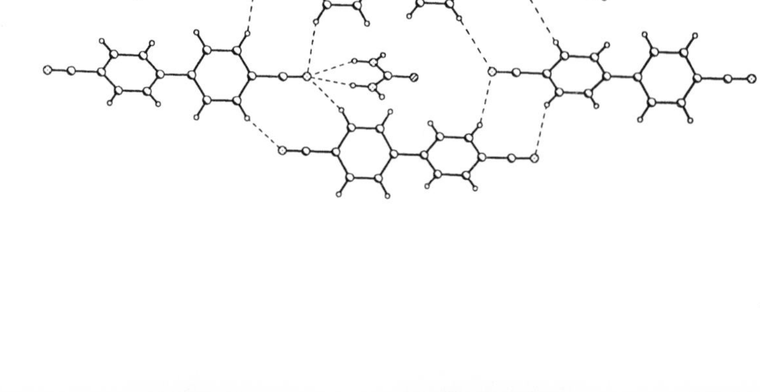

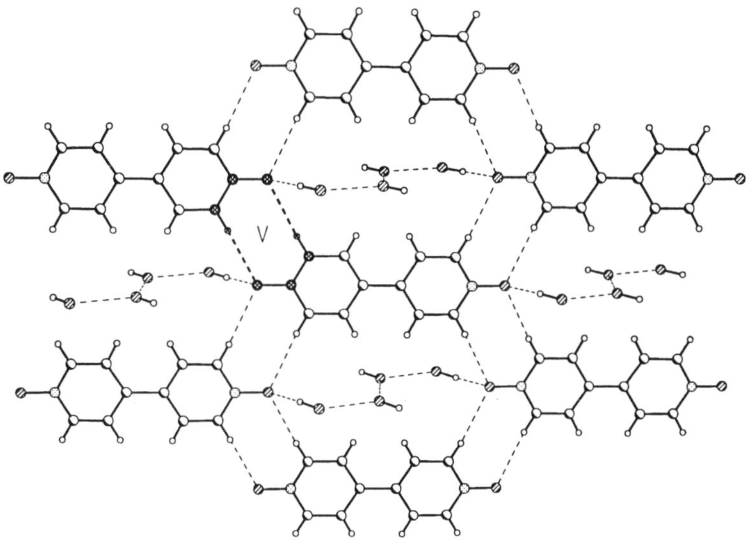

Figure 3. C–H⋯O synthon **V** in the 1:1 hydrate of 4,4'-bipyridine *N,N'*-dioxide. Notice the hexagonal channels.

and it was felt that 4,4'-dinitrobiphenyl, **8**, would be an appropriate host substance. If formed, such a framework would be mediated by the equivalent C–H⋯O based synthon, **VI**. These expectations were fully borne out in the structure of the 1:1 complex of biphenyl **8** with urea (Figure 4). This structure is very similar to that of complex **5** in that the urea is completely enclosed by the hexagonal channels formed by **8**.

It is well-known, that in most carboxylic acid crystal structures, the conformation of the carboxyl group is synplanar and that the most frequent and dominant interlink is the *syn-syn* centrosymmetric dimer synthon, **II**. This synthon is found in nearly a third of all crystalline carboxylic acids, with or without any other functionality being present in the molecule, signifying robustness.[21] Against such a background, the crystal structures of 1,4-cubane-

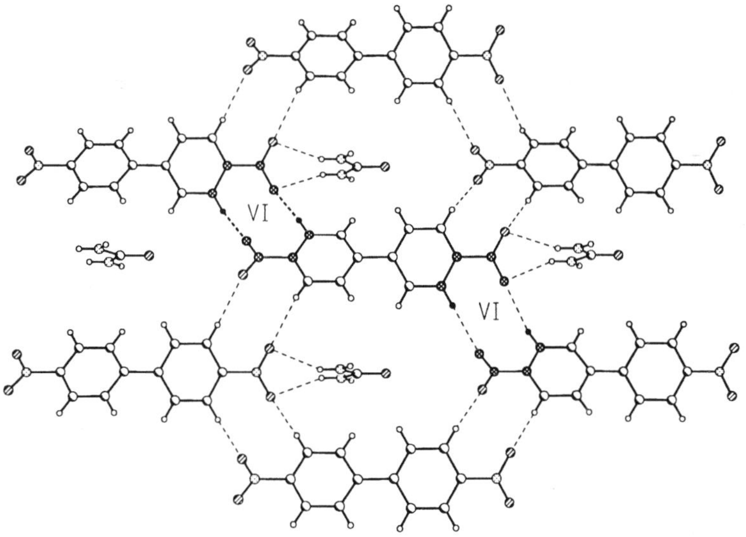

Figure 4. Hexagonal C–H⋯O channels in the synthon **VI** based structure of the 1:1 complex 4,4'-dinitrobiphenyl:urea. The similarity to Figure 1 is obvious.

One learns from these molecular complexes that equivalent synthons can lead to virtually identical crystal structures. Synthons **III**, **V** and **VI** are chemically and geometrically equivalent though they originate from different molecules, a nitrile, an *N*-oxide and a nitro compound. These three synthons are used in crystal design in almost the same way. So, different molecules may yield similar crystal structures if they are capable of forming equivalent synthons. This is a powerful concept because it establishes a many-to-one correspondence between molecular and crystal structures.

dicarboxylic acid, **9**, and 1,3,5,7-cubanetetracarboxylic acid are exceptional in that they contain the rare *syn-anti* catemer, **VII**.[22–24] Given the robustness of synthon **II**, and the fact that it is not formed, but rather a very uncommon packing is observed in the only two reported crystal structures of cubanecarboxylic acids, a more detailed study of this family of compounds was undertaken. An uncommon molecular feature leads to an uncommon structural feature. Is there any possible connection here?

Figure 5 shows the crystal structures of 4-chloro, 4-bromo, 4-iodo and 4-carbomethoxy-1-cubane car-

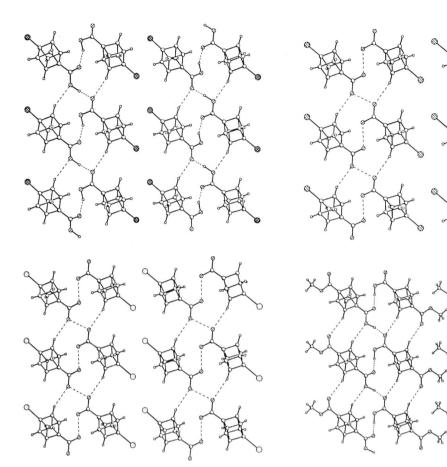

Figure 5. Crystal structures of 4-chloro-1-cubanecarboxylic acid (top left), 4-bromo-1-cubanecarboxylic acid (bottom left), 4-iodo-1-cubanecarboxylic acid (top right) and 4-carbomethoxy-1-cubanecarboxylic acid (bottom right) to show the layer with zig-zag arrangement of synthon **VIII**.

Strong and weak hydrogen bonds are indicated with dashed lines. Notice the close similarities between these structures and the insensitivity of the catemer pattern to the nature of the 4-substituent.

boxylic acids.[25] These structures are related in that all contain catemer **VII** as the common hydrogen bond pattern. Alternating *syn* and *anti* carboxylic acid groups are necessarily symmetry-independent and constitute the O–H···O catemer. Of some significance are the C–H···O hydrogen bonds formed by the cubyl H-atoms to the carbonyl/carboxyl O-atoms of the translation-related molecules (2.37–2.65 Å, 153–158 °). These weak hydrogen bonds may be taken, along with catemer **VII**, to derive the larger synthon **VIII**. This larger pattern **VIII**, contains within it, the smaller **VII**. The crystal structure of diacid **9** (Figure 6) then stands rationalized in terms of the symmetrical occurrence of synthon **VIII** on either side of the cubyl skeleton. This last observation hints incidentally that the catemer structure for these acids is independent of the 4-halogen substituent.

The repeated occurrence of the otherwise rare catemer **VII** in the family of cubanecarboxylic acids under study here is quite likely the result of fortification of the catemer by the C–H···O hydrogen bond formed by the acidic cubyl C–H group.[26] This

functional group may be considered to be the cause of *structural interference* – in other words it is able to perturb a normally robust scenario (formation of dimer synthon **II**) to a critical enough extent that the packing type changes completely. According to such an argument, the catemer structure might not have been observed, if such an interfering functionality were not present. However, once the nature of the interference has been understood it may well be used in further design efforts. Inspection of the structure of the 4-chloro acid shows that the catemer tapes are well-insulated via close-packing of the halogen atoms and it is not so difficult at this stage to anticipate that the 4-bromo and 4-iodo derivatives will have similar crystal structures. That the 4-carbomethoxy derivative adopts the same structure is more serendipitous but then, crystal engineering – as does any other form of synthesis – benefits not only from design but also from chance.

It has been noted above that synthon **VIII** incorporates within it synthon **VII**. Which of these is the more useful representation? The efficacy of a supramolecular synthon as an indicator of crystal pack-

Figure 6. Synthon **VIII** based structure of 1,4-cubanedicarboxylic acid. O-atoms are shaded. All hydrogen bonds are indicated.

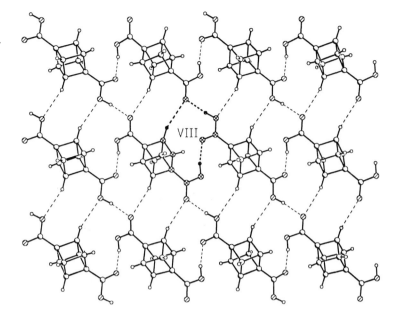

ing arises not only from its frequency of occurrence but also from its size. Synthons represent a carry-over of structural information from molecules to crystals and also between crystal structures; as their size increases, so does their information content. So, larger synthons are potentially more useful than smaller ones. In the present case, **VIII** is more useful than **VII** because it contains additional and vital information about a C–H···O hydrogen bond formed by the cubyl group. Without this information, it would be very difficult to appreciate why **VII** is even formed at all. Clearly then, **VIII** is chemically more complete than **VII**. However, there is a general dilemma here – as the size of a synthon increases, its occurrence becomes less frequent and, in this sense, the two criteria for identifying useful synthons appear to be contradictory. Still, it should be noted that, between these extremes of small size and numerous occurrences and large size and infrequent occurrences, there lies an optimal region wherein the maximum structural information is contained in a synthon of minimum size. It is in this intermediate region that the visualization of supramolecular synthons and crystal design is most effectively accomplished.

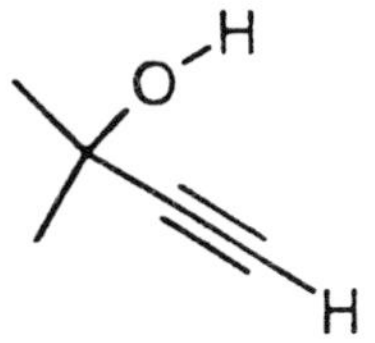

The previous example shows that, as a group of crystal structures is better understood, structural interference may be tamed to the extent that it forms the basis of a new family which may be developed and analyzed in terms of repetitive and structurally insulated synthons. Sometimes, however, a large number of crystal structures of related molecules fail to yield repeating supramolecular patterns. This is the case for the 90 or so compounds with a geminal hydroxy ethynyl moiety, **10**. These structures contain many diverse hydrogen-bond arrangements made up of all manner of permutations of O–H···O, C–H···O, O–H···π and C–H···π hydrogen bonds (Scheme 3). The lack of structural

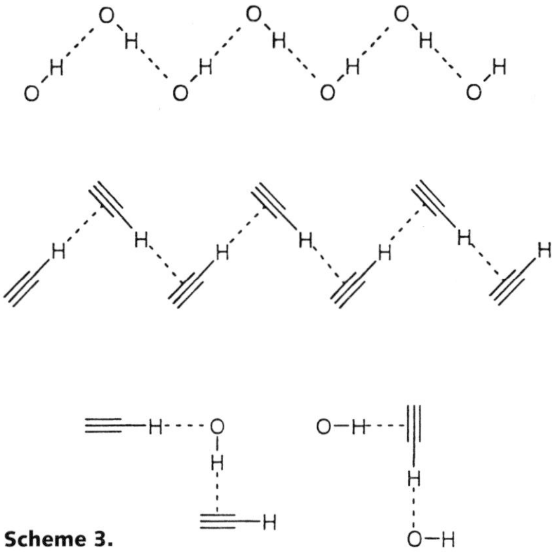

Scheme 3.

repetitivity among these compounds may arise from the close juxtaposition of two hydrogen bond donors and two acceptors within moiety **10**. In this sterically hindered situation, and also with their incorporation into cooperative networks, the four possible interactions O–H···O, C–H···O, O–H···π and C–H···π become competitive.[27] Thus, the packing adopted by any particular *gem*-alkynol becomes extremely sensitive to other molecular features. In

Figure 7. Crystal structure of alkynol **11** in the (100) layer. C–H···O and C–H···π hydrogen bonds are indicated. The hydrocarbon and hydrogen bonded regions both lie in the plane of the drawing signifying heavy structural interference between these domains.

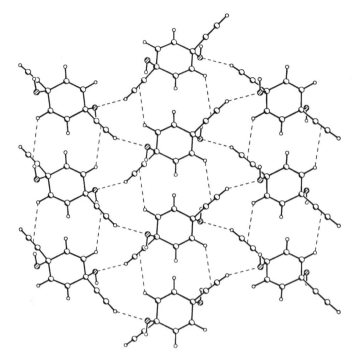

practice, the very high levels of interaction interference among these compounds lead to several quite different hydrogen-bonded networks so that it is not at all easy to establish the molecule-to-crystal correspondences that are so important in crystal engineering. How does one manipulate such systems?

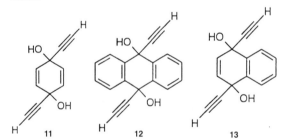

Figures 7 and 8 show the crystal structures of the *trans gem*-alkynols, **11** and **12** respectively. Let us consider the former structure.[28] A number of C–H···O and C–H···π interactions are found in the (100) layer whilst the O–H groups form an infinite cooperative O–H···O–H···O–H··· network along (100), not shown in Figure 7. Though the structure is straightforward, it is hardly what might be termed predictable. The cyclohexadiene ring lies approximately in (100) and a substitution say, at any of the four alkenic positions by the simplest of groups would be expected to change the crystal structure drastically because one or more hydrogen bonds (strong or weak) would be disturbed. The various molecular functionalities – hydroxy, ethynyl, alkene – are intimately involved with one another and they also interact with the hydrocarbon residues. This structure is therefore an example of heavy structural

interference and does not lend itself well to extrapolations. Now let us consider the latter structure – i.e, that of **12** in Figure 8. Here, there are two symmetry independent half molecules each lying on a distinct inversion centre. The alkynol groups from these two sets of molecules result in the centrosymmetric cooperative synthons, **IX** and **X**. Both these overlapping synthons involve both the symmetry independent molecules (A and B in Scheme 1), but while **IX** is constituted with O–H···O and C–H···O hydrogen bonds, **X** is constituted with C–H···O and C–H···π bonds. The encouraging feature in the *trans-gem*-alkynol **12** vis-à-vis alkynol **11** is that the strong

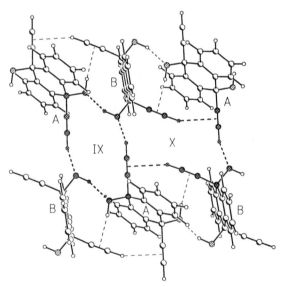

Figure 8. Crystal structure of alkynol **12** to show synthons **IX** and **X**. There are two symmetry independent molecules and each lies on an inversion centre. Hydrogen bonds are indicated. Notice here that the hydrogen-bonded and hydrocarbon regions are well insulated.

and weak hydrogen bonds form a two-dimensional sheet structure *that does not generally include the hydrocarbon residues.* Instead, the fused benzo rings protrude from either side of the hydrogen bonded sheet and interdigitate with the aromatic rings from adjacent sheets. Effectively, the two major domains of the crystal, namely the hydrophobic aromatic rings and the hydrogen-bonded network, are in a mutually orthogonal disposition and as such, well-insulated.

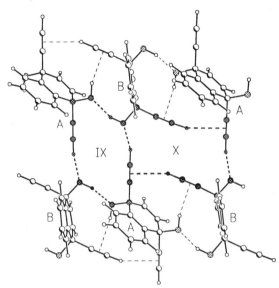

Figure 9. Crystal structure of alkynol **13**. Notice the similarity to Figure 8. The crystal structure is essentially the same as that of alkynol **12**.

Such orthogonality is of course, a most valuable attribute in crystal engineering and it is not hard to extrapolate to the unsymmetrical *trans* alkynol, **13**, the crystal structure of which is shown in Figure 9. The extended hydrogen-bonded sheet observed in alkynol **12** is repeated here and the aromatic rings interdigitate between adjacent layers in the same way as they do in **12**. In general, it may be expected that the benzo rings in **12** and **13** may be further substituted or annelated without much change in the hydrogen-bonded sheet, leading to structural repetitivity. A key element, therefore, in the control of crystal structures, in systems where severe interference is likely, is to seek out orthogonal and therefore non-interfering patterns of supramolecular synthons.

Supramolecular Synthons and Crystallization

The crystallization of an organic molecule is a complex and yet highly efficient process in which a number of molecular functionalities compete with each other as recognition sites. During the early stages of crystallization prior to nucleation, all functional groups present in the molecule sample numerous possible trajectories towards alternative intermolecular interactions and patterns, even though only some of these putative recognition events will be ultimately fruitful. Two or more molecules may come together to form, in principle, several competing and mutually exclusive supramolecular synthons. However, there is a simplifying feature here – only a few of the interaction combinations leading to these synthons are thermodynamically or kinetically favored, and so the alternative possibilities are quickly and efficiently excluded.[29] Accordingly, a molecule with more recognition sites may cascade into a stable crystal structure more easily, in a manner reminiscent of protein-folding. The fact that robust supramolecular synthons are found in so many structural families (in other words, that crystal structure patterns are not totally incomprehensible) hints that crystallization proceeds through these intermediate, and perhaps kinetically stabilized states. These intermediate states – that is, the most robust synthons – could result in a considerable increase in crystallization efficiency.[30,31]

An understanding of crystallization is important for the systematic development of crystal engineering, but it is not a simple phenomenon and many would agree that it is still far too difficult to study in a rigorous way, either experimentally or theoretically. However, indirect approaches to the study of crystallization are evolving. Three possible types of crystals that may be pertinent to this endeavor are: (1) polymorphs – these represent cases of alternative crystallization, (2) pseudosymmetric structures with multiple molecules in the asymmetric unit – these could represent cases of incomplete crystallization, and (3) solvated crystals or pseudopolymorphs – these may represent cases of interrupted crystallization. These three scenarios are now sketched very briefly and the treatment given is necessarily selective.

Given the complementary nature of molecular recognition, it would appear that, when a compound crystallizes, the crystallization pathway, and hence the crystal structure obtained, should be quite specific to the molecule in question. However, the very existence of the phenomenon of polymorphism indicates that, under certain conditions, alternative crystallization pathways are feasible. In the special circumstance of concomitant polymorphism, or the simultaneous appearance of polymorphic forms in the same crystallization batch, these pathways even co-exist.[32] So, in general, the study of polymorphic systems has a bearing on a better

Scheme 4.

understanding of crystallization itself. Here, we discuss only a particular aspect of this subject, and one which is relevant in the context of synthon theory.[33] Let us consider 3-nitro-4-hydroxy-β-nitrostyrene, **14**. This compound is dimorphic and both structures contain the C–H···O based synthon, **XI**. The difference in the two forms lies in the completely different ways in which **XI** is generated (Scheme 4). In one of the polymorphs (Form A), it arises from the aromatic nitro group, whereas, in the other (Form B), it is constituted with the aliphatic nitro group. The roles of the alkenyl and aromatic H-atoms are switched appropriately in the two cases. Considering that it is manifested in both polymorphs, **XI** is clearly an important synthon for this compound. Because of the multiple and non-equivalent occurrences of various functionalities in the molecular structure, this synthon is non-uniquely defined – hence, polymorphism follows. Implicit in this discussion, of course, is that crystallization proceeds through the more robust synthons.

The occurrence of multiple molecules in the asymmetric unit in pseudosymmetric crystal structures is also pertinent to studies of crystallization. Figure 10 shows the crystal structures of quinoxaline, or 1,4-diazanaphthalene.[34] The figure on the top is the experimentally observed one. The space group is $P2_12_12_1$ and there are five molecules in the asymmetric unit, labelled 1 through 5. The structure may be analyzed in terms of a C–H···N catemer synthon and a general packing arrangement which in itself is unremarkable. However, the structure is pseudosymmetric; the environment around each of

the five symmetry-independent molecules is nearly the same. It is not really difficult to model a simpler structure in the same space group but with just one

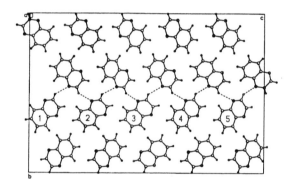

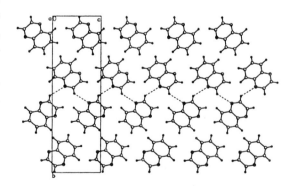

Figure 10. Crystal structure of quinoxaline showing the unit cell and the C–H···N hydrogen bonded catemer. Top: Experimental structure with $Z'= 5$. The symmetry independent molecules are numbered. Bottom: Structure determined with the Polymorph Predictor (*Cerius²*) software and having $Z'=1$. The two structures are nearly equivalent. Would one find other related structures with different values of Z'?

symmetry-independent molecule.[35] Such a simulated structure is shown in the bottom figure. It may be noted that the experimental (Z'=5) and simulated (Z' = 1) structures have calculated X-ray powder patterns and crystal energies that are nearly identical.

What is the significance then of the experimental structure? The occurrence of five symmetry-independent molecules in a crystal is very unusual and molecular dynamics studies on this structure shows that the preferred displacements about the atomic positions are such that they tend to remove the minor differences between the orientations of the five symmetry-independent molecules. All this hints that the experimental structure is kinetically locked in because of the low temperature (120 K) and method of crystal growth (crystallization from the neat liquid of the compound, m.p. 30 °C, followed by quick cooling with a stream of cold nitrogen after mounting on the diffractometer). We have not attempted to anneal the experimental structure into the simulated structure. Is it possible that during the (as yet unobserved) crystallization of the simulated structure, the C–H···N catemer synthon is initially formed as in the experimental structure while structural fine-tuning from pseudosymmetry to full symmetry comes later? If so, this might be evidence that synthons are of relevance, not only in the equilibrium crystal structure, but also in the evolving structure. One might then term the experimental structure an example of incomplete crystallization.[36]

15

Finally, let us turn to the phenomenon of solvation. Most non-ionic organic crystals (85 %) are unsolvated because of the entropic advantage in eliminating solvent molecules from solute-solvent aggregates into the bulk solvent, prior to enthalpically driven synthon mediated nucleation. However, if multipoint recognition between solvent and solute is unusually robust, the solvent might persist within the crystal nucleus leading to solvated or pseudopolymorphic structures.[37] In this context, the crystal structure of morellic acid, **15**, is of interest because it yields gums, resins and powders rather than well-defined crystals from many solvents. Such behavior is not unknown in molecules with conformationally flexible side-chains. However, it was also observed that good crystals of acid **15** are obtained from pyridine.[38] X-ray analysis of these crystals showed that they correspond to a 1:1 acid:pyridine solvate. A section of the structure of this solvate is shown in Figure 11, from which it may be noted that two adjacent acid molecules are hydrogen bonded to the pyridine with both strong O–H···N and weak C–H···O interactions (**a, b, c** and **d** in the figure). The role of the pyridine molecule in bringing the acid molecules together is noteworthy. The carboxy-pyridine synthon **XII** is surrounded by a more diffuse C–H···O pattern (interactions **e, f** and **g**) and the whole arrangement guides the morellic acid molecules into a hairpin shape and a close-packed geometry shown as a stereoview in Figure 12.

Since the expulsion of solvent from the nucleating crystal seems to be enthalpically disfavored for acid **15**, the formation of a solvated crystal may be likened to an interruption of the sequence of events that accompany normal crystallization. As for the failure of this acid to form crystals from other solvents, it has been noted earlier in this chapter that most crystalline carboxylic acids contain dimers or catemers – in other words, that carboxyl–carboxyl recognition is important. In the present instance, the formation of dimers and catemers may just not be enough to ensure crystallization because of the flexible nature of the molecule. The entropic penalty that has to be paid upon the loss of conformational freedom that would necessarily accompany crystallization is perhaps not compensated for adequately by enthalpic factors such as O–H···O hydrogen bonding between carboxylic groups. A solvent like pyridine, which can act as a multi-point hydrogen bond nucleator, might immobilize distant portions of the solute molecule to just an extent that is critically sufficient to induce crystallization.

Our knowledge of crystallization mechanisms is still quite rudimentary. However, crystal structures such as 3-nitro-4-hydroxy-β-nitrostyrene, quinoxaline and morellic acid provide snapshot views of some of the events that may occur during crystallization of more 'normal' compounds. One attempts to get a better appreciation of the dynamic processes that occur during crystallization events by studying crystal structures which though static, as are all crystal structures, are still suggestive.

Figure 11. Crystal structure of morellic acid, **15** in its 1:1 complex with pyridine. Hydrogen bonds **a** (strong) and **b** through **g** (weak) are marked. The pyridine molecule assists in assembling the structure.

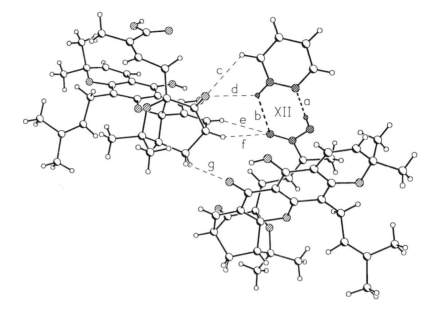

Figure 12. Stereoview of the morellic acid:pyridine complex showing more explicitly the function of the pyridine molecule. Hydrogen bonds are indicated. Notice the hairpin shape of the acid molecules.

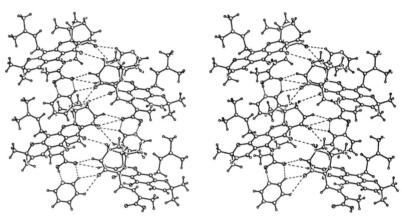

Epilogue

Molecules and crystals differ fundamentally with respect to their descriptors and the ways in which their structures are related to their properties. This distinction is the essential philosophical one between the molecular and the supramolecular paradigm. The significance of supramolecular synthons is that they constitute a bridge between molecules and crystals. These compact structural units are formed because of geometrical and chemical recognition between molecules. Specificity of recognition leads to structural robustness and synthons may be said to be implicated in the entire sequence of events: molecules → recognition → nucleation → crystallization → growth → crystal. Since synthons reflect the building up of crystals from molecules they may be dissected retrosynthetically to define crystal engineering strategies. Yet, the significance of the synthon goes beyond the purely structural to the mechanistic. As our knowledge of the crystallization event grows, the involvement and implication of these representative structural units will surely continue to be increasingly appreciated.

References and Notes

1. G. R. Desiraju, *Crystal Engineering. The Design of Organic Solids*, Elsevier, Amsterdam, **1989**.
2. The term *synthon* has been used in the manner originally intended by Corey (E. J. Corey, *Pure Appl. Chem.*, "General methods for the construction of complex molecules", **1967**, *14*, 19–37), and refers to a sub-structural unit.
3. G. R. Desiraju, "Supramolecular synthons in crystal engineering – a new organic synthesis", *Angew. Chem.*, **1995**, *107*, 2541–2557; *Angew. Chem. Int. Ed. Engl.*, **1995**, *34*, 2311–2327.
4. E. J. Corey, "Retrosynthetic thinking – essentials and examples", *Chem. Soc. Rev.*, **1988**, *17*, 111–133.
5. A. Nangia, G. R. Desiraju, "Supramolecular structures – reason and imagination", *Acta Crystallogr., Section A*, **1998**, *54*, 934–944.
6. V. R. Thalladi, B. S. Goud, V. J. Hoy, F. H. Allen, J. A. K. Howard, G. R. Desiraju, "Supramolecular synthons in crystal engineering. Structure simplification, synthon robustness and supramolecular retrosynthesis", *Chem.Commun.*, **1996**, 401–402.
7. No special significance need be ascribed to the use of the word *ribbon*, as opposed to say, *tape*. By ribbon is meant a generally flat one-dimensional motif.
8. F. H. Allen, B. S. Goud, V. J. Hoy, J. A. K. Howard, G. R. Desiraju, "Molecular recognition via iodo⋯nitro and

iodo⋯cyano interactions: crystal structures of the 1:1 complexes of 1,4-diiodobenzene with 1,4-dinitrobenzene and 7,7,8,8-tetracyanoquinodimethane (TCNQ)", *J. Chem. Soc., Chem. Commun.,* **1994**, 2729–2730.

9. J. A. R. P. Sarma, F. H. Allen, V. J. Hoy, J. A. K. Howard, R. Thaimattam, K. Biradha, G. R. Desiraju, "Design of a SHG-active crystal, 4-iodo-4'-nitrobiphenyl. The role of supramolecular synthons", *Chem. Commun.,* **1997**, 101–102.

10. Some subsequent discussion on compound **3** followed our original paper on this subject (N. Masciocchi, M. Bergamo, A. Sironi, "Comments on the elusive crystal structure of 4-iodo-4'-nitrobiphenyl" *Chem. Comm.,* **1998**, 1347–1348; J. Hulliger, P. J. Langley, "On intrinsic and extrinsic defect-forming mechanisms determining the disordered structure of 4-iodo-4'-nitrobiphenyl" *Chem. Comm.,* **1998**, 2557–2558), but these papers, which describe the effects of small amounts of 4,4'-dinitrobiphenyl impurity in **3**, only reinforce the idea of structures based on NO₂⋯I synthons.

11. J.-M. Lehn, *Supramolecular Chemistry,* VCH, Weinheim, **1995**.

12. This author has always distinguished between the terms *synthon* and *building block* and has rarely used the latter, because it implies modularity and a chemical insensitivity that is scarcely known in molecular crystals.

13. F. H. Allen, J. P. M. Lommerse, V. J. Hoy, J. A. K. Howard, G. R. Desiraju, "The halogen⋯O(nitro) supramolecular synthon in crystal engineering: a combined crystallographic database and *ab initio* molecular orbital study", *Acta Crystallogr., Section B,* **1997**, *53*, 1006–1016.

14. Insulation is effectively achieved if one particular set of interactions is very much stronger than the rest. This is the situation that arises in many coordination polymers (S. R. Batten, R. Robson, "Interpenetrating nets: ordered, periodic entanglement", *Angew. Chem. Int. Ed. Engl.,* **1998**, *37*, 1469–1494; *Angew. Chem.,* **1998**, *110*, 1558–1595). But coordination polymers are not molecular in all three dimensions, and as such are not within the scope of this essay.

15. L. Pauling, M. Delbrück, "Nature of the intermolecular forces operative in biological processes", *Science,* **1940**, *92*, 77–79.

16. Fischer's lock-and-key analogy expresses the same idea of complementarity in molecular recognition (E. Fischer, "Einfluss der Configuration auf die Wirkung der Enzyme", *Ber. Dtsch. Chem. Ges.,* **1894**, *27*, 2985–2993) as does Kitaigorodskii's close-packing principle (*Molecular Crystals and Molecules,* Academic Press, New York, **1973**). In a more recent context, strategies of rational drug design attempt to exploit such recognition complementarities.

17. An important distinction between molecular and supramolecular chemistry is that, in the latter, hydrocarbon residues are functionalities.

18. The identification of the set of interactions that constitute a synthon is, in the end, subjective. In the limit, any possible combination of interactions may be defined as a supramolecular synthon but then there is the risk that the term will be degraded or fall into disuse much like its molecular sibling. Interactions or groups of interactions that are needlessly identified as supramolecular synthons, but are unable subsequently to sustain a predictive role in crystal engineering, will

drop out from practical usage. In this way, one would hope that the scope and meaning of the term *synthon* will be refined during the evolution of crystal engineering into a mature subject. The collection of 75 cyclic hydrogen bonded synthons in the recent paper by Allen *et al.* (F. H. Allen, W. D. S. Motherwell, P. R. Raithby, G. P. Shields, R. Taylor, "Systematic analysis of the probabilities of formation of bimolecular hydrogen-bonded ring motifs in organic crystal structures", *New. J. Chem.,* **1999**, 25–34) provides a starting point for such refinement.

19. R. Thaimattam, D. S. Reddy, F. Xue, T. C. W. Mak, A. Nangia, G. R. Desiraju, "Interplay of strong and weak hydrogen bonding in molecular complexes of some 4,4'-disubstituted biphenyls with urea, thiourea and water", *J. Chem. Soc., Perkin Trans. 2,* **1998**,1783–1789.

20. The novel feature of these inclusion compounds is that the species which is linked with weak hydrogen bonds completely surrounds the species which employs strong hydrogen bonding in crystal assembly. Using topological conventions, the former is the *host,* and the latter the *guest.* All this discussion shows is that the terms *host* and *guest* are somewhat subjective, and that their definition is partly chemical and partly linguistic.

21. That the dimer synthon is found in as many as a third of all carboxylic acids is quite impressive, considering the very large variety of other functional groups that are present in the acids contained in the CSD. If, however, one were to consider only simple carboxylic acids – that is, with no other functionality present other than carboxyl and hydrocarbon residues – then the proportion of acids that contain dimer synthon **II** rises to around 85 %. Hence, the qualifier *robust* is justified in this case.

22. L. Leiserowitz, "Molecular packing modes. Carboxylic acids", *Acta Crystallogr. Section B.,* **1976**, *32*, 775–802.

23. O. Ermer, J. Lex, "Shortened C–C bonds and antiplanar O=C–O–H torsion angles in 1,4-cubanecarboxylic acid", *Angew. Chem. Int. Ed. Engl.,* **1987**, *26*, 447–449; *Angew. Chem.,* **1987**, *99*, 455–457.

24. R. J. Butcher, A. Bashir-Hashemi, R. Gilardi, "Network hydrogen bonding: the crystal and molecular structure of cubane-1,3,5,7-tetracarboxylic acid dihydrate", *J. Chem. Cryst.,* **1997**, *27*, 99–107.

25. S. S. Kuduva, D. C. Craig, A. Nangia, G. R. Desiraju, "Cubane carboxylic acids. Crystal engineering considerations and the role of C–H⋯O hydrogen bonds in determining O–H⋯O networks", *J. Am. Chem. Soc.,* **1999**, *121*, 1936–1944.

26. The acidity of the unactivated cubyl-H is comparable to NH₃ (pK_a ~ 38). Cubyl H-atoms are at least 10^5–10^6 times more acidic than vinyl and phenyl hydrogens (K. A. Lukin, J. Li, P. E. Eaton, N. Kanomata, J. Hain, E. Punzalan, R. Gilardi, "Synthesis and chemistry of 1,3,5,7-tetranitrocubane including measurement of its acidity, formation of *o*-nitro anions, and the first preparations of pentanitrocubane and hexanitrocubane", *J. Am. Chem. Soc.,* **1997**, *119*, 9591–9602 and the references cited therein).

27. G. R. Desiraju, T. Steiner, *The Weak Hydrogen Bond in Structural Chemistry and Biology* by Oxford University Press, Oxford, **1999**.

28. N. N. L. Madhavi, C. Bilton, J. A. K. Howard, F. H. Allen, A. Nangia, G. R. Desiraju, "Establishing structural repeti-

tivity in systems with interaction interference: crystal engineering in the *gem*-alkynol family", *New J. Chem.*, **2000**, *24*, 1–4.

29. Both thermodynamic and kinetic factors need to be considered. Take, for instance, acetic acid. The liquid contains mostly dimer but the crystal contains the catemer and no (polymorphic) dimer crystal has ever been obtained. Various computations (R. S. Payne, R. J. Roberts, R. C. Rowe, R. Docherty, "Generation of crystal structures of acetic acid and its halogenated analogs", *J. Comput. Chem.*, **1998**, *19*, 1–20; W. T. M. Mooij, B. P. van Eijck, S. L. Price, P. Verwer, J. Kroon, "Crystal structure predictions for acetic acid", *J. Comput. Chem.*, **1998**, *19*, 459–474) show the relative stability of the dimer. Perhaps the dimer is not formed in the crystal because it is 0-dimensional and as such, not able to propagate so easily to the bulk crystal as say, the 1-dimensional catemer.

30. G. R. Desiraju, "Crystal gazing: structure prediction and polymorphism", *Science*, **1997**, *278*, 404–405.

31. The correction of mistakes is an important aspect of crystal nucleation and is achieved by the inherent weakness of most interactions in the typical molecular crystal. Such correction could be facile because only around 15 % of all crystal structures in the CSD are disordered.

32. J. Bernstein, R. J. Davey, J.-O. Henck, "Concomitant polymorphs", *Angew. Chem. Int. Ed.*, **1999**, *38*, 3440–3461; *Angew. Chem.*, **1999**, *111*, 3646–3669.

33. In general, polymorphism can be enthalpically and entropically driven and, in general, is susceptible to both thermodynamic and kinetic factors.

34. A. Anthony, G. R. Desiraju, R. K. R. Jetti, S. S. Kuduva, N. N. L.Madhavi, A. Nangia, R. Thaimattam, V. R. Thalladi, "Crystal engineering: some further strategies", *Crystal Engineering*, **1998**, *1*, 1–18.

35. This simulation and other computational studies were carried out with the Polymorph Predictor and Crystal Packer modules in the *Cerius²* suite of programs from Molecular Simulations. We thank MSI (Cambridge and San Diego) for their continuing cooperation and assistance.

36. As an aside, it may be noted that the crystal structure of isoquinoline was reported recently (K. Hensen, R. Mayr-Stein, M. Bolte, "Isoquinoline", *Acta Crystallogr., Section C*, **55**, 1565–1567). It is disordered and isostructural to naphthalene. Clearly, the introduction of a second heteroatom into the naphthalene ring brings with it interesting consequences.

37. Solvent, in general, can affect crystallization during any of three stages – nucleation, growth and transformation. Here, we are concerned only with the first of these stages.

38. As a caveat, this discussion assumes that the 'good' crystals obtained from pyridine have a better degree of long-range order when compared to the powders obtained from other solvents. If the crystalline domains in the powders are already sufficiently large, one cannot really invoke molecule-level mechanisms to explain 'difficulties' in crystallization from the other solvents. Still, it may be noted that morellic acid and a number of other gamboge pigments are crystallized only with difficulty (K. V. N. Rao, P. L. N. Rao, "α- and β-Guttiferins", *Experientia*, **1961**, *17*, 213–214).

Luminescent Logic and Sensing

A. Prasanna de Silva, David B. Fox, and Thomas S. Moody

*School of Chemistry,
Queen's University
of Belfast BT9 5AG,
Northern Ireland*

*Phone: +44 28 90 335418,
Fax: +44 28 90 382117,
e-mail: a.desilva@qub.ac.uk*

Concept: A set of chemical inputs can generate a particular light output from a light-powered molecular-level device. Such output patterns correspond to various members of the logic vocabulary. Besides offering opportunities in information processing at the Angstrom scale, these systems also enable the continuous monitoring of key players in a whole spectrum of disciplines.

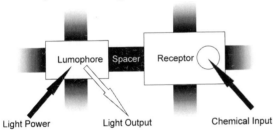

Abstract: Luminescent molecular sensors and switches can be classified according to their logical truth tables. In this chapter, a few examples illustrating YES (PASS), NOT, AND, XOR and INHIBIT are discussed. We emphasis the formatting of these molecular devices in terms of lumophores for photonic communication, receptors for chemical transactions and spacers for segregation. Photoinduced electron transfer (PET) is an important mechanistic thread running through most of the examples. These molecules connect chemistry to the life sciences, informatics and engineering.

Keywords:
- Luminescence
- Fluorescence
- Phosphorescence
- Sensors
- Switches
- Logic Gates
- Supramolecular Systems
- Truth Tables
- Photoinduced Electron Transfer
- Molecular-Level Devices

Prologue

Light can readily bridge the gap between the world of molecules and ours. While we chemists make molecules our business, these objects can be visualized most readily if they are empowered with light-emitting capabilities. The information transfer between the two worlds can be enriched if the light-emitting capability can be controlled or – in the extreme – switched 'on' or 'off'.[1–7] Then, molecular encounters fall under the glare of the chemists' spotlight. The fact that some of these molecular or atomic species are critical cogs in life processes or in high-value technological operations only serves to highlight the importance of such visualization.

Many biologically or technologically interesting species are not blessed with light-emitting properties because they don't have to be. Covalent tagging with a lumophore would enable emission but would probably destroy their primary function. However, we can tag them reversibly with a lumophore via an attached receptor. The encounter of the species of interest with the lumophore-appended receptor creates the luminescent supramolecular complex. For such a strategy to be successful, we must ensure that the lumophore-appended receptor is itself non-emissive. Then the objects of our interest will be the only points of light clearly contrasted against a dark backcloth. Similar, even if less appealing, contrast can be arranged if a brightly luminescent lumophore-appended receptor system was to be switched 'off' upon encountering the active species.

From a different perspective, the lumophore-appended receptor represents a molecular-scale light switch, which is triggered by a chemical species (whether or not it is of biological or technological consequence). Thus, photonic signals can be chemically generated with spatial, temporal, colour and other forms of resolution. It is encouraging to note that the opposite situation of photonically generated signals underlies our gift of vision. Molecular photonic logic gates can be viewed as switches with non-simplistic behaviour. Logic gates drive the information technology revolution. Suddenly, rudimentary computation at the molecular-scale emerges over the horizon.

Appending Lumophores to Receptors

The coordination, supramolecular and biological branches of chemistry provide us with a wealth of receptors, which seems to grow with each passing week.[8] A large variety of binding selectivity profiles are available with some bordering on guest-specificity under certain conditions. The binding constant for the receptor-guest association is particularly valuable to designers since its reciprocal gives the guest concentration on which the binding process hinges. If we go about one order of magnitude higher, the association is virtually complete: about one order of magnitude lower and the association is essentially nil. Thus, receptors can be chosen rationally with a given application in mind. However, there is a long way to go before a practically suitable receptor will be available for every need.

A similarly wide choice of lumophores can be found within photophysics and photochemistry.[9] The excitation (or absorption) and emission bands of luminescence come in a variety of wavelength positions, intensities and shapes. Another parameter, which is gaining in popularity among designers, is the luminescence lifetime. Time-resolved observation is a neat way of dissecting out the response of the luminescent device from the emissive noise of real-life matrices. Photostability of lumophores is a parameter which perhaps deserves more attention as more and more demanding applications are being tackled.

Now that receptors and lumophores have been sourced, we can focus on how they can be joined to create the devices we are seeking. This objective can be achieved in two fundamentally different ways: directly or via a connector (Figure 1).[10] The former approach integrates the receptor and lumophore so seamlessly that the individual 'components' can hardly be recognised. On the other hand, the latter approach clearly preserves the properties of the receptor and lumophore on account of the isolatory role of the connector. Isolation is relative, however. The connector offers spatial separation, which cuts off all inter-component communication channels except those of sufficiently long range. 'Spacer' is therefore a more informative descriptor of the connector. The integrated approach can lead to such strong mutual perturbations of the receptor and lumophore that very interesting systems result. Indeed, perhaps the most popular chemical sensors in physiology arose from this route in the hands of Roger Tsien (now at San Diego).[11] However, strong switching of light signals from 'off' to 'on' or vice versa is easier to arrange with the spaced approach. So this is the direction our story will take.

The long-range interaction central to our argument is the very same one which fuels our planet via photosynthesis – photoinduced electron transfer (PET).[12] PET processes are rather rapid across spacers composed of a few atoms if thermodynamically allowed and if kinetic barriers are absent. The receptor and lumophore pair need to be chosen with the latter points in mind. In which case, the 'lumophore-spacer-receptor' system[10] will have its luminescence emission quenched by the rapid PET process,[13] i.e., the luminescence will be switched 'off'. However, when the guest species of interest is captured by the receptor component the situation can be swung rather sharply, mostly by electrostatic influences. The PET process can be retarded by several orders of magnitude, in which case luminescence can reassert itself, i.e., it will be switched 'on'. It is equally feasible to start with a 'lumophore-spacer-receptor' system which is rationally arranged to have its luminescence switched 'on' and then to organize an acceleration of a PET process upon arrival of the guest so that the luminescence is switched 'off'.

We can now use these PET processes and relatives thereof as a mechanistic foundation on which a small set of illustrative examples will be placed. The examples will be presented in order of increasing logical complexity. The first two cases, in spite of their beauty and power, are relatively simple logic devices,[14] but they clearly show how far beyond chemistry the application areas can lie. They underline the message that luminescent sensors are beneficial molecular devices, right here, right now. The later examples cannot enjoy the above claim, but it is our belief that they will be in the vanguard for the molecular technologies of the future.

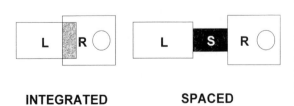

INTEGRATED   **SPACED**

Figure 1. Lumophore-receptor connection formats.

YES (PASS)

The YES logic operation is so trivial in an electronic context that it hardly bears serious discussion. The idea of an arrival of a voltage pulse triggering the release of a similar pulse is commonplace and can be seen at all points along an electrical conductor. Molecular versions of wires which can be rigorously tested in an electrical sense are now available. Thus, all-electronic YES operations are demonstrable at the molecular-scale. In contrast, YES logic operations with chemical input and light output are non-trivial. As pointed out in the prologue, the triggering of a light output upon arrival of a chemical input forms the basis of many fluorescent molecular sensors in analytical, bio- and clinical chemistry which are involved in reversible interactions with their target analyte chemical. We focus our attention upon one of these which has illuminated a particularly important problem. How does a cell in all its complexity respond to its environment? The realisation that a simple atomic species is a key player in this life process led Roger Tsien to apply the accumulated wisdom of coordination chemistry to this problem. Ca^{2+} is the atomic species in question and therefore a suitably modified EGTA [ethylenebis(oxyethylenenitrilo)tetraacetate] serves as a selective receptor. The latter can be discerned as the tetracarboxylate unit in YES gate **1** (Figure 2).[15]

calcium's private intracellular life in truly cinematic fashion. These pictures, moving in more than one way, have unearthed a wealth of information for life scientists to ponder.

On examining structure **1**, we notice the steric crowding at the link between the tetracarboxylate receptor and the linearly fused tricyclic system. Thus, orthogonalization occurs at this link, making it a virtual spacer. The spatial isolation of the two π-electron systems allows us to describe the tricyclic unit as the lumophore. The tetracarboxylate receptor also coexists as a clearly distinguishable module.

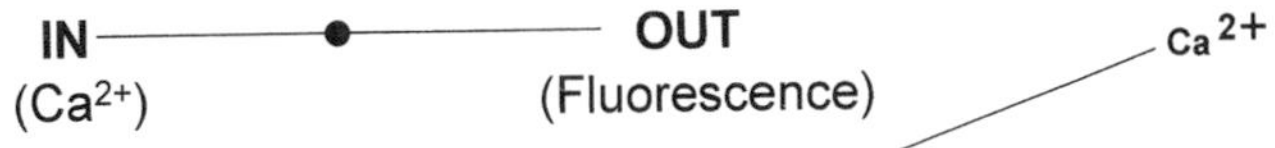

Ca^{2+} sensing by **1** can be understood by considering a PET process originating in the Ca^{2+}-free tetracarboxylate receptor and the linearly fused tricyclic system. The rapidity of the process may be ascribed to the low oxidation potential of the electron rich alkoxyaniline donor and to the short distance of separation between the two transfer components. The minimal separation arranged by a virtual spacer of zero carbon atoms is worthy of note. A low level of luminescence is seen owing to the fast PET.

The switching 'on' of luminescence upon Ca^{2+} binding to **1** has a special cause in addition to the general arguments advanced earlier in this essay. Ca^{2+} demands an optimal spatial draping of ligating atoms around it, in spite of its s-block character. This arrangement is best attained by rotation about the aryl C–N bonds in the *N*-aryl iminodiacetate moie-

Figure 2. (**a**) Electronic representation, (**b**) truth table, and (**c**) molecular implementation of YES (PASS) logic. 0 and 1 are digital representations of low and high signal levels respectively in all Figures.

b)

IN (Ca^{2+})	OUT $(Fluor^n)$
0	0
1	1

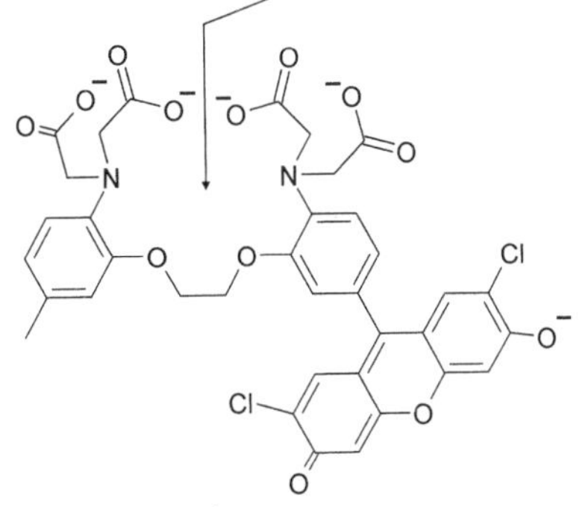

1

In particular, the binding of H^+ and Mg^{2+} are suppressed to the point that these ions have little or no influence on **1** within the physiological regime. This arrangement opens up the way for **1** to spy on Ca^{2+} as it goes about its business in the cytosol. Initial experiments conducted on cell populations with **1** and its forerunners revealed heightened levels of intracellular Ca^{2+} during the stimulation of various cell types with many different agents. However, it was left to later fluorescence microscopy studies on single live cells to open the window on

ties, among other things. In so doing, the alkoxyaniline π-electron system is split into alkoxybenzene and amine systems. The oxidation potential of the Ca^{2+}-bound receptor shoots up accordingly, making the PET process thermodynamically unfavourable and kinetically slow. Luminescence emission therefore reasserts itself emphatically.

NOT

Unlike its YES counterpart, the electronic NOT logic operation is anything but trivial. The NOT logic gate inverts any signal that is received e.g. a voltage input pulse results in no output for that period whereas a voltage output is generated the rest of the time.

As with YES operations, molecular NOT logic can be discerned in many previously published luminescence phenomena. This occurrence is especially so because of the ease with which luminescence quenching can be set up. NOT logic in such contexts can be easily interpreted since the arrival of the chemical input causes luminescence output to be extinguished to a greater or lesser extent. We have chosen to highlight the case highlighted in Figure 3 because of its wide applicability in situations unanticipated by most chemical scientists.

through. They realised that air pressure at a given point on the aerofoil surface is proportional to the concentration of O_2 there. In practice, the O_2 concentration is measured within a thin layer of polymer painted on the aerofoil surface. How is this done? Simply by dissolving phosphor **2** in the polymer layer and measuring its phosphorescence intensity. The triplet excited state of **2** is quite long-lived and is quenched upon encountering O_2 in its triplet ground state. In this instance, the encounter is collisional whereas all the other cases, but one, discussed in this chapter involve the formation of complexes of easily measurable lifetimes. Nevertheless, if we view the receptor as the site of interaction with the input species we see that all the non-σ electrons in **2** make up both the receptor and the lumophore. If **2** is to succeed in its present task, rapid diffusion of O_2 is essential and simple quenching (i.e. Stern–Volmer) kinetics is desirable. On the

Figure 3. (**a**) Electronic representation, (**b**) truth table, and (**c**) molecular implementation of NOT logic.

a) IN (O₂) — OUT (Phosphorescence)

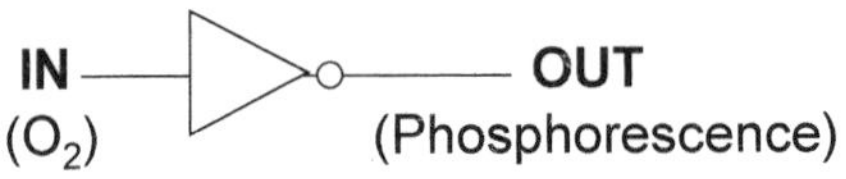

b) IN (O_2)	OUT $(Phos^n)$
0	1
1	0

c) [structure of **2**, Pt(II) porphyrin with O_2]

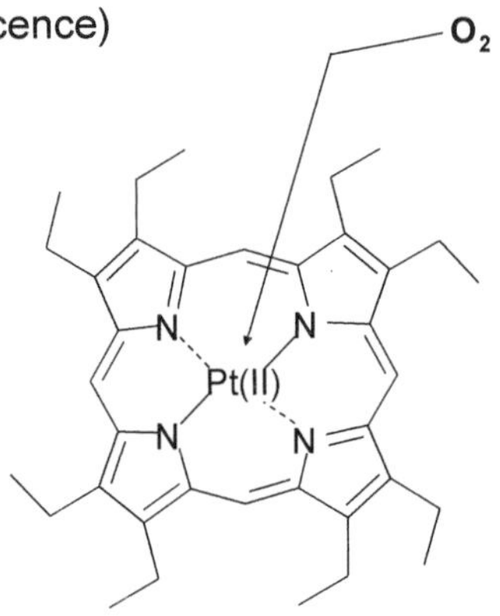

2

The story starts with aeroplanes, birds and helicopters. Each of these defy gravity via the lift created by differential airflow across upper and lower wing surfaces. While the design principles behind the two artificial cases are (comfortingly) well established since the time of Bernouilli, engineers continue to strive for improved performance. Related, but less critical, situations arise during the streamlining of various moving objects in our technological world. The need then arises for the continuous mapping of air pressure at aerofoil surfaces as a function of wind velocity and other engineering design variables. This need represents a call for remote monitoring in real-space and real-time – an ideal application for luminescent molecular sensors. However, this crossover from chemistry to the very different culture of aeronautical engineering required a leap of imagination. A group at the Central Aerohydrodynamic Institute[16] in Moscow and Martin Gouterman[17] in Seattle provided the break-

non-chemical side, intensity data spewing out of every pixel of the wing luminescence image needs to be acquired, maintained in-register with the corresponding zero-time data and processed.

Development issues aside, this approach to molecular luminescent barometry is sheer elegance, especially when compared to the older engineering approach of riddling the wing surface with a myriad of individual pressure sensors. It is heartening to realise that virtually any journey we make by air or land in the very near future will be underpinned by a luminescent NOT gate molecule.

AND

Now we step towards logic operations, which require more than one input. The AND operation has a significance that stretches beyond electronics or chemistry. It is the AND logic operation among people that leads to friendships, families, societies and other organisations which critically depend on co-operation. The truth table for AND logic (Figure 4b) can be very humanly described by the slogan 'United we stand, divided we fall'. At the far simpler level of electronics two voltage pulses arriving simultaneously along two input wires is the only condition under which the AND logic gate will deliver a voltage pulse along the output wire. Simpler still, H$^+$ and Na$^+$ are required to arrive simultaneously and to be nearly coincident on molecules 3[18] or 4[19] in order to elicit a bright fluorescence response (Figure 4a). Notably the amine and crown ether receptors handle their chosen input (H$^+$ and Na$^+$ respectively) nearly specifically, thus removing the need for wiring.

concentration blocks the lone electron pair on the amine, thus stopping the PET process associated with it. Similarly, Na$^+$ loading into the benzocrown ether cavity would cause a raised oxidation potential of the 1,2-dioxybenzene group and curtail its PET activity.

AND gate 4 differs from 3 in several aspects of performance and design. Experimentally, 4 produces a virtually perfect truth table. The output logic 1 state has a fluorescence quantum yield (ϕ_f) of 0.22 and the three logic 0 states do not rise above a ϕ_f value of 0.009. A fluorescence enhancement (FE) factor exceeding an order of magnitude such as this is a joy to work with since the switching phenomena are so clearly visible.

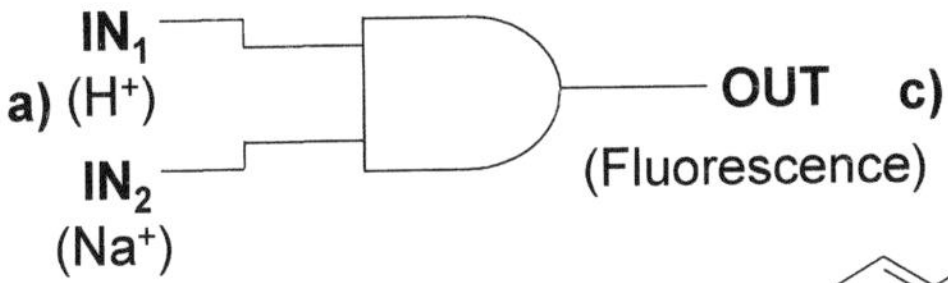

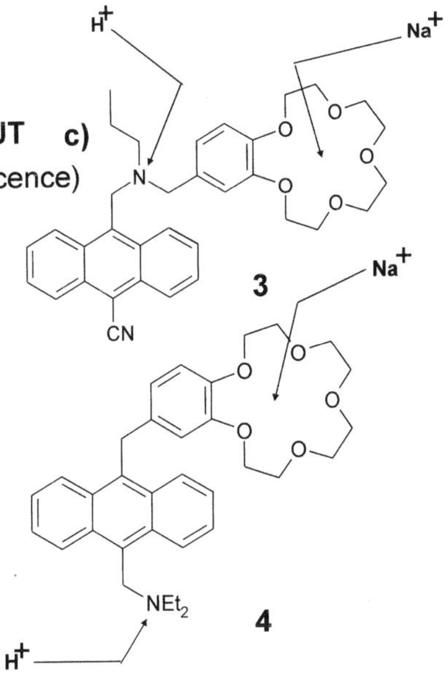

Figure 4. (**a**) Electronic representation, (**b**) truth table, and (**c**) molecular implementation of AND logic.

IN$_1$ (H$^+$)	IN$_2$ (Na$^+$)	OUT (Fluor^n)
0	0	0
0	1	0
1	0	0
1	1	1

The mechanistic basis of 3 and 4 is a double application of the argument used for the YES operation with 1. Each receptor in 3 is capable of launching a PET process if the lumophore is powered up by ultraviolet excitation. In other words, the fluorescence of the 9-cyanoanthracene unit is efficiently quenched by either the amine or the 1,2-dioxybenzene group within the benzo-15-crown-5 ether unit. These quenching processes are predictable from thermodynamic calculations or from related bimolecular quenching experiments in the literature. The small separation of the amine from the lumophore ensures rapid PET kinetics. On the other hand, the presence of four bonds between the benzo-15-crown-5 ether and the lumophore is probably responsible for the incomplete quenching seen between this pair. No information is yet available regarding possible folded conformations of 3.

Each of these PET processes can be arrested upon ionic command. Provision of H$^+$ at high enough

The obvious design difference in 4 versus 3 is the formatting of components. Since the lumophore in 4 is flanked by the amine and benzo-15-crown-5 ether receptors, it is best described as a 'receptor$_1$-spacer$_1$-lumophore-spacer$_2$-receptor$_2$' system. In contrast, 3 is built according to the 'lumophore-spacer$_1$-receptor$_1$-spacer$_2$-receptor$_2$' format. The two distances over which the PET processes operate in 4 are thus the minimum commensurate with a real spacer, i.e., a methylene group. Accordingly, the PET processes are very efficient at quenching the anthracene fluorescence, resulting in exquisitely low ϕ_f values.

However, there is an important subtlety in 4. While amine receptor-anthracene lumophore pairs produce excellent PET based quenching, a benzo-15-crown-5 ether receptor is demonstrably incapable of transferring an electron to an excited anthracene lumophore unless the latter carries an electron withdrawing substituent. Then, what is the secret of 4's success? First, only one rapid PET process is required to produce an output logic 0 state. In fact only one PET process can occur at a given time, even in multi-component systems under the conditions of low excitation intensity employed here. Second, a protonated amine group serves as an electron withdrawing group with regard to the anthracene lumophore, especially because it is positioned only a methylene group away. Thus, the PET process from the benzo-15-crown-5 ether comes into action just when we need it to satisfy the truth table, i.e., under the input conditions of high H^+ and low Na^+ concentrations.

XOR

The XOR logic operation is particularly interesting because it uses two inputs, no more and no less. In fact it is vital for comparing the digital state of two signals, whether they are similar or different. The output delivers logic state 1 only if the two input signals are logically different, i.e., 0 and 1 or 1 and 0.

The first molecular emulation of an XOR gate (Figure 5) can be found in the work by Vincenzo Balzani, Fraser Stoddart and their co-workers in Bologna and Birmingham.[20] The molecular com-

electron donor dioxynaphthalene unit within **6** and the electron acceptor diazapyrenium unit within **5**. Complexation is clearly signalled by not one but three optical channels – the emergence of a red colour (absorbing at $\lambda > 450nm$) and the disappearance of strong fluorescence in the ultraviolet (due to **6**) and in the blue-green (due to **5**).

Let's focus on the ultraviolet emission of **6** at 343 nm to see XOR logic in action. The first input is H^+ and the second is tri(n-butyl)amine. In the absence of either input, the 343 nm emission is virtually dead, i.e., the output is in logic state 0. Provision of H^+ alone at sufficiently high concentration causes protonation of **6** (at perhaps the aliphatic ether oxygens) and concurrent collapse of the **5·6** complex. The protonated form of **6** still emits at 343 nm. If tri(n-butyl)amine was provided instead, the **5·6** complex collapses again due to the stronger charge transfer interaction between the aliphatic amine and **5**. Fluorophore **6** thus freed displays the 343 nm emission. So the output achieves logic state 1 in the two situations where one or the other input is present. The final row of the truth table (Figure 5b) can be arranged by providing both inputs H^+ and tri(n-butyl)amine in stoichiometric quantities. Mutual neutralization occurs between

Figure 5. (**a**) Electronic representation, (**b**) truth table and (**c**) molecular implementation of XOR logic.

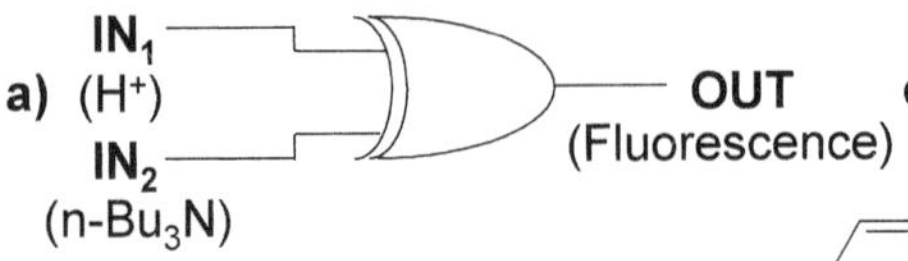

IN₁ (H⁺)	IN₂ (n-Bu₃N)	OUT (Fluorⁿ)
0	0	0
0	1	1
1	0	1
1	1	0

plex between **5** and **6** represents a 'ring on a string' situation. The use of such ideas and systems within an information handling framework takes us back to the imperial dynasties of ancient China when the abacus emerged as a computing machine. The casting of such an enduring device in molecular terms had to wait until rotaxanes and pseudorotaxanes could be constructed consistently and conveniently. The pseudorotaxane **5·6** hangs together due to the charge-transfer complexation between the

the inputs, leaving the complex **5·6** untouched. The 343 nm emission remains extinguished.

Besides the logic aspect, the chemically induced behavior of complex **5·6** connects with other macro-scale experiences such as threading/dethreading of a needle or piston motion in a cylinder. And so it turns out that the abacus will continue to be a particular inspiration for the design of controllable molecular motions for computational purposes. Complex **5·6** illuminates this path.

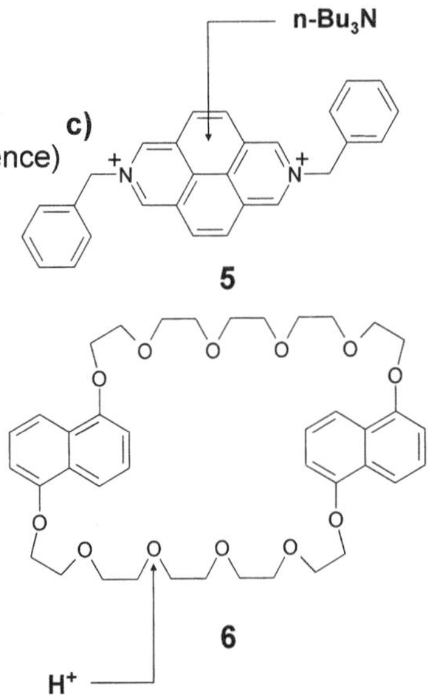

INHIBIT

One key reason for the rise and rise of electronics technology is the ease of integrating logic gates. At the simplest level, integration is as straightforward as wiring the output of one gate to the input of another. This trafficking is possible because electronic logic gates employ qualitatively identical inputs and outputs of voltage pulses. The molecular logic gates discussed here do not have this luxury. In fact, the only examples known so far would have died at birth if they did not have qualitatively different inputs, outputs and power supplies. The current state of knowledge would not allow the intelligent handling of all these types of traffic in and out of a molecule unless they are all distinguishable to the molecule and to the human operator.

In that case we need to think laterally if higher-level logic operations are to be achieved at the molecular-scale. One approach is functional integration within a single molecular system. This design can avoid the 'wiring' necessary during physical integration. For example, the INHIBIT logic operation as usually represented is the first molecular-scale example which requires a minimum of three inputs. Its electronic symbol (Figure 6a) clearly shows its constituents: a three-input AND gate with an inverter inserted into the third input line. In the electronic case, the arrival of a voltage pulse along the third line disables the entire system for that time period. The output achieves the logic state 1 only when two inputs are both live and when the third input is absent. In other words, the system acts as a normal two input AND gate if the third line is signal-free. Molecule 7[21] allows us to set up the truth table (Figure 6b) which looks rather daunting with regard to the current state of play in molecular-scale logic. Here, we combine three quite different photochemical processes. PET is one of these of course. The electron donor is the alkoxyaniline moiety within the tetracarboxylate receptor as was observed previously with 1. The electron acceptor is the bromonaphthalene lumophore whose excitation to its triplet excited state provides the necessary driving force. Triplet states possess less energy than their corresponding singlet excited states and hence are poorer drivers of PET processes. In compensation however, the relatively slow phosphorescence emission from triplet excited states provides poorer competition to PET processes. So the phosphorescence output from 7 is efficiently suppressed. Ca^{2+} input to the tetracarboxylate receptor raises its oxidation potential – as noted in the case of 1 and blocks this PET channel.

The second photochemical process engineered within 7 is the propensity of triplet excited states to linger long enough to meet a kindred species and annihilate one another. Such mutual destruction usually prevents us from observing strong phosphorescence from organic species in fluid solution at room temperature. However, there is help at hand. Enveloping the lumophore in a transparent sleeve such as the donut-like β-cyclodextrin shields the phosphor from its suicidal tendency, thus setting the stage for observable emission.

The photochemical process built into 7 was encountered previously with regard to 2, i.e., the capability of O_2 in quenching excited states of sufficiently long lifetimes. In the case of 7, the process is so efficient that ambient levels of O_2 completely kill off phosphorescence, even if the phosphor is enveloped by β-cyclodextrin.

Now it is clear that O_2 serves as our disabling third input, since it achieves an overall quenching

Figure 6. (a) Electronic representation, (b) truth table, and (c) molecular implementation of INHIBIT logic.

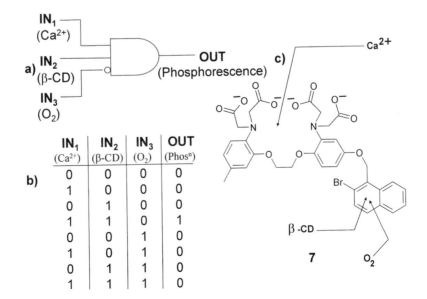

IN₁ (Ca²⁺)	IN₂ (β-CD)	IN₃ (O₂)	OUT (Phosⁿ)
0	0	0	0
1	0	0	0
0	1	0	0
1	1	0	1
0	0	1	0
1	0	1	0
0	1	1	0
1	1	1	0

of phosphorescence. The first and second inputs are Ca^{2+} and β-cyclodextrin respectively. It is only when Ca^{2+} blocks the PET channel and when β-cyclodextrin disrupts the triplet-triplet annihilation pathway that strong phosphorescence output is observed, provided that O_2 is kept at sufficiently low levels. This is INHIBIT logic action at the molecular-scale.

The relatively long lifetime of phosphorescence displayed by INHIBIT gate **7** leads to two distinguishing features vis-à-vis its fluorescent cousins such as AND gate **3**. As we saw above, the long lifetime causes high sensitivity to quenching by O_2 which fluorophores are largely free of. A more physical feature is the possibility of time-delayed observation following pulse excitation. A delay of the order of milliseconds is sufficient to pick out phosphorescence clearly after any fluorescence has subsided. Armed with these two distinguishing features we can choose to tune-in to any one of these two types of gates co-existing in a medium. Sequential addressing of logic gates thus becomes available, at least at a prototype level.

Epilogue

The previous pages of this chapter advanced the thesis that chemical species, either singly or in combination, can be persuaded by design to produce physical signals which impact human senses. The key supramolecular systems which arrange such signalling can be profitably classified and viewed as logic devices in all their diversity. Increasing demands from clinicians, biologists and engineers for monitoring and imaging situations involving particular sets of chemical players will provide one impetus for researchers in luminescent logic and sensing. Another driver will be the growing capability of designed single molecules to perform humanly useful functions in a variety of contexts. In all probability, the future will demonstrate that light and molecules form an even more potent mix than before.

Acknowledgements

We appreciate those fellow scientists who have been courageous enough to examine chemical problems from the perspective of other sciences and vice versa. Our efforts would have been weaker without the support of the Department of Education in Northern Ireland, the Engineering and Physical Sciences Research Council (UK), the European Social Fund and the ERASMUS Program. We specifically thank Nimal Gunaratne, Colin McCoy, Isabelle Dixon, Thorfinnur Gunnlaugsson, Pamela Maxwell and Terence Rice for their valuable contributions.

References and Notes

1. A.P. de Silva, H.Q.N. Gunaratne, T. Gunnlaugsson, A.J.M. Huxley, C.P. McCoy, J.T. Rademacher, T.E. Rice, "Signalling recognition events with fluorescent sensors and switches", *Chem. Rev.* **1997**, *97*, 1515–1566.
2. A.W. Czarnik, Ed., *Fluorescent Chemosensors for Ion and Molecule Recognition*, ACS Books, Washington, **1993**. This collection contains the contributions of 12 laboratories.
3. A.W. Czarnik, J.-P. Desvergne, Eds., *Chemosensors of Ion and Molecule Recognition*, Kluwer, Dordrecht, **1997**. This collects the contributions of 16 laboratories.
4. B. Valeur, "Principles of fluorescent probe design for ion recognition" In *Topics in Fluorescence Spectroscopy. Vol. 4. Probe Design and Chemical Sensing* (Ed. J.R. Lakowicz), Plenum, New York, **1994**, pp 21–48.
5. L. Fabbrizzi, A. Poggi, "Sensors and switches from supramolecular chemistry", *Chem. Soc. Rev.* **1995**, *24*, 197–202.
6. T.D. James, K.R.A.S. Sandanayake, S. Shinkai, "Saccharide sensing with molecular receptors based on boronic acid", *Angew. Chem., Int. Ed. Engl.* **1996**, *35*, 1910–1922.
7. A. Fernandez-Gutierrez, A. Munoz de la Pena, "Determination of inorganic substances by luminescence methods", In *Molecular Luminescence Spectroscopy. Methods and Applications. Part 1* (Ed. S.G. Schulman), Wiley, New York, **1985**, pp 371–546.
8. J.-M. Lehn, *Supramolecular Chemistry*, VCH, Weinheim, **1995**.
9. B.M. Krasovitskii, B.M. Bolotin, *Organic Luminescent Materials*, VCH, Weinheim, **1989**.
10. R.A. Bissell, A.P. de Silva, H.Q.N. Gunaratne, P.L.M. Lynch, G.E.M. Maguire, K.R.A.S. Sandanayake, "Molecular fluorescent signalling with 'fluor-spacer-receptor' systems – approaches to sensing and switching devices via supramolecular photophysics", *Chem. Soc. Rev.* **1992**, *21*, 187–195.
11. R.Y. Tsien, "Intracellular signal transduction in 4 dimensions – from molecular design to physiology", *Am. J. Physiol.* **1992**, *263*, C723-C728.
12. G.J. Kavarnos, *Fundamentals of Photoinduced Electron Transfer*, VCH, Weinheim, New York, **1993**.
13. R.A. Bissell, A.P. de Silva, H.Q.N. Gunaratne, P.L.M. Lynch, G.E.M. Maguire, C.P. McCoy, K.R.A.S. Sandanayake, "Fluorescent PET (photoinduced electron transfer) sensors", *Top. Curr. Chem.* **1993**, *168*, 223–264.
14. A.P. Malvino, J.A. Brown, *Digital Computer Electronics* 3rd Ed., Glencoe, Lake Forest, **1993**.
15. A. Minta, J.P.Y. Kao, R.Y. Tsien, "Fluorescent indicators for cytosolic calcium based on rhodamine and fluorescein chromophores" *J. Biol. Chem.* **1989**, *264*, 8171–8178.
16. V. Moshasrov, V. Radchenko, S. Fonov, *Luminescent pressure sensors in aerodynamic experiments*, Central Aerohydrodynamic Institute, Moscow, **1998**.
17. M. Gouterman, "Oxygen quenching of luminescence of pressure sensitive paint for wind tunnel research", *J. Chem. Educ.* **1997**, *74*, 697–702.

18. A.P. de Silva, H.Q.N. Gunaratne, C.P. McCoy, "A molecular photoionic AND gate based on fluorescence signalling", *Nature* **1993**, *364*, 42–44.

19. A.P. de Silva, H.Q.N. Gunaratne, C.P. McCoy, "Molecular photoionic AND logic gates with bright fluorescence and "off-on" digital action", *J. Am. Chem. Soc.* **1997**, *119*, 7891–7892.

20. A. Credi, V. Balzani, S.J. Langford, J.F. Stoddart, "Logic operations at the molecular level. An XOR gate based on a molecular machine", *J. Am. Chem. Soc.* **1997**, *119*, 2679–2681.

21. A.P. de Silva, I.M. Dixon, H.Q.N. Gunaratne, T. Gunnlaugsson, P.R.S. Maxwell, T.E. Rice, "Integration of logic functions and sequential operation of gates at the molecular-scale", *J. Am. Chem. Soc.* **1999**, *121*, 1393–1394.

Nanochemistry – Architecture at the Mesoscale

**Stefan Becker
and Klaus Müllen**

*Max-Planck-Institut
für Polymerforschung,
Ackermannweg 10,
D-55128 Mainz, Germany*

Keywords ■ *Polycyclic aromatic hydrocarbons* ■ *Conjugated polymers* ■ *Self-assembly* ■ *Nanotechnology* ■ *Dendrimers* ■ *Dyes* ■ *Hydrogen bonding*

Concepts: Think small! Chemists have for a long time been making and visualizing nanoparticles such as pigments or lattices. Novel methods in physics allow the direct visualization and manipulation of single molecules and thus the establishment of the nanosciences. This extreme case of miniaturization and of moving along length scales creates new opportunities for chemistry since molecular characteristics are obtained which no longer represent an ensemble average. The task for nanochemistry then is to make and assemble optimal target molecules which provide an insight into structure and function at a single molecule level.

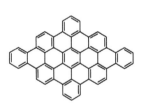

Abstract: Chemists may find it difficult to admit that their concepts and opportunities have always been strongly influenced by the available methods for characterization and analysis. Physics, has, of course, the lead when it comes to the visualization of single molecules in real space and to the detection of their specific, not ensemble-averaged properties. The challenge for chemistry is to provide molecules as objects of study which really disclose new concepts of structure and function. This chapter presents a chemical approach toward nanosciences which comprises: (i) design and synthesis, (ii) immobilization, often using principles of self-assembly, (iii) visualization, e.g. by scanning probe techniques, and (iv) manipulation and function. Obviously, the latter aspect has tremendous implications: "seeing" a single molecule by its light emission or recording its current-potential characteristics would ultimately lead to as efficient use of materials as one could imagine. For the sake of clarity the title molecular structures are classified according to their dimensionality, i.e., one proceeds from rods to discs and spheres with electronic properties as a key issue. Use of the benzene ring as a synthetic, modular block offers a great degree of structural versatility.

Prologue

Miniaturization plays a key role in the future development of modern technologies. Currently the main process for the fabrication of nanostructures is lithography, which nowadays faces tremendous challenges in the attempt of patterning in the sub-100 nm range. The conceptual and practical problems associated with this downscaling might suggest investigation of an alternative approach: the use of arrays of single atoms or single molecules as fundamental electronic devices. Such a bottom-up instead of the conventional top-down approach would have profound consequences for device operation, because single electron phenomena could come into play.

When a chemist writes down a molecular formula, it normally represents an ensemble of many molecules. The new challenge is to handle *single* molecules or defined aggregates of a few molecules. One would then want to visualize these nano-objects, to move them around in real space, to record physical functions such as light emission and their response to a stimulus, and even to perform chemical reactions on them.

There are now physical methods available to achieve such goals. These methods are often complex. So one might argue that the emerging nanosciences will be restricted to physics and physical chemistry. It appears, however, that the molecular and supramolecular structures of molecules and their aggregates must fulfill a series of requirements to qualify for such experiments, especially if one wants not only to record nice pictures, but also to measure complex functions of a single molecule. Doing chemistry at the level of a single molecule is an exciting prospect for a chemist, for it is a way of "taking each molecule seriously" and of making the most efficient use of materials.

The study of principles relevant to building nanoscale structures of, for example, 1–100 nm in size has recently been referred to as nanoscience, and it is expected that methods for the fabrication of nanostructures will evolve into nanotechnologies. The present approach toward a "nanochemistry" is of more fundamental character since it is restricted to single molecules of a few nm in size or even smaller and focuses on the molecular pattern.

The concept of nanochemistry which we are outlining herein implies four steps:

– (i) *design and synthesis of suitable molecules*: as long as we are not subjecting single molecules to chemical transformations, synthesis is still a macroscopic experiment performed in the bulk. Since the size and shape of the molecules are crucial, the chemical design often requires transition from small repeat units to oligomers and polymers.

– (ii) *immobilization*: this can include incorporation into an amorphous matrix, such as in single molecule spectroscopy, but also deposition as a film, in an extreme case as a monolayer on a surface. This process will often include supramolecular order occurring under the influence of weak intermolecular forces, such as π-π-interaction, alkyl chain packing, hydrogen bonding or Coulomb effects. It would be advantageous to already have encoded these supramolecular features in the molecular structure.

one can chemically transform single molecules under direct control.

The above four-step procedure (i)–(iv) will now be outlined for different classes of compounds where the dimensionality of the molecular objects (chains, discs, spheres) serves as a useful guideline. The present text is concerned only with *synthetic organic molecules and macromolecules*. It is clear that an analogous concept can be established for biologically relevant molecules and their aggregates as well as for inorganic materials such as metals or semiconductors at different length scales. Further, since electronic function plays a major role, non-conjugated molecules such as molecular "worms" or "bottle brushes" are not considered.

Chains

The prototypes of electronically active molecules are linear conjugated chains made from olefinic and aromatic repeat units (Scheme 1).

(i) The synthesis of such chains by repetitive carbon-carbon bond formation processes often requires organometallic intermediates and transition metal catalysis.[1] Examples of rigid rod-type structures are oligo- and polyphenylenes **4** as well as oligo- and polyphenyleneethynylenes **5**. Suitable alkyl substitution is often mandatory to obtain solution processable compounds (see below).

Scheme 1. Chemical Structures of some important linear π-conjugated polymers. Polyenes **1**, polyynes **2**, polyenynes **3**, poly(*p*-phenylene)s (PPP) **4**, polyphenylenevinylenes (PPV) **5**, polyphenylene-ethynylenes (PPE) **6**.

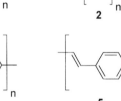

– (iii) *visualization*: this is the actual physical measurement which can, e.g., consist of single molecule spectroscopy (SMS), scanning tunneling microscopy (STM), or atomic force microscopy (AFM). Unlike STM, AFM does not normally give a molecular resolution. It is, however, experimentally less demanding. STM rests upon the occurrence of a tunnel current through molecules and requires their deposition on a conducting substrate.

– (iv) *function and manipulation*: this is certainly the most exciting step leading to questions of whether one can move molecules around in space, deform them mechanically, or use their electronic function such as charge-transport or electroluminescence, and in particular, whether

Another key point is selective chemical functionalization at one or both ends, or inside the chain (see scheme 2).[2] Thus, thiolo functions can serve as clips to create contact with metal surfaces or particles. Quantitative end functionalization of the rigid-rod on one end is a key step toward rod-coil copolymer synthesis (see scheme 3),[3] and such a covalent coupling of incompatible polymer blocks is relevant for supramolecular organization.[4]

(ii) The chain stiffness of rigid rods has served as a major structure-forming principle in supramolecular organization in the bulk, at the surface of solid substrates or at an air-water interface whereby the side-by-side packing of the chains is a typical motif and the supramolecular structure formation can be

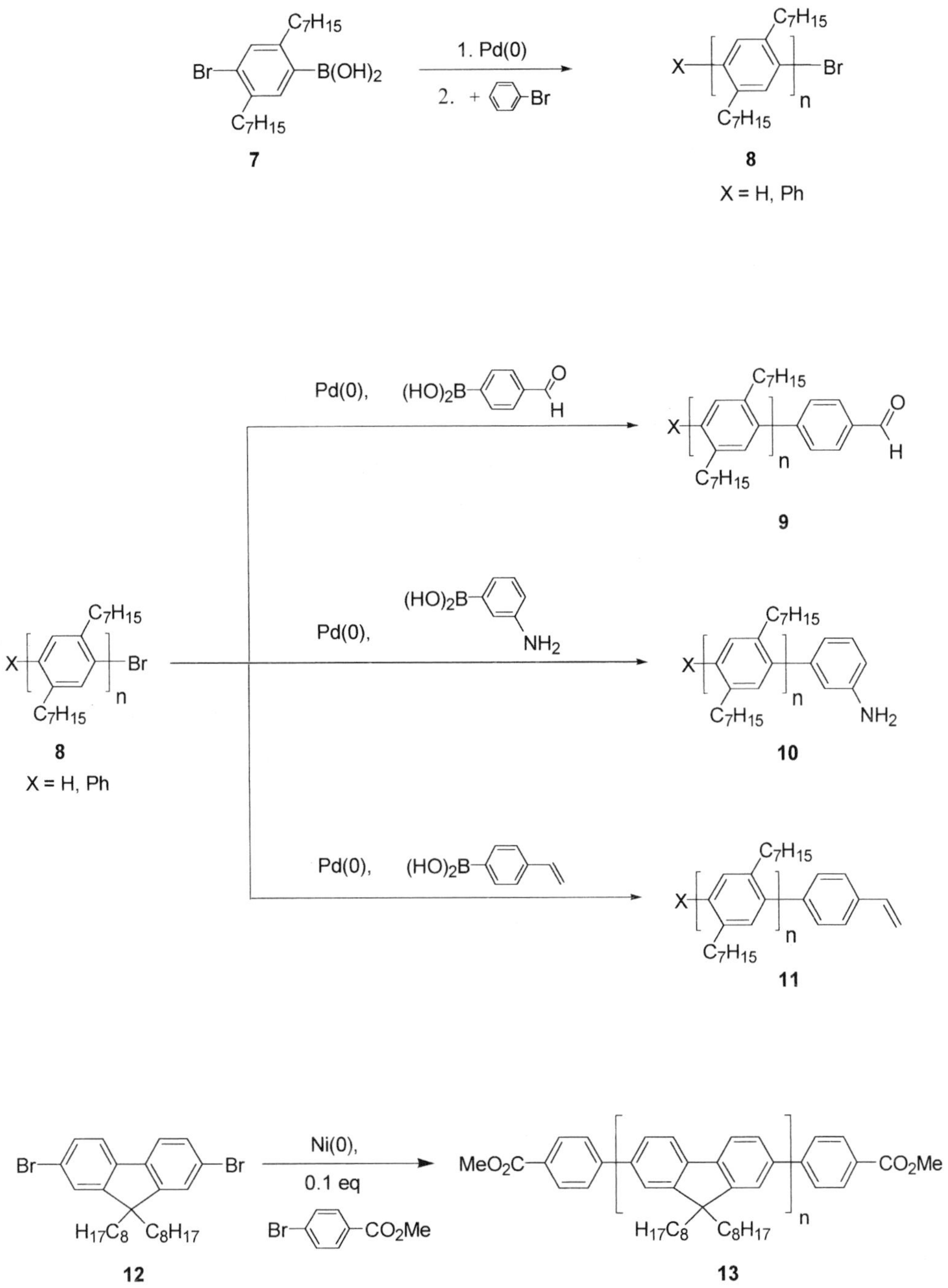

Scheme 2. Synthesis of endfunctionalized conjugated polymers. The bromoendfunctionalized poly(p-phenylene) **8** is the starting material for the synthesis of various endfunctionalized polymers **9**, **10**, **11**. Two-fold carboxy- phenyl terminated polyfluorene **13** is afforded by adding the endcapper 4-bromobenzoic acid methyl ester during the nickel catalyzed coupling reaction of monomer **12**.

assisted by the interdigitation of alkyl chains (Figure 1).[5]

The concept of a molecular wire is a remarkable case; in this a conjugated rod-like molecule bridges the gap between nanoelectrodes with molecular clips as crucial components.[6]

(iii) Oligophenyleneethynylenes can be deposited by physical adsorption (physisorption) from solution onto graphite surfaces,[7] and the resulting pattern can be visualized by scanning tunneling microscopy.

A typical lamellar arrangement of the rods is formed in which the benzenoid π-systems lie flat on the surface. The nature of the substrate surface is, however, crucial. If the corresponding polyphenyleneethynylenes **21** are deposited onto mica surfaces and the resulting patterns visualized by AFM, one detects micrometer long needles, whose width corresponds to the length of the polymers (Figure 2).[8]

The height of the needles suggests the formation of mono- and double layers in which the polymers

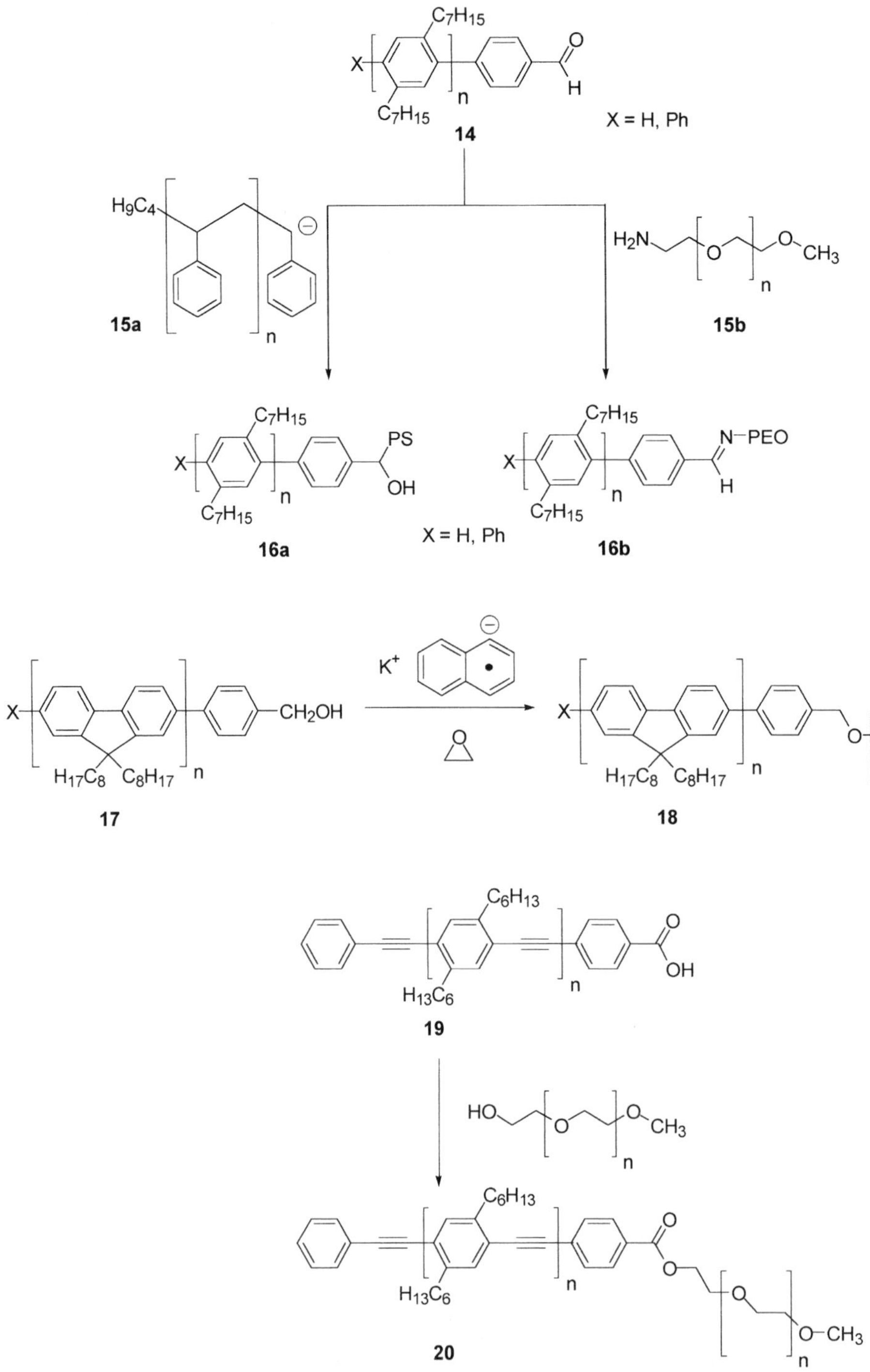

Scheme 3. Synthetic routes leading to various rod-coil block copolymers via "grafting onto" (**16a**, **16b**, **20**) or "grafting from" (**18**) reactions.

"stand" on their alkyl chains. Instead of optimizing the π–π interaction with the substrate surface, they create a face-to-face arrangement among themselves. It is a remarkable feature of this organization process that polymers of different lengths cluster into very similar sized areas on the surface.

(iv) The role of conjugated oligomers and polymers as potential molecular wires between electrodes has attracted much attention as a first step toward molecular electronics.[9] The concept of conductivity of a single molecule has been subject to much dispute.[10] The problem of short oligophenylenes, e.g. between break-junctions, is the significant role of tunnel currents; further, semiconductor-type oligophenylenes should require formation of charge carriers, e.g. by doping. A more feasible approach toward testing electrical conductivity as a function of single molecules is the use of metal-like

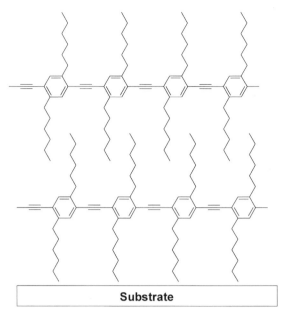

Figure 1. Schematic representation of the molecular packing of a polyphenylenethynylene.

nanotubes.[11] Indeed, nanoelectrodes with about 100 nm distance have been bridged by nanotubes and currents through single objects measured.[12]

Discs

A transition from 1D- to 2D π-conjugated structures has important consequences for the π–electron structure.[13] Furthermore, it is challenging to investigate supramolecular architectures formed from disc-type structures.

(i) The prototypes of disc structures are polycyclic aromatic hydrocarbons (PAH), which for a long time have attracted the attention of synthetic, theoreti-

cal and industrial chemists. The π-topology is crucial for both the stability and the electronic properties.[14] We have recently proposed a simple two-step procedure for the synthesis of PAHs of hitherto unprecedented size (scheme 4).[15]

The key feature of the molecular design is a variation of size and shape as well as of the perimeter type. Thus, benzene-homologous hexagonal structures from $C_{42}H_{18}$ (hexabenzocoronene, HBC) **27** to $C_{222}H_{42}$ **28** (scheme 5), but also the homologues with D_{2h} symmetry (scheme 6) $C_{60}H_{22}$ **29**, $C_{78}H_{26}$ **30**, $C_{96}H_{30}$ **31** and the "supertriphenylene" **32** (scheme 7) with D_{3h} symmetry are available. One anticipates that the giant PAHs will also achieve significance as molecularly defined subunits of graphite.

(ii) As with the rigid-rod structures, alkyl substitution is important to bring about solubility and assist in supramolecular ordering. Two motifs are important: one is the formation of discotic mesophases with a stack-type arrangement of the π-systems, and the other is the deposition of discs into monomolecular adsorption layers on substrate surfaces. The size of the discs is crucial because it stabilizes the discotic mesophases over a broad temperature range and induces very high charge-carrier mobilities within the stacks in photoconduction experiments (Figure 3).[16]

A particularly useful case is the hexaphenyl HBC **34** because it forms a mesophase even at room temperature and exhibits a constant charge-carrier mobility over a large temperature range.[17] It is possible to draw fibers from the anisotropic melts in which, according to X-ray diffractometry, the fiber axis is oriented along the stacking axis, the fiber thus constituting a macroscopic analog of multilayered cyclophanes.[18] A mechanical stabilization of

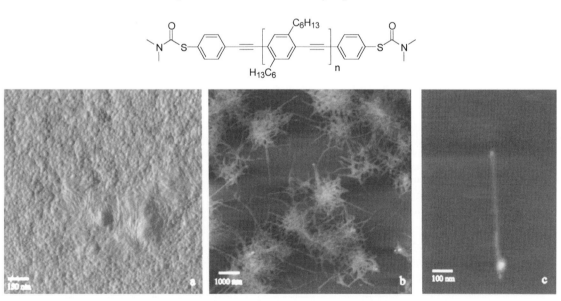

Figure 2. Topographical tapping mode SFM images of polyphenylenethynylene **21** on mica prepared by solution casting from (**a**) 2.0 g/l polymer in MeOH, (**b**) 0.14 g/l

polymer in THF, (**c**) 0.07 g/l polymer in THF. Pictures taken from ref. [8].

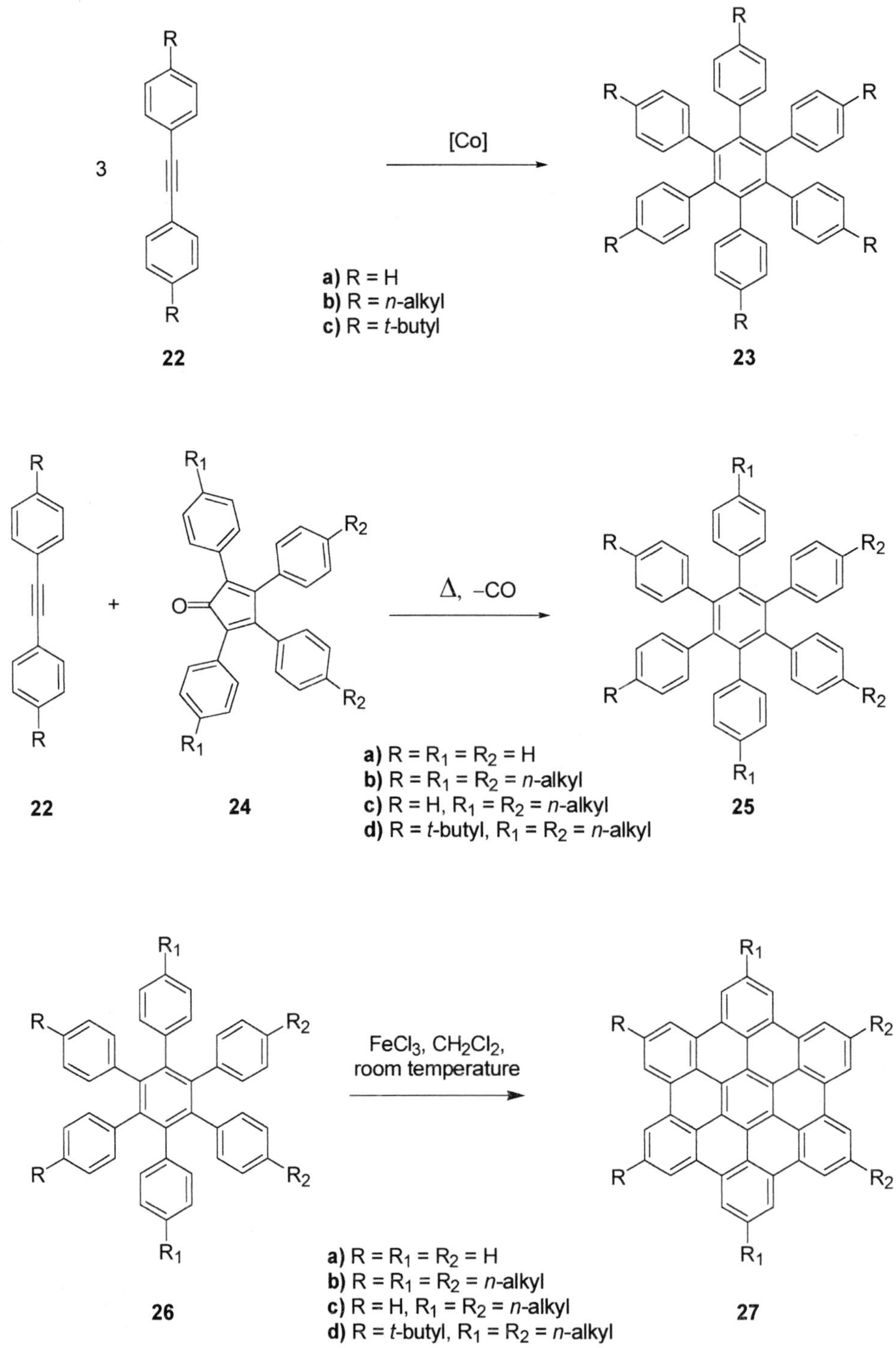

Scheme 4. The two-step procedure for the synthesis of PAHs (HBC is given as an example). Hexaphenylbenzenes are afforded in the first step either by cobalt-catalyzed cyclotrimerization of suitable diphenylethynylenes **22** or by Diels-Alder reaction of **22** and cyclopentadienones **24**. The second step is an oxidative cyclodehydrogenation to yield HBCs **27**.

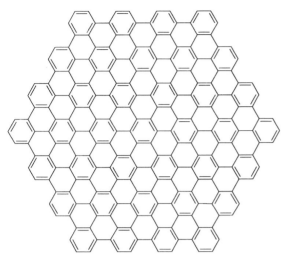

Scheme 5. Giant PAH **28**. [28]

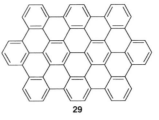

Scheme 6. Examples for PAH homologues with D_{2h} symmetry $C_{60}H_{22}$ **29**, $C_{78}H_{26}$ **30** and $C_{96}H_{30}$ **31**.

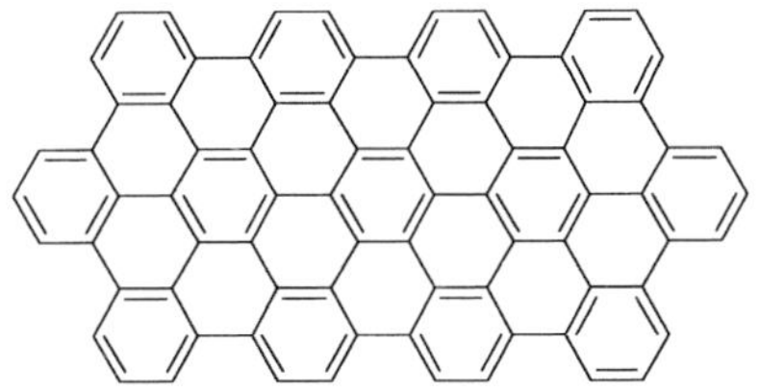

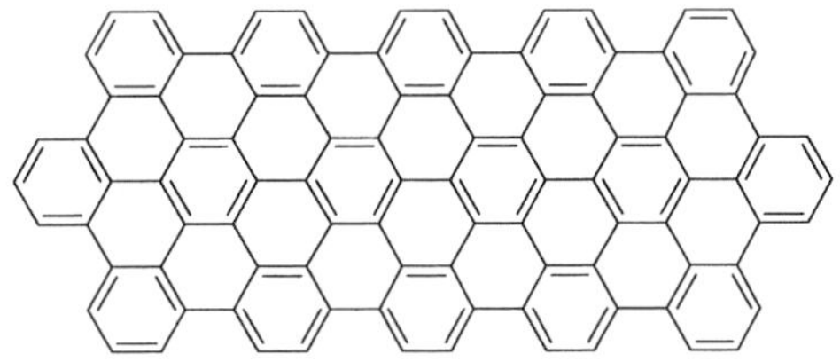

mesophases is possible by forming 1:1-complexes **37** between the monocarboxylic acid derivative **36** and a polyethylene iminium salt **35** (scheme 8).[19] This complex can be regarded as a true molecular composite since the liquid crystalline order persists and the mechanical stability is enhanced.

Fibers from disc-type HBCs are also obtained upon precipitation in various organic solvents. According to electron microscopy, the precipitation process also gives rise to the formation of belt-type structures with a diameter of 0.5–1µm.

(iii) The deposition of monomolecular adsorption layers of, e.g., HBC on graphite surfaces can be achieved under ultrahigh vacuum. Here the coincidence of substrate and adsorbate symmetry is important for the formation of an ordered motif. Similar experiments can be made using the rhombus-type PAH **38** and the resulting order visualized by both low-energy energy diffraction (LEED) and STM (Figure 4).[20] Desorption measurements point toward a binding energy of PAH **38** on the surface of about 2 eV/mol.[21]

More complex structures are also accessible; thus one can use an HBC monolayer as the "ground floor" for the deposition of a second layer of perylenediimides.[22] The electron-rich character of the hydrocarbon and the electron poor character of the dye create a pair of single molecules suitable for measuring currents in both, forward and reverse direction.

Deposition of PAHs on surfaces and the formation of monolayers by UHV-techniques is experimentally demanding. Therefore it is important that monolayers of soluble alkyl substituted HBCs can also be obtained by physisorption from solution.

Scheme 7. The "super-triphenylene" **32**.

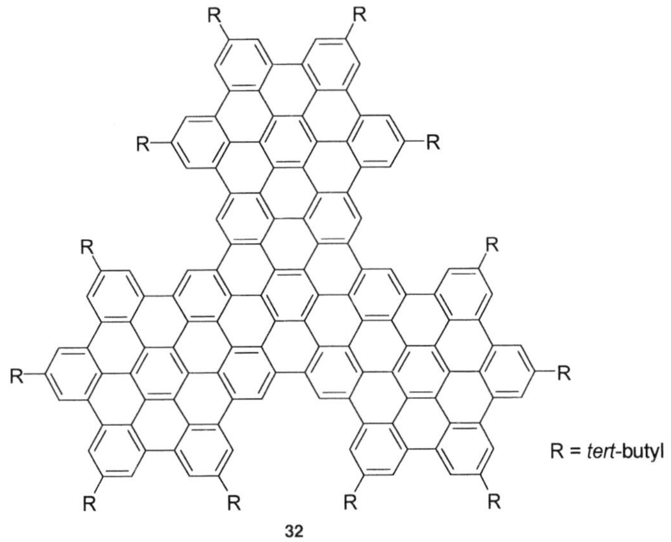

R = *tert*-butyl

32

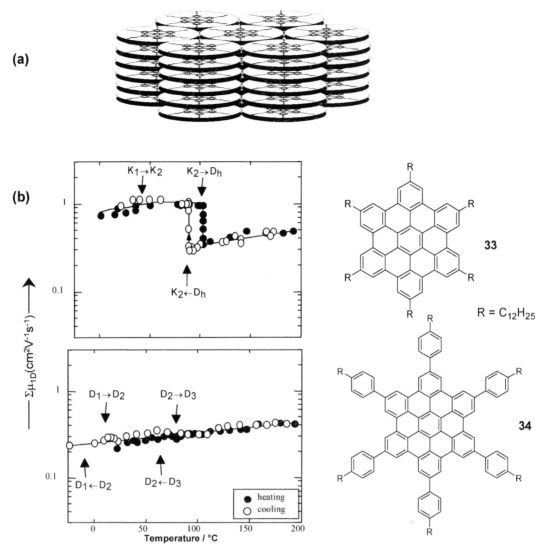

Figure 3. (a) Representation of the hexagonal discotic mesophase formed by hexadodecyl substituted HBC **33**. (b) The temperature dependence of the intracolumnar charge carrier mobilities for **33** and **34**. Phase transition temperatures as indicated by DSC are shown by vertical arrows. Hexaphenyl HBC **34** forms a mesophase even at room temperature and exhibits a constant charge carrier mobility.

Scheme 8. 1:1 complex **37** formed from a hydrophobic modified polyethyleneimine **35** and HBC monocarboxylic acid derivative **36**.

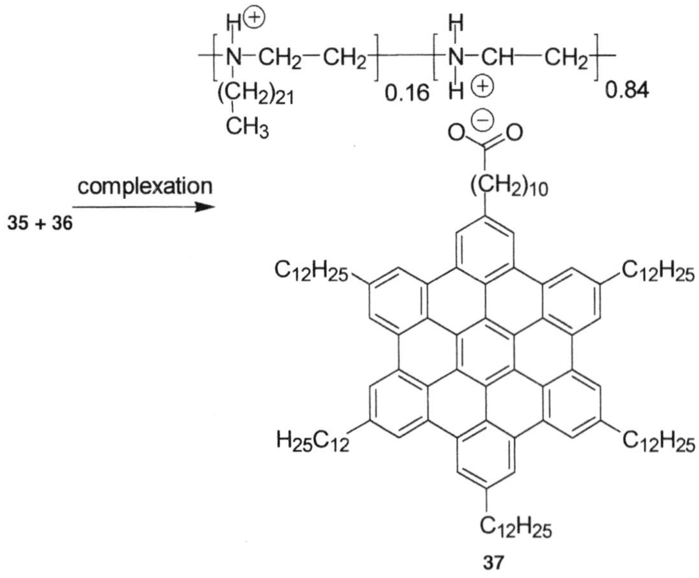

Figure 4. STM picture of the rhombus type PAH **38**. Picture taken from ref. [20].

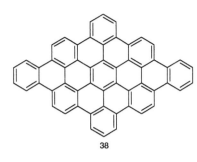

38

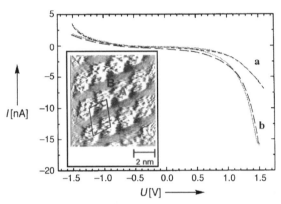

Figure 5. Current voltage curve of hexadodecylsubstituted HBC **33** on graphite. Symmetric curve **a** for the aliphatic part. Diode-like curve **b** for the aromatic part of the molecule. The inset shows STM image of **33** with unit cell depicted. A and B indicate the aliphatic and the aromatic part. Picture taken from ref. [24].

Next to the π-interactions, the arrangement of alkyl chains along the axis of symmetry of the graphite seems to be crucial. While over the years chemists have accumulated ample empirical evidence on the formation of the 3D structures in single crystals, the knowledge of 2D-crystal formation, i.e., of monolayers on surfaces, is still in its infancy. We have seen that, reducing the symmetry of hexaalkyl HBCs by going to a functionalized bromoderivative, such as

39 (scheme 9), physisorption from solution gives rise to different patterns consisting of both trimeric ensembles and single molecules.[23]

(iv) The direct visualization of single HBC molecules suggests a closer look into their electronic function. Interestingly, it is possible to combine STM with scanning tunneling spectroscopy because a submolecular resolution can be obtained. As a consequence one can record current-potential curves for single molecules with different characteristics for the aliphatic and aromatic domains.[24] While curves recorded for the aliphatic part are symmetric and correspond to those of aliphatic alcohols on graphite surfaces, those for the aromatic part show a strong asymmetry, and thus display diode-type behavior (Figure 5). The origin of this asymmetry is still a matter of dispute.

Similar current-potential measurements have not been successful for smaller PAHs, which is not surprising in view of the spatial resolution of the experiment.

An interesting variant of the deposition of disc-type molecules on surfaces and of the single molecule detection is to increase the distance between the adsorbate molecule and the surface by introducing bulky 3,5-di-*tert.*-butylphenyl substituents. This has been done for the tetraphenylporphyrin

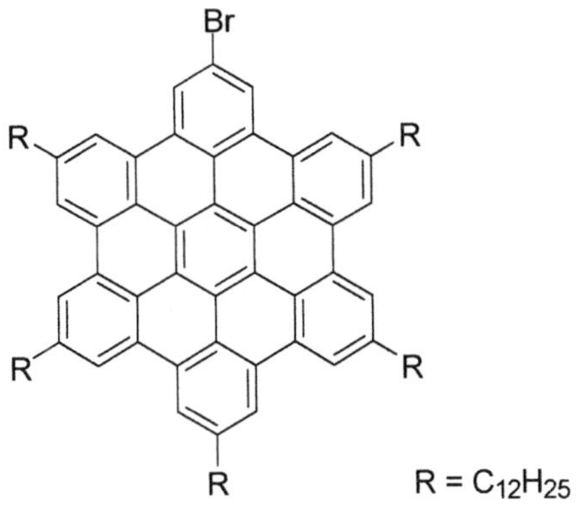

39

Scheme 9. Monobromo derivative of HBC **39**.

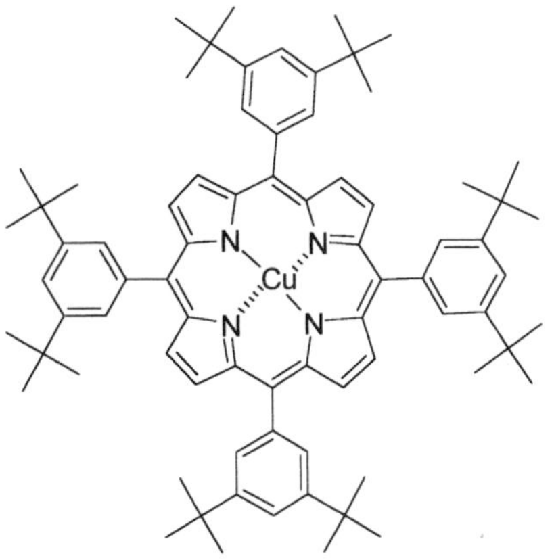

R = C₁₂H₂₅

40

Scheme 10. Porphyrin derivative bearing bulky 3,5-di-tert.-butylphenyl substituents **40**.

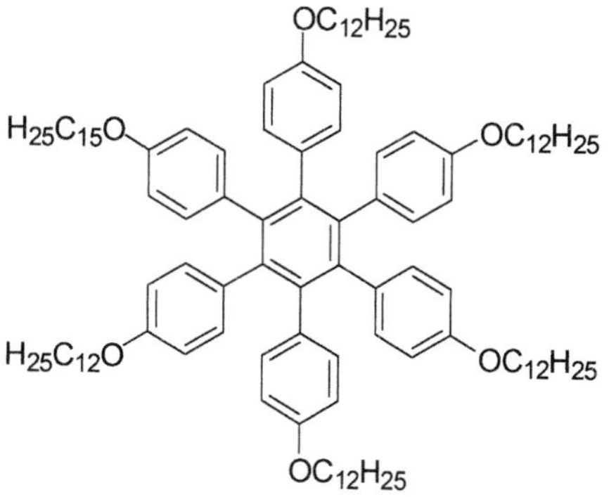

Scheme 11. Hexa(dodecyloxy)hexaphenylbenzene **41** is used as a precursor in a template-mediated synthesis of PAHs.

face whereby a carbon–oxygen cleavage under removal of the ether alkyl groups and an intramolecular dehydrogenation occurs, finally forming the fully fused HBC structure **27** (figure 6).

The decrease in thickness of the monolayer upon chemical "flattening" of the molecules can be shown by STM. The importance of the defined graphitic structures suggests the transfer of this novel strategy, namely the surface-induced PAH synthesis, to considerably larger oligophenylene precursors.

Dyes as Functional Discs

Dye stuffs are among the oldest topics of synthetic organic and industrial chemistry, and it is exciting to follow their role in nanosciences and single molecule spectroscopy.

(i) A particularly remarkable class of compounds in view of their optical properties and their stability has been derived from perylene-3,4,9,10-tetracarboxdiimide **42** (scheme 12).[27] Partial hydrolysis and decarboxylation has provided perylene-3,4-dicarboximide **43** which can be functionalized to yield a broad variety of derivatives serving as building blocks for further chemistry.

A typical aryl-aryl coupling and subsequent cyclodehydrogenation have afforded higher homologues of **42**, such as the terrylene-**47** and quaterry-

derivative **40** (scheme 10) and related cases utilizing UHV-deposition.[25] As a result of the weaker binding between substrate and adsorbate it is possible to shift molecules around on surfaces under single molecule detection.

Increasing the size of PAHs makes their deposition on surfaces difficult, because they can neither be sublimed nor made sufficiently soluble for solution processing. A precursor route has thus been designed according to which molecules are deposited on a surface and transformed into the final disc-type adsorbate structures in a thermal solid-state reaction with the substrate surface acting as a template.[26] An exciting example is the hexaether **41** (scheme 11) which is sublimed onto a Cu-(111) sur-

Figure 6. STM images of hexaether **41** adsorbed on Cu(111) and heated for 10–15 min to temperatures of (**a**) 570 K, (**b**) 620 K, (**c**) 670 K, and (**d**) 720 K. The line profiles on the right were taken along the white lines indicated in the images. The decrease of the height with increasing temperatures is strong evidence for the stepwise planarization of **41** and formation of the fully fused HBC structure. Pictures taken from ref. [26].

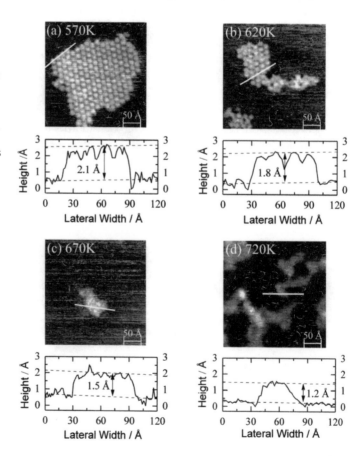

R = alkyl, aryl

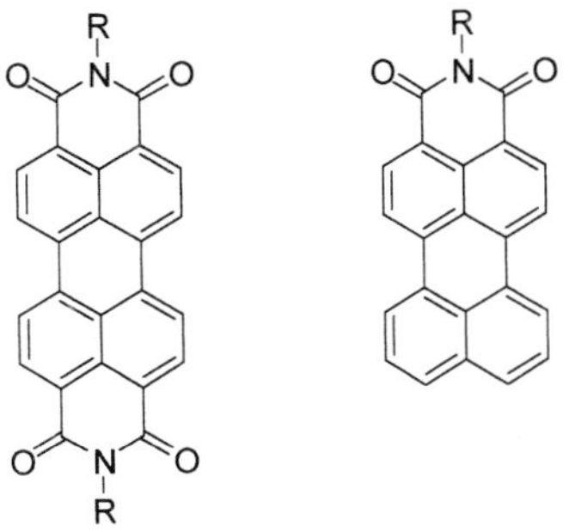

42   43

Scheme 12. Perylene-3,4,9,10-tetracarboxdiimide **42** and perylen-3,4-dicarboximide **43** are typical representatives for organic colorants.

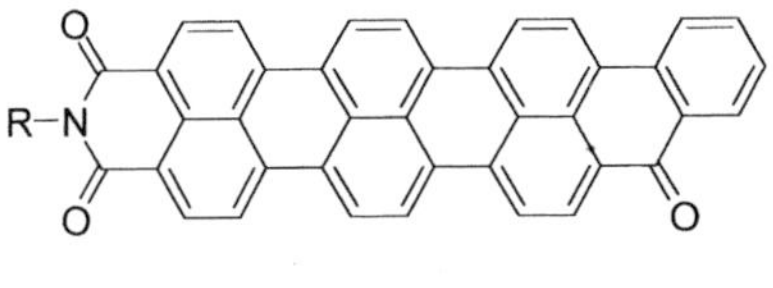

R = alkyl, aryl   **50**

Scheme 14. Benzoylterrylenmonoimide **50**: A dye with a long wavelength absorbance maximim at 676 nm.

lenetetracarboxdiimides **49** and benzoylterrylene-dicarboximide **50** (schemes 13 and 14).[28]

Remarkably enough, **47** is a blue dye with a high photoluminescence quantum yield and, nevertheless, a high photochemical stability. Figure 7 shows the spectra of the different rylenetetracarboxdiimide oligomers.

(ii) Soluble dyes and insoluble pigments differ significantly in their optical properties and in their methods of processing. It is important that suitable chemical modification of pigment structures should not only produce soluble, but also phase-forming materials.[29] Thus, coronenetetracarboximide derivatives **51** and the monoimide **52** form liquid crystalline phases,[30] where for the latter an anisotropic melt even exists at room temperature (scheme 15).

This is important from the perspective of the above-mentioned photoconductivity of liquid crystalline materials, since the inclusion of dyes allows

Scheme 13. Synthetic pathway including aryl-aryl coupling and subsequent dehydrogenation to terrylene tetracarboxdiimide **47** and quaterrylene tetracarboxdiimide **49**.

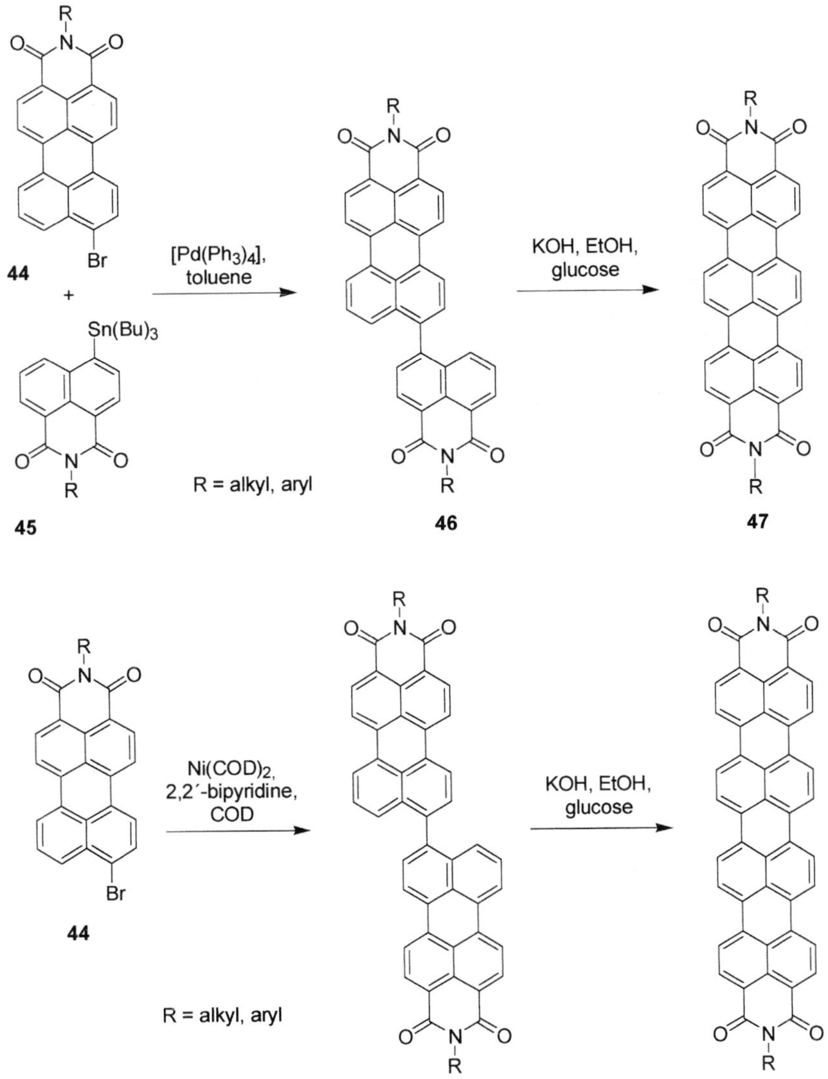

Figure 7. UV-VIS-NIR spectra of perylene-, terrylene-, and quaterrylene tetracarboxdiimidees **42**, **47**, **49**.

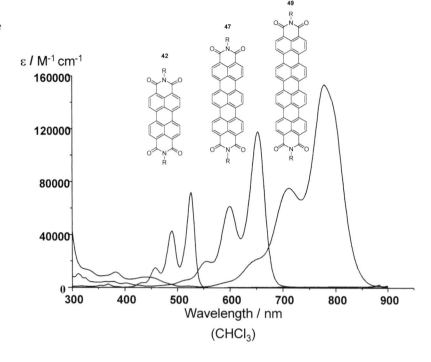

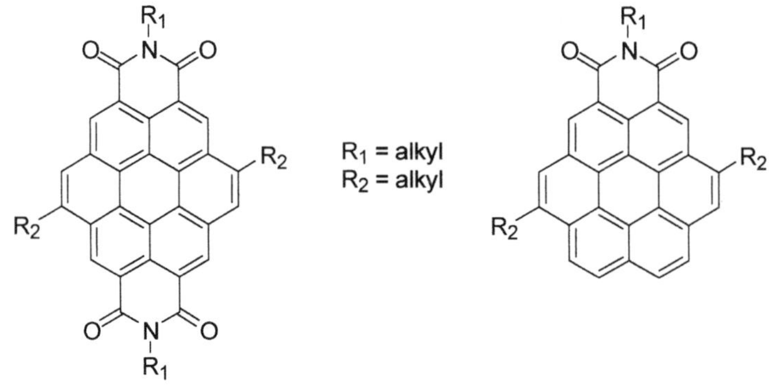

Scheme 15. Coronene-tetracarboximide **51** and –dicarboximide **52** with alkyl substituents form liquid crystalline phases.

R_1 = alkyl
R_2 = alkyl

51

52

one to modify the wavelength of the exciting light. In alkyl substituted quinacridones such as **53** (scheme 16), one can detect supramolecular ordering in 3D under the influence of intermolecular hydrogen bonding and alkyl chain packing.[31]

Incorporation of the dyes into host matrices raises concerns about achievable concentration, aggregate formation, and migration fastness.

(iii) Perylene-3,4,9,10-tetracarboxdianhydrides and related structures have been deposited on substrates and the resulting patterns visualized by scanning tunneling microscopy. A key issue is the correlation of tunnel current and the properties of the relevant π-electron orbitals. A review about "surface achitecture" with large organic molecules, e.g. perylene-3,4,9,10-tetracarboxdianhydride, was recently presented.[32]

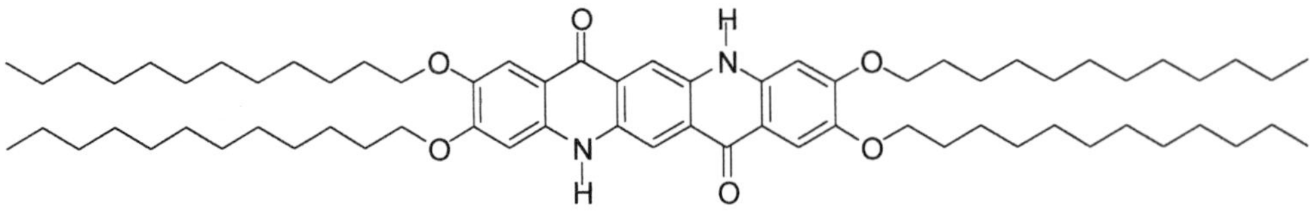

53

Scheme 16. 2,3,9,10-Tetra(dodecyloxy)quinacridone **53** is an attractive example for supramolecular ordering of chromophores.

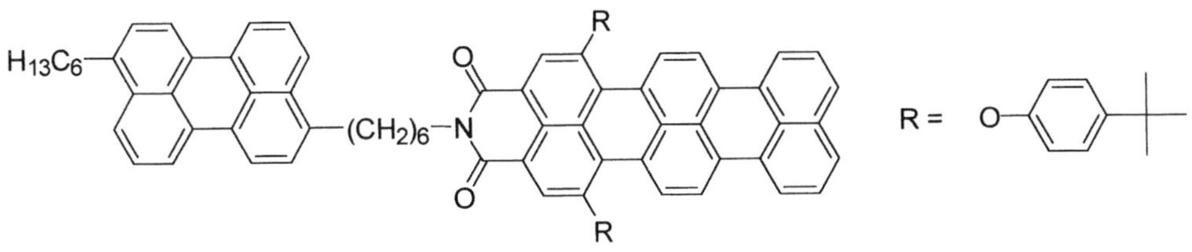

Scheme 17. Bis-chromophore **54** serves as an example for the detection of fluorescence resonance energy transfer at the single molecule level.

(iv) The fact, that single molecules incorporated into highly diluted matrices can be seen by means of their fluorescence, presents exciting opportunities for new research in single molecule detection. The relevant photophysics is beyond the scope of this text, but it is clear that synthetic chemistry comes into play since the dyes, in order to qualify for single molecule spectroscopy, have to fulfil a series of requirements. Terrylene **47** has played an important role since its single molecule detection was possible even at room temperature.[33] Just as with the STM visualization of single molecules, it is the outstanding challenge of single molecule spectroscopy to proceed from just the visualization of molecules to the recording of more complex functions. An example is the detection of energy transfer for a single diad which has, indeed, become possible for the structure **54** (scheme 17).[34]

Along the same lines, it is exciting to monitor chemical transformations of single molecules. The perylene dicarboximide derivative **55** has a high practical potential as a thermotropic dye for writ-

ing on plastics (Figure 8). It should be possible to detect this process, which goes along with a change in both emission wavelength and fluorescence quantum yield, by single molecule spectroscopy. The ultimate goal in this research, which is currently in progress, is the application of heat to a single molecule and thereby the initiation of an optically detectable chemical reaction of a single molecule. The examination of such defined chemical reactions accompanied by a change of fluorescence properties may lead to optical data storage on the single molecule level.

Hydrogen-Bonded Aggregates

It has been emphasized that in the construction of complex molecular and supramolecular architectures the use of non-covalent bonding can be more efficient and more economic than the conventional covalent approach.[35] We outline here, how the interplay of a hydrogen bonding and alkyl chain

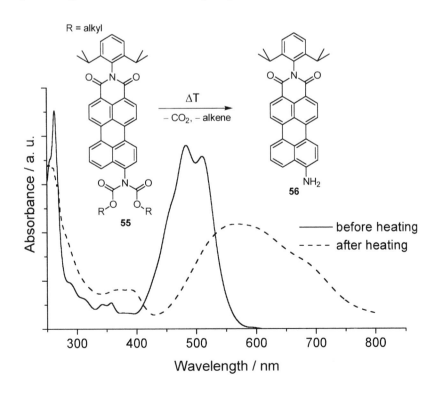

Figure 8. Perylenedicarboximide derivative **55** has a high practical potential for laser writing on plastics. Upon heating the alkoxycarbonyl groups are cleaved and **56** is yielded. This reaction is accompanied by an eye-catching color change. The UV-VIS spectra are recorded from a polystyrene film containing **55** before and after heat exposure.

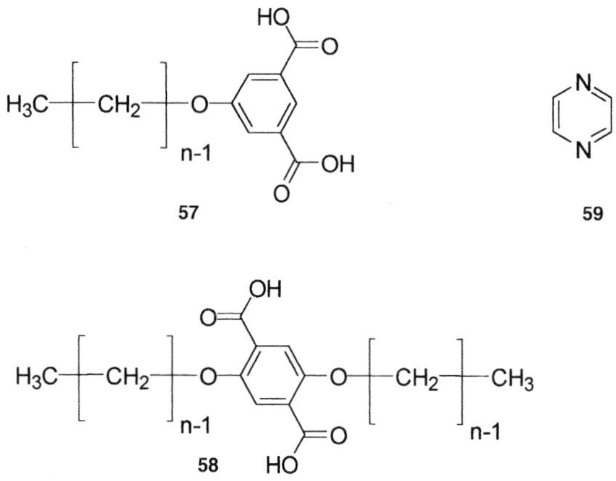

Scheme 18. Structures of 5-alkoxyisophthalic acid **57** (C_nISA, *n*: number of C-atoms in the alkoxy chain), 2,5-dialkoxyterephthalic acid **58** and pyrazine **59**.

packing can be used to control supramolecular structure formation in *different phases*. The key concern is the visualization of monomolecular adsorbate layers by STM and a comparison of the resulting 2D crystals with conventional 3D crystals.

(i) The building blocks used are 5-alkoxyisophthalic acid derivatives **57** (C_nISA, *n*: number of C atoms in the alkoxy chain; see scheme 18). In that case synthesis is, of course, trivial although it should be noted that different moieties must be incorporated into the alkyl chain because C_nISA is supposed to serve as a self-organizing carrier for chemical and physical functions.

(ii) Crystal structure analysis reveals that depending upon the alkyl chain length, C_nISA can either form cyclic hexamers ($6 < n < 12$) or lamella-type ($n > 12$) structures (Figure 9).[36] The latter are created from hydrogen bonded strands and interdigitated alkyl chains.

For a 1:1 complex of C_{12}ISA and pyrazine **59** (as hydrogen bond acceptor) the resulting crystal structure reveals the supramolecular alignment of hydrogen bonded strands by intercalated pyrazine molecules (Figure 10a).[37] The resulting extended sheet structures are planar, and it is then typical that the very same motif can be built by physisorption from solution on graphite and the resulting pattern visualized by STM (Figure 10b). While the coincidence of 2D and 3D molecular motifs is remarkable, this should not be understood as evidence that the knowledge of 2D crystals is generally sufficient to predict 3D crystal structures.

Remarkably enough, dialkoxyterephthalic acids, both in 2D and 3D, form linear hydrogen bonded strands which are reminiscent of the rigid rods formed from alkyl substituted poly-para-phenylenes.

(iii) As in the PAH case, STM patterns of physisorbed ISAs help to build up a broader knowledge of the formation of 2D crystals. The deposition of ISAs with stereogenic centers in the chains is particularly exciting, because, when depositing racemic mixtures, one can detect the formation of enantiomorphous domains of the pure enantiomers which can be regarded as kind of Pasteur experiment in 2D.[38] A good case can also be made when synthesizing ISAs, in which the alkyl chains contain (CF_2)-

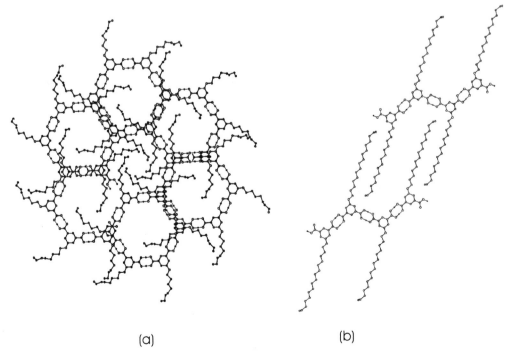

(a)   (b)

Figure 9. Crystal structure of C_nISA: Depending on the length of the alkoxy chain, C_nISA forms either cyclic hexamers (**a**: C_8ISA) or lamella type structures (b: C_{16}ISA).

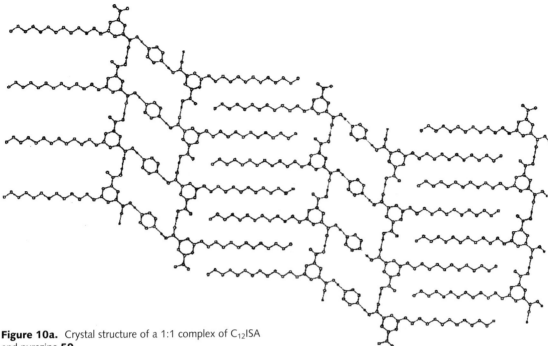

Figure 10a. Crystal structure of a 1:1 complex of $C_{12}ISA$ and pyrazine **59**.

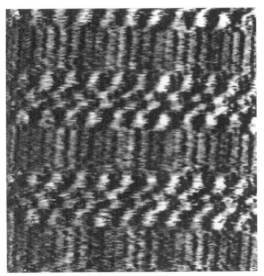

Figure 10b. The STM image proves the coincidence of 2D and 3D structure motifs for the pyrazine-$C_{12}ISA$ complex. Picture taken from ref. [37].

segments. The question of incompatibility between alkyl and perfluoroalkyl chains can now be studied, both, in 3D and 2D.[39]

(iv) The quest for the function of single molecules can be documented for the light driven modification of the monolayers. The first example is $C_{12}(AZO)C_{12}ISA$ **60**, in which an azobenzene unit is incorporated into the alkyl chain (scheme 19). By irradiating the 2D monolayer formed upon physisorption from solution, one can now promote transformation from the *trans-* to the *cis*-isomer, which disturbs the lattice and leads to a different, although less stable pattern containing only *cis*-azobenzene units, indeed, possessing a kink in the chains (Figure 11).[40]

However, an interpretation of this experiment is not straightforward in view of the experimental conditions, which imply a monolayer on the substrate surface covered by the solution of molecules. *Cis-trans* isomerization for an observed molecule is supposed to promote desorption, so that a pattern change includes absorption-desorption equilibria. It would be important therefore, to observe a light induced change of a 2D pattern only for adsorbed molecules. This is, indeed, possible for $C_9(DIA)C_8ISA$

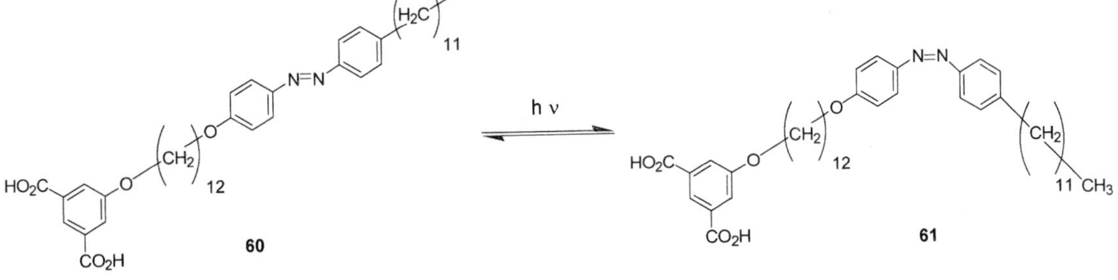

Scheme 19. Photoinduced *cis-trans* isomerization of $C_{12}(AZO)C_{12}ISAs$ **60** and **61**.

Figure 11. STM image of a monolayer of (**a**) *cis*-C$_{12}$(AZO)C$_{12}$ISA **61** and (**c**) *trans*-C$_{12}$(AZO)C$_{12}$ISA **60** on graphite. The molecular models for the two-dimensional packing of **61** and **60** with proposed unit cells are represented in (**b**) and (**d**). Pictures taken from ref. [40].

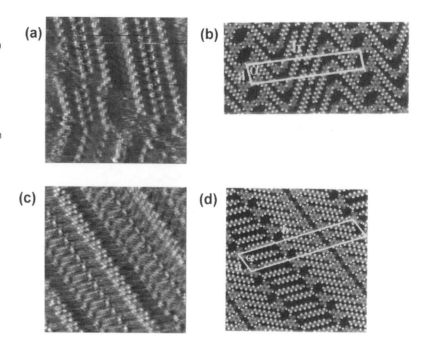

63 which contain diacetylene units.[41] The lamella, upon physisorption from solution, produce a side-by-side arrangement of diacetylenes, which, depending on the distance between neighboring diacetylenes and the twist of diacetylenes with respect to the packing axis, fulfills the requirements for a topochemical transformation as seen for crystalline diacetylenes. Accordingly, one can now transform diacetylene molecules, which have been self-assembled upon physisorption onto a surface, into polymer chains under direct STM control, representing another way of information storage at a single molecule level (Figure 12).

Spheres Made From Dendritic (3D)Polyphenylenes

A challenging class of compounds, when it comes to the single molecule detection of molecules with a more spherical shape, are the dendrimers.

(i) In general, dendrimers are synthesized by the repetition of a sequence of reaction steps leading after each repetition to a dendrimer one generation higher. Two general methods for dendrimer synthesis are known: a divergent route, building the dendrimer layer-by-layer from a central core to the periphery and a convergent one, where the dendrimer grows from the periphery toward the central core. Of course, a review of the synthesis of all the

Figure 12. STM image of a monolayer of C$_9$(DIA)C$_8$ISA **63** (**a**) before and (**b**) after photoinduced polymerization of the diacetylene units. Pictures taken from ref. [41].

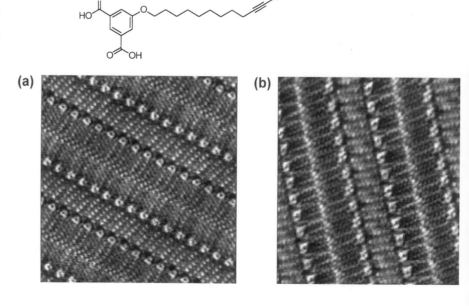

different types of dendrimers which have been prepared is beyond the scope of this text.[42] In view of the nanochemistry concept outlined above, polyphenylene-based dendrimers are of particular interest because they are shape persistent and, therefore, offer the opportunity of building a well defined three-dimensional architecture. Moreover, it has been demonstrated that different three-dimensional structures of these dendrimers are easily accessible by choosing a suitable core from a "building block kit" (Figure 13).[43]

It is also possible to decorate the dendrimer surfaces with a variety of functional groups such as halo, cyano, carboxy, amino, hydroxy and thiomethyl.[44] It stands to reason that surface functional-

ization will be the key step towards applications of dendrimers.

(ii) *and* (iii) Alkyl chain substitution, for example, allows the deposition of polyphenylene dendrimers on HOPG by spin coating from solution. Using AFM measurements it has recently been proven that a clear distinction between an inner phenylene core and an outer alkyl shell is crucial for the formation of self assembled supramolecular structures on graphite.[45] Second generation tetrahedral core polyphenylene dendrimers bearing dodecyl substituents, which create this type of an alkyl shell, form rod-like structures of 6 nm diameter.

In order to obtain isolated molecules on a surface, fourth generation polyphenylene dendrimers with-

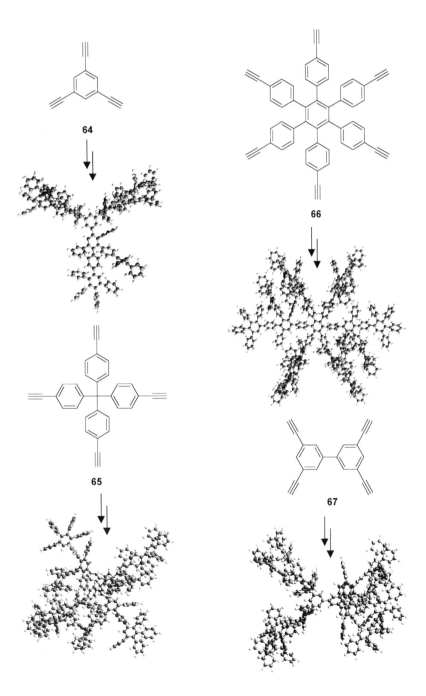

Figure 13. Various cores (**64, 65, 66, 67**) and ball stick models of the second-generation dendrimers resulting from them.

out any further functionalization have been deposited via spin-coating from highly dilute solution onto a MICA. Also using AFM spatially separated particles of 5 nm size, as expected for a single dendrimer molecule,[46] were observed moreover, the size of the dendrimers has also been confirmed by TEM.

(iv) Experiments are underway using an AFM tip to investigate the compressibility of single nanoparticles. This is relevant in assessing their role as potential nanolubricants which can be further investigated by surface force measurements. Another challenging question concerns the possible changes in STM tunnel currents during compression.

Furthermore, dendrimers can serve advantageously as scaffolds in molecular architecture in order to build well defined 3D functional units. Outstanding examples of this are the strongly emitting nanoparticles **68** and **70** that are obtained after attaching fluorescent perylene imide chromophores to dendrimer surfaces (scheme 20).[47]

Optical properties of dendrimers bearing eight chromophores have been examined by single molecule spectroscopy techniques. It is especially exciting that variations in the spectra are recorded if one of these dendrimers is observed for a period of time under continuous irradiation.[48] The fluorescence intensity of the dendrimer nanoparticle also jumps between discrete emissive levels. All these findings suggest the existence of strong electronic interactions between several perylene imide chromophores within one dendrimer and provide new

Scheme 20. Polyphenylene dendrimers in the 1st **68** and in the 2nd generation **70** which are decorated with fluorescent perylene imide chromophores on the surface. Perylenedicarboximide derivative **69** serves as a model compound for spectroscopic investigations.

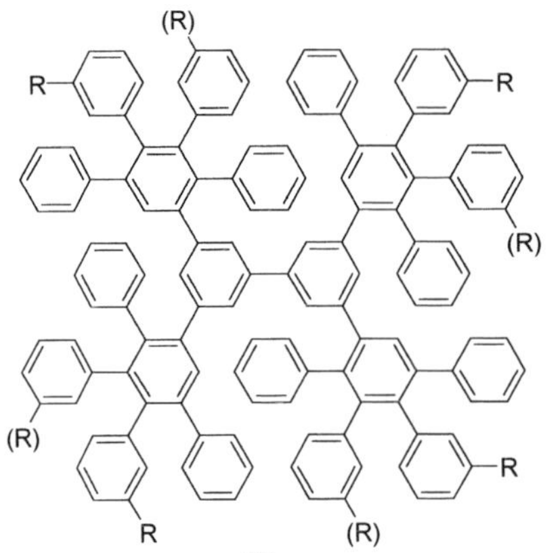

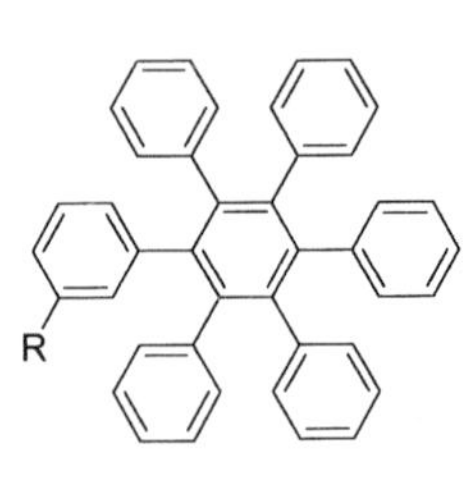

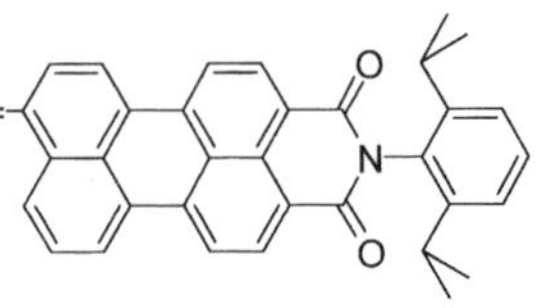

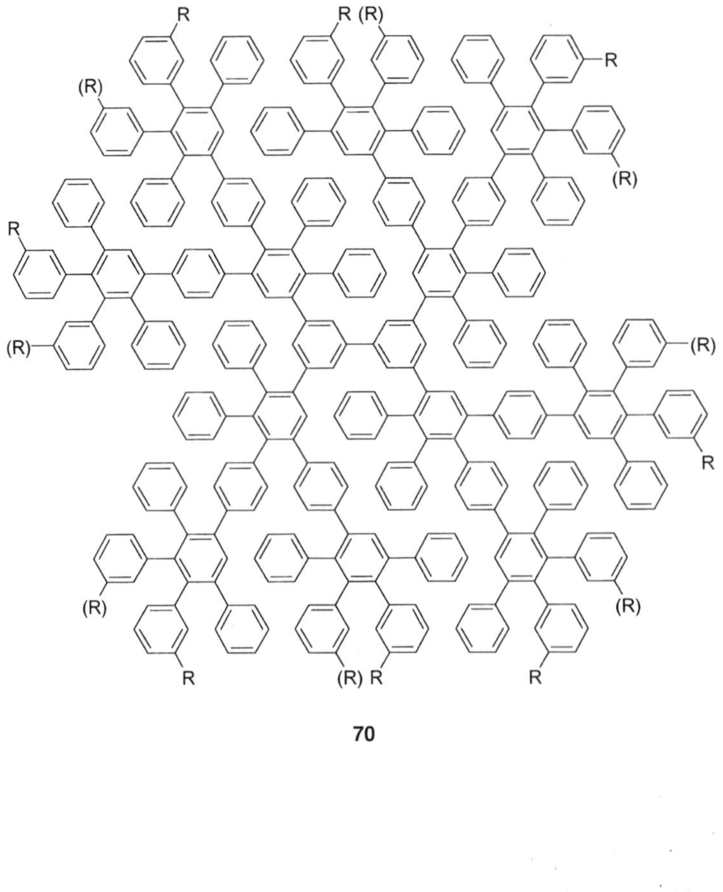

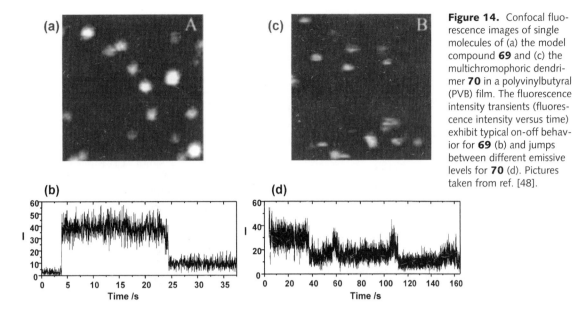

Figure 14. Confocal fluorescence images of single molecules of (a) the model compound **69** and (c) the multichromophoric dendrimer **70** in a polyvinylbutyral (PVB) film. The fluorescence intensity transients (fluorescence intensity versus time) exhibit typical on-off behavior for **69** (b) and jumps between different emissive levels for **70** (d). Pictures taken from ref. [48].

insight into both molecular photophysics and molecular dynamics. Such an investigation of a well defined multichromophoric system on the single molecule scale might open a way to understand the parameters that determine the transition from the single molecule, with its typical behavior of "on-off" fluorescence under continuous irradiation, to ensemble behavior where the emission is constant and continuous.

Nanotechnology?

The grand vision of an applicable nanotechnology has been pointed out by R. P. Feynman as early as 1960: He emphasized that all the books ever written could be stored in a cube with 6 mm sized edges, if 100 atoms were sufficient to store one bit of information.[49] While the above single molecule operations are certainly relevant for processes such as pattern formation or information storage on one hand and conduction or rectification on the other hand, the present approach toward nanochemistry is of a purely fundamental character. As has been mentioned above, semiconductor industries over the years have produced successively higher device performance at lower costs whereby microlithography has been a key technological stimulus. Nowadays one searches for concepts of nanofabrication including writing, that is making "masters" and replication, i.e. transferring patterns.

Nanochemistry has not yet advanced to a level where it can meet crucial technological requirements. The above examples, however, convincingly support the view of molecules as potential basic functional units. For a long time, so-called molecular electronics has been looked at as remote since there was no way of creating a contact between molecules and the outside world. This interfacing is now possible. This, together with progress in synthesis and self-assembly of molecules, has paved the way to increasingly complex single molecule operations.

The key questions are: When will "intelligent" molecular systems be applicable? And, in particular, will it be possible to create memory and logic circuits on the level of a few molecules? To this end, devices like supramolecular atom relay transistors (SMART) and molecular single electron switching transistors (MOSES) have been proposed in the literature and evaluated as promising candidates with regard to practical utilization.[50] Synthesis of appropriate molecular structures with branches, dots and gates as well as the molecular design of multiple output wiring are the key concerns towards realization of nanoscale information processing.

To answer the above mentioned questions more directly: both, a variety of necessary molecular structures (e.g. insulating and conducting polymers for MOSES) and many kinds of molecular engineering tools for the future electronic information processing circuit, are already available from state of the art chemical knowledge. Therefore, several authors believe that the goal of an applicable nanotechnology will be achieved within the next few decades. One has to bear in mind that nowadays an interdisciplinary effort involving organic chemistry as well as inorganic chemistry, physics and the engineering sciences is directed towards the fabrication of nanostructured molecular devices. Finally, one should also be aware of the fact that nature has already evolved a working biological nanotechnology, including the features of reading and writing information, by means of the DNA/RNA system in molecular genetics. Thus, it is not unreasonable to believe that we will be able to achieve viable nanotechnologies in the not too distant future.

Acknowledgements

It is a pleasure to gratefully acknowledge the contributions of all our colleagues, who have done the experimental work. First, we would like to give our sincere thanks to all our partners in a variety of cooperations. Among them are Prof. T. Basché, Prof. C. Bräuchle, Prof. H.-J. Butt, Prof. F. De Schryver, Prof. W. Eisenmenger, Dr. V. Enkelmann, Prof. N. Karl, Prof. K. Leo, Dr. G. Lieser, Prof. A. J. Meixner, Prof. J. P. Rabe, Dr. P. Samori, Dr. W. Strohmeier, Prof. J. Warman, Prof. G. Wegner, and Prof. C. Wöll. We know that, without them, the advances in our research would not have been possible. Secondly we wish to record our thanks for internal efforts of all our colleagues within our group at the Max-Planck-Institute for Polymer Research.

Furthermore, financial support by the BASF AG, the Bundesministerium für Bildung und Forschung, the European Commission (TMR-Program SISITO-MAS), and the Volkswagen Stiftung is gratefully acknowledged.

Last, but by no means least, we thank all the people who have helped us in the preparation of this review. We would like to commend Dr. A. Grimsdale who has kindly read the entire manuscript and made many valuable comments and M. Sieffert for his kind assistance in preparing the figures for this text. Also we would like to thank D. Stiep, Dr. D. Marsitzky, A. Fechtenkötter, M. Harbison, and Dr. M. Watson for their valuable insight towards the preparation of this paper.

References and Notes

1. a) W. J. Feast, J. Tsibouklis, K. L. Pouwer, L. Groenendaal, E. W. Meyer, *Polymer* **1996**, *37*, 5017; b) Y. Geerts, G. Klärner, K. Müllen in *Electronic Materials – The Oligomer Approach* (Eds.: K. Müllen and G. Wegner), Wiley-VCH, Weinheim, **1998**, pp. 1 –103 and references therein.

2. a) M. Müller, F. Morgenroth, U. Scherf, T. Soczkaguth, G. Klärner, K. Müllen, *Phil. Trans. Royal Soc. Ser. A – Phys. Sc. Eng.* **1997**, *355*, 715. b) V. Francke, T. Mangel, K. Müllen, *Macromolecules* **1998**, *31*, 2447; c) D. Marsitzky, T. Brand, Y. Geerts, M. Klapper, K. Müllen, *Macromol. Rapid Commun.* **1998**, *19*, 385.

3. a) V. Francke, H. J. Räder, Y. Geerts, K. Müllen, *Macromol. Rapid Commun.* **1998**, *19*, 275; b) D. Marsitzky, M. Klapper, K. Müllen, *Macromolecules* **1999**, *32*, 8685.

4. a) G. Widawski, M. Rawiso, B. Francois, *Nature* **1994**, *369*, 387; b) B. Francois, G. Widawski, M. Rawiso, B. Cesar, *Syn. Met.* **1995**, *69*, 463.

5. a) T. Vahlenkamp, G. Wegner, *Macromol. Chem. Phys.* **1994**, *195*, 1933; b) U. Lauter, W. H. Meyer, G. Wegner, *Macromolecules* **1997**, *30*, 2092; c) U. Lauter, W. H. Meyer, V. Enkelmann, G. Wegner, *Macromol. Chem. Phys.* **1998**, *2*, 242; d) V. Cimrová, M. Remmes, D. Neher, G. Wegner, *Adv. Mater.* **1996**, *8*, 146; e) J. C. Wittmann, P. Smith, *Nature* **1991**, *352*, 414; f) K. L. Prime, G. M. Whitesides, *Science* **1991**, *252*, 1164; g) P. E. Laibini, J. J. Hickmann, M. S. Wrighton, G. M. Whitesides, *Science* **1991**, *245*, 854; h) P. E. Laibinis, G. M. Whitesides, *J. Am. Chem. Soc.* **1992**, *114*, 1990; i) C. A. Alves, E. L. Smith, M. D. Porter, *J. Am. Chem. Soc.* **1992**, *114*, 1222; j) T. J. Lenk, V. M. Hallmark, J. F. Rabolt, L. Häussling, H. Ringsdorf, *Macromolecules* **1993**, *26*, 1230; k) O. B. Steiner, M. Rehahn, W. R. Caseri, U. W. Suter, *Macromolecules* **1994**, *27*, 1983; l) G. Wegner, *Mol. Cryst. Liq. Cryst.* **1993**, *235,1*; m) U. Lauter, W. H. Meyer, G. Wegner, *Macromolecules* **1997**, *30*, 2092.

6. a) M. A. Reed, C. Zhou, C. J. Müller, T. P. Burgin, J. M. Tour, *Science* **1997**, *278*, 252; b) J. M. Tour, L. Jones, D. L. Pearson, J. J. S. Lamba, T. P. Burgin, G. M. Whitesides, D. L. Allara, A. N. Parikh, S. V. Atre, *J. Am. Chem. Soc.* **1995**, *117*, 9529; c) J. S. Schumm, D. L. Pearson, J. M. Tour, *Angew. Chem. Int. Ed. Engl.* **1994**, *33*, 1360; d) K. Müllen, S. Valiyaveettil, V. Francke, V. S. Iyer, *NATO ASI Series E: Applied Sciences* (Eds. C. Joachim, S. Roth), Kluwer Academic Publishers, Dordrecht, Boston, London, **1997**, *Vol. 341*, 61.

7. P. Samori, V. Francke, T. Mangel, K. Müllen, J. P. Rabe, *Opt. Mater.* **1998**, *9*, 390.

8. P. Samori, V. Francke, K. Müllen, J. P. Rabe, *Thin Solid Films* **1998**, *336*, 13.

9. a) A. Aviram, *Molecular Electronics: Science and Technology*, Conference Proceedings No. 262, American Institute of Physics, New York **1992**; b) J. M. Tour, *Trends Polym. Sci.* **1994**, *2*, 332.

10. L. A. Bumm, J. J. Arnold, M. T. Cygan, T. D. Bunbar, T. P. Burgin, L. Jones, D. L. Allara, J. M. Tour, P. S. Weiss, *Science* **1996**, *271*, 1705.

11. a) S. J. Tans, M. H. Devoret, H. Dai, A. Thess, R. E. Smalley, L. J. Geerligs, C. Dekker, *Nature* **1997**, *386*, 474; b) M. Bockrath, D. H. Cobden, P. L. McEuen, N. G. Chopra, A. Zettl, A. Thess, R. E. Smalley, *Science* **1997**, *275*, 1922.

12. M. Burghard, V. Krstic, G. S. Duesberg, G. Philipp, J. Muster, S. Roth, *Syn. Met.* **1999**, *193*, 2540.

13. A. J. Berresheim, M. Müller, K. Müllen, *Chem. Rev.* **1999**, *99*, 1747.

14. a) E. Clar, *The Aromatic Sextet*, 1st ed., Wiley, London, **1972**; b) H. Hosoya, *Top. Curr. Chem.* **1990**, *153*, 255.

15. a) A. Stabel, P. Herwig, K. Müllen, J. P. Rabe, *Angew. Chem. Int. Ed. Eng.* **1995**, *34*, 1609; b) V. S. Iyer, M. Wehmeier, J. D. Brand, M. A. Keegstra, K. Müllen, *Angew. Chem. Int. Ed. Eng.* **1997**, *36*, 1604; c) M. Müller, V. S. Iyer, C. Kübel, V. Enkelmann, K. Müllen, *Angew. Chem. Int. Ed. Eng.* **1997**, *36*, 1607.

16. a) D. Demus, J. Goodby, G. Gray, H. W. Spiess, V. Vill, *Handbook of Liquid Crystals*, Vol. 2B, Wiley VCH, Weinheim, **1998**; b) A. M. van de Craats, J. M. Warman, K. Müllen, Y. Geerts, J.-D. Brand, *Adv. Mater.* **1998**, *10*, 36; c) J. Simmerer, B. Glüsen, W. Paulus, A. Kettner, P. Schuhmacher, D. Adam, K. H. Etzbach, K. Siemensmeyer, J. Wendorff, H. Ringsdorff, D. Haarer, *Adv. Mater.* **1996**, *8*, 815; d) A. M. van de Craats, M. P. de Haas, J. M. Warman, *Synth. Met.* **1997**, *86*, 2125.

17. A. Fechtenkötter, K. Saalwächter, M. A. Harbison, K. Müllen, H. W. Spiess, *Angew. Chem. Int. Ed. Eng.* **1999**, *38*, 3039.

18. S. Ito, M. Wehmeier, J. D. Brand, C. Kübel, R. Epsch, J. P. Rabe, K. Müllen, *Chem. Eur. J.*, in press.

19. A. F. Thüneman, D. Ruppelt, S. Ito, K. Müllen, *J. Mater. Chem.* **1999**, *9*, 1055.

20. a) M. Müller, *PhD Thesis*, Johannes Gutenberg-Universität, Mainz, **1998**; b) C. Ludwig, B. Gompf, J. Petersen, R. Strohmeier, W. Eisenmenger, *Z. Phys. B* **1994**, *93*, 365.

21. M. Müller, J. Petersen, R. Strohmeier, C. Günther, N. Karl, K. Müllen, *Angew. Chem. Int. Ed. Eng.* **1996**, *35*, 886.

22. C. Kübel, *PhD Thesis*, Johannes Gutenberg-Universität Mainz, Mainz, **1998**.

23. S. Ito, M. Wehmeier, J. D. Brand, C. Kübel, R. Epsch, J. P. Rabe, K. Müllen, *Chem. Eur. J.*, in press.

24. A. Stabel, P. Herwig, K. Müllen, J. P. Rabe, *Angew. Chem. Int. Ed. Eng.* **1995**, *34*, 1609;

25. a) For a review see: J. K. Gimzewski, C. Joachim, *Science* **1999**, *283*, 1683; b) T. A. Jung, R. R. Schlittler, J. K. Gimzewski, H. Tang, C. Joachim, *Science* **1996**, *271*, 181.

26. K. Weiss, G. Beernink, F. Dötz, A. Birkner, K. Müllen, C. H. Wöll, *Angew. Chem. Int. Ed. Eng.* **1999**, *38*, 3748.

27. a) For a review see: H. Langhals, *Heterocycles* **1995**, *40*, 477; b) K. Hunger, W. Herbst, *Industrielle Organische Pigmente: Herstellung, Eigenschaften, Anwendungen*, 1st ed., VCH, Weinheim **1987**; c) W. S. Czajkowski in *Modern Colorants, Synthesis and Structure* (Eds. A. T. Peters, H. S. Freeman), Chapman & Hall, New York, **1995**; d) H. Zollinger, *Color Chemistry*, 2nd ed., VCH, Weinheim, **1991**.

28. a) F. O. Holtrup, G. R. J. Müller, H. Quante, S. De Feyter, F. C. De Schryver, K. Müllen, *Chem. Eur. J.* **1997**, *3*, 219; b) H. Quante, K. Müllen, *Angew. Chem. Int. Ed. Engl.* **1995**, *34*, 1323; c) Y. Geerts, H. Quante, H. Platz, R. Mahrt, M. Hopmeier, A. Böhm, K. Müllen, *J. Mater. Chem.* **1998**, *8*, 2357; d) S. K. Lee, Y. B. Zu, A. Herrmann, Y. Geerts, K. Müllen, A. J. Bard, *J. Am. Chem. Soc.* **1999**, *121*, 3513; e) F. O. Holtrup, G. R. J. Müller, J. Uebe, K. Müllen, *Tetrahedron* **1997**, *53*, 6847.

29. a) C. Göltner, D. Pressner, K. Müllen, H.-W. Spiess, *Angew. Chem. Int. Ed. Eng.* **1993**, *32*, 1660; b) P. Schlichting, U. Rohr, K. Müllen, *J. Mater. Chem.* **1998**, *8*, 2651; c) G. R. J. Müller, C. Meiners, V. Enkelmann, Y. Geerts, K. Müllen, *J. Mater. Chem.* **1998**, *8*, 61; d) A. M. van de Craats, J. M. Warman, P. Schlichting, U. Rohr, Y. Geerts, K. Müllen, *Synthetic Metals* **1999**, *102*, 1550; e) R. A. Cormier, B. A. Gregg, *Chem. Mater.* **1998**, *10*, 1309.

30. U. Rohr, P. Schlichting, A. Böhm, M. Gross, K. Meerholz, C. Bräuchle, K. Müllen, *Angew. Chem. Int. Ed. Engl.* **1998**, *37*, 1434.

31. U. Keller, K. Müllen, D. De Feyter, F. C. De Schryver, *Adv. Mater.* **1996**, *8*, 490.

32. E. Umbach, K. Glockler, M. Sokolowski, *Surf. Sci.* **1998**, *404*, 20.

33. a) L. Fleury, B. Sick, G. Zumhofen, B. Hecht, U. P. Wild, *Mol. Phys.* **1998**, *95*, 1333; b) F. Kulzer, F. Koberling, T. Christ, A. Mews, T. Basché, *Chem. Phys.* **1999**, *247*, 23.

34. a) P. Schlichting, B. Duchscherer, G. Seisenberger, T. Basché, C. Bräuchle, K. Müllen, *Chem. Eur. J.* **1999**, *121*, 3513.

35. For a review see: a) J. M. Lehn, *Supramolecular Chemistry*, VCH, Weinheim **1995**; b) J. P. Glusker, *Top. Curr. Chem.* **1998**, *198*, 1; c) G. R. Desiraju, *Angew. Chem.* **1995**, *107*, 254.

36. V. Enkelmann, S. Valiyaveettil, G. Moessner, K. Müllen, *Supramol. Science* **1995**, *2*, 3.

37. K. Eichhorst-Gerner, A. Stabel, G. Moessner, D. Declerq, S. Valiyaveettil, V. Enkelmann, K. Müllen, J. P. Rabe, *Angew. Chem.* **1996**, *108*, 1599.

38. a) S. De Feyter, P. C. M. Grim, M. Rücker, P. Vanoppen, C. Meiners, M. Sieffert, S. Valiyaveettil, K. Müllen, F. C. De Schryver, *Angew. Chem.* **1998**, *110*, 1281; b) for similar experiments using terephthailic acid derivatives see: S. De Feyter, A. Gesquière, P. C. M. Grim, F. C. De Schryver, S. Valiyaveettil, C. Meiners, M. Sieffert, K. Müllen, *Langmuir* **1999**, *15*, 2817.

39. A. Gesquière, M. M. Abdel-Mottaleb, F. C. De Schryver, M. Sieffert, K. Müllen, *Langmuir* **1999**, *15*, 6821.

40. P. Vanoppen, P. C. M. Grim, M. Rücker, S. De Feyter, G. Moessner, S. Valiyaveettil, K. Müllen, F. C. De Schryver, *J. Phys. Chem.* **1996**, *100*, 19636.

41. P. C. M. Grim, S. De Feyter, A. Gesquière, P. Vanopen, M. Rücker, S. Valiyaveettil, G. Moessner, K. Müllen, F. C. De Schryver, *Angew. Chem.* **1997**, *109*, 2713.

42. For a review see: a) F. Vögtle, *Topics in Current Chemistry* **1998**, *197*, 1; b) G. Newcome, *Dendritic Molecules*, Wiley-VCH, Weinheim, **1996**.

43. a) F. Morgenroth, K. Müllen, *Tetrahedron* **1997**, *53*, 15349; b) A. J. Berresheim, F. Morgenroth, U.-M. Wiesler, K. Müllen, *Polym. Prepr.* **1998**, *39*, 721; c) F. Morgenroth, C. Kübel, K. Müllen, *J. Mater. Chem.* **1998**, *7*, 1207.

44. U.-M. Wiesler, T. Weil, K. Müllen, unpublished results.

45. H.-J. Butt, unpublished results.

46. F. De Schryver, unpublished results.

47. a) J. Hofkens, L. Latterini, G. De Belder, T. Gensch, M. Maus, T. Vosch, Y. Karni, G. Schweitzer, F. C. De Schryver, A. Herrmann, K. Müllen, *Chem. Phys. Lett.* **1999**, *304*, 1; b) Y. Karni, S. Jordens, G. De Belder, G. Schweitzer, J. Hofkens, T. Gensch, M. Maus, F. C. De Schryver, A. Herrmann, K. Müllen, *Chem. Phys. Lett.* **1999**, *310*, 73.

48. T. Gensch, J. Hofkens, A. Herrmann, K. Tsuda, W. Verheijen, T. Vosch, T. Christ, T. Basché, K. Müllen, F. C. De Schryver, *Angew. Chem. Int. Ed. Eng.* **1999**, *38*, 3752.

49. R. P. Feynman, *Eng. Sci.* **1960**, *23*, 22.

50. Y. Wada, *Molecular Electronics: Science and Technology* (ed. A. Aviram, M. Ratner), The New York Academy of Science, New York **1998**, 257.

IV Biological Aspects

Enzyme Mimics

Anthony J. Kirby

University Chemical Laboratory, Cambridge CB2 1EW, England

Phone: +1223 336370, Fax: +1223 336913, e-mail: ajk1@cam.ac.uk

Keywords ■ Catalysis ■ Enzyme ■ Transition State Analog ■ RNA Catalysis ■ DNA catalysis ■ Catalytic Antibodies ■ Host-Guest

Concept: "One of the great intellectual challenges presented to Science by Nature is a proper understanding of how enzymes work. At one level we can 'explain' enzyme catalysis – what an enzyme does is bind, and thus stabilise, selectively the transition state for a particular reaction. But our current level of understanding fails the more severe, practical test – that of designing and making artificial enzyme systems with catalytic efficiencies which rival those of natural enzymes."[1]

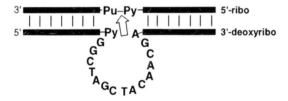

Abstract: This chapter updates but mostly supplements the author's Angewandte Review,[1] setting in context recent advances based on protein and nucleic acid engineering. Systems qualify as a true enzyme mimics if there is experimental evidence for both the initial binding interaction and catalysis with turnover, generally in the shape of saturation kinetics. They are discussed under five broad headings: mimics based on natural enzymes, on other proteins, on other biopolymers, on synthetic macromolecules and on small-molecule host–guest interactions.

Prologue

The chemist's fascination for enzymes shows no sign of abating. The more we understand about biological catalysis the more remarkable proves to be the chemistry that enzymes do. It has long been accepted that enzymes use normal mechanisms, but in special ways and environments that support their extraordinary catalytic efficiencies. Just how extraordinary these are is underlined by the work of Wolfenden, who has painstakingly measured rates of typical biological reactions in the absence of catalysis. Reactions that proceed in enzyme active sites typically with rate constants k_{cat} in the range 1–1000 s^{-1} may have half-lives in solution at pH 7 of hundreds of thousands of years. [2-4] Enzymes catalyzing such very slow reactions stabilise and thus – at least formally – bind the rate determining transition states for the reactions more strongly than does the solvent by well over 100 kJ mol^{-1}.[1] Molecular recognition at this level, in a powerful solvent like water, presents a tremendous challenge to chemists interested in molecular recognition even for the passive binding[5] of stable molecules (particularly small molecules!). Mimicking the *dynamic binding*[5] of transition states is the ultimate challenge.

This chapter will consider some of the most interesting of current approaches to the evolution of enzyme mimics, in the context of continuing dramatic progress in protein and nucleotide engineering. There are excellent practical as well as intellectual reasons for the broad interest in this topic. Catalysis is a major preoccupation of the chemical industry: if the application of the principles of biocatalysis can lead to robust and efficient catalysts tailor-made for reactions of economic importance the area will become even more a focus of intense activity and investment.

Definitions – and Problems

The groundwork for this article was laid in an Angewandte Review,[1] and it is convenient to start with the definition adopted there.

"Enzyme mimics catalyze reactions by mechanisms which are demonstrably enzyme-like. The minimum requirement is that the reactions concerned should involve an initial binding interaction between the substrate and the catalyst. This gives rise to Michaelis–Menten kinetics: reactivity is measured in terms of the familiar parameters k_{cat} and K_m, and we use E to denote enzyme mimic as well as enzyme."

Thus we will consider only systems where there is experimental evidence for both catalysis with turnover and the initial binding interaction, generally in the shape of saturation kinetics (and reserve the use of the unmodified term enzyme specifically to mean a protein enzyme).

The minimalist reaction of interest is thus:

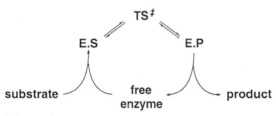

Scheme 1.

Scheme 1 is a gross over-simplification for almost any enzyme-catalyzed reaction of a specific substrate, based as it is on a one-step reaction with a single, rate-determining transition state; but it is appropriate for many, if not most reactions catalyzed by simple enzyme mimics. Most important for present purposes, it emphasises the most important properties of enzyme reactions which the design of mimics, or artificial enzymes, must address, namely:

(i) **E** must bind the substrate (but not too strongly[6]),

(ii) The transition state **TS**$^+$ must be efficiently stabilized (bound more strongly than the substrate) for catalysis to be significant, and

(iii) The product **P** must be released rapidly to regenerate the catalyst.

This third requirement is a particular problem for synthetic processes, where the reaction catalyzed is the formation of a new covalent bond or bonds. In the general case binding of the product will inevitably be stronger than that of the of the reactant fragments (the chelate effect), and product inhibition is to be expected.

Classifying Enzyme Mimics

Since the best natural enzymes have solved all the problems – and more – listed above it it no surprise that some of the most stimulating and effective approaches to the design of mimics start from real enzymes. Thus there arise immediately questions of definition. An enzyme which has been "improved" by protein engineering – having for example a higher k_{cat} or k_{cat}/K_m than the wild type as a result of site-directed mutagenesis [7] – is clearly not reasonably defined as an enzyme mimic: even – though getting closer – when the mutant has been designed to favor the reverse reaction (a classic example is subtiloligase [8]). Nor, to be strict, are the very interesting systems where a deleted active site group can be replaced by an equivalent functional group in the substrate,[9] or even free in solution.[10] On the other hand an enzyme which has been more or less completely redesigned – to catalyze an entirely different reaction, or by combining binding and catalytic domains from different enzymes to catalyze a reaction of a different substrate, will qualify. In this article the author's decision – however arbitrary – is final!

We consider systems under the following headings:

Mimics based on natural enzymes
Mimics based on other proteins
Mimics based on other biopolymers
Mimics based on synthetic macromolecules
Mimics based on small-molecule host–guest interactions

Mimics Based on Natural Enzymes

A remarkable example which appears to steer well clear of such common pitfalls as contamination by enzyme impurities is the redesign as a peptidase of a cyclophilin from *E. coli*. The reaction is significant because peptide cleavage is one of the most popular targets in the enzyme mimics area, and one of the most difficult. The cleavage of an ordinary amide at pH 7 is very slow reaction, with a half-life of the order of hundreds of years,[2] and thus a highly ambitious target, well beyond the range of ordinary enzyme mimics.

The design of this successful system started with an enzyme with a ready-made peptide binding site, with well established specificity. The cyclophilins catalyze the *cis-trans* isomerization of peptide bonds to proline:

passing through a conformation in which – significantly – the amide bond is rotated out of plane, and thus stereoelectronically destabilized as the π-type delocalization of the nitrogen lone pair is turned off. In this conformation an amide bond is highly reactive, as shown by the recent demonstration that the "most twisted amide" **1** is hydrolyzed effectively instantaneously at pH 7 to the amino-acid **2**.[11]

Quemeneur, Moutiez, Charbonnier and Menez[12] started by generating four different mutants in which serine replaced residues – Arg-48, Gln-56, Ala-91 and Thr-93 – known from crystal structures to be within possible striking distance of the carbonyl group of a bound substrate. Of these mutants one, A91S, proved highly effective, with k_{cat} near 4×10^{-2} s^{-1} for the hydrolysis of an acyl-Ala-Pro peptide – some 10^7 faster than the uncatalyzed reaction. Further logical modifications of this successful lead compound involved the systematic introduction of histidine and then aspartate residues, thus constructing, at least in principle, the catalytic triad familiar from the serine proteases. The finished article, named cyproase 1 (final mutations A91S, F104H, N106D), is a respectable endopeptidase, with k_{cat}/k_{uncat} close to 10^9, and no residual *cis-trans* isomerase activity.

Cyproase 1 is in effect a new enzyme, produced by systematic protein engineering firmly based on sound chemical principles. Its status as an enzyme mimic may be debatable: its efficiency is not. It was shown to hydrolyze 25% of bonds to proline in a (denatured) peptide toxin in two hours at pH 7.0, with some 400 turnovers. It is one of the two most efficient enzyme mimics we will encounter in this article.[13]

It is a reasonable supposition that step-by-step improvements of this sort, involving mutations of individual amino-acid residues, have contributed to enzyme evolution. A more radical method of interest in this context is "gene shuffling," whereby new genes, encoding new proteins, may be produced by the combination of exons of unrelated genes. The technology involved in creating a new enzyme by combining catalytic and binding domains from different proteins is now fairly readily available, and

this is a rapidly expanding area of interest. The problems should not be underestimated: protein domains do not normally behave as independent "modules" which can simply be plugged into each other, and certainly not where such a high-specification function as catalysis is involved. Thus most published examples of chimeric proteins have been designed to combine separate binding functions.

Successful catalysts have been created in this way in two types of system, both of which are particularly favourable, though for quite different reasons. Nixon, Warren and Benkovic[14] created hybrid proteins by combining the N-terminal, substrate binding region of an enzyme (glycinamide ribonucleotide transformylase, which transfers the formyl group from the cofactor N^{10}-formyltetrahydrofolate to the substrate) with the C-terminal cofactor-binding domain of a related enzyme which catalyzes its hydrolysis to formate. In a resulting protein the formyl group was also transferred successfully from the cofactor to the substrate, though 100–1000 times more slowly across the new inter-domain space. This work confirms that creating reasonably precise positioning in three dimensions is possible; in a case where the reactants and the reaction catalysed are "familiar" and the two proteins involved rather closely related.

The second example is a different sort of hybrid which uses a DNA-binding motif as the recognition element. The recognition and binding of *single-strand* DNA by the complementary sequence is the basis of many different "antisense" projects, including several enzyme mimics discussed below under small-molecule host–guest interactions. Here we discuss specifically chimeric enzymes, which use (*duplex*) DNA-binding motifs for substrate recognition. Chimeric restriction enzymes[15] offer the possibility of targeting a specific site in the genome, a fundamental requirement of one form of gene therapy – an area of much current interest and activity.

One of the most successful chimeric (hybrid) systems is based on a non-specific DNA cleavage domain, fused to selected DNA-binding motifs from other proteins. If the recognition sequence is an integral part of the active catalytic apparatus, tailoring the sequence specificity will also affect the catalytic efficiency, almost certainly adversely. So a favored approach is to base the design on one of a class IIS bacterial restriction endonucleases, which cleaves DNA some 10 nucleotides downstream – thus well away from – the recognition site (in duplex DNA). (This strategy is more or less essential for the design of any *artificial* nuclease with sequence specificity.) Such enzymes evidently have separate recognition and catalytic sites: the geometry of the interaction between the two is clearly crucial, and recognition and catalytic activity can be uncoupled by a single amino-acid substitution.[16]

For the recognition sequence to be unique within a genome it needs to be relatively extensive (of the order of 16–18 base pairs in length), so the procedure was to exchange the recognition site of a selected class IIS enzyme for suitable DNA-binding regions from other proteins. Of these the most interesting are those based on the zinc-finger motif.[15] These are specific DNA recognition elements from gene regulatory proteins, and have in common tandemly arranged segments of α-helix and β-sheet pointing outwards like fingers, held together by coordination to Zn^{++} ions of cysteine and histidine side-chains. The α-helical regions bind in the major groove of the DNA and the primary amino-acid sequence gives each finger specificity for a particular base-pair triplet. Since the zinc fingers appear to bind as independent modules it is in principle possible to construct DNA-binding proteins for any desired sequence.

A number of such combinations have been shown to be effective site-specific endonucleases, able to cleave DNA *in vitro* at least close to the intended target sequences.[15] We can look forward to further exciting developments in this general area.

Mimics Based on Other Proteins

In the present context proteins not (directly) related to enzymes fall into two distinct classes: those having *accidental* catalytic properties, and the generally more interesting systems, particularly catalytic antibodies, raised specifically as potential catalysts.

Work linking the two classes has been reported by the Kirby and Hilvert groups. The report of Thorn and Hilvert[17] of antibodies that catalysed the Kemp elimination (**3** → **4**) remarkably efficiently led Hollfelder *et al.*[18] to screen a small but carefully selected set of common proteins for evidence of similar catalysis.

The simple selection criteria were a hydrophobic binding site and a general base within reasonably close proximity. It was found that the serum albumins, common transport proteins which meet these simple criteria, catalyse the Kemp elimina-

tion, in reactions showing Michaelis–Menten kinetics and with efficiencies comparable to the antibodies of interest. Similar results were reported by Kikuchi et al.,[19] and the two groups concerned joined forces to try to set their results in the general context of catalysis by enzyme mimics.[20]

The Kemp elimination is of special interest because it is known to be extraordinarily sensitive to the medium, and particularly well suited as a test reaction for potential enzyme mimics because it is a simple, one-step process. The joint conclusions from this work were that catalysis involves a combination of a number of different factors, even for this simple reaction by these simple catalysts.

One of these factors is undoubtedly that the energetic requirements for catalysis of the single-transition state reaction **3** → **4** are modest, compared with more demanding proton transfers in nature. Another is likely to be the convenient geometry of the process – an extended, almost linear transition state with at one end a delocalized anion developing in a favorable, hydrophobic, stabilizing environment, generated by the removal of the proton by an adjacent catalytic general base. At the other end the (lysine amine) general base remains largely solvated and so can take advantage of hydrogen-bonding solvation as the ammonium cation develops. Such "compartmentalization" of environments is not possible in homogeneous solution but pays a crucial role in catalysis by enzymes.

Even in such a simple system it is not possible to dissect the observed catalysis quantitatively into contributions from specific and non-specific medium effects and from positioning of the general base. However, the relatively small shifts of the pK_as from normal solution values suggest that hydrogen-bonding desolvation of the general base in the active site is not the major factor, and is worth no more than perhaps an order of magnitude in rate for antibody 34E4, which uses carboxylate as the general base.[17] For the serum albumins this factor is smaller still (it is in the opposite direction for an amino-group).[20] The most convincing candidate for the largest single factor, certainly in the case of BSA and HSA and probably in the case of 34E4, is thought to be specific solvation of the delocalizing π-system in the transition state.

Catalytic antibodies are proteins designed and produced specifically to act as enzyme mimics, and depend on natural enzymes for neither binding specificity nor catalytic activity. Their binding specificity can be tailored, generally successfully, to select a specific substrate, and even a single enantiomer of a chiral substrate.[21] This desirable result is achieved by recruiting the immense resources of the immune system, which can be manipulated to produce (usually monoclonal) antibodies capable of binding a specific antigen (the hapten) exceedingly tightly. The trick which can add catalytic capability is to use a hapten which is a good transition state analog. If the result is an antibody which can bind a particular transition state well it will catalyse the target reaction, acting by definition as an enzyme mimic.[1,22–24]

in principle – gives access to abzymes with both these advantages.

The most interesting developments involve catalysis of simple aldol reactions. The key to reactive immunisation is the use of a hapten that is chemically reactive, rather than a passive template. This means that (i) relevant chemistry is going on during the course of antibody induction, which thus happens in the presence of intermediates involved in the reaction, and so may be modified to favor the formation of antibodies which bind these intermediates (and perhaps transition states leading to them). Furthermore (ii) it becomes possible to select for antibodies that react with, rather than just bind, to the hapten. The system used for the development of aldolase antibodies is outlined in Scheme 2.

Scheme 2. Reactive immunization using a 1,3-diketone as hapten. (R (= –(CH$_2$)$_3$CO–) makes the connection to the carrier protein).

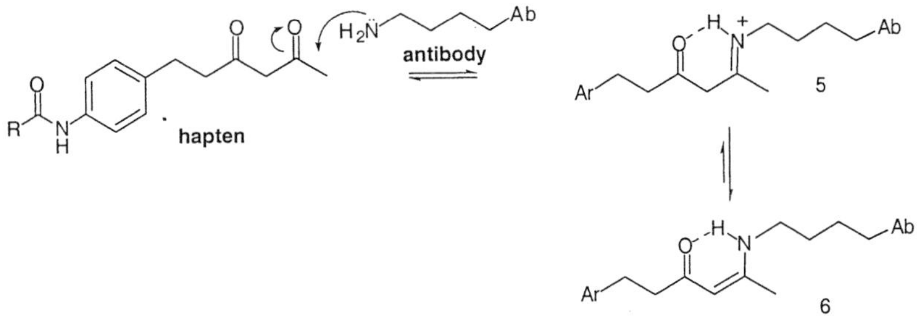

Developments in the area have been rapid. The initial proof of the (definitely stimulating!) concept attracted a host of enthusiastic new practitioners, most of whom soon discovered that the development of a new abzyme from scratch is not in practice a simple operation: this for two main reasons. (i) Success rates are generally low: it is relatively simple to produce antibodies that will bind haptens well but few of these turn out to be catalytic. This is in part because (ii) catalytic efficiency is typically low, especially when compared with natural enzymes.[21] As a result it is not currently possible to produce a catalytic antibody capable of cleaving a simple amide bond at a useful rate (though progress in this direction has been reported[25]), still less a phosphodiester. However, useful synthetic reactions are not typically very slow and recent advances have been geared to potential applications in synthesis. The first catalytic antibody is now commercially available,[26] thus by-passing the chancy development stages for the reaction concerned (see below). So direct comparison with commercially available enzymes is relevant: the experimental protocols are much the same.

"Made-to-measure" protein catalysts can in principle offer two significant advantages over enzymes: the ability to catalyse reactions of unnatural substrates, and substrate specificities broad enough to cover a given reaction for a good range of structures, as with classical synthetic methods. Reactive immunization[27] is a new technique which – at least

Here the hapten (Scheme 2) is a 1,3-diketone, which incorporates structural features of both reactants – ketone donor and aldehyde acceptor (see below, Scheme 3) – in the aldol reaction of interest. In favorable cases the hapten reacts with the primary amino-group of a lysine residue in the complementary-determining region of an antibody to form a Schiff base 5, which readily tautomerises to the more stable vinylogous amide 6.

Antibodies produced by this procedure were screened for their ability to react with the hapten to form the vinylogous amide 6, which has a convenient UV chromophore near 318nm, clear of the main protein absorption. Two antibodies selected in this way catalysed the expected aldol reaction of acetone with aldehyde 7 by way of the enamine 8 (Scheme 3): the remainder did not. These two effective aldolase mimics have been studied in some detail, and a crystal structure is available for (a *Fab'* fragment of) one of them.[26, 28]

Antibodies 38C2 and 33F12 turn out to have very broad, though not unlimited substrate specificity, catalyzing reactions involving 20-odd different donors and a similar number of acceptors, generally in good yields, with enantiomeric excesses better than 95 % and, where appropriate, good diastereoselectivity.[26, 28] The reverse, retroaldol reaction is also catalyzed, and since equilibrium favors cleavage a large excess of the donor ketone is used to drive reaction in the forward direction. Typical conditions for preparative-scale synthesis are 1M ketone donor, 1mM aldehyde acceptor and several micromolar antibody, in predominantly aqueous buffer pH 7.4 for several days at room temperature.

This work represents a definite practical advance, and points the way towards solutions of some major problems with the applications of protein enzyme mimics. Enzymes are most efficient at concentrations close to their K_Ms, which mirror the low concentrations of natural substrates in the cell. 38C2 and 33F12 have high K_Ms, up to 1M for typical substrates, suitable for the higher concentrations of interest for synthesis. This holds also for retroaldolisation, so product inhibition is not observed and hundreds of turnovers are achieved. However, absolute levels of catalytic efficiency remain a problem: the high molecular weight of the abzyme protein (~75 000 per active site) means that 2–3 micromoles of catalyst weigh more than a millimole of a small molecule substrate. Enzymes solve this problem by their high catalytic efficiency. Abzymes 38C2 and 33F12 catalyze the chemistry shown in Scheme 3 relatively efficiently, not least by controlling the pK_a of the catalytic lysine amino-group (lowered from around 10 to about 6, in common with class I aldolases) so that it is present in the reactive free base at pH 7. There is clearly some way to go before antibody catalysis becomes a primary method of choice for the synthetic chemist, but the way forward is being energetically explored.[29, 30]

Mimics Based on Other Biopolymers

For a poly- or oligonucleotide catalyst–substrate interaction the problem of binding-specificity is in principle solved simply by using the complementary sequence, and chemistry based on classical Watson–Crick base-pairing has attracted much recent interest. This is basically what ribozymes do: specific binding interactions bring the 5'-hydroxyl group of a guanosine residue in close proximity to a phosphodiester linkage far removed along the polynucleotide chain, and with the help of one or more metal cations catalyse a nucleophilic displacement at phosphorus which effects chain transfer. We will not discuss ribozymes as such, though their chemistry is very relevant to the problems of synthetic enzyme mimics,[31, 32] and there is good evidence that they can fold to set up the sort of three-dimensional active site familiar from crystal structures of enzymes.[33] However, one result with a chimeric deoxyribozyme is of particular interest because it represents a successful attempt to augment the functionality of DNA.

Catalysis by DNA

Roth and Breaker[33] used *in vitro* selection to evolve DNA molecules of the type illustrated in Scheme 4, which self-cleave *in the presence of histidine*. (This is in principle mimicking a disabled mutant enzyme, of the sort mentioned in the introduction.[10]) In this case a pool of 2×10^{13} modified DNAs was attached at the 5'-end, tightly but reversibly, to a solid support. The polynucleotide was made up of a randomised sequence of 40 deoxynucleotides (N_{40} in Scheme 4) flanked by two regions of base-pairing complementarity to the sequences on either side of a single upstream RNA linkage (rA).

Cleavage in the presence of histidine releases the oligonucleotide fragment containing the random sequence, and this is amplified by PCR and the cycle repeated. This selection procedure produced deoxyribozymes that require no metal cofactor but have a specific requirement for L-histidine, which is presumed (from pH-rate profiles) to act as a general

Scheme 4.

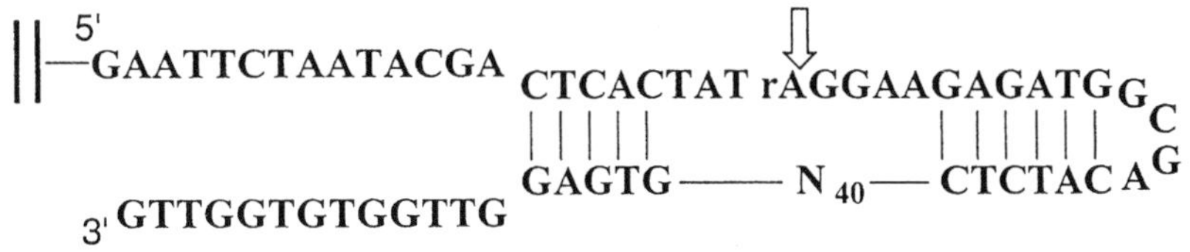

base as in an RNAseA-type mechanism. The rate enhancement, for what is only formally a cyclization reaction, is of the order of 10^6.

This system illustrates the preferred strategy for catalysis by nucleic acids: DNA is the potential catalyst, because of its stability, and RNA is the preferred substrate because of its higher reactivity. However, the longer the oligonucleotide, for improved specificity, the stronger will be the multiple hydrogen-bonding interactions, and the less the chance of useful turnover. So this chemistry has mostly involved shorter, oligonucleotides as hosts – usually referred to in this context as templates.[35] This work is discussed below. A useful general discussion of templating is given by the Andersons.[36]

DNA binds and reacts with carcinogenic and similar compounds which alkylate it through cationic intermediates, in some cases extraordinarily fast:[37] and can in the process catalyse the hydrolysis of some substrates, like the bay-region diol epoxides derived from benzpyrene.[38] In the context of enzyme mimics these reactions are primarily of curiosity value: DNA lacks the conformational flexibility and the chemical functionality to offer the prospect of efficient catalysis for ordinary reactions.

As might be expected, this judgement represents a challenge taken up by a number of groups, for example by Li and Breaker,[39] whose optimized deoxyribozyme-kinase catalyzes its own selective phosphorylation by ATP with a rate enhancement in k_{cat} of 6×10^9. In a complementary approach progress is being made in equipping DNA with additional catalytic functionality: thus Sakthivel and Barbas have shown that dTTP can be replaced in the polymerase chain reaction by synthetic nucleotides with amino, imidazole and carboxy groups in an extended side-chain attached at the 5'-position.[40]

Catalysis by RNA

The potential availability of vast numbers of potential catalytic systems of varying length and sequence, coupled with the development of efficient methods of *in vitro* selection and growing interest in the implications of an RNA-world, inspire increasingly sophisticated searches for catalytic activities of RNA.[41] It is estimated [42] that the chance that a given random sequence of RNA will catalyse a suitable phosphoryl transfer reaction is of the order of 1 in 10^{13}, odds that give a reasonable probability of success if combinatorial libraries of potential catalysts can be screened efficiently. The numbers involved suggest that the transition state analog approach on which the design of catalytic antibodies is based could be fruitful, but only one modest success has been reported: Prudent *et al.* using a planar hapten to elicit an RNA molecule that catalysed the isomerization[43] of bridged biphenyl **10**, with multiple turnover and k_{cat}/k_{uncat} = 88.

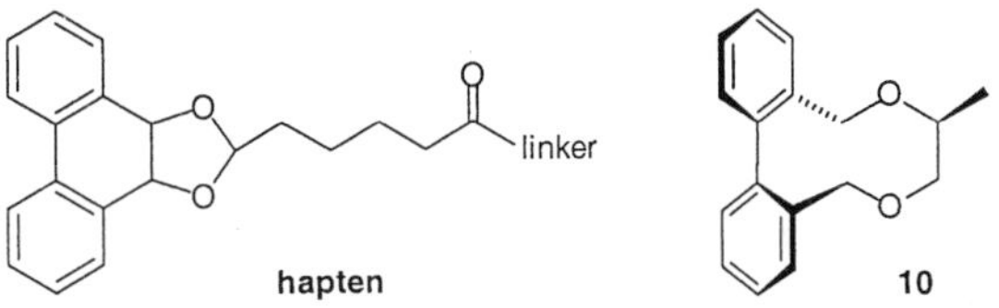

hapten   **10**

Of a (small) number of examples in which RNA molecules have been generated to catalyse simple organic reactions[44] perhaps the most interesting for the organic chemist is the *in vitro* selection (from a library of ~10^{14} unique RNA sequences) of catalysts for the Diels–Alder reaction.[45] The RNA molecules were made up of constant sequence elements flanking a 100-nucleotide randomised region (Scheme 5), modified by substituting for UTP in the transcription reaction a nucleotide triphosphate (**11**) in which uridine carries a pyridine-containing side chain. The diene was attached to this RNA by a long

Scheme 5.

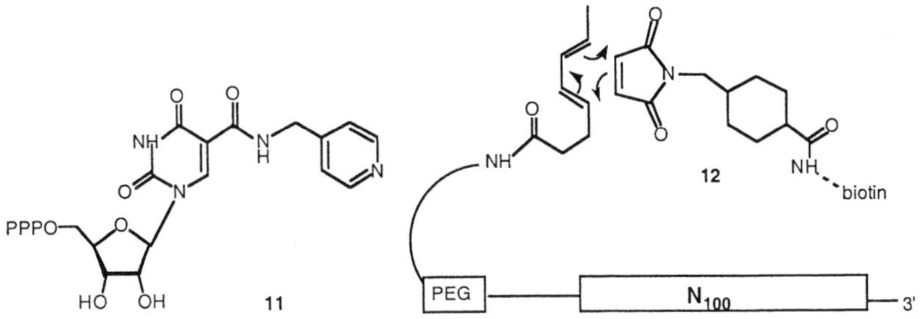

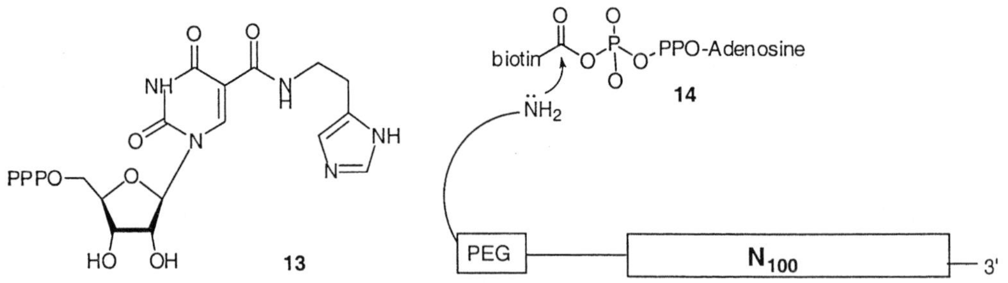

Scheme 6.

polyethyleneglycol chain (not to boost intramolecular reactivity but to allow selection). The dienophile **12** was too insoluble for a full Michaelis–Menten analysis, but the binding step is not in doubt in view of the absence of catalysis in the general case and the observation of product inhibition. The reaction depends absolutely on the presence of Cu^{++} ions, and in the best case is some 800 times faster than in the presence of non-catalytic RNA molecules.

Larger accelerations were observed, using the same principle, for the geometrically less demanding formation of an amide from a primary aliphatic amine and an acyl phosphate (**14**, Scheme 6). In this case the uridine bases carry an imidazole side-chain (**13**). After 16 rounds of increasing selection pressure the best catalytic species accelerated the reaction of Scheme 6 over 10^5-fold.

Finally, the most complex synthetic reaction clearly catalysed by RNA molecules generated by *in vitro* selection is the formation of the C–N bond of a nucleoside (Scheme 7), from 4-thiouracil and most of the natural substrate for the natural (uracil phosphoribotransferase) reaction.[46] (Thiouracil was used because it is easily tagged by alkylation on sulfur.) The catalytic RNAs produced by 11 rounds of selection required Mg^{++} cations and had k_{cat} as high as 0.13 min^{-1}, with k_{cat}/K_M at least 10^7 times greater than the (undetectable) uncatalyzed reaction. Once again these systems are convincing, rather efficient enzyme mimics.

Mimics Based on Synthetic Macromolecules

Oligonucleotides

As might be expected, most oligonucleotide-based enzyme mimics catalyze reactions involving phosphate transfer, as do the natural ribozymes.[47, 48] There is an understandable trend towards shorter, synthetically accessible oligonucleotide catalysts, with made-to-order sequence-specific restriction endonucleases an objective which seems eminently achievable. Thus Santoro and Joyce describe a procedure for the development of "a DNA enzyme that can be made to cleave almost any targeted RNA substrate under simulated physiological conditions"[49]. These "DNA enzymes" are made up of some 30 deoxynucleotides, with a central catalytic domain of 15 lying between outer, substrate recognition, sequences. Two catalytic domains emerged from the selection procedures, one of which has the sequence shown in Scheme 8 (the lower strand is the "DNA enzyme", the upper strand the oligoribonucleotide substrate). It is effective against a wide range of substrate sequences complementary to its

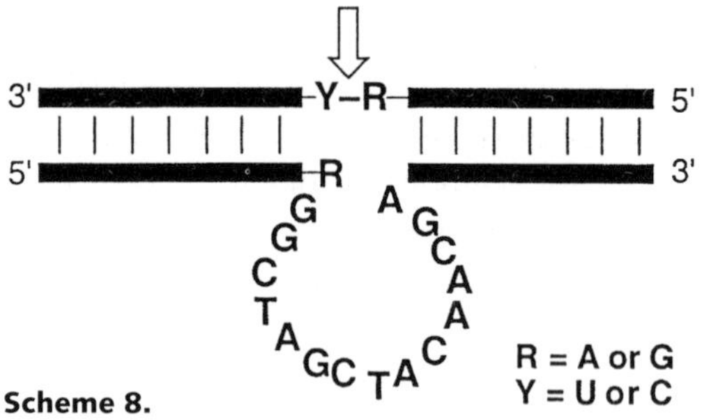

Scheme 8.

Scheme 7.

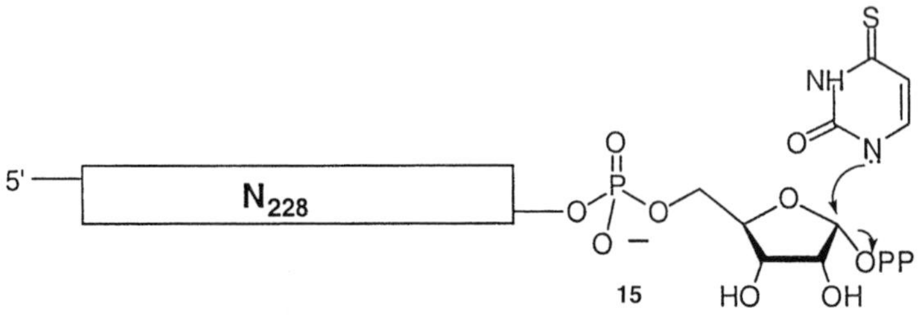

outer, recognition sequences, typically cleaving the P–O(3') bond of a purine-pyrimidine linkage (**Y–R**). A–U linkages are cleaved fastest, with k_{cat} of the order of minutes. This compares with 10^{4-5} min^{-1} for catalysis by ribonuclease A, but K_M is of the order of nanomolar for the DNA enzyme, much lower than for the protein. This makes the "DNA enzyme" faster in terms of k_{cat}/K_M ($\sim 10^9$M^{-1}min^{-1}) than ribonuclease A for some substrates, and comparable with the endopeptidase cyproase discussed above. At this rate substrate-binding is likely to be rate determining at $[S] < K_M$, and multiple turnover is observed.

This highly efficient enzyme mimic, like natural ribozymes, requires Mg^{++} ions, and a requirement for a metal cation was thought to be general for DNA-based mimics in particular, adding potential catalytic functionality in the simplest way to the parent structure – and making it possible to extend the range of catalysis to the much slower DNA cleavage. Thus Breaker and his coworkers used an *in vitro* selection protocol similar to that used by Santoro and Joyce to develop a 46-nucleotide deoxyribozyme that cleaves complementary DNA substrates in the presence of Cu^{++} ions, using simultaneous duplex and triplex binding.[50] In this case however the cleavage mechanism is oxidative, and selectivity is not high enough to prevent concurrent cleavage of the catalytic DNA. (In practice high sequence specificity *requires* P–O cleavage – in what can be seen as another manifestation of the reactivity-selectivity principle!) Lanthanide cations in particular, with three or even four positive charges, are very effective promoters of phosphate cleavage which have been used as cofactors for "DNAzymes".[51] However it appears that reasonably efficient DNA catalysis is possible even in the absence of the usual catalytic metal cations, at least for RNA cleavage.[52, 53]

A logical extension of the work on phosphoryl cleavage reactions is the study of the reverse, ligation reaction. The principle of microscopic reversibility offers the comforting thought that the requirements for catalysis are basically the same, and Watson–Crick base-pairing provides a simple system for bringing reacting centres together. This templating approach works well with oligonucleotides when a nucleophilic group on one reactant is brought into reasonably close proximity with a suitable electrophile. For example, Cuenoud and Szostak[54] reported a 47-mer DNAzyme which in the presence of Zn^{++} catalyzes the formation of a new phosphodiester bond between the terminal 5'-hydroxyl of one oligodeoxynucleotide and the activated phosphoroimidazolide attached to the 3'-position of a second. Product release was rate limiting, as might be expected in the circumstances, but some turnover was observed. Furthermore, given the appropriate template it is possible to add mononucleotides sequentially to an RNA primer[55] and to design self-replicating DNA-based systems based on this principle.[56]

However, there is a second aspect to reactions controlled by templating. Simple proximity effects accelerate bond formation between activated groups on bound substrates held together closely enough in favorable orientations, but an enzyme – or a well designed enzyme mimic – will also *catalyze* the chemical "ligation" reaction. Few current enzyme mimics achieve this more challenging function: most examples involve oligoribonucleotide ligation, partly because of its relevance to RNA-world chemistry but probably also because the 2'-OH group may help to preorganize the metal cations involved in the simplest sort of catalysis.[57, 58] One remarkable result is of particular interest in the context of oligonucleotide templating – and of templating in general.

The problem of product release in ligation reactions has been addressed in several ways: only with luck can it be ignored. It has been addressed for example by incorporating a special heating step in the cycle, or by building in features designed to weaken templating.[35] Binding can be remarkably sensitive to apparently minor changes in conformation, as shown by a series of experiments from the Lynn group. The formation of the iminium system **18** (Scheme 9) from the trinucleotide amine and aldehyde **16** and **17** was, as expected, strongly

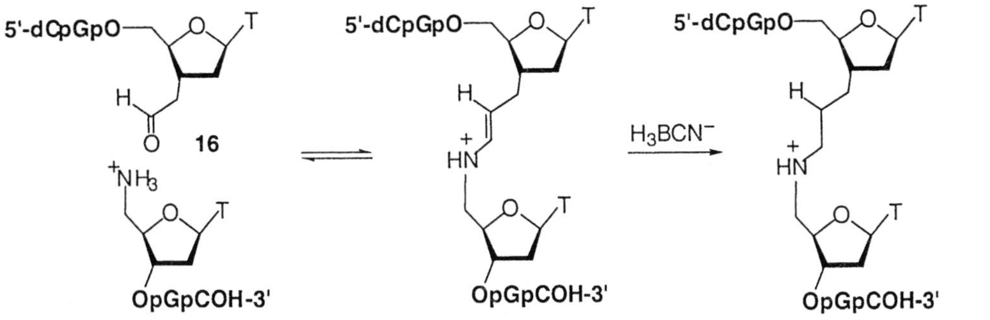

Scheme 9.

favored in the presence of the complementary hexanucleotide template 5'-HOd**GCAACG**OH. The condensation reaction is reversible, so this is an example of thermodynamic templating[36] (as opposed to the various kinetically controlled systems discussed so far). Templating favors the equilibrium of Scheme 9 by a factor of at least 10^5, but simply reducing the imine double bond to give the more conformationally flexible **19** weakens binding to the template by a factor of over 10^6. In the presence of cyanoborohydride 1 mol% of the hexanucleotide (over 90% saturated under these conditions) catalyzes the reductive amination of **16** by **17** by trapping the bound iminium cation as **19**, which is then readily released (allowing multiple turnover).[59] Here too the catalyst acts simply as a template, but this sort of chemistry offers the prospect of incorporating catalytic functionality into potential enzyme mimics which have solved the problems of substrate binding and product release.

Oligopeptides

Templating by oligopeptides is a more complicated proposition than the essentially linear solution available for nucleotide systems. Regular structures can be designed, but conformations typically depend on pH, and template aggregation is potentially more of a problem. Two groups have achieved success in the sort of template-mediated ligation reactions we have discussed for nucleotide systems, and their results illustrate the problems.[60, 61]

The approach in both cases is based on activated substrate groups (compare Scheme 9, above). The templates are 32–36-residue oligopeptides based on systems known to form coiled coils (assemblies of two (or more) α-helical peptides). The substrates are in effect the cleavage products of the templates, one having an amino-terminal cysteine with a free NH_2 group, the other with its carboxyl-terminus activated as a thiolester. The ligation reaction, designed to take place on the hydrophilic surface of the cleaved coiled coil, involves two stages: thiolester exchange – rapid enough that protecting groups are not needed – and a second, intramolecular acyl transfer to NH_2 (Scheme 10).

The ligation product is a copy of the template, so this represents a self-replication procedure for **P1–Cys–P2**. Initial rates depend on the square root of the template concentration (a common observation, readily explained where the template is active as the monomer but present mostly as the dimer), but rates fall off, as expected, as the product accumulates.

"Electronic mismatches" in the sequences of the substrate peptides slow the coupling reaction,[60] templates act better at lower or higher pH depending on the ionizing side-chains present,[61, 62] and yields and diastereoselectivity depend strongly on the ionic strength, all consistent with a critical dependence on the three-dimensional structure of the template. These systems are relatively inefficient, with $(k_{cat}/K_M)/k_{uncat}$ about 10^4 at best, but they should be regarded as important proofs of concept rather than optimized catalytic systems. Again, no catalysis of the ligation reaction is involved.

Baltzer has reported catalytic systems based on synthetic oligopeptides designed, successfully, to fold into a hairpin helix-loop-helix structure that brings side-chain functional groups into interesting proximity, either along or between helices.[63, 64] For example, KO-42, with 42 amino acid residues, which dimerizes to a four-helix bundle, forming a cluster of six histidine residues, catalyzes the hydrolysis of mono-*p*-nitrophenyl fumarate and its transesterification to form the corresponding trifluoroethyl ester in 10 vol% trifluoroethanol. At pH 4.1 the reactions are some 10^3 times faster than the reaction catalyzed by 4-methylimidazole.

Mimics Based on Synthetic Macromolecules

The transition state analog (TSA) approach[65] which has proved so successful in the design of enzyme inhibitors and catalytic antibodies lends itself nicely, at least in principle, to the molecular imprinting of polymers. Polymerization carried out in the presence of the TSA, or with the TSA covalently but readily reversibly bound to a monomer, produces a polymer with a number of embedded TSA molecules. If these can be removed under rea-

Scheme 10.

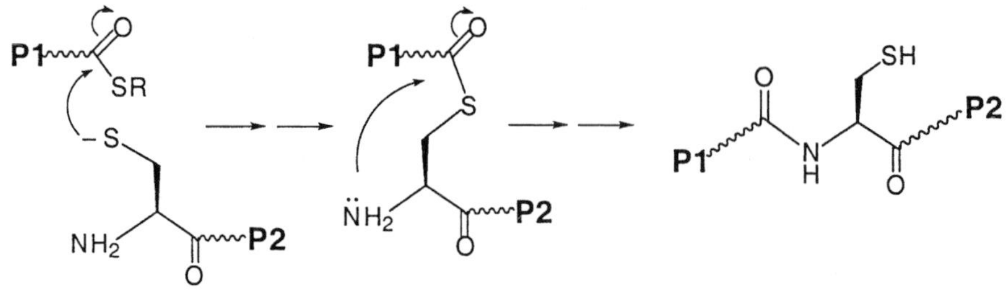

sonably mild conditions the polymer produced contains tailor-made binding sites for the transition state concerned, so should be capable of catalyzing the reaction of interest. Synthetic polymers can be much more robust than proteins and the approach has attracted considerable interest.

Results have generally been disappointing. It can be difficult to remove the TSA from the polymer, but a more fundamental problem concerns the efficiency of the catalysis observed. The most efficient systems catalyze the hydrolysis of carboxylate and reactive phosphate esters with Michaelis–Menten kinetics and accelerations $[(k_{cat}/K_M)/k_{uncat}]$ approaching 10^3,[66] but the prospects for useful catalysis of more complex reactions look unpromising. Apart from the usual difficulties the "active sites" produced are relatively inflexible, and the balance between substrate binding and product inhibition is particularly acute.

An attempt to generate more flexible catalysts has been reported by Hollfelder at al.,[67] using structures based on polyethyleneimine. The polymer was randomly alkylated with systematically varied proportions of methyl, benzyl and dodecyl side-chains, the latter two designed to create hydrophobic binding regions necessarily close to the ubiquitous alkylamino-groups of the polymer, which are available to act as potential general bases. The design criteria are are similar to those used to identify BSA catalysis (discussed above) and the test reaction, the Kemp elimination, was the same. Most of the many hundreds of catalysts produced showed low activity, but the best showed rate accelerations near pH 6 (the pK_as of some of the amino-groups are lowered by 4–5 units compared with simple alkylamines) of up to 10^6, and no problems with product inhibition. These "synzymes" provide the most efficient artificial catalytic systems for proton transfer from carbon. The authors consider that they work by stabilising the negatively charged transition state TS (for the conversion of **3** to **4**, above) primarily through a specific medium effect involving dispersion interactions with the delocalized π-system, with an additional contribution from through-space electrostatic stabilization by the predominantly cationic polymer backbone.

Both these examples are designed to act in an aqueous medium. A recent report describes the construction of dendrimers with an outer envelope of tetradecyl side-chains surrounding an inner core containing many benzyl alcohol groups.[68] In organic solvents these molecules form reverse micelles, with a central cluster of hydroxyl groups creating a "protected" polar environment, capable of stabilizing the transition states of suitable ionic reactions. In cyclohexane in the presence of sodium bicarbonate the system catalyzed the E1 reactions of Scheme 11 with apparently unlimited turnover, though there is no evidence as yet for Michaelis–Menten kinetics.

Mimics Based on Small-Molecule Host–Guest Interactions

This is a mature field, and some of the most interesting new systems have the advantage of a long (in single investigator terms) evolutionary history. One of the most recent and successful of Breslow's many contributions to the field of enzyme mimics[69] is illustrated in Scheme 12.[70]

The catalyst **20** is a tetraarylporphyrin with β-cyclodextrins attached to the para-positions of each aromatic ring by a convenient sulfur linkage. The reaction is the oxidation by iodosobenzene, PhI=O, of the steroid **21** to a single alcohol and is a model for oxidations catalyzed by the cytochrome P-450 group of enzymes. This class of reactions has been a popular target for enzyme modellers, because the radical mechanism involved supports easy oxidations of unactivated alkane CH bonds: also the stereoelectronic requirements of the hydrogen-atom transfer are relatively flexible, allowing greater geometric tolerance than many ionic mechanisms. In this instance the reaction observed is the completely selective oxidation of the 6α hydrogen atom (shown) of the steroid to give the 6α-OH derivative: in 95 % yield when the reaction is run with 1.46mM substrate **21** and 1 mol% of catalyst **20** at room temperature in water for a few hours.

The highly evolved catalyst **20** combines several features that have proved successful in simpler cases. The ionic sulfonate groups make the substrate sufficiently soluble for the reaction to be run in water. (The four hydrophilic cyclodextrins perform the same service for the catalyst.) The target reaction, the selective oxidation of the steroid skeleton, goes back to the early days of enzyme models,[71] and the choice of porphyrin and of manganese as the metal cation are based on many years' experience. The aryl groups are perfluorinated because an earlier version of the catalyst suffered self-oxidation.

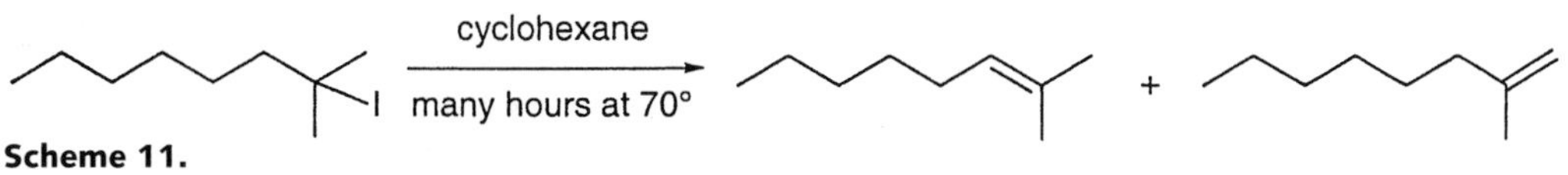

Scheme 11.

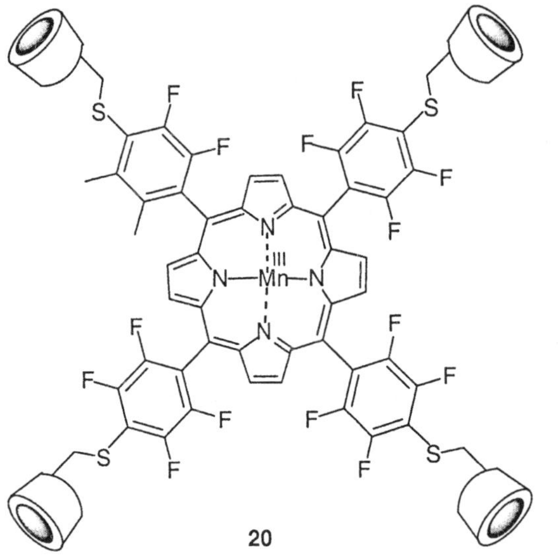

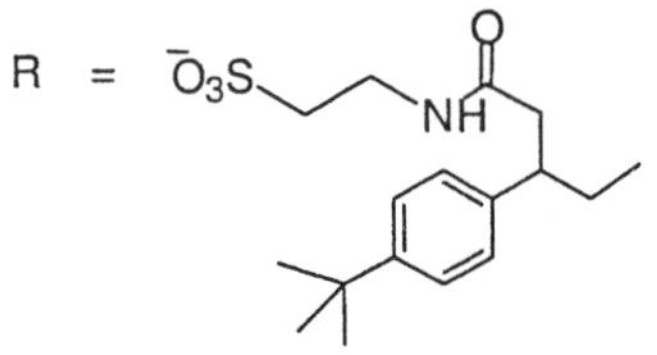

Scheme 12.

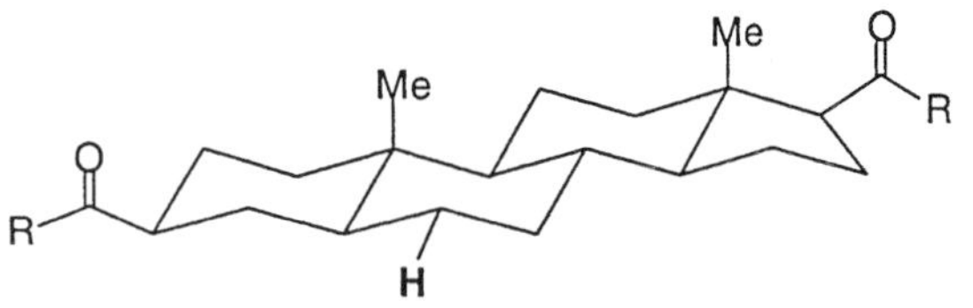

And the binding preferences of the cyclodextrins are well understood, after many years of study by many groups. Most important is the incorporation of multiple binding functionality. The *p-t*-butylphenyl groups of the substrate are known to bind well to β-cyclodextrins: binding two groups means that binding is more than twice as strong, but also that the substrate is stretched across the face of the porphyrin, bringing a specific region of the steroid skeleton into close proximity with the active Mn=O oxidant. This simple strategy has proved reliably successful in a number of contexts,[72] and we can expect more examples to appear.

Major advances in supramolecular chemistry in recent years have produced systems capable of effective and sophisticated molecular recognition, but attempts to translate this into effective catalysis have been generally disappointing. One of the more successful examples is Diederich's pyruvate oxidase mimic, a cyclophane with thiazolium and flavin attached, in the course of an 18-step synthesis, on opposite sides of the central cavity.[73] 2-Naphthaldehyde is bound in the cavity and its oxidation in basic methanol to the methyl ester is catalyzed with saturation kinetics and over 100 turnovers. The catalytic cycle is completed by electrochemical oxidation of the reduced flavin, and the reaction can be carried out on a reasonable preparative scale. However, attaching a catalytic group to a host known to cater for a sensitive guest does not generally work well, and the expertise needed to optimise this sort of system is still accumulating. This area has recently been discussed by Sanders.[74]

References and Notes

1. A. J. Kirby, *Angew. Chem., Intl. Ed. Engl. 35* (**1996**) 707–724.
2. R. Wolfenden, *Science 267* (**1994**) 90–93.
3. R. Wolfenden, C. Ridgway, G. Young, *J. Am. Chem. Soc. 120* (**1998**) 833–834.
4. R. Wolfenden, X. Lu, G. Young, *J. Am. Chem. Soc. 120* (**1998**) 6814–6815.4
5. A. J. Kirby, *Phil. Trans. Roy. Soc. (London), Series A 345* (**1993**) 67.
6. A. R. Fersht, *Enzyme Structure and Mechanism*, Freeman, New York **1985**.
7. P. Berglund, M. R. S. Grace DeSantis, Xiao Shang, Marvin Gold, Richard R. Bott, Thomas P. Graycar, Tony Hing Lau, Colin Mitchinson, J. B. Jones, *J. Am. Chem. Soc. 119* (**1997**) 5265–5266.
8. L. Abrahamsen, J. Tom, J. Burnier, K. A. Butcher, A. Kossiakoff, J. A. Wells, *Biochemistry 30* (**1991**) 4151.
9. P. Carter, J. A. Wells, *Science 237* (**1987**) 394–399.
10. Q. Wang, R. W. Graham, D. Trimbur, R. A. J. Warren, S. G. Withers, *J. Am. Chem. Soc., 116* (**1994**) 11594–11595.
11. A. J. Kirby, I. V. Komarov, P. D. Wothers, N. Feeder, *Angew. Chem. Intern. Ed. 37* (**1998**) 785–786.
12. E. Quemeneur, M. Moutiez, J. B. Charbonnier, A. Menez, *Nature 391* (**1998**) 301–304.
13. For an important recent development in this area ("Directed evolution of new catalytic activity using the alpha/beta-barrel scaffold") see M. M. Altamirano, J. M. Blackburn, C. Aguayo, A. R. Fersht, *Nature 403* (**2000**) 617–622.
14. A. E. Nixon, M. S. Warren, S. J. Benkovic, *Proc. Natl. Acad. Sci. USA 94* (**1997**) 1069–1073.
15. S. Chandrasegaran, J. Smith, *Biol. Chem. 380* (**1999**) 841–848.
16. D. S. Waugh, R. T. Sauer, *Proc. Natl. Acad. Sci. USA 90* (**1993**) 9596–9660.
17. S. N. Thorn, R. G. Daniels, M.-T. M. Auditor, D. Hilvert, *Nature 373* (**1995**) 228–230.
18. F. Hollfelder, A. J. Kirby, D. S. Tawfik, *Nature 383* (**1996**) 60–63.
19. K. Kikuchi, S. Thorn, D. Hilvert, *J. Am. Chem. Soc. 118* (**1996**) 8184–8185.
20. F. Hollfelder, A. J. Kirby, D. S. Tawfik, K. Kikuchi, D. Hilvert, *J. Am. Chem. Soc. 122* (**2000**) 1022–1029.

21. A. J. Kirby, *Acta Chem. Scand. 50* (**1996**) 203–210.

22. R. A. Lerner, A. Tramontano, *Scientific American 258* (**1988**) 58.

23. D. B. Smithrud, S. J. Benkovic, *Curr. Op. Biotechnol. 8* (**1997**) 459–466.

24. H. Wade, T. S. Scanlan, *Ann. Rev. Biophys. Biomol Str. 26* (**1997**) 461–493.

25. C. Gao, B. J. Lavey, C.-H. L. Lo, A. Datta, J. P. Wentworth, K. D. Janda, *J. Am. Chem. Soc. 120* (**1998**) 2211–2217.

26. T. Hoffmann, G. Zhong, B. List, D. Shabat, J. Anderson, S. Gramatikova, R. A. Lerner, C. F. B. III, *J. Am. Chem. Soc. 120* (**1998**) 2768–2779.

27. P. Wirsching, J. A. Ashley, C.-H. L. Lo, K. D. Janda, R. A. Lerner, *Science 270* (**1995**) 1775–1782.

28. C. F. Barbas III, A. Heine, G. Zhong, T. Hoffmann, S. Gramatikova, R. Björnestedt, B. List, J. Anderson, E. A. Stura, I. A. Wilson, R.A. Lerner, *Science 278* (**1997**) 2085–2092.

29. B. List, D. Shabat, C. F. Barbas, R. A. Lerner, *Chem. Eur. J. 4* (**1998**) 881–885.

30. D. Shabat, B. List, R. A. Lerner, C. F. Barbas, *Tetrahedron Lett.* (**1999**) 1437–1440.

31. G. J. Narlikar, D. Herschlag, *Ann. Rev. Biochem.* (**1997**) 19–59.

32. D. Herschlag, *Nature 395* (**1998**) 548–549.

33. K. J. Hertel, A. Peracchi, O. C. Uhlenbeck, D. Herschlag, *Proc. Natl. Acad. Sci. USA 94* (**1997**) 8497–8502.

34. A. Roth, R. R. Breaker, *Proc. Natl. Acad. Sci. USA 95* (**1998**) 6207–6231.

35. Y. Gat, D. G. Lynn, in F. Diederich, P. J. Stang (Eds.): *Templated Organic Synthesis*, Wiley-VCH, Weinheim **2000**, p. 133–157.

36. S. Anderson, H. L. Anderson, in F. Diederich, P. J. Stang (Eds.): *Templated Organic Synthesis*, Wiley-VCH, Weinheim **2000**, p. 1–38.

37. M. A. Warpehoski, D. E. Harper, *J. Am. Chem. Soc. 117* (**1995**) 2951–2952.

38. N. B. Islam, D. L. Whalen, H. Tagi, D. M. Jerina, *J. Am. Chem. Soc. 109* (**1987**) 2108–2111.

39. Y. Li, R. R. Breaker, *Proc. Natl. Acad. Sci. USA 96* (**1999**) 2746–2751.

40. K. Sakthivel, C. F. Barbas, *Angew. Chem., Intl. Ed. 37* (**1998**) 2872–2875.

41. For proponents of an „RNA-world", predating protein catalysis, to regard RNA catalysts as mimicking enzymes – rather than vice versa – may seem perverse. We use this classification for consistency, and without prejudice...

42. J. R. Lorsch, J. W. Szostak, *Accts. Chem. Res. 29* (**1996**) 103–110.

43. J. R. Prudent, T. Uno, P. G. Schultz, *Science 264* (**1994**) 1924–1927.

44. C. Frauendorf, A. Jäschke, *Angew. Chem. Intern. Ed. 37* (**1998**) 1378–1380.

45. T. M. Tarasow, S. L. Tarasow, B. E. Eaton, *Nature 389* (**1997**) 54–57.

46. P. J. Unrau, D. P. Bartel, *Nature 395* (**1998**) 260–263.

47. R. R. Breaker, *Chem. Rev. 97* (**1997**) 371–390.

48. B. N. Trawick, A. T. Danifer, J. K. Bashkin, *Chem. Rev. 98* (**1998**) 939–960.

49. S. W. Santoro, G. F. Joyce, *Proc. Natl. Acad. Sci. (USA) 94* (**1997**) 4262–4266.

50. N. Carmi, S. R. Balkhi, R. R. Breaker, *Proc. Natl. Acad. Sci. USA 95* (**1998**) 2233–2237.

51. D. Sen, C. R. Geyer, *Curr. Opinion Chem. Biol.* (**1998**) 680–687.

52. D. Faulhammer, M. Famulok, *J. Mol. Biol. 269* (**1997**) 188–202.

53. C. R. Geyer, D. Sen, *Chem. & Biol. 4* (**1997**) 579–593.

54. B. Cuenoud, J. W. Szostak, *Nature 375* (**1995**) 611.

55. E. H. Ekland, D. P. Bartel, *Nature 382* (**1996**) 373–376.

56. A. Luther, R. Brandsch, G. von Kiedrowski, *Nature 396* (**1998**) 245–248.

57. R. Rohatgi, D. P. Bartel, J. W. Szostak, *J. Am. Chem. Soc. 118* (**1996**) 3332–3339.

58. R. Rohatgi, D. P. Bartel, J. W. Szostak, *J. Am. Chem. Soc. 118* (**1996**) 3340–3344.

59. Z.-Y. J. Zhan, D. G. Lynn, *J. Am. Chem. Soc. 119* (**1997**) 12420–12421.

60. K. Severin, D. H. Lee, J. A. Martinez, M. R. Ghadiri, *Nature 389* (**1997**) 706–709.

61. S. Yao, I. Ghosh, R. Zutshi, J. Chmielewski, *J. Am. Chem. Soc. 119* (**1997**) 10559–10560.

62. S. Yao, I. Ghosh, R. Zutshi, J. Chmielewski, *Angew. Chem. Intern. Ed. 37* (**1998**) 478–481.

63. L. Baltzer, *Topics in Current Chemistry 202* (**1999**) 39–76.

64. L. Baltzer, K. S. Broo, H. Nilsson, J. Nilsson, *Bioorg Med. Chem. 7* (**1999**) 83–91.

65. M. M. Mader, P. A. Bartlett, *Chem. Rev. 97* (**1997**) 1281–1301.

66. G. Wulff, in F. Diederich, P. J. Stang (Eds.): *Templated Organic Synthesis*, Wiley-VCH, Weinheim **2000**, p. 39–73.

67. F. Hollfelder, A. J. Kirby, D. S. Tawfik, *J. Am. Chem. Soc. 119* (**1997**) 9578–9579.

68. M. E. Piotti, F. Rivera, R. Bond, C. J. Hawker, J. M. J. Fréchet, *J. Am. Chem. Soc. 121* (**1999**) 9471–9472.

69. R. Breslow, *Acc. Chem. Res. 28* (**1995**) 146–153.

70. R. Breslow, B. Gabriele, J. Yang, *Tetrahedron Lett.* (**1998**) 2887–2890.

71. R. Breslow, *Chem. Soc. Rev. 1* (**1972**) 553–580.

72. B. Zhang, R. Breslow, *J. Am. Chem. Soc. 119* (**1997**) 1676–1681.

73. P. Mattei, F. Diederich, *Helv. Chim. Acta 80* (**1997**) 1555–1588.

74. J. K. M. Sanders, *Chem. Eur. J. 4* (**1998**) 1378–1383.

Enzyme Inhibitors

Kevin D. Kreutter and Chi-Huey Wong

*Department of Chemistry
and the Skaggs Institute
for Chemical Biology
The Scripps Research
Institute, La Jolla,
CA 92037, USA*

*Phone: (858) 784-2487
Fax: (858) 784-2409
e-mail: wong@scripps.edu*

Concept: A new approach to the rational design of enzyme inhibitors has emerged in the last ten to fifteen years that incorporates a substrate (or transition state) analog "core" molecule with additional binding determinants spanning beyond the immediate active site area. This approach, which is made more robust by the application of structure-based and computer-aided drug design principles, has been successfully used to generate HIV protease inhibitors that combat the emergence of resistant forms of the protease.

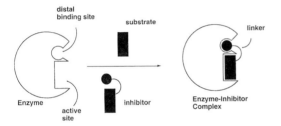

Abstract: Traditional approaches to the rational design of enzyme inhibitors include the use of transition state analogs, suicide substrates, "quiescent" affinity labels, and bisubstrate analogs. Although each of these classes of inhibitor includes at least one (man-made or Nature-made) FDA-approved drug, many enzyme inhibitors bear only marginal resemblance to their target enzymes' substrates, and often have large functional groups (aromatics, heterocycles, alkyl chains, etc) extending from the "core" active site-bound moiety. This method of drug design exploits binding pockets that may have no catalytic function and are often relatively far from the active site. Termed "distal binding analogs", the optimization of these inhibitors is made easier by the existence of X-ray crystallographic data for the enzyme-(parent) inhibitor complex (structure-based design), molecular modeling to create more potent virtual candidate inhibitors (computer-aided drug design), and the use of relatively small, focused combinatorial chemistry libraries to synthesize next-generation inhibitors. A potentially important recent advance with respect to distal binding analog design (including bisubstrate analog design) is the quantification of the entropies of configuration (S_{config}) of linkers as a function of linker identity and length.

Prologue

Enzyme inhibitors are useful not only in the treatment of diseases but also as probes of enzyme mechanism and cell processes. Inhibitors "rationally" designed from enzyme-mechanism principles have traditionally exploited one or more motifs – transition state analogs, suicide substrates, "quiescent" affinity labels, bisubstrate analogs, etc. – that target the core active site binding pocket and have a well-understood basis for delivering potency. We first discuss the characteristics of transition state analogs and the irreversible inactivators before detailing recent developments in bisubstrate analog design and in the more general case of distal binding analog design.

Transition State Analogs

When Pauling surmised that enzymes bind their substrates in the transition state (en route to product) much more strongly than in the ground state,[1] it became apparent that "transition state analogs" might be potent inhibitors of enzymes. Wolfenden extended the theory behind this approach[2] when he showed that substrate-saturated enzymes accelerate their reactions by factors of 10^6–10^{17} over the nonenzymatic rate (k_{cat}/k_{uncat}), and bind their substrate(s) in the transition state with calculated dissociation constants (K_{TS}) of 10^{-7} M (carbonic anhydrase) to 10^{-24} M (orotate monophosphate decarboxylase). Transition state affinities in the range of 10^{-18} – 10^{-20} M are typical (Fig. 1).

Because a stable transition state analog can only approximate the actual transition state in terms of bond lengths, angles, partial charges, etc., inhibitors using this approach have usually fallen many orders of magnitude short of their theoretical maxi-

Keywords ■ Enzyme Inhibitors ■ Transition State Analogs ■ Suicide Substrates ■ "Quiescent" Affinity Labels
■ Bisubstrate Analogs ■ Distal Binding Analogs ■ Linker Design ■ Entropy of Configuration
■ Structure-Based Design ■ Library Synthesis ■ Glycosyltransfer ■ Proteases

Figure 1. Enzymes bind the transition state (ES)$^+$ more tightly than the ground state (ES) by a factor approximately equal to the rate of acceleration (i.e., $K_{TS}/K_S \sim k_{uncat}/k_{cat}$).

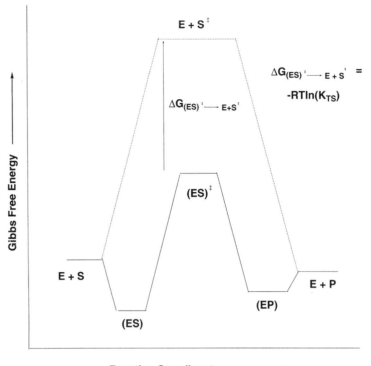

$E + S^{\ddagger}$

$\Delta G_{(ES)' \longrightarrow E+S'} = -RTln(K_{TS})$

$\Delta G_{(ES)' \longrightarrow E+S'}$

$(ES)^{\ddagger}$

$E + S$

$E + P$

(EP)

(ES)

Gibbs Free Energy

Reaction Coordinant ⟶

mum potency. However, many have been found to bind their target enzymes with dissociation constants of 10^{-8}–10^{-12} M, with typical examples including inhibition of adenosine deaminase[3] by coformycin ($K_i = 1 \times 10^{-11}$ M) and inhibition of sucrase by 1-deoxynojirimycin ($K_i = 3.2 \times 10^{-8}$ M)[4]. The tripeptide phosphonate ZFV^P(O)F, inhibitor of carboxypeptidase A,[5] is one of the strongest binding transition state analogs to date ($K_i = 1$–2×10^{-14} M) (Fig. 2).

Interestingly, although many transition state analogs bind noncovalently to the target enzyme's active site via a one-step kinetic mechanism (Scheme 1a) and would therefore be expected to exhibit no time-dependent properties of inhibition, inhibitors with K_i values of $\leq 10^{-10}$ M (like coformycin) usually have a slow onset of inhibition $k_{observed} \leq 10^{-2}\ s^{-1}$ (i.e., an approach to equilibrium inhibition of ≥ 1 min).[6] This is merely an assay artifact due to

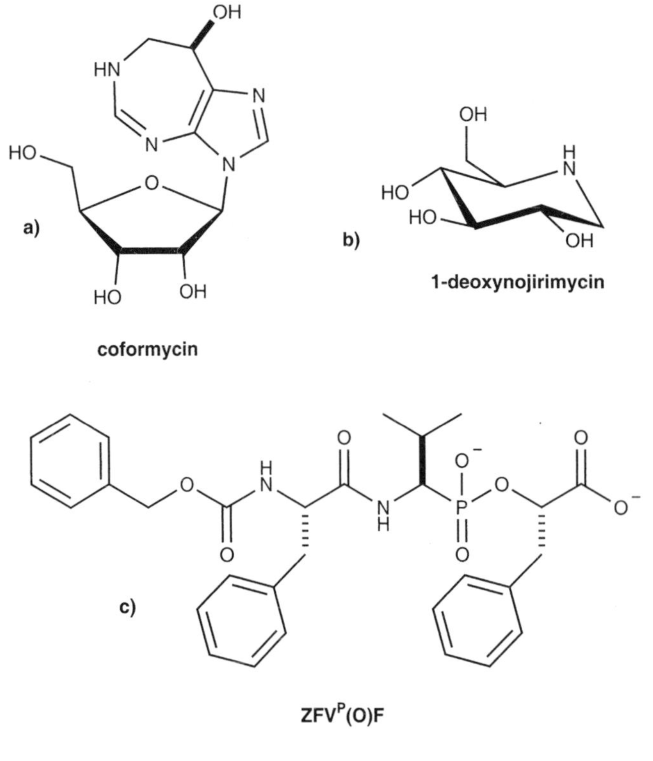

Figure 2. Transition-state analogs of **a**) adenosine deaminase, **b**) sucrase (and other glucosidases), and **c**) carboxypeptidase A.

a)

coformycin

b)

1-deoxynojirimycin

c)

ZFV^P(O)F

a) $E + I \rightleftharpoons E \cdot I$

b) $E + I \rightleftharpoons (E \cdot I)^* \rightleftharpoons E \cdot I$

Scheme 1. The kinetic scheme for **a)** "classic" reversible inhibition, and **b)** "slow-binding" inhibition.

the fact that such potent inhibitors must be assayed at concentrations around their K_i values, and because the upper rate limit for diffusion of inhibitor into an enzyme active site is $\sim 10^8$ M^{-1}s^{-1}. In other cases, however, such "slow-binding inhibition" is due to formation of an initial relatively weak enzyme-inhibitor complex (Scheme 1b). For example, 1-deoxynojirimycin initially binds to sucrase with $K_i^* = 1.9 \times 10^{-6}$ M, and this complex then slowly becomes 98 % isomerized[4] to give an equilibrium (i.e., final) $K_i = 3.2 \times 10^{-8}$ M. The chemical basis for this kinetic behavior may be that the conjugate acid of this iminocyclitol has a $pK_a = 6.4$, and is therefore predominantly in the uncharged form upon binding sucrase at neutral pH. It has been suggested that the slow isomerization step may reflect slow protonation of the azasugar by the active site general acid responsible for activation of the fructose leaving group. Only upon protonation of the ring nitrogen would 1-deoxynojirimycin mimic the presumed positively charged oxocarbenium ion transition state for sucrose cleavage (Fig. 3).

Bartlett has derived a method[8] for proving that a putative transition state analog exerts its inhibitory power from successfully mimicking the transition state. If a series of structurally-related inhibitors (all containing the identical core chemical structure meant to simulate the transition state) bind to the target enzyme with log (K_i) values that linearly correlate (slope = 1) with the log (K_M/k_{cat}) values of the same series of structurally-related substrates, then

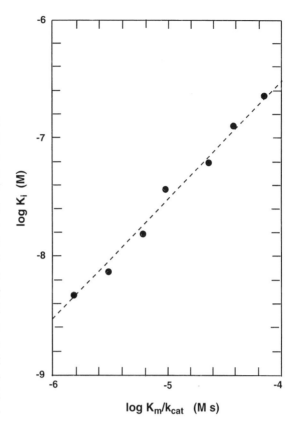

Figure 4. A diagnostic test for proving that an inhibitor is a true transition state analog (data points are representative only).

those inhibitors are true transition state analogs (Fig. 4). This analysis is based on equations 1–2 which follow from the thermodynamic cycle for enzyme E -catalyzed versus uncatalyzed turnover of substrate S (Fig. 5). Equations 3–4 (based on transition-state theory) then assume that K_i is proportional to K_{TS} if the inhibitor mimics the transition state and that ES is in rapid pre-equilibrium with E+S such that $K_S \sim K_m$ (the Michaelis constant).

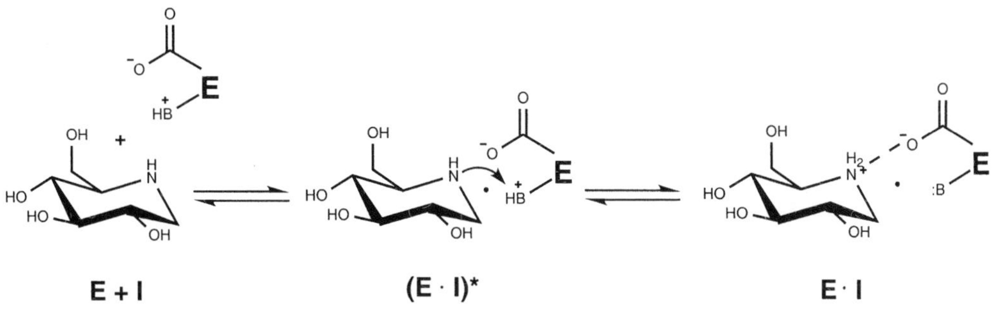

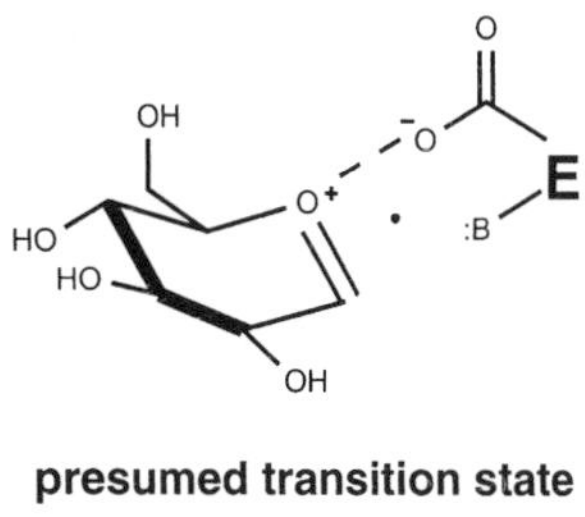

Figure 3. Slow-onset of inhibition by 1-deoxynojirimycin may be due to slow protonation of the inhibitor.

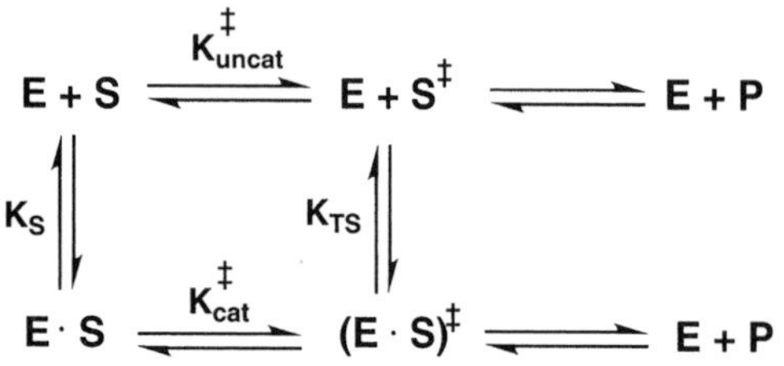

$$E + S \xrightleftharpoons{K_{uncat}^{\ddagger}} E + S^{\ddagger} \rightleftharpoons E + P$$

$$K_S \updownarrow \qquad K_{TS} \updownarrow$$

$$E \cdot S \xrightleftharpoons{K_{cat}^{\ddagger}} (E \cdot S)^{\ddagger} \rightleftharpoons E + P$$

Eqn. 1 $(K_{cat}^{\ddagger})(1/K_S) = (K_{uncat}^{\ddagger})(1/K_{TS})$

Eqn. 2 $K_{TS} = (K_{uncat}^{\ddagger}/K_{cat}^{\ddagger})(K_S)$

Eqn. 3 $K_i = (c) K_{TS} = (c)(K_m)(k_{uncat}/k_{cat})$

Eqn. 4 $\log(K_i) = \log(K_m/k_{cat}) + \log(C)(k_{uncat})$

constant of proportionality

Figure 5. The theoretical basis for the diagnostic test given in Fig. 4.

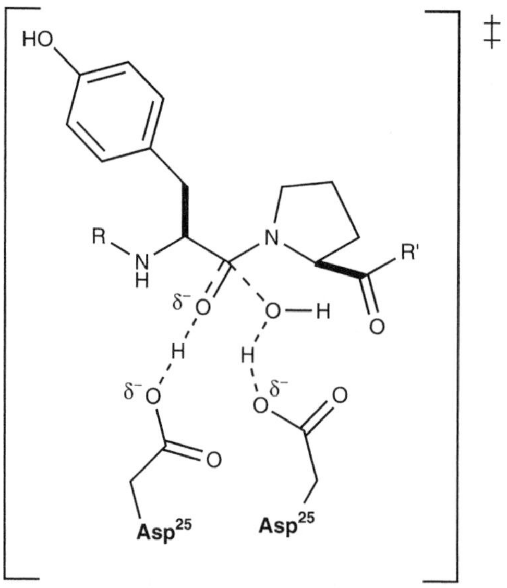

Figure 6. The putative transition state for HIV protease-catalyzed protein cleavage.

Table 1. FDA approved enzyme inhibitors

Drug/ Enzyme Target/ Mode of Inhibition	Structure
Norvir™, Ritonavir, ABT-538 Abbot HIV protease Transition state analog	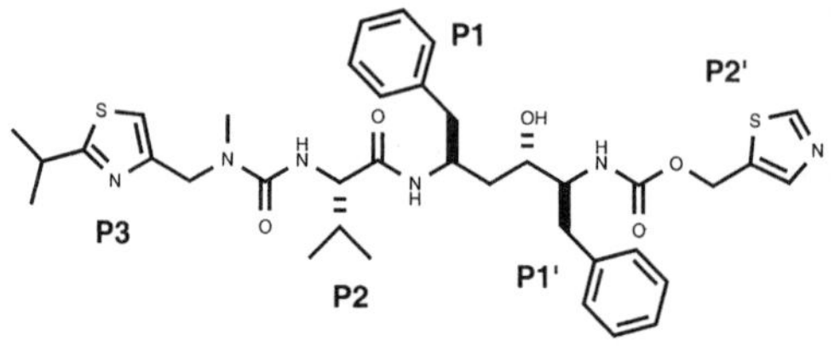
Invirase™, Saquinavir, Ro 31-8959 Hoffmann-La Roche HIV protease Transition state analog	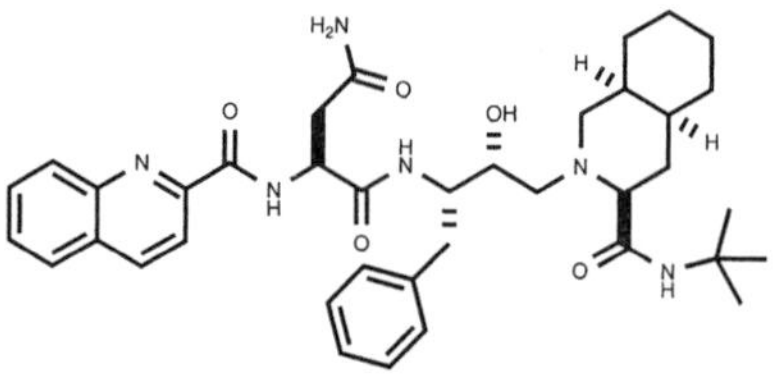
Nelfinavir, Viracept, AG1343 Agouron HIV protease Transition state analog	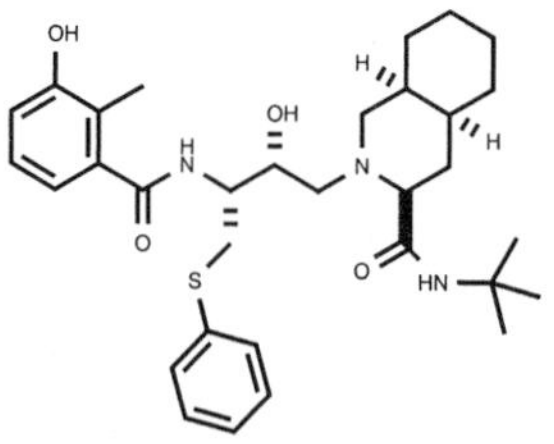
Crixivan™, MK-639, L-735524 Merck HIV protease Transition state analog	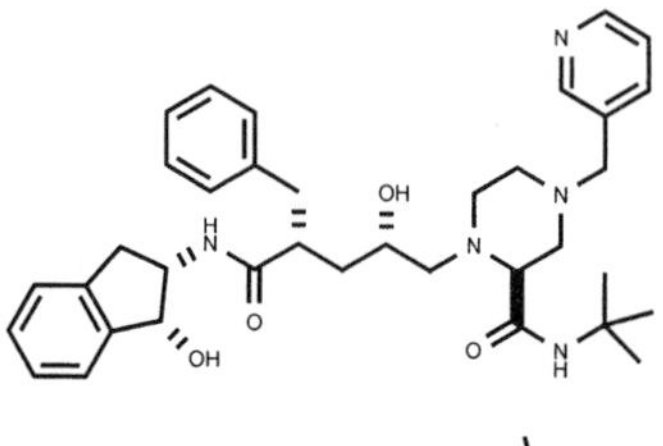
Amprenavir, Agenerase, VX278 Glaxo-Wellcome HIV protease Transition state analog	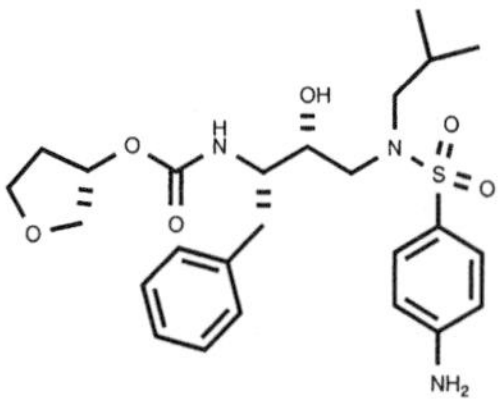

Table 1. (cont.)

Drug/ Enzyme Target/ Mode of Inhibition	Structure
Clavulanic acid **β-Lactamase** **Suicide substrate**	
Eflornithine, DL-α-difluoromethylornithine **Ornithine decarboxylase** **Suicide substrate**	
Aspirin, acetylsalicylic acid **Cyclooxygenase-2** **"Quiescent" affinity label**	
Celebrex **Cyclooxygenase-2**	
Vioxx **Merck** **Cyclooxygenase-2**	
Vasotec, enalapril **Merck** **Angiotensin converting enzyme (ACE)** **$770 million**	
Mevacor, lovastatin (R=CH₃, R"=H) **Zocor, simvastatin (R=CH₃, R"=CH₃)** **Compactin (R=R'=H; not FDA-approved)** **Merck** **HMG-CoA reductase** **Distal binding analogs**	
Viagra **Pfizer** **cGMP hydrolase** **Distal binding analog?**	

This analysis reveals that enzymes bind the transition state more tightly than the ground state by a factor approximately equal to the rate of acceleration (i.e., $K_{TS}/K_S \sim k_{uncat}/k_{cat}$). This method has been used to show, for example, that the peptide phosphonate inhibitors of carboxypeptidase A are true transition state analogs.

Several FDA-approved drugs had their genesis from initial transition state analog leads. Notably, five HIV protease inhibitors recently put on the market (Table 1) all share a hydroxyethylene core motif with the same relative hydroxyl group stereochemistry, which is meant to mimic the tetrahedral intermediate for substrate cleavage between the P1 tyrosine residue and the P1' proline residue (Fig. 6). It would be interesting to determine whether these drugs remain true transition state analogs (using

Bartlett's method) after having been extensively optimized at points distal from the hydroxyethylene core.

Suicide Substrates

Suicide substrates and "quiescent" affinity labels, unlike the other types of inhibitors discussed in this chapter, form covalent bonds with active site nucleophiles and thereby irreversibly inactivate their target enzymes. A suicide substrate,[9] also described by Silverman in a comprehensive review[10] as a mechanism-based inactivator, is a molecule that resembles its target enzyme's true substrate but contains a latent (relatively unreactive) electrophile. When the target enzyme attempts to turn over the suicide "substrate" to "product", it unmasks the latent electrophile, thereby generating a potent acylating or alkylating agent in the active site. For example, during the mechanism of inactivation of β-hydroxy-decanoyl thioester dehydrase by 3-decynoyl-*N*-acetyl-cysteamine, Bloch and coworkers showed[11] that the unmasked allene electrophile (I*) is formed (Scheme 2). There is then a competition between attack of the electrophile by an active site histidine, leading to irreversible inactivation of the enzyme (k_5), versus escape of the electrophile from the active site (k_6) where it may solvolyze or react indiscriminately with other macromolecules or buffer components.

is not uncommon) may cause these molecules to sometimes be quite toxic to living tissue. Nonetheless, several suicide substrates are FDA-approved drugs (Table 1).

"Quiescent" Affinity Labels

One approach that avoids the release of potent electrophiles into the general cellular milieu is use of the "quiescent" affinity label.[13] These evolved from the classic (non-quiescent!) affinity labels such as L-1-chloro-3-tosylamido-4-phenyl-2-butanone (Fig. 7a)

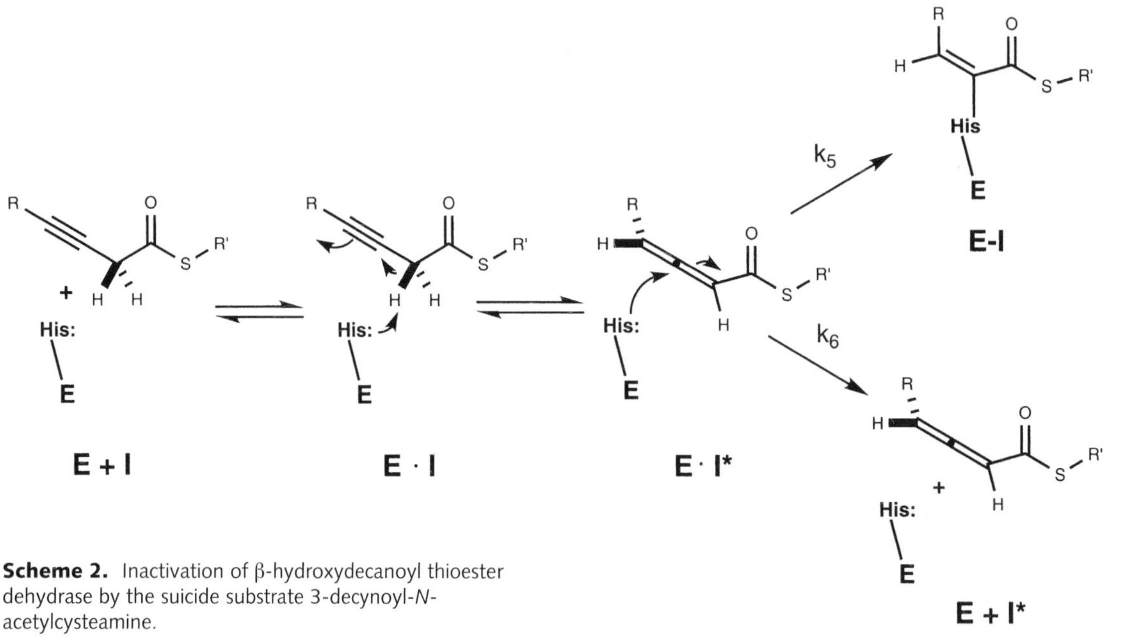

Scheme 2. Inactivation of β-hydroxydecanoyl thioester dehydrase by the suicide substrate 3-decynoyl-*N*-acetylcysteamine.

Silverman has pointed out that several criteria must be met to demonstrate that a compound is a true suicide substrate:[10] (1) Loss of enzyme activity must be time-dependent, and it must be first-order in [inactivator] at low concentrations and zero-order at higher concentrations (saturation kinetics), (2) substrate must protect the enzyme from inactivation (by blocking the active site), (3) the enzyme must be irreversibly inactivated and be shown to have a 1:1 stoichiometry of suicide substrate:active site (dialysis of enzyme previously treated with radiolabeled suicide substrate must not release radiolabel into the buffer), (4) the enzyme must unmask the suicide substrate's potent electrophile via a catalytic step,[12] and (5) the enzyme must not be covalently labeled with the activated form of the suicide substrate following its escape from the active site (the presence of bulky scavenging thiol nucleophiles in the buffer must not decrease the observed rate of inactivation).

The pronounced tendency of suicide substrates to have non-negligible partitioning ratios ($k_6/k_5 > 100$

that resemble the target enzyme's natural substrate but (unlike suicide substrates) carry an exposed potent electrophile.[14,15] Originally used to identify the catalytic serine and histidine in the serine protease family, they were far too intrinsically reactive for use as drugs.[16] However, Krantz has shown[13] that if dramatic reductions in the reactivity of the electrophile are coupled with compensating increases in the favorable non-covalent interactions between the inactivator and enzyme, the synthesis of extremely potent inactivators – termed "quiescent" affinity labels – is possible. A typical example is the peptidyl (acyloxy)methane cathepsin B (cysteine protease) inactivator shown in Fig. 7b ($k_{inact}/K_I = 2.3 \times 10^5$ M^{-1}s^{-1}). Their kinetic behavior is identical to that of the classic affinity labels (Fig. 7c). Two interesting natural products that seem to be "quiescent" affinity labels are microcystin, a potent depsipeptide inactivator that contains a dehydroalanyl moiety which is attacked by the active site cysteine of protein phosphatases, and aspirin (Table 1), which selectively acetylates an active site serine of cyclooxygenase-2.

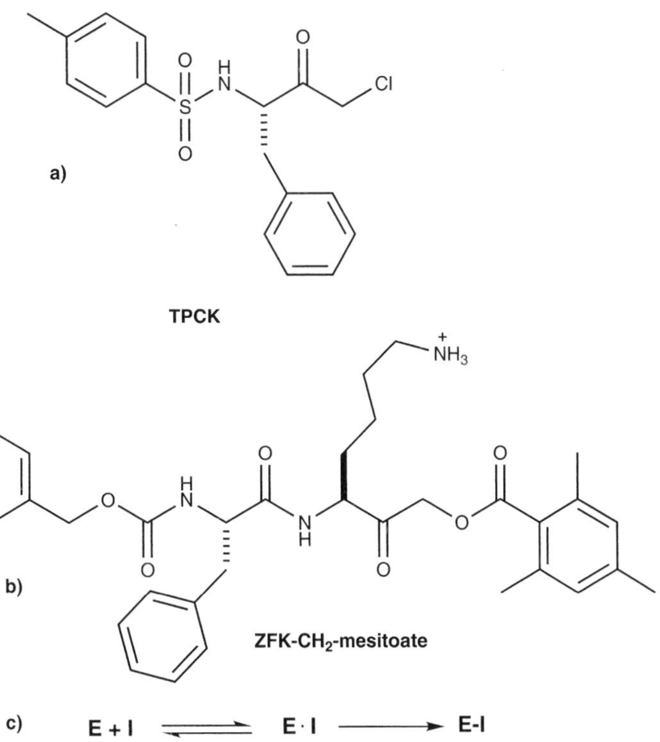

Figure 7. Two examples of irreversible inactivators that are not suicide substrates: **a**) TPCK, a "classic" affinity label of the serine protease chymotrypsin, **b**) ZFK-CH$_2$-mesitoate, a "quiescent" affinity label of the cysteine protease cathepsin B, and **c**) the kinetic scheme for both forms of affinity label-inactivation.

TPCK

ZFK-CH$_2$-mesitoate

c)   E + I $\rightleftharpoons$ E · I $\longrightarrow$ E-I

These inactivators typically have negligible reactivity toward cellular nucleophiles, in contrast to the classic affinity labels and the activated (escaped) form of suicide substrates (I*). However, all classes of irreversible inactivators – even in the ideal case of covalently labeling only their target enzymes – suffer from the possibility of eliciting an undesired immune response against the inactivator-derivatized protein following protein denaturation and degradation.[17]

Multi-Site Binding

Bisubstrate (or Multisubstrate) Analogues

The "chelate effect" in inorganic chemistry is the ability of two or more covalently linked metal ligands (e.g., thiols, amines, carboxylates, etc) to bind a metal cation much more strongly than the individual (unlinked) metal ligands (Fig. 8). This phe-

nomenon holds for any system with more than one binding site.[18] Early (and continuing) efforts at applying the "chelate effect" to enzyme inhibitor design focused on those enzymes that utilized two substrates, and merely involved covalently linking those two substrates. This approach can also provide very potent inhibitors (K_i values < 10^{-8} M). For example, β-thioglycinamide ribonucleotide didea-zafolate (Fig. 9) binds to glycinamide ribonucleotide transformylase[19] with a K_i = 2.5 × 10^{-10} M.

The free energy of binding of A-linker-B to its target enzyme, $\Delta G_{\text{A-linker-B}}$, is given by equation 5:

$$\Delta G_{\text{A-linker-B}} = \Delta G_A + \Delta G_B + \Delta G^S_{\text{linker}} \qquad \text{(Eq. 5)}$$

where ΔG_A and ΔG_B are the observed free energies of binding of A and B to the enzyme (Fig. 10).[18,20] $\Delta G^S_{\text{linker}}$, the "observed effective concentration" term, includes the "intrinsic effective concentration" brought about by having lost only one (not two) sets of translational and rotational entropies upon binding A-linker-B (relative to binding A and B

Figure 8. The chelate effect: **a**) relatively weak binding of pyridine to ferrous ion, and **b**) relatively strong binding of 1,10-phenanthroline to ferrous ion.

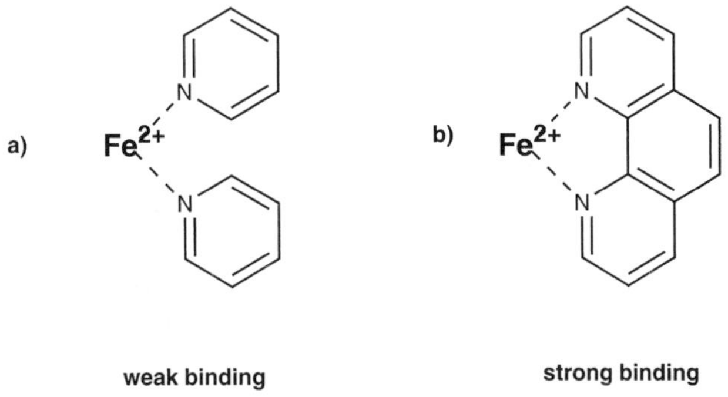

a)   Fe^{2+}

b)   Fe^{2+}

weak binding   **strong binding**

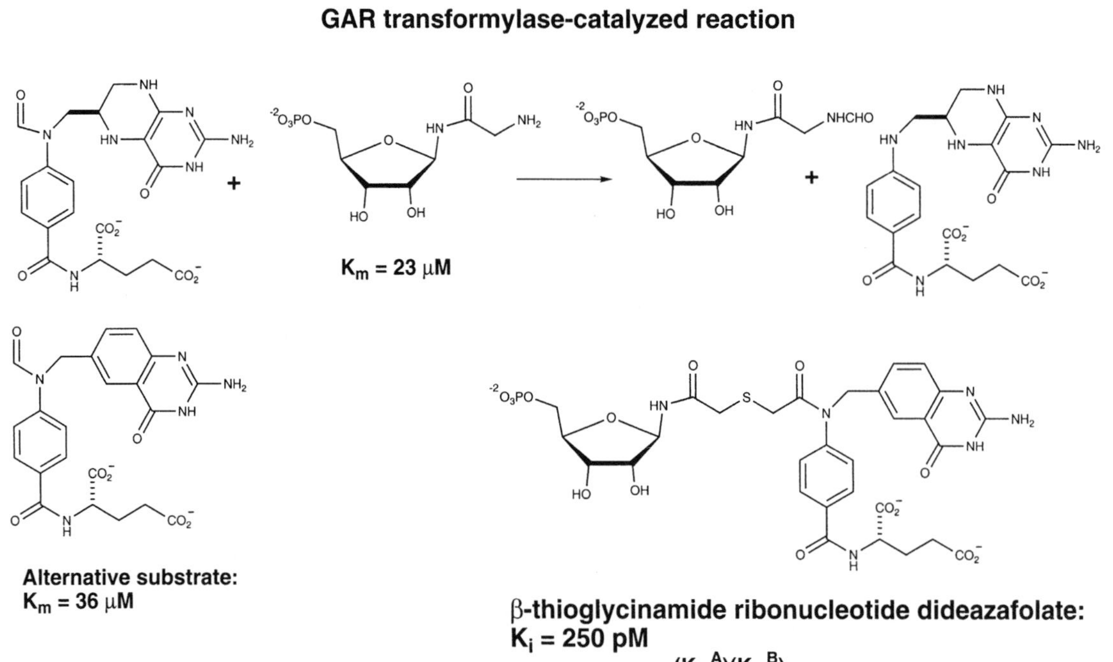

GAR transformylase-catalyzed reaction

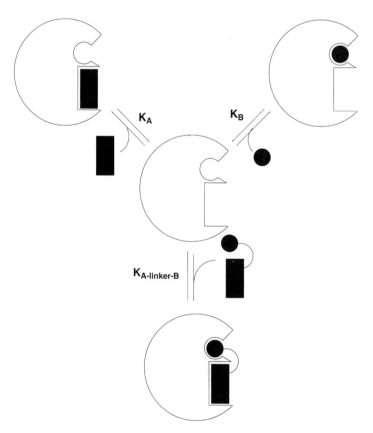

$$K_m = 23 \ \mu M$$

Alternative substrate:
$$K_m = 36 \ \mu M$$

β-thioglycinamide ribonucleotide dideazafolate:
$$K_i = 250 \ pM$$

$$\frac{(K_m^A)(K_m^B)}{K_i} = 3.3 \ M$$

Figure 9. β-Thioglycinamide ribonucleotide dideazafolate: a typical bisubstrate inhibitor.

independently)[21], as well as the free energy of binding of the linker to the enzyme. The latter may encompass favorable or unfavorable enthalpic interactions with the enzyme, as well as an unfavorable loss of entropy of configuration (except in the case of a truly rigid linker).

Page and Jencks have pointed out that the "intrinsic effective concentration" (i.e., the theoreti-

cally maximum effective concentration) of the second substrate covalently linked to the first is 10^8 M.[21] Whitesides and coworkers have recently calculated the entropies of configuration of different chemical linkers,[22] and they vary from the extremely stiff perfluoroalkyl chain (~0.02 kcal mol^{-1} C–C bond^{-1}) to the "typical" alkane, ether, or aminoalkane chain (~0.4 – 0.5 kcal mol^{-1} per C–C,

Figure 10. The binding energies of A and B to the enzyme are not additive, since entropic effects and binding energy of the linker itself also contribute to the binding energy of A-linker-B.

C–O, or C–N bond) to the quite floppy thioether chain (~1.0 kcal mol^{-1} C-S bond^{-1}).

"Observed effective concentrations" as high as 10^8 M have (to the best of our knowledge) not yet been seen in a bisubstrate inhibitor. Possible reasons for this include the potential unfavorable enthalpic interactions between the linker and the enzyme (e.g., threading a polymethylene linker through a pyrophosphate binding pocket) and the loss of configurational entropy in the linker upon binding to the enzyme mentioned earlier. In addition, the tethered ligands may not be as perfectly positioned in their binding pockets as the free ligands A and B,[23] and the tether itself may decrease the affinity of one or both ligands for their binding pockets (by masking a favorable ligand-enzyme interaction at the point of attachment of the tether to the ligand, for example.) Conversely, *favorable* enthalpic interactions between the linker and enzyme may partially offset these sources of lost binding energy.[20]

In practice, these considerations usually reduce ΔG^S_{linker} by $\geq$ three orders of magnitude ($\geq \sim 4$ kcal mol^{-1}) since "observed effective concentrations" $> 10^5$ M are rarely if ever seen in bisubstrate inhibitors. For example, N-phosphonoacetyl-L-aspartate (PALA),[24] in clinical trials in the 1980s, is a "typical" bisubstrate analog[25] with $K_i = 27$ nM (Fig. 11) and an "observed effective concentration" of > 17 M (equations 6–9):

$$K_i^{A\text{-}linker\text{-}B} = (K_i^A)(K_i^B)(K^S_{linker}) \tag{Eq. 6}$$

$$\Delta G^S_{linker} = -RT\ln(K^S_{linker}) \tag{Eq. 7}$$

$$K^S_{linker} = (2.7 \times 10^{-8} \text{ M})/(> 1.7 \times 10^{-2} \text{ M})$$
$$(> 2.7 \times 10^{-5} \text{ M}) = \, < 5.9 \times 10^{-2} \text{ M}^{-1} \tag{Eq. 8}$$

"Observed effective concentration" =
$$1/K^S_{linker} = (K_i^A)(K_i^B)/(K_i^{A\text{-}linker\text{-}B}) = \, > 17 \text{ M} \tag{Eq. 9}$$

Distal-Binding Analogs

The bisubstrate (multisubstrate) analog approach is in fact a special case for a more general method of inhibitor design whereby a substrate (or transition state) analog spans not only the core active site region but also one or more other binding pockets on the enzyme. Abeles has pointed out that these additional binding pockets need not be catalytically important.[26] An important early successful application of this approach stems from the discovery of the toxic fungal metabolite compactin, which potently binds HMG-CoA reductase ($K_i = 1.1$ nM) with an "observed effective concentration" term of $\sim 5 \times 10^4$ M.[26] Close analogs of compactin such as lovastatin (Mevacor) have met with FDA approval (Table 1).

Proteases, whose substrates have an easily discerned modular nature (i.e., at the unit of the individual amino acids within the peptide chain), are obvious candidates for this approach whereby S and S' binding sites (or pseudosites) remote from the catalytic machinery are filled with P and P'[27] moieties linked end to end via sequential amide (or other functional group) bonds. Each P or P' ligand would have to interact favorably with its respective S or S' site with a binding energy of at least 1.4 kcal mol^{-1} simply to pay for the loss of entropy of configuration caused by the cessation of rotation about the C–C, C–N (amide), and N–C bonds connecting successive P or P' ligands (0.86 + 0.06 + 0.49 kcal mol, respectively, = ~ 1.4 kcal mol^{-1}).[22] Fortunately, this is not difficult for two reasons. First, the measured "intrinsic" binding energies even of small groups like methyl, methylthio, amino, ammonium, and hydroxyl can be substantial (2.0–3.9, 4.9, 4.5, 6.7, 8 kcal mol, respectively),[18] and second, the three-bond linker between successive P or P' ligands often pays for its own loss of configurational entropy

Aspartate transcarbamylase-catalyzed reaction

$K_d > 17$ mM $K_d > 27$ µM

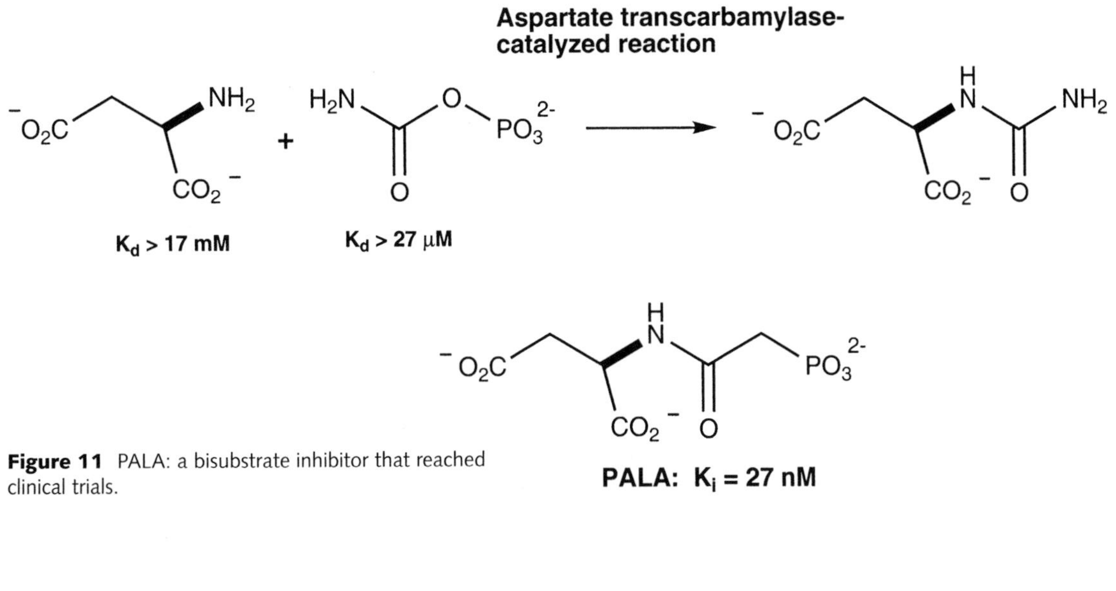

Figure 11 PALA: a bisubstrate inhibitor that reached clinical trials.

PALA: $K_i = 27$ nM

upon binding the protease target using favorable enthalpic interactions between the NH or CO groups and the protein side-chains.

This second effect has been observed to occur[20] with a series of oligoglycine-derived para-substituted benzenesulfonamides bound to carbonic anhydrase II (Fig. 12). With the benzenesulfonamide moiety strongly anchored to the active site Zn atom, it

n = 1: **inhibitor A**
2: **inhibitor B**
.
.
.
6: **inhibitor F**

Figure 12. Oligoglycine-linked para-substituted benzene-sulfonamides as inhibitors of carbonic anhydrase II.

was found that, although the first three glycyl moieties closest to the phenyl ring were restrained from moving due to the closeness of the surrounding protein residues, and the last three glycyl moieties had relative freedom of movement, inhibitors A–F had no difference in observed dissociation constant from the enzyme within experimental error [K_d = 0.33 ± 0.04 μM ($\pm$ 1 SD)].

The two influenza A neuraminidase inhibitors (Fig. 13) recently approved by the FDA can also be considered as distal binding analogs as in both cases the transition-state core is linked to a distal binding determinant (guanidino and isopentyl, respectively) to interact with a new group in the enzyme (carboxyl and hydrophobic, respectively) not involved in catalysis to further improve affinity and specificity[28,29]. Current efforts in our own lab in development of distal binding analog inhibitors of glycosyltransfer enzymes (e.g., fucosidases and fucosyltransferases) involve the use of iminocyclitols as transition-state mimics of the reacting sugar moiety to exploit distal binding determinants.[30–32] In addition, a hydroxamic acid moiety is linked to the iminocyclitol to occupy the pyrophosphate binding site and to chelate the metal ion (Fig. 14).

We also have an interest in exploiting the distal binding analog approach to design inhibitors of HIV protease. These might prove effective against mutant forms of the enzyme that have lost some of their ability to bind some of the drugs listed at the top of Table 1. In particular, mutations glycine-to-valine at position 48 (G48V) and valine-to-phenylalanine at position 82 (V82F) of the HIV pro-

tease homodimer substantially reduce the size of the S3 and S3' pockets. The former mutation causes the protease drug Invirase (Ro 31–8959) to bind with an IC_{50} value 24-fold higher than wild-type (4.4 nM vs 106 nM); the latter mutation raises the IC_{50} value six-fold (4.4 nM vs 26.5 nM). Even greater reductions in binding affinity are seen with Norvir (ABT-538): the K_i increases 17-fold for the G48V mutation and 90-fold for the V82F mutation.[33–35]

Use of the feline immunodeficiency virus (FIV) protease, whose active site is largely superimposable with that of the HIV protease, yet has wild-type S3 and S3' pockets that are as constricted as the mutant HIV protease S3 and S3' pockets discussed above, provides a good model for designing inhibitors to counteract this mode of resistance (Fig. 15).[36] In fact, replacement of the bulky napthyl and ^tbutylsulfonyl groups with methyl and benzyloxycarbonyl, respectively, in the P3/P3' and P4/P4' positions of the drug candidate HOE/BAY793 (to give TL-3, Table 2)

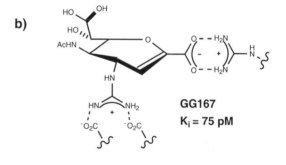

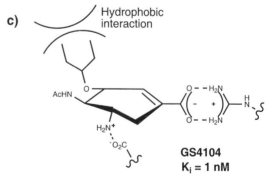

Figure 13 **a**) The putative transition state for the neuraminidase-catalyzed reaction. **b**) and **c**) Two FDA-approved distal binding analog inhibitors of neuraminidase.

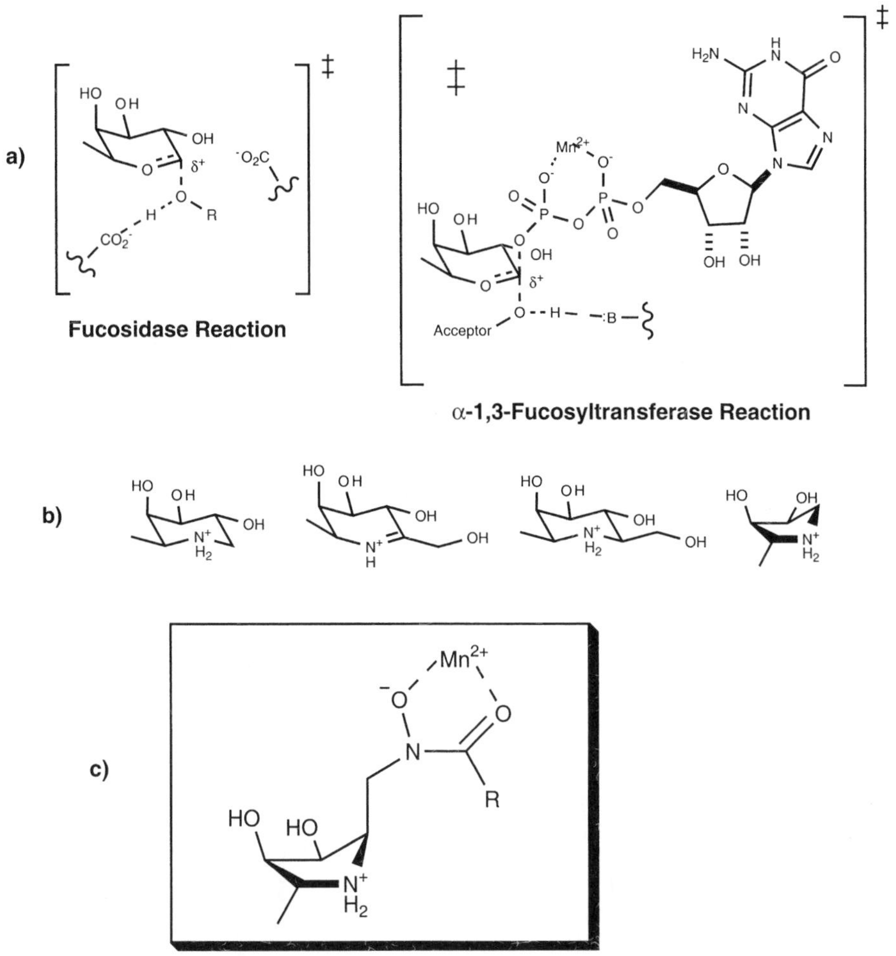

Figure 14. a) The putative transition states for the fucosidase- and fucosyltransferase-catalyzed reactions. **b**) Iminocyclitol transition state analogs which may be used as building blocks for exploiting enzyme pockets distal from the core active site region. **c**) Hydroxamic acid moiety as a pyrophosphate mimic.

Table 2. Interaction of HIV and FIV proteases with protease inhibitors

Inhibitor	WT HIV protease	G48V HIV	V82F HIV	WT FIV protease
	IC$_{50}$ (or K$_i^*$) values; data from references 35 – 37			
Invirase™ 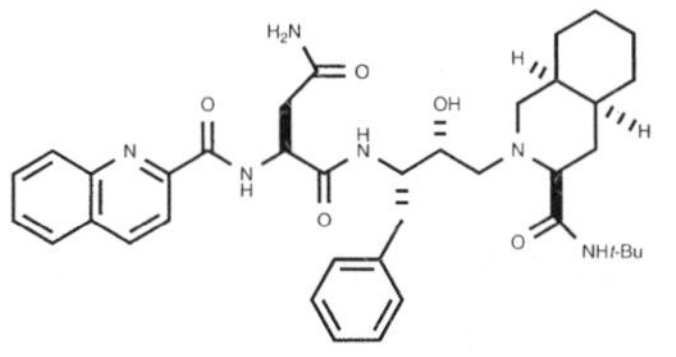	4.4 nM	106 nM	26.5 nM	76,000 nM*
HOE/BAY793 (not FDA-approved) 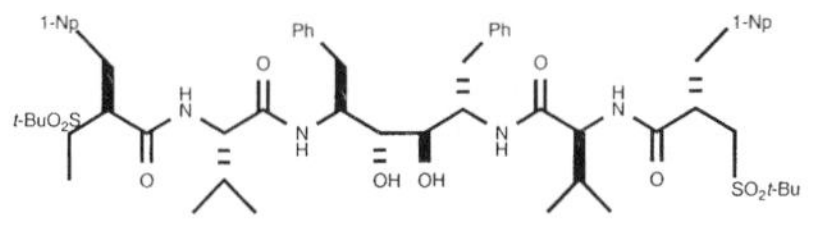	0.3 nM			20,000 nM
TL-3 (not FDA-approved) 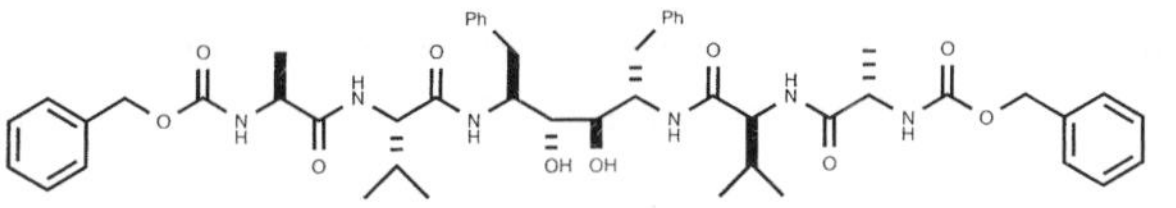	1.5 nM*	20.5 nM	14.9 nM	41 nM*

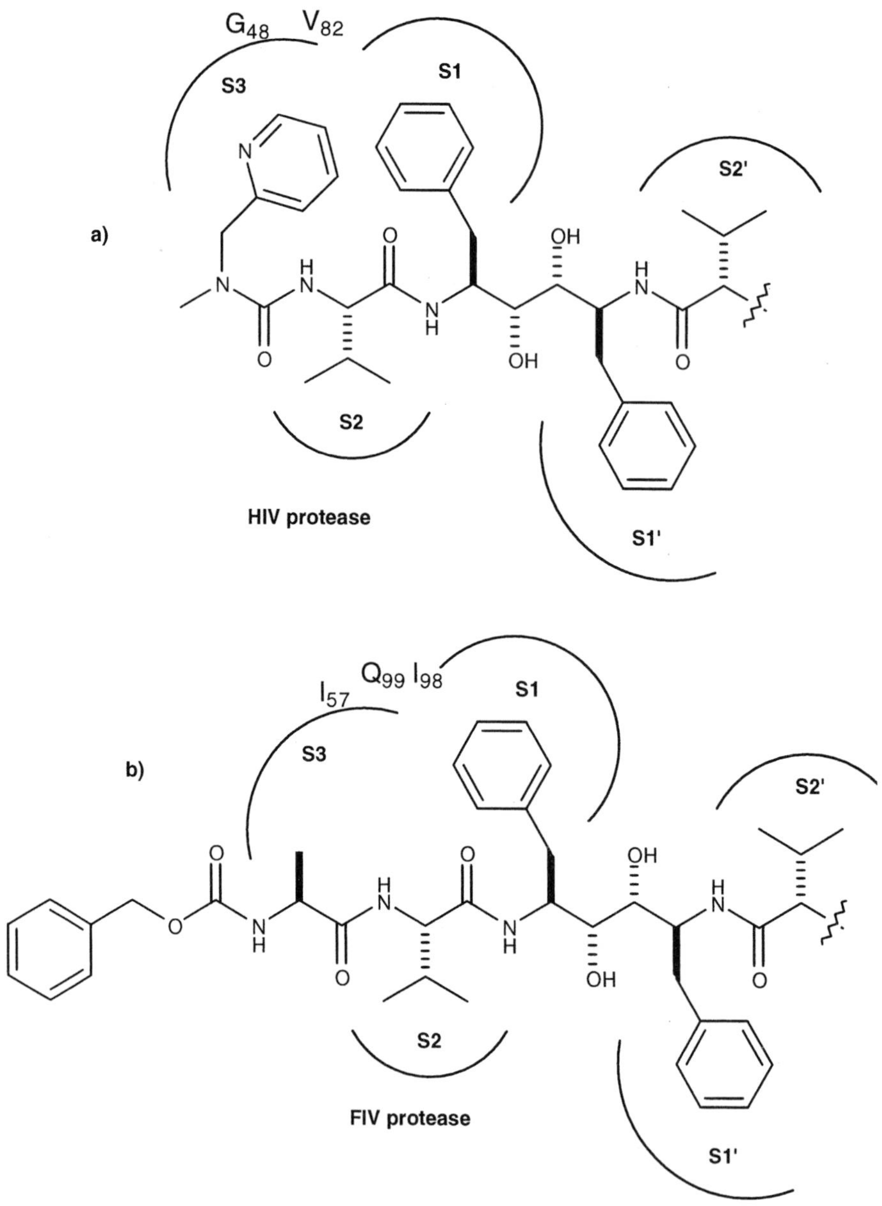

Figure 15. a) The active site of WT HIV protease complexed with A-76889, an inhibitor with a bulky P3 residue. Note the relatively spacious S3 pocket made possible by the presence of the small glycine and valine residues. **b**) The active site of WT FIV protease complexed with TL-3, an inhibitor with a small P3 residue. Note the relatively small constrained S3 pocket due to the presence of the relatively large isoleucine and glutamine residues.

HIV protease

FIV protease

led to greatly improved inhibition of FIV protease (IC_{50} 20 μM vs K_i 41 nM), and virtually no change in HIV wild type inhibition (IC_{50} 0.3 nM vs K_i 1.5 nM). Importantly, the improved inhibitor TL-3 is two- to five-fold more potent than Invirase against the G48V and V82F HIV protease mutants (IC_{50} 15–20 nM) (Table 2).[35,37]

Epilogue

The distal binding analog approach to drug design has yielded clinically approved drugs, (e.g., lovastatin, simvastatin), as well as a large number of exceedingly potent inhibitors, particularly of proteases. Nature in fact has provided us with the distal binding analog hirudin, an inhibitor of thrombin. More recently, parallel synthesis of a series of purine analogs substituted at the 2-, 6-, and 9-positions yielded a 6 nM inhibitor of the cyclin-dependent kinase 2 (CDK2)-cyclin A complex. X-ray crystallography revealed it to make a variety of novel contacts with the enzyme not made by the substrate adenosine triphosphate.[38] Similarly, the determination of structure-activity relationships using nmr spectroscopy ("SAR by NMR") has proven useful in discovering a 19 nM distal binding analog inhibitor of FK506 binding protein (FKBP).[39] Parallel and combinatorial synthesis, high resolution structure determination of enzyme-inhibitor complexes, and/or molecular modeling have played prominent roles in generating the recently invented distal binding inhibitors, and these methods will continue to be used in tandem to extend the number and kinds of enzymes potently inhibited by this approach. With advances in the development of functional genomics and proteomics, many new enzymes will be identified as targets for drug discovery and new tools (e.g., high throughput screening and analysis of inhibitors in a mixture of gene products) will be developed to facilitate the discovery and evaluation of enzyme inhibitors and their effect on other proteins.

References and Notes

1. L. Pauling, *Chem. Eng. News.* **1946**, *24*, 1375.

2. R. Wolfenden, "Conformational Aspects of Inhibitor Design: Enzyme-Substrate Interactions in the Transition State", *Bioorg. Med. Chem.* **1999**, *7*, 647–652.

3. V. L. Schramm, D. C. Baker, "Spontaneous Epimerization of (*S*)-Deoxycoformycin and Interaction of (*R*)-Deoxycoformycin, (*S*)-Deoxycoformycin, and 8-Ketodeoxycoformycin with Adenosine Deaminase", *Biochemistry.* **1985**, *24*, 641–646.

4. G. Hanozet, H.-P. Pircher, P. Vanni, B. Oesch, and G. Semenza, "An Example of Enzyme Hysteresis: The Slow and Tight Interaction of Some Fully Competitive Inhibitors with Small Intestinal Sucrase", *J. Biol. Chem.* **1981**, *256*, 3703–3711.

5. A. P. Kaplan, P. A. Bartlett, "Synthesis and Evaluation of an Inhibitor of Carboxypeptidase A with a K_i Value in the Femtomolar Range", *Biochemistry.* **1991**, *30*, 8165–8170.

6. J. F. Morrison, C. T. Walsh, "The Behavior and Significance of Slow-Binding Enzyme Inhibitors", *Adv. In Enzymol. Relat. Areas Mol. Biol.* **1987**, *61*, 201–301.

7. T. D. Heightman, A. T. Vasella, "Recent Insights into Inhibition, Structure, and Mechanism of Configuration-Retaining Glycosidases", *Angew. Chem. Int. Ed.* **1999**, *38*, 750–770.

8. M. M. Mader, P. A. Bartlett, "Binding Energy and Catalysis: The Implications for Transition-State Analogs and Catalytic Antibodies", *Chem. Rev.* **1997**, *97*, 1281–1301.

9. R. H. Abeles, A. L. Maycock, "Suicide Enzyme Inactivators", *Acc. Chem. Res.* **1976**, *9*, 313–319.

10. R. B. Silverman, *Mechanism-Based Enzyme Inactivation: Chemistry and Enzymology, 2 Vols,* CRC Press, Inc., Boca Raton, Florida, **1988**.

11. M. Morisaki, K. Bloch, "Inhibition of β-hydroxydecanoyl thioester dehydrase by some allenic acids and their thioesters", *Bioorg. Chem.* **1971**, *1*, 188.

12. An interesting clinically useful prodrug is 5-fluorouracil, which is converted *in vivo* to 5-fluoro-2'-deoxyuridine 5'-monophosphate, a potent irreversible inactivator of thymidylate synthase. It is sometimes characterized as a "dead end inactivator" rather than a "suicide substrate" since no electrophile is unmasked during attempted catalytic turnover. Rather, since a fluorine atom replaces the proton found on the normal substrate, enzyme-catalyzed deprotonation at the 5'-position of uracil cannot occur. The enzyme-inactivator covalent adduct (analogous to the normal enzyme-substrate covalent intermediate) therefore cannot break down and has reached a "dead end" (R. R. Rando, "Mechanism-Based Enzyme Inactivators", *Pharm. Rev.* **1984**, *36*, 111–142).

13. A. Krantz, "Peptidyl (Acyloxy)methanes as Quiescent Affinity Labels for Cysteine Proteases", *Methods Enzymol.* **1994**, *244*, 656–671.

14. M. Prorok, A. Albeck, B. M. Foxman, R. H. Abeles, "Chloroketone Hydrolysis by Chymotrypsin and N-Methylhistidyl-57-chymotrypsin: Implications for the Mechanism of Chymotrypsin Inactivation by Chloroketones", *Biochemistry.* **1994**, *33*, 9784–9790.

15. K. Kreutter, A. C. U.Steinmetz, T.-C. Liang, M. Prorok, R. H. Abeles, D. Ringe, "Three-Dimensional Structure of Chymotrypsin Inactivated with (2*S*)-N-Acetyl-L-alanyl-L-phenylalanyl α-chloroethane: Implications for the Mechanism of Inactivation of Serine Proteases by Chloroketones", *Biochemistry,* **1994**, *33*, 13792–13800.

16. H. Angliker, P. Wikstrom, E. Shaw, C. Brenner, R. S. Fuller, "The synthesis of inhibitors for processing proteinases and their action on the Kex2 proteinase of yeast", *Biochem. J.* **1993**, *293*, 75–81.

17. H. E. Amos, B. K. Park, R. Dixon, *Immunotoxicology and Immunopharmacology,* Raven Press, New York, **1985**.

18. W. P. Jencks, "On the attribution and additivity of binding energies", *Proc. Natl. Acad. Sci. USA.* **1981**, *78*, 4046–4050.

19. J. Inglese, R. A. Blatchly, S. J. Benkovic, "A Multisubstrate Adduct Inhibitor of a Purine Biosynthetic Enzyme with a Picomolar Dissociation Constant", *J. Med. Chem.* **1989**, *32*, 937–940.

20. A. Jain, S. G. Huang, G. M. Whitesides, "Lack of Effect of the Length of Oligoglycine- and Oligo(ethylene glycol)-Derived para-Substituents on the Affinity of Benzenesulfonamides for Carbonic Anhydrase II in Solution", *J. Am. Chem. Soc.* **1994**, *116*, 5057–5062.

21. M. I. Page, W. P. Jencks, "Entropic Contributions to Rate Accelerations in Enzymic and Intramolecular Reactions and the Chelate Effect", *Proc. Nat. Acad. Sci. USA.* **1971**, *68*, 1678–1683.

22. M. Mammen, E. I. Shakhnovich, G. M. Whitesides, "Using a Convenient, Quantitative Model for Torsional Entropy To Establish Qualitative Trends for Molecular Processes That Restrict Conformational Freedom", *J. Org. Chem.* **1998**, *63*, 3168–3175.

23. M. Mammen, S.-K. Choi, G. M. Whitesides, "Polyvalent Interactions in Biological Systems: Implications for Design and Use of Multivalent Ligands and Inhibitors", *Angew. Chem. Int. Ed.* **1998**, *37*, 2754–2794.

24. K. D. Collins, G. R. Stark, "Aspartate Transcarbamylase: Interaction With the Transition State Analogue N-(Phosphonacetyl)-L-Aspartate", *J. Biol. Chem.* **1971**, *246*, 6599–6605.

25. A. J. Broom, "Rational Design of Enzyme Inhibitors: Multisubstrate Analogue Inhibitors", *J. Med. Chem.* **1989**, *32*, 2–7.

26. R. H. Abeles, "Enzyme Inhibitors: Ground State/Transition-State Analogs", *Drug, Dev. Res.* **1987**, *10*, 221–234.

27. According to the nomenclature of Schechter and Berger, P and P' moieties refer to the amino acids at the amino terminal side and the carboxy terminal side, respectively, of the site of peptide bond cleavage. Hence, following cleavage the P1 residue has its carboxylate group exposed and the P1' residue has its amino group exposed. These P and P' residues fit in their respective S and S' binding pockets on the protease (I. Schechter, A. Berger, "On the Size of the Active Site in Proteases. I. Papain", *Biochem. Biophys. Res. Comm.* **1967**, *27*, 157–162).

28. C. U. Kim, W. Lew, M. A. Williams, H. Liu, L. Zhang, S. Swaminathan, N. Bischofberger, M. S. Chen, D. B. Men-

del, C. Y. Tai, W. G. Laver, R. C. Stevens, "Influenza Neuraminidase Inhibitors Possessing a Novel Hydrophobic Interaction in the Enzyme Active Site: Design, Synthesis, and Structural Analysis of Carbocyclic Sialic Acid Analogues with Potent Anti-Influenza Activity", *J. Am. Chem. Soc.* **1997**, *119*, 681–690.

29. M. von Itzstein, W.- Y. Wu, G. B. Kok, M. S. Pegg, J. C. Dyason, B. Jin, T. V. Phan, M. L. Smythe, H. F. White, S. W. Oliver, P. M. Colman, J. N. Varghese, D. M. Ryan, J. M. Woods, R. C. Bethell, V. J. Hotham, J. M. Cameron, C. R. Penn, "Rational design of potent sialidase-based inhibitors of influenza virus replication", *Nature*, **1993**, *363*, 418–423.

30. R. Wischnat, R. Martin, S. Takayama, C.-H. Wong, "Chemoenzymatic Synthesis of Iminocyclitol Derivatives: A Useful Library Strategy for the Development of Selective Fucosyltransfer Enzymes Inhibitors", *Bioorg. Med. Chem. Lett.* **1998**, *8*, 3353–3358.

31. R. Wischnat, R. Martin, C.-H. Wong, "Synthesis of a New Class of N-Linked Lewis and LacNAc Analogues as Potential Inhibitors of Human Fucosyltransferases: A General Method for the Incorporation of an Iminocyclitol as a Transition-State Mimetic of the Donor Sugar to the Acceptor", *J. Org. Chem.* **1997**, *63*, 8361–8365.

32. P. Sears, C.-H. Wong, "Carbohydrate Mimetics: A New Strategy for Tackling the Problem of Carbohydrate-Mediated Biological Recognition", *Angew. Chem. Int. Ed.* **1999**, *38*, 2300–2324.

33. S. V. Gulnik, L. I. Suvorov, B. Liu, B. Yu, B. Anderson, H. Mitsuya, J. W. Erickson, "Kinetic Characterization and Cross-Resistance Patterns of HIV-1 Protease Mutants Selected under Drug Pressure", *Biochemisry*, **1995**, *34*, 9282–9287.

34. S. I. Wilson, L. H. Phylip, J. S. Mills, S. V. Gulnik, J. W. Erickson, B. M. Dunn, J. Kay, "Escape mutants of HIV-1 proteinase: enzymic efficiency and susceptibility to inhibition", *Biochim. Biophys. Acta* **1997**, *1339*, 113–125.

35. T. Lee, V.-D. Le, D. Lim, Y.-C. Lin, G. M. Morris, A. L. Wong, A. J. Olson, J. H. Elder, C.-H. Wong, "Development of a New Type of Protease Inhibitors, Efficacious against FIV and HIV Variants", *J. Am. Chem. Soc.* **1999**, *121*, 1145–1155.

36. T. Lee, G. S. Laco, B. E. Torbett, H. S. Fox, D. L. Lerner, J. H. Elder, C.-H. Wong, "Analysis of the S3 and S3' subsite specificities of feline immunodeficiency virus (FIV) protease: Development of a broad-based protease inhibitor efficacious against FIV, SIV, and HIV *in vitro* and *ex vivo*", *Proc. Natl. Acad. Sci. USA.* **1998**, *95*, 939–944.

37. M. Li, G. M. Morris, T. Lee, G.S. Laco, C.-H. Wong, A. J. Olson, J. H. Elder, A. Wlodawer, and A. Gustchina, "Structural Studies of FIV and HIV-1 Proteases Complexed With an Efficient Inhibitor of FIV Protease", *Proteins: Structure, Function, and Genetics*, **2000**, *38*, 29–40.

38. N. S. Gray, L. Wodicka, A.-M. W. H. Thunnissen, T. C. Norman, S. Kwon, F. H. Espinoza, D. O. Morgan, G. Barnes, S. LeClerc, L. Meijer, S.-H. Kim, D. J. Lockhart, P. G. Schultz, "Exploiting Chemical Libraries, Structure, and Genomics in the Search for Kinase Inhibitors", *Science*, **1998**, *281*, 533–538.

39. S. B. Shuker, P. J. Hajduk, R. P. Meadows, S. W. Fesik, "Discovering High-Affinity Ligands for Proteins: SAR by NMR", *Science*, **1996**, *274*, 1531–1534.

Organic Synthesis and Cell Biology

Dieter Kadereit, Jürgen Kuhlmann, and Herbert Waldmann

Max-Planck-Institut
für molekulare Physiologie,
Otto-Hahn-Straße 11,
D-44227 Dortmund,
Germany,

Phone:
Intl. (+) 231-133-2400
Fax: Intl. (+) 231-133-2499
e-mail:
herbert.waldmann@mpi-dortmund.mpg.de

Keywords ■ Bioorganic Chemistry ■ Biological Signal Transduction ■ Cell Biology
■ Biophysics ■ Lipidated Proteins

Concept: The basic concept is to study cell biological phenomena with an approach originating from organic chemistry. Based on structural information available for a given biological phenomenon unsolved chemical problems are identified and new synthetic pathways and methods are developed. This new chemistry then is used to synthesize tools for subsequent biological investigations. If required, the biological properties of these molecular probes are determined and then biological experiments are carried out. This combination of organic synthesis, biological chemistry and cell biology may open up new and alternative opportunities to gain knowledge about the biological phenomenon that could not be obtained by employing biological techniques alone.

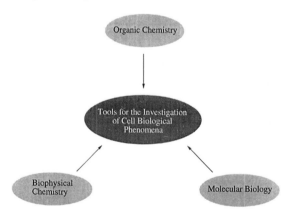

Abstract: The interplay between three disciplines, organic synthesis, biophysics and cell biology in the study of protein lipidation and its relevance to targeting proteins towards the plasma membrane of cells in precise molecular detail is described in this concept. This interplay is highlighted using the Ras protein as a representative example. Included herein is: the development of methods for the synthesis of Ras-derived peptides and fully functional Ras proteins, the determination of the biophysical properties, in particular, the ability to bind to model membranes, and finally the use of synthetic Ras peptides and Ras proteins in cell biological experiments.

Prologue

During the past decades, the biological sciences have undergone a fundamental and dramatic change. Whereas biology for a long time was more phenomenologically oriented, today many biological processes are investigated and understood in molecular detail. The tools for the study of various biological phenomena provided by classical biological techniques often are not sufficient to address the prevailing issues in precise molecular detail. Due to the fundamental chemical nature of biological problems, chemical expertise is urgently required, and the strengths of both chemical and biological methodology have to be used. Such highly interdisciplinary research enterprises in the field of "Bioorganic Chemistry" or "Chemical Biology" may follow a cycle of investigation that begins with the structural analysis of a biological phenomenon, in particular the structural information available for the individual biomacromolecules (i.e. proteins or protein conjugates) or low molecular weight compounds (i.e. natural products or drugs) influencing it (Scheme 1).[1,2] Based on these structural data, unsolved chemical problems are identified and solved by developing new synthetic methods and

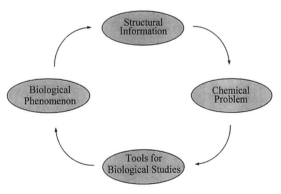

Scheme 1. Interplay between organic chemistry and biology

techniques or by devising new pathways to a desired product. If required, the biophysical properties of the synthesized compounds are determined and used for designing new syntheses and planning subsequent biological experiments. Finally, these compounds are used in biological studies aimed at gaining new insights into the biological phenomenon of interest.

As an illustrative example for the successful realization of this concept, here the interplay between three disciplines, organic synthesis, biophysics and cell biology in the study of protein lipidation and its relevance to targeting of proteins to the plasma membrane of cells in precise molecular detail is described. The interplay is highlighted using the Ras protein as a representative example. Included herein is: the development of methods for the synthesis of Ras-derived peptides and fully functional Ras proteins, the determination of the biophysical properties, in particular the ability to bind to model membranes, and finally the use of synthetic Ras peptides and Ras proteins in cell biological experiments.

Protein Lipidation

Lipidation of proteins may involve N-terminal myristoylation,[3] S-prenylation (farnesyl- and geranylgeranyl groups) of cysteine residues at or close to the C-terminus, and S-palmitoylation[3] of cysteines throughout proteins (Scheme 2).

Lipid modified proteins are often attached to cell membranes. In many cases, they play crucial roles in the transduction of extracellular signals across the plasma membrane and into the nucleus. A particularly important example are the N-, K-, and H-Ras proteins. All Ras proteins terminate in a farne-

Myristoyl   Palmitoyl   n:1 Farnesyl
                                n:2 Geranylgeranyl

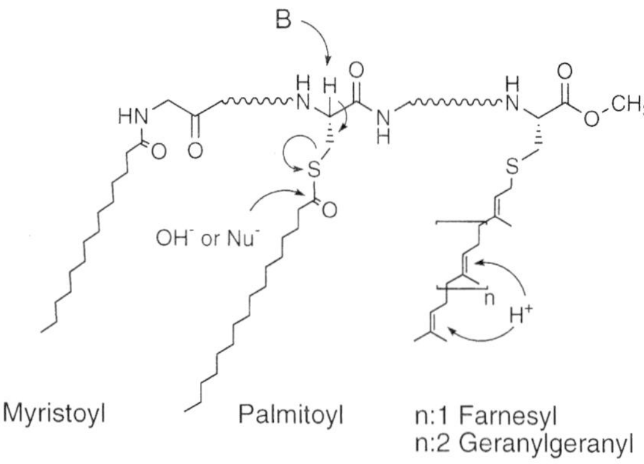

Scheme 2. Lipid modifications of proteins

sylated cysteine methyl ester. In addition, fully modified N-Ras and K-Ras$_B$ are palmitoylated at a cysteine close to the C-terminus while H-Ras is palmitoylated twice (Scheme 3). Lipid modification is essential for both membrane association and biological function of all Ras proteins.

Signal Transduction via Ras Proteins

Ras proteins influence numerous signal transduction processes.[4] In the Ras signal transduction cascade, growth promoting signals e.g. given by a growth factor are passed on to the Ras protein via receptor phosphorylation, recruitment of adaptor molecules and catalysis of GDP/GTP exchange on Ras (Scheme 4). Activated GTP-Ras then binds to further proteins like Raf thereby passing the signal on.

The Ras signal transduction cascade is of extreme physiological importance. It is central to the regulation of cell growth and differentiation, and false regulation of this signal pathway can be one of the critical steps leading to cell transformation.[5] A

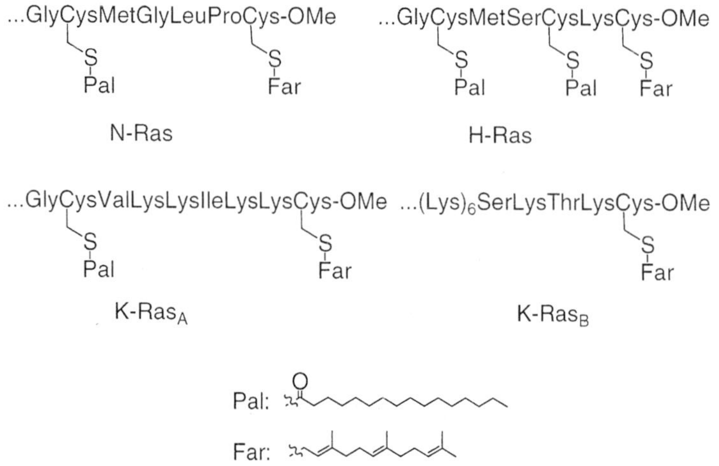

...GlyCysMetGlyLeuProCys-OMe
        |S |S
        Pal   Far
            N-Ras

...GlyCysMetSerCysLysCys-OMe
        |S |S |S
        Pal   Pal Far
            H-Ras

...GlyCysValLysLysIleLysLysCys-OMe
        |S |S
        Pal   Far
            K-Ras$_A$

...(Lys)$_6$SerLysThrLysCys-OMe
                        |S
                        Far
            K-Ras$_B$

Pal: [structure]

Far: [structure]

Scheme 3. Structure of lipid-modified C-termini of the Ras proteins

Scheme 4. Ras activation by receptor tyrosine kinases

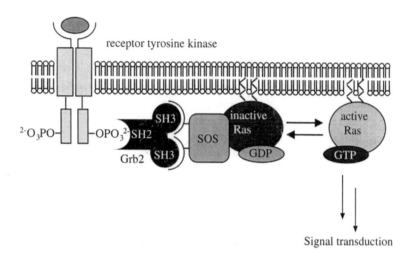

mutation in ras genes can lead to continuously active Ras proteins emitting a permanent growth signal that can result in tumor formation. This malfunction is found in approximately 30 % of all human cancers and in some of the major malignancies it reaches 80 %.[6]

During the last decades extensive research activities were initiated aiming at the precise understanding of signal transduction processes and particularly, the biological role and molecular details of Ras mediated signaling. In general, these investigations were based on genetic and cell-biological

Scheme 5. An interdisciplinary approach toward the investigation of biological phenomena

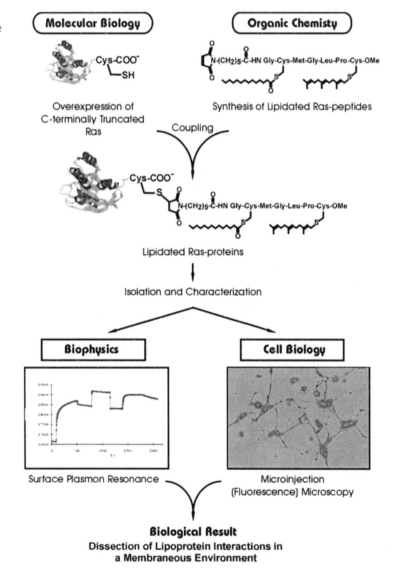

approaches. The Ras proteins employed in these studies were bacterially synthesized proteins lacking the C-terminal lipid modifications. Furthermore, these biological techniques did not allow the introduction of modified lipid groups. Consequently, the function of the lipid groups in signaling via Ras remained largely unclear, in particular their role in the selective targeting of Ras to the plasma membrane and the possible involvement in interactions with upstream or downstream effectors. For studying these problems, a flexible access to differently lipidated and biologically functional Ras proteins to be used subsequently as molecular probes was needed. This demand only could be met by an approach that combined techniques of organic synthesis, molecular biology, biophysics and cell biology.[7] It includes: the development of methods for the synthesis of sensitive Ras-peptides, the coupling to appropriately designed and expressed Ras mutants, the determination of the biophysical properties of the synthetic neo-Ras proteins, and applications in cell-biological experiments. By combination of organic synthesis, biophysics and cell biology, the precise role of the lipid-modifications of farnesylated and palmitoylated Ras proteins in their selective targeting to the plasma membrane could be better understood (Scheme 5).

Synthesis of Lipidated Ras Peptides

The synthesis of lipidated peptides is implicated by the base-lability of the thioester and the acid sensitivity of the prenyl group double bonds. Thus, new protecting groups are required which can be removed under extremely mild, preferably neutral, conditions.

An efficient solution of this synthetic problem consists in the use of enzyme labile protecting groups[8] since enzymatic transformations often can be carried out under characteristically mild reaction conditions (pH 6–8, room temperature to 40 °C). In addition, enzymes often combine a high specificity for the recognized substrates with a large tolerance for secondary structure. Alternatively, noble metal transformations offer reaction conditions that are also mild enough to be compatible with sensitive, doubly lipidated peptides.

Enzyme Labile Amine Protecting Groups

Enzyme-labile protecting groups have to embody a functional group which is specifically recognized by the enzyme, and most importantly, an urethane structure to avoid racemization upon amino acid activation. Unfortunately, most enzymes available today do not attack urethanes. This might be due to diminished reactivity of the urethane carbonyl group preventing a direct attack on the carbonyl C=O. An alternative strategy would be to employ a biocatalyst which attacks a different bond, e.g. an *O*-alkyl bond in an appropriately designed urethane.

Following this concept, the design of a carbohydrate based urethane protecting group was based on the ability of glycosidases to cleave the glycosidic bond to the corresponding sugar, i.e. the bond between C-1 and O-1.[9] In this case, the enzyme (here: a glucosidase) initially cleaves the glycosidic C–O bond to the carbohydrate thereby rendering an attack on the less reactive urethane unnecessary (Scheme 6).

In an alternative general strategy the urethane is linked via a spacer to a functional group which is specifically recognized by the enzyme. Upon cleavage of the enzyme labile bond, the spacer undergoes spontaneous fragmentation and liberates the desired peptide or peptide conjugate (Scheme 7).

The principle of the enzymatic deprotection depicted in Scheme 7 is general. Depending on the acyl group chosen, the fragmentation of the resulting *p*-acyloxybenzyl urethane can be initiated with an appropriate enzyme.

Scheme 6. Carbohydrate based urethane protecting groups

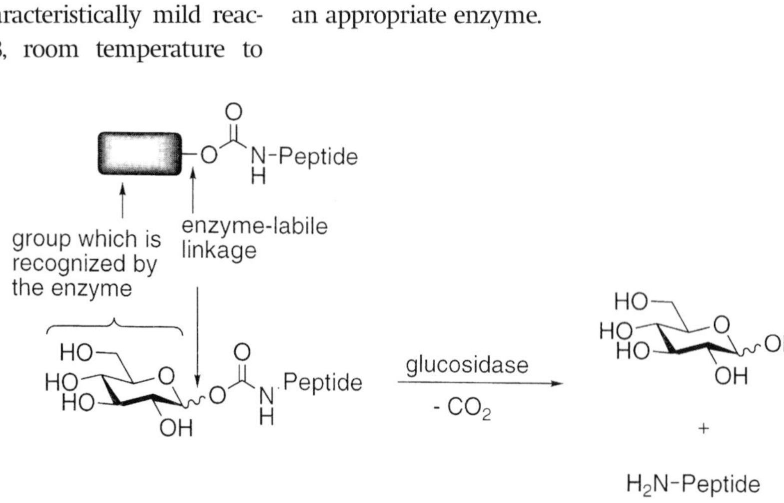

group which is
recognized by
the enzyme

enzyme-labile
linkage

group which undergoes spontaneous
fragmentation upon cleavage of the
enzyme-labile linkage

enzymatic cleavage

fragmentation

$$O= \quad = \quad + \quad CO_2 + H_2N\text{-Peptide}$$

H_2O

Scheme 7. Spacer-based enzymatically removable protecting groups

The *p*-Acetoxybenzyloxycarbonyl Group (AcOZ)

The *p*-acetoxybenzyloxycarbonyl (AcOZ) group can be removed efficiently by a lipase from *Mucor miehei* and an acetyl esterase from the *flavedo* of oranges under exceptionally mild conditions (pH 5–6).[10,11,12] Acetyl esterase discriminated between acetyl and longer acyl side chains (Scheme 8).

In the course of the enzymatic reactions (Scheme 8) neither the C-terminal methyl ester nor the base labile thioester was effected.[11]

Enzyme Labile Carboxyl Protecting Groups; Choline Ester (OCho)

Acid- and base-sensitive lipidated peptides can be selectively deprotected by enzymatic hydrolysis of choline esters.[13a] Choline esters of simple peptides, but also of sensitive peptide conjugates like phosphorylated and glycosylated peptides,[14] nucleopeptides[15] and lipidated peptides,[13,16a] can be cleaved with acetyl choline esterase (AChE) and butyryl choline esterase (BChE) under virtually neutral conditions with complete chemoselectivity. Acid-labile farnesyl groups and base-sensitive thioesters are not attacked.

For instance, in a synthesis of N-Ras lipopeptide **8**, the choline ester in the palmitoylated tripeptide **5** was removed selectively and in high yield by means of the butyryl choline esterase (BChE). Efficient cou-

Scheme 8. AcOZ strategy for the synthesis of palmitoylated and farnesylated N-Ras heptapeptide

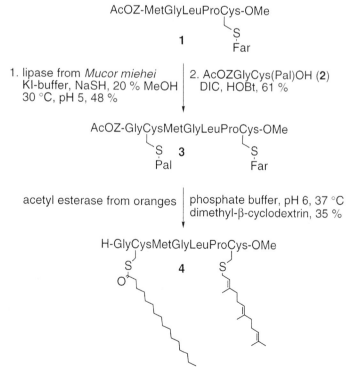

AcOZ-MetGlyLeuProCys-OMe

1

S
|
Far

1. lipase from *Mucor miehei*
   KI-buffer, NaSH, 20 % MeOH
   30 °C, pH 5, 48 %

2. AcOZGlyCys(Pal)OH (**2**)
   DIC, HOBt, 61 %

AcOZ-GlyCysMetGlyLeuProCys-OMe

S **3**   S
| |
Pal   Far

acetyl esterase from oranges | phosphate buffer, pH 6, 37 °C
dimethyl-β-cyclodextrin, 35 %

H-GlyCysMetGlyLeuProCys-OMe

S **4**   S
| |
O

Scheme 9. N-Ras peptide synthesis employing choline esters as C-terminal protecting group. EDC: 1-ethyl-3 (dimethylamino)propyl-carbodiimide hydrochloride, HOBt: 1-hydroxybenzotriazole

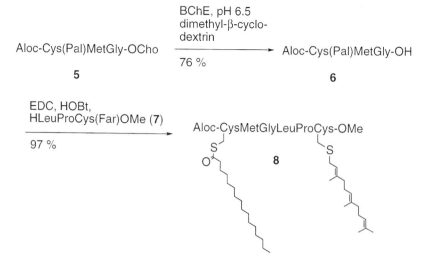

pling of both lipid-modified tripeptides **6** and **7** in high yield completed the synthesis of the target peptide **8**.

A Noble Metal-Sensitive Alternative: Allyl Ester (All) and the Allyloxycarbonyl (Aloc) Group

The selectivity and mildness of the Pd(0)-catalyzed deprotection of allyl (All) esters and the allyloxycarbonyl (Aloc) urethanes[17, 18] reaction also allowed for the successful and efficient application of this blocking group technology in the synthesis of acid- and base-labile lipidated peptides.[7,12,13b,19,20,21]

The suitability of the Aloc group for the construction of lipidated peptides is emphasized by the synthesis of the maleimidocaproyl-modified, S-palmitoylated and farnesylated heptapeptide **16** which corresponds to the N-Ras C-terminus (Scheme 10).[21] In contrast to classical urethane-type protecting groups, the Aloc group can be removed in the presence of additional functional groups and under neutral conditions. It is therefore a very convenient protecting group for the synthesis of very hydrophobic lipid-modified peptides, which are not soluble in the aqueous media required for enzyme catalyzed transformations.

Pd⁰-catalyzed deprotection of S-palmitoylated dipeptide **9** yielded the corresponding selectively deprotected peptide **10**. Condensation of **10** with farnesylated pentapeptide **15**, which was readily accessible via the Aloc methodology[20] as depicted in Scheme 10 and alternatively using AcOZ[11] as the protecting group, resulted in the formation of target peptide **16**.[7,25]

Synthesis of Lipidated Peptides for Biological Investigations

For the study of biological phenomena, analogues with modified lipid or peptide structure may be required. In addition, the introduction of reporter groups, which allow for monitoring the intracellular fate of the peptide conjugates, may be necessary. Depending on the nature of the lipid group, several problems have been studied. Under physiological conditions S-palmitoylation is a reversible process. Thus, the regulation of palmitoylation/depalmitoylation processes may be involved in the steering of biological phenomena like regulated membrane trapping mechanisms. To investigate such mechanisms, peptides are needed which either can not be palmitoylated (i.e. Cys→Ser or Cys→Ala mutants) or which are irreversibly modified, i.e. embodying a cysteine hexadecyl thioether instead of the corresponding palmitoyl thioester (Scheme 11).

Similarly, the farnesyl group in proteins may participate in protein-protein interactions[22,23] and the existence of farnesyl receptors in membranes was proposed but not proven.[24] On the other hand, only the hydrophobicity of that lipid group might account for its physiological consequences. Thus, for the investigation of the biological importance of protein farnesylation, analogs are needed which display a similar hydrophobicity but different structure (e.g. *n*-alkyl ethers). Other analogs, especially in the case of palmitoylated Ras proteins, contain a serine instead of a cysteine.[7,25]

For biological assays, lipidated peptides embodying a fluorescent label like the bimanyl- and the NBD-group, are required for determining membrane binding or subcellular distribution by fluorescence spectroscopy and fluorescence microscopy. Also, attachment of a biotin group allows research-

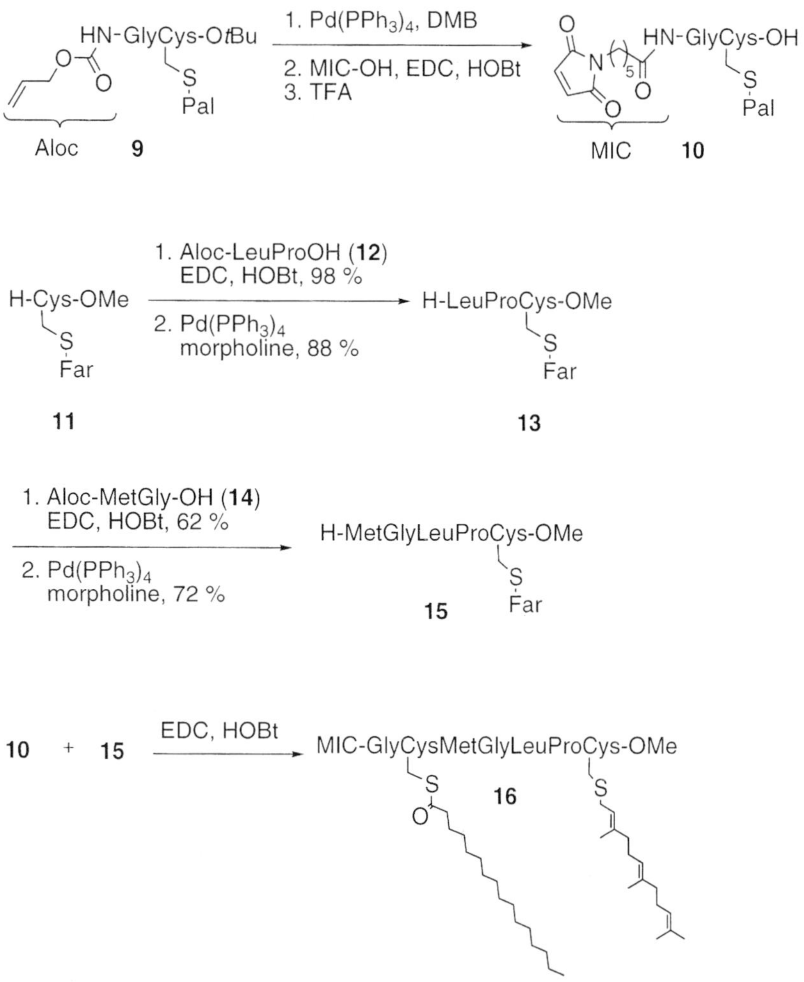

Scheme 10. Synthesis of maleimidocaproyl (MIC) modified heptapeptide using the Aloc strategy. TFA: trifluoroacetic acid, EDC: 1-ethyl-3(dimethylamino) propylcarbodiimide hydrochloride, HOBt: 1-hydroxybenzotriazole, DMB: dimethylbarbituric acid

HN-GlyCys-O*t*Bu
Aloc, Pal — **9**

1. Pd(PPh$_3$)$_4$, DMB
2. MIC-OH, EDC, HOBt
3. TFA

→ HN-GlyCys-OH
MIC, Pal — **10**

H-Cys-OMe
S, Far — **11**

1. Aloc-LeuProOH (**12**) EDC, HOBt, 98 %
2. Pd(PPh$_3$)$_4$ morpholine, 88 %

→ H-LeuProCys-OMe
S, Far — **13**

1. Aloc-MetGly-OH (**14**) EDC, HOBt, 62 %
2. Pd(PPh$_3$)$_4$ morpholine, 72 %

→ H-MetGlyLeuProCys-OMe
15, S Far

10 + **15** →[EDC, HOBt] MIC-GlyCysMetGlyLeuProCys-OMe
16

ers to trace modified peptides by means of the protein streptavidin, which may carry a fluorescent label or gold clusters.[26] In general, such functional groups can be attached to the amino group of selectively N-terminal deprotected peptides like **19** (Scheme 12). The peptides **20a-c** were employed in the synthesis of labeled C-terminal N-Ras peptides.[12,27]

For the synthesis of the peptide–protein conjugates, peptides with a reactive linker group, i.e. a maleimido group at the N-terminal amino function, were required. The synthesis of these peptides

O-palmitoylation, *S*-hexadecylation, free thiol group
↓
Pal   serine instead of cysteine
S   ↓
GlyCysMetGlyLeuProCys-OMe
↑   S
serine instead of cysteine   Far
↑
S-hexadecylation

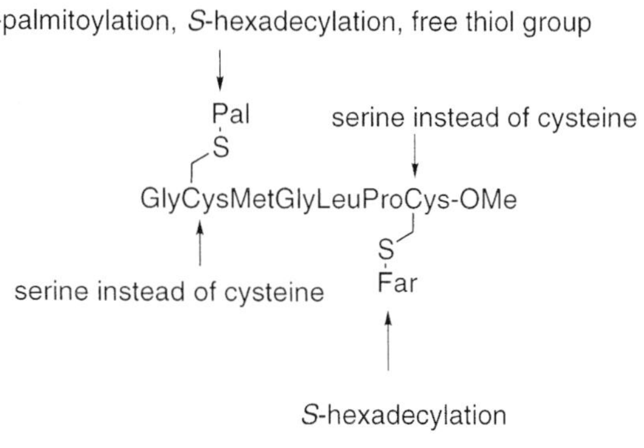

Scheme 11. Variation of *S*-lipidation of palmitoylatable amino acid residues for the investigation of the biological function of protein lipidation

was achieved by selective introduction of the maleimidocaproyl (MIC) linker at the N-terminal peptide amino group as described in Scheme 12.

By appropriate combination of the lipid modifications, linkers, and markers depicted in Schemes 11 and 12, a variety of peptides was synthesized and employed as molecular tools in biological investigations, the details of which are outlined in later sections.

Synthesis of Lipidated Proteins

Lipidated peptides embodying the characteristic linkage region found in the parent lipoproteins and bearing additional functional groups, which could be traced in biological systems or which allowed for their use in biophysical experiments, were used successfully in model studies. However, such model studies only provide a limited amount of information. In order to approximate the situation in a biological system more precisely, experiments with differently lipidated proteins are required.

As mentioned above, at least in the case of the Ras proteins, fully and correctly lipidated proteins

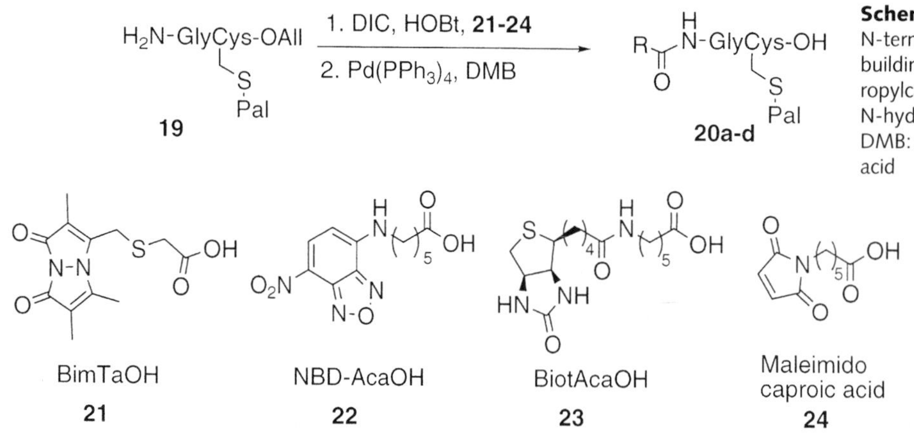

Scheme 12. Synthesis of N-terminal labeled peptide building blocks. DIC: diisopropylcarbodiimide, HOBt: N-hydroxybenzotriazole, DMB: dimethylbarbituric acid

BimTaOH **21**   NBD-AcaOH **22**   BiotAcaOH **23**   Maleimido caproic acid **24**

can not be obtained from yeast or baculoviral expression systems. In addition, such biological techniques are not suitable for the introduction of modified lipid groups into the proteins. Thus, methods for the synthesis of differently lipidated proteins were required. Such a method was developed by combining the methodology of molecular biology with the techniques of organic synthesis (Scheme 5).[7] It consisted of the expression of a suitable Ras mutant lacking the lipidatable C-terminus and subsequent coupling with differently lipidated peptides corresponding to the C-terminus of Ras using maleimido-chemistry.

To this end, on the one hand, C-terminally truncated wildtype and mutated H-Ras proteins were generated. These mutants carried a cysteine at the C-terminal position 181,[28] which lies on the surface of the protein and is the only cysteine residue accessible to external reagents. On the other hand, differently lipidated peptides carrying a maleimidocaproyl (MIC) group at the N-terminus were synthesized by means of the enzyme labile and Pd(0)-sensitive protecting group strategies described above.[7]

The Ras proteins were then allowed to react with the MIC-modified peptides in stoichiometric amounts. The maleimido group[26] reacts specifically with mercapto groups of proteins by conjugate addition of the thiol to the α,β-unsaturated carbonyl compound. The Ras mutants reacted smoothly with the MIC modified peptides and in

high yield. Mass spectroscopic analysis of the proteins both without and after protease digestion revealed that only one lipopeptide was introduced and that only the C-terminal cysteine had been modified.

By means of this method, a variety of Ras proteins with different lipidation patterns could be synthesized in multimilligram amounts. For instance, proteins were generated with the natural lipid combination, i.e. a farnesyl thioether and a palmitoyl thioester. Furthermore, analogous proteins were synthesized embodying only one lipid residue in which either the farnesyl- or the palmitoyl group was replaced by a stable hexadecyl thioether. In addition, proteins were built up containing a serine instead of a cysteine residue at the critical sites which normally are lipidated. In a further series of experiments, lipidated Ras proteins which carry a fluorescent Mant group incorporated into the farnesyl-type modification were synthesized.[25]

Synthetic Lipidated Peptides and Proteins: Biophysical Properties

The insertion of proteins into intracellular membranes has incising effects upon the kinetic and thermodynamic properties of the corresponding biological interactions. Although diffusion in membranes is approx. 100-fold slower than in aqueous solution the probability for two molecules to meet

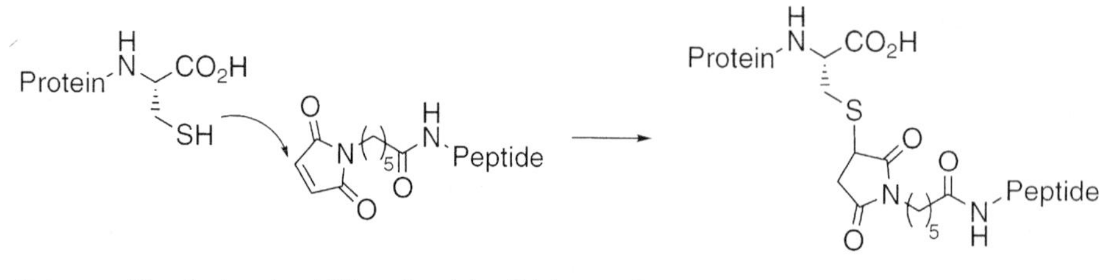

Scheme 13. Conjugate addition of cysteine thiol group to the maleiimido functionality for the synthesis of protein–peptide constructs

in a lateral matrix can accelerate the overall reaction dramatically. As a second effect the local concentrations of the reacting molecules may exceed a critical value, if two binding partners are translocated to the same membrane. In the case of signal transduction via Ras proteins, the affinity of the Ras nucleotide exchange factor protein Sos towards Ras is too small to allow sufficient binding events for the activation of Ras as long as Sos is in the cytoplasm. In response to an extracellular signal Sos is adressed towards the plasma membrane of the cell via an adaptor protein. This membrane adressing generates a local concentration of Ras and Sos molecules in the plasma membrane which is sufficiently above the apparent K_D-value of the binding partners and subsequently allows initiation of nucleotide exchange reactions for a biological response (*vide supra*).

Additionally the membrane itself can contribute to further modifications of the protein-protein interactions. It can provide additional electrostatic and hydrophobic interactions distinct from the lipid anchorage and thereby affect conformation and/or activity of membrane associated proteins.

The principle interactions of lipoproteins with a single hydrophobic modification and membranes has been studied with isoprenylated di-, tri- and tetrapeptides representing the C-terminus of GTP-binding proteins involved in signal transduction (K-Ras, Ral1, Rac2, and RhoC).[29] The binding of these lipopeptides with phospholipid vesicles was monitored utilizing a fluorescent bimanyl label and revealed that a single isoprenyl group is sufficient for membrane association only if supported by carboxymethylation of the C-terminal cysteine.

To achieve stable membrane binding some proteins utilize hydrophobic and electrostatic interactions in concert. The hydrophobic contribution can be realized by a myristoylation (e.g. Src, HIV-1 Gag, MARCKS), farnesylation (e.g. K-Ras4B) or geranylgeranylation (e.g. G25K). The effect of a combination of positively charged side groups and an isoprenoid modification upon membrane binding has been studied with the *C*-termini of K-Ras4B and G25K.[30]

Hydrophobic and electrostatic properties of these lipopeptides show synergistic effects upon binding with membranes.[31] Due to the long stretch of basic amino acids the electrostatic interaction of the K-Ras4B peptide with negatively charged vesicles results in an approx. 10^3-fold increase in binding compared with a neutral membrane.

Further investigations with bimanyl labeled K-Ras4B peptides demonstrated that relatively small differences in membrane charging (approx. 10 mol%) are sufficient for an electrostatic enrichment in the more negative environment.[32] With the farnesyl group as a hydrophobic anchor the peptide is still mobile and can swap between vesicles but may find its target membrane with the sensitive surface potential-sensing function of its lysine residues.

The second class of stable membrane anchoring motives does not rely on electrostatic interactions but supports the first (often isoprenoid) hydrophobic modification by additional thioester formation with fatty acids (e.g. the H- and N-isoforms of Ras or in the α subunits of heterotrimeric G-proteins) or a second isoprenoid moiety (e.g. Rab proteins).[33]

At physiological temperature (37 °C) the dissociation of doubly modified lipopeptides with an isoprenyl thioether and a palmitoyl thioester is rather slow and characterized by half times in the order of 50 h. Here, the relative effect of the carboxymethylation is significantly reduced. Palmitoyl groups with their C_{16} alkane chain due to their length contribute more efficiently to membrane anchoring than farnesyl- or geranylgeranyl-modifications. This findings led to the conclusion that the regulation of membrane anchored proteins has to be achieved by other mechanisms than spontaneous dissociation. In principal binding to an "escort protein" or de-*S*-acylation may induce dissociation of the lipoproteins out of the membrane structure.

The results summarized above were obtained by using fluorescence based assays employing phospholipid vesicles and fluorescent labeled lipopeptides. Recently, surface plasmon resonance (SPR) was developed as new a technique for the study of membrane association of lipidated peptides. Thus, artificial membranes on the surface of biosensors offered new tools for the study of lipopeptides. In SPR (surface plasmon resonance) systems[7,13b] changes of the refractive index (RI) in the proximity of the sensor layer are monitored. In a commercial BIAcore™ system[34] the resonance signal is proportional to the mass of macromolecules bound to the membrane and allows analysis with a time resolution of seconds. Vesicles of defined size distribution were prepared from mixtures of lipids and biotinylated lipopeptides by extruder technique and fused with a alkane thiol surface of a hydrophobic SPR sensor.

The insertion stability of several lipopeptides of the C-terminal sequence of N-Ras in such an artificial membrane is shown in Scheme 14. Here streptavidine is applied to indicate biotinylated lipopetides on the membrane surface.

Again there is a clear difference between single and double hydrophobic modified peptides in their ability to persist in the lipid layer. A farnesylated and palmitoylated heptapeptide dissociates rather

Scheme 14. Binding and dissociation of streptavidine to biotinylated lipopeptides inserted into a SPR membrane

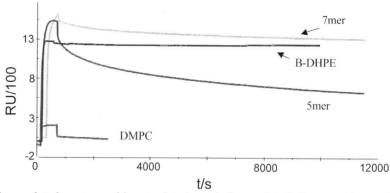

slowly, whereas an only farnesylated pentapeptide has an observed half time in the matrix of less than two hours. While these findings agree qualitatively with the results from vesicle experiments in solution[35] their values differ by two orders of magnitude compared to data derived from intervesicle transfer.[27] Despite of other experimental parameters (e.g. temperature) in the SPR system dissociated lipopeptides have a high probability to bind back to the sensor surface because they are not caught by vesicles in solution.

A new quality in the analysis of hydrophobically post-translational modified proteins could be achieved by the construction of lipidated proteins in a combination of bioorganic synthesis of activated lipopeptides and bacterial expression of the protein backbone as described before. The physico-chemical properties of such artificial lipoproteins differ substantially from those of the corresponding lipopeptides. The pronounced dominance of the hydrophilic protein moiety (e.g. for the Ras protein 181 amino acids) over a short lipopeptide with one or two hydrophobic modifications keeps the construct soluble up to 10^{-4} M, while the biotinylated or fluorescence labeled lipopeptides exhibit low solubility in aqueous solutions and can be applied in the biophysical experiments only in vesicle integrated form or dissolved in organic solvent.

Thus, lipoproteins could be injected over the surface of a lipid covered SPR sensor in a detergent free buffer solution and showed spontaneous insertion into the artificial membrane.[7] Again two hydrophobic modifications are necessary for stable insertion into the lipid layer, whereas lipoproteins with a farnesyl group only dissociate significantly faster out of the membrane. Therefore the isoprenylation of a protein is sufficient to allow interaction with membraneous structures, while trapping of the molecule at a particular location requires a second hydrophobic anchor. Interaction between the Ras protein and its effector Raf-kinase depends on complex formation of Ras with GTP (instead of the Ras*GDP complex, present in the resting cell). If a synthetically modified Ras protein with a palmi-

toylated and farnesylated lipopeptide at its C-terminus is inserted into an artificial membrane of a BIAcore™ sensor a GST-fusion construct with the Ras-binding domain (RBD) of Raf-kinase only shows weak non-specific binding (mostly due to the GST-domain). This binding increases specifically, if the Ras-complexed GDP is exchanged on lane for the non-hydrolyzable GTP-analog guanosine-5'-(β,γ-imido)-triphosphate (GppNHp). The SPR setup can now be applied for the study of interactions between membrane associated proteins and their effectors and regulators in a membrane environment mimicking the situation in the living cell.

Synthetic Lipidated Peptides and Proteins: Function in Living Cells

Eukaryotic cells utilize an efficient transport system that delivers macromolecules fast and secure to their destination. In the case of the small GTP binding proteins of the Ras family the modified C-terminus seems to be sufficient for addressing the polypeptide to its target membrane (in the case of Ras itself the plasma membrane). Lipopeptides with the *C*-terminal structure of N-Ras (either a pentamer with a C-terminal carboxymethylation and farnesylation or a heptapeptide with a palmitoyl thioester in addition) and a N-terminal 7-nitrobenz-2-oxa-1,3-diazolyl (NBD) fluorophore were microinjected into NIH3T3 fibroblast cells and the distribution of the fluorophore was monitored by confocal laser fluorescence microscopy. Enrichment of the protein in the plasma membrane was efficient only for peptides with two hydrophobic modification sites, while the farnesylated but not palmitoylated peptide was distributed in the cytosol.[12]

In a related experiment CV-1 fibroblasts were incubated with fluorescent N-Ras lipopeptides bearing a free palmitoylation site. These peptides cause staining of the CV-1 plasma membrane and efficient *S*-acylation even if the farnesyl group was replaced by a *n*-octyl group.[27] The association of the N-Ras lipopeptides with the plasma membrane was not effected by brefeldin A (which blocks endosomal transport) or reduced temperatures, which inhibit vesicular transport by a different mechanism. These findings support the kinetic targeting model, where a singly lipid-modified protein bearing an *S*-acylation site near the isoprenylated residue can switch between different membrane surfaces inside a cell. If the protein enters a membrane with an acylating enzyme (e.g. the still putative prenylprotein specific palmitoyltransferase) the protein aquires its second hydrophobic modification and is now trapped in the distinct membrane as long as its thioester is not hydrolyzed (Scheme 15).[33]

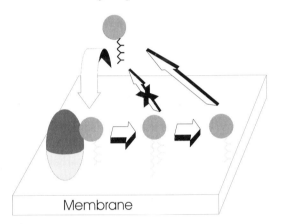

Scheme 15. Kinetic trapping model for the enrichment of isoprenylated proteins with a free acylation site in specific membranes

Replacement of the farnesyl group by lipid analogues could be performed for full length Ras proteins *in vitro* by means of the enzyme farnesyltransferase. When such partially modified Ras constructs were applied in *Xenopus* oocytes the cellular machinery completed modification (endoprotease activity, carboxymethylation and palmitoylation). In these cases the H-Ras farnesyl group could be stripped off most of its isoprenoid features that distinguish it from a fatty acid without any apparent effect on its ability to induce oocyte maturation and activation of mitogen-activated protein kinase. In contrast, replacement by the less hydrophobic isoprenoid geranyl causes severely delayed oocyte activation.

The powerful tool of molecular genetics allows the modification of each single amino acid in the peptide chain of a protein, e.g. deletion of side residues necessary for isoprenylation or palmitoylation[36] or introduction of additional charged amino acids for electrostatic interaction with the plasma membrane.[37] Even some artificial modifications can be introduced by means of recombinant enzymes as shown above.

Limitations of molecular biology become obvious if hydrophobic modifications do not match the specificity of the enzymes available or if the nature of the chemical bond should be changed. This is for example necessary if the labile palmitoyl thioester of membrane anchored proteins should be replaced by stable thioethers with the corresponding alkane chains.

These restrictions could be overcome with the coupling of C-terminally truncated protein and activated lipopeptides as described earlier. These neo-lipoproteins could be synthesized in large amounts and proved to be efficient tools for biochemical, biophysical and biological experiments. Their biological activity has been demonstrated in experiments with the rat pheochromocytoma cell line PC12. This cell line can be induced to differentiate by oncogenic Ras proteins[38] and this effect can be correlated to the transforming potential of these mutants. If oncogenic Ras protein (substitution of glycine by valine at codon 12, RasG12V) from bacterial synthesis is microinjected into PC12 cells the enzymatic machinery of the cell performs all modification steps (as for endogenous Ras) to generate active oncogenic protein. As a consequence most of the cells develop neurite-like outgrowths.[39]

Protein–lipopeptide constructs with the oncogenic mutation and the natural C-terminal modification introduced by the lipopeptide also induce neurite outgrowth in the same manner as full length non-modified RasG12V (Scheme 16).[7] Cells microinjected with the truncated RasG12VΔ181 do not respond to the oncogenic protein since the protein can no longer be post-translationally modified in the cell. A RasG12V construct with a C-terminal farnesyl thioether and carboxy methylation but without palmitoylation is nearly inactive. This is in agreement with transfection experiments using Ras constructs with mutations in the palmitoylatable cysteine residues, which had no effect on farnesylation but dramatically reduced transforming activity and plasma membrane localisation,[7] indicating that one hydrophobic modification is not sufficient for biological activity of H-Ras.

If the farnesyl moiety of the lipopeptide is replaced by a linear unbranched alkane chain, the corresponding coupling product displays the same biological activity as the farnesylated analogue. This finding is in line with experiments utilizing structural analogues of the farnesyl group enzymatically incorporated in the H-Ras protein which showed biological activity even with reduced isoprenoid character.[40] Therefore, no specific isoprenylation receptor seems to be involved in the localiation of Ras to the plasma membrane – a hydropobic alkane chain and a palmitate are sufficient.

One of the major advantages of the strategy for the synthesis of hybrid proteins is the ability to

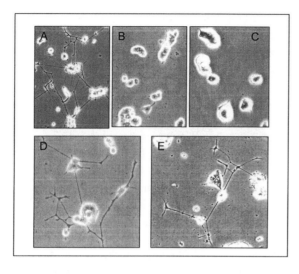

Scheme 16. Biological activity of Ras–lipopeptide constructs in microinjection experiments

A: RasG12V, fl

B: RasG12VD181

C: RasG12V-5Far

D: RasG12V-7PalFar

E: RasG12V-7PalHD

design libraries of C-terminal modifications relying on the efficiency of chemical lipopeptide synthesis. If the labile palmitoyl thioester is replaced by a stable hexadecyl thioether the hybrid protein Ras(G12V)-HDFar is also biologically active but the efficiency of neurite formation is significantly reduced. The reduced biological readout of the 7HDFar-hybrid additionally indicates that the palmitoylation step of H-Ras occurs in the plasma membrane. The farnesylated and hexadecylated protein will insert into any cellular membrane with a low probability of detaching from it. In contrast the reversibility of thioester formation may allow palmitoylated constructs to switch between several membrane structures after induced or spontaneous hydrolysis of the thioester bond. If a corresponding protein specific palmitoyltransferase would be located exclusively in the plasma membrane, Ras constructs which may become palmitoylated would enrich there automatically.

By combining bacterial expression and chemical synthesis Ras constructs with the properties of the post-translationally modified protein can be generated. These hybrid proteins can insert into artifical and biological membranes, have been proven to be efficient tools for biochemical, biophysical and biological experiments and can be synthesized in large amounts. Principally the same method is applicable to many of the Ras-related GTP-binding proteins or the γ-subunit of heterotrimeric G proteins.

Epilogue

In this concept we have described the successful interplay between organic synthesis, biophysics and cell biology in the study of protein lipidation and its role in the selective targeting of proteins like Ras to the plasma membrane. The development of new methods for the synthesis of sensitive Ras peptides and entire Ras proteins, the analysis of their membrane binding properties by means of vesicle based assays and surface plasmon resonance techniques and the use of synthetic Ras peptides and Ras proteins in microinjection experiments led to a better understanding of the molecular details that govern plasma membrane binding of lipidated proteins and the mechanisms by which selective plasma membrane trapping is achieved.

This highly interdisciplinary research provides an illustrative and representative example of what we define as "Bioorganic Chemistry" or "Chemical Biology". In this field, research has to be carried out in both chemistry and biology. The researcher has to cross the barrier and bridge the undoubtedly existing gap in research culture between the two disciplines. The researcher will be rewarded by experiencing the excitement that is created in both disciplines, by the ability to describe a biological phenomenon in the precise molecular language of chemistry and by gaining insights that could not have been obtained by employing either discipline alone.

References and Notes

1. Bioorganic Chemistry: Organic Chemistry meets Biology, H. Waldmann, M. Famulok, submitted for publication.

2. Organic Synthesis and Biological Signal Transduction, K. Hinterding, D. Alonso-Díaz, H. Waldmann, *Angew. Chem.* **1998**, *110*, 716–780, *Angew. Chem. Int. Ed.* **1998**, *37*, 688–749.

3. (a) Fatty Acylation of Proteins: New Insights into Membrane Targeting of Myristoylated and Palmitoylated Proteins, M. D. Resh, *Biochim. Biophys. Acta* **1999**, *1451*, 1–16; (b) Signalling Functions of Protein Palmitoylation, J. T. Dunphy, M. E. Linder, *Biochim. Biophys. Acta* **1998**, *1436*, 245–261.

4. For recent reviews see (a) GEFs, GAPs, GDIs and Effectors: Taking a Closer (3D) Look at the Regulation of Ras-Related GTP-Binding proteins, M. Geyer, A. Wittinghofer, *Curr. Op. Struct. Biol.* **1997**, *7*, 786–792; (b) Ras – a Versatile Cellular Switch, C. Rommel, E. Hafen, *Curr. Op. Gen. Dev.* **1998**, *8*, 412–418; (c) Increasing Complexity of the Ras Signaling Pathway, A. B. Vojtek, C. J. Der, *J. Biol. Chem.* **1998**, *273*, 19925–19928.

5. H. Waldmann, A. Wittinghofer, *Angew. Chem.*, submitted for publication.

6. Ras Genes, M. Barbacid, *Annu. Rev. Biochem.* **1987**, *56*, 779–289.

7. Bioorganic Synthesis of Ras Proteins for the Study of Signal Transduction, B. Bader, K. Kuhn, D. J. Owen, H. Waldmann, A. Wittinghofer, J. Kuhlmann, *Nature* **2000**, *403*, 223–226.

8. (a) Enzymatic Protecting Group Techniques, H. Waldmann, D. Sebastian, *Chem. Rev.* **1994**, *94*, 911–937; (b) Protecting Group Strategies in Organic Synthesis, M. Schelhaas, H. Waldmann, *Angew. Chem.* **1996**, *108*, 2192–2219, *Angew. Chem. Int. Ed.* **1996**, *35*, 2056–2083 (c) Enzymes and Protecting Group Chemistry, T. Pathak, H. Waldmann, *Curr. Op. Chem. Biol.* **1998**, *2*, 112–120.

9. The Tetrabenzylglycosyloxycarbonyl(BGloc)-Group – An Enzyme Labile Carbohydrate Derived Urethane Blocking Group, T. Kappes, H. Waldmann, *Carbohydr. Res.* **1998**, *305*, 341–349.

10. Synthesis of the Palmitoylated and Farnesylated C-Terminal Lipohexapeptide of the Human N-Ras Protein by Employing an Enzymatically Removable Urethane Protecting Group, H. Waldmann, E. Nägele, *Angew. Chem.* **1995**, *107*, 2425–2428, *Angew. Chem. Int. Ed.* **1995**, *34*, 2259–2262.

11. Chemoenzymatic Synthesis of N-Ras Lipopeptides, E. Nägele, M. Schelhaas, N. Kuder, H. Waldmann, *J. Am. Chem. Soc.* **1998**, *120*, 6889–6902.

12. Chemoenzymatic Synthesis of Fluorescent N-Ras Lipopeptides and Their Use in Membrane Localization Studies In Vivo, H. Waldmann, M. Schelhaas, E. Nägele, J. Kuhlmann, A. Wittinghofer, H. Schroeder, J. R. Silvius, *Angew. Chem.* **1997**, *109*, 2334–2337, *Angew. Chem. Int. Ed.* **1997**, *36*, 2238–2241.

13. (a) Enzymatic Synthesis of Peptides and Ras Lipopeptides Employing Choline Ester as a Solubilizing, Protecting, and Activating Group, M. Schelhaas, S. Glomsda, M. Hänsler, H.-D. Jakubke, H. Waldmann, *Angew. Chem.* **1996**, *108*, 82–85, *Angew. Chem. Int. Ed.* **1996**, *35*, 106–109; (b) Chemoenzymatic Synthesis of Biotinylated Ras Peptides and Their Use in Membrane Binding Studies of Lipidated Model Proteins by Surface Plasmon Resonance, M. Schelhaas, E. Nägele, N. Kuder, B. Bader, J. Kuhlmann, A. Wittinghofer, H. Waldmann, *Chem. Eur. J.* **1999**, *5*, 1239–1252.

14. Chemoenzymatic Synthesis of a Characteristic Glycophosphopeptide from the Transactivation Domain of the Serum Response Factor, J. Sander, H. Waldmann, *Angew. Chem.* **1999**, *111*, 1337–1339, *Angew. Chem. Int. Ed.* **1999**, *38*, 1250–1252.

15. Chemoenzymatic Synthesis of Nucleopeptides, S. Flohr, V. Jungmann, H. Waldmann, *Chem. Eur. J.* **1999**, *5*, 669–681.

16. (a) Synthesis of the *N*-terminal Lipohexapeptide of Human $G_{\alpha O}$-Protein and Fluorescent Labeled Analogues for Biological Studies, A. Cotté, B. Bader, J. Kuhlmann, H. Waldmann, *Chem. Eur. J.* **1999**, *5*, 922–936; (b)

Synthesis of Characteristic Lipopeptides of Lipid Modefied Proteins Employing the Allyl Ester as Protecting Group, T. Schmittberger, A. Cotté, H. Waldmann, *Chem. Commun.* **1998**, 937–940.

17. (a) Die Allylgruppe als mild und selektiv abspaltbare Carboxy-Schutzgruppe zur Synthese empfindlicher *O*-Glycopeptide, H. Kunz, H. Waldmann, *Angew. Chem.* **1984**, *96*, 49–50, *Angew. Chem. Int. Ed.* **1984**, *23*, 71–72; (b) Der Allyloxycarbonyl(Aloc)-Rest – die Verwandlung einer untauglichen in eine wertvolle Aminoschutzgruppe für die Peptidsynthese, H. Kunz, C. Unverzagt, *Angew. Chem.* **1984**, *96*, 426–427, *Angew. Chem. Int. Ed.* **1984**, *23*, 436–437.

18. (a) Allylic Protecting Groups and Their Use in a Complex Environment Part I: Allylic Protection of Alkohols, F. Guibé, *Tetrahedron* **1997**, *53*, 13509–13556; (b) Allylic Protecting Groups and Their Use in a Complex Environment Part II: Allylic Protecting Groups and their Removal through Catalytic Palladium π-Allyl Methodology, F. Guibé, *Tetrahedron* **1998**, *54*, 2967–3042.

19. Synthesis of the Palmitoylated and Prenylated C-terminal Lipopeptides of the Human R- and N-Ras Proteins, T. Schmittberger, H. Waldmann, *Bioorg. & Med. Chem.* **1999**, *7*, 749–762.

20. Synthesis of Characteristic Lipopeptides of the Human N-Ras Protein and Their Evaluation as Possible Inhibitors of Protein Farnesyl Transferase, P. Stöber, M. Schelhaas, E. Nägele, P. Hagenbuch, J. Retey, H. Waldmann, *Bioorg. & Med. Chem.* **1997**, *5*, 75–83.

21. Farnesylation of Ras is Important for the Interaction with Phosphoinositide 3-Kinase γ, I. Rubio, U. Wittig, C. Meyer, R. Heinze, D. Kadereit, H. Waldmann, J. Downward, R. Wetzker, *Eur. J. Biochem.* **1999**, *266*, 70–82.

22. (a) Cystein-Rich Region of Raf-1 Interacts with Activator Domain of Post-Translationally Modified Ha-Ras, C.-D. Hu, K.-I. Kariya, M. Tamaoka, K. Akasaka, M. Shirouzu, S. Yokoyama, T. Kataoka, *J. Biol. Chem.* **1995**, *270*, 30274–30277; (b) Ras Interaction with Two Distinct Binding Domains in Raf-1 May Be Required for Ras Transformation, J. D. Drugan, R. Khosravi-Far, M. A. White, C. J. Der, Y.-J. Sung, Y.-W. Hwang, S. L. Campbell, *J. Biol. Chem.* **1996**, *271*, 233–237; (c) Post-Translational Modification of H-Ras Is Required for Activation of, but Not for Association with, B-Raf, T. Okada, T. Masuda, M. Shinkai, K.-I. Kariya, T. Kataoka, *J. Biol. Chem.* **1996**, *271*, 4671–4678; (d) An Intact Raf Zinc Finger Is Required for Optimal Binding to Processed Ras and for Ras-Dependent Raf Activation In Situ, Z. Luo, B. Diaz, M. S. Marshall, J. Avruch, *Mol. Cell. Biol.* **1997**, *17*, 46–53; (e) Activation of c-Raf-1 by Ras and Src Through Different Mechanisms: Activation in vivo and in vitro, D. Stokoe, F. McCormick, *EMBO J.* **1997**, *16*, 2384–2396.

23. The Farnesyl Group of H-Ras Facilitates the Activation of a Soluble Upstream Activator of Mitogen-activated Protein Kinase, P. McGeady, S. Kuroda, K. Shimizu, Y. Takai, M. H. Gelb, *J. Biol. Chem.* **1995**, *270*, 26347–26351.

24. (a) Dislodgement and Accelerated Degradation of Ras, R. Haklai, M. Gana-Weisz, G. Elad, A. Paz, D. Marciano, Y. Egozi, G. Ben-Baruch, Y. Kloog, *Biochemistry* **1998**, *37*, 1306–1314; (b) Evidence for a High Affinity, Saturable, Prenylation-dependent p21^{Ha-ras} Binding Site in Plasma Membranes, A. A. Siddiqui, J. R. Garland, M. B. Dalton, M. Simensky, *J. Biol. Chem.* **1998**, *273*, 3712–3717.

25. Synthesis of Fluorescent Functional Ras-Proteins by a Combined Organic Synthesis and Molecular Biology

Approach, D. J. Owen, K. Kuhn, B. Bader, A. Wittingho-fer, J. Kuhlmann, H. Waldmann, submitted for publication.

26. G. T. Hermanson, Bioconjugate Techniques, Academic Press, London **1996**.

27. *S*-Acylation and Plasma Membrane Targeting of the Farnesylated Carboxyl-Terminal Peptide of N-Ras in Mammalian Fibroblasts, H. Schroeder, R. Leventis, S. Rex, M. Schelhaas, E. Nägele, H. Waldmann, J. R. Silvius, *Biochemistry* **1997**, *36*, 13102–13109.

28. The Ras-RasGAP-Complex: Structural Basis for GTPase Activation and Its Loss in Oncogenic Ras Mutants, K. Scheffzek, M. R. Ahmadian, W. Kabsch, L. Wiesmüller, A. Lantwein, F. Schmitz, A. Wittinghofer, *Science* **1997**, *277(5324)*, 333–338.

29. (a) Interbilayer Transfer of Phospholipid-Anchored Macromolecules via Monomer Diffusion, J. R. Silvius, M. J. Zuckermann, *Biochemistry* **1993**, *32*, 3153–3161; (b) Spontaneous Interbilayer Transfer of Phospholipids: Dependence on Acyl Chain Composition, J. R. Silvius, R. Leventis, *Biochemistry* **1993**, *32*, 13318–13326.

30. Binding of Prenylated and Polybasic Peptides to Membranes: Affinities and Intervesicle Exchange, F. Ghomashchi, X. Zhang, L. Liu, M. H. Gelb, *Biochemistry* **1995**, *34*, 11910–11918.

31. Electrostatic Interaction of Myristoylated Proteins with Membranes: Simple Physics, Complicated Biology, D. Murray, N. Ben-Tal, B. Honig, S. McLaughlin, *Structure* **1997**, *5*, 985–989.

32. Lipid-Binding Characteristics of the Polybasic Carboxy-Terminal Sequence of K-Ras4B, R. Leventis, J. R. Silvius, *Biochemistry* **1998**, *37*, 7640–7648.

33. Doubly-Lipid Modified Protein Sequence Motifs Exhibit Long-Lived Anchorage to Lipid Bilayer Membranes, S. Shahinian, J. R. Silvius, *Biochemistry* **1995**, *34*, 3813–3822.

34. Analysis of Macromolecular Interactions Using Immobilized Ligands, I. Chaiken, S. Rose, R. Karlsson, *Analytical Biochemistry* **1992**, *201*, 197–210.

35. Fluorimetric Evaluation of the Affinities of Isoprenylated Peptides for Lipid Bilayers, J. R. Silvius, F. l'Heureux, *Biochemistry* **1994**, *33*, 3014–3022.

36. Novel Determination of H-Ras Plasma Membrane Localization and Transformation, B. M. Willumsen, A. D. Cox, P. A. Solski, C. J. Der, J. E. Buss, *Oncogene* **1996**, *13*, 1901–1909.

37. A Non-farnesylated Ha-Ras Protein Can Be Palmitoylated and Trigger Potent Differentiation and Transformation, M. A. Booden, T. L. Baker, P. A. Solski, C. J. Der, S. G. Punke, J. E. Buss, *J. Biol. Chem.* **1999**, *274*, 1423–1431.

38. Microinjection of the Ras Oncogene Protein into PC12 Cells Induces Morphological Differentiation, D. Bar-Sagi, J. R. Feramisco, *Cell* **1985**, *42*, 841–848.

39. Biochemical and Biological Consequences of Changing the Specificity of p21(Ras) from Guanosine to Xanthosine Nucleotides, G. Schmidt, C. Lenzen, I. Simon, R. Deuter, R. H. Cool, R. S. Goody, A. Wittinghofer, *Oncogene* **1996**, *12*, 87–96.

40. Replacement of the H-Ras Farnesyl Group by Lipid Analogues: Implication for Downstream Processing and Effector Activation in Xenopus Oocytes, T. Dudler, M. H. Gelb, *Biochemistry* **1997**, *36*, 12434–12441.

Subject Index